International Series in Operations Research & Management Science

Volume 208

Series Editor
Frederick S. Hillier
Stanford University, CA, USA

Special Editorial Consultant
Camille C. Price
Stephen F. Austin State University, TX, USA

For further volumes:
http://www.springer.com/series/6161

Wade D. Cook • Joe Zhu
Editors

Data Envelopment Analysis

A Handbook on the Modeling of Internal Structures and Networks

Springer

Editors
Wade D. Cook
Schulich School of Business
York University
Toronto, ON, Canada

Joe Zhu
School of Business
Worcester Polytechnic Institute
Worcester, MA, USA

ISSN 0884-8289 ISSN 2214-7934 (electronic)
ISBN 978-1-4899-7846-2 ISBN 978-1-4899-8068-7 (eBook)
DOI 10.1007/978-1-4899-8068-7
Springer New York Heidelberg Dordrecht London

Printed on acid-free paper

Springer is part of Springer Science+Business Media (www.springer.com)

In Memory of William Wager Cooper
(July 23, 1914–June 20, 2012)

Preface

Data envelopment analysis (DEA) is a linear programming based approach for measuring relative efficiencies or performances of peer decision making units (DMUs). The performance or efficiency of a DMU is expressed in terms of a set of measures which are classified or coined as DEA inputs and outputs. In conventional DEA, each DMU is treated as a black-box and its internal structures and operations are ignored.

With the publication of the 2nd edition of *Handbook on Data Envelopment Analysis* (eds, Cooper et al. 2011), DEA models for treating DMUs that have internal or network structures have been identified as being on the research frontier (see, for example, Cook and Seiford 2009, and Liu et al. 2013). In fact, there already exists a significant amount of research on both the theory and applications of the network DEA approach. A significant number of researchers and scholars have started to look into the internal structures of DMUs.

Färe and Grosskopf (1996) are the first to propose DEA models when inputs and outputs of DMUs form a network structure. Castelli et al. (2004) study several types of DMU internal structures and develop DEA-type models to measure the overall and component efficiencies. In a different line of research, Kao and Hwang (2008) and Liang et al. (2008) model a specific type of internal structure where DMUs are composed of a two-stage process, namely the output measures from the first stage become input measures to the second stage. Tone and Tsutsui (2009) develop slacks-based network DEA model. There are other variations or extensions to the above earlier work on network DEA models, depending on the particular DMU network structures. Some are based upon the DEA envelopment form and some on the DEA multiplier form.

The current handbook serves as a complement to the *Handbook on Data Envelopment Analysis* (eds, Cooper et al. 2011) in an effort to extend the frontier of DEA research. It provides a comprehensive source for the state-of-the art DEA modeling on internal structures and network DEA.

Chapter 1 by Cook and Zhu provides a survey on two-stage network performance decomposition and modeling techniques. Chapter 2 by Chen et al. discusses the pitfalls in network DEA modeling. The authors point out that caution should be paid when models are developed based upon the envelopment or multiplier forms, because the usual duality (or equivalence) between the DEA envelopment and multiplier linear models is no longer true. Chapter 3 by Kao discusses efficiency decompositions in network DEA under three types of structures, namely series, parallel, and dynamic.

Chapter 4 by Chen, Cook and Zhu studies the determination of the network DEA frontier. In Chap. 5 the same authors then discuss additive efficiency decomposition in network DEA. Kao and Hwang present an approach in scale efficiency measurement in two-stage networks in Chap. 6. Sahoo, Zhu and Tone further discuss the scale efficiency decomposition in two stage networks in Chap. 7.

Chapter 8 by Du et al. offers a bargaining game approach to modeling two-stage networks. Chen et al. in Chap. 9 study shared resources and efficiency decomposition in two-stage networks. Chapter 10 by Chen introduces an approach to computing the technical efficiency scores for a dynamic production network and its sub-processes.

In Chap. 11 Tone and Tsutsui present a slacks-based network DEA. Chapter 12 by Li et al. discusses a DEA modeling technique for a two-stage network process where the inputs of the second stage include both the outputs from the first stage and additional inputs to the second stage.

Chapter 13 by Golany, Hackman and Passy presents an efficiency measurement methodology for multi-stage production systems. Färe' Grosskopf, and Whittaker in Chap. 14 discuss network DEA models, both static and dynamic. The discussion also explores various useful objective functions that can be applied to the models to find the optimal allocation of resources for processes within the black box that are normally invisible to DEA. Chapter 15 by Castelli and Pesenti provides a comprehensive review of various types of network DEA modeling techniques.

In Chap. 16, Cook et al. present shared resources models for deriving aggregate measures of bank-branch performance, with accompanying component measures that make up that aggregate value.

In Chap. 17, Cook et al. examine a set of manufacturing plants operating under a single umbrella, with the objective being to use the component or function measures to decide what might be considered as each plant's core business.

Chapter 18 by Cook et al. considers problem settings where there may be clusters or groups of DMUs that form a hierarchy. The specific case of a set of electric power plants is examined in this context.

Chapter 19 by Fukuyama and Weber models bad outputs in two-stage network DEA. Chapter 20 by Lewis presents an application of network DEA to performance measurement of Major League Baseball (MLB) teams. Lu et al. in Chap. 21 present an application of a two-stage network DEA model for examining the performance of 30 U.S. airline companies. Chapter 22 by Triantis presents two distinct network efficiency models that are applied to engineering systems.

Toronto, ON, Canada Wade D. Cook
Worcester, MA, USA Joe Zhu

References

Castelli, L., Pesenti, R., & Ukovich, W. (2004). DEA-like models for the efficiency evaluation of hierarchically structured units. *European Journal of Operational Research*, *154*(2), 465–476.

Cook, W. D., & Seiford, L. M. (2009). Data envelopment analysis (DEA) – Thirty years on. *European Journal of Operational Research*, *192*(1), 1–17.

Cooper, W. W., Seiford, L. M., & Zhu, J. (2011). *Handbook on data envelopment analysis* (2nd ed.). New York: Springer.

Färe, R., & Grosskopf, S. (1996). Productivity and intermediate products: A frontier approach. *Economics Letters*, *50*, 65–70.

Kao, C., & Hwang, S. N. (2008). Efficiency decomposition in two-stage data envelopment analysis: an application to non-life insurance companies in Taiwan. *European Journal of Operational Research*, *185*(1), 418–429.

Liang, L. Cook, W.D., & Zhu, J. (2008). DEA models for two-stage processes: Game approach and efficiency decomposition. *Naval Research Logistics*, *55*, 643–653.

Liu, J. S., Lu, L. Y., Lu, W.-M., & Lin, B. J. (2013). Data envelopment analysis 1978–2010: A citation based literature survey. *Omega*, *41*(1):3–15.

Tone, K., & Tsutsui, M. (2009), Network DEA: A slacks-based measure approach. *European Journal of Operational Research*, *197*(1), 243–252.

Contents

Chapter 1
DEA for Two-Stage Networks: Efficiency Decompositions and Modeling Techniques

Wade D. Cook and Joe Zhu

Abstract Data envelopment analysis (DEA) is a method for identifying best practices among peer decision making units (DMUs). An important area of development in recent years has been that devoted to applications wherein DMUs represent network processes. One particular subset of such processes is those in which all the outputs from the first stage become inputs to the second stage. We call these types of DMU structures "two-stage networks". Existing approaches in modeling efficiency of two-stage networks can be categorized as using either Stackelberg (leader-follower), or cooperative game concepts. There are two types of efficiency decomposition; multiplicative and additive. In multiplicative efficiency decomposition, the overall efficiency is defined as a product of the two individual stages' efficiency scores, whereas in additive efficiency decomposition, the overall efficiency is defined as a weighted average of the two individual stages' efficiency scores. We discuss modeling techniques used for solving two-stage network DEA models in linear programs.

Keywords Data envelopment analysis (DEA) • Efficiency • Decomposition • Game • Intermediate measure • Network • Cooperative • Two-stage

1.1 Introduction

Data envelopment analysis (DEA), introduced by Charnes et al. (1978), is an approach for identifying best practices among peer decision making units (DMUs) in the presence of multiple inputs and outputs. In many cases DMUs may consist of

W.D. Cook (✉)
Schulich School of Business, York University, Toronto, ON M3J 1P3, Canada
e-mail: wcook@schulich.yorku.ca

J. Zhu
School of Business, Worcester Polytechnic Institute, Worcester, MA 01609, USA
e-mail: jzhu@wpi.edu

W.D. Cook and J. Zhu (eds.), *Data Envelopment Analysis*,
International Series in Operations Research & Management Science 208,
DOI 10.1007/978-1-4899-8068-7_1, © Springer Science+Business Media New York 2014

two-stage network structures with intermediate measures. In other words, DMUs under evaluation share a common feature found in many two-stage network structures, namely that outputs from the first stage become the inputs to the second stage. We refer to these as intermediate measures. For example, Seiford and Zhu (1999) use a two-stage network structure to measure the profitability and marketability of US commercial banks. In their study, profitability is measured relative to labor and assets as inputs, and the outputs are profits and revenues. In the second stage, for marketability, the profits and revenue are then used as inputs, while market value, returns and earnings per share constitute the outputs. Zhu (2000) applies the same two-stage network structure to the Fortune Global 500 companies.

Seiford and Zhu (1999) use the standard DEA approach which does not address potential conflicts between the two stages arising from the intermediate measures. Namely, the second stage may have to reduce its inputs (intermediate measures) in order to achieve an 'efficient' status. Such an action would, however, imply a reduction in the first stage outputs, thereby reducing the efficiency of that stage.

Note that these types of DMUs have not only inputs and outputs, but also intermediate measures that flow from one stage to the other. Each stage may also have its own inputs and outputs. Recently, a number of studies have focused on DMUs that appear as two-stage processes. Kao and Hwang (2008) describe a two-stage process where 24 non-life insurance companies use operating and insurance expenses to generate premiums in the first stage, and then underwriting and investment profits in the second stage. Other examples include the impact of information technology use on bank branch performance (Chen and Zhu 2004), two stage Major League Baseball performance (Sexton and Lewis 2003), health care applications (Chilingerian and Sherman 2004), and many others.

Kao and Hwang (2008) define the overall efficiency of the DMU as the product of the efficiencies of the two stages. Such *multiplicative* efficiency decomposition is also studied in Liang et al. (2008), where three DEA models/efficiency decompositions are developed using game theory concepts. More recently, Chen et al. (2009b) present a methodology for representing overall radial efficiency of a DMU as an *additive* weighted average of the radial efficiencies of the individual stages or components that make up the DMU. Cook et al. (2010) extend the additive decomposition approach of Chen et al. (2009b) into more general network structures.

In a review study done by Cook et al. (2010), the authors classify various existing DEA models for measuring efficiency in the aforementioned two-stage network structures or processes. The models fall into four categories: standard DEA approach; efficiency decomposition approach; network DEA approach; and game theoretic approach. Except for the standard DEA approach, all other approaches attempt to correct for the above-referenced conflict issue existing between the two stages.

The rest of the chapter is organized as follows. Section 1.2 presents the generic two-stage process and a general literature review and classification of papers dealing with DMUs having such processes.

In Sects. 1.3 and 1.4, we discuss the efficiency decomposition methodology and game-theoretic approaches. We begin with the work by Liang et al. (2006) where DEA models are developed to measure the performance of supply chains with two members. In their study, because some of the inputs to the second stage are not from the first stage, one of the DEA models is non-linear. However, if we apply their approach to our two-stage processes, and use the overall efficiency definition from Kao and Hwang (2008), we can obtain linear DEA models as in Liang et al. (2008). This establishes the relationships among the works of Liang et al. (2006), Castelli et al. (2004), Kao and Hwang (2008) and Liang et al. (2008). These approaches are then re-categorized as (1.1) the centralized models of Kao and Hwang (2008) and Liang et al. (2008), and (1.2) the non-cooperative (or leader-follower) model. It is shown how to test for uniqueness of the efficiency decomposition.

We then proceed to the network DEA approach in Sect. 1.5. We show that the Kao and Hwang (2008) model and the centralized model of Liang et al. (2008) are equivalent to the network DEA approach of Färe and Grosskopf (1996). Note the fact that, as demonstrated in Chen et al. (2009a), Chen and Zhu's (2004) model under the CRS assumption is equivalent to the Kao and Hwang (2008) model. As a result, we establish the equivalence among these models in dealing with two-stage processes. We discuss as well the determination of the efficient frontier of the two-stage process. Since it is possible that no single DMU is efficient, the standard DEA projections can no longer be used to generate the frontier. See Chen et al. (2010a, 2013) on issues related to DEA frontier identification under network DEA models.

Section 1.6 presents a technique for solving non-linear network DEA models via linear programming problems. Such a technique is often used in additive efficiency decompositions (see, e.g., Liang et al. 2006, 2011, 2013). Section 1.7 discusses a two-stage network structure where outputs from the second stage can be fed back as inputs to the first stage (Liang et al. 2011). Conclusions appear in Sect. 1.8.

1.2 Classification of Network DEA Modeling

Consider a generic two-stage network structure or process as shown in Fig. 1.1. Using the notation of Chen and Zhu (2004), we assume each DMU_j ($j = 1, 2, \ldots, n$) has m inputs x_{ij}, ($i = 1, 2, \ldots, m$) to the first stage, and D outputs z_{dj}, ($d = 1, 2, \ldots, D$)

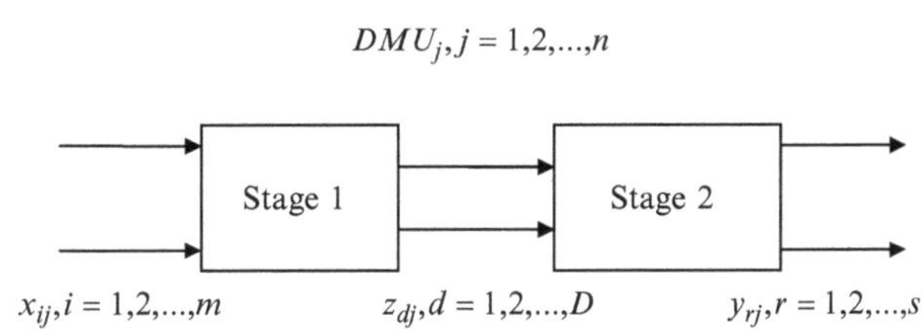

Fig. 1.1 Two-stage process

from that stage. These D outputs then become the inputs to the second stage and will be referred to as intermediate measures. The outputs from the second stage are y_{rj}, $(r = 1, 2, \ldots, s)$.

We denote the efficiency for the first stage as e_j^1 and second stage as e_j^2, for each DMU_j. Using the Constant Returns to Scale (CRS) DEA model of Charnes et al. (1978), we define

$$e_j^1 = \frac{\sum_{d=1}^{D} w_d z_{dj}}{\sum_{i=1}^{m} v_i x_{ij}} \quad \text{and} \quad e_j^2 = \frac{\sum_{r=1}^{s} u_r y_{rj}}{\sum_{d=1}^{D} \widetilde{w}_d z_{dj}} \tag{1.1}$$

where v_i, w_d, $\widetilde{w}_d$, and u_r are unknown non-negative weights. Note that w_d can be equal to $\widetilde{w}_d$.

There are four types of papers that use various approaches to the modeling of efficiency of DMUs with two-stage processes. Some approaches are equivalent.

1.2.1 Standard DEA Methodology

The first type simply uses the standard DEA model. i.e. two separate DEA runs are applied to the two stages to calculate e_j^1 and e_j^2, respectively. For example, Chilingerian and Sherman (2004) describe a two-stage process in measuring physician care. Their first stage is a manager-controlled process with inputs including registered nurses, medical supplies, and capital and fixed costs. These inputs generate the outputs or intermediate measures (inputs to the second stage), including patient days, quality of treatment, drugs dispensed, among others. The outputs of the second (physician controlled) stage include research grants, quality of patients, and quantity of individuals trained, by specialty. Other examples include Fortune 500 companies performance (Seiford and Zhu 1999; Zhu 2000). Similar to Seiford and Zhu (1999), Sexton and Lewis (2003) also use the standard DEA approach where in one of their standard DEA models, projected (efficient) intermediate measures are used in the second stage efficiency calculation.

However, as discussed earlier, such an approach does not treat z_{dj} in a coordinated manner. For example, suppose the first stage is DEA efficient and the second stage is not. When the second stage improves its performance, by reducing the inputs z_{dj} via an input-oriented DEA model, the reduced z_{dj} may render the first stage inefficient.

1.2.2 Efficiency Decomposition Methodology

It is useful to point out that given individual efficiency measures e_j^1 and e_j^2, for stages 1 and 2, respectively, it is reasonable to define the efficiency of the overall two-stage process either as $\frac{1}{2}\left(e_j^1 + e_j^2\right)$ or $e_j^1 \bullet e_j^2$. If the input-oriented DEA model is used, then we should as well require that $e_j^1 \le 1$ and $e_j^2 \le 1$. The above definition ensures that the two-stage process is efficient if and only if $e_j^1 = e_j^2 = 1$.

If we define $e_j = \frac{\sum_{r=1}^{s} u_r y_{ro}}{\sum_{i=1}^{m} v_i x_{io}}$ as the two-stage overall efficiency, then we arrive at another type of research, as in Kao and Hwang (2008) who describe a two-stage process where 24 non-life insurance companies use operating and insurance expenses to generate premiums in the first stage, and then underwriting and investment profits in the second stage. As in Kao and Hwang (2008), we have $e_j = e_j^1 \bullet e_j^2$ at optimality provided we assume $w_d = \widetilde{w}_d$. Note that such a decomposition of efficiency is not available in the standard DEA approach, and the network DEA approaches.

1.2.3 Network DEA

We point out that in these above examples, it is the case that the intermediate measures are the *only* inputs to the second stage, i.e., there are no additional independent inputs to that stage. There are, of course, other types of two-stage processes and even DMUs with network structures that may have inputs to the second stage in addition to the intermediate measures. In a more general situation than two-stage processes, Castelli et al. (2004) discuss DMUs with two-stage and two-layer structures. The network DEA approach of Färe and Whittaker (1995) and Färe and Grosskopf (1996), and the slacks-based network DEA approach of Tone and Tsutsui (2009, 2010) may involve more than two stages. Fukuyama and Weber (2010) considers a slacks-based measure for a two-stage process with bad outputs. More recently, Chen (2009) developed a network DEA model incorporating dynamic effects in production networks. A number of empirical studies have used this type of DEA technique, see, e.g., Avkiran (2009), and Yu and Lin (2008), among others. We call these network DEA approaches.

Similar network DEA approaches are used in two-stage processes described in Fig. 1.1. For example, Chen and Zhu (2004) study the impact of information technology use on bank branches performance (Wang et al. 1997). Under the assumption of variable returns to scale (VRS), Chen and Zhu (2004) and

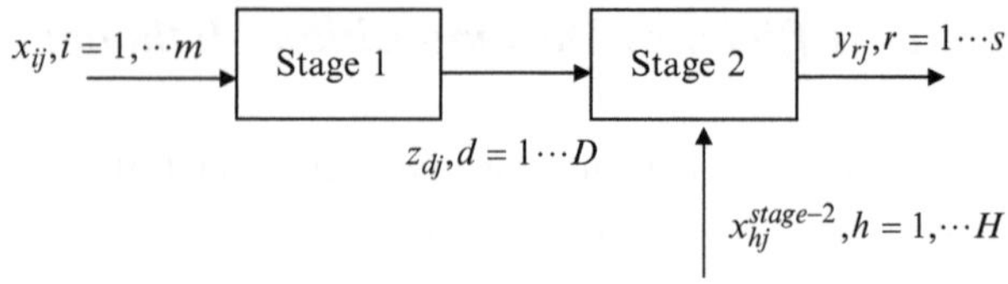

Fig. 1.2 Two-stage process with additional inputs to the second stage

Chen et al. (2006) develop linear and non-linear models for measuring the impact of information technology on the firm performance via a two-stage process. However, their individual stage efficiency scores do not provide information on the overall performance and best-practices of the two-stage process.

1.2.4 Game-Theoretic Approaches

The fourth type of approach uses game theory concepts. It originates from the work of Liang et al. (2006) who use DEA to measure the performance of supply chains with two members (as in a manufacturer-retailer setting, for example). In Liang et al. (2006), the concepts of the Stackelberg game (or leader-follower) and the cooperative game are used to develop models for measuring performance in supply chain settings. We should point out that in their paper, the second stage(retailer) has not only the inputs from the first stage (manufacturer), but also its own inputs not linked with the first stage, i.e. additional inputs to the second stage are introduced (see, for example, Fig. 1.2 above). As a results, $e_j^2 = \dfrac{\sum_{r=1}^{s} u_r y_{rj}}{\sum_{d=1}^{D} \widetilde{w}_d z_{dj} + \sum_{h=1}^{H} Q_h x_{hj}^2}$, where x_{hj}^2 $(h = 1, \ldots, H)$ are inputs to the second stage that are not related to the first stage. In this case, it may be more convenient and tractable to express the overall efficiency as $\frac{1}{2}\left(e_j^1 + e_j^2\right)$, since the alternative, namely $e_j^1 \bullet e_j^2$, results in a highly non-linear problem.

We note that their models can actually be directly applied to the two-stage process described in Fig. 1.1, since if there are no additional inputs x_{hj}^2 $(h = 1, \ldots, H)$, the structure of their two-member supply chain is identical to the two-stage process shown. Liang et al. (2008) provide detailed models for the two-stage process using the same modeling principle as in Liang et al. (2006).

While the current chapter focuses on the two-stage processes that have only the intermediate measures linking the stages, we will discuss the relations among DEA models for specific two-stage processes, and for the more general network structures.

1.3 Centralized Model

Liang et al. (2006) show that using the concept of cooperative game theory, or centralized control, the two stage process can be viewed as one where the stages jointly determine a set of optimal weights on the intermediate factors to maximize their efficiency scores. This would be the case in situations where the manufacturer and retailer jointly determine prices, order quantities, etc., to achieve maximum profit (Huang and Li 2001). In other words, the cooperative or centralized approach is characterized by letting $w_d = \widetilde{w}_d$ in (1.1), and the efficiency scores of both stages are optimized simultaneously. The optimization can be based upon maximizing the average of e_o^1 and e_o^2 in a non-linear program as in Liang et al. (2006), Kao and Hwang (2008), and Liang et al. (2008). However, it is noted that because of the assumption $w_d = \widetilde{w}_d$ in (1.1), $e_o^1 \bullet e_o^2$ becomes $\frac{\sum_{r=1}^{s} u_r y_{ro}}{\sum_{i=1}^{m} v_i x_{io}}$. Therefore, instead of maximizing the average of e_o^1 and e_o^2, we have

$$e_o^{centralized} = Max\ e_o^1 \bullet e_o^2 = \frac{\sum_{r=1}^{s} u_r y_{ro}}{\sum_{i=1}^{m} v_i x_{io}} \tag{1.2}$$
$$s.t.\ \ e_j^1 < 1 \quad \text{and} \quad e_j^2 < 1 \quad \text{and} \quad w_d = \widetilde{w}_d.$$

Model (1.2) can be converted into the following linear program format:

$$\begin{aligned} e_o^{centralized} = \text{Max} & \sum_{r=1}^{s} u_r y_{ro} \\ \text{s.t.} \quad & \sum_{r=1}^{s} u_r y_{rj} - \sum_{d=1}^{D} w_d z_{dj} \le 0 \quad j = 1, 2, \ldots., n \\ & \sum_{d=1}^{D} w_d z_{dj} - \sum_{i=1}^{m} v_i x_{ij} \le 0 \quad j = 1, 2, \ldots., n \\ & \sum_{i=1}^{m} v_i x_{io} = 1 \\ & w_d \ge 0, \quad d = 1, 2, \ldots, D; \quad v_i \ge, \quad i = 1, 2, \ldots, m; \quad u_r \ge, \\ & \quad r = 1, 2, \ldots, s \end{aligned} \tag{1.3}$$

Model (1.3) is the Kao and Hwang (2008) model and the centralized model developed in Liang et al. (2008). Note that constraints $\sum_{r=1}^{s} u_r y_{rj} - \sum_{i=1}^{m} v_i x_{ij} \le 0$ are redundant in Kao and Hwang's (2008) model, since $\sum_{r=1}^{s} u_r y_{rj} - \sum_{d=1}^{D} w_d z_{dj} \leq 0$ and $\sum_{d=1}^{D} w_d z_{dj} - \sum_{i=1}^{m} v_i x_{ij} \leq 0$ imply $\sum_{r=1}^{s} u_r y_{rj} - \sum_{i=1}^{m} v_i x_{ij} \leq 0$.

Model (1.3) gives the overall efficiency of the two-stage process. Assume the above model (1.3) yields a unique solution. We can then obtain

$$e_o^{1,centralized} = \frac{\sum_{d=1}^{D} w_d^* z_{do}}{\sum_{i=1}^{m} v_i^* x_{io}} = \sum_{d=1}^{D} w_d^* z_{do} \quad \text{and} \quad e_o^{2,centralized} = \frac{\sum_{r=1}^{s} u_r^* y_{ro}}{\sum_{d=1}^{D} w_d^* z_{do}} \tag{1.4}$$

as the efficiencies for the first and second stages, respectively. If we denote the optimal value to model (1.3) as $e_o^{centralized}$, then we have $e_o^{centralized} = e_o^{1,centralized} \bullet e_o^{2,centralized}$.

If only one layer is considered in the internal structure of Castelli et al. (2004), then the same above efficiency decomposition can be obtained. Therefore, the approaches of Castelli et al. (2004) and Kao and Hwang (2008) can be viewed as cooperative game models.

As noted in Kao and Hwang (2008), optimal multipliers from model (1.3) may not be unique. They propose deriving the maximum achievable value of $e_o^{1,centralized}$ or $e_o^{2,centralized}$. In fact, as shown in Liang et al. (2008), their models can also be used to test whether $e_o^{1,centralized}$ and $e_o^{2,centralized}$, obtained from model (1.3), are unique. The maximum achievable value of $e_o^{1,centralized}$ can be determined via

$$\begin{aligned}
e_o^{1+} = \; & \text{Max} \sum_{d=1}^{D} w_d z_{do} \\
s.t. \quad & \sum_{r=1}^{s} u_r y_{ro} = e_o^{centralized} \\
& \sum_{d=1}^{D} w_d z_{dj} - \sum_{i=1}^{m} v_i x_{ij} \leq 0 \quad j = 1, 2, \ldots, n \\
& \sum_{r=1}^{s} u_r y_{rj} - \sum_{d=1}^{D} w_d z_{dj} \leq 0 \quad j = 1, 2, \ldots, n \\
& \sum_{i=1}^{m} v_i x_{io} = 1 \\
& w_d \geq 0, \quad d = 1, 2, \ldots, D; \quad v_i \geq 0, \quad i = 1, 2, \ldots, m; \quad u_r \geq 0, \\
& \quad r = 1, 2, \ldots, s
\end{aligned} \tag{1.5}$$

This yields the minimum of $e_o^{2,centralized}$, namely, $e_o^{2-} = \frac{e_o^{centralized}}{e_o^{1+}}$. The maximum of $e_o^{2,centralized}$ can be calculated via the following linear program,

$$
\begin{aligned}
e_o^{2+} = \text{Max} & \sum_{r=1}^{s} u_r y_{ro} \\
s.t. \quad & \sum_{r=1}^{s} u_r y_{ro} - e_o^{centralized} \bullet \sum_{i=1}^{m} v_i x_{io} = 0 \\
& \sum_{r=1}^{s} u_r y_{rj} - \sum_{d=1}^{D} w_d z_{dj} \leq 0 \quad j = 1, 2, \ldots., n \\
& \sum_{d=1}^{D} w_d z_{dj} - \sum_{i=1}^{m} v_i x_{ij} \leq 0 \quad j = 1, 2, \ldots., n \\
& \sum_{d=1}^{D} w_d z_{do} = 1 \\
& w_d \geq 0, \; d = 1, 2, \ldots, D; \; v_i \geq 0, i = 1, 2, \ldots, m; \; u_r \geq 0, \; r = 1, 2, \ldots, s,
\end{aligned}
\tag{1.6}
$$

and the minimum of $e_o^{1,centralized}$ is then calculated as $e_k^{1-} = e_o^{centralized}/e_o^{2+}$ Note that $e_o^{1-} = e_o^{1+}$ if and only of $e_o^{2-} = e_o^{2+}$. Note also if $e_o^{1-} = e_o^{1+}$ or $e_o^{2-} = e_o^{2+}$, then $e_o^{1,centralized}$ and $e_o^{2,centralized}$ are uniquely determined via model (1.3). If $e_o^{1-} \neq e_o^{1+}$ or $e_o^{2-} \neq e_o^{2+}$, Liang et al. (2008) develop a procedure to obtain an alternative decomposition of $e_o^{1,centralized}$ and $e_o^{2,centralized}$.

Table 1.1 presents data on 24 non-life insurance companies in Taiwan where there are two intermediate measures (Kao and Hwang 2008). The two inputs to the first stage (premium acquisition) are Operating expenses and Insurance expenses. The intermediate measures (or the outputs from the first stage) are Direct written premiums and Reinsurance premiums. The outputs of the second stage (profit generation) are Underwriting profit and Investment profit.

The efficiency scores for the two individual stages are calculated based upon (1.4) via a set of optimal solutions from model (1.3) (see the 2nd, 3rd and 4th columns of Table 1.2). Note that the efficiency decompositions are identical to those in Kao and Hwang (2008). In fact, the use of models (1.5) and (1.6) indicates that $e_o^{1-} = e_o^{1+}$ and $e_o^{2-} = e_o^{2+}$ for all the DMUs. Therefore, the $e_o^{1,centralized}$ and $e_o^{2,centralized}$ defined in (1.4), or the efficiency decompositions in Kao and Hwang (2008), are uniquely determined via model (1.3).

Table 1.1 Non-life insurance companies in Taiwan

DMU	Company	Operation expenses (x1)	Insurance expenses (x2)	Direct written premiums (z1)	Reinsurance premiums (z2)	Underwriting profit (y1)	Investment profit (y2)
1	Taiwan Fire	1,178,744	673,512	7,451,757	856,735	984,143	681,687
2	Chung Kuo	1,381,822	1,352,755	10,020,274	1,812,894	1,228,502	834,754
3	Tai Ping	1,177,494	592,790	4,776,548	560,244	293,613	658,428
4	China Mariners	601,320	594,259	3,174,851	371,863	248,709	177,331
5	Fubon	6,699,063	3,531,614	37,392,862	1,753,794	7,851,229	3,925,272
6	Zurich	2,627,707	668,363	9,747,908	952,326	1,713,598	415,058
7	Taian	1,942,833	1,443,100	10,685,457	643,412	2,239,593	439,039
8	Ming Tai	3,789,001	1,873,530	17,267,266	1,134,600	3,899,530	622,868
9	Central	1,567,746	950,432	11,473,162	546,337	1,043,778	264,098
10	The First	1,303,249	1,298,470	8,210,389	504,528	1,697,941	554,806
11	Kuo Hua	1,962,448	672,414	7,222,378	643,178	1,486,014	18,259
12	Union	2,592,790	650,952	9,434,406	1,118,489	1,574,191	909,295
13	Shing kong	2,609,941	1,368,802	13,921,464	811,343	3,609,236	223,047
14	South China	1,396,002	988,888	7,396,396	465,509	1,401,200	332,283
15	Cathay Century	2,184,944	651,063	10,422,297	749,893	3,355,197	555,482
16	Allianz President	1,211,716	415,071	5,606,013	402,881	854,054	197,947
17	Newa	1,453,797	1,085,019	7,695,461	342,489	3,144,484	371,984
18	AIU	757,515	547,997	3,631,484	995,620	692,731	163,927
19	North America	159,422	182,338	1,141,950	483,291	519,121	46,857
20	Federal	145,442	53,518	316,829	131,920	355,624	26,537
21	Royal & Sunalliance	84,171	26,224	225,888	40,542	51,950	6,491
22	Asia	15,993	10,502	52,063	14,574	82,141	4,181
23	AXA	54,693	28,408	245,910	49,864	0.1	18,980
24	Mitsui Sumitomo	163,297	235,094	476,419	644,816	142,370	16,976

Table 1.2 Results for non-life insurance companies in Taiwan

	Centralized model			Stage 1 as the leader			Stage 2 as the leader		
	$e_o^{1,Cooperative}$	$e_o^{2,Cooperative}$	$e_o^{Cooperative}$	$\theta_o^1(=e_o^{1*})$	e_o^{2*}	$e_o^{1*} \cdot e_o^{2*}$	$e_o^{1^o}$	$\theta_o^2\left(=e_o^{2^o}\right)$	$e_o^{1^o} \cdot e_o^{2^o}$
1	0.99257	0.70447	0.69923	0.99257	0.70447	0.69923	0.92622	0.71337	0.66074
2	0.9985	0.62571	0.62477	0.9985	0.62571	0.62477	0.99086	0.62748	0.62175
3	0.69002	1	0.69002	0.69002	1	0.69002	0.69002	1	0.69002
4	0.72435	0.41999	0.30422	0.72435	0.41999	0.30422	0.49792	0.43231	0.21526
5	0.83066	0.92334	0.76698	0.83752	0.8057	0.67478	0.73759	1	0.73759
6	0.96062	0.40566	0.38968	0.96369	0.40101	0.38645	0.96062	0.40566	0.38968
7	0.67064	0.41241	0.27658	0.75208	0.35216	0.26485	0.29991	0.53784	0.1613
8	0.66302	0.41503	0.27517	0.7256	0.37803	0.2743	0.38992	0.51135	0.19939
9	1	0.22328	0.22328	1	0.22328	0.22328	0.43904	0.29196	0.12818
10	0.86154	0.54084	0.46596	0.86154	0.54084	0.46596	0.25868	0.6736	0.17425
11	0.64679	0.25344	0.16392	0.74055	0.16753	0.12407	0.47185	0.32667	0.15414
12	1	0.75958	0.75958	1	0.75958	0.75958	1	0.75958	0.75958
13	0.67198	0.30925	0.20781	0.81068	0.24306	0.19705	0.33839	0.54349	0.18391
14	0.66992	0.43086	0.28864	0.72462	0.37396	0.27098	0.30964	0.51782	0.16034
15	1	0.61383	0.61383	1	0.61383	0.61383	0.71007	0.70473	0.50041
16	0.88558	0.36152	0.32015	0.9072	0.33557	0.30443	0.59872	0.38475	0.23035
17	0.62761	0.57363	0.36001	0.72331	0.45958	0.33242	0.2507	1	0.2507
18	0.79354	0.32619	0.25884	0.79354	0.32619	0.25884	0.65507	0.37366	0.24477
19	1	0.4112	0.4112	1	0.4112	0.4112	0.97884	0.41578	0.40697
20	0.93322	0.58566	0.54655	0.93322	0.58566	0.54655	0.40728	0.90137	0.36711
21	0.7321	0.27425	0.20078	0.75052	0.26232	0.19688	0.69178	0.27951	0.19336
22	0.58952	1	0.58952	0.58952	1	0.58952	0.58952	1	0.58952
23	0.84256	0.49889	0.42034	0.85005	0.45124	0.38358	0.68119	0.55992	0.38141
24	0.42869	0.31447	0.13481	1	0.08703	0.08703	0.39866	0.33509	0.13359

1.4 Stackelberg Game

In the previous section we examined the cooperative or centralized game approach to the two stage problem. In this section we look at the two-stage process from the perspective of the non-cooperative game. The non-cooperative approach is characterized by the leader-follower, or Stackelberg game. For example, consider a case of a supply chain where there is non-cooperative advertising on the part of the manufacture (leader) and the retailer (follower). The manufacturer determines its optimal brand name investment and local advertising allowance based on an estimation of the local advertisement by the retailer to maximize its profit. The retailer, as a follower on the other hand, based on the information from the manufacturer, determines the optimal local advertisement cost to maximize its profit (Huang and Li 2001).

In a similar manner, if we assume that the first stage is the leader, then the first stage performance is more important, and the efficiency of the second stage is computed subject to the requirement that the efficiency of the first stage is to stay fixed. We first calculate the efficiency for the first stage. Based upon the CRS model, we have for a specific DMU_o

$$
\begin{aligned}
e_o^{1*} &= \text{Max} \sum_{d=1}^{D} w_d z_{do} \\
\text{s.t.} \quad & \sum_{d=1}^{D} w_d z_{dj} - \sum_{i=1}^{m} v_i x_{ij} \leq 0 \quad j = 1, 2, ..., n \\
& \sum_{i=1}^{m} v_i x_{io} = 1 \\
& w_d \geq 0, \quad d = 1, 2, \ldots, D; \quad v_i \geq 0, \quad i = 1, 2, \ldots, m.
\end{aligned}
\tag{1.7}
$$

Note that model (1.7) is in fact the standard (CCR) DEA model. i.e., e_o^{1*} is the regular DEA efficiency score.

Once we obtain the efficiency for the first stage, the second stage will only consider w_d that maintains $e_o^1 = e_o^{1*}$. Or, in other words, the second stage now treats $\sum_{d=1}^{D} w_d z_{dj}$ as the "single" input subject to the restriction that the efficiency score of the first stage remains at e_o^{1*}. The model for computing e_o^2, the second stage's efficiency, can be calculated as (Liang et al. 2008)

$$
\begin{aligned}
e_o^{2*} = \mathrm{Max} & \frac{\sum_{r=1}^{s} U_r y_{ro}}{Q\sum_{d=1}^{D} w_d z_{do}} \\
s.t. \quad & \frac{\sum_{r=1}^{s} U_r y_{rj}}{Q\sum_{d=1}^{D} w_d z_{dj}} \le 1 \quad j = 1, 2, \ldots.., n \\
& \sum_{d=1}^{D} w_d z_{dj} - \sum_{i=1}^{m} v_i x_{ij} \le 0 \quad j = 1, 2, \ldots.., n \\
& \sum_{i=1}^{m} v_i x_{io} = 1 \\
& \sum_{d=1}^{D} w_d z_{do} = e_o^{1*} \\
& U_r, Q, w_d, v_i \ge 0, \quad r = 1, 2, \ldots, s; \quad d = 1, 2, \ldots, D; \quad i = 1, 2, \ldots, m
\end{aligned}
\tag{1.8}
$$

Note that in model (1.8), the efficiency of the first stage is set equal to e_o^{1*}. Let $u_r = \frac{U_r}{Q}, \quad r = 1, 2, \ldots, s$. Model (1.8) is then equivalent to the following linear model

$$
\begin{aligned}
e_o^{2*} = \mathrm{Max} & \left(\sum_{r=1}^{s} u_r y_{ro}\right) / e_o^{1*} \\
s.t. \quad & \sum_{r=1}^{s} u_r y_{rj} - \sum_{d=1}^{D} w_d z_{dj} \le 0 \quad j = 1, 2, \ldots.., n \\
& \sum_{d=1}^{D} w_d z_{dj} - \sum_{i=1}^{m} v_i x_{ij} \le 0 \quad j = 1, 2, \ldots.., n \\
& \sum_{i=1}^{m} v_i x_{io} = 1 \\
& \sum_{d=1}^{D} w_d z_{do} = e_o^{1*} \\
& w_d \ge 0, \; d = 1, 2, \ldots, D; \; v_i \ge 0, \; i = 1, 2, \ldots, m; \; u_r \ge 0, \; r = 1, 2, \ldots, s
\end{aligned}
\tag{1.9}
$$

In a similar manner, if we take the second stage as the leader, we then calculate the regular DEA efficiency ($e_o^{2^o}$) for the second stage first using the CCR model.

Once we obtain the second stage efficiency, the efficiency for the first stage, namely $e_o^{1^o}$, is calculated via the following linear program (see Liang et al. 2008)

$$
\begin{aligned}
\frac{1}{e_o^{1^o}} &= \text{Min} \sum_{i=1}^{m} v_i x_{io} \\
s.t. \quad & \sum_{d=1}^{D} w_d z_{dj} - \sum_{i=1}^{m} v_i x_{ij} \leq 0 \quad j = 1, 2, \ldots, n \\
& \sum_{r=1}^{s} u_r y_{rj} - \sum_{d=1}^{D} w_d z_{dj} \leq 0 \quad j = 1, 2, \ldots, n \\
& \sum_{d=1}^{D} w_d z_{do} = 1 \\
& \sum_{r=1}^{s} u_r y_{ro} = e_o^{2^o} \\
& w_d \geq 0, \;\; d = 1, 2, \ldots, D; \;\; v_i \geq 0, \;\; i = 1, 2, \ldots, m; \;\; u_r \geq 0, \;\; r = 1, 2, \ldots, s
\end{aligned}
\tag{1.10}
$$

We note that in (1.9), $e_o^{1*} \bullet e_o^{2*} = \sum_{r=1}^{s} u_r^* y_{ro}$ at optimality, with $\sum_{i=1}^{m} v_i^* x_{io} = 1$. i.e., $e_o^{1*} \bullet e_o^{2*} = \frac{\sum_{r=1}^{s} u_r^* y_{ro}}{\sum_{i=1}^{m} v_i^* x_{io}}$. Note also that at optimality, $\frac{\sum_{r=1}^{s} u_r^* y_{ro}}{\sum_{i=1}^{m} v_i^* x_{io}} = e_o^{1^o} \bullet e_o^{2^o}$ in model (1.10). This indicates that the leader-follower approach also implies an efficiency decomposition for the two-stage process. i.e., the overall efficiency is a product of efficiencies of individual stages. Further, note that in the first-stage leader case, e_o^{1*} and e_o^{2*}, and in the second-stage leader case, $e_o^{1^o}$ and $e_o^{2^o}$, are optimal values to linear programs. Therefore, such efficiency decomposition is unique, and is not affected by possible multiple optimal solutions. However, the two approaches may not yield the same efficiency decomposition.

Note that ultimately, a common set of weights is used at both stages in both centralized and Stackelberg game approaches. However, in the Stackelberg game approach, the efficiency scores of two stages, e_o^1 and e_o^2, are not optimized simultaneously.

Liang et al. (2008) also study the relationships among non-cooperative and centralized models and the standard DEA approach. We here summarize their findings.

Let θ_o^1 and θ_o^2 be the standard CRS efficiency scores for the two stages.

Theorem 1 *If there is only one intermediate measure, then* $e_o^{1*} = \theta_o^1$ *and* $e_o^{2*} = \theta_o^2$ *regardless of the assumption of whether the first stage is a leader or follower, where* e_o^{1*} *and* e_o^{2*} *are obtained via the non-cooperative approach.*

Theorem 1 indicates that when there is only one intermediate measure, the non-cooperative approach yields the same result as applying the standard DEA model to each stage.

Under the condition of multiple intermediate measures, we have

Theorem 2 *For a specific* DMU_o, $e_o^{centralized} \geq e_o^{1*} \bullet e_o^{2*}$, *where* $e_o^{centralized}$ *is the optimal value to model (1.3), and* e_o^{1*} *and* e_o^{2*} *are obtained via the non-cooperative (leader-follower) approach.*

Based upon Theorems 1 and 2, we must have

Theorem 3 *If there is only one intermediate measure, then* $e_o^{centralized} = \theta_o^1 \bullet \theta_o^2$ *with* $\theta_o^1 = e_o^{1,centralized}$ *and* $\theta_o^2 = e_o^{2,centralized}$, *where* θ_o^1 *and* θ_o^2 *are the CRS efficiency scores for the two stages, respectively, and* $e_o^{1,centralized}$ *and* $e_o^{2,centralized}$ *are defined in (1.4).*

When there is only one intermediate measure, Theorem 3 indicates that (i) both the non-cooperative and centralized models yield the same result as applying the standard DEA model to each stage, and (ii) the efficiency decomposition is unique.

We finally note that the following is true with respect to the relations between the non-cooperative and centralized approaches.

Theorem 4

(i) $e_o^{1,centralized} \geq e_0^{1*}$ *and* $\theta_o^2 (= e_o^{2*}) \geq e_o^{2,centralized}$ *when the second stage is the leader,*

(ii) $e_o^{2,centralized} \geq e_o^{2*}$ *and* $\theta_o^1 (= e_o^{1*}) > e_o^{1,centralized}$ *when the first stage is the leader.*

The results in Table 1.2 also verify Theorems 2 and 4. We finally note that $e_o^{centralized} = e_o^{1*} \bullet e_o^{2*}$ holds for 12 DMUs (50 % of the companies), where e_o^{1*} and e_o^{2*} represent the efficiency scores for the two stages when the first stage is treated as the leader. Note also that $e_o^{centralized} = e_o^{1^o} \bullet e_o^{2^o}$ holds for only one DMU, namely DMU 6, where $e_o^{1^o}$ and $e_o^{2^o}$ represent the efficiency scores for the two stages when the second stage is treated as the leader. This may indicate that the first stage or the premium-generating stage is more important.

1.5 Network DEA

If we model the two-stage process shown in Fig. 1.1 using the network approach of Färe and Grosskopf (1996), we have

$$
\begin{aligned}
&\min_{\Theta,\lambda_j,\mu_j,\widetilde{z}} \Theta \\
&\textit{subject to} \\
&(\text{stage } 1) \\
&\sum_{j=1}^{n} \lambda_j x_{ij} \le \Theta x_{ij_o} \quad i = 1, \ldots, \mathrm{m} \\
&\sum_{j=1}^{n} \lambda_j z_{dj} \ge \widetilde{z}_{dj_o} \quad d = 1, \ldots, \mathrm{D} \\
&\lambda_j \ge 0, \quad j = 1, \ldots, n \\
&(\text{stage } 2) \\
&\sum_{j=1}^{n} \mu_j z_{dj} \le \widetilde{z}_{dj_o} \quad d = 1, \ldots, \mathrm{D} \\
&\sum_{j=1}^{n} \mu_j y_{rj} \ge y_{rj_o} \quad r = 1, \ldots, \mathrm{s} \\
&\mu_j \ge 0, \quad j = 1, \ldots, n
\end{aligned} \tag{1.11}
$$

where $\widetilde{z}_{dj_o}$ are set as decision variables related to the intermediate measures.

Model (1.11) is equivalent to the following model

$$
\begin{aligned}
&\min_{\Theta,\lambda_j,\mu_j,\widetilde{z}} \Theta \\
&\textit{subject to} \\
&\sum_{j=1}^{n} \lambda_j x_{ij} \le \Theta x_{ij_o} \quad i = 1, \ldots, \mathrm{m} \\
&\sum_{j=1}^{n} (\lambda_j - \mu_j) z_{dj} \ge 0 \quad d = 1, \ldots, \mathrm{D} \\
&\sum_{j=1}^{n} \mu_j y_{rj} \ge y_{rj_o} \quad r = 1, \ldots, \mathrm{s} \\
&\lambda_j, \mu_j \ge 0, \quad j = 1, \ldots, n
\end{aligned} \tag{1.12}
$$

Model (1.12) is the dual to the centralized model (1.3). Therefore, the network DEA approach of Färe and Grosskopf (1996) yields results equivalent to the centralized model (1.3) of Liang et al. (2008) and Kao and Hwang (2008).

Chen et al. (2009a) show that the following CRS version of the Chen and Zhu's (2004) model is equivalent to model (1.3). (If we add the convexity constraints $\sum \lambda_j = \sum \mu_j = 1$ into model (1.13), then model (1.13) becomes the original Chen and Zhu (2004) model under the variable returns to scale assumption.)

$$
\begin{aligned}
&\min_{\alpha,\beta,\lambda_j,\mu_j,\tilde{z}} \alpha - \beta \\
&\textit{subject to} \\
&\text{(stage 1)} \\
&\sum_{j=1}^{n} \lambda_j x_{ij} \le \alpha x_{ij_o} \quad i = 1, \ldots, \mathrm{m} \\
&\sum_{j=1}^{n} \lambda_j z_{dj} \ge \tilde{z}_{dj_o} \quad d = 1, \ldots, \mathrm{D} \\
&\lambda_j \ge 0, \quad j = 1, \ldots, n \\
&\alpha \le 1 \\
&\text{(stage 2)} \\
&\sum_{j=1}^{n} \mu_j z_{dj} \le \tilde{z}_{dj_o} \quad d = 1, \ldots, \mathrm{D} \\
&\sum_{j=1}^{n} \mu_j y_{rj} \ge \beta y_{rj_o} \quad r = 1, \ldots, \mathrm{s} \\
&\mu_j \ge 0, \quad j = 1, \ldots, n \\
&\beta \ge 1
\end{aligned}
\tag{1.13}
$$

Thus, since both the network DEA model (1.11) and model (1.13) are equivalent to model (1.3), they ((1.11) and (1.13)) must then be equivalent to each other. This implies that $\beta = 1$ at optimality in model (1.13).

Chen et al. (2010a) demonstrate that the centralized model (1.3) may not yield information on the efficient frontier of the two-stage process in Fig. 1.1. In other words, due to the existence of intermediate measures, the usual procedure of adjusting the inputs or outputs by the efficiency scores obtained from model (1.3), as in the standard DEA approach, does not necessarily yield a frontier projection.

We note that the network DEA approach only provides information on the overall efficiency of the two-stages, and does not yield information on the individual stages. However, the equivalence between models (1.11) and (1.13) indicates that the network DEA approach generates an efficient frontier point, since model (1.13) ensures that a frontier point is obtained if $\alpha < 1$ in optimality. See Chen et al. (2010a).

1.6 Searching for the Global Optimal Solution

While in the previous sections, the DEA models can be converted into linear programs due to the specific nature of two-stage network processes depicted in Fig. 1.1. A slight modification to Fig. 1.1, for example, by introducing additional (independent) inputs to the second stage, the resulting models are not necessarily linear.

For example, Li et al. (2012) considers a situation depicted in Fig. 1.2, where inputs to the second stage are denoted as $x_{hj}^{stage-2}$ (h = 1, 2, …, H).

Model (1.2) now becomes

$$\begin{aligned}
&\max \theta_1^{o*} \theta_2^o = \max \frac{\sum_{d=1}^{D} w_d z_{do}}{\sum_{i=1}^{m} v_i x_{io}} * \frac{\sum_{r=1}^{s} u_r y_{ro}}{\sum_{d=1}^{D} w_d z_{do} + \sum_{h=1}^{H} Q_h x_{ho}^{stage-2}} \\
&s.t. \quad \frac{\sum_{d=1}^{D} w_d z_{dj}}{\sum_{i=1}^{m} v_i x_{ij}} \le 1 \quad \forall j \\
&\qquad \frac{\sum_{r=1}^{s} u_r y_{rj}}{\sum_{d=1}^{D} w_d z_{dj} + \sum_{h=1}^{H} Q_h x_{hj}^{stage-2}} \le 1 \quad \forall j \\
&\qquad v_i, w_d, Q_h, u_r \ge 0, \forall i, d, h, r
\end{aligned} \tag{1.14}$$

where θ_1^o and θ_2^o represent the ratio efficiencies for stages 1 and 2, respectively. Due to the additional inputs to the second stage $\left(\sum_{h=1}^{H} Q_h x_{ho}^{stage-2}\right)$, model (1.13) cannot be converted into a linear program. Li et al. (2012) introduce a heuristic method to solve this problem. In fact, such a heuristic method can be found in Liang et al. (2006, 2011), and Du et al. (2011). Further, such a heuristic method can be used when an additive form of objective function of model (1.2) is used, namely, Lim and Zhu (2013)

$$\begin{aligned}
&\max \quad \frac{\sum_{d=1}^{D} w_d z_{do}}{\sum_{i=1}^{m} v_i x_{io}} + \frac{\sum_{r=1}^{s} u_r y_{ro}}{\sum_{d=1}^{D} w_d z_{do}} \\
&\text{s.t.} \quad \frac{\sum_{d=1}^{D} w_d z_{dj}}{\sum_{i=1}^{m} v_i x_{ij}} \le 1, \quad \text{j} = 1, \ldots, \text{n} \\
&\qquad \frac{\sum_{r=1}^{s} u_r y_{rj}}{\sum_{d=1}^{D} w_d z_{dj}} \le 1 \quad \text{j} = 1, \ldots, \text{n} \\
&\qquad w_d \ge 0, \ d = 1, 2, \ldots, D; \ v_i \ge, \ i = 1, 2, \ldots, m; \ u_r \ge, \ r = 1, 2, \ldots, s
\end{aligned} \tag{1.15}$$

Next we apply the following transformation to model (1.15):

$$t_1 = \frac{1}{\sum v_i x_{io}}, \quad t_2 = \frac{1}{\sum w_d z_{do}}$$

$$\omega_i = t_1 v_i, c_d^1 = t_1 w_d, \quad \mu_r = t_2 u_r, \quad c_d^2 = t_2 w_d$$

Note that in the above transformation, $c_d^1 = t_1 w_d$ and $c_d^2 = t_2 w_d$ imply a linear relationship between c_d^1 and c_d^2. Therefore, we can assume $c_d^2 = hc_d^1$, where h is a positive real number. Then model (1.15) (Cook and Hababou 2001) can be transformed into:

$$\begin{aligned}
\max \ & \sum_{d=1}^{D} c_d^1 z_{do} + \sum_{r=1}^{s} \mu_r y_{ro} \\
\text{s.t.} \ & \sum_{d-1}^{D} c_d^1 z_{ij} \leq \sum_{i=1}^{m} \omega_i x_{ij} \\
& \sum_{r=1}^{s} \mu_r y_{rj} \leq \sum_{d=1}^{D} c_d^2 z_{ij} \\
& \sum_{i=1}^{m} \omega_i x_{io} = 1 \\
& \sum_{d=1}^{D} c_d^2 z_{io} = 1 \\
& c_d^2 = hc_d^1 \\
& c_d^1, c_d^2 \geq 0, \ d = 1, 2, \ldots, D; \ \omega_i \geq , \ i = 1, 2, \ldots, m; \\
& \quad \mu_r \geq , \ r = 1, 2, \ldots, s, h > 0
\end{aligned} \tag{1.16}$$

Taking h as a parameter, model (1.16) can be considered a parametric linear program, and it can be solved via a simple line search method over a certain range of h, as is done in Liang et al. (2006).

To secure a search range of h, it is needed to calculate a lower bound and an upper bound on h, which can be done similarly with Liang et al. (2006). See Lim and Zhu (2013) for a detailed discussion and a numerical example.

1.7 Two Stage Network System with Feedbacks

Liang et al. (2011) extend the basic two-stage network system as shown in Fig. 1.1 to include situations where outputs from the second stage can be fed back as inputs to the first stage. Such feed-back variables thus serve a dual role. Liang et al (2011) develop models for examining performance under such feedback setting (see Fig. 1.3).

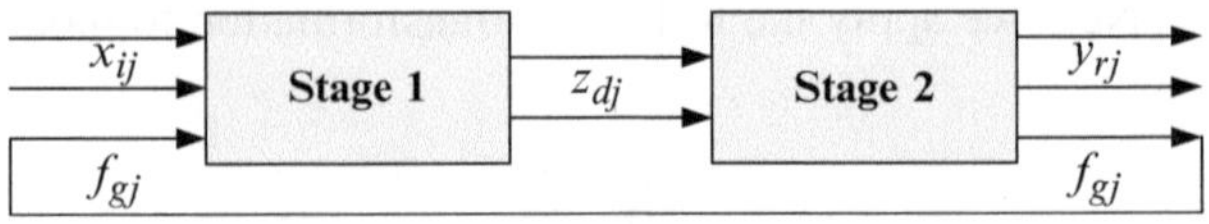

Fig. 1.3 Two-stage process with feedback

Note that both intermediate variables and the feed-back variables constitute what Cook et al. (2006) and Cook and Zhu (2007) call dual-role variables, in that they play a role of both output and input simultaneously.

To gain a sense of how feedback, as pictured in Fig. 1.3, might materialize in a practical setting, consider an example involving highway resurfacing. Liang et al. (2011) point out many highway resurfacing operations can be thought of as constituting a multi stage process. The way it often works is as follows: A machine passes over the old pavement and planes or grinds off the top damaged layer of asphalt. Generally, a truck follows behind the grinding machine, and the used asphalt is deposited into that truck, which then takes it off to a storage area. At a somewhat later time (the same day or even much later), a paving machine passes over the piece of highway, and the resurfacing process takes place. That is, asphalt is brought from a processing plant and deposited into a paving machine that runs over the road and puts the new surface on.

The resurfacing operation described, might be viewed as a two-stage process as follows:

Stage 1 might be regarded as the asphalt plant where the raw material (asphalt) is prepared. Note that in most cases new asphalt is prepared and then mixed with used asphalt (from a storage site).Using reclaimed asphalt has become a common practice. This is a less expensive way to resurface highways, and the resulting quality is nearly as good as using all new materials. There are recommended percentages regarding the mix of old and new asphalt to get varying life spans from the resurfaced structure. The inputs to this first stage are then the materials, including the old asphalt from the storage area. The output is the prepared asphalt ready to be taken to the resurfacing site.

Stage 2 is the actual resurfacing operation. Inputs include the prepared asphalt from Stage 1, together with a grinding machine, paving machine, manpower, etc. The output from this stage is the reclaimed old pavement from the grinding process, and the resulting resurfaced piece of road created by the paving machine. Clearly, one might choose to call the grinding process Stage 2 and the paving process that follows, Stage 3.

Note that in this problem one does not normally use the reclaimed pavement from stage 2 (that is taken off to a storage area) as the input to stage 1, but rather old pavement that went to the storage area earlier, from **some other** paving job. This is because all processes where finished goods become part of the input to make more finished goods (feedback) appear as a time series. Nothing is really fed **back** to the past but fed **forward** to the future.

Another example is bank branch operations where employees and assets are used to generate deposits in the first stage, which become inputs to a second stage which generates loans and profits as outputs. Part of the profits generated in the second stage is then re-invested in the first stage, as capital to finance the ongoing operation, and constituting a form of self-regenerative process.

Liang et al. (2011) develop a base model where the (additive) average of two stages' efficiency scores is defined as the overall efficiency as in Chen et al. (2009b). This base average model can be regarded as a centralized model (Liang et al. 2008). To adopt the concept of leader-follower approach, Liang et al. (2011) then develop a bi-level model.

In addition to the notation used in the previous sections related to Fig. 1.1, we have outputs f_{gj} $(g = 1, 2, \ldots, p)$ that flow back to stage 1 and become part of the set of inputs to the first stage. We refer to the f_{gj} as feedback variables. The z_{dj} and f_{gj} can then be regarded as dual-role variables in the sense of Cook et al. (2006) and Cook and Zhu (2007).

Using additive efficiency decomposition, model (1.2) becomes

$$
\begin{aligned}
& Max\ \frac{1}{2}\left(\frac{\sum_{d=1}^{D} h_d z_{d0}}{\sum_{i=1}^{m} v_i x_{i0} + \sum_{g=1}^{p} w_g f_{g0}} + \frac{\sum_{g=1}^{p} w_g f_{g0} + \sum_{r=1}^{s} u_r y_{r0}}{\sum_{d=1}^{D} h_d z_{d0}}\right) \\
& s.t.\ \frac{\sum_{d=1}^{D} h_d z_{dj}}{\sum_{i=1}^{m} v_i x_{ij} + \sum_{g=1}^{p} w_g f_{gj}} \le 1 \\
& \quad\ \frac{\sum_{g=1}^{p} w_g f_{gj} + \sum_{r=1}^{s} u_r y_{rj}}{\sum_{d=1}^{D} h_d z_{dj}} \le 1 \\
& \quad\ v_i, w_g, h_d, u_r \ge 0
\end{aligned}
\tag{1.17}
$$

where v_i, u_r, w_g and h_d are unknown weights. Note that it is assumed each stage applies the same weights h_d on the intermediate measures and w_g on the feedback variables f_{gj}.

To create a more tractable version of (1.17) (which, in its current form, is highly nonlinear), we propose the following change of variables in the spirit of Charnes and Cooper (1962). Specifically, let $t_1 = \frac{1}{\sum_{i=1}^{m} v_i x_{i0} + \sum_{g=1}^{p} w_g f_{g0}}$, $t_2 = \frac{1}{\sum_{d=1}^{D} h_d z_{d0}}$, and $\beta = \frac{t_1}{t_2}$, $\nu_i = t_1 v_i$, $\varsigma_d^1 = t_1 h_d$, $\omega_g^1 = t_1 w_g$, $\mu_r = t_2 u_r$, $\varsigma_d^2 = t_2 h_d$, $\omega_g^2 = t_2 w_g$. Note that $\frac{\varsigma_d^1}{\varsigma_d^2} = \frac{\omega_g^1}{\omega_g^2} = \frac{t_1}{t_2} = \beta$, and then model (1.17) is equivalent to the following problem:

$$
\begin{aligned}
Max\ & \frac{1}{2}\left(\beta + \sum_{g=1}^{p} \omega_g^2 f_{g0} + \sum_{r=1}^{s} \mu_r y_{r0}\right) \\
s.t.\ & \beta \sum_{d=1}^{D} \varsigma_d^2 z_{dj} - \sum_{i=1}^{m} v_i x_{ij} - \beta \sum_{g=1}^{p} \omega_g^2 f_{gj} \le 0 \\
& \sum_{g=1}^{p} \omega_g^2 f_{gj} + \sum_{r=1}^{s} \mu_r y_{rj} - \sum_{d=1}^{D} \varsigma_d^2 z_{dj} \le 0 \\
& \sum_{i=1}^{m} v_i x_{i0} + \beta \sum_{g=1}^{p} \omega_g^2 f_{g0} = 1 \\
& \sum_{d=1}^{D} \varsigma_d^2 z_{d0} = 1 \\
& \nu_i, \varsigma_d^2, \omega_g^2, \mu_r \ge 0
\end{aligned}
\tag{1.18}
$$

Note that $\beta = \frac{t_1}{t_2} = \frac{\sum_{d=1}^{D} h_d z_{d0}}{\sum_{i=1}^{m} v_i x_{i0} + \sum_{g=1}^{p} w_g f_{g0}} \le 1$, and $\beta > 0$, i.e. $\beta \in (0,\ 1]$. Now, let $\nu_i = \beta \upsilon_i$, $\varsigma_d = \varsigma_d^2$, $\omega_g = \omega_g^2$. Model (1.18) is then equivalent to the following model:

$$
\begin{aligned}
Max\ & \frac{1}{2}\left(\beta + \sum_{g=1}^{p} \omega_g f_{g0} + \sum_{r=1}^{s} \mu_r y_{r0}\right) \\
s.t.\ & \sum_{d=1}^{D} \varsigma_d z_{dj} - \sum_{i=1}^{m} \upsilon_i x_{ij} - \sum_{g=1}^{p} \omega_g f_{gj} \le 0 \\
& \sum_{g=1}^{p} \omega_g f_{gj} + \sum_{r=1}^{s} \mu_r y_{rj} - \sum_{d=1}^{D} \varsigma_d z_{dj} \le 0 \\
& \beta\left(\sum_{i=1}^{m} \upsilon_i x_{i0} + \sum_{g=1}^{p} \omega_g f_{g0}\right) = 1 \\
& \sum_{d=1}^{D} \varsigma_d z_{d0} = 1 \\
& \nu_i, \varsigma_d, \omega_g, \mu_r \ge 0
\end{aligned}
\tag{1.19}
$$

Given that model (1.19) is still a nonlinear problem, we can view it as a parametric linear programming program, with $\beta \in (0,\ 1]$ as the parameter. One can search over β, and pick values that are optimal to model (1.19).

In model (1.19), at optimality, the chosen β and the resulting $\sum_{g=1}^{p} \omega_g^* f_{g0} + \sum_{r=1}^{s} \mu_r^* y_{r0}$ are efficiency scores for stages 1 and 2, respectively. Yet, it is likely that model (1.19) has alternative optimal solutions, meaning that efficiency scores for the two stages may not be uniquely determined. Due to our search approach, we actually obtain all possible pairs of β and $\sum_{g=1}^{p} \omega_g^* f_{g0} + \sum_{r=1}^{s} \mu_r^* y_{r0}$ that are optimal solutions to model (1.19). Therefore, we can select one pair as our efficiency scores. For example, we can choose a pair β and $\sum_{g=1}^{p} \omega_g^* f_{g0} + \sum_{r=1}^{s} \mu_r^* y_{r0}$ that minimizes the efficiency gap between the two stages. Alternatively, we can choose a pair of efficiency scores that has the maximum β value (stage 1's efficiency is maximized), or maximum $\sum_{g=1}^{p} \omega_g^* f_{g0} + \sum_{r=1}^{s} \mu_r^* y_{r0}$ (stage 2's efficiency is maximized).

Note that in fact, for a given β, model (1.19) is equivalent to

$$
\begin{aligned}
\psi = Max\ & \sum_{g=1}^{p} \omega_g f_{g0} + \sum_{r=1}^{s} \mu_r y_{r0} \\
s.t.\ & \sum_{d=1}^{D} \varsigma_d z_{dj} - \sum_{i=1}^{m} \upsilon_i x_{ij} - \sum_{g=1}^{p} \omega_g f_{gj} \le 0 \\
& \sum_{g=1}^{p} \omega_g f_{gj} + \sum_{r=1}^{s} \mu_r y_{rj} - \sum_{d=1}^{D} \varsigma_d z_{dj} \le 0 \\
& \beta \left(\sum_{i=1}^{m} \upsilon_i x_{i0} + \sum_{g=1}^{p} \omega_g f_{g0} \right) = 1 \\
& \sum_{d=1}^{D} \varsigma_d z_{d0} = 1 \\
& \nu_i, \varsigma_d, \omega_g, \mu_r \ge 0
\end{aligned}
\tag{1.20}
$$

For some chosen β, model (1.20) (or model (1.19)) may be infeasible. Letting $\beta_{\max}$ be the largest β for which model (1.20) is feasible, we have

Theorem 5 *For any $\beta \in (0, \beta_{max}]$, β is feasible for model (1.20) and model (1.19) has optimal solutions and optimal value.*

Proof See Liang et al. (2011).

To find optimal solutions to model (1.19), we can start with a small β (say 0.001) as the initial point and set a small increment Δs ($=$ 0.001). Theorem 5 indicates that we solve model (1.20) for each $\beta_i = \beta_0 + \Delta s \times i$ ($i = 0, 1, 2, \ldots$) until $\beta = 1$ or model (1.20) is infeasible.

The previous model is based upon the arithmetic mean of the efficiency scores of the two individual stages. In other words, the relative importance of the two stages are assumed to be equal. Liang et al. (2011) also develop a bi-level model in which they model the performance of the two-stage process by viewing the stages consecutively rather than simultaneously. For example, stage 1 can be considered more important or a dominant stage.

Following the bi-level logic, it becomes necessary to deal with the presence of variables on both the output and input sides. Liang et al. (2011) adopt the assumption that regardless of which stage, first or second, is given priority, if a variable appears as an input in one stage and an output in another, we determine the weighting (importance) of that variable wherever it plays the role of **output**. Hence, the weights 'h' for the intermediate variables 'z' will be determined in the stage 1 problem, where they (the z-variables) play the role of outputs; those for the feedback variables 'f' will be determined in the stage 2 problem, where they (the f-variables) assume an output role. This requirement becomes clear in Liang et al.'s (2011) empirical application involving Chinese Universities. For example, the number of SCI papers published and awards received are two outputs from stage 1, and research funds received is an output from stage 2, and is the feedback measure. Each University will want to maximize these measures regardless of whether the first or second stage is given the first priority.

Given the above discussion, if stage 1 is more important, then the efficiency of stage 1 is given first priority and the efficiency of stage 2 depends on that of stage 1. We therefore establish the following bi-level programming model:

$$
\begin{aligned}
&(P1:)\ \theta^A = \underset{v_i, h_d}{Max} \frac{\sum_{d=1}^{D} h_d z_{d0}}{\sum_{i=1}^{m} v_i x_{i0} + \sum_{g=1}^{p} w_g f_{g0}} \\
&\textit{where } w_g \textit{ solve}: \\
&(P2:)\ \theta^B = \underset{w_g, u_r}{Max} \frac{\sum_{g=1}^{p} w_g f_{g0} + \sum_{r=1}^{s} u_r y_{r0}}{\sum_{d=1}^{D} h_d z_{d0}} \\
&\quad s.t. \quad \frac{\sum_{d=1}^{D} h_d z_{dj}}{\sum_{i=1}^{m} v_i x_{ij} + \sum_{g=1}^{p} w_g f_{gj}} \le 1 \\
&\qquad\qquad \frac{\sum_{g=1}^{p} w_g f_{gj} + \sum_{r=1}^{s} u_r y_{rj}}{\sum_{d=1}^{D} h_d z_{dj}} \le 1 \\
&\qquad\qquad v_i, w_g, h_d, u_r \ge 0
\end{aligned}
\tag{1.21}
$$

Model (1.21) treats the first stage as the dominant stage. However, the second stage still gets the first priority to improve the feedback measures.

Letting $t_1 = \frac{1}{\sum_{i=1}^{m} v_i x_{i0} + \sum_{g=1}^{p} w_g f_{g0}}$, $t_2 = \frac{1}{\sum_{d=1}^{D} h_d z_{d0}}$, model (1.21) is equivalent to the following model:

$$
\begin{aligned}
&(P1:) \quad \theta^A = \underset{v_i, h_d}{Max} \quad \sum_{d=1}^{D} t_1 h_d z_{d0} \\
&\textit{where } w_g \textit{ solve}: \\
&(P2:) \quad \theta^B = \underset{w_g, u_r}{Max} \quad \sum_{g=1}^{p} t_2 w_g f_{g0} + \sum_{r=1}^{s} t_2 u_r y_{r0} \\
&\qquad s.t. \quad \sum_{d=1}^{D} t_1 h_d z_{dj} - \left(\sum_{i=1}^{m} t_1 v_i x_{ij} + \sum_{g=1}^{p} t_1 w_g f_{gj} \right) \le 0 \\
&\qquad\qquad \left(\sum_{g=1}^{p} t_2 w_g f_{gj} + \sum_{r=1}^{s} t_2 u_r y_{rj} \right) - \sum_{d=1}^{D} t_2 h_d z_{dj} \le 0 \\
&\qquad\qquad \sum_{i=1}^{m} t_1 v_i x_{i0} + \sum_{g=1}^{p} t_1 w_g f_{g0} = 1 \\
&\qquad\qquad \sum_{d=1}^{D} t_2 h_d z_{d0} = 1 \\
&\qquad\qquad v_i, h_d, w_g, u_r \ge 0
\end{aligned}
\tag{1.22}
$$

Let $\beta = \frac{t_1}{t_2}$ and note that $0 < \frac{t_1}{t_2} = \frac{\sum_{d=1}^{D} h_d z_{d0}}{\sum_{i=1}^{m} v_i x_{i0} + \sum_{g=1}^{p} w_g f_{g0}} \le 1$, thus $\beta \in (0, 1]$.

Substitute $t_1 = \beta t_2$ and model (1.22) becomes:

$$
\begin{aligned}
&(P1:) \ \theta^A = \underset{v_i, h_d}{Max} \quad \sum_{d=1}^{D} \beta t_2 h_d z_{d0} \\
&\textit{where } w_g \textit{ solve}: \\
&(P2:) \ \theta^B = \underset{w_g, u_r}{Max} \quad \sum_{g=1}^{p} t_2 w_g f_{g0} + \sum_{r=1}^{s} t_2 u_r y_{r0} \\
&\qquad s.t. \quad \sum_{d=1}^{D} \beta t_2 h_d z_{dj} - \left(\sum_{i=1}^{m} \beta t_2 v_i x_{ij} + \sum_{g=1}^{p} \beta t_2 w_g f_{gj} \right) \le 0 \\
&\qquad\qquad \left(\sum_{g=1}^{p} t_2 w_g f_{gj} + \sum_{r=1}^{s} t_2 u_r y_{rj} \right) - \sum_{d=1}^{D} t_2 h_d z_{dj} \le 0 \\
&\qquad\qquad \sum_{i=1}^{m} \beta t_2 v_i x_{i0} + \sum_{g=1}^{p} \beta t_2 w_g f_{g0} = 1 \\
&\qquad\qquad \sum_{d=1}^{D} t_2 h_d z_{d0} = 1 \\
&\qquad\qquad v_i, h_d, w_g, u_r \ge 0
\end{aligned}
\tag{1.23}
$$

Let $\upsilon_i = v_i t_2$, $\varsigma_d = h_d t_2$, $\omega_g = w_g t_2$, $\mu_r = u_r t_2$, model (1.23) is now equivalent to the following model:

$$
\begin{aligned}
&(P1:)\ \theta^A = \underset{\upsilon_i,\varsigma_d}{Max} \quad \beta \\
&\textit{where } \omega_g \textit{ solve}: \\
&(P2:)\ \theta^B = \underset{\omega_g,\mu_r}{Max} \quad \sum_{g=1}^{p} \omega_g f_{g0} + \sum_{r=1}^{s} \mu_r y_{r0} \\
&\qquad s.t. \quad \sum_{d=1}^{D} \varsigma_d z_{dj} - \left(\sum_{i=1}^{m} \upsilon_i x_{ij} + \sum_{g=1}^{p} \omega_g f_{gj} \right) \le 0 \\
&\qquad\qquad \left(\sum_{g=1}^{p} \omega_g f_{gj} + \sum_{r=1}^{s} \mu_r y_{rj} \right) - \sum_{d=1}^{D} \varsigma_d z_{dj} \le 0 \\
&\qquad\qquad \beta \left(\sum_{i=1}^{m} \upsilon_i x_{i0} + \sum_{g=1}^{p} \omega_g f_{g0} \right) = 1 \\
&\qquad\qquad \sum_{d=1}^{D} \varsigma_d z_{d0} = 1 \\
&\qquad\qquad \upsilon_i, \varsigma_d, \omega_g, \mu_r \ge 0
\end{aligned}
\tag{1.24}
$$

For model (1.24), the first goal is to maximize the efficiency score of stage 1, θ^{A}, and the second goal is to maximize the efficiency score of stage 2, θ^{B}, when the first goal is achieved.

Similarly, we can develop the following bi-level programming problem when stage 2 is more important.

Models (1.24) are bi-level programming problems where some of the constraints are nonlinear. Liang et al. (2011) point out the model has its own unique features such that the level-two program uses neither its optimal solutions nor its optimal objective function value as the optimal feedback to the objective function of the level-1 program.

Note that when β is a given value, solving the sub-programming problem (P2) of model (1.24) is equivalent to solving model (1.20). Liang et al. (2011) show that the optimal value to model (1.20) is a non-increasing continuous function of β.

Recall that model (1.19) is solved by searching over $\beta_i = \beta_0 + \Delta s \times i$ ($i = 0$, 1, 2, ...). For each β_i we denote the optimal value to model (1.20) as θ_i^B. Namely, when model (1.19) is solved, we already have obtained a set of optimal (β_i, θ_i^B) for each β_i. Therefore, we can obtain the solutions to model (1.24) based upon the following procedures.

For the solution to model (1.24), we select the largest β_i, $\max_i \beta_i = \beta_{\max}^A$, and its corresponding optimal value to model (1.20), denoted as θ^{B^*}, from set $\{(\beta_i, \theta_i^B)\}$. Liang et al. (2011) show that this pair of ($\beta_{\max}^A$, θ^{B^*}) is uniquely determined. Thus, the efficiency score for stage 1 is $\theta^A = \beta_{\max}^A$ and efficiency score for stage 2 is $\theta^B = \theta^{B^*}$.

1.8 Conclusions

In the current chapter, we only focus on a sub-set of modeling approaches and techniques in dealing with DMUs that have two stage network structures. We note that there are variations to the simple two-stage process as shown in Fig. 1.1. For example, Du et al. (2011) use a Nash bargaining game model to model Fig. 1.1. Chen et al. (2010b) study a modified version of Fig. 1.1 where shared input resources are used in both stages of operations. For example, in hospital operations, some of the input resources such as equipment, personnel, and information technology are used in the first stage to generate medical records to track treatments, tests, drug dosages, and costs. Premachandra et al. (2012) study another modification of Fig. 1.1 where there are additional independent inputs to the second stage. While the underlying two stage network process is very similar in Premachandra et al. (2012) and Li et al. (2012), their modeling techniques are very different.

The following chapters provide a more in-depth discussion regarding DEA methodologies on network structures. Some recent applications include Avkiran (2009) and Chen et al. (2012).

References

Avkiran, N. K. (2009). Opening the black box of efficiency analysis: An illustration with UAE banks. *Omega, 37*, 930–941.

Castelli, L., Pesenti, R., & Ukovich, W. (2004). DEA-like models for the efficiency evaluation of hierarchically structured units. *European Journal of Operational Research, 154*(2), 465–476.

Charnes, A., & Cooper, W.W. (1962). Programming with linear fractional functionals, *Naval Research Logistics Quarterly, 9*, 181–186.

Charnes, A., Cooper, W. W., & Rhodes, E. (1978). Measuring the efficiency of decision making units. *European Journal of Operational Research, 2*, 429–444.

Chen, C.-M. (2009). A network-DEA model with new efficiency measures to incorporate the dynamic effect in production networks. *European Journal of Operational Research, 194*(3), 687–699.

Chen, Y., & Zhu, J. (2004). Measuring information technology's indirect impact on firm performance. *Information Technology & Management Journal, 5*(1–2), 9–22.

Chen, Y., Liang, L., Yang, F., & Zhu, J. (2006). Evaluation of information technology investment: A data envelopment analysis approach. *Computers & Operations Research, 33*(5), 1368–1379.

Chen, Y., Liang, L., & Zhu, J. (2009a). Equivalence in two-stage DEA approaches. *European Journal of Operational Research, 193*(2), 600–604.

Chen, Y., Cook, W. D., Li, N., & Zhu, J. (2009b). Additive efficiency decomposition in two-stage DEA. *European Journal of Operational Research, 196*, 1170–1176.

Chen, Y., Cook, W. D., & Zhu, J. (2010a). Deriving the DEA frontier for two-stage processes. *European Journal of Operational Research, 202*, 138–142.

Chen, Y., Du, J., Sherman, H. D., & Zhu, J. (2010b). DEA model with shared resources and efficiency decomposition. *European Journal of Operational Research, 207*, 339–349.

Chen, C.-L., Zhu, J., Yu, J.-Y., & Noori, H. (2012). A new methodology for evaluating sustainable product design performance with two-stage network data envelopment analysis. *European Journal of Operational Research, 221*(2), 348–359.

Chen, Y., Cook, W. D., Kao, C., & Zhu, J. (2013). Network DEA pitfalls: Divisional efficiency and frontier projection under general network structures. *European Journal of Operational Research, 226*, 507–515.

Chilingerian, J., & Sherman, H. D. (2004). Health care applications: From Hospitals to Physician, from productive efficiency to quality frontiers. In W. W. Cooper, L. M. Seiford, & J. Zhu (Eds.), *Handbook on data envelopment analysis*. Boston: Springer.

Cook, W. D., & Hababou, M. (2001). Sales performance measurement in bank branches. *Omega, 29*, 299–307.

Cook, W. D., & Zhu, J. (2007). Classifying inputs and outputs in data envelopment analysis. *European Journal of Operational Research, 180*(2), 692–699.

Cook, W. D., Green, R., & Zhu, J. (2006). Dual-role factors in data envelopment analysis. *IIE Transactions, 38*(2), 105–115.

Cook, W. D., Zhu, J., Yang, F., & Bi, G.-B. (2010). Network DEA: Additive efficiency decomposition. *European Journal of Operational Research, 207*(2), 1122–1129.

Du, J., Liang, L., Chen, Y., Cook, W. D., & Zhu, J. (2011). A bargaining game model for measuring performance of two-stage network structures. *European Journal of Operational Research, 210*(2), 390–397.

Färe, R., & Grosskopf, S. (1996). Productivity and intermediate products: A frontier approach. *Economics Letters, 50*, 65–70.

Färe, R., & Whittaker, G. (1995). An intermediate input model of dairy production using complex survey data. *Journal of Agricultural Economics, 46*(2), 201–213.

Fukuyama, H., & Weber, W. L. (2010). A slacks-based inefficiency measure for a two-stage system with bad outputs. *Omega 38*(5), 398–409.

Huang, Z. M., & Li, S. X. (2001). Co-op advertising models in a manufacturing-retailing supply chain: A game theory approach. *European Journal of Operational Research, 135*, 527–544.

Kao, C., & Hwang, S. N. (2008). Efficiency decomposition in two-stage data envelopment analysis: An application to non-life insurance companies in Taiwan. *European Journal of Operational Research, 185*(1), 418–429.

Li, Y., Chen, Y., Liang, L., & Xie, J. (2012). DEA models for extended two-stage network structures. *Omega, 40*(5), 611–618.

Liang, L., Yang, F., Cook, W. D., & Zhu, J. (2006). DEA models for supply chain efficiency evaluation. *Annals of Operations Research, 145*(1), 35–49.

Liang, L., Cook, W. D., & Zhu, J. (2008). DEA models for two-stage processes: Game approach and efficiency decomposition. *Naval Research Logistics, 55*, 643–653.

Liang, L., Li, Z.-Q., Cook, W. D., & Zhu, J. (2011). DEA Efficiency in two-stage networks with feed back. *IIE Transactions, 43*, 309–322.

Lim, S., & Zhu, J. (2013). Integrated data envelopment analysis: Global vs local optimum. *European Journal of Operational Research, 229*(1), 276–278.

Premachandra, I. M., Zhu, J., Watson, J., & Galagedera, D. U. A. (2012). Best-performing US mutual fund families from 1993 to 2008: Evidence from a novel two-stage DEA model for efficiency decomposition. *Journal of Banking and Finance, 36*(12), 3302–3317.

Seiford, L. M., & Zhu, J. (1999). Profitability and marketability of the top 55 US commercial banks. *Management Science, 45*(9), 1270–1288.

Sexton, T. R., & Lewis, H. F. (2003). Two-stage DEA: An application to major league baseball. *Journal of Productivity Analysis, 19*(2–3), 227–249.

Tone, K., & Tsutsui, M. (2009). Network DEA: A slacks-based measure approach. *European Journal of Operational Research, 197*(1), 243–252.

Tone, K., & Tsutsui, M. (2010). Dynamic DEA: A slacks-based measure approach. *Omega, 38*(3–4), 145–156.

Wang, C. H., Gopal, R., & Zionts, S. (1997). Use of data envelopment analysis in assessing information technology impact on firm performance. *Annals of Operations Research, 73*, 191–213.
Yu, M.-M., & Lin, E. T. J. (2008). Efficiency and effectiveness in railway performance using a multi-activity network DEA model. *Omega, 36*, 1005–1017.
Zhu, J. (2000). Multi-factor performance measure model with an application to Fortune 500 companies. *European Journal of Operational Research, 123*(1), 105–124.

Chapter 2
Network DEA Pitfalls: Divisional Efficiency and Frontier Projection

Yao Chen, Wade D. Cook, Chiang Kao, and Joe Zhu

Abstract Recently network DEA models have been developed to examine the efficiency of DMUs with internal structures. The internal network structures range from a simple two-stage process to a complex system where multiple divisions are linked together with intermediate measures. In general, there are two types of network DEA models. One is developed under the standard multiplier DEA models based upon the DEA ratio efficiency, and the other under the envelopment DEA models based upon production possibility sets. While the multiplier and envelopment DEA models are dual models and equivalent under the standard DEA, such is not necessarily true for the two types of network DEA models. Pitfalls in network DEA are discussed with respect to the determination of divisional efficiency, frontier type, and projections. We point out that the envelopment-based network DEA model should be used for determining the frontier projection for inefficient

---

Part of this chapter is based upon Chen, Y., Cook, W.D., Kao, C. and Zhu, Joe, Network DEA pitfalls: Divisional efficiency and frontier projection under general network structures, *European Journal of Operational Research*, Vol. 226 (2013), 507–515, with permissions from Elsevier Science.

Y. Chen (✉)
Manning School of Business, University of Massachusetts at Lowell, Lowell, MA 01845, USA
e-mail: Yao_Chen@uml.edu

W.D. Cook
Schulich School of Business, York University, Toronto, ON M3J 1P3, Canada
e-mail: wcook@schulich.yorku.ca

C. Kao
Department of Industrial and Information Management, National Cheng Kung University, Tainan 70101, Taiwan, ROC
e-mail: ckao@mail.ncku.edu.tw

J. Zhu
School of Business, Worcester Polytechnic Institute, Worcester, MA 01609, USA
e-mail: jzhu@wpi.edu

W.D. Cook and J. Zhu (eds.), *Data Envelopment Analysis*,
International Series in Operations Research & Management Science 208,
DOI 10.1007/978-1-4899-8068-7_2, © Springer Science+Business Media New York 2014

DMUs while the multiplier-based network DEA model should be used for determining the divisional efficiency. Finally, we demonstrate that under general network structures, the multiplier and envelopment network DEA models are two different approaches. The divisional efficiency obtained from the multiplier network DEA model can be infeasible in the envelopment network DEA model. This indicates that these two types of network DEA models use different concepts of efficiency. We further demonstrate that the envelopment model's divisional efficiency may actually be the overall efficiency.

Keywords Data envelopment analysis (DEA) • Efficiency • Network • Intermediate measure • Link • Frontier

2.1 Introduction

Data envelopment analysis (DEA) is used to identify best practices or (efficient) frontier decision making units (DMUs), in the presence of multiple inputs and outputs (Charnes et al. 1978). DEA provides not only efficiency scores for inefficient DMUs, but also provides for frontier projections for such units onto an efficient frontier. In recent years, a number of DEA studies have focused on DMUs with internal network structures. For example, Cook et al. (2010) review DEA models for treating two-stage network structures. Others have developed DEA-based models for more complicated network structures (see Färe and Grosskopf (2000) and Tone and Tsutsui (2009)). While the focus of the current study is not to review all the existing network DEA approaches, we note that many of these approaches require significant modifications to the standard DEA structures. Therefore, a rational question to ask is whether the network DEA model retains the property of the standard DEA model, namely that it yields both (divisional) efficiency scores and a frontier projection in a single model.

Following a thorough review of the existing network DEA approaches, Chen et al. (2013) conclude that there are two types of structures based upon the standard DEA models used. One type is the multiplier-based network DEA models which calculate the overall network efficiency by integrating the ratio efficiency of each division in the network via geometric or arithmetic averages. Such a network model is then converted into a linear program that looks like the DEA multiplier model. The other type is developed by using the production possibility set for each division in the network. The resulting model takes on the appearance of the DEA envelopment model.

In the standard DEA context, the multiplier model is equivalent to the envelopment model which yields the DEA projection and the efficiency due to the linear programming duality. However, under the network structure, such duality may not lead to a particular pair of network multiplier and envelopment models, where frontier projections and divisional efficiency scores are generated in a single network DEA model.

The current chapter first uses a simple two-stage network structure to demonstrate that under the condition of constant returns to scale (CRS), the envelopment network DEA model does not necessarily provide information on divisional efficiency and only provides information on the frontier projection. This can be a pitfall when we use the envelopment network DEA approach. Such a pitfall is caused by the fact that the envelopment-based network DEA approach does not account for the intermediate measures (or links) in calculating the divisional efficiency. While the multiplier-based network DEA models provide both overall and divisional efficiency scores, their duals may not yield correct information on frontier projections without proper adjustments to those dual models.

Note in the standard DEA approach, variable returns to scale (VRS) is achieved by adding a convexity constraint into the CRS envelopment model or equivalently a free variable into the multiplier model. We demonstrate that under the network DEA model, the above equivalence no longer holds. We further show that envelopment and multiplier network DEA models are two very different approaches using different efficiency concepts. Further, divisional efficiency obtained from the multiplier network DEA model can be infeasible on the envelopment side. We also demonstrate that the envelopment model's divisional efficiency may actually be the overall efficiency.

The rest of the chapter is organized as follows. Section 2.2 briefly introduces the multiplier and envelopment-based network DEA models under a simple two-stage network structure, where outputs from the first stage (division) are the only inputs to the second stage (division). Sections 2.3 and 2.4 then discuss the pitfalls for determining divisional efficiencies and frontier projections. In Sect. 2.5 we examine the VRS case. Section 2.6 is devoted to discussing network DEA models under general network structures. Conclusions follow in Sect. 2.7.

2.2 Two-Stage Network DEA

For simplicity, we consider a generic two-stage process as shown in Fig. 2.1, for each of a set of n DMUs. We assume each $DMU_j(j = 1, 2, \ldots, n)$ has m inputs x_{ij}, $(i = 1, 2, \ldots, m)$ to the first stage, and D outputs z_{dj}, $(d = 1, 2, \ldots, D)$ from that stage. These D outputs then become the inputs to the second stage, hence behaving as intermediate measures. The outputs from the second stage are y_{rj}, $(r = 1, 2, \ldots, s)$.

For DMU_j we denote the efficiency ratios for the first stage (division) as θ_j^1 and the second as θ_j^2. Based upon the *input-oriented* DEA model of Charnes et al. (1978), we have the following standard DEA models for each stage (division):

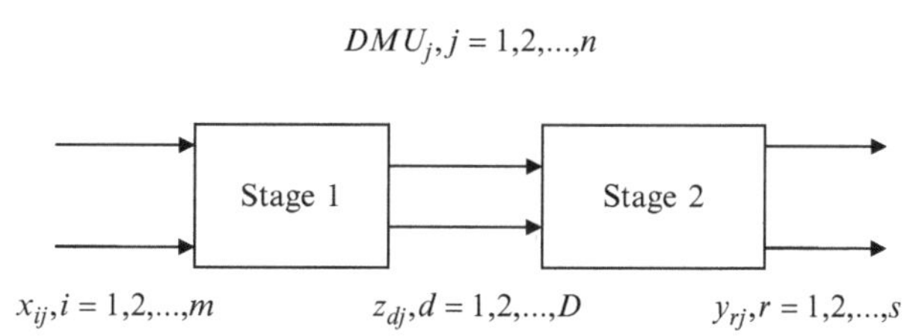

Fig. 2.1 Two-stage process

$$\theta_j^1 = \max \frac{\sum_{d=1}^{D} w_d z_{dj}}{\sum_{i=1}^{m} v_i x_{ij}} \quad \text{and} \quad \theta_j^2 = \max \frac{\sum_{r=1}^{s} u_r y_{rj}}{\sum_{d=1}^{D} \widetilde{w}_d z_{dj}} \tag{2.1}$$

$$\text{Subject to} \qquad\qquad \text{Subject to}$$

$$\frac{\sum_{d=1}^{D} w_d z_{dj}}{\sum_{i=1}^{m} v_i x_{ij}} \leq 1 \qquad\qquad \frac{\sum_{r=1}^{s} u_r y_{rj}}{\sum_{d=1}^{D} \widetilde{w}_d z_{dj}} \leq 1$$

where v_i, w_d, $\widetilde{w}_d$, and u_r are unknown non-negative weights. In order to model the two-stage network based upon the two efficiency ratios defined in (2.1) the variables w_d are set equal to $\widetilde{w}_d$ as in Kao and Hwang (2008) and in Liang et al. (2008). As a result, the two-stage overall efficiency ratio can be defined as $\theta_j^1 \bullet \theta_j^2$ which is equal to $\theta_j = \frac{\sum_{r=1}^{s} u_r y_{ro}}{\sum_{i=1}^{m} v_i x_{io}}$. To calculate the overall efficiency of θ_j, Kao and Hwang (2008) present the following model (this model is called centralized model in Liang et al. (2008))

$$Max\ \theta_j^1 \bullet \theta_j^2 = \frac{\sum_{r=1}^{s} u_r y_{ro}}{\sum_{i=1}^{m} v_i x_{io}} \tag{2.2}$$
$$s.t. \quad \theta_j^1 \leq 1 \quad \text{and} \quad \theta_j^2 \leq 1 \quad \text{and} \quad w_d = \widetilde{w}_d$$

Model (2.2) can be converted into the following linear program

$$\begin{aligned}
&\text{Max} \sum_{r=1}^{s} u_r y_{ro} \\
&\text{s.t.} \quad \sum_{r=1}^{s} u_r y_{rj} - \sum_{d=1}^{D} w_d z_{dj} \leq 0 \quad j = 1, 2, \ldots, n \\
&\sum_{d=1}^{D} w_d z_{dj} - \sum_{i=1}^{m} v_i x_{ij} \leq 0 \quad j = 1, 2, \ldots, n \\
&\sum_{i=1}^{m} v_i x_{io} = 1 \\
&w_d \geq 0, \ d = 1, 2, \ldots, D; \ v_i \geq, \ i = 1, 2, \ldots, m; \ u_r \geq, \ r = 1, 2, \ldots, s
\end{aligned} \tag{2.3}$$

In a similar manner, we can develop output-oriented models. Model (2.3) yields the overall efficiency. After the overall efficiency is obtained, divisional efficiency can be obtained via efficiency decomposition (see Kao and Hwang (2008)). Specifically,

If we denote the optimal value to model (2.3) as θ_o^*, then we have $\theta_o^* = \theta_o^{1*} \bullet \theta_o^{2*}$. Note that optimal multipliers from model (2.3) may not be unique, meaning that θ_o^{1*} and θ_o^{2*} may not be unique. To test for uniqueness, we can first determine the maximum achievable value of θ_o^{1*} via

$$
\begin{aligned}
\theta_o^{1+} = \text{Max} & \sum_{d=1}^{D} w_d z_{do} \\
s.t. \quad & \sum_{r=1}^{s} u_r y_{ro} = \theta_o^* \\
& \sum_{d=1}^{D} w_d z_{dj} - \sum_{i=1}^{m} v_i x_{ij} \leq 0 \quad j = 1, 2, \ldots, n \\
& \sum_{r=1}^{s} u_r y_{rj} - \sum_{d=1}^{D} w_d z_{dj} \leq 0 \quad j = 1, 2, \ldots, n \\
& \sum_{i=1}^{m} v_i x_{io} = 1 \\
& w_d \geq 0, \;\; d = 1, 2, \ldots, D; \;\; v_i \geq 0, \;\; i = 1, 2, \ldots, m; \;\; u_r \geq 0, \;\; r = 1, 2, \ldots, s
\end{aligned}
$$

It then follows that the minimum of θ_o^{2*} is given by $\theta_o^{2-} = \frac{\theta_o^*}{\theta_o^{1+}}$. This also gives an efficiency decomposition of $\theta_o^* = \theta_o^{1+} \bullet \theta_o^{2-}$.

Liang et al. (2008) provide a procedure for testing the uniqueness of efficiency decomposition. The maximum of θ_o^{2*}, which we denote by θ_o^{2+}, can be calculated in a manner similar to the above, and the minimum of θ_o^{1*} is then calculated as $\theta_o^{1-} = \theta_o^*/\theta_o^{2+}$. Note that $\theta_o^{1-} = \theta_o^{1+}$ if and only of $\theta_o^{2-} = \theta_o^{2+}$. Note also that if $\theta_o^{1-} = \theta_o^{1+}$ or $\theta_o^{2-} = \theta_o^{2+}$, then θ_o^{1*} and θ_o^{2*} are uniquely determined via model (2.3).

Model (2.3) is based upon the ratio DEA efficiency and then is converted into a DEA multiplier-type linear program. Therefore, we can refer to this type of network DEA approach as multiplier-based. On the other hand, Tone and Tsutsui (2009) develop a slacks-based network DEA model by using the production possibility sets, where the intermediate measures $z_{dj}(d = 1, \ldots, \text{D})$ are called links. Relative to Fig. 2.1, the constraints for the slacks-based model take the form:

$$
\begin{aligned}
& \sum_{j=1}^{n} \lambda_j x_{ij} + s_i^- = x_{io} \qquad i = 1, \ldots, \text{m} \\
& \sum_{j=1}^{n} \mu_j y_{rj} - s_r^+ = y_{ro} \qquad r = 1, \ldots, \text{s} \\
& \sum_{j=1}^{n} \lambda_j z_{dj} = \sum_{j=1}^{n} \mu_j z_{dj} \qquad d = 1, \ldots, \text{D}
\end{aligned}
\tag{2.4}
$$

In Tone and Tsutsui (2009), models based upon (2.4) are referred to as the "fixed link" case.

Note that z_{do} are outputs from the first stage and are inputs to the second stage. Therefore, based upon the standard DEA model, the production possibility set can be defined as

$$\begin{aligned}
&\sum_{j=1}^{n} \lambda_j x_{ij} + s_i^- = x_{io} && i = 1, \ldots, \text{m} \\
&\sum_{j=1}^{n} \mu_j y_{rj} - s_r^+ = y_{ro} && r = 1, \ldots, \text{s} \\
&\sum_{j=1}^{n} \lambda_j z_{dj} \geq \tilde{z}_{do} && d = 1, \ldots, \text{D} \\
&\sum_{j=1}^{n} \mu_j z_{dj} \leq \tilde{z}_{do} && d = 1, \ldots, \text{D}
\end{aligned} \tag{2.5}$$

where we use $\tilde{z}_{do}$ to denote unknown decision variables for the intermediate measures (or links as referred to in Tone and Tsutsui (2009)). As in Tone and Tsutsui (2009), these measures can be increased or decreased in the optimal solution of a network DEA model based upon (2.5). Tone and Tsutsui (2009) refer to (2.5) as the "free link" case.

The difference between (2.4) and (2.5) (or between "fixed link" and "free link") is very minor. The slacks-based (envelopment) network DEA models based upon (2.4) and (2.5) will yield identical optimal slack values if the constraints related to $\tilde{z}_{do}$ become binding at optimality. Otherwise, if these constraints are not binding, then the two models based upon (2.4) and (2.5) will yield different optimal slack values. If one uses a radial measure, the difference between (2.4) and (2.5) is negligible with respect to the radial efficiency scores. Therefore, in the discussion to follow, we will not specifically distinguish "fixed link" and "free link" for the intermediate measures.

A version of the input-oriented envelopment network DEA model under "free link" intermediate measures can be written as

$$\begin{gathered}
\text{Max} \sum_{i=1}^{m} \frac{s_i^-}{x_{io}} \\
\text{Subject to (5)}
\end{gathered} \tag{2.6}$$

Since the intermediate measures are the only outputs from stage-1, and the only inputs to stage-2, Tone and Tsutsui's (2009) input-oriented slacks-based network DEA model will not have the divisional efficiency for stage 2. In other words, the divisional efficiency for stage 1 should be regarded as the overall efficiency, based upon Tone and Tsutsui's (2009) definition under either "fixed link" or "free link". In a similar manner, Tone and Tsutsui's (2009) output-oriented slacks-based network DEA model will not have the divisional efficiency for stage 1, namely the divisional efficiency for stage 2 is the overall efficiency.

Moreover, as shown in Chen et al. (2009b), model (2.3) is equivalent to the following linear program, where the intermediate measures are treated as "free link" defined in Tone and Tsutsui (2009).

$$
\begin{aligned}
& \min \widetilde{\theta} \\
& s.t. \quad \sum_{j=1}^{n} \lambda_j x_{ij} \leq \widetilde{\theta} x_{io} \qquad i = 1, \ldots, \mathrm{m} \\
& \quad \sum_{j=1}^{n} \mu_j y_{rj} \geq y_{ro} \qquad r = 1, \ldots, \mathrm{s} \\
& \quad \sum_{j=1}^{n} \lambda_j z_{dj} \geq \widetilde{z}_{do} \qquad d = 1, \ldots, \mathrm{D} \\
& \quad \sum_{j=1}^{n} \mu_j z_{dj} \leq \widetilde{z}_{do} \qquad d = 1, \ldots, \mathrm{D} \\
& \quad \dot{z}_{do} \geq 0, \qquad d = 1, \ldots, \mathrm{D} \\
& \quad \lambda_j \geq 0, \quad j = 1, \ldots, n \\
& \quad \mu_j \geq 0, \quad j = 1, \ldots, n \\
& \quad \widetilde{\theta} \leq 1
\end{aligned}
\tag{2.7}
$$

Model (2.7) can be viewed as the radial version of the two-stage network DEA model of Tone and Tsutsui's (2009) based upon (2.5). In this case, model (2.7), like the input-oriented slacks-based network DEA model (2.6), can only generate the overall efficiency. In this regard, θ^* cannot be treated as the divisional efficiency of stage 1.

Models developed based upon (2.4) or (2.5) can be called envelopment DEA network models, as they are similar to the standard envelopment DEA model format. If we add $\sum \lambda_j = \sum \mu_j = 1$ into (2.4) or (2.5), the existing DEA literature claims that the VRS envelopment network DEA model is obtained, since this is how VRS envelopment model is obtained under the standard DEA model. However, we believe that this issue needs to be further examined.

2.3 Two-Stage Network: Divisional Efficiency Pitfall

Consider the numerical example given in Table 2.1 where we have five DMUs and two intermediate measures. Table 2.2 reports the optimal slacks and intermediate measures when (2.4) and (2.5) are used. The last three columns report the overall efficiency based upon (2.2), and its efficiency decomposition for divisional efficiency scores based upon Kao and Hwang (2008) and Liang et al. (2008).

It can be seen that the envelopment-based network DEA model can generate a score for overall efficiency and frontier projections, and the multiplier-based

Table 2.1 Numerical example for two-stage network

DMU	X1	X2	Z1	Z2	Y1	Y2
1	2	4	3	4	7	8
2	12	9	4	3	9	12
3	3	4	5	4	10	12
4	7	9	6	12	21	16
5	4	8	10	11	18	16

Table 2.2 Optimal slacks and intermediate measures

DMU	s1	s2	z1	z2	s1	s2	Overall	Stage 1	Stage 2
Equation (2.5)									
1	0.9037	1.8074	2.7407	3.0148	0	0	0.54815	0.60000	0.91358
2	10.4	5.8	4	4.4	0	0	0.35556	0.35556	1.00000
3	1.3704	0.7407	4.0740	4.4814	0	0	0.81481	1.00000	0.81481
4	3.7692	2.5384	8.0769	8.8846	0	7.4769	0.71795	0.71795	1.00000
5	1.2308	2.4615	6.9231	7.6154	0	4.1231	0.69231	1.00000	0.69231
Equation (2.4)									
1	0.8991	1.7982	2.7523	3.0275	0.1560	0			
2	10.348	5.6972	4.1284	4.5412	1.7339	0			
3	1.3486	0.6972	4.1284	4.5412	0.7339	0			
4	3.7692	2.5384	8.0769	8.8846	0	7.4769			
5	1.2308	2.4615	6.9231	7.6153	0	4.1231			

two-stage network DEA model (2.2) or (2.4) is able to generate the divisional efficiency scores for both stages. In other words, under the network DEA approach, both multiplier and envelopment-based models are needed to generate (i) overall efficiency, (ii) divisional efficiency, and (iii) frontier projections.

Under (2.5), model (2.2) or its equivalent model (2.7) yields the identical optimal intermediate measures as model (2.6). This also implies that the input-oriented slacks-based network DEA model does not yield information on divisional efficiency.

For the output-oriented situation, we can obtain the similar conclusion that while the multiplier network DEA model can decompose the overall efficiency into divisional efficiency scores, the slacks-based or envelopment network DEA model only yields information on the overall efficiency along with the frontier projection.

We finally consider the non-oriented case. For example, for (2.4) or (2.5), we can use sum of slacks or the ratio form from Tone and Tsutsui (2009). For example, we can have the following non-oriented envelopment network DEA model.

$$\begin{aligned} &\text{Max} \sum_{i=1}^{m} \frac{s_i^-}{x_{io}} + \sum_{r=1}^{s} \frac{s_r^+}{y_{ro}} \\ &\text{Subject to (4) or (5)} \end{aligned} \tag{2.8}$$

Table 2.3 Slacks and intermediate measures based on non-oriented model

DMU	s_1^{-*}	s_2^{-*}	z1*	z2*	s_1^{+*}	s_2^{+*}
1	0	0	5	5.5	6	6.533333
2	7.5	0	11.25	12.375	20.25	20.7
3	1	0	5	5.5	3	2.533333
4	2.5	0	11.25	12.375	8.25	16.7
5	0	0	10	11	8	13.06667

Or, it appears that we can build a radial version of (2.8), that is

$$
\begin{aligned}
&\min \alpha - \beta \\
&s.t. \quad \sum_{j=1}^{n} \lambda_j x_{ij} \le \alpha x_{io} \qquad i = 1, \ldots, \mathrm{m} \\
&\sum_{j=1}^{n} \mu_j y_{rj} \ge \beta y_{ro} \qquad r = 1, \ldots, \mathrm{s} \\
&\sum_{j=1}^{n} \lambda_j z_{dj} \ge \tilde{z}_{do} \qquad d - 1, \ldots, \mathrm{D} \\
&\sum_{j=1}^{n} \mu_j z_{dj} \le \tilde{z}_{do} \qquad d = 1, \ldots, \mathrm{D} \\
&\tilde{z}_{do} \ge 0, \qquad d = 1, \ldots, \mathrm{D} \\
&\lambda_j \ge 0, \quad j = 1, \ldots, n \\
&\mu_j \ge 0, \quad j = 1, \ldots, n \\
&\alpha \le 1, \beta \ge 1
\end{aligned}
\tag{2.9}
$$

where α and β represent the divisional efficiency scores for stages 1 and 2 respectively. Model (2.9) assumes "free link". If "fixed link" is assumed, we use $\sum_{j=1}^{n} \lambda_j z_{dj} = \sum_{j=1}^{n} \mu_j z_{dj}$ in model (2.9).

However, as shown in Chen et al. (2009b), $\alpha^* = 1$ and $1/\beta^*$ is equal to the overall efficiency obtained from model (2.2) at optimality. This indicates that α and β actually do not represent the divisional efficiency scores. This implies that the slack based measures cannot be used to represent the divisional efficiency scores in model (2.8).

Table 2.3 reports the results from (2.8). Both (2.4) ("fixed link") and (2.5) ("free link") yield identical optimal slacks and intermediate measures, namely, the inequality constraints in (2.5) are binding at optimality. It can be seen from the input slacks that DMUs 1 and 5 are efficient and DMUs 2, 3, and 4 are weakly efficient. This corresponds to the situation $\alpha^* = 1$. Based upon the last two columns of Table 2.2, stage 1 in DMU5 is not efficient, for example.

The above phenomenon can be regarded as a two-stage network DEA pitfall in calculating divisional efficiency. It is recommended that the envelopment-based network DEA model, for example, Tone and Tsutsui's (2009), be used to calculate the frontier projection, and the multiplier-based approach is used to calculate the overall and divisional efficiencies.

This pitfall can be due to the fact that the envelopment-based network DEA model does not consider the optimal intermediate measures in its calculation of the divisional efficiency. In the current study, we argue that divisional efficiency should be based upon the DEA ratio efficiency as defined in (2.1) where (optimal) intermediate measures are considered. This is due to the fact that once the optimal intermediate measures are determined, they become inputs (or outputs) to a division.

2.4 Two-Stage Network: Frontier Projection Pitfall

While the envelopment-based network DEA model provides a frontier projection for inefficient DMUs, the dual to the multiplier-based network DEA model does not necessarily provide the frontier projection. For example, the dual to model (2.3) is

$$\begin{aligned}
& Min\,\theta \\
& s.t. \quad \sum_{j=1}^{n} \lambda_j x_{ij} \le \theta x_{io} \quad i = 1, 2, \ldots, m \\
& \qquad \sum_{j=1}^{n} \mu_j y_{rj} \ge y_{ro} \quad r = 1, 2, \ldots, m \\
& \qquad \sum_{j=1}^{n} (\lambda_j - \mu_j) z_{dj} \ge 0 \quad d = 1, 2, \ldots, D \\
& \qquad \lambda_j, \mu_j \ge 0 \quad \theta \le 1
\end{aligned} \tag{2.10}$$

As shown in Chen et al. (2010) or Chap. 4, ($\theta^* x_{io}$, z_{do}, y_{ro}) is not on the frontier as the projection generated by model (2.10), and we have to determine an optimal z_{do}. Based upon the discussion in the previous section, we know that for an inefficient DMU to be projected onto the network DEA frontier, its intermediate measures will have to be adjusted (increased or decreased). In fact, Chen et al. (2010) show that model (2.10) is equivalent to model (2.7). Therefore, the dual to the multiplier-based network DEA model (namely model (2.10)) has to be adjusted as model (2.7) in order to calculate the optimal intermediate measures so that we obtain the correct frontier projection as $(\theta^* x_{io}, \tilde{z}^*_{do}, y_{ro})$ or ($\theta^* x_{io}$, $\sum \lambda_j^* z_{ij}$, y_{ro}).

Since Fig. 2.1 presents a simple network structure, we are able to modify the model (2.10) to model (2.7). In a complicated network structure, such task may not be possible.

Such frontier projections have an interesting aspect. We consider the two-stage network structure involving 24 Taiwanese non-life insurance companies studied in Kao and Hwang (2008). The two stages represent premium acquisition and profit generation respectively. The inputs to the first stage are operational expenses and insurance expenses, and the outputs from the second stage are underwriting profit and investment profit. There are two intermediate measures between the two stages, namely direct written premiums and reinsurance premiums.

Table 2.4 reports a set frontier projection points based upon model (2.7). Table 2.5 reports the overall efficiency and its decomposition based upon model (2.3). Note that none of the DMUs are efficient.

Because of the existence of intermediate measures, we cannot apply the standard DEA to each stage separately. However, since we now have the frontier, we should be able to apply the standard DEA to each stage after we include the projected DMUs in Table 2.4. In other words, we now have a set of 48 DMUs, of which 24 are projections of the original DMUs.

The last two columns of Table 2.5 report the CRS efficiency scores. Interestingly, these scores are equal to the standard CRS scores when the original 24 DMUs are evaluated. That is,, the added projected DMUs do not change the CRS efficiency scores. Note that the projected DMUs are obtained from the network DEA model and represent the frontier for the two-stage process. Such a frontier cannot be obtained from the standard CRS model.

The above discussion may indicate that the network DEA model behaves very differently from the standard DEA model, although it is built upon the standard DEA model.

Finally, we point out that one may argue that the dual variables to the multiplier model (2.3) could be used to obtain the frontier projections. However, without the help of transforming the model (2.10) (which is the dual to model (2.3)) to model (2.7), we cannot obtain the frontier projections directly based upon the dual variables. The same is true that we cannot obtain the divisional efficiency directly based upon the dual variables to the envelopment model (2.7). In other words, both models (2.3) and (2.7) are needed to calculate the divisional efficiency and frontier projections. We will further demonstrate this point in the next section.

2.5 Two-Stage Network: Variable Returns to Scale

Discussions in Sects. 2.3 and 2.4 are based upon CRS. Under the standard DEA approach, by adding the convexity constraint (e.g. $\sum \lambda_j = 1$), we obtain the envelopment model under VRS. Note that under the standard DEA approach, the VRS multiplier model is obtained by introducing a free variable. The issue here is whether the multiplier network model is equivalent to the envelopment network model under the VRS condition. To address such an issue is computationally difficult because model (2.2) for VRS version cannot be converted into a linear program. An alternative approach is to use an additive form of weighed average of

Table 2.4 Frontier projection

DMU		Operation expenses (X1)	Insurance expenses (X2)	Direct written premiums (Z1)	Reinsurance premiums (Z2)	Underwriting profit (Y1)	Investment profit (Y2)
1	Taiwan Fire	824,177.805	470,919.59	5,129,409	673,373.7	984,143	681,687
2	Chung Kuo	863,362.386	845,201.324	6,287,502	827,782.2	1,228,502	834,754
3	Tai Ping	812,470.86	409,025.1	4,776,548	560,244	293,613	658,428
4	China Mariners	182,921.544	180,773.588	1,332,365	174,166.4	248,709	177,331
5	Fubon	5,138,181.32	2,708,747.94	30,127,364	4,177,166	7,851,229	3,925,272
6	Zurich	1,024,017.42	260,461.061	3,807,167	435,393.8	1,713,598	415,058
7	Taian	537,387.608	399,161.46	3,738,287	654,045.2	2,239,593	439,039
8	Ming Tai	1,042,733.08	515,595.456	5,553,015	1,009,007	3,899,530	622,868
9	Central	350,077.682	212,231.466	2,166,576	351,793.5	1,043,778	264,098
10	The First	607,314.034	605,087.02	4,417,507	671,133.1	1,697,941	554,806
11	Kuo Hua	321,645.227	110,208.655	941,872.5	263,658.4	1,486,014	18,259
12	Union	1,969,483.28	494,463.139	7,166,191	849,582.4	1,574,191	909,295
13	Shingkong	542,345.74	284,437.056	2,649,297	644,399.8	3,609,236	223,047
14	South China	402,886.177	285,393.077	2,749,750	454,687.7	1,401,200	332,283
15	Cathay Century	1,341,118.63	399,622.469	5,663,750	634,667.6	3,355,197	555,482
16	Allianz President	387,991.463	132,905.734	1,899,396	188,102.5	854,054	197,947
17	Newa	523,366.92	390,606.84	3,504,900	725,272.3	3,144,484	371,984
18	AIU	196,044.882	141,821.624	1,356,947	224,515.2	692,731	163,927
19	North America	65,554.3264	74,977.3856	474,800	108,242	519,121	46,857
20	Federal	79,498.5972	29,252.9388	302,964.9	63,960.38	355,624	26,537
21	Royal & Sunalliance	16,901.5368	5,265.7792	68,296	9,610.947	51,950	6,491
22	Aisa	9,427.8735	6,190.929	52,063	14,574	82,141	4,181
23	AXA	22,987.4679	11,939.8824	137,689.9	16,149.72	0.1	18,980
24	Mitsui Sumitomo	22,012.4356	31,690.6712	159,644.2	32,943.34	142,370	16,976

Table 2.5 Overall efficiency and its decomposition

DMU	Overall efficiency	Stage 1	Stage 2	New 1	New 2
1	0.69915	0.99246	0.70446	0.99248	0.71337
2	0.62473	0.99845	0.62570	0.99845	0.62748
3	0.68995	0.68995	1.00000	0.68996	1.00000
4	0.30420	0.72430	0.41999	0.72431	0.43232
5	0.76691	0.83043	0.92351	0.83752	1.00000
6	0.38968	0.96062	0.40566	0.96369	0.40566
7	0.27654	0.67093	0.41217	0.75208	0.53784
8	0.27517	0.66302	0.41502	0.72559	0.51135
9	0.22326	0.99966	0.22334	1.00000	0.29196
10	0.46593	0.86146	0.54086	0.86153	0.67360
11	0.16390	0.64637	0.25357	0.74053	0.32667
12	0.75958	1.00000	0.75958	1.00000	0.75958
13	0.20780	0.67142	0.30949	0.81068	0.54349
14	0.28859	0.66968	0.43094	0.72461	0.51782
15	0.61380	0.99990	0.61386	1.00000	0.70473
16	0.32015	0.88558	0.36152	0.90720	0.38475
17	0.35997	0.62734	0.57380	0.72331	1.00000
18	0.25880	0.79353	0.32614	0.79353	0.37366
19	0.41118	1.00000	0.41118	1.00000	0.41578
20	0.54653	0.93301	0.58577	0.93321	0.90137
21	0.20078	0.73211	0.27424	0.75049	0.27951
22	0.58950	0.58950	1.00000	0.58950	1.00000
23	0.42030	0.84245	0.49890	0.85005	0.55992
24	0.13480	0.42883	0.31434	1.00000	0.33509

divisional efficiency scores (Chen et al. 2009a). However, Chen et al. (2009a) discover that CRS scores are greater than the related VRS scores for several DMUs in the case of 24 Taiwanese non-life insurance companies. This may indicate that the properties related to returns to scale in the standard DEA model do not apply in network DEA.

Kao and Hwang (2011) recently developed an alternative approach to study efficiency decomposition under both CRS and VRS conditions. Based upon model (2.3), we denote E_0, E_0^1, and E_0^2 to be the system, stage 1 , and stage 2 CRS efficiencies of the two-stage system, respectively. Now, let T_0, T_0^1, and T_0^2 be the respective technical efficiencies (under VRS), and S_0, S_0^1, and S_0^2 the respective scale efficiencies. Since the outputs of the first stage are the inputs of the second, if one wants to improve the efficiency of the first stage via increasing its outputs, then the efficiency of the second stage will be affected. Therefore, Kao and Hwang (2011) used the input-oriented VRS model to calculate T_0^1 and the output-oriented VRS model to calculate T_0^2, so that the intermediate products can remain intact. As in the conventional case, $S_0^1 = E_0^1/T_0^1$ and $S_0^2 = E_0^2/T_0^2$. Note that the former is an input-oriented scale efficiency and the latter output-oriented one. The technical and scale efficiencies of the overall system are the products of those of the first and second stages, respectively, i.e., $T_0 = T_0^1 \times T_0^2$ and $S_0 = S_0^1 \times S_0^2$.

To calculate the input-oriented VRS technical efficiency of stage 1 and the output-oriented VRS technical efficiency of stage 2, the production possibility based envelopment models are:

$$\begin{aligned} \min.\ & \theta \\ \text{s.t.}\ & \sum\nolimits_{j=1}^{n} \lambda_j X_{ij} \leq \theta X_{i0} \\ & \sum\nolimits_{j=1}^{n} \lambda_j Z_{dj} \geq Z_{d0} \\ & \sum\nolimits_{j=1}^{n} \lambda_j = 1 \\ & \sum\nolimits_{j=1}^{n} \mu_j Z_{dj} \leq Z_{d0} \\ & \sum\nolimits_{j=1}^{n} \mu_j Y_{rj} \geq Y_{r0} \end{aligned} \tag{2.11}$$

$$\begin{aligned} \min.\ & \eta \\ \text{s.t.}\ & \sum\nolimits_{j=1}^{n} \lambda_j X_{ij} \leq X_{i0} \\ & \sum\nolimits_{j=1}^{n} \lambda_j Z_{dj} \geq Z_{d0} \\ & \sum\nolimits_{j=1}^{n} \mu_j Z_{dj} \leq Z_{d0} \\ & \sum\nolimits_{j=1}^{n} \mu_j Y_{rj} \geq Y_{r0}/\eta \\ & \sum\nolimits_{j=1}^{n} \mu_j = 1 \end{aligned} \tag{2.12}$$

Based on Kao and Hwang (2011), T_0^1 and T_0^2 are calculated as follows:

$$\begin{aligned} T_0^1 = \max.\ & \left(\sum\nolimits_{d=1}^{D} \widetilde{w}_d Z_{d0} - \widetilde{w}_0\right) \Big/ \sum\nolimits_{i=1}^{m} v_i \widetilde{X}_{i0} \\ \text{s.t.}\ & \sum\nolimits_{r=1}^{s} u_r Y_{r0} \Big/ \sum\nolimits_{i=1}^{m} v_i X_{i0} = E_0 \\ & \sum\nolimits_{d=1}^{D} w_d Z_{dj} - \sum\nolimits_{i=1}^{m} v_i X_{ij} \leq 0 \\ & \sum\nolimits_{r=1}^{s} u_r Y_{rj} - \sum\nolimits_{d=1}^{D} w_d Z_{dj} \leq 0 \\ & \left(\sum\nolimits_{d=1}^{D} \widetilde{w}_d Z_{dj} - \widetilde{w}_0\right) - \sum\nolimits_{i=1}^{m} v_i \widetilde{X}_{ij} \leq 0 \end{aligned}$$

$$\begin{aligned} T_0^1 = \max.\ & \sum\nolimits_{r=1}^{s} u_r \hat{Y}_{r0} \Big/ \left(\sum\nolimits_{d=1}^{D} \hat{w}_d Z_{d0} + \hat{w}_0\right) \\ \text{s.t.}\ & \sum\nolimits_{r=1}^{s} u_r Y_{r0} \Big/ \sum\nolimits_{i=1}^{m} v_i X_{i0} = E_0 \\ & \sum\nolimits_{d=1}^{D} w_d Z_{dj} - \sum\nolimits_{i=1}^{m} v_i X_{ij} \leq 0 \\ & \sum\nolimits_{r=1}^{s} u_r Y_{rj} - \sum\nolimits_{d=1}^{D} w_d Z_{dj} \leq 0 \\ & \sum\nolimits_{r=1}^{s} u_r \hat{Y}_{rj} - \left(\sum\nolimits_{d=1}^{D} \hat{w}_d Z_{dj} + \hat{w}_0 \leq 0\right) \end{aligned}$$

The linearized forms for calculating T_0^1 and its dual are:

$$\begin{aligned} T_0^1 = \max.\ & \sum\nolimits_{d=1}^{D} \widetilde{w}_d Z_{d0} - \widetilde{w}_0 && \text{dual} \\ \text{s.t.}\ & \sum\nolimits_{i=1}^{m} v_i \widetilde{X}_{i0} = 1 && \theta_1 \\ & E_0 \sum\nolimits_{i=1}^{m} v_i X_{i0} - \sum\nolimits_{r=1}^{s} u_r Y_{r0} = 0 && \eta \\ & \sum\nolimits_{d=1}^{D} w_d Z_{dj} - \sum\nolimits_{i=1}^{m} v_i X_{ij} \leq 0 && \lambda_j \\ & \sum\nolimits_{r=1}^{s} u_r Y_{rj} - \sum\nolimits_{d=1}^{D} w_d Z_{dj} \leq 0 && \mu_j \\ & \sum\nolimits_{d=1}^{D} \widetilde{w}_d Z_{dj} - \widetilde{w}_0 - \sum\nolimits_{i=1}^{m} v_i \widetilde{X}_{ij} \leq 0 && \alpha_j \end{aligned}$$

$$\begin{aligned} \min.\ & \theta_1 \\ & \sum\nolimits_{j=1}^{n} \alpha_j X_{ij} \leq \theta_1 X_{i0} \\ & \sum\nolimits_{j=1}^{n} \alpha_j = 1 \\ & \sum\nolimits_{j=1}^{n} \alpha_j Z_{dj} \geq Z_{d0} \\ & \sum\nolimits_{j=1}^{n} \lambda_j X_{ij} \leq \eta E_0 X_{i0} \\ & \sum\nolimits_{j=1}^{n} \mu_j Y_{rj} \geq \eta Y_{r0} \\ & \sum\nolimits_{j=1}^{n} (\lambda_j - \mu_j) Z_{dj} \geq 0 \end{aligned}$$

From the dual, which is an envelopment model, it is clear that $\sum_{j=1}^{n}\alpha_j Z_{dj}$, $\sum_{j=1}^{n}\alpha_j X_{ij}$, $\sum_{j=1}^{n}\lambda_j X_{ij}$, $\sum_{j=1}^{n}\mu_j Y_{rj}$, and $\sum_{j=1}^{n}\lambda_j Z_{pj}$ (or $\sum_{j=1}^{n}\mu_j Z_{pj}$) are the projections of VRS-Z_{d0}, VRS-X_{i0}, CRS-X_{i0}, CRS-Y_{r0}, and CRS-Z_{d0}, respectively. The CRS efficiency of stage two is η, which is equal to $1/E_0^2$. The CRS efficiency of stage one is E_0^1, or $\eta E_0 (=\eta \times E_0^1 \times E_0^2)$.

Obviously, the dual model is very different from the envelopment-based model (2.11). Under careful inspection, one finds that the last three sets of constraints have nothing to do with calculating θ_1, and can thus be deleted. In other words, VRS technical efficiency can be calculated independently of the CRS efficiencies.

Similarly, the linearized forms for T_0^2 and its dual are:

$$
\begin{array}{lll}
T_0^2 = \max. \sum_{r=1}^{s} \hat{u}_r Y_{r0} & \text{dual} & \min.\ \theta_2 \\
\text{s.t.}\ \ \sum_{d=1}^{D} \hat{w}_d Z_{d0} + \hat{w}_0 = 1 & \theta_2 & \text{s.t.}\ \ \sum_{j=1}^{n} \beta_j Z_{dj} \le \theta_2 Z_{d0} \\
E_0 \sum_{i=1}^{m} v_i X_{i0} - \sum_{r=1}^{s} u_r Y_{r0} = 0 & \eta & \sum_{j=1}^{n} \beta_j = \theta_2 \\
\sum_{d=1}^{D} w_d Z_{dj} - \sum_{i=1}^{m} v_i X_{ij} \le 0 & \lambda_j & \sum_{j=1}^{n} \beta_j Y_{rj} \ge Y_{r0} \\
\sum_{r=1}^{s} u_r Y_{rj} - \sum_{d=1}^{D} w_d Z_{dj} \le 0 & \mu_j & \sum_{j=1}^{n} \lambda_j X_{ij} \le \eta E_0 X_{i0} \\
\sum_{r=1}^{s} \hat{u}_r Y_{rj} - \sum_{d=1}^{D} \hat{w}_d Z_{dj} - \hat{w}_0 \le 0 & \beta_j & \sum_{j=1}^{n} \mu_j Y_{rj} \ge \eta Y_{r0} \\
 & & \sum_{j=1}^{n} (\lambda_j - \mu_j) Z_{dj} \ge 0
\end{array}
$$

The projections for VRS-Z_{d0} and VRS-Y_{r0} are $\sum_{j=1}^{n}(\beta_j/\theta_2)Z_{dj}$ and $\sum_{j=1}^{n}(\beta_j/\theta_2)$ Y_{rj}, respectively. Other interpretations are similar to that of the first stage. Again, the VRS technical efficiency can be calculated independently of the CRS efficiencies.

The models for stages one and two can be combined as:

$$
\begin{array}{lll}
\text{Max.} \left(\sum_{d=1}^{D} \tilde{w}_d Z_{d0} - \tilde{w}_0\right) + \left(\sum_{r=1}^{s} \hat{u}_r Y_{r0}\right) & \text{dual} & \min.\ \theta_1 + \theta_2 \\
\text{s.t.}\ \ \sum_{i=1}^{m} v_i \tilde{X}_{i0} = 1 & \theta_1 & \text{s.t.}\ \ \sum_{j=1}^{n} \alpha_j X_{ij} \le \theta_1 X_{i0}, \sum_{j=1}^{n} \beta_j Z_{dj} \le \theta_2 Z_{d0} \\
\sum_{d=1}^{D} \hat{w}_d Z_{d0} + \hat{w}_0 = 1 & \theta_2 & \sum_{j=1}^{n} \alpha_j = 1,\quad \sum_{j=1}^{n} \beta_j = \theta_2 \\
E_0 \sum_{i=1}^{m} v_i X_{i0} - \sum_{r=1}^{s} u_r Y_{r0} = 0 & \eta & \sum_{j=1}^{n} \alpha_j Z_{dj} \ge Z_{d0},\quad \sum_{j=1}^{n} \beta_j Y_{rj} \ge Y_{r0} \\
\sum_{d=1}^{D} w_d Z_{dj} - \sum_{i=1}^{m} v_i X_{ij} \le 0 & \lambda_j & \sum_{j=1}^{n} \lambda_j X_{ij} \le \eta E_0 X_{i0} \\
\sum_{r=1}^{s} u_r Y_{rj} - \sum_{d=1}^{D} w_d Z_{dj} \le 0 & \mu_j & \sum_{j=1}^{n} \mu_j Y_{rj} \ge \eta Y_{r0} \\
\sum_{d=1}^{D} \tilde{w}_d Z_{dj} - \tilde{w}_0 - \sum_{i=1}^{m} v_i \tilde{X}_{ij} \le 0 & \alpha_j & \sum_{j=1}^{n} (\lambda_j - \mu_j) Z_{dj} \ge 0 \\
\sum_{r=1}^{s} \hat{u}_r Y_{rj} - \sum_{d=1}^{D} \hat{w}_d Z_{dj} - \hat{w}_0 \le 0 & \beta_j &
\end{array}
$$

The interpretations are straightforward. Note that θ_1 and θ_2 are independent, and can thus be calculated separately. If the model is developed under the envelopment form, then it will be something similar to model (2.9) with the convexity constraints of $\sum \lambda_j = 1$ and $\sum \mu_j = 1$, which is obviously different from the dual of the VRS multiplier model derived here.

If there are multiple solutions, such that the decomposition of $E_0 = E_0^1 \times E_0^2$ is not unique, then stage one (or stage 2) must be calculated first, then calculate the second stage by requiring the CRS efficiency of stage one is equal to E_0^1, i.e. $\sum_{D=1}^{D} w_d Z_{d0} \Big/ \sum_{i=1}^{m} v_i X_{i0} = E_0^1$.

The above discussion reveals the following interesting observation. While in the standard DEA model, the multiplier and envelopment models are equivalent, in the two-stage network DEA, such equivalence does not exist. This indicates that the multiplier-based and envelopment-based network DEA models are two different approaches. We will further illustrate this point in the next section. The next section will show that under general network structures, the multiplier and envelopment network models not only use different efficiency concepts, but also do not correspond with each other.

2.6 Multiplier Versus Envelopment Network DEA: Pitfall

We now assume that in addition to the intermediate measures (z_{dj}, (d =1, 2, . . ., D)), there are inputs to the second stage pictured in Fig. 2.1. We denote these inputs to the second stage shown as $x_{hj}^{stage-2}$ (h =1, 2, . . ., H), using the same notations from Li et al. (2012). Figure 2.1 then becomes

Model (2.2) now becomes

$$\max \theta_1^o * \theta_2^o = \max \frac{\sum_{d=1}^{D} w_d z_{do}}{\sum_{i=1}^{m} v_i x_{io}} * \frac{\sum_{r=1}^{s} u_r y_{ro}}{\sum_{d=1}^{D} w_d z_{do} + \sum_{h=1}^{H} Q_h x_{ho}^{stage-2}}$$

$$\begin{aligned} s.t. \quad & \frac{\sum_{d=1}^{D} w_d z_{dj}}{\sum_{i=1}^{m} v_i x_{ij}} \le 1 \quad \forall j \\ & \frac{\sum_{r=1}^{s} u_r y_{rj}}{\sum_{d=1}^{D} w_d z_{dj} + \sum_{h=1}^{H} Q_h x_{hj}^{stage-2}} \le 1 \quad \forall j \\ & v_i, w_d, Q_h, u_r \ge 0, \forall i, d, h, r \end{aligned} \tag{2.13}$$

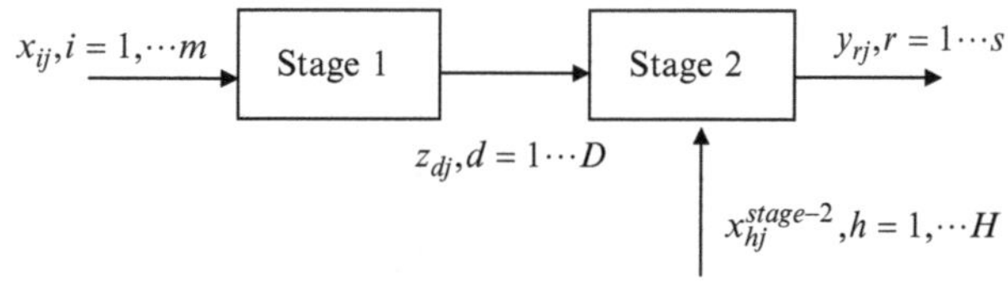

Fig. 2.2 Two-stage process with additional inputs to the second stage

where θ_1^o and θ_2^o represent the ratio efficiencies for stages 1 and 2, respectively. Note that the "link" between the two stages is indicated by the same weights w_d for the intermediate measures. Due to the additional inputs to the second stage $\left(\sum_{h=1}^{H} Q_h x_{ho}^{stage-2}\right)$, model (2.13) cannot be converted into a linear program. Li et al. (2012) introduce a heuristic method to solve this problem.

Model (2.13) is regarded as the multiplier model. To establish the envelopment DEA network model for Fig. 2.2, we follow Tone and Tsutsui (2009) and provide the following model based upon the concept of the production possibility set. Note that in this case, the intermediate measures are treated as "fixed link".

$$
\begin{aligned}
&\min \theta^1 + \theta^2 \\
&s.t. \quad \sum_{j=1}^{n} \lambda_j x_{ij} \le \theta^1 x_{io} \quad i = 1, \ldots, \mathrm{m} \\
&\sum_{j=1}^{n} \lambda_j z_{dj} = \sum_{j=1}^{n} \mu_j z_{dj} \quad d = 1, \ldots, \mathrm{D} \\
&\sum_{j=1}^{n} \mu_j x_{hj}^{stage-2} \le \theta^2 x_{ho}^{stage-2} \quad h = 1, \ldots, \mathrm{H} \\
&\sum_{j=1}^{n} \mu_j y_{rj} \ge y_{ro} \quad r = 1, \ldots, \mathrm{s} \\
&\theta^1, \theta^2 \le 1 \\
&\lambda_j \mu_j \ge 0
\end{aligned}
\tag{2.14}
$$

We use radial measures θ^1, θ^2 rather than slacks-based measures in model (2.14), because in the standard DEA approach, the radial measure in the envelopment model is equivalent to the ratio efficiency defined in the multiplier model.

The issue here is whether θ^{1*} and θ^{2*} obtained from model (2.14) represent the efficiency scores for stages 1 and 2. To address this issue, we need to compare models (2.13) and (2.14). Since as demonstrated in Li et al. (2012), model (2.13) can only be converted into a nonlinear program, we are not able to compare model (2.13) and the dual to model (2.14). Note however that the data set used in Li et al. (2012) yields a unique efficiency decomposition (divisional efficiency). Therefore, we can compare the divisional efficiency scores obtained from models (2.13) and (2.14).

Table 2.6 provides the data for the R&D system for the 30 Provincial Level Regions in China used in Li et al. (2012). The two stages are technology

Table 2.6 Inputs and outputs of R& D for 30 provincial level regions in mainland China

DMU	Region	R&DP	R&DE	S&TF/TFE	Patent	Paper	CV	GDP	TE	UPCDAI	GOHI
1	Beijing	10.34786	668.6351	5.445765	9,157	65,951	1,236.245	12,153.03	483.7932	26,738.48	2,757.14
2	Chongqing	2.00665	79.45994	1.203814	834	13,737	38.31581	6,530.01	42.80071	15,748.67	352.84
3	Shanghai	6.46163	423.3774	7.201873	5,997	32,733	435.4108	15,046.45	1,417.96027	28,837.78	5,557.45
4	Tianjin	2.8783	178.4661	3.023746	1,889	12,472	105.4611	7,521.85	298.92719	21,402.01	1,901.07
5	Anhui	3.01654	135.9535	1.702644	795	13,699	35.61736	10,062.82	88.86487	14,085.74	460.31
6	Fujian	2.27886	135.3819	1.975486	824	9,075	23.25944	12,236.53	533.1911	19,576.83	1,972.01
7	Gansu	1.27445	37.26124	0.817095	227	7,856	35.62869	3,387.56	7.35512	11,929.78	67.39
8	Guangdong	12.97681	652.982	3.887568	11,355	35,773	170.985	39,482.56	3,589.54893	21,574.72	17,161.94
9	Guizhou	0.77328	26.41343	1.040003	322	4,946	1.780611	3,912.68	13.56612	12,862.53	293.64
10	Hainan	0.17583	5.7806	1.249058	84	2,726	0.555627	1,654.21	13.08632	13,750.85	54.75
11	Hebei	3.8808	134.8446	1.125763	691	17,970	17.21118	17,235.48	156.88902	14,718.25	629.17
12	Heilongjiang	3.70197	109.1704	1.062859	1,142	14,553	48.855	8,587	100.82127	12,565.98	311.4
13	Henan	4.79963	174.7599	1.222261	1,129	21,188	26.30461	19,480.46	73.45376	14,371.56	953.23
14	Hubei	5.12124	213.449	1.211472	1,478	25,268	77.03287	12,961.1	99.78796	14,367.48	1,039.52
15	Hunan	3.49591	153.4995	1.339839	1,752	21,042	44.04324	13,059.69	54.92034	15,084.31	648.75
16	Jiangsu	10.67826	701.9529	2.912858	5,322	47,441	108.2184	34,457.3	1,991.9919	20,551.72	13,015.35
17	Jiangxi	1.83522	75.8936	0.857803	386	6,811	9.78927	7,655.18	73.68488	14,021.54	755.65

18	Jilin	2.60875	81.36019	1.28305	719	8,987	19.75983	7,278.75	31.24935	14,006.27	537.66
19	Liaoning	5.43947	232.3687	2.143081	1,993	20,801	119.7095	15,212.49	334.14928	15,761.38	1,313.84
20	Qinghai	0.30013	7.59379	0.982114	35	1,240	8.496721	1,081.27	2.51876	12,691.85	19.22
21	Shandong	8.33303	519.592	1.924254	2,865	26,941	71.9391	33,896.65	794.90706	17,811.04	4,555.71
22	Shanxi	2.52624	80.85633	1.127415	603	6,757	16.20675	7,358.31	28.37455	13,996.55	196.47
23	Shanxi	4.23465	189.5063	1.131443	1,342	26,403	69.80741	8,169.8	39.88149	14,128.76	717.04
24	Sichuan	4.87863	214.459	0.79759	1,596	22,568	54.59769	14,151.28	141.69447	13,839.4	1,766.76
25	Yunnan	1.22051	37.23044	0.972869	476	7,101	10.24687	6,169.75	45.13252	14,423.93	147.17
26	Zhejiang	5.90844	398.8367	3.74258	4,818	25,638	56.45805	22,990.35	1,330.12954	24,610.81	2,672.09
27	Guangxi	1.56993	47.20277	1.114432	326	9,982	1.766182	7,759.16	83.7537	15,451.48	273.67
28	Inner Mongolia	1.27057	52.07259	0.937557	178	3,214	14.76515	9,740.25	23.15476	15,849.19	236.61
29	Ningxia	0.33954	10.44221	1.018058	52	1,365	0.898229	1,353.31	7.4293	14,024.7	32.89
30	Xinjiang	0.82683	21.80426	1.198296	120	5,688	1.207767	4,277.05	109.34563	12,257.52	23.74

Table 2.7 Divisional efficiency scores

	Multiplier model (2.13)		Envelopment model (2.14)		
DMU	Stage 1	Stage 2	Stage 1	Stage 2	Envelopment model feasibility[a]
1	1	0.1598	0.1598	0.0238	
2	1	0.2489	0.2337	0.2595	
3	0.8950	0.5365	0.4802	0.1447	
4	0.6774	0.5704	0.3864	0.1756	
5	0.6697	0.3895	0.1949	0.4453	
6	0.5668	1	0.5668	1	
7	1	0.2207	0.2213	0.1266	
8	1	1	1	1	
9	0.9398	1	0.9398	1	
10	1	1	1	1	
11	0.8885	0.8351	1	0.4819	
12	0.9328	0.2648	0.2271	0.2818	
13	0.8493	0.7373	0.9878	0.2529	
14	0.9060	0.2816	0.2551	0.2815	
15	1	0.3685	0.2421	0.4707	
16	0.9225	1	0.9225	1	
17	0.5644	0.9914	0.8329	0.5330	
18	0.7152	0.4947	0.6137	0.2059	
19	0.6671	0.3668	0.2447	0.2159	
20	0.4573	1	0.4706	0.0957	Infeasible
21	0.7101	0.8176	0.6612	0.7281	
22	0.5708	0.5156	0.6670	0.1194	
23	1	0.1941	0.1896	0.1970	
24	1	0.4566	0.5071	0.4055	
25	1	0.5846	0.9922	0.1486	
26	0.7293	0.9171	0.5854	1	
27	1	1	1	1	
28	0.3599	1	0.3626	1	Infeasible
29	0.4300	1	0.4300	1	
30	1	1	1	1	

[a]We test for whether the divisional efficiency scores based upon multiplier model (2.13) are feasible solution for θ^1, θ^2 in model (2.14)

development process, and economic application process. The inputs to the first stage are: R&D expenditure (R&DE), R&D personnel (R&DP) and the proportion of regional science and technology funds in regional total financial expenditure (S&TF/TFE). The intermediate measures are the number of patents and the number of papers. The second stage also has an input of contract value (CV) in technology market. The outputs from the second stage are GDP, total exports (TE), urban per capita disposable annual income (UPCDAI), and gross output of high-tech industry (GOHI).

Columns 2 and 3 of Table 2.7 report the unique divisional efficiency scores obtained from model (2.13) under Li et al.'s (2012) algorithm. Columns 4 and 5 report the divisional efficiency scores from model (2.14). We observe that except for 8 DMUs (6, 8, 9, 10, 16, 27, 29, 30), divisional efficiency scores from models (2.13) and (2.14) are very different. This indicates that models (2.13) and (2.14) produce different results under general network structures.

We next test whether the divisional efficiency scores for each DMU obtained from the multiplier model (2.13) are feasible solutions for θ^1 and θ^2 in the envelopment model (2.14). The last column in Table 2.7 indicates that DMUs 20 and 28's model (2.13) divisional efficiency scores are infeasible under model (2.14). However, all DMUs' divisional efficiency scores obtained from model (2.14) are feasible scores under model (2.13). In particular, model (2.14) yields projection points. If we apply model (2.13) to the projection points obtained from model (2.14), each DMU is efficient.

The above study indicates that (i) the multiplier and envelopment network DEA models are different with respect to defining divisional efficiency, and (ii) the projection points based upon the envelopment network DEA model are efficient under the multiplier network DEA model.

To further illustrate the above points, we modify the objective function of model (2.14) from $(\theta^1 + \theta^2)$ to $(\theta^1 + \theta^2)/2$, which will not affect the solution. Then the dual of model (2.14) becomes:

$\text{Min.}(\theta^1+\theta^1)/2$	dual	$\max.\sum_{r=1}^{s} u_r Y_{r0}$	
s.t $\sum_{j=1}^{n} \lambda_j X_{ij} \le \theta^1 X_{i0}$	v_i	s.t. $\sum_{i=1}^{m} v_i X_{i0} = 1/2$	
$\sum_{j=1}^{n} \mu_j X_{hj}^{stage-2} \le \theta^2 X_{h0}^{stage-2}$	q_h	$\sum_{h=1}^{H} q_h X_{h0}^{stage-2} = 1/2$	
$\sum_{j=1}^{n} \mu_j Y_{rj} \ge Y_{r0}$	u_r	$\sum_{d=1}^{D} w_d Z_{dj} - \sum_{i=1}^{m} v_i X_{ij} \le 0$	
$\sum_{j=1}^{n} (\lambda_j - \mu_j) Z_{dj} = 0$	w_d	$\sum_{r=1}^{s} u_r Y_{rj} - \sum_{d=1}^{D} w_d Z_{dj}$ $- \sum_{h=1}^{H} q_h X_{hj}^{stage-2} \le 0$	
θ^1, θ^2 free in sign.		w_d free in sign.	

Since the system efficiency can be defined as $\sum_{r=1}^{s} u_r Y_{r0} / \left(\sum_{i=1}^{m} v_i X_{i0} + \sum_{h=1}^{H} q_h X_{h0}^{stage-2}\right)$, the constraint of the dual should be $\left(\sum_{i=1}^{m} v_i X_{i0} + \sum_{h=1}^{H} q_h X_{h0}^{stage-2}\right) = 1$, rather than $\sum_{i=1}^{m} v_i X_{i0} = 1/2$ and $\sum_{h=1}^{H} q_h X_{h0}^{stage-2} = 1/2$. This indicates that model (2.14) is too restrictive to provide correct solutions. (The objective function, $(\theta^1 + \theta^2)/2$, intends to represent the system efficiency.) Moreover, the multiplier w_d should be positive. If we take these conditions into consideration, we would require that $\left(\sum_{i=1}^{m} v_i X_{i0} + \sum_{h=1}^{H} q_h X_{h0}^{stage-2}\right) = 1$, or $\sum_{i=1}^{m} v_i X_{i0} = \pi$ and $\sum_{h=1}^{H} q_h X_{h0}^{stage-2} = 1 - \pi$, and w_d to be positive. Then the dual becomes:

$$
\begin{array}{llll}
\max \sum_{r=1}^{s} u_r Y_{r0} & & \text{dual} & \min \theta^2 \\
\text{s.t. } \sum_{i=1}^{m} v_i X_{i0} = \pi & & \theta^1 & \text{s.t. } \sum_{j=1}^{n} \lambda_j X_{ij} \le \theta^1 X_{i0} \\
\sum_{h=1}^{H} q_h X_{h0}^{stage-2} = 1 - \pi & & \theta^2 & \sum_{j=1}^{n} \mu_j X_{hj}^{2} \le \theta^2 X_{h0}^{stage-2} \\
\sum_{d=1}^{D} w_d Z_{dj} - \sum_{i=1}^{m} v_i X_{ij} \le 0 & & \lambda_j & \theta^1 = \theta^2 \\
\sum_{r=1}^{s} u_r Y_{rj} - \sum_{d=1}^{D} w_d Z_{dj} & & & \\
- \sum_{h=1}^{H} q_h X_{hj}^{stage-2} \le 0 & \mu_j & & \sum_{j=1}^{n} \mu_j Y_{rj} \ge Y_{r0} \\
\pi \text{ free in sign} & & & \sum_{j=1}^{n} (\lambda_j - \mu_j) Z_{dj} \ge 0 \\
 & & & \theta^1, \theta^2 \text{ free in sign}
\end{array}
$$

Although, theoretically, π is free in sign, the constraint of $\sum_{i=1}^{m} v_i X_{i0} = \pi$ ensures it to be positive (and so is $1 - \pi$). The dual indicates that θ^1 and θ^2 are equal and do not represent divisional efficiencies, but rather they represent overall system efficiency.

The above discussion indicates that the so-called divisional efficiency scores in the envelopment model are not efficiency scores for divisions under the concept of ratio DEA efficiency, whether the network structure is a simple two-stage process or a general one.

2.7 Conclusions

The current chapter presents several pitfalls in network DEA modeling. We start the discussion with a simple two-stage network structure where only intermediate measures exist between the two stages and the first stage has inputs only and the second stage outputs only. This simple structure allows one to (i) establish an equivalence between the multiplier-based and envelopment-based network DEA models, and (ii) demonstrate the difference between the multiplier-based and envelopment-based network DEA models.

Under a general network structure, we demonstrate that the outcomes from the multiplier and envelopment models are not necessarily equivalent. The divisional efficiency scores obtained from the multiplier model can be infeasible under the envelopment model under the condition of CRS. We demonstrate that the divisional efficiency scores based upon the envelopment model do not necessarily represent divisional efficiencies, and may actually be the overall efficiency. This indicates that cautions needs to be taken when developing a network DEA model using production possibility sets.

It is our view that overall efficiency along with divisional efficiencies should be defined under the DEA multiplier (ratio) model, as in Kao and Hwang (2008) and

Liang et al. (2008), for example. Such a definition is related to other definitions of efficiency used in engineering and science, as well as in business and economics. For example, the CCR efficiency was modeled on the definition from "combustion" engineering, where efficiency is defined as "the ratio of actual amount of heat liberated ... to the maximum amount which could be liberated" (Charnes et al. (1978)). Many of the business efficiency measures appear in the form of ratios, such as earnings per share and profit per employee.

While in conventional DEA, the envelopment model or the distance function-based efficiency is equivalent to multiplier (ratio) efficiency, in the case of network DEA, the distance function-based envelopment models do not necessarily yield information on divisional efficiency. Although the envelopment network DEA models might give the appearance of providing optimal divisional efficiency, we show that in reality the envelopment efficiency is a measure of overall efficiency.

As a result of the current study, many existing production possibility set-based network DEA models including Tone and Tsutsui's (2009) slacks-based approach need to be re-examined with respect to their rationale for the (divisional) efficiency definition. Our study indicates that current envelopment models are not able to calculate divisional efficiencies. However, this does not mean that it is impossible to calculate divisional efficiencies by using envelopment models; rather there would appear to be a need to develop new envelopment-based models for accomplishing this task.

Finally, due to the fact that we are not able to obtain multiplier divisional efficiency scores under the condition of VRS (because the resulting model cannot be solved as a linear program), we cannot perform such a comparison under the condition of VRS. Therefore, it is important to develop algorithms that will enable one to derive multiplier-based divisional efficiency under VRS.

References

Charnes, A., Cooper, W. W., & Rhodes, E. (1978). Measuring the efficiency of decision making units. *European Journal of Operational Research, 2*, 429–444.

Chen, Y., Cook, W. D., Li, N., & Zhu, J. (2009a). Additive efficiency decomposition in two-stage DEA. *European Journal of Operational Research, 196*, 1170–1176.

Chen, Y., Liang, L., & Zhu, J. (2009b). Equivalence in two-stage DEA approaches. *European Journal of Operational Research, 193*(2), 600–604.

Chen, Y., Cook, W. D., & Zhu, J. (2010). Deriving the DEA frontier for two-stage processes. *European Journal of Operational Research, 202*, 138–142.

Chen, Y., Cook, W. D., Kao, C., & Zhu, J. (2013). Network DEA pitfalls: Divisional efficiency and frontier projection under general network structures. *European Journal of Operational Research, 226*, 507–515.

Cook, W. D., Liang, L., & Zhu, J. (2010). Measuring performance of two-stage network structures by DEA: A review and future perspective. *Omega, 38*, 423–430.

Färe, R., & Grosskopf, S. (2000). Network DEA. *Socio-Economic Planning Sciences, 34*, 35–49.

Kao, C., & Hwang, S.-N. (2008). Efficiency decomposition in two-stage data envelopment analysis: An application to non-life insurance companies in Taiwan. *European Journal of Operational Research, 185*(1), 418–429.

Kao, C., & Hwang, S.-N. (2011). Decomposition of technical and scale efficiencies in two-stage production systems. *European Journal of Operational Research, 211*, 515–519.

Li, Y., Chen, Y., Liang, L., & Xie, J. (2012). DEA models for extended two-stage network structures. *Omega, 40*(5), 611–618.

Liang, L., Cook, W. D., & Zhu, J. (2008). DEA models for two-stage processes: Game approach and efficiency decomposition. *Naval Research Logistics, 55*, 643–653.

Tone, K., & Tsutsui, M. (2009). Network DEA: A slacks-based measure approach. *European Journal of Operational Research, 197*(1), 243–252.

Chapter 3
Efficiency Decomposition in Network Data Envelopment Analysis

Chiang Kao

Abstract In order to measure the efficiency of systems composed of several processes more appropriately, various network data envelopment analysis (DEA) models have been developed. One type of the model, which is able to calculate the system and process efficiencies at the same time, is relational. This paper discusses the relationship between the system and process efficiencies measured from this model, and derives five properties. The first is general to all types of network structure, which states that the efficiency slack of the system is the sum of those of the component processes. This implies that a system is efficient if and only if all its component processes are. The second to fourth correspond to three types of structure, series, parallel, and dynamic. The last states that any unstructured system can be transformed into a series of parallel structures for efficiency decomposition. Numerical examples are used to help explain the idea of each type of decomposition. Efficiency decomposition enables decision makers to identify the processes that cause the inefficiency of a system, and thus to make effective changes to it.

Keywords Data envelopment analysis • Network • Efficiency decomposition • Relational model

3.1 Introduction

Data envelopment analysis (DEA), first developed by Charnes et al. (1978), is a technique for measuring the relative efficiency of a set of decision making units (DMUs) that apply multiple inputs to produce multiple outputs. In its original settings, only the inputs supplied to the system and the outputs produced from it

---

C. Kao (✉)
Department of Industrial and Information Management, National Cheng Kung University, Tainan, Taiwan
e-mail: ckao@mail.ncku.edu.tw

W.D. Cook and J. Zhu (eds.), *Data Envelopment Analysis*,
International Series in Operations Research & Management Science 208,
DOI 10.1007/978-1-4899-8068-7_3,

Fig. 3.1 Block-box system

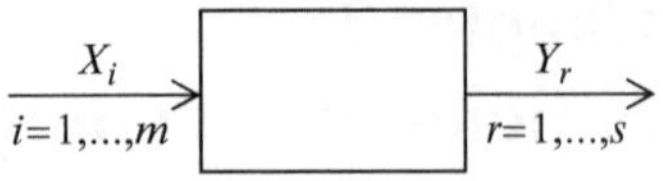

are considered, neglecting the operations and interrelations of the processes within the system. The system is thus called a black-box system, and the associated model a black-box one.

Figure 3.1 shows a black-box system, where inputs X_{ij}, $i = 1, \ldots, m$, are utilized to produce outputs Y_{rj}, $r = 1, \ldots, s$, for each DMU j, $j = 1, \ldots, n$. The input-oriented model for measuring the efficiency of DMU k under the assumption of constant returns to scale can be formulated as (Charnes et al. 1978):

$$\begin{aligned} E_k^{CCR} = \max. & \sum_{r=1}^{s} u_r Y_{rk} \\ \text{s.t.} \ & \sum_{i=1}^{m} v_i X_{ik} = 1 \\ & \sum_{r=1}^{s} u_r Y_{rj} - \sum_{i=1}^{m} v_i X_{ij} \leq 0, \quad j = 1, \ldots, n \\ & u_r, v_i \geq \varepsilon, \quad r = 1, \ldots, s, \quad i = 1, \ldots, m \end{aligned} \tag{3.1}$$

where u_r and v_i are virtual multipliers, and ε is a small non-Archimedean number imposed to avoid ignoring any factor in calculating efficiency (Charnes and Cooper 1984). Larger values of E_k^{CCR} imply a better performance, and a value of 1 is considered as efficient.

Usually a production system is composed of several interrelated processes. If the operations of the component processes are ignored, then Model (3.1) may produce efficiency scores that are misleading. When the internal structure is considered, one faces a network system. Figure 3.2 shows a general network system, where each process p, $p = 1, \ldots, q$, uses exogenous inputs $X_i^{(p)}$, $i \in I^{(p)}$, supplied from outside the system, and endogenous inputs $Z_f^{(p)}$, $f \in M_I^{(p)}$, produced by other processes, to produce outputs $Y_r^{(p)}$, $r \in O^{(p)}$, as the final outputs of the system, and endogenous outputs $Z_g^{(p)}$, $g \in M_O^{(p)}$, to be utilized by other processes. $Z_f^{(p)}$ (or $Z_g^{(p)}$) are also called intermediate products. The whole system has m inputs, s outputs, and h intermediate products. The sets $I^{(p)}$, $O^{(p)}$, $M_I^{(p)}$, and $M_O^{(p)}$ contain the indices of the inputs, outputs, intermediate products used, and intermediate products produced, respectively, for process p.

Comparing the black-box system of Fig. 3.1 with the network system of Fig. 3.2, it is noted that, for each DMU j, the sum of the exogenous inputs of all q processes is equal to the inputs of the system, $\sum_{p=1}^{q} X_{ij}^{(p)} = X_{ij}$, $i = 1, \ldots, m$, and the sum of the exogenous outputs of all q processes is equal to the outputs of the

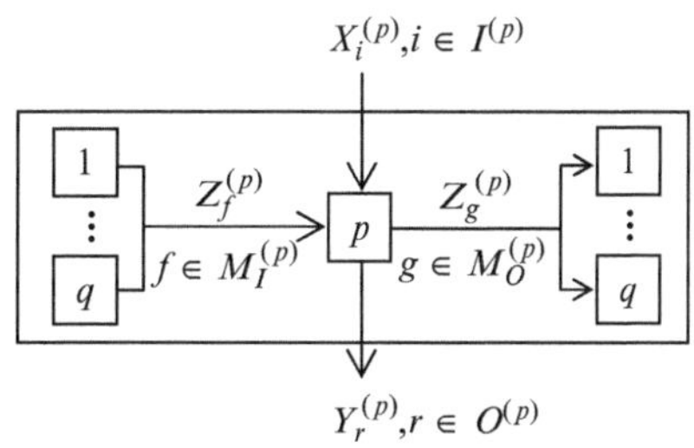

Fig. 3.2 General network system

system, $\sum_{p=1}^{q} Y_{rj}^{(p)} = Y_{rj}$, $r = 1, \ldots, s$. Moreover, the sum of the intermediate products used by all q processes is equal to the sum of the intermediate products produced by all q processes; that is, all intermediate products are produced and consumed within the system.

Several models for measuring the efficiency of a network system have been developed (see, for example, the reviews and classifications in Castelli et al. (2010) and Kao and Hwang (2010)). Some models can measure the system and process efficiencies at the same time, and derive mathematical relationships between them, based on which the most effective way to improve the efficiency of a DMU can be identified. This paper applies the relational model of Kao (2009a) for network systems to decompose the system efficiency into process efficiencies. For structured systems, including series, parallel, and dynamic ones, mathematical relationships exist between the system and process efficiencies. Unstructured systems can thus be transformed into structured ones, and the relationships within them derived. In the next section, we will briefly review the relational DEA model for network systems. The process of efficiency decomposition for series, parallel, dynamic, and general unstructured network systems is then discussed.

3.2 Relational Network DEA Model

In a network system, the same input X_i may be used by several processes, and the same output Y_r may be produced by different processes. The relational model requires the same factor, either input X_i or output Y_r, to have the same multiplier v_i or u_r. The rationale is that the same factor should have the same market value, no matter which process it is associated with. This concept also applies to the intermediate product, and so the same intermediate product Z_f should have the same multiplier w_f associated with it, no matter whether it plays the role of input or output, and no matter which process it is associated with. If different multipliers are used for the same factor, then all processes are actually working independently. In contrast, using the same multiplier relates one process with another.

The difference between the network and black-box models is that the former takes the operations of the component processes into account. Referring to Fig. 3.2, the operation of each process p is described by the following constraint:

$$\left(\sum_{r\in O^{(p)}} u_r Y_{rj}^{(p)} + \sum_{g\in M_O^{(p)}} w_g Z_{gj}^{(p)}\right) - \left(\sum_{i\in I^{(p)}} v_i X_{ij}^{(p)} + \sum_{f\in M_I^{(p)}} w_f Z_{fj}^{(p)}\right) \le 0,$$

$$p = 1, \ldots, q, \quad j = 1, \ldots, n$$

That is, the aggregate output should be less than or equal to the aggregate input. Adding this set of constraints to Model (3.1) and inserting slack variables obtains the following relational network DEA model:

$$\begin{aligned}
E_k = \max.& \sum_{r=1}^{s} u_r Y_{rk} \\
\text{s.t.}& \sum_{i=1}^{m} v_i X_{ik} = 1 \\
& \sum_{r=1}^{s} u_r Y_{rj} - \sum_{i=1}^{m} v_i X_{ij} + s_j = 0, \quad j = 1, \ldots, n \\
& \sum_{r\in O^{(p)}} u_r Y_{rj}^{(p)} + \sum_{g\in M_O^{(p)}} w_g Z_{gj}^{(p)} - \left(\sum_{i\in I^{(p)}} v_i X_{ij}^{(p)} + \sum_{f\in M_I^{(p)}} w_f Z_{fj}^{(p)}\right) \\
& \quad + s_j^{(p)} = 0, \quad p = 1, \ldots, q, \quad j = 1, \ldots, n \\
& u_r, v_i, w_f \ge \varepsilon, \quad r = 1, \ldots, s, \quad i = 1, \ldots, m, \quad f = 1, \ldots, h
\end{aligned} \tag{3.2}$$

Let $(u_r^*, v_i^*, w_f^*, s_k^*, s_k^{(p)*})$ be an optimal solution to Model (3.2). The system and process efficiencies for DMU k are:

$$E_k = \frac{\sum_{r=1}^{s} u_r^* Y_{rk}}{\sum_{i=1}^{m} v_i^* X_{ik}} = \sum_{r=1}^{s} u_r^* Y_{rk}$$

$$E_k^{(p)} = \frac{\sum_{r\in O^{(p)}} u_r^* Y_{rk}^{(p)} + \sum_{g\in M_O^{(p)}} w_g^* Z_{gk}^{(p)}}{\sum_{i\in I^{(p)}} v_i^* X_{ik}^{(p)} + \sum_{f\in M_I^{(p)}} w_f^* Z_{fk}^{(p)}}, \quad p = 1, \ldots, q$$

For each DMU j, the sum of the constraints associated with the q component processes is:

$$\sum_{p=1}^{q}\left[\left(\sum_{r\in O^{(p)}} u_r Y_{rj}^{(p)} + \sum_{g\in M_O^{(p)}} w_g Z_{gj}^{(p)}\right) - \left(\sum_{i\in I^{(p)}} v_i X_{ij}^{(p)} + \sum_{f\in M_I^{(p)}} w_f Z_{fj}^{(p)}\right)\right]$$
$$= \sum_{r=1}^{s} u_r Y_{rj} - \sum_{i=1}^{m} v_i X_{ij}$$

because $\sum_{p=1}^{q} X_{ij}^{(p)} = X_{ij}$, $\sum_{p=1}^{q} Y_{rj}^{(p)} = Y_{rj}$, and all of the intermediate products produced in the system have been consumed. This implies that the constraints associated with the system in Model (3.2) for all DMUs are redundant. Moreover, the slack variable associated with the system of a DMU is equal to the sum of those associated with its processes, i.e., $s_k^* = \sum_{p=1}^{q} s_k^{(p)*}$. Therefore, we have the following property.

Property 1 For a network system, the efficiency slack corresponding to the system is equal to the sum of those corresponding to the process.

This property also implies that a system is efficient if and only if all its component processes are.

Cook et al. (2010) proposed a way to decompose the system efficiency into a weighted average of the process efficiencies. The key point is that their system efficiency is defined differently from the conventional one, so that decomposition is possible. Kao (2013b) also proposed a way to achieve the same goal, and it was slacks-based measures (Tone and Tsutsui 2009), rather than the conventional radial ones.

In the following sections, we will discuss the efficiency decomposition for systems with specific structures, including series, parallel, and dynamic, and then general unstructured systems, based on Model (3.2).

3.3 Series Systems

The simplest network system, which is also the most widely discussed in the literature, is the two-stage one, where the first stage (process) utilizes all the inputs supplied from outside of the system to produce some intermediate products for the second stage (process) to produce the final outputs of the system. Two types of efficiency decomposition have been proposed, the multiplicative one of Kao and Hwang (2008) and the additive one of Chen et al. (2009). Both can be extended to series systems of more than two stages. However, it should be noted that the system efficiency in Chen et al. (2009) is defined differently from the conventional one, in order to make the additive decomposition possible.

Figure 3.3 shows a series system of q processes, where all of the intermediate products produced by a process are utilized by its successor. The intermediate products used by the first process, $Z_f^{(0)}$, are the inputs of the system, X_i, and the

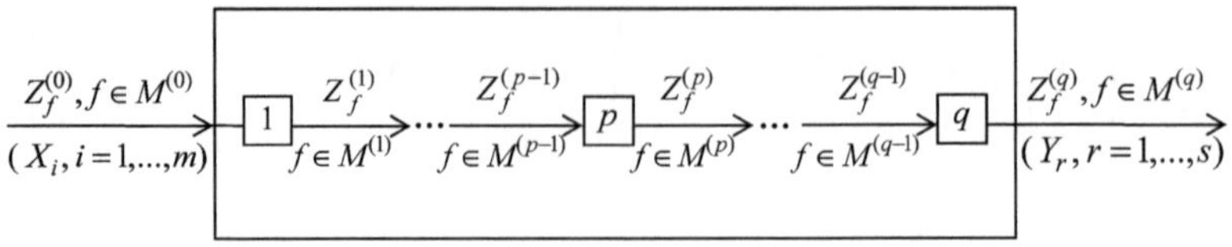

Fig. 3.3 Series system

intermediate products produced by the last process, $Z_f^{(q)}$, are the final outputs of the system, Y_r. Each process, except the first, does not consume exogenous inputs, and all processes, except the last, do not produce exogenous outputs. If we treat X_i as $Z_f^{(0)}$ and Y_r as $Z_f^{(q)}$, then the series system is a special network system, where each process p does not have inputs $X_i^{(p)}$, nor outputs $Y_r^{(p)}$. For this well-structured system, the general network model (3.2) can be formulated more compactly as:

$$
\begin{aligned}
E_k = \max. & \sum_{f \in M^{(q)}} w_f Z_{fk}^{(q)} \\
\text{s.t.} & \sum_{f \in M^{(0)}} w_f Z_{fk}^{(0)} = 1 \\
& \sum_{f \in M^{(q)}} w_f Z_{fj}^{(q)} - \sum_{f \in M^{(0)}} w_f Z_{fj}^{(0)} + s_j = 0, \quad j = 1, \ldots, n \\
& \sum_{f \in M^{(p)}} w_f Z_{fj}^{(p)} - \sum_{f \in M^{(p-1)}} w_f Z_{fj}^{(p-1)} + s_j^{(p)} = 0, \quad p = 1, \ldots, q \\
& u_r, v_i, w_f \geq \varepsilon, \quad r = 1, \ldots, s, \quad i = 1, \ldots, m, \quad f = 1, \ldots, h
\end{aligned}
\tag{3.3}
$$

Denote (u_r^*, v_i^*, w_f^*, s_k^*, $s_k^{(p)*}$) as an optimal solution. The system and process efficiencies are:

$$
E_k = \frac{\sum_{f \in M^{(q)}} w_f^* Z_{fk}^{(q)}}{\sum_{f \in M^{(0)}} w_f^* Z_{fk}^{(0)}}
$$

$$
E_k^{(p)} = \frac{\sum_{f \in M^{(p)}} w_f^* Z_{fk}^{(p)}}{\sum_{f \in M^{(p-1)}} w_f^* Z_{fk}^{(p-1)}}, \quad p = 1, \ldots, q
$$

When all q process efficiencies are multiplied together, one obtains:

$$
\prod_{p=1}^{q} E_k^{(p)} = \prod_{p=1}^{q} \left[\frac{\sum_{f \in M^{(p)}} w_f^* Z_{fk}^{(p)}}{\sum_{f \in M^{(p-1)}} w_f^* Z_{fk}^{(p-1)}} \right] = \frac{\sum_{f \in M^{(q)}} w_f^* Z_{fk}^{(q)}}{\sum_{f \in M^{(0)}} w_f^* Z_{fk}^{(0)}}
$$

Table 3.1 Data for the series system example

	Input		Intermediate product		Output	
DMU	$X_1(Z_1^{(0)})$	$X_2(Z_2^{(0)})$	$Z_3^{(1)}$	$Z_4^{(2)}$	$Y_1(Z_5^{(3)})$	$Y_2(Z_6^{(3)})$
A	1	3	2	3	2	2
B	2	3	3	5	2	2
C	4	2	3	4	3	2
D	4	3	3	5	3	3
E	3	5	4	6	3	3

Table 3.2 Results of the series system example

	Black-box	Network system		Process 1		Process 2		Process 3	
DMU	E_k^{CCR}	E_k	(s_k)	$E_k^{(1)}$	$(s_k^{(1)})$	$E_k^{(2)}$	$(s_k^{(2)})$	$E_k^{(3)}$	$(s_k^{(3)})$
A	1	0.9	(0.1)	1	(0)	0.9	(0.1)	1	(0)
B	0.8571	0.6667	(0.3333)	1	(0)	1	(0)	0.6667	(0.3333)
C	1	0.8	(0.2)	1	(0)	0.8	(0.2)	1	(0)
D	1	0.8	(0.2)	0.8	(0.2)	1	(0)	1	(0)
E	0.7941	0.6666	(0.3334)	0.8888	(0.1112)	0.9	(0.0889)	0.8333	(0.1333)

which is the system efficiency. Thus, for series systems, one has the following property, in addition to the general efficiency decomposition of $s_k^* = \sum_{p=1}^{q} s_k^{(p)*}$, stated in Property 1.

Property 2 The system efficiency in series systems is the product of all process efficiencies.

Consider a series system of three processes, where the first process consumes inputs $X_1(Z_1^{(0)})$ and $X_2(Z_2^{(0)})$ to produce intermediate product $Z_3^{(1)}$, the second process consumes $Z_3^{(1)}$ to produce intermediate product $Z_4^{(2)}$, and the third process consumes $Z_4^{(2)}$ to produce the final outputs $Y_1(Z_5^{(3)})$ and $Y_2(Z_6^{(3)})$. Table 3.1 shows the data of an example of five DMUs, labeled as $A \sim E$. By setting the non-Archimedean number ε to 0.0001, the efficiencies of the black-box system for the five DMUs are calculated via Model (3.1), with the results shown in the second column of Table 3.2. Three DMUs, *A*, *C*, and *D* are efficient. By taking into account the operations of the three processes, that is, applying Model (3.2), none of the five DMUs is efficient, as is seen in the results shown in column three of Table 3.2. Clearly, the network model is more discriminative, as it identifies DMU *A* (from *A*, *C*, and *D*) as the one with the best performance.

The last three double-columns show the efficiencies of the three processes and their associated efficiency slacks. The results verify Property 1, that the efficiency slack of the system is equal to the sum of those of the process. Using DMU *E* to explain this, one has 0.3334 = 0.1112 + 0.0889 + 0.1333. Property 2, which states that the system efficiency is the product of the process efficiencies, is also verified:

$0.6666 = 0.8888 \times 0.9 \times 0.8333$. From the efficiency scores or the efficiency slacks of a DMU, one can identify the process that affects the system efficiency the most. For example, Process 3 has the smallest efficiency score and the largest efficiency slack, and thus improving the efficiency of this process will be the most effective way to improve the system efficiency of this DMU.

3.4 Parallel Systems

Series and parallel are the two basic structures of network systems. Surprisingly, parallel systems are not as widely discussed as series systems in the literature (refer to the review of Kao and Hwang (2010)). Kao (2009b, 2012) followed the relational model of Kao (2009a) to derive a relationship between the system and process efficiencies, in that the former is a weighted average of the latter, either from the input or output side. Here, we will only discuss these from the input side.

Consider a parallel system with q processes operating independently, as shown in Fig. 3.4, where process p uses inputs $X_i^{(p)}$, $i \in I^{(p)}$ to produce outputs $Y_r^{(p)}$, $r \in O^{(p)}$. The sum of the inputs used by all q processes is equal to the system inputs, $\sum_{p=1}^{q} X_{ij}^{(p)} = X_{ij}$. Similarly, the sum of the outputs produced by all q processes is equal to the system outputs, $\sum_{p=1}^{q} Y_{rj}^{(p)} = Y_{rj}$. The parallel system does not have intermediate products to connect different processes. Some studies (Beasley 1995; Färe et al. 1992, 1997) discuss the best way of allocating the total inputs X_i to different processes so that the system will be more efficient. The processes in this case are interrelated. When the operations of the processes are considered, the following constraints, one corresponding to one process, are added to Model (3.1):

$$\sum_{r \in O^{(p)}} u_r Y_{rj}^{(p)} - \sum_{i \in I^{(p)}} v_i X_{ij}^{(p)} \leq 0, \quad p = 1, \ldots, q$$

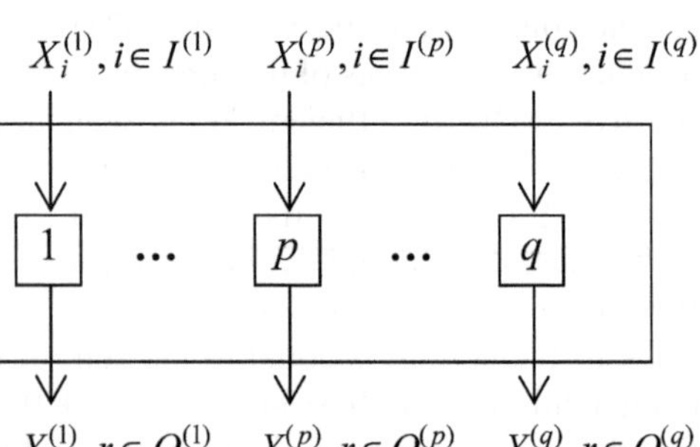

Fig. 3.4 Parallel system

The network model for parallel systems thus becomes:

$$
\begin{aligned}
E_k = \max.&\sum_{r=1}^{s} u_r Y_{rk} \\
\text{s.t.}\ &\sum_{i=1}^{m} v_i X_{ik} = 1 \\
&\sum_{r=1}^{s} u_r Y_{rj} - \sum_{i=1}^{m} v_i X_{ij} + s_j = 0, \quad j = 1, \ldots, n \\
&\sum_{r \in O^{(p)}} u_r Y_{rj}^{(p)} - \sum_{i \in I^{(p)}} v_i X_{ij}^{(p)} + s_j^{(p)} = 0, \ p = 1, \ldots, q, \ j = 1, \ldots, n \\
&u_r, v_i \geq \varepsilon, \quad r = 1, \ldots, s, \quad i = 1, \ldots, m
\end{aligned}
\tag{3.4}
$$

Let $(u_r^*, v_i^*, s_k^*, s_k^{(p)*})$ be an optimal solution. The system and process efficiencies are:

$$
\begin{aligned}
E_k &= \frac{\sum_{r=1}^{s} u_r^* Y_{rk}}{\sum_{i=1}^{m} v_i^* X_{ik}} = \sum_{r=1}^{s} u_r^* Y_{rk} \\
E_k^{(p)} &= \frac{\sum_{r \in O^{(p)}} u_r^* Y_{rk}^{(p)}}{\sum_{i \in I^{(p)}} v_i^* X_{ik}^{(p)}}, \quad p = 1, \ldots, q
\end{aligned}
$$

If we define the weight $\omega^{(p)}$ for process p as the aggregate input consumed by process p in that consumed by all q processes, i.e. $\omega^{(p)} = \sum_{i \in I^{(p)}} v_i^* X_{ik}^{(p)} / \sum_{i=1}^{m} v_i^* X_{ik}$, then the average of the process efficiencies weighted by $\omega^{(p)}$ can be derived as:

$$
\begin{aligned}
\sum_{p=1}^{q} \omega^{(p)} E_k^{(p)} &= \sum_{p=1}^{q} \left[\left(\frac{\sum_{i \in I^{(p)}} v_i^* X_{ik}^{(p)}}{\sum_{i=1}^{m} v_i^* X_{ik}} \right) \left(\frac{\sum_{r \in O^{(p)}} u_r^* Y_{rk}^{(p)}}{\sum_{i \in I^{(p)}} v_i^* X_{ik}^{(p)}} \right) \right] \\
&= \sum_{p=1}^{q} \left(\frac{\sum_{r \in O^{(p)}} u_r^* Y_{rk}^{(p)}}{\sum_{i=1}^{m} v_i^* X_{ik}} \right) = \frac{\sum_{r=1}^{s} u_r^* Y_{rk}}{\sum_{i=1}^{m} v_i^* X_{ik}}
\end{aligned}
$$

which is the system efficiency. Thus, in addition to Property 1, the efficiencies in a parallel system also possess the following property.

Property 3 The system efficiency for a parallel system is equal to the weighted average of the process efficiencies, where the weight associated with process p is the aggregate input consumed by process p in that consumed by all q processes.

Suppose there is a parallel system of three processes. The first process applies input X_1 to produce output Y_1, the second process applies inputs X_1 and X_2 to produce outputs Y_1 and Y_2, and the third process applies input X_2 to produce output Y_2.

Table 3.3 Data for the parallel system example

	Process 1		Process 2				Process 3	
DMU	$X_1^{(1)}$	$Y_1^{(1)}$	$X_1^{(2)}$	$X_2^{(2)}$	$Y_1^{(2)}$	$Y_2^{(2)}$	$X_2^{(3)}$	$Y_2^{(3)}$
A	3	2	1	2	2	3	2	2
B	2	3	1	4	2	4	2	1
C	4	3	3	2	2	4	3	2
D	3	3	3	3	3	2	3	3
E	3	4	4	3	4	3	2	3

There are five DMUs to be evaluated, with the data shown in Table 3.3. By applying Model (3.1), the conventional black-box CCR efficiencies are calculated as shown in the second column of Table 3.4, where three DMUs, *A*, *B*, and *E*, are efficient. The network model (3.4), on the other hand, finds no efficient DMUs. However, it is able to distinguish DMU *B* as the best, *A* the second, and *E* the third. The two inefficient DMUs evaluated by the black-box model are still evaluated as the worst by the network model.

The efficiencies of the three processes, efficiency slacks (in parentheses), and weights (in square brackets) are shown in the last three double-columns of Table 3.4. Based on Property 1, the efficiency slack of the system must be equal to the sum of those of the three processes, and this is easily verified. For example, DMU *A* has an efficiency slack of 0.24 for the system, which is exactly the sum of those of the three processes, 0.1, 0, and 0.14. Furthermore, according to Property 3, the system efficiency is a weighted average of the process efficiencies. For DMU *A*, the weights for the three processes are $3v_1 = 0.18$, $v_1 + 2v_2 = 0.44$, and $2v_2 = 0.38$. The total weight is $4v_1 + 4v_2 = 1$. The weighted average of the three process efficiencies is $0.18(0.4444) + 0.44(1) + 0.38(0.6316) = 0.76$, which is exactly the value of the system efficiency.

3.5 Dynamic Systems

Another type of network systems that also can be described systematically is dynamic systems (Chen and Dalen 2010; Jaenicke 2000; Nemoto and Goto 1999; Tone and Tsutsui 2010), where a DMU repeats the same operation from period to period, and two consecutive periods are connected by carryovers.

Figure 3.5 shows the structure of the dynamic system, where each period (process) *p* consumes inputs $X_i^{(p)}$ and carryover $Z_f^{(p-1)}$ to produce outputs $Y_r^{(p)}$ and carryover $Z_f^{(p)}$. If we examine the structure carefully, we will find that it is the structure of the series system in Fig. 3.3 overlapped with that of the parallel system in Fig. 3.4. Horizontally, it is a series system where each process consumes intermediate products (carryovers) produced by the proceeding process to produce the same ones for the succeeding process to use. Vertically, it is a parallel system where each process consumes the same exogenous inputs to produce the same

Table 3.4 Results of the parallel system example

	Black-box	Network system		Process 1		Process 2		Process 3	
DMU	E_k^{CCR}	E_k	(s_k)	$E_k^{(1)}$ $[\omega^{(1)}]$	$(s_k^{(1)})$	$E_k^{(2)}$ $[\omega^{(2)}]$	$(s_k^{(2)})$	$E_k^{(3)}$ $[\omega^{(3)}]$	$(s_k^{(3)})$
A	1	0.76	(0.24)	0.4444 [0.1800]	(0.1)	1 [0.4400]	(0)	0.6316 [0.3800]	(0.14)
B	1	0.8331	(0.1669)	0.75 [0.6661]	(0.1665)	0.9994 [0.3336]	(0)	0.3333 [0.0003]	(0.0002)
C	0.9623	0.6715	(0.3285)	0.5 [0.1752]	(0.0876)	1 [0.4088]	(0)	0.4211 [0.4161]	(0.2409)
D	0.7989	0.5714	(0.4286)	0.6667 [0.4282]	(0.1427)	0.5713 [0.5000]	(0.2143)	0.0042 [0.0715]	(0.0715)
E	1	0.7591	(0.2409)	0.8889 [0.1314]	(0.0146)	0.6420 [0.5912]	(0.2117)	0.9474 [0.2774]	(0.0146)

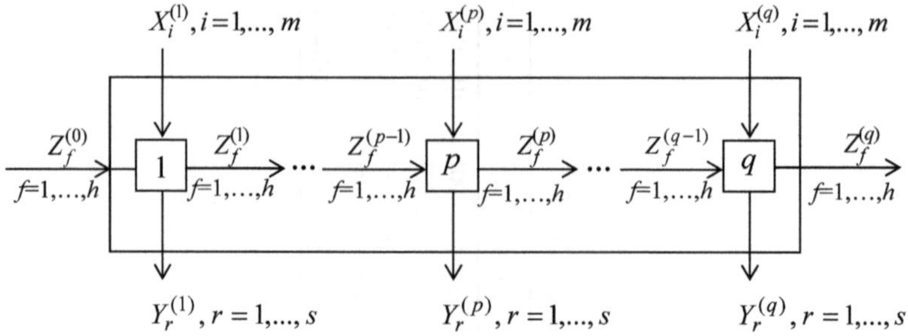

Fig. 3.5 Dynamic system

exogenous outputs. Kao (2013a) developed a relational model to calculate the system and process efficiencies of this system.

Referring to Fig. 3.5, the inputs to the system are a little different from those of the conventional black-box one. In addition to the conventional inputs of $\sum_{p=1}^{q} X_{ij}^{(p)} = X_{ij}$, the dynamic system has initial carryovers of $Z_f^{(0)}$. Similarly, in addition to the conventional outputs of $\sum_{p=1}^{q} Y_{rj}^{(p)} = Y_{rj}$, it has terminal carryovers of $Z_f^{(q)}$. The constraint corresponding to each process p is:

$$\left(\sum_{r=1}^{s} u_r Y_{rj}^{(p)} + \sum_{f=1}^{h} w_f Z_{fj}^{(p)}\right) - \left(\sum_{i=1}^{m} v_i X_{ij}^{(p)} + \sum_{f=1}^{h} w_f Z_{fj}^{(p-1)}\right) \leq 0, \quad p = 1, \ldots, q.$$

The complete model is:

$$\begin{aligned}
E_k = \max.\ & \sum_{r=1}^{s} u_r Y_{rk} + \sum_{f=1}^{h} w_f Z_{fk}^{(q)} \\
\text{s.t.}\ & \sum_{i=1}^{m} v_i X_{ik} + \sum_{f=1}^{h} w_f Z_{fk}^{(0)} = 1 \\
& \sum_{r=1}^{s} u_r Y_{rj} + \sum_{f=1}^{h} w_f Z_{fj}^{(q)} - \left(\sum_{i=1}^{m} v_i X_{ij} + \sum_{f=1}^{h} w_f Z_{fj}^{(0)}\right) \\
& \quad + s_j = 0, \quad j = 1, \ldots, n \\
& \sum_{r=1}^{s} u_r Y_{rj}^{(p)} + \sum_{f=1}^{h} w_f Z_{fj}^{(p)} - \left(\sum_{i=1}^{m} v_i X_{ij}^{(p)} + \sum_{f=1}^{h} w_f Z_{fj}^{(p-1)}\right) \\
& \quad + s_j^{(p)} = 0, \quad p = 1, \ldots, q, \quad j = 1, \ldots, n \\
& u_r, v_i, w_f \geq \varepsilon, \quad r = 1, \ldots, s, \quad i = 1, \ldots, m, \quad f = 1, \ldots, h
\end{aligned} \tag{3.5}$$

For each DMU j, the sum of the constraints associated with the q processes is:

$$\sum_{p=1}^{q}\left[\left(\sum_{r=1}^{s}u_rY_{rj}^{(p)}+\sum_{f=1}^{h}w_fZ_{fj}^{(p)}\right)-\left(\sum_{i=1}^{m}v_iX_{ij}^{(p)}+\sum_{f=1}^{h}w_fZ_{fj}^{(p-1)}\right)\right]$$
$$=\left(\sum_{r=1}^{s}u_rY_{rj}+\sum_{f=1}^{h}w_fZ_{fj}^{(q)}\right)-\left(\sum_{i=1}^{m}v_iX_{ij}+\sum_{f=1}^{h}w_fZ_{fj}^{(0)}\right) \tag{3.6}$$

which is just the constraint corresponding to the system. Therefore, similar to the previous network models, the constraints corresponding to the system are redundant, and the efficiency slack of the system, s_j, is the sum of those of the processes, $s_j^{(p)}$. The system and process efficiencies, according to Model (3.5) are:

$$E_k=\frac{\sum_{r=1}^{s}u_r^*Y_{rk}+\sum_{f=1}^{h}w_f^*Z_{fk}^{(q)}}{\sum_{i=1}^{m}v_i^*X_{ik}+\sum_{f=1}^{h}w_f^*Z_{fk}^{(0)}}=\sum_{r=1}^{s}u_r^*Y_{rk}+\sum_{f=1}^{h}w_f^*Z_{fk}^{(q)}=1-s_k^*$$
$$E_k^{(p)}=\frac{\sum_{r=1}^{s}u_r^*Y_{rk}^{(p)}+\sum_{f=1}^{h}w_f^*Z_{fk}^{(p)}}{\sum_{i=1}^{m}v_i^*X_{ik}^{(p)}+\sum_{f=1}^{h}w_f^*Z_{fk}^{(p-1)}},\quad p=1,\ldots,q$$

where $(u_r^*,\ v_i^*,\ w_f^*,\ s_k^*,\ s_k^{(p)*})$ is an optimal solution. Dividing both sides of equation (3.6) by $(\sum_{i=1}^{m}v_i^*X_{ik}+\sum_{f=1}^{h}w_f^*Z_{fk}^{(0)})$ and multiplying the left-hand side by an identity of $(\sum_{i=1}^{m}v_i^*X_{ik}^{(p)}+\sum_{f=1}^{h}w_f^*Z_{fk}^{(p-1)})/(\sum_{i=1}^{m}v_i^*X_{ik}^{(p)}+\sum_{f=1}^{h}w_f^*Z_{fk}^{(p-1)})$ for period p results in:

$$\sum_{p=1}^{q}\left[\left(E_k^{(p)}-1\right)\left(\sum_{i=1}^{m}v_i^*X_{ik}^{(p)}+\sum_{f=1}^{h}w_f^*Z_{fk}^{(p-1)}\right)\Big/\left(\sum_{i=1}^{m}v_i^*X_{ik}+\sum_{f=1}^{h}w_f^*Z_{fk}^{(0)}\right)\right]=E_k-1$$

Denoting $\omega^{(p)}=(\sum_{i=1}^{m}v_i^*X_{ik}^{(p)}+\sum_{f=1}^{h}w_f^*Z_{fk}^{(p-1)})/(\sum_{i=1}^{m}v_i^*X_{ik}+\sum_{f=1}^{h}w_f^*Z_{fk}^{(0)})$, the above equation becomes:

$$\sum_{p=1}^{q}\left[\omega^{(p)}\left(1-E_k^{(p)}\right)\right]=1-E_k$$

Note that the sum of $\omega^{(p)}$ over $p=1,\ldots,q$ is: $\sum_{p=1}^{q}\omega^{(p)}=1+\sum_{p=2}^{q}\sum_{f=1}^{h}w_f^*Z_{fk}^{(p-1)}/(\sum_{i=1}^{m}v_i^*X_{ik}+\sum_{f=1}^{h}w_f^*Z_{fk}^{(0)})$, which is greater than 1. We thus obtain the following property.

Property 4 The complement of the system efficiency for a dynamic system is a linear, and not necessarily a convex, combination of those of the process efficiencies.

Consider a dynamic system of three periods, where each DMU applies two inputs, X_1 and X_2, to produce two outputs, Y_1 and Y_2, with a carryover Z. Table 3.5 contains the data of five hypothetical DMUs. When the black-box Model (3.1) is applied (with $Z^{(0)}$ as another input and $Z^{(3)}$ as another output), four DMUs, $A \sim D$, are evaluated as efficient, as shown in the second column of Table 3.6. If the network Model (3.5) is applied, then no DMU is efficient; however, the system efficiencies, as shown in the third column of Table 3.6, are able to distinguish their performance. The last three double-columns show the results for the three periods, including efficiency scores, efficiency slacks (in parentheses), and weights (in square brackets).

To check whether Property 1 holds or not for this dynamic system, we simply add the efficiency slacks of the three periods to see if the sum is equal to that of the system. Using DMU E to explain this, the sum of the three periods is 0.2960+ 0.0368 + 0.0007, or 0.3335, which is just the efficiency slack of the system. Other DMUs can be checked similarly.

According to Property 4, the complement of the system efficiency is a linear combination of those of the three processes. To see this, the complement of the system efficiency for DMU E is $1 - E_E = 0.3335$, and the sum of those of the three processes multiplied by their respective weights is 0.4448(1 − 0.3347) + 0.3339 (1 − 0.8897) + 0.2233(1 − 0.9969), or 0.3334 (=1 − 0.6666). (Note that there is a slight difference caused by truncation.) The number under the system efficiency, in square brackets, is the total weight of the three processes. Only DMUs A and E have a total weight close to 1, indicating that their complement of the system efficiency is close to the weighted average of those of the three processes.

3.6 Unstructured Systems

Network systems that cannot be described systematically, as the preceding three structures can be, are considered as unstructured systems. Since the decomposition of this type of system depends on the related structure, we will use an example to explain this. The application of this idea to systems with other structures should then be straightforward.

Consider the structure shown in Fig. 3.6, which has been discussed in the literature (Färe and Grosskopf 1996, 2000; Färe et al. 2007). The system has three processes, where Process 1 consumes inputs X_1 and X_2 to produce output Y_1, Process 2 consumes inputs X_1 and X_2 to produce output Y_2, and Process 3 consumes a part of Y_1, denoted as $Y_1^{(I)}$, and a part of Y_2, denoted as $Y_2^{(I)}$, to produce output Y_3. The outputs of the system are the remaining parts of Y_1 and Y_2, denoted as

Table 3.5 Data for the example dynamic system

	Period 1					Period 2					Period 3					System				
DMU	$Z^{(0)}$	$X_1^{(1)}$	$X_2^{(1)}$	$Y_1^{(1)}$	$Y_2^{(1)}$	$Z^{(1)}$	$X_1^{(2)}$	$X_2^{(2)}$	$Y_1^{(2)}$	$Y_2^{(2)}$	$Z^{(2)}$	$X_1^{(3)}$	$X_2^{(3)}$	$Y_1^{(3)}$	$Y_2^{(3)}$	$Z^{(3)}$	X_1	X_2	Y_1	Y_2
A	3	1	2	3	3	2	3	2	2	2	3	1	2	2	2	2	5	6	7	7
B	3	2	2	3	4	3	2	4	3	1	5	1	2	2	2	2	6	8	8	7
C	2	2	3	2	4	3	2	2	3	2	4	2	3	2	2	3	6	8	7	8
D	3	3	3	3	2	3	3	3	3	3	5	1	3	3	3	3	7	9	9	8
E	5	3	4	2	3	4	3	3	4	3	6	2	2	3	3	3	8	9	9	9

Table 3.6 Results for the example dynamic system

	Black-box	Network system		Period 1		Period 2		Period 3	
DMU	E_k^{CCR}	$E_k\ [\sum_{p=1}^{3}\omega^{(p)}]$	(s_k)	$E_k^{(1)}\ [\omega^{(1)}]$	$(s_k^{(1)})$	$E_k^{(2)}\ [\omega^{(2)}]$	$(s_k^{(2)})$	$E_k^{(3)}\ [\omega^{(3)}]$	$(s_k^{(3)})$
A	1	0.7777 [1.0010]	(0.2223)	1 [0.3335]	(0)	0.6669 [0.3339]	(0.1112)	0.6671 [0.3335]	(0.1110)
B	1	0.75 [1.5263]	(0.25)	0.9189 [0.4868]	(0.0395)	1 [0.5395]	(0)	0.5789 [0.5]	(0.2105)
C	1	0.7352 [1.6171]	(0.2648)	0.9441 [0.5293]	(0.0296)	0.9413 [0.4998]	(0.0294)	0.6499 [0.5880]	(0.2059)
D	1	0.7128 [1.5333]	(0.2872)	0.6381 [0.5385]	(0.1949)	0.9619 [0.5385]	(0.0205)	0.8427 [0.4564]	(0.0718)
E	0.8571	0.6665 [1.0020]	(0.3335)	0.3347 [0.4448]	(0.2960)	0.8897 [0.3339]	(0.0368)	0.9969 [0.2233]	(0.0007)

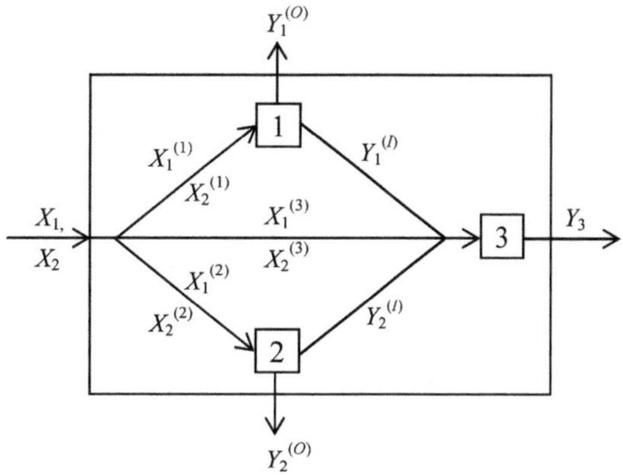

Fig. 3.6 Example of unstructured network system

$Y_1^{(O)}$ and $Y_2^{(O)}$, respectively, and Y_3. Following Model (3.2), the model for measuring the system and process efficiencies is formulated as:

$$
\begin{aligned}
E_k = \max.\ & u_1 Y_{1k}^{(O)} + u_2 Y_{2k}^{(O)} + u_3 Y_{3k} \\
\text{s.t.}\ & v_1 X_{1k} + v_2 X_{2k} = 1 \\
& \left(u_1 Y_{1j}^{(O)} + u_2 Y_{2j}^{(O)} + u_3 Y_{3j}\right) - \left(v_1 X_{1j} + v_2 X_{2j}\right) + s_j = 0, \quad j = 1, \ldots, n \\
& u_1 Y_{1j} - \left(v_1 X_{1j}^{(1)} + v_2 X_{2j}^{(1)}\right) + s_j^{(1)} = 0 \quad j = 1, \ldots, n \\
& u_2 Y_{2j} - \left(v_1 X_{1j}^{(2)} + v_2 X_{2j}^{(2)}\right) + s_j^{(2)} = 0, \quad j = 1, \ldots, n \\
& u_3 Y_{3j} - \left(v_1 X_{1j}^{(3)} + v_2 X_{2j}^{(3)} + u_1 Y_{1j}^{(I)} + u_2 Y_{2j}^{(I)} + s_j^{(3)} = 0, \quad j = 1, \ldots, n\right) \\
& u_1, u_{21}, u_3, v_1, v_2 \geq \varepsilon
\end{aligned}
\tag{3.7}
$$

After the optimal multipliers u_1^*, u_2^*, u_3^*, v_1^*, and v_2^* are calculated from this model, the efficiencies of the system and three processes are obtained as:

$$
\begin{aligned}
E_k &= \frac{u_1^* Y_{1k}^{(O)} + u_2^* Y_{2k}^{(O)} + u_3^* Y_{3k}}{v_1^* X_{1k} + v_2^* X_{2k}} \\
E_k^{(1)} &= \frac{u_1^* Y_{1k}}{v_1^* X_{1k}^{(1)} + v_2^* X_{2k}^{(1)}} \\
E_k^{(2)} &= \frac{u_2^* Y_{2k}}{v_1^* X_{1k}^{(2)} + v_2^* X_{2k}^{(2)}} \\
E_k^{(3)} &= \frac{u_3^* Y_{3k}}{v_1^* X_{1k}^{(3)} + v_2^* X_{2k}^{(3)} + u_1^* Y_{1k}^{(I)} + u_2^* Y_{2k}^{(I)}}
\end{aligned}
$$

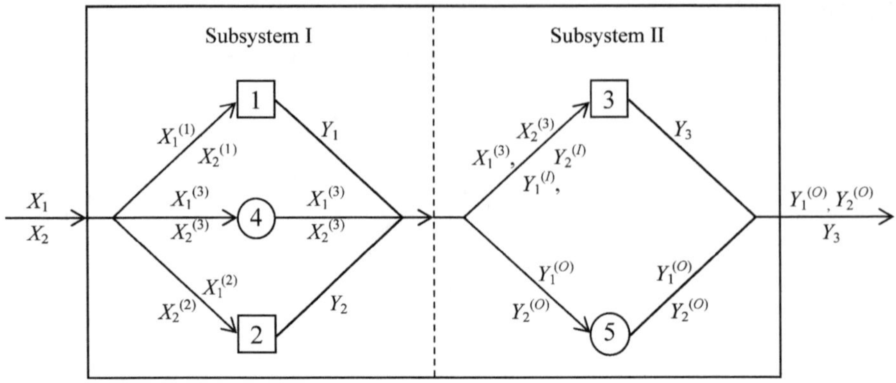

Fig. 3.7 Series-parallel transformation of the example unstructured system

Since the constraint corresponding to the system is the sum of those corresponding to the three processes, the efficiency slack of the system is the sum of those of the process, $s_k^* = s_k^{(1)*} + s_k^{(2)*} + s_k^{(3)*}$, as stated in Property 1.

Another way to decompose the system efficiency, proposed by Kao (2009a), is to express the system as a series of subsystems by introducing dummy processes, where each subsystem is composed of processes connected in parallel. Then, by applying Property 2 for decomposing efficiencies for series systems, and Property 3 for parallel systems, the system efficiency can be expressed as a function of process efficiencies. Figure 3.7 shows a transformation of this system, which is composed of two subsystems connected in series and each subsystem is composed of a set of processes connected in parallel, where squares and circles represent real and dummy processes, respectively. A dummy process does not exist in the original system, and is used solely for representation. The system has two inputs, X_1 and X_2. The first subsystem has three processes, real 1 and 2 and dummy 4, where Process 1 consumes $X_1^{(1)}$ and $X_2^{(1)}$ to produce Y_1, Process 2 consumes $X_1^{(2)}$ and $X_2^{(2)}$ to produce Y_2, and Process 4 consumes $X_1^{(3)}$ and $X_2^{(3)}$ to produce the same items $X_1^{(3)}$ and $X_2^{(3)}$ as outputs. The outputs of this subsystem, Y_1, Y_2, $X_1^{(3)}$, and $X_2^{(3)}$, are then used by the second subsystem, which also has two processes, real 3 and dummy 5. In this subsystem, Process 3 consumes $X_1^{(3)}$, $X_2^{(3)}$, $Y_1^{(I)}$, and $Y_2^{(I)}$ to produce Y_3, and Process 5 consumes $Y_1^{(O)}$ and $Y_2^{(O)}$ to produce the same items $Y_1^{(O)}$ and $Y_2^{(O)}$. The outputs of this subsystem, $Y_1^{(O)}$, $Y_2^{(O)}$, and Y_3, are the outputs of the system.

Since the outputs and inputs are the same for each dummy process, the corresponding constraint is redundant, and the process is always efficient, with an efficiency score of 1. For example, the constraint corresponding to dummy Process 4 is $(v_1X_{1j}^{(3)} + v_2X_{2j}^{(3)}) - (v_1X_{1j}^{(3)} + v_2X_{2j}^{(3)}) \leq 0$, which certainly holds, and the efficiency for DMU k is $(v_1^*X_{1k}^{(3)} + v_2^*X_{2k}^{(3)})/(v_1^*X_{1k}^{(3)} + v_2^*X_{2k}^{(3)})$, which is definitely equal to 1. The efficiency of each subsystem is the aggregate output divided by the aggregate input:

$$E_k^I = \frac{u_1^*Y_{1k} + u_2^*Y_{2k} + v_1^*X_{1k}^{(3)} + v_2^*X_{2k}^{(3)}}{v_1^*X_{1k} + v_2^*X_{2k}}$$

$$E_k^{II} = \frac{u_1^*Y_{1k}^{(O)} + u_2^*Y_{2k}^{(O)} + u_3^*Y_{3k}}{u_1^*Y_{1k} + u_2^*Y_{2k} + v_1^*X_{1k}^{(3)} + v_2^*X_{2k}^{(3)}}$$

To restrict the subsystem efficiency to be less than or equal to one, the aggregate output must be no greater than the aggregate input, which for the two subsystems are:

$$\left(u_1Y_{1j} + u_2Y_{2j} + v_1X_{1j}^{(3)} + v_2X_{2j}^{(3)}\right) - \left(v_1X_{1j} + v_2X_{2j}\right) \le 0, \qquad j = 1, \ldots, n$$
$$\left(u_1Y_{1j}^{(O)} + u_2Y_{2j}^{(O)} + u_3Y_{3j}\right) - \left(u_1Y_{1j} + u_2Y_{2j} + v_1X_{1j}^{(3)} + v_2X_{2j}^{(3)}\right) \le 0, \quad j = 1, \ldots, n$$

By deleting identical terms, they become:

$$u_1Y_{1j} + u_2Y_{2j} - \left(v_1X_{1j}^{(1)} + v_2X_{2j}^{(1)} + v_1X_{1j}^{(2)} + v_2X_{2j}^{(2)}\right) \le 0, \quad j = 1, \ldots, n$$
$$u_3Y_{3j} - \left(v_1X_{1j}^{(3)} + v_2X_{2j}^{(3)} + u_1Y_{1j}^{(I)} + u_2Y_{2j}^{(I)}\right) \le 0, \qquad j = 1, \ldots, n$$

which are exactly the sum of the two constraints corresponding to Processes 1 and 2, and that corresponding to Process 3 formulated in Model (3.7), respectively. Therefore, they are redundant, and need not repeat in the model.

Clearly, the product of the two subsystem efficiencies is equal to the system efficiency, $E_k^I \times E_k^{II} = E_k$, as required by Property 2 for series systems. Furthermore, since each subsystem has a parallel structure, according to Property 3, the subsystem efficiency is a weighted average of those of the component processes:

$$E_k^I = \omega^{(1)}E_k^{(1)} + \omega^{(2)}E_k^{(2)} + \omega^{(4)}E_k^{(4)} = \omega^{(1)}E_k^{(1)} + \omega^{(2)}E_k^{(2)} + \left(1 - \omega^{(1)} - \omega^{(2)}\right)$$
$$E_k^{II} = \omega^{(3)}E_k^{(3)} + \omega^{(5)}E_k^{(5)} = \omega^{(3)}E_k^{(3)} + \left(1 - \omega^{(3)}\right)$$

where $\omega^{(1)} = (v_1^*X_{1k}^{(1)} + v_2^*X_{2k}^{(1)})/(v_1^*X_{1k} + v_2^*X_{2k})$, $\omega^{(2)} = (v_1^*X_{1k}^{(2)} + v_2^*X_{2k}^{(2)})/(v_1^*X_{1k} + v_2^*X_{2k})$, and $\omega^{(4)} = (v_1^*X_{1k}^{(3)} + v_2^*X_{2k}^{(3)})/(v_1^*X_{1k} + v_2^*X_{2k})$ for Subsystem I, and $\omega^{(3)} = (v_1^*X_{1k}^{(3)} + v_2^*X_{2k}^{(3)} + u_1^*Y_{1k}^{(I)} + u_2^*Y_{2k}^{(I)})/(v_1^*X_{1k}^{(3)} + v_2^*X_{2k}^{(3)} + u_1^*Y_{1k} + u_2^*Y_{2k})$ and $\omega^{(5)} = (u_1^*Y_{1k}^{(O)} + u_2^*Y_{2k}^{(O)})/(v_1^*X_{1k}^{(3)} + v_2^*X_{2k}^{(3)} + u_1^*Y_{1k} + u_2^*Y_{2k}) = 1 - \omega^{(3)}$ for Subsystem II. In sum, we have $E[\omega^{(1)}E_k^{(1)} + \omega^{(2)}E_k^{(2)} + (1 - \omega^{(1)} - \omega^{(2)})][\omega^{(3)}E_k^{(3)} + (1 - \omega^{(3)})]$.

Table 3.7 Data for the example unstructured system

DMU	X_1	$(X_1^{(1)}X_1^{(2)}X_1^{(3)})$	X_2	$(X_2^{(1)}X_2^{(2)}X_2^{(3)})$	Y_1	$(Y_1^{(I)}Y_1^{(O)})$	Y_2	$(Y_2^{(I)}Y_2^{(O)})$	Y_3
A	6	(1 2 3)	7	(2 3 2)	4	(2 2)	4	(2 2)	2
B	9	(2 3 4)	5	(1 2 2)	3	(1 2)	3	(2 1)	3
C	9	(2 3 4)	8	(2 3 3)	5	(3 2)	5	(2 3)	4
D	8	(2 3 3)	6	(2 1 3)	5	(3 2)	4	(1 3)	4
E	7	(2 3 2)	9	(3 3 3)	5	(3 2)	4	(3 1)	3

This series-parallel type of representation can be applied to other network systems. The idea is to find the longest path of the system being considered. The number of processes in the path will be the number of subsystems, and each subsystem will be composed of a number of processes plus one dummy process, where the latter is used to carry the inputs to be used in later stages and the outputs already produced. With this representation, the system efficiency can be decomposed into the process efficiencies according to the following property.

Property 5 For general network systems, the system efficiency is the product of a set of subsystem efficiencies, where the subsystem efficiency is a weighted average of some process efficiencies and 1.

Table 3.7 contains a set of data taken from Färe et al. (2007). To make it more complicated, two more DMUs, *D* and *E*, are added. By applying Model (3.1), the black-box efficiencies for the five DMUs are calculated, with the results shown in column two of Table 3.8. Of the five DMUs, three are efficient. The network Model (3.7), on the other hand, identifies DMU *D* as the best, followed by *B* and *A*. Interestingly, this model ranks DMU *C* the second, in contrast to the worst, as ranked by the black-box model. The results are shown in the third column of Table 3.8.

The last two sets of columns show the results of the efficiencies of the two subsystems. First, it is noted that the product of the two subsystem efficiencies is equal to the system efficiency. Using DMU *E* to explain this, we have $0.8261 \times 0.7401 = 0.6114$. Second, the efficiency slack of the system is equal to the sum of those of the three processes. For DMU *E*, this is $0.3886 = 0.0870 + 0.0870 + 0.2147$. Third, the efficiency of the subsystem is the average of those of the component processes weighted by $\omega^{(p)}$. For Subsystem I of DMU *E*, this is $0.8261 = 0.3043 \times 0.7143 + 0.3913 \times 0.7778 + (1 - 0.3043 - 0.3913)$ and for Subsystem II it is $0.7401 = 0.8026 \times 0.6762 + (1 - 0.8026)$. Combining these two results, the system efficiency is decomposed as: $0.6114 = [0.3043 \times 0.7143 + 0.3913 \times 0.7778 + (1 - 0.3043 - 0.3913)] \times [0.8026 \times 0.6762 + (1 - 0.8026)]$. The other four DMUs can be verified similarly.

Table 3.8 Results for the example unstructured system

	Black-box	Network system		Subsystem I					Subsystem II		
DMU	E_k^{CCR}	E_k	(s_k)	E_k^I	$E_k^{(1)}$ $[\omega^{(1)}]$	$(s_k^{(1)})$	$E_k^{(2)}$ $[\omega^{(2)}]$	$(s_k^{(2)})$	E_k^{II}	$E_k^{(3)}$ $[\omega^{(3)}]$	$(s_k^{(3)})$
A	1	0.6513	(0.3487)	1	1 [0.2105]	(0)	1 [0.3684]	(0)	0.6513	0.5093 [0.7105]	(0.3487)
B	1	0.7926	(0.2074)	0.7926	1 [0.2057]	(0)	0.4585 [0.3829]	(0.2074)	1	1 [0.7532]	(0)
C	0.8888	0.8095	(0.1905)	0.9107	1 [0.2381]	(0)	0.75 [0.3571]	(0.0893)	0.8889	0.8455 [0.7190]	(0.1012)
D	1	1	(0)	1	1 [0.3125]	(0)	1 [0.2188]	(0)	1	1 [0.7190]	(0)
E	0.9996	0.6114	(0.3886)	0.8261	0.7143 [0.3043]	(0.0870)	0.7778 [0.3913]	(0.0870)	0.7401	0.6762 [0.8026]	(0.2147)

3.7 Conclusion

Network production systems are very common in the real world. Ignoring the operations of their internal processes by applying conventional black-box models may overstate the efficiency of such systems. More seriously, some results may also be misleading. This paper measures the efficiency of network systems using the relational model of Kao (2009a), and decomposes the system efficiency into process efficiencies. Since the operations of the processes have been taken into account, the network models have stronger discriminating power than the black-box one in ranking the performance of a set of DMUs. The decomposition is expressed by mathematical relationships. Based on the decomposition, the processes that cause the inefficiency of the system are identified, and the most effective way to improve the efficiency of a DMU is obtained.

Several properties are obtained in decomposing the system efficiency of different structures. First, for any network system, the efficiency slack of the system is equal to the sum of those of the component processes. A consequence of this is that a system is efficient if and only if all its component processes are. Second, for series systems, the system efficiency is the product of the process efficiencies. Third, for parallel systems, the system efficiency is a weighted average of the process efficiencies. Fourth, the system efficiency is a linear, and not necessarily convex, combination of the process efficiencies.

Series and parallel are the two basic structures of network systems. Any network system can be expressed as a series of subsystems, where each subsystem is composed of a set of processes connected in parallel. Based on the properties of efficiency decomposition for series and parallel structures, a relationship between the system and process efficiencies is then obtained. A decomposition is thus possible for other unstructured network systems, but it is dependent on the structure of the system. Property 5 states that the system efficiency is the product of a set of subsystem efficiencies, where the subsystem efficiency is a weighted average of some process efficiencies and 1. How many processes are included in each subsystem depends on the structure of the network system.

The efficiency decomposition discussed in this paper is based on the relational model of Kao (2009a). In that paper, the models for network systems are classified as independent, connected, and relational. Whether there are different ways of decomposition for the other two types of models is a direction for future studies. In addition to radial measures of efficiency, there is the slacks-based measure (SBM) proposed by Tone and Tsutsui (2009). In addition, Kao (2013b) recently decomposed the system efficiency of general network systems measured from an SBM model into a weighted average of the process efficiencies. Whether there are different ways of decomposition for other non-radial efficiency measures is another avenue for future research.

References

Beasley, J. E. (1995). Determining teaching and research efficiencies. *Journal of Operational Research Society, 46*, 441–452.

Castelli, L., Pesenti, R., & Ukovich, W. (2010). A classification of DEA models when the internal structure of the Decision Making Units is considered. *Annals of Operations Research, 173*, 207–235.

Charnes, A., & Cooper, W. W. (1984). The non-Archimedean CCR ratio for efficiency analysis: A rejoinder to Boyd and Färe. *European Journal of Operational Research, 15*, 333–334.

Charnes, A., Cooper, W. W., & Rhodes, E. (1978). Measuring the efficiency of decision making units. *European Journal of Operational Research, 2*, 429–444.

Chen, C. M., & van Dalen, J. (2010). Measuring dynamic efficiency: Theories and an integrated methodology. *European Journal of Operational Research, 203*, 749–760.

Chen, Y., Cook, W. D., Ning, L., & Zhu, J. (2009). Additive efficiency decomposition in two-stage DEA. *European Journal of Operational Research, 196*, 1170–1176.

Cook, W. D., Zhu, J., Bi, G., & Yang, F. (2010). Network DEA: Additive efficiency decomposition. *European Journal of Operational Research, 207*, 1122–1129.

Färe, R., & Grosskopf, S. (1996). Productivity and intermediate products: A frontier approach. *Economics Letters, 50*, 65–70.

Färe, R., & Grosskopf, S. (2000). Network DEA. *Socio-Economic Planning Sciences, 34*, 35–49.

Färe, R., Grosskopf, S., & Li, S. K. (1992). Linear programming models for firm and industry performance. *The Scandinavian Journal of Economics, 94*, 599–608.

Färe, R., Grabowski, R., Grosskopf, S., & Kraft, S. (1997). Efficiency of a fixed but allocatable input: A non-parametric approach. *Economics Letters, 56*, 187–193.

Färe, R., Grosskopf, S., & Whittaker, G. (2007). Network DEA. In J. Zhu & W. D. Cook (Eds.), *Modeling data irregularities and structural complexities in data envelopment analysis* (pp. 209–240). New York: Springer.

Jaenicke, E. C. (2000). Testing for intermediate outputs in dynamic DEA models: Accounting for soil capital in rotational crop production and productivity measures. *Journal of Productivity Analysis, 14*, 247–266.

Kao, C. (2009a). Efficiency decomposition in network data envelopment analysis: A relational model. *European Journal of Operational Research, 192*, 949–962.

Kao, C. (2009b). Efficiency measurement for parallel production systems. *European Journal of Operational Research, 196*, 1107–1112.

Kao, C. (2012). Efficiency decomposition for parallel production systems. *Journal of Operational Research Society, 63*, 64–71.

Kao, C. (2013a). Dynamic data envelopment analysis: A relational analysis. *European Journal of Operational Research, 227*, 325–330.

Kao, C. (2013b). Efficiency decomposition in network data envelopment analysis with slacks-based measures. *European Journal of Operational Research, 92*, 949–962.

Kao, C., & Hwang, S. N. (2008). Efficiency decomposition in two-stage data envelopment analysis: An application to non-life insurance companies in Taiwan. *European Journal of Operational Research, 185*, 418–429.

Kao, C., & Hwang, S. N. (2010). Efficiency measurement for network systems: IT impact on firm performance. *Decision Support Systems, 48*, 437–446.

Nemoto, J., & Goto, M. (1999). Dynamic data envelopment analysis modeling intertemporal behavior of a firm in the presence of productive inefficiencies. *Economics Letters, 64*, 51–56.

Tone, K., & Tsutsui, M. (2009). Network DEA: A slacks-based measure approach. *European Journal of Operational Research, 197*, 243–252.

Tone, K., & Tsutsui, M. (2010). Dynamic DEA: A slacks-based measure approach. *Omega, International Journal of Management Science, 38*, 145–156.

Chapter 4
Two-Stage Network Processes: DEA Frontier Identification

Yao Chen, Wade D. Cook, and Joe Zhu

Abstract The current chapter focuses on how to identify DEA frontier when decision making units (DMUs) are in forms of two-stage network processes. In these two stage network processes, all the outputs from the first stage are intermediate measures that make up the inputs to the second stage. Due to the existence of intermediate measures, the usual procedure of adjusting the inputs or outputs by the efficiency scores, as in the standard DEA approach, does not necessarily yield a frontier projection. The current chapter presents an approach for determining the frontier points for inefficient DMUs within the framework of two-stage network processes.

Keywords Data envelopment analysis (DEA) • Efficiency • Two-stage • Intermediate measure • Frontier • Network

---

Part of this chapter is based upon Chen, Y., Cook, W. D. and Zhu, Joe, Deriving the DEA Frontier for Two-Stage Processes, European Journal of Operational Research, Vol. 202 (2010), 138–142, with permissions from Elsevier Science.

Y. Chen (✉)
Manning School of Business, University of Massachusetts at Lowell, Lowell, MA 01845, USA
e-mail: Yao_Chen@uml.edu

W.D. Cook
Schulich School of Business, York University, Toronto, ON M3J 1P3, Canada
e-mail: wcook@schulich.yorku.ca

J. Zhu
School of Business, Worcester Polytechnic Institute, Worcester, MA 01609, USA
e-mail: jzhu@wpi.edu

W.D. Cook and J. Zhu (eds.), *Data Envelopment Analysis*,
International Series in Operations Research & Management Science 208,
DOI 10.1007/978-1-4899-8068-7_4, © Springer Science+Business Media New York 2014

4.1 Introduction

Data envelopment analysis (DEA) is an approach for identifying best practices of peer decision making units (DMUs), in the presence of multiple inputs and outputs (Charnes et al. 1978). DEA provides not only efficiency scores for inefficient DMUs, but also provides for efficient projections for inefficient units onto the best-practice frontier. In recent years, a number of DEA studies have focused on two-stage network processes (see Chap. 1).

Consider a generic two-stage process as shown in Fig. 4.1, for each of a set of n DMUs. We assume each DMU_j $(j = 1, 2, \ldots, n)$ has m inputs x_{ij}, $(i = 1, 2, \ldots, m)$ to the first stage, and D outputs z_{dj}, $(d = 1, 2, \ldots, D)$ from that stage. These D outputs then become the inputs to the second stage, hence behaving as intermediate measures. The outputs from the second stage are y_{rj}, $(r = 1, 2, \ldots, s)$.

As pointed out by a number of authors, including Kao and Huang (2008), Lewis and Sexton (2004), and Castelli et al. (2010), adjusting the inputs and outputs by the efficiency scores in a two-stage process is generally not sufficient to yield a frontier projection. Chen et al. (2009) present a model similar to that of Kao and Huang (2008), but in an additive format. However, as with the multiplicative setup, the usual input and output adjustments do not yield the efficient frontier here either. For many of the cases addressed in the DEA literature dealing with structured DMUs, this will be the situation.

In the sections to follow we show that the overall efficiency scores resulting from Kao and Hwang (2008) and Liang et al. (2008) are not direct indicators of potential input reductions or output increases not realized by the inefficient DMUs, e.g. how much more output each DMU can produce given its present inputs, or how much each DMU could reduce its input-use while still producing the same output. In other words, the resulting DEA scores do not provide complete information on how to project inefficient DMUs onto the DEA frontier for a specific two-stage network process. Although we know the efficiency scores, we still do not know where the DEA frontier is.

This chapter presents the work of Chen et al. (2010) who develop an approach for determining the DEA frontier or DEA projections for inefficient DMUs under the framework of Kao and Hwang (2008) and Liang et al. (2008). We revisit Kao and Hwang's (2008) application involving Taiwanese non-life insurance companies and illustrate the fact that none of the DMUs are efficient, meaning that the usual DEA projections fail to identify the frontier. We then present the approach of Chen et al. (2009) for determining the DEA frontier for the two-stage processes, and illustrate this using the aforementioned data set. Conclusions are given in the last section.

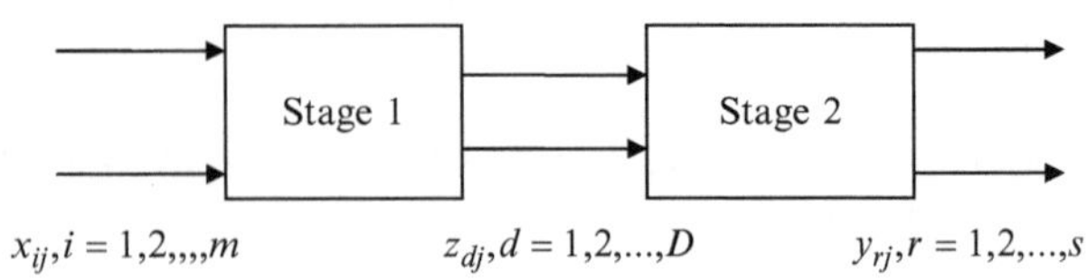

Fig. 4.1 Two-stage process

4.2 Input- and Output-Oriented DEA Models for Two-Stage Network Processes

4.2.1 *Input Orientation*

For DMU_j we denote the efficiency ratios for the first stage as θ_j^1 and the second as θ_j^2. Based upon the *input-oriented* DEA model of Charnes et al. (1978), we define

$$\theta_j^1 = \frac{\sum_{d=1}^{D} w_d z_{dj}}{\sum_{i=1}^{m} v_i x_{ij}} \quad \text{and} \quad \theta_j^2 = \frac{\sum_{r=1}^{s} u_r y_{rj}}{\sum_{d=1}^{D} \widetilde{w}_d z_{dj}} \tag{4.1}$$

where v_i, w_d, $\widetilde{w}_d$, and u_r are unknown non-negative weights. It is assumed that w_d are set equal to $\widetilde{w}_d$ as in Kao and Hwang (2008) and Liang et al. (2008). As a result, the two-stage overall efficiency ratio is defined as $\theta_j^1 \bullet \theta_j^2$ which is equal to $\theta_j = \frac{\sum_{r=1}^{s} u_r y_{ro}}{\sum_{i=1}^{m} v_i x_{io}}$. To calculate the overall efficiency of θ_j, Kao and Hwang (2008) and Liang et al. (2008) present the following (centralized) model

$$\begin{aligned} &Max \ \theta_j^1 \bullet \theta_j^2 = \frac{\sum_{r=1}^{s} u_r y_{ro}}{\sum_{i=1}^{m} v_i x_{io}} \\ &s.t. \ \ \theta_j^1 \leq 1 \text{ and } \theta_j^2 \leq \text{ and } w_d = \widetilde{w}_d. \end{aligned} \tag{4.2}$$

Applying the usual Charnes-Cooper transformation, model (4.2) can be converted into the following linear program

$$\begin{aligned} &\text{Max} \sum_{r=1}^{s} u_r y_{ro} \\ &\text{s.t.} \sum_{r=1}^{s} u_r y_{rj} - \sum_{d=1}^{D} w_d z_{dj} \leq 0 \ j = 1, 2, \ldots., n \\ &\sum_{d=1}^{D} w_d z_{dj} - \sum_{i=1}^{m} v_i x_{ij} \leq 0 \ j = 1, 2, \ldots., n \\ &\sum_{i=1}^{m} v_i x_{io} = 1 \\ &w_d \geq 0, \ d = 1, 2, \ldots, D; \ v_i \geq , \ i = 1, 2, \ldots, m; \ u_r \geq , \ r = 1, 2, \ldots, s \end{aligned} \tag{4.3}$$

The dual to model (4.3) can be expressed as

$$\begin{aligned}
&Min\theta \\
&s.t.\ \sum_{j=1}^{n}\lambda_j x_{ij} \le \theta x_{io} \quad i=1,2,\ldots,m \\
&\sum_{j=1}^{n}\mu_j y_{rj} \ge y_{ro} \quad r=1,2,\ldots,m \\
&\sum_{j=1}^{n}(\lambda_j-\mu_j)z_{dj} \ge 0 \quad d=1,2,\ldots,D \\
&\lambda_j,\mu_j \ge 0, \quad \theta \le 1
\end{aligned} \tag{4.4}$$

4.2.2 Output Orientation

The *output-oriented* version of the above model is given by:

$$\begin{aligned}
&\text{Min}\ \frac{\sum_{i=1}^{m} v_i x_{ij_0}}{\sum_{r=1}^{s} u_r y_{rj_0}} \\
&s.t.\ \theta_j^1 \le 1 \quad \text{and} \quad \theta_j^2 \le 1 \quad \text{for all } j \\
&w_d = \widetilde{w}_d \quad \text{for all } d,
\end{aligned} \tag{4.5}$$

which is equivalent to the linear programming formulation

$$\begin{aligned}
&\text{Min} \ \sum_{i=1}^{m} v_i x_{io} \\
&\text{s.t.} \ \sum_{r=1}^{s} u_r y_{rj} - \sum_{d=1}^{D} w_d z_{dj} \le 0 \quad j=1,2,....,n \\
&\sum_{d=1}^{D} w_d z_{dj} - \sum_{i=1}^{m} v_i x_{ij} \le 0 \quad j=1,2,....,n \\
&\sum_{r=1}^{s} u_r y_{ro} = 1 \\
&w_d, \quad d=1,2,\ldots,D; \quad v_i, \quad i=1,2,\ldots,m; \quad u_r, \quad r=1,2,\ldots,s \ge 0
\end{aligned} \tag{4.6}$$

The dual to model (4.6) can be expressed as

$$
\begin{aligned}
& Max\phi \\
& s.t. \ \sum_{j=1}^{n} \lambda_j x_{ij} \le x_{io} && i = 1, 2, \ldots, m \\
& \sum_{j=1}^{n} \mu_j y_{rj} \ge \phi y_{ro} && r = 1, 2, \ldots, m \\
& \sum_{j=1}^{n} (\lambda_j - \mu_j) z_{dj} \ge 0 && d = 1, 2, \ldots, D \\
& \lambda_j, \mu_j \ge 0 \quad \phi \ge 1
\end{aligned}
\tag{4.7}
$$

Note that both the input- and output-oriented two-stage DEA models (4.4) and (4.7) are identical to the standard DEA envelopment model if $\lambda_j = \mu_j$.

To illustrate their model, Kao and Hwang (2008) present a numerical example with three DMUs A, B, and C which use 2, 4, and 5 units of input X to produce 1.5, 4, and 4 units of intermediate product Z at stage 1, which in turn become inputs to stage 2 to produce 1.5, 5, and 6 units of output Y, respectively. It is shown that DMU A has an overall score of 0.5, indicating that this DMU would be able to produce twice as much if it used its inputs efficiently. In this example the overall DEA efficiency scores show how much more output each DMU can produce given its present inputs, or how much each DMU could reduce its input-use while still producing the same output. i.e., the overall efficiency scores obtained from model (4.4) (or (4.7)) are intended to identify the DEA frontier points. The following discussion shows that this can be not true and that we, therefore, need alternative models to determine the frontier points for two-stage processes.

4.3 DEA Frontier

We here revisit the two-stage application involving 24 Taiwanese non-life insurance companies studied in Kao and Hwang (2008). The two stages represent premium acquisition and profit generation respectively. The inputs to the first stage are operational expenses and insurance expenses, and the outputs from the second stage are underwriting profit and investment profit. There are two intermediate measures between the two stages, namely direct written premiums and reinsurance premiums. The data appear in Table 4.1.

The second and third columns of Table 4.2 report the overall efficiency scores obtained from models (4.4) and (4.7), respectively. It can be seen that $\theta_j^* = 1/\phi_j^*$ for all the DMUs, as expected under the condition of constant returns to scale (CRS). As in the standard DEA approach, we first calculate the DEA projections by multiplying the current inputs (outputs) with the related optimal values θ^* (ϕ^*) while keeping the intermediate measures constant. We then apply these DEA projections to models (4.4) and (4.7) to examine whether they are efficient. Note that the current study does not consider potential DEA slacks in the projections.

Table 4.1 Data set

DMU		Operation expenses (X1)	Insurance expenses (X2)	Direct written premiums (Z1)	Reinsurance premiums (Z2)	Underwriting profit (Y1)	Investment profit (Y2)
1	Taiwan Fire	1,178,744	673,512	7,451,757	856,735	984,143	681,687
2	Chung Kuo	1,381,822	1,352,755	10,020,274	1,812,894	1,228,502	834,754
3	Tai Ping	1,177,494	592,790	4,776,548	560,244	293,613	658,428
4	China Mariners	601,320	594,259	3,174,851	371,863	248,709	177,331
5	Fubon	6,699,063	3,531,614	37,392,862	1,753,794	7,851,229	3,925,272
6	Zurich	2,627,707	668,363	9,747,908	952,326	1,713,598	415,058
7	Taian	1,942,833	1,443,100	10,685,457	643,412	2,239,593	439,039
8	Ming Tai	3,789,001	1,873,530	17,267,266	1,134,600	3,899,530	622,868
9	Central	1,567,746	950,432	11,473,162	546,337	1,043,778	264,098
10	The First	1,303,249	1,298,470	8,210,389	504,528	1,697,941	554,806
11	Kuo Hua	1,962,448	672,414	7,222,378	643,178	1,486,014	18,259
12	Union	2,592,790	650,952	9,434,406	1,118,489	1,574,191	909,295
13	Shingkong	2,609,941	1,368,802	13,921,464	811,343	3,609,236	223,047
14	South China	1,396,002	988,888	7,396,396	465,509	1,401,200	332,283
15	Cathay Century	2,184,944	651,063	10,422,297	749,893	3,355,197	555,482
16	Allianz President	1,211,716	415,071	5,606,013	402,881	854,054	197,947
17	Newa	1,453,797	1,085,019	7,695,461	342,489	3,144,484	371,984
18	AIU	757,515	547,997	3,631,484	995,620	692,731	163,927
19	North America	159,422	182,338	1,141,951	483,291	519,121	46,857
20	Federal	145,442	53,518	316,829	131,920	355,624	26,537
21	Royal & Sunalliance	84,171	26,224	225,888	40,542	51,950	6,491
22	Aisa	15,993	10,502	52,063	14,574	82,141	4,181
23	AXA	54,693	28,408	245,910	49,864	0.1	18,980
24	Mitsui Sumitomo	163,297	235,094	476,419	644,816	142,370	16,976

Source: Kao and Hwang (2008)

Table 4.2 DEA scores

DMU	Input-oriented	Output-oriented	Efficiency scores for projected DMUs: Case I[a] Input-oriented	Case I Output-oriented	Case I 1/Output-oriented score	Case II[b] Input-oriented	Case II Output-oriented	Case II 1/Input-oriented score
1	0.6992	1.4301	0.2202	1.4722	0.6792	0.7134	1.0075	1.4018
2	0.6248	1.6006	0.2296	1.5634	0.6396	0.6275	1.0015	1.5937
3	0.6900	1.4492	0.2206	1.4492	0.6900	1	1.4492	1
4	0.3042	3.2871	0.2295	1.5614	0.6405	0.4323	1.3805	2.3131
5	0.7670	1.3038	0.2199	1.4843	0.6737	0.9738	1.1940	1.0269
6	0.3897	2.5662	0.2364	1.6220	0.6165	0.4057	1.0377	2.4651
7	0.2766	3.6156	0.2229	1.5405	0.6491	0.5378	1.3296	1.8593
8	0.2752	3.6341	0.2177	1.5665	0.6384	0.5113	1.3782	1.9556
9	0.2233	4.4787	0.2192	1.5226	0.6568	0.2920	1	3.4251
10	0.4660	2.1461	0.2311	1.6117	0.6205	0.6627	1.1607	1.5090
11	0.1639	6.1005	0.1965	1.7292	0.5783	0.3267	1.3504	3.0612
12	0.7596	1.3165	0.2366	1.5135	0.6607	0.7596	1	1.3165
13	0.2078	4.8121	0.2053	1.7447	0.5732	0.5435	1.2335	1.8400
14	0.2886	3.4645	0.2192	1.5261	0.6553	0.5178	1.3800	1.9312
15	0.6138	1.6291	0.2295	1.7451	0.5730	0.7047	1	1.4190
16	0.3202	3.1235	0.2233	1.6082	0.6218	0.3847	1.1023	2.5991
17	0.3600	2.7777	0.2181	1.5901	0.6289	0.8066	1.3825	1.2397
18	0.2588	3.8634	0.2209	1.5260	0.6553	0.3737	1.2602	2.6762
19	0.4112	2.4319	0.2367	1.7495	0.5716	0.4158	1	2.4051
20	0.5466	1.8297	0.2093	1.8341	0.5452	0.7891	1.0716	1.2673
21	0.2008	4.9806	0.2251	1.8015	0.5551	0.2795	1.3324	3.5777
22	0.5895	1.6963	0.2122	1.6963	0.5895	1	1.6963	1
23	0.4203	2.3790	0.2206	1.4492	0.6900	0.5599	1.1764	1.7860
24	0.1348	7.4178	0.2351	1.7297	0.5781	0.3351	1.2652	2.9843

[a]Case I: all DMUs are replaced with their projections

[b]Case II: only the DMU under evaluation is replaced with its projection, and the data for other DMUs remain unchanged

We now consider two cases. In Case I, all the DMUs are first replaced by the projected DMUs and then models (4.4) and (4.7) are applied. Namely, in model (4.4), for all j, x_{1j} and x_{2j} are replaced with $\theta_j^* x_{1j}$ and $\theta_j^* x_{2j}$ and all the intermediate measures and outputs are kept unchanged, and in model (4.7), for all j, y_{1j} and y_{2j} are replaced with $\phi_j^* y_{1j}$ and $\phi_j^* y_{2j}$, and all the intermediate measures and inputs are kept unchanged. The fourth column of Table 4.2 reports the input-oriented scores obtained from model (4.4) and the fifth column of Table 4.2 displays the output-oriented scores obtained from model (4.7). It can be seen that none of the projected DMUs are efficient. It can also be seen that the newly obtained θ_j^* is no longer equal to $1/\varphi_j^*$.

In Case II, we only replace the DMU under evaluation by its projection in models (4.4) and (4.7). Namely, we apply models (4.4) and (4.7) to the projected DMU_o, and the data for other DMU_j $(j \neq o)$ remain unchanged. The seventh column of Table 4.2

reports the input-oriented scores obtained from model (4.4) and the eighth column of Table 4.2 displays the output-oriented scores obtained from model (4.7). It can be seen that except for DMUs 3 and 22 (under the input-orientation), and DMUs 9, 12, 15 and 19 (under the output-orientation), none of the projected DMUs are efficient. Also, models (4.4) and (4.7) do not identify the same efficient DMUs. It can also be seen that the output-oriented score is not equal to the reciprocal of the input-oriented score for all the DMUs, as expected under the condition of CRS.

Under the standard DEA approach, Case I and Case II should yield identical results. However, Case I and Case II yield very different results under both models (4.4) and (4.7). Thus, models (4.4) and (4.7) only yield an overall efficiency score, but fail to provide the complete information on how to project inefficient DMUs on to the DEA frontier.

To address this deficiency, Chen et al. (2010) propose a model that is equivalent to model (4.4) (or model (4.7)) and generates a set of new inputs, outputs and intermediate measures that constitute an efficient point (projection) under model (4.4) (or (4.7)).

To this end, consider the input-oriented model (4.4). For each DMU_o, we introduce $\widetilde{z}_{do}$ (d = 1, …, D), representing a set of new intermediate measures to be determined. We then break the constraints $\sum_{j=1}^{n}(\lambda_j - \mu_j)z_{dj} \geq 0$ into two new sets of constraints

$$\sum_{j=1}^{n}\lambda_j z_{dj} \geq \widetilde{z}_{do} \qquad d = 1, \ldots, \mathrm{D}$$
$$\sum_{j=1}^{n}\mu_j z_{dj} \leq \widetilde{z}_{do} \qquad d = 1, \ldots, \mathrm{D}$$

The first new set of constraints treats the $\widetilde{z}_{do}$ as "outputs", and the second set treats the $\widetilde{z}_{do}$ as "inputs". We now propose the following DEA type model:

$$\begin{aligned}
&\min \widetilde{\theta} \\
&s.t. \quad \sum_{j=1}^{n}\lambda_j x_{ij} \leq \widetilde{\theta} x_{io} \quad i = 1, \ldots, \mathrm{m} \\
&\sum_{j=1}^{n}\mu_j y_{rj} \geq y_{ro} \quad r = 1, \ldots, \mathrm{s} \\
&\sum_{j=1}^{n}\lambda_j z_{dj} \geq \widetilde{z}_{do} \quad d = 1, \ldots, \mathrm{D} \\
&\sum_{j=1}^{n}\mu_j z_{dj} \leq \widetilde{z}_{do} \quad d = 1, \ldots, \mathrm{D} \\
&\widetilde{z}_{do} \geq 0, \quad d = 1, \ldots, \mathrm{D} \\
&\lambda_j \geq 0, \quad j = 1, \ldots, n \\
&\mu_j \geq 0, \quad j = 1, \ldots, n \\
&\widetilde{\theta} \leq 1
\end{aligned} \tag{4.8}$$

Chen et al. (2010) show that model (4.8) and model (4.4) yield the same efficiency score, and model (4.8) provides an efficient projection.

In fact, the dual to model (4.8) can be expressed as:

$$
\begin{aligned}
&\text{Max} \sum_{r=1}^{s} u_r y_{ro} \\
&\text{s.t.} \quad \sum_{r=1}^{s} u_r y_{rj} - \sum_{d=1}^{D} w_d^2 z_{dj} \le 0 \quad j = 1, 2, \ldots, n \\
&\sum_{d=1}^{D} w_d^1 z_{dj} - \sum_{i=1}^{m} v_i x_{ij} \le 0 \quad j = 1, 2, \ldots., n \\
&\sum_{i=1}^{m} v_i x_{io} = 1 \\
&w_d^2 - w_d^1 \le 0, \quad d = 1, 2, \ldots, D \\
&w_d^1, w_d^2 \ge 0, \quad d = 1, 2, \ldots, D; \quad v_i \ge, \quad i = 1, 2, \ldots, m; \quad u_r \ge, \quad r = 1, 2, \ldots, s
\end{aligned}
\tag{4.9}
$$

Note that in model (4.8), the constraints $\widetilde{z}_{do} \ge 0$ are redundant, hence we can omit $\widetilde{z}_{do} \ge 0$, and let the $\widetilde{z}_{do}$ be unrestricted variables. Then, the constraint $w_d^2 - w_d^1 \le 0$, $d = 1, 2, \ldots, D$ in model (4.9) becomes $w_d^2 = w_d^1$ meaning that we put the same weights w_d on the z_{dj} in both settings (as outputs from stage 1 and as inputs to stage 2), as in model (4.2). Thus, problem (4.9) above is identical to problem (4.3). This further indicates that model (4.8) yields the same overall efficiency score as model (4.4).

Based upon model (4.8), the projection point for DMU_o is given by $(\widetilde{\theta}^* x_{io}, \widetilde{z}_{dj}^*, y_{ro})$ which is efficient under models (4.8) and (4.4), namely the optimal objective function value for model (4.8) is equal to unity for this projection.

In a similar manner, we can show that in the output-oriented case, model (4.7) is equivalent to the following model

$$
\begin{aligned}
&\max \widetilde{\phi} \\
&s.t. \quad \sum_{j=1}^{n} \lambda_j x_{ij} \le x_{io} \qquad i = 1, \ldots, \mathrm{m} \\
&\sum_{j=1}^{n} \mu_j y_{rj} \ge \widetilde{\phi} y_{ro} \quad r = 1, \ldots, \mathrm{s} \\
&\sum_{j=1}^{n} \lambda_j z_{dj} \ge \widetilde{z}_{do} \qquad d = 1, \ldots, \mathrm{D} \\
&\sum_{j=1}^{n} \mu_j z_{dj} \le \widetilde{z}_{do} \qquad d = 1, \ldots, \mathrm{D} \\
&\widetilde{z}_{do} \ge 0, \quad d = 1, \ldots, \mathrm{D} \\
&\lambda_j \ge 0, \quad j = 1, \ldots, n \\
&\mu_j \ge 0, \quad j = 1, \ldots, n \\
&\phi \ge 1
\end{aligned}
\tag{4.10}
$$

Table 4.3 Optimal intermediate measures for DEA frontier

	Input-oriented (model 8)		Output-oriented (model 10)	
DMU	z1	z2	z1	z2
1	5,129,409	673,373.7	7,335,749	963,015.6
2	6,287,502	827,782.2	10,063,742	1,324,944
3	4,776,548	560,244	6,922,331	811,924.1
4	1,332,365	174,166.4	4,379,619	572,502.8
5	30,127,364	4,177,166	39,280,374	5,446,233
6	3,807,167	435,393.8	9,769,924	1,117,304
7	3,738,287	654,045.2	13,516,290	2,364,791
8	5,553,015	1,009,007	20,180,210	3,666,832
9	2,166,576	351,793.5	9,703,191	1,575,536
10	4,417,507	671,133.1	9,480,468	1,440,327
11	941,872.5	263,658.4	5,745,796	1,608,421
12	7,166,191	849,582.4	9,434,406	1,118,489
13	2,649,297	644,399.8	12,748,722	3,100,926
14	2,749,750	454,687.7	9,526,595	1,575,280
15	5,663,750	634,667.6	9,226,915	1,033,948
16	1,899,396	188,102.5	5,932,758	587,537.6
17	3,504,900	725,272.3	9,735,516	2,014,580
18	1,356,947	224,515.2	5,242,330	867,375.7
19	474,800	108,242	1,154,679	263,236.6
20	302,964.9	63,960.38	554,326.4	117,026.5
21	68,296	9,610.947	340,151.7	47,867.8
22	52,063	14,574	88,313.65	24,721.65
23	137,689.9	16,149.72	327,564.3	38,420.2
24	159,644.2	32,943.34	1,184,206	244,366.5

Model (4.10) yields a set of new inputs, outputs and intermediate measures that render the DMU efficient under model (4.7).

From model (4.8) or (4.10), it can be seen that in addition to the overall efficiency scores, we have to obtain a set of optimal intermediate measures (z). Models (4.4) and (4.7) do not immediately yield the set of optimal z values that are on the DEA frontier, rather we have to use model (4.8) or (4.10) to determine the frontier points for two-stage processes.

Table 4.3 reports the optimal intermediate measures for both orientations, obtained from models (4.8) and (4.10). If we apply the new (projected) DMUs with the intermediate measures reported in Table 4.3 and $\theta_j^* x_{1j}$ and $\theta_j^* x_{2j}$ (or $\phi_j^* y_{1j}$ and $\phi_j^* y_{2j}$) to model (4.4) (or (4.7)) under either Case I or Case II, the overall efficiency scores are all equal to one.

Note that under the input-oriented models, for DMUs 3 and 22, model (4.8) indicates the newly obtained intermediate measures are equal to the original values. This is confirmed by the fact that model (4.4) identifies DMUs 3 and 22 as efficient under the Case II. The same situation is found for DMU12 under the output-oriented models.

While DMUs 9, 15 and 19 are efficient under model (4.7), model (4.10) identifies a set of different values on intermediate measures for DMUs 9, 15 and 19 (see the last two columns of Table 4.3). This indicates that multiple optimal solutions exist for intermediate measures in models (4.8) and (4.10). In fact, if we calculate model (4.10) using the original intermediate values and $\phi_j^* y_{1j}$ and $\phi_j^* y_{2j}$ for DMUs 9, 15, and 19, model (4.10) yields an efficiency score of 1.

4.4 Conclusions

This chapter presents the approach of Chen et al. (2010) for determining the DEA frontier points (projections) for inefficient DMUs under the framework of the DEA model for two-stage network processes. The current study is based upon the assumption of CRS. Chen et al. (2009) develop a two-stage DEA model under the condition of variable returns to scale (VRS) wherein the overall efficiency is expressed as a (weighted) sum of the efficiencies of the individual stages. Although overall VRS efficiency scores as well as scores for individual stages can be obtained by using Chen et al. (2009), adjusting the inputs or outputs by the efficiency scores is not sufficient to yield VRS frontier projections. Because Chen et al. (2009) focus on the additive efficiency decomposition, the newly developed approach in the current study cannot be (directly) applied. Further study is then needed to develop models for determining the DEA frontier points for VRS inefficient DMUs.

References

Castelli, L., Pesenti, R., & Ukovich, W. (2010). A classification of DEA models when the internal structure of the Decision Making Units is considered. *Annals of Operations Research, 173*(1), 207–235.

Charnes, A., Cooper, W. W., & Rhodes, E. (1978). Measuring the efficiency of decision making units. *European Journal of Operational Research, 2*, 429–444.

Chen, Y., Cook, W. D., Li, N., & Zhu, J. (2009). Additive efficiency decomposition in two-stage DEA. *European Journal of Operational Research, 196*(2009), 1170–1176.

Chen, Y., Cook, W. D., & Zhu, J. (2010). Deriving the DEA frontier for two-stage processes. *European Journal of Operational Research, 202*(2010), 138–142.

Kao, C., & Hwang, S.-N. (2008). Efficiency decomposition in two-stage data envelopment analysis: An application to non-life insurance companies in Taiwan. *European Journal of Operational Research, 185*(1), 418–429.

Lewis, H. F., Sexton, T. R. (2004). Network DEA: Efficiency analysis of organizations with complex internal structures. *Computers and Operations Research, 31*(9), 1365–1410.

Liang, L., Cook, W. D., & Zhu, J. (2008). DEA models for two-stage processes: Game approach and efficiency decomposition. *Naval Research Logistics, 55*, 643–653.

Chapter 5
Additive Efficiency Decomposition in Network DEA

Yao Chen, Wade D. Cook, and Joe Zhu

Abstract In conventional data envelopment analysis (DEA), decision making units (DMUs) are generally treated as a black-box in the sense that internal structures are ignored, and the performance of a DMU is assumed to be a function of a set of chosen inputs and outputs. A significant body of work has been directed at problem settings where the DMU is characterized by a multistage process; supply chains and many manufacturing processes take this form. The current chapter presents DEA modeling approaches for network DEA where additive efficiency decompositions are assumed for sub-units/processes/stages. In the additive efficiency decomposition approach, the overall efficiency is expressed as a (weighted) sum of the efficiencies of the individual stages. This approach can be applied under both constant returns to scale (CRS) and variable returns to scale (VRS) assumptions.

Keywords Data envelopment analysis (DEA) • Efficiency • Intermediate measure • Two-stage • Multistage • Serial systems • Additive decomposition

---

Part of this chapter is based upon (i) Chen, Y., Cook, W.D., Li, N. and Zhu, Joe, Additive Efficiency Decomposition in Two-Stage DEA, *European Journal of Operational Research*, Vol. 196 (2009), 1170–1176 and (ii) Cook, W.D., Zhu, Joe, Yang, F. and Bi, G-B, Network DEA: Additive Efficiency Decomposition, *European Journal of Operational Research*, Vol. 207, Issue 2 (2010), 1122–1129, with permissions from Elsevier Science.

Y. Chen (✉)
Manning School of Business, University of Massachusetts at Lowell, Lowell, MA 01845, USA
e-mail: Yao_Chen@uml.edu

W.D. Cook
Schulich School of Business, York University, Toronto, ON M3J 1P3, Canada
e-mail: wcook@schulich.yorku.ca

J. Zhu
School of Business, Worcester Polytechnic Institute, Worcester, MA 01609, USA
e-mail: jzhu@wpi.edu

W.D. Cook and J. Zhu (eds.), *Data Envelopment Analysis*,
International Series in Operations Research & Management Science 208,
DOI 10.1007/978-1-4899-8068-7_5,

5.1 Introduction

Data envelopment analysis (DEA) is a tool for measuring the relative efficiency of peer decision making units (DMUs) that have multiple inputs and outputs. In many cases, DMUs may have internal or network structures; see for example, Färe and Grosskopf (1996), Castelli et al. (2004) and Tone and Tsutsui (2009). In the latter case, the authors provide a slacks-based model that captures the overall efficiency of the DMU, and provides, as well, measures for the components (referred to as divisions) or stages that make up the DMU. The overall efficiency is expressed as a weighted average of the component efficiencies, where weights are exogenously imposed to reflect the perceived importance of the components. (see Chap. 11 for detailed discussions.)

Based upon the work of Chen et al. (2009) and Cook et al. (2010), the current chapter focuses on the derivation of a radial measure of efficiency that can be decomposed into a convex combination of radial measures for the individual components that make up the DMU. We note that in these two work, the weights used for individual stage's efficiency aggregation are variables, and not imposed exogenously.

Chen et al. (2009) present a methodology for representing overall radial efficiency of a DMU as an *additive* weighted average of the radial efficiencies of the individual stages or components that make up the DMU. While the approach of Chen et al. (2009) can be extended to DMUs that have more than two stages, such an extension requires that the multi-stage processes share the unique feature that all outputs from any stage represent the only inputs to the next stage. In other words, except for the first stage, all other stages do not have their own independent inputs (and/or outputs), that enter (exit) the process at that point. While these *closed* systems do exist, the more prevalent case is that where each stage is *open*, that is it has its own inputs (and/or outputs) in addition to the intermediate measures (that exist in-between two stages).

Such open multistage structures are relatively common, particularly in processing industries. Consider, for example, the situation in which a coal mining company wishes to evaluate the efficiency of a set of collieries (mining operations) in a large coal field. Typically, the process of delivering finished products to the customer is multistage in nature. In crude terms, *Stage 1* would involve the *extraction* of the raw or run-of-mine (ROM) coal from underground or open pit coal reserves. At the mine site, the ROM is generally put through a process where screens separate the product into different size categories; e.g. a 'more than one inch in diameter' category, and a 'less than one inch' category. The resulting 'size grades', representing the outputs from this first stage, are then transported to an on-site *washing facility*, which might be deemed *Stage 2*. The washing process filters out any material below a certain specific gravity; this portion is unsuitable for sale and is discarded. A portion of the remaining usable coal (outputs from Stage 2) is sold to the open market as a finished product, and at management's discretion (based on estimates of the demand), the remaining product is sent to *Stage 3*, the *crusher*.

The crushing process also produces waste or discard, with the remaining material, sometimes referred to as 'middlings', being sold or blended with other materials to make such products as briquettes. This latter process might be thought of as *Stage 4*.

Numerous such examples from processing industries exist. In many cases a portion of the outputs from one stage may be in 'finished' form and go to the consumer market, with the remainder being reprocessed at the next stage to get a more pure form of the product. The petrochemical industry, perfume manufacturing and so on, are examples.

It is important to note that the models of Kao and Hwang (2008), Liang et al. (2008) and Chen et al. (2009) concentrate specifically on pure serial processes. Cook et al. (2010) develop linear models for DMUs that have multiple stages, with each stage being open, having its own inputs and outputs. Cook et al. (2010) also obtain an additive efficiency decomposition of the overall efficiency score. The advantage of additive efficiency decomposition is that we can also study performance under assumptions of both constant returns to scale (CRS) and variable returns to scale (VRS).

The current chapter starts with the approach of Chen et al. (2009) where a simple two-stage network process is studies. We then present the work of Cook et al. (2010) where additive efficiency decomposition approach is applied to general network structures. For ease of notation, we begin in Sect. 5.5 by examining open serial systems. We then present a model for measuring the overall radial efficiency of the general serial multi-stage process, and show that this measure can be decomposed into radial measures of efficiency for the components or stages making up the overall process. Section 5.6 then extends this model structure to include more complex multistage processes. Our approach is illustrated in Sect. 5.7 with the supply chain data set in Liang et al. (2006). As well, we re-evaluate the data set provided in Tone and Tsutsui (2009).

5.2 A Two-Stage Network Process: Constant Returns to Scale

Consider a two-stage process shown in Fig. 5.1. Suppose we have n DMUs, and that each $DMU_j (j = 1, 2, \ldots, n)$ has m inputs to the first stage, $x_{ij} (i = 1, 2, \ldots, m)$, and D outputs from this stage, z_{dj}, $(d = 1, 2, \ldots, D)$. These D outputs then become the inputs to the second stage, and are referred to as intermediate measures. The outputs from the second stage are denoted y_{rj}, $(r = 1, 2, \ldots, s)$. Based upon the CRS model (Charnes et al. 1978), the (CRS) efficiency scores for DMU_{j_o} in the first and second stages can be calculated in the following two CRS models (5.1) and (5.2), respectively:

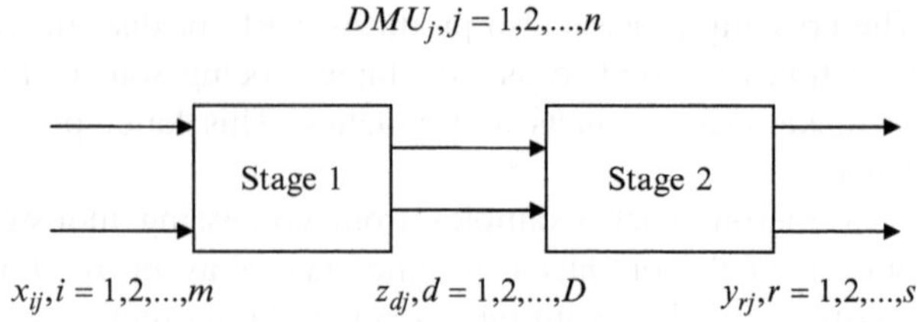

Fig. 5.1 Two-stage process

$$\theta_{j_o}^1 = \max \frac{\sum_{d=1}^{D} \eta_d^A z_{dj_o}}{\sum_{i=1}^{m} v_i x_{ij_o}}$$

$$\text{s.t.} \quad \frac{\sum_{d=1}^{D} \eta_d^A z_{dj}}{\sum_{i=1}^{m} v_i x_{ij}} \le 1 \quad \text{j} = 1, \ldots, \text{n} \tag{5.1}$$

$$\eta_d^A, v_i \ge 0$$

$$\theta_j^2 = \max \frac{\sum_{r=1}^{s} u_r y_{rj_o}}{\sum_{d=1}^{D} \eta_d^B z_{dj_o}}$$

$$\text{s.t.} \quad \frac{\sum_{r=1}^{s} u_r y_{rj}}{\sum_{d=1}^{D} \eta_d^B z_{dj}} < 1, \quad \text{j} = 1, \ldots, \text{n} \tag{5.2}$$

$$\eta_d^B, u_r \ge 0$$

The overall CRS efficiency score can be calculated from the following CRS model (5.3)

$$\max \frac{\sum_{r=1}^{s} u_r y_{rj_o}}{\sum_{i=1}^{m} v_i x_{ij_o}}$$
$$\text{s.t.} \quad \frac{\sum_{r=1}^{s} u_r y_{rj}}{\sum_{i=1}^{m} v_i x_{ij}} \leq 1, \quad \mathrm{j} = 1, \ldots \mathrm{n} \tag{5.3}$$
$$v_i, u_r \geq 0$$

In Kao and Hwang's (2008) and Liang et al. (2008) two-stage network DEA approach, it is required that the input of the second stage to be the expected output of the first stage, i.e., given the inputs to the first stage x_{ij}, that stage yields the optimal intermediate measure $\sum_{d=1}^{D} \eta_d^* z_{dj}$ which is then used as the (aggregated) input in the second stage. Thus, it is assumed that $\eta_d^A = \eta_d^B = \eta_d$, and the overall efficiency of a DMU is given by:

$$\theta_{j_o} = Max \frac{\sum_{d=1}^{D} \eta_d z_{dj_o}}{\sum_{i=1}^{m} v_i x_{ij_o}} \bullet \frac{\sum_{r=1}^{s} u_r y_{rj}}{\sum_{d=1}^{D} \eta_d z_{dj}} = \frac{\sum_{r=1}^{s} u_r y_{rj_O}}{\sum_{i=1}^{m} v_i x_{ij_O}}$$
$$s.t. \quad \frac{\sum_{d=1}^{D} \eta_d z_{dj}}{\sum_{i=1}^{m} v_i x_{ij}} \leq 1, \quad \mathrm{j} = 1, \ldots, \mathrm{n} \tag{5.4}$$
$$\frac{\sum_{r=1}^{s} u_r y_{rj}}{\sum_{d=1}^{D} \eta_d z_{dj}} \leq 1, \quad \mathrm{j} = 1, \ldots, \mathrm{n}$$
$$v_i, u_r, \eta_d \geq 0$$

It can be seen from the objective function of model (5.4) that the overall efficiency is the product of the efficiencies of the two stages, i.e., $\theta_{j_0}^1 \bullet \theta_{j_0}^2 = \frac{\sum_{r=1}^{s} u_o^* y_{rj_0}}{\sum_{i=1}^{m} v_o^* x_{ij_0}} = \theta_{j_o}$, where $\theta_{j_0}^1 = \frac{\sum_{d=1}^{D} \eta_d^* z_{dj_o}}{\sum_{i=1}^{m} v_i^* x_{ij_o}}$ and $\theta_{j_0}^2 = \frac{\sum_{r=1}^{s} u_r^* y_{rj}}{\sum_{d=1}^{D} \eta_d^* z_{dj}}$ and (*) denotes optimal value from model (5.4).

Note $\eta_d^A = \eta_d^B$ is a key and rational assumption in that the value accorded the outputs from the first stage should reasonably be assumed as their value when they assume the additional role as inputs to the second stage. Without this assumption, model (5.4) becomes a non-linear program, as the terms $\sum_{d=1}^{D} \eta_d^A z_{do}$ and $\sum_{d=1}^{D} \eta_d^B z_{do}$ cannot be cancelled in the objective function. Also, without this assumption, solving model (5.4) is equivalent to applying the CRS model to stages 1 and 2 independently, and then taking the geometric mean of the two CCR efficiency scores. Throughout the chapter we therefore maintain the assumption that $\sum_{d=1}^{D} \eta_d z_{do}$ is to be the same for the two stages.

In the interest of modeling two-stage processes in a more general way, and specifically to allow for VRS settings, we propose that rather than combine the stages in a multiplicative (geometric) manner as in Kao and Hwang (2008) and Liang et al. (2008), we use a weighted additive (arithmetic mean) approach.

As will be explained below, the multiplicative and additive models are two different, but equally valid ways of aggregating the components of a two-stage process. Thus, we propose to define overall efficiency of the two stage process as

$$w_1 \bullet \frac{\sum_{d=1}^{D} \eta_d z_{dj_0}}{\sum_{i=1}^{m} v_i x_{ij_0}} + w_2 \bullet \frac{\sum_{r=1}^{s} u_r y_{rj_0}}{\sum_{d=1}^{D} \eta_d z_{dj_0}}, \tag{5.5}$$

Where w_1 and w_2 are user-specified weights such that $w_1 + w_2 = 1$. These weights are not optimization variables, but rather are functions of the optimization variables.

We thus propose deriving the overall efficiency of the process by solving the following problem:

$$Max\left[w_1 \bullet \frac{\sum_{d=1}^{D} \eta_d z_{dj_0}}{\sum_{i=1}^{m} v_i x_{ij_0}} + w_2 \bullet \frac{\sum_{r=1}^{s} u_r y_{rj_0}}{\sum_{d=1}^{D} \eta_d z_{dj_0}}\right]$$

$$\begin{aligned} s.t. \quad & \frac{\sum_{d=1}^{D} \eta_d z_{dj}}{\sum_{i=1}^{m} v_i x_{ij}} \leq 1 \\ & \frac{\sum_{r=1}^{s} u_r y_{rj}}{\sum_{d=1}^{D} \eta_d z_{dj}} \leq 1 \\ & \eta_d, u_r, v_i \geq 0, \quad j = 1, 2, \ldots, n \end{aligned} \tag{5.6}$$

It is observed that model (5.6) cannot be turned into a linear program using the usual Charnes and Cooper (1962) transformation. For example, if we let $t_1 = \frac{1}{\sum_{i=1}^{m} v_i x_{ij_0}}$, $t_2 = \frac{1}{\sum_{d=1}^{D} \eta_d z_{dj_0}}$, and set $\pi_d^1 = t_1 \bullet \eta_d$, $\omega_i = t_1 \bullet v_i$, $\mu_r = t_2 \bullet u_r$, $\pi_d^2 = t_2 \bullet \eta_d$, then the transformations $\pi_d^1 = t_1 \bullet \eta_d$ and $\pi_d^2 = t_2 \bullet \eta_d$ imply a linear relationship between π_d^1 and π_d^2, namely, $\pi_d^1 = \frac{\sum_i \omega_i x_{ij_o}}{\sum_k \pi_k^1 z_{kj_o}} \bullet \pi_d^2$. Then, model (5.6) becomes

$$Max\left[w_1 \bullet \sum_{d=1}^{D} \pi_d^1 z_{dj_0} + w_2 \bullet \sum_{r=1}^{s} \mu_r y_{rj_0}\right]$$

$$\begin{aligned} s.t. \quad & \sum_{i=1}^{m} \omega_i x_{ij} - \sum_{d=1}^{D} \pi_d^1 z_{dj} \geq 0 \\ & \sum_{d=1}^{D} \pi_d^2 z_{dj} - \sum_{r=1}^{s} \mu_r y_{rj} \geq 0 \\ & \sum_{i=1}^{m} \omega_i x_{ij_o} = 1 \\ & \sum_{d=1}^{D} \pi_d^2 z_{dj_o} = 1 \\ & \pi_d^1 = \frac{\sum_i \omega_i \bullet x_{i,j_0}}{\sum_k \pi_k^1 \bullet z_{k,j_0}} \bullet \pi_d^2 \\ & \pi_d^1, \pi_d^2, \mu_r, \omega_i \geq 0, \quad j = 1, 2, \ldots, n \end{aligned} \tag{5.7}$$

which is a non-linear program. We, therefore, seek an alternative way to convert model (5.6) into a linear form, by appropriate choice of the w_1 and w_2.

Note that w_1 and w_2 are intended to represent the relative importance or contribution of the performances of stages 1 and 2, respectively, to the overall performance of the DMU. One argument is that the 'size' of a stage reflects its importance, (as measured by its weight). One reasonable representation of size is the portion of total resources devoted to each stage. Letting $\sum_{i=1}^{m} v_i x_{ij_0} + \sum_{d=1}^{D} \eta_d z_{dj_0}$ represent the total size of (amount of resources consumed by) the two-stage process, and $\sum_{i=1}^{m} v_i x_{ij_0}$ and $\sum_{d=1}^{D} \eta_d z_{dj_0}$, the sizes of the stages 1 and 2 respectively, we define

$$w_1 = \frac{\sum_{i=1}^{m} v_i x_{ij_0}}{\sum_{i=1}^{m} v_i x_{ij_0} + \sum_{d=1}^{D} \eta_d z_{dj_0}} \quad \text{and}$$
$$w_2 = \frac{\sum_{d=1}^{D} \eta_d z_{dj_0}}{\sum_{i=1}^{m} v_i x_{ij_0} + \sum_{d=1}^{D} \eta_d z_{dj_0}} \tag{5.8}$$

Then, the objective function of model (5.6) becomes:

$$\frac{\sum_{d=1}^{D} \eta_d z_{dj_0} + \sum_{r=1}^{s} u_r y_{rj_0}}{\sum_{i=1}^{m} v_i x_{ij_0} + \sum_{d=1}^{D} \eta_d z_{dj_0}} \tag{5.9}$$

Under the CRS case, model (5.6) becomes

$$\begin{aligned} & Max \frac{\sum_{d=1}^{D} \eta_d z_{dj_0} + \sum_{r=1}^{s} u_r y_{rj_0}}{\sum_{i=1}^{m} v_i x_{ij_0} + \sum_{d=1}^{D} \eta_d z_{dj_0}} \\ & s.t. \ \frac{\sum_{d=1}^{D} \eta_d z_{dj}}{\sum_{i=1}^{m} v_i x_{ij}} \leq 1 \\ & \frac{\sum_{r=1}^{s} u_r y_{rj}}{\sum_{d=1}^{D} \eta_d z_{dj}} \leq 1 \\ & \eta_d, u_r, v_i \geq 0, \quad j = 1, 2, \ldots, n \end{aligned} \tag{5.10}$$

Using the Charnes-Cooper transformation, model (5.10) is equivalent to

$$\begin{aligned} & Max \sum_{r=1}^{s} \mu_r y_{rj_o} + \sum_{d=1}^{D} \pi_d z_{dj_o} \\ & s.t. \ \sum_{d=1}^{D} \pi_d z_{dj} - \sum_{i=1}^{m} \omega_i x_{ij} \leq 0 \\ & \sum_{r=1}^{s} \mu_r y_{rj} - \sum_{d=1}^{D} \pi_d z_{dj} \leq 0 \\ & \sum_{i=1}^{m} \omega_i x_{ij_o} + \sum_{d=1}^{D} \pi_d z_{dj_o} = 1 \\ & \pi_d, \mu_r, \omega_i \geq 0, \quad j = 1, 2, \ldots, n \end{aligned} \tag{5.11}$$

Once we obtain an optimal solution to (5.11), we can calculate efficiency scores for the two individual stages. However, model (5.11) can have alternative optimal solutions. As a result, the decomposition of the overall efficiency defined in (5.5) may not be unique. We here follow Kao and Hwang's (2008) approach to find a set of multipliers which produces the largest first (or second) stage efficiency score while maintaining the overall efficiency score.

We therefore propose the following procedure. Given the overall efficiency obtained from (5.11) (denoted as θ_o), we calculate either the first stage's efficiency (θ_j^{1*}) or the second stage's efficiency (θ_j^{2*}) first, and then derive from that the efficiency of the other stage.

In case the first stage is to be given pre-emptive priority, the following model determines its efficiency (θ_o^{1*}), while maintaining the overall efficiency score at θ_o calculated from model (5.11).

$$
\begin{aligned}
\theta_o^{1*} &= Max \frac{\sum_{d=1}^{D} \eta_d z_{dj_o}}{\sum_{i=1}^{m} v_i x_{ij_o}} \\
s.t. \quad & \frac{\sum_{d=1}^{D} \eta_d z_{dj}}{\sum_{i=1}^{m} v_i x_{ij}} \le 1 \\
& \frac{\sum_{r=1}^{s} u_r y_{rj}}{\sum_{d=1}^{D} \eta_d z_{dj}} \le 1 \\
& \frac{\sum_{d=1}^{D} \eta_d z_{dj_o} + \sum_{r=1}^{s} u_r y_{rj_o}}{\sum_{i=1}^{m} v_i x_{ij_o} + \sum_{d=1}^{D} \eta_d z_{dj_o}} = \theta_0 \\
& \eta_d, u_r, v_i \ge 0, \quad j = 1, 2, \ldots, n
\end{aligned}
\tag{5.12}
$$

or equivalently,

$$
\begin{aligned}
\theta_o^{1*} &= Max \sum_{d=1}^{D} \pi_d z_{dj_o} \\
s.t. \quad & \sum_{d=1}^{D} \pi_d z_{dj} - \sum_{i=1}^{m} \omega_i x_{ij} \le 0 \\
& \sum_{r=1}^{s} \mu_r y_{rj} - \sum_{d=1}^{D} \pi_d z_{dj} \le 0 \\
& (1 - \theta_o) \sum_{d=1}^{D} \pi_d z_{dj_o} + \sum_{r=1}^{s} \mu_r y_{rj_o} = \theta_o \\
& \sum_{i=1}^{m} \omega_i x_{ij_o} = 1 \\
& \pi_d, \mu_r, \omega_i \ge 0, \quad j = 1, 2, \ldots, n
\end{aligned}
\tag{5.13}
$$

The efficiency for the second stage is then calculated as

$$\theta_o^2 = \frac{\theta_o - w_1^* \cdot \theta_o^{1*}}{w_2^*}$$

where w_1^* and w_2^* represent optimal weights obtained from model (5.11) by way of (5.8).

Note that we here use (*) in θ_o^{1*} to indicate that the efficiency of the first stage is given the pre-emptive priority and is optimized first. In this case, the resulting second stage efficiency score is denoted as θ_o^2.

In case the second stage is to be given pre-emptive priority, the following model determines the second stage's efficiency (θ_o^{2*}) while maintaining the overall efficiency score at θ_o calculated from model (5.11).

$$\begin{aligned} \theta_o^{2*} &= Max \frac{\sum_{r=1}^{s} u_r y_{rj_o}}{\sum_{d=1}^{D} \eta_d z_{dj_o}} \\ s.t. \quad & \frac{\sum_{d=1}^{D} \eta_d z_{dj}}{\sum_{i=1}^{m} v_i x_{ij}} \le 1 \\ & \frac{\sum_{r=1}^{s} u_r y_{rj}}{\sum_{d=1}^{D} \eta_d z_{dj}} \le 1 \\ & \frac{\sum_{d=1}^{D} \eta_d z_{dj_o} + \sum_{r=1}^{s} u_r y_{rj_o}}{\sum_{i=1}^{m} v_i x_{ij_o} + \sum_{d=1}^{D} \eta_d z_{dj_o}} = \theta_0 \\ & \eta_d, u_r, v_i \ge 0, \quad j = 1, 2, \ldots, n \end{aligned} \tag{5.14}$$

Model (5.14) is equivalent to

$$\begin{aligned} \theta_o^{2*} &= Max \sum_{r=1}^{s} \mu_r y_{rj_o} \\ s.t. \quad & \sum_{d=1}^{D} \pi_d z_{dj} - \sum_{i=1}^{m} \omega_i x_{ij} \le 0 \\ & \sum_{r=1}^{s} \mu_r y_{rj} - \sum_{d=1}^{D} \pi_d z_{dj} \le 0 \\ & \sum_{d=1}^{D} \pi_d z_{dj_o} + \sum_{r=1}^{s} \mu_r y_{rj_o} - \theta_o \sum_{i=1}^{m} \omega_i x_{ij_o} = \theta_o \\ & \sum_{d=1}^{D} \pi_d z_{dj_o} = 1 \\ & \pi_d, \mu_r, \omega_i \ge 0, \quad j = 1, 2, \ldots, n \end{aligned} \tag{5.15}$$

and the efficiency for the first stage is calculated as

$$\theta_o^1 = \frac{\theta_o - w_2^* \cdot \theta_o^{2*}}{w_1^*}.$$

Note that we here use (*) in θ_o^{2*} to indicate that second stage is given pre-emptive priority in terms of its efficiency being optimized first. In this case, the resulting first stage efficiency score is denoted as θ_o^1.

Finally, note that if $\theta_o^{1*} = \theta_o^1$ or $\theta_o^{2*} = \theta_o^2$, then this indicates that we have a unique efficiency decomposition.

5.3 Two-Stage Network DEA: Variable Returns to Scale

While the discussion in the previous section is based upon the assumption of CRS, the above approach enables us to study the efficiency of two-stage processes under VRS. The VRS efficiency scores for the two stages can be determined by the following VRS models (Banker et al. 1984):

$$\max E_{j_o}^1 = \frac{\sum_{d=1}^{D} \eta_d^A z_{dj_o} + u^A}{\sum_{i=1}^{m} v_i x_{ij_o}}$$

$$\text{s.t.} \quad \frac{\sum_{d=1}^{D} \eta_d^A z_{dj} + u^A}{\sum_{i=1}^{m} v_i x_{ij}} \leq 1, \quad \text{j} = 1, \ldots, \text{n}$$

$$\eta_d^A, v_i \geq 0 \quad u^A \text{ free in sign}$$

and

$$\max E_{j_o}^2 = \frac{\sum_{r=1}^{s} u_r y_{rj_o} + u^B}{\sum_{d=1}^{D} \eta_d^B z_{dj_o}}$$

$$\text{s.t.} \quad \frac{\sum_{r=1}^{s} u_r y_{rj} + u^B}{\sum_{d=1}^{D} \eta_d^B z_{dj}} < 1, \quad \text{j} = 1, \ldots, \text{n}$$

$$\eta_d^B, u_r \geq 0 \quad \text{and} \quad u^B \text{ free in sign}$$

Note that the approach of Kao and Hwang (2008) and Liang et al. (2008) cannot be extended to the VRS assumption, because $E_{j_o}^1 \bullet E_{j_o}^2$ cannot be converted into a linear form under the condition of $\eta_d^A = \eta_d^B$, due to the free variable u^A in the numerator of $E_{j_o}^1$. On the other hand, using our approach, we have the VRS overall efficiency as using the weights defined under the CRS assumption

$$
\begin{aligned}
& Max \frac{\sum_{d=1}^{D} \eta_d z_{dj_0} + u^A + \sum_{r=1}^{s} u_r y_{rj_0} + u^B}{\sum_{i=1}^{m} v_i x_{ij_0} + \sum_{d=1}^{D} \eta_d z_{dj_0}} \\
& s.t. \quad \frac{\sum_{d=1}^{D} \eta_d z_{dj} + u^A}{\sum_{i=1}^{m} v_i x_{ij}} \le 1 \\
& \frac{\sum_{r=1}^{s} u_r y_{rj} + u^B}{\sum_{d=1}^{D} \eta_d z_{dj}} \le 1 \\
& \eta_d, u_r, v_i \ge 0, \quad j = 1, 2, \ldots, n \\
& u^A, u^B, \text{free in sign}
\end{aligned}
\tag{5.16}
$$

Note that this is an input-oriented model. If we use output-oriented VRS models, the weights will be defined as $w_1 = \frac{\sum_{d=1}^{D} \eta_d z_{dj_0}}{\sum_{r=1}^{s} u_r y_{rj_0} + \sum_{d=1}^{D} \eta_d z_{dj_0}}$ and $w_2 = \frac{\sum_{r=1}^{s} u_i y_{rj_0}}{\sum_{r=1}^{s} u_i y_{rj_0} + \sum_{d=1}^{D} \eta_d z_{dj_0}}$.

Model (5.16) is equivalent to the following linear programming program

$$
\begin{aligned}
& Max \sum_{r=1}^{s} \mu_r y_{rj_o} + u^1 + \sum_{d=1}^{D} \pi_d z_{dj_o} + u^2 \\
& s.t. \sum_{d=1}^{D} \pi_d z_{dj} - \sum_{i=1}^{m} \omega_i x_{ij} + u^1 \le 0 \\
& \sum_{r=1}^{s} \mu_r y_{rj} - \sum_{d=1}^{D} \pi_d z_{dj} + u^2 \le 0 \\
& \sum_{i=1}^{m} \omega_i x_{ij_o} + \sum_{d=1}^{D} \pi_d z_{dj_o} = 1 \\
& \pi_d, \mu_r, \omega_i \ge 0, \quad j = 1, 2, \ldots, n \\
& u^1, u^2 \ \text{free in sign}
\end{aligned}
\tag{5.17}
$$

Once we obtain the overall efficiency, models similar to (5.13) and (5.15) can be developed to determine the efficiency of each stage. Specifically, assuming pre-emptive priority for stage 1, the following model determines that stage's efficiency (E_o^{1*}), while maintaining the overall efficiency score at E_o calculated from model (5.17).

$$
\begin{aligned}
E_o^{1*} &= Max\sum_{d=1}^{D} \pi_d z_{dj_o} + u^1 \\
\text{s.t.} &\sum_{d=1}^{D} \pi_d z_{dj} + u^1 - \sum_{i=1}^{m} \omega_i x_{ij} \leq 0 \\
&\sum_{r=1}^{s} \mu_r y_{rj} + u^2 - \sum_{d=1}^{D} \pi_d z_{dj} \leq 0 \\
&(1 - E_o)\sum_{d=1}^{D} \pi_d z_{dj_o} + \sum_{r=1}^{s} \mu_r y_{rj_o} + u^1 + u^2 = E_o \\
&\sum_{i=1}^{m} \omega_i x_{ij_o} = 1 \\
&\pi_d, \mu_r, \omega_i \geq 0, \quad j = 1, 2, \ldots, n \\
&u^1, u^2 \text{ free in sign}
\end{aligned}
\tag{5.18}
$$

Similarly, if stage 2 is to be given pre-emptive priority, the following model determines the efficiency (E_j^{2*}) for that stage, while maintaining the overall efficiency score at E_o calculated from model (5.17).

$$
\begin{aligned}
E_o^{2*} &= Max\sum_{r=1}^{s} \mu_r y_{rj_o} + u^2 \\
s.t. &\sum_{d=1}^{D} \pi_d z_{dj} + u^1 - \sum_{i=1}^{m} \omega_i x_{ij} \leq 0 \\
&\sum_{r=1}^{s} \mu_r y_{rj} + u^2 - \sum_{d=1}^{D} \pi_d z_{dj} \leq 0 \\
&\sum_{d=1}^{D} \pi_d z_{dj_o} + \sum_{r=1}^{s} \mu_r y_{rj_o} - E_o\sum_{i=1}^{m} \omega_i x_{ij_o} + u^1 + u^2 = E_o \\
&\sum_{d=1}^{D} \pi_d z_{dj_o} = 1 \\
&\pi_d, \mu_r, \omega_i \geq 0, \quad j = 1, 2, \ldots, n \\
&u^1, u^2 \text{ free in sign}
\end{aligned}
\tag{5.19}
$$

Once the efficiency score for one of the stages is calculated using (5.18) or (5.19), the score for the other stage can be derived in the similar manner as in the CRS case.

5.4 Two-Stage Network DEA: Application of Additive Efficiency Decomposition

We here apply the above approach to the 24 Taiwanese non-life insurance companies studied in Kao and Hwang (2008). The two-stage process consists of premium acquisition and profit generation. There are two inputs to the first stage which is characterized by marketing of the insurance and generation of premiums, and two outputs from the second stage which is characterized by investment and generation of profit. The two inputs are operational expenses and insurance expenses, and the outputs are underwriting profit and investment profit. There are also two intermediate measures between the two stages, namely direct written premiums and reinsurance premiums. The data are provided in Table 5.1.

The CRS results from models (5.11), (5.13) and (5.15) are reported in Table 5.2. The third column reports the overall CRS efficiency obtained from model (5.11). The optimal weights from model (5.11) for each DMU are reported under columns 4 and 5. The rest of the columns report the efficiency score for each individual stage based upon models (5.13) and (5.15).

It can be seen from Table 5.2 that we have unique efficiency decompositions for all DMUs. This arises from the fact that models (5.13) and (5.15) yield identical efficiency scores for the two stages. (Note that the uniqueness result is only true to this specific data set.)

Since the overall efficiency definition presented herein is different from that assumed by Kao and Hwang (2008), the overall efficiency scores from the two approaches cannot be directly compared. The last three columns of Table 5.3 report the CRS scores based upon Kao and Hwang's (2008) approach. We, however, note that except for 8 DMUs (7, 8, 11, 13, 14, 17, 21, and 24), our first and second stage's efficiency scores are identical to those of Kao and Hwang (2008). This indicates that Kao and Hwang's approach also yields unique efficiency decompositions for the remaining 16 DMUs.

Table 5.3 reports the rankings of the CRS scores based upon our new approach and Kao and Hwang's (2008). It can be seen they do not yield the same exact ranking. DMUs 9 and 14 show a big ranking difference. In fact, if we apply the average to Kao and Hwang's (2008) first and second stage scores, a different ranking is obtained. However, the Spearman Rank Correlation coefficient for the rankings in Table 5.3 is 0.971 which is significant at the 0.01 level, indicating an approximately equal ranking based upon the two different approaches. It is also the case that the Pearson Correlation Coefficient for the two sets of raw CRS scores is 98 %.

We next turn to the case of VRS reported in Table 5.4. Two DMUs (5.5 and 5.22) are VRS overall efficient. Also, we have unique VRS efficiency decompositions for all DMUs, as the results obtained from models (5.18) and (5.19) are identical.

Under the standard DEA approach, the scores under the VRS assumption are not less than the ones under CRS assumption. This is true as well for the overall efficiency scores in our models. However, we note that this is not the case for

Table 5.1 Data set

DMU		Operation expenses (X1)	Insurance expenses (*X*2)	Direct written premiums (Z1)	Reinsurance premiums (Z2)	Underwriting profit (Y1)	Investment profit (Y2)
1	Taiwan Fire	1,178,744	673,512	7,451,757	856,735	984,143	681,687
2	Chung Kuo	1,381,822	1,352,755	10,020,274	1,812,894	1,228,502	834,754
3	Tai Ping	1,177,494	592,790	4,776,548	560,244	293,613	658,428
4	China Mariners	601,320	594,259	3,174,851	371,863	248,709	177,331
5	Fubon	6,699,063	3,531,614	37,392,862	1,753,794	7,851,229	3,925,272
6	Zurich	2,627,707	668,363	9,747,908	952,326	1,713,598	415,058
7	Taian	1,942,833	1,443,100	10,685,457	643,412	2,239,593	439,039
8	Ming Tai	3,789,001	1,873,530	17,267,266	1,134,600	3,899,530	622,868
9	Central	1,567,746	950,432	11,473,162	546,337	1,043,778	264,098
10	The First	1,303,249	1,298,470	8,210,389	504,528	1,697,941	554,806
11	Kuo Hua	1,962,448	672,414	7,222,378	643,178	1,486,014	18,259
12	Union	2,592,790	650,952	9,434,406	1,118,489	1,574,191	909,295
13	Shingkong	2,609,941	1,368,802	13,921,464	811,343	3,609,236	223,047
14	South China	1,396,002	988,888	7,396,396	465,509	1,401,200	332,283
15	Cathay Century	2,184,944	651,063	10,422,297	749,893	3,355,197	555,482
16	Allianz President	1,211,716	415,071	5,606,013	402,881	854,054	197,947
17	Newa	1,453,797	1,085,019	7,695,461	342,489	3,144,484	371,984
18	AIU	757,515	547,997	3,631,484	995,620	692,731	163,927
19	North America	159,422	182,338	1,141,951	483,291	519,121	46,857
20	Federal	145,442	53,518	316,829	131,920	355,624	26,537
21	Royal & Sunalliance	84,171	26,224	225,888	40,542	51,950	6,491
22	Aisa	15,993	10,502	52,063	14,574	82,141	4,181
23	AXA	54,693	28,408	245,910	49,864	0.1	18,980
24	Mitsui Sumitomo	163,297	235,094	476,419	644,816	142,370	16,976

Source: Kao and Hwang (2008)

Table 5.2 CRS results

DMU		CRS overall efficiency	w_1	w_2	θ_o^{1*}	θ_o^2	θ_o^1	θ_o^{2*}
1	Taiwan Fire	0.849	0.502	0.498	0.993	0.704	0.993	0.704
2	Chung Kuo	0.812	0.500	0.500	0.998	0.626	0.998	0.626
3	Tai Ping	0.817	0.592	0.408	0.690	1	0.690	1
4	China Mariners	0.596	0.580	0.420	0.724	0.420	0.724	0.420
5	Fubon	0.873	0.546	0.454	0.831	0.923	0.831	0.923
6	Zurich	0.689	0.510	0.490	0.961	0.406	0.961	0.406
7	Taian	0.580	0.571	0.429	0.752	0.352	0.752	0.352
8	Ming Tai	0.579	0.580	0.420	0.726	0.378	0.726	0.378
9	Central	0.612	0.500	0.500	1	0.223	1	0.223
10	The First	0.713	0.537	0.463	0.862	0.541	0.862	0.541
11	Kuo Hua	0.509	0.578	0.422	0.729	0.207	0.729	0.207
12	Union	0.880	0.500	0.500	1	0.760	1	0.760
13	Shingkong	0.557	0.552	0.448	0.811	0.243	0.811	0.243
14	South China	0.577	0.580	0.420	0.725	0.374	0.725	0.374
15	Cathay Century	0.807	0.500	0.500	1	0.614	1	0.614
16	Allianz President	0.639	0.530	0.470	0.886	0.362	0.886	0.362
17	Newa	0.613	0.580	0.420	0.723	0.460	0.723	0.460
18	AIU	0.587	0.558	0.442	0.794	0.326	0.794	0.326
19	North America	0.706	0.500	0.500	1	0.411	1	0.411
20	Federal	0.765	0.517	0.483	0.933	0.586	0.933	0.586
21	Royal & Sunalliance	0.541	0.571	0.429	0.751	0.262	0.751	0.262
22	Aisa	0.742	0.629	0.371	0.590	1	0.590	1
23	AXA	0.685	0.543	0.457	0.843	0.499	0.843	0.499
24	Mitsui Sumitomo	0.544	0.500	0.500	1	0.087	1	0.087

DMUs 1, 12 and 20 for the first stage scores. This may be attributed to the fact that the constraint spaces for (5.13) and (5.18) are not the same, and hence the intermediate scores may not obey the conventional principles.

We finally note that w_1 and w_2 as defined in the current chapter, are variables related to the inputs and the intermediate measures. By virtue of the optimization process, it can turn out that either $w_1 = 1$ and $w_2 = 0$ or $w_1 = 0$ and $w_2 = 1$ at optimality. To overcome this problem, we can require that $w_1 \geq \alpha$ and $w_2 \geq \alpha$ in model (5.6), where α is a selected constant and $0\ \% < \alpha \leq 50\ \%$. Such additional constraints can also be viewed as user's preference regarding the relative importance of the two stages. If such additional constraints are need, we can then study the sensitivity of the overall efficiency scores relative to changes in this parameter α.

In the current chapter, however, there is no need to add additional constraints of $w_1 \geq \alpha$ and $w_2 \geq \alpha$ into models (5.11) and (5.17), because non-zero weights are obtained for both stages. We point out, however, that it is likely that model (5.11) (or model (5.17)) can be infeasible with certain α values. For example, when

Table 5.3 Ranking of CRS scores

		Kao and Hwang's (2008) results			
DMU	Our ranking	Ranking	First stage	Second stage	Overall efficiency
1	3	3	0.993	0.704	0.699
2	5	5	0.998	0.626	0.625
3	4	4	0.690	1	0.690
4	16	15	0.724	0.420	0.304
5	2	1	0.831	0.923	0.767
6	11	12	0.961	0.406	0.390
7	18	17	0.671	0.412	0.277
8	19	18	0.663	0.415	0.275
9	**15**	**20**	1	0.223	0.223
10	9	9	0.862	0.541	0.466
11	24	23	0.647	0.253	0.164
12	1	2	1	0.760	0.760
13	21	21	0.672	0.309	0.208
14	**20**	**16**	0.670	0.431	0.289
15	6	6	1	0.614	0.614
16	13	14	0.886	0.362	0.320
17	14	13	0.628	0.574	0.360
18	17	19	0.794	0.326	0.259
19	10	11	1	0.411	0.411
20	7	8	0.933	0.586	0.547
21	23	22	0.732	0.274	0.201
22	8	7	0.590	1	0.590
23	12	10	0.843	0.499	0.420
24	22	24	0.429	0.314	0.135

$\alpha = 40$ %, model (5.11) is infeasible for DMU22 and when $\alpha = 50$ %, model (5.11) is infeasible for DMUs 1, 2, 5, 6, 10, 16, 20 and 23. This indicates that the input mixes for these DMUs do not allow such weighting structures.

5.5 General Multi-stage Serial Processes

Consider the P-stage process pictured in Fig. 5.2. We denote the input vector to stage 1 by z_o. The output vectors from stage p (p = 1,. . .,P) take two forms, namely z_p^1 and z_p^2. Here, z_p^1 represents that output that leaves the process at this stage and is not passed on as input to the next stage. The vector z_p^2 represents the amount of output that becomes input to the next (p + 1) stage. These types of intermediate measures are called *links* in Tone and Tsutsui (2009). In addition, there is the provision for new inputs z_p^3 to enter the process at the beginning of stage p + 1. Specifically, when p = 2,3,. . ., we define

Table 5.4 VRS results

DMU		VRS overall efficiency	w_1	w_2	E_o^{1*}	E_o^2	E_o^1	E_o^{2*}
1	Taiwan Fire	0.867	0.503	0.497	0.990	0.743	0.990	0.743
2	Chung Kuo	0.856	0.500	0.500	1	0.711	1	0.711
3	Tai Ping	0.818	0.587	0.413	0.690	1	0.690	1
4	China Mariners	0.599	0.581	0.419	0.726	0.424	0.726	0.424
5	Fubon	1	0.483	0.517	1	1	1	1
6	Zurich	0.732	0.511	0.489	0.964	0.490	0.964	0.490
7	Taian	0.684	0.571	0.429	0.752	0.593	0.752	0.593
8	Ming Tai	0.754	0.523	0.477	0.783	0.722	0.783	0.722
9	Central	0.639	0.501	0.499	1	0.276	1	0.276
10	The First	0.780	0.538	0.462	0.862	0.727	0.862	0.727
11	Kuo Hua	0.614	0.576	0.424	0.741	0.443	0.741	0.443
12	Union	0.887	0.511	0.489	0.968	0.803	0.968	0.803
13	Shingkong	0.804	0.494	0.506	0.846	0.763	0.846	0.763
14	South China	0.654	0.581	0.419	0.725	0.555	0.725	0.555
15	Cathay Century	0.940	0.503	0.497	1	0.880	1	0.880
16	Allianz President	0.676	0.526	0.474	0.911	0.417	0.911	0.417
17	Newa	0.840	0.581	0.419	0.724	1	0.724	1
18	AIU	0.618	0.517	0.483	0.850	0.369	0.850	0.369
19	North America	0.833	0.515	0.485	1	0.657	1	0.657
20	Federal	0.946	0.548	0.452	0.902	1	0.902	1
21	Royal & Sunalliance	0.679	0.575	0.425	0.913	0.362	0.913	0.362
22	Aisa	1	0.634	0.366	1	1	1	1
23	AXA	0.815	0.547	0.453	0.976	0.620	0.976	0.620
24	Mitsui Sumitomo	0.564	0.517	0.483	1	0.098	1	0.098

(i) z_{pr}^{j1} the rth component ($r = 1,\ldots. R_p$) of the R_p -dimensional **output** vector for DMU j flowing from stage p, that *leaves* the process at that stage, and is not passed on as an input to stage p + 1.
(ii) z_{pk}^{j2} the kth component ($k = 1, \ldots S_p$) of the S_p -dimensional **output** vector for DMU j flowing from stage p, and is passed on as a portion of the **inputs** to stage p + 1.
(iii) z_{pi}^{j3} the ith component ($i = 1, \ldots I_p$) of the I_p -dimensional **input** vector for DMU j at the stage p + 1, that enters the process at the beginning of that stage.

Note that in the last stage P, all the outputs are viewed as z_{pr}^{j1}, as they leave the process.

We denote the multipliers (weights) for the above factors as

(i) u_{pr} is the multiplier for the output component z_{pr}^{j1} flowing from stage p.
(ii) η_{pk} is the multiplier for the output component z_{pk}^{j2} at stage p, and is as well the multiplier for that same component as it becomes an input to stage p + 1.

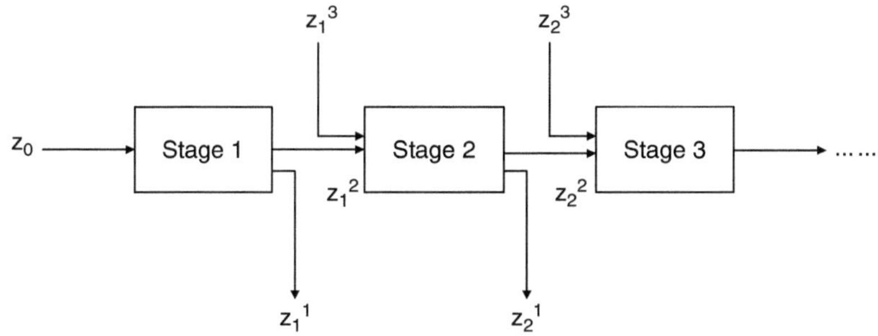

Fig. 5.2 Serial multi-stage DMU

(iii) ν_{pi} is the multiplier for the input component z_{pi}^{j3} entering the process at the beginning of stage p + 1.

Thus, when p = 2, 3, …, the efficiency ratio for DMU j (for a given set of multipliers) would be expressed as:

$$\theta_p = \left(\sum_{r=1}^{R_p} u_{pr} z_{pr}^{j1} + \sum_{k=1}^{S_p} \eta_{pk} z_{pk}^{j2}\right) \Big/ \left(\sum_{k=1}^{S_{p-1}} \eta_{p-1k} z_{p-1k}^{j2} + \sum_{i=1}^{I_p} \nu_{p-1i} z_{p-1i}^{j3}\right) \tag{5.20}$$

Note that there are no outputs flowing into stage 1. The efficiency measure for stage 1 of the process (namely, p = 1), for DMU_j becomes

$$\theta_1 = \left(\sum_{r=1}^{R_1} u_{1r} z_{1r}^{j1} + \sum_{k=1}^{S_1} \eta_{1k} z_{1k}^{j2}\right) \Big/ \sum_{i=1}^{I_0} \nu_{0i} z_{0i}^{j} \tag{5.21}$$

where z_{0i}^{j} are the (only) inputs to the first stage represented by the input vector z_o.

We claim that the overall efficiency measure of the multistage process can reasonably be represented as a convex linear combination of the P (stage-level) measures, namely

$$\theta = \sum_{p=1}^{P} w_p \theta_p \text{ where } \sum_{p=1}^{P} w_p = 1.$$

As in Sect. 5.3, we use $\sum_{i=1}^{I_0} \nu_{0i} z_{0i}^{j} + \sum_{p=2}^{P} \left(\sum_{k=1}^{S_{p-1}} \eta_{p-1k} z_{p-1k}^{j2} + \sum_{i=1}^{I_p} \nu_{p-1i} z_{p-1i}^{j3}\right)$ to represent the total size of or total amount of resources consumed by the entire process, and we define the weight w_p to be the proportion of the total input used at the pth stage. We then have

$$w_1 = \sum_{i=1}^{I_0} \nu_{0i} z_{0i}^{j} \Big/ \left\{ \sum_{i=1}^{I_0} \nu_{0i} z_{0i}^{j} + \sum_{p=2}^{P} \left(\sum_{k=1}^{S_{p-1}} \eta_{p-1k} z_{p-1k}^{j2} + \sum_{i=1}^{I_p} \nu_{p-1i} z_{p-1i}^{j3} \right) \right\},$$

$$w_p = \left(\sum_{k=1}^{S_{p-1}} \eta_{p-1k} z_{p-1k}^{j2} + \sum_{i=1}^{I_p} \nu_{p-1i} z_{p-1i}^{j3} \right) \Big/ \left\{ \sum_{i=1}^{I_0} \nu_{0i} z_{0i}^{j} \right.$$

$$\left. + \sum_{p=2}^{P} \left(\sum_{k=1}^{S_{p-1}} \eta_{p-1k} z_{p-1k}^{j2} + \sum_{i=1}^{I_p} \nu_{p-1i} z_{p-1i}^{j3} \right) \right\}, p > 1$$

Thus, we can write the overall efficiency θ in the form

$$\theta = \sum_{p=1}^{P} \left(\sum_{r=1}^{R_p} u_{pr} z_{pr}^{j1} + \sum_{k=1}^{S_p} \eta_{pk} z_{pk}^{j2} \right) \Big/ \left\{ \sum_{i=1}^{I_0} \nu_{0i} z_{0i}^{j} \right.$$

$$\left. + \sum_{p=2}^{P} \left(\sum_{k=1}^{S_{p-1}} \eta_{p-1k} z_{p-1k}^{j2} + \sum_{i=1}^{I_p} \nu_{p-1i} z_{p-1i}^{j3} \right) \right\}, \tag{5.22}$$

We then set out to optimize the overall efficiency θ of the multistage process, subject to the restrictions that the individual measures θ_p must not exceed unity, or in the linear programming format, after making the usual Charnes and Cooper transformation,

$$\max \sum_{p=1}^{P} \left(\sum_{r=1}^{R_p} u_{pr} z_{pr}^{o1} + \sum_{k=1}^{S_p} \eta_{pk} z_{pk}^{o2} \right)$$

subject to

$$\begin{aligned}
&\left\{ \sum_{i=1}^{I_0} \nu_{0i} z_{0i}^{o} + \sum_{p=2}^{P} \left(\sum_{k=1}^{S_{p-1}} \eta_{p-1k} z_{p-1k}^{o2} + \sum_{i=1}^{I_p} \nu_{p-1i} z_{p-1i}^{o3} \right) \right\} = 1 \\
&\left(\sum_{r=1}^{R_1} u_{1r} z_{1r}^{j1} + \sum_{k=1}^{S_1} \eta_{1k} z_{1k}^{j2} \right) \le \sum_{i=1}^{I_0} \nu_{0i} z_{0i}^{j} \\
&\left(\sum_{r=1}^{R_p} u_{pr} z_{pr}^{j1} + \sum_{k=1}^{S_p} \eta_{pk} z_{pk}^{j2} \right) \le \left(\sum_{k=1}^{S_{p-1}} \eta_{p-1k} z_{p-1k}^{j2} + \sum_{i=1}^{I_p} \nu_{p-1i} z_{p-1i}^{j3} \right) \forall j \\
&u_{pr}, \eta_{pk}, \nu_{pi}, \nu_{0i} \ge 0
\end{aligned} \tag{5.23}$$

Note that we should impose the restriction that the overall efficiency scores for each j should not exceed unity, but since these are redundant, this is unnecessary.

Note again that the w_p, as defined above, are variables related to the inputs and the intermediate measures. By virtue of the optimization process, it can turn out that some $w_p = 0$ at optimality. To overcome this problem, one can impose bounding restrictions $w_p \ge \beta$, where β is a selected constant.

5.6 General Multi-stage Processes

In the process discussed in the previous section it is assumed that the components of a DMU are arranged in series as depicted in Fig. 5.2. There, at each stage p, the inputs took one of two forms, namely (1) those that are outputs from the previous stage p-1, and (2) new inputs that enter the process at the start of stage p. On the output side, those (outputs) emanating from stage p take two forms as well, namely (1) those that leave the system as finished 'products', and (2) those that are passed on as inputs to the *immediate* next stage $p + 1$.

The model presented to handle such strict serial processes is easily adapted to more general network structures. Specifically, the efficiency ratio for an overall process can be expressed as the weighted average of the efficiencies of the individual components. The efficiency of any given component is the ratio of the total output to the total input corresponding to that component. Again, the weight w_p to be applied to any component p is expressed as

$$w_p = (\text{component } p \,\text{input})/(\text{total input across all components}).$$

There is no convenient way to represent a network structure that would lend itself to a generic mathematical representation analogous to model (5.23) above. The sequencing of activities and the source of inputs and outputs for any given component will differ from one type of process to another. However, as a simple illustration, consider the following two examples of network structures:

5.6.1 Parallel Processes

Consider the process depicted in Fig. 5.3. Here, an initial input vector z_o enters component 1. Three output vectors exit this component, that is z_1^1 leaves the process, z_1^2 is passed on as an input to component 2, and z_1^3 as an input to component 3. Additional inputs z_1^4 and z_1^5 enter components 2 and 3 respectively, from outside the process. Components 2 and 3 have z_2^1 and z_3^1, respectively as output vectors which are passed on as inputs to component 4, where a final output vector z_4^1 is the result.

Component Efficiencies
Component 1 efficiency ratio: $\theta_1 = (u_1 z_1^1 + \eta_1^2 z_1^2 + \eta_1^3 z_1^3)/\nu_o z_o$
Component 2 efficiency ratio: $\theta_2 = \eta_2^1 z_2^1/(\eta_1^2 z_1^2 + \nu_1 z_1^4)$
Component 3 efficiency ratio: $\theta_3 = \eta_3^1 z_3^1/(\eta_1^3 z_1^3 + \nu_2 z_1^5)$
Component 4 efficiency ratio: $\theta_4 = u_4 z_4^1/(\eta_2^1 z_2^1 + \eta_3^1 z_3^1)$

Component Weights
Note that the total (weighted) input across all components is given by the sum of the denominators of θ_1 through θ_4, namely

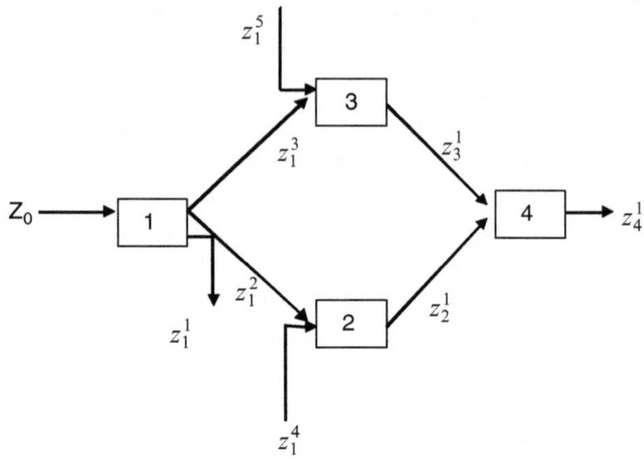

Fig. 5.3 Multi-stage DMU with parallel processes

$$I = \nu_o z_o + \eta_1^2 z_1^2 + \nu_1 z_1^4 + \eta_1^3 z_1^3 + \nu_2 z_1^5 + \eta_2^1 z_2^1 + \eta_3^1 z_3^1.$$

Now express the w_p as:

$$\begin{aligned} w_1 &= \nu_o z_o / I \\ w_2 &= \left(\eta_1^2 z_1^2 + \nu_1 z_1^4\right)/I \\ w_3 &= \left(\eta_1^3 z_1^3 + \nu_2 z_1^5\right)/I \\ w_4 &= \left(\eta_2^1 z_2^1 + \eta_3^1 z_3^1\right)/I \end{aligned}$$

With this, the overall network efficiency ratio is given by

$$\theta = \sum_{p=1}^{4} w_p \theta_p = \left(u_1 z_1^1 + \eta_1^2 z_1^2 + \eta_1^3 z_1^3 + \eta_2^1 z_2^1 + \eta_3^1 z_3^1 + u_4 z_4^1\right)/I,$$

And one then proceeds, as in (5.4) above, to derive the efficiency of each DMU and its components.

5.6.2 *Non-immediate Successor Flows*

In the previous example all flows of outputs from a stage or component either leave the process entirely or enter as an input to an *immediate successor* stage. In Fig. 5.2, stage p outputs flow to stage $p + 1$. In Fig. 5.3, the same is true except that there is more than one immediate successor of stage 1.

Consider Fig. 5.4. Here, the inputs to stage 3 are of three types, namely outputs from stage 2, inputs coming from outside the process, and outputs from a previous, but not immediately previous stage. Again the above rationale for deriving weights w_p can be applied and a model equivalent to (5.23) solved to determine the decomposition of an overall efficiency score into scores for each of the components in the process.

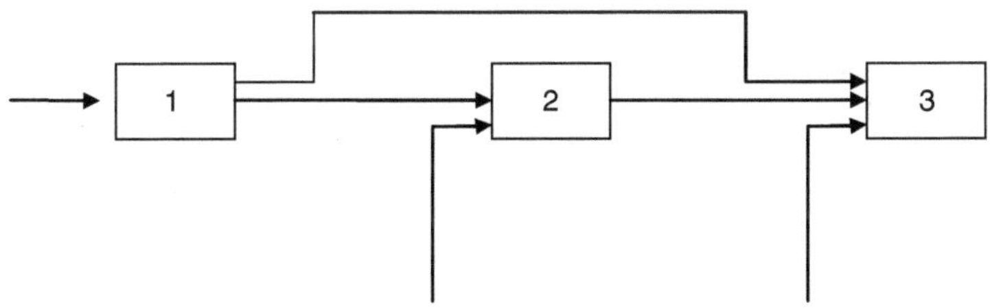

Fig. 5.4 Non-immediate successor flows

5.7 General Multi-stage Processes: An Illustrative Application

We here re-visit the supply chain data set used in Liang et al. (2006). This data set consists of a two-stage process, or a seller-buyer supply chain. The inputs to the first stage (seller) are labor (z^j_{01}), operating cost (z^j_{02}) and shipping cost (z^j_{03}). The outputs from the first stage are number of product A shipped (z^{j2}_{11}), number of product B shipped (z^{j2}_{12}) and number of product C shipped (z^{j2}_{13}). This data set assumes that all outputs from the first stage become inputs to the second stage, i.e., there is no z^1_1. There is one input to the second stage (buyer), labor (z^{j3}_{11}), and two outputs from the second stage, sales (z^{j1}_{21}) and profits (z^{j1}_{22}). Table 5.5 provides the data set.

In this case, we have, for DMU_o

$$w_1 = \sum_{i=1}^{3} \nu_{0i} z^o_{0i} \Big/ \left(\sum_{i=1}^{3} \nu_{0i} z^o_{0i} + \sum_{k=1}^{3} \eta_{1k} z^{o2}_{1k} + \nu_{11} z^{o3}_{11} \right),$$
$$w_2 = \left(\sum_{k=1}^{3} \eta_{1k} z^{o2}_{1k} + \nu_{11} z^{o3}_{11} \right) \Big/ \left(\sum_{i=1}^{3} \nu_{0i} z^o_{0i} + \sum_{k=1}^{3} \eta_{1k} z^{o2}_{1k} + \nu_{11} z^{o3}_{11} \right)$$

$$\begin{aligned}
&\text{Max} \sum_{k=1}^{3} \eta_{2k} z^{o2}_{1k} + \sum_{r=1}^{2} u_{2r} z^{o1}_{2r} \\
&\text{subject to} \\
&\quad \sum_{i=1}^{3} \nu_{0i} z^o_{0i} + \sum_{k=1}^{3} \eta_{1k} z^{o2}_{1k} + \nu_{11} z^{o3}_{11} = 1 \\
&\quad \sum_{k=1}^{3} \eta_{1k} z^{j2}_{1k} \le \sum_{i=1}^{3} \nu_{0i} z^j_{0i}, \text{j} = 1, \ldots, 10 \text{ (for stage 1)} \\
&\quad \sum_{r=1}^{2} u_{2r} z^{j1}_{2r} \le \sum_{k=1}^{3} \eta_{1k} z^{j2}_{1k} + \nu_{11} z^{j3}_{11}, \text{j} = 1, \ldots, 10 \text{ (for stage 2)}
\end{aligned} \tag{5.24}$$

Table 5.5 Data set

	Labor	Operating cost	Shipping cost	Product A	Product B	Product C	Labor	Sales	Profits
DMU	z_{01}^{j}	z_{02}^{j}	z_{03}^{j}	z_{11}^{j2}	z_{12}^{j2}	z_{13}^{j2}	z_{11}^{j3}	z_{21}^{j1}	z_{22}^{j1}
1	9	50	1	20	10	5	8	100	25
2	10	18	10	10	15	7	10	70	20
3	9	30	3	8	20	2	8	96	30
4	8	25	1	20	20	10	10	80	20
5	10	40	5	15	20	5	15	85	15
6	7	35	2	35	10	5	5	90	35
7	7	30	3	10	25	8	10	100	30
8	12	40	4	20	25	4	8	120	10
9	9	25	2	10	10	5	15	110	15
10	10	50	1	20	15	9	10	80	20

where efficiency scores for DMU_o in stages 1 and 2 can be expressed as

$$\theta_1 = \sum_{k=1}^{3} \eta_{1k} z_{1k}^{o2} \Big/ \sum_{i=1}^{3} \nu_{0i} z_{0i}^{o}$$

$$\theta_2 = \sum_{r=1}^{2} u_{2r} z_{2r}^{o1} \Big/ \left(\sum_{k=1}^{3} \eta_{1k} z_{1k}^{o2} + \nu_{11} z_{11}^{o3} \right)$$

Table 5.6 reports the results from model (5.24) where the last two columns display the efficiency scores derived from the cooperative model of Liang et al. (2006). Note that the differences between the two approaches are not significant. For example, the two approaches yield identical efficiency scores for the two stages for DMUs, 2, 5, 6, and 9. The Liang et al. (2006) approach is based upon a non-linear program and its solution is obtained by using heuristic search. While the current approach uses a linear program and guarantees a global optimal solution.

Note that the average of the two stages' efficiency scores is used as the objective function in Liang et al. (2006) non-linear model, namely, the weights for the two individual efficiency scores are equal, $w_1 = w_2$. The current approach yields $w_1 = w_2 = 0.5$ for DMUs 4 and 7. Yet, our results are different from those obtained from Liang et al. (2006). For example, in DMU 7, the efficiency score for the second stage is 0.54762 compared to 0.81888 from Liang et al. (2006). This is due to the fact that our choice of weights actually introduces some sort of value judgment into the DEA model, and restricts the multiplier values in model (5.24). This is why Liang et al. (2006) score is larger than ours when $w_1 = w_2 = 0.5$ in optimality.

Table 5.6 Results

	Our results (model (5.5))					Liang et al. (2006)	
DMU	Overall score	w_1	w_2	θ_1	θ_2	θ_1	θ_2
1	0.92495	0.30843	0.69157	0.75666	1	1	0.89394
2	0.86486	0.51974	0.48026	0.92403	0.80082	0.92403	0.80082
3	0.85898	0.34817	0.65183	0.59497	1	0.69106	1
4	0.77381	0.5	0.5	1	0.54762	1	0.62786
5	0.62073	0.46194	0.53806	0.67595	0.57332	0.67595	0.57332
6	1	0.27992	0.72008	1	1	1	1
7	0.90405	0.5	0.5	1	0.80811	1	0.81888
8	0.92886	0.21477	0.78523	0.66875	1	0.74667	1
9	0.78091	0.43817	0.56183	0.5	1	0.5	1
10	0.75444	0.54281	0.45719	0.84226	0.65018	1	0.59596

Table 5.7 Results with $\beta_1 = 0.5, \beta_2 = 0.5$

DMU	Overall score	w_1	w_2	θ_1	θ_2
1	0.86323	0.5	0.5	0.72645	1
2	0.85303	0.5	0.5	0.9222	0.78386
3	0.83629	0.5	0.5	0.67258	1
4	0.77381	0.5	0.5	1	0.54762
5	0.61749	0.5	0.5	0.67595	0.55903
6	0.99678	0.5	0.5	0.99357	1
7	0.90405	0.5	0.5	1	0.80811
8	0.81756	0.5	0.5	0.72772	0.9074
9	0.75	0.5	0.5	0.5	1
10	0.75435	0.5	0.5	0.85137	0.65732

Note that weights $w_p(p = 1, 2, \ldots, P)$ defined are actually variables related to the multiplier decision variables. We next, therefore, impose additional restrictions on w_1 and w_2 in model (5.24) via

$$w_1 = \left\{ \sum_{i=1}^{3} \nu_{0i} z_{0i}^{o} \Big/ \left(\sum_{i=1}^{3} \nu_{0i} z_{0i}^{o} + \sum_{k=1}^{3} \eta_{1k} z_{1k}^{o2} + \nu_{11} z_{11}^{o3} \right) \right\} \geq \beta_1$$

$$w_2 = \left\{ \left(\sum_{k=1}^{3} \eta_{1k} z_{1k}^{o2} + \nu_{11} z_{11}^{o3} \right) \Big/ \left(\sum_{i=1}^{3} \nu_{0i} z_{0i}^{o} + \sum_{k=1}^{3} \eta_{1k} z_{1k}^{o2} + \nu_{11} z_{11}^{o3} \right) \right\} \geq \beta_2$$

where β_1 and β_2 are user-specified parameters. In this way, we can perform sensitivity analysis on w_1 and w_2.

We first impose $\beta_1 = \beta_2$ and change β_1 and β_2 0.1–0.5 with a 0.1 increment each time. Note that when $\beta_1 = \beta_2 = 0.5$, we explicitly require that $w_1 = w_2 = 0.5$ as in Liang et al. (2006). Table 5.7 reports the results when $\beta_1 = \beta_2 = 0.5$. Both our approach and Liang et al. (2006) yield identical efficiency scores for DMU9. Except for DMU1, Liang et al. (2006) score is larger than ours when $w_1 = w_2 = 0.5$ in optimality. For DMU1, the definition of our weights and restrictions on our weights

Table 5.8 Results with $\beta_1 = \beta_2 = 0.1$ (0.2, 0.3, 0.4)

DMU	Overall score	w_1	w_2	θ_1	θ_2
2	0.86486	0.51974	0.48026	0.92403	0.80082
4	0.77381	0.5	0.5	1	0.54762
5	0.62073	0.46194	0.53806	0.67595	0.57332
6	1	0.31591	0.68409	1	1
7	0.90405	0.5	0.5	1	0.80811
9	0.78091	0.43817	0.56183	0.5	1
10	0.75444	0.54281	0.45719	0.84226	0.65018

Table 5.9 Results for DMUs 1, 3, and 8

DMU	Overall score	w_1	w_2	θ_1	θ_2	
1	0.92495	0.30843	0.69157	0.75666	1	$\beta_1 = \beta_2 = 0.1, 0.2, 0.3$
1	0.90182	0.4	0.6	0.75455	1	$\beta_1 = \beta_2 = 0.4$
3	0.85898	0.34817	0.65183	0.59497	1	$\beta_1 = \beta_2 = 0.1, 0.2, 0.3$
3	0.85186	0.4	0.6	0.62966	1	$\beta_1 = \beta_2 = 0.4$
8	0.92886	0.21477	0.78523	0.66875	1	$\beta_1 = \beta_2 = 0.1, 0.2$
8	0.91627	0.3	0.7	0.72091	1	$\beta_1 = \beta_2 = 0.3$
8	0.89238	0.4	0.6	0.73095	1	$\beta_1 = \beta_2 = 0.4$

turn the efficient stage 1 under Liang et al. (2006) approach into an inefficient stage, and the inefficient stage 2 under Liang et al. (2006) approach into efficient.

Table 5.8 reports the results for DMUs 2, 4, 5, 6, 7, 9 and 10 whose efficiency scores along with the optimized weights remain unchanged when $\beta_1 = \beta_2 = 0.1$, 0.2, 0.3 and 0.4, respectively.

Table 5.9 reports the results for DMUs 1, 3 and 8 whose efficiency scores changed when β_1 and β_2 are changed (see the last column of Table 5.5). For DMUs 1 and 3, change in the efficiency scores does not occur until $\beta_1 = \beta_2 = 0.4$. For DMU 8, a change in the efficiency score for the first stage is observed when $\beta_1 = \beta_2 = 0.3$ and 0.4.

It can be seen that up to $\beta_1 = \beta_2 = 0.3$, most of the DMUs have the same weights and efficiency scores with respect to different values of β_1 and β_2. As expected, when $\beta_1 = \beta_2 = 0.4$, some of the resulting weights are different from the previous cases. However, we note that the efficiency scores do not change significantly. We also note that the efficiency scores for the second stage do not change when β_1 and β_2 are increased from 0.1 to 0.4.

We also performed calculations when β_1 is fixed at 0.2 and β_2 is changed from 0.3 to 0.8 with an increment of 0.1 each time (results are not reported here). In overall, the efficiency scores do not change significantly.

The above sensitivity analysis indicates that efficiency scores obtained based upon our approach are robust with respect to our choice of weights.

We finally apply our approach to the numerical example used in Tone and Tsutsui (2009). Table 5.10 provides the data. We have two intermediate measures or outputs flow from one stage to the other. Table 5.11 reports the results. In this case, if we do not impose a lower bound for the w_p(p = 1, 2, 3), we have some

Table 5.10 Data set in Tone and Tsutsui (2009)

	Stage 1	Stage 2		Stage 3		Intermediate measure	
	Input 1	Input 2	Output 2	Input 3	Output 3	Link12	Link23
A	0.838	0.277	0.879	0.962	0.337	0.894	0.362
B	1.233	0.132	0.538	0.443	0.18	0.678	0.188
C	0.321	0.045	0.911	0.482	0.198	0.836	0.207
D	1.483	0.111	0.57	0.467	0.491	0.869	0.516
E	1.592	0.208	1.086	1.073	0.372	0.693	0.407
F	0.79	0.139	0.722	0.545	0.253	0.966	0.269
G	0.451	0.075	0.509	0.366	0.241	0.647	0.257
H	0.408	0.074	0.619	0.229	0.097	0.756	0.103
I	1.864	0.061	1.023	0.691	0.38	1.191	0.402
J	1.222	0.149	0.769	0.337	0.178	0.792	0.187

Table 5.11 Results on three-stage process

	Overall	Stage 1	Stage 2	Stage 3	w_1	w_2	w_3
A	0.579	0.410	0.646	0.971	0.46	0.41	0.13
B	0.386	0.211	0.339	0.414	0.10	0.10	0.80
C	1.000	1.000	1.000	0.999	0.42	0.48	0.10
D	0.917	0.225	0.942	1.000	0.10	0.10	0.80
E	0.478	0.167	0.501	0.953	0.36	0.42	0.22
F	0.598	0.470	0.656	0.984	0.51	0.37	0.11
G	0.762	0.551	0.717	0.983	0.24	0.44	0.32
H	0.675	0.711	0.599	0.843	0.46	0.44	0.10
I	0.922	0.245	1.000	0.990	0.10	0.64	0.26
J	0.476	0.249	0.423	0.511	0.10	0.10	0.80

$w_p = 1$ at optimality (for DMUs B, D, I and J). Therefore, we impose $w_p > 0.1$ ($p = 1, 2, 3$) in model (5.23). Because our approach is different from Tone and Tsutsui's (2009) and our choice of weights introduces restrictions on the multipliers, our results are different from theirs.

5.8 Conclusions

The current chapter introduces the DEA approaches of Chen et al. (2009) and Cook et al. (2010) for DMUs that have a general multi-stage or network structure. We first study the simple two-stage network processes where outputs form the first stage become the only inputs to the second stage. We then examine pure serial networks where each stage has its own inputs and two types of outputs. One type of output from any given stage p is passed on as an input to the next stage, and the other type exits the process at stage p. Work closely related to the current chapter is the non-linear approach of Liang et al. (2006) where a two-member supply chain structure is studied. While Liang et al. (2006) developed a heuristic search algorithm after converting the non-linear model into a parametric linear model, their approach cannot be generalized into cases where supply chains have more than two members. The approach of Cook et al. (2010) can, however, handle via a linear model, situations where more than two stages are present.

In general, the intermediate measures are those that exist between two members of the network. In many cases, the intermediate measures are obvious, as indicated in our examples mentioned in the Introduction. Tone and Tsutsui (2009) provides other good examples. Sometimes, the selection of intermediate measures is not so obvious. The important thing is that intermediate measures are neither "inputs" (to be reduced) nor "outputs" (to be increased), rather these measures need to be "coordinated" to determine their efficient levels.

Note that models under Sects. 5.5 and 5.6 are developed under the assumption of CRS. We should point out that these models can directly be applied to VRS by adding the free-in-sign variable in our ratio efficiency definition, just as in the standard VRS DEA model and the two stage network DEA approach discussed in Sect. 5.3.

References

Banker, R. D., Charnes, A., & Cooper, W. W. (1984). Some models for the estimation of technical and scale inefficiencies in data envelopment analysis. *Management Science, 30*, 1078–1092.

Castelli, L., Pesenti, R., & Ukovich, W. (2004). DEA-like models for the efficiency evaluation of hierarchically structured units. *European Journal of Operational Research, 154*(2), 465–476.

Charnes, A., & Cooper, W. W. (1962). Programming with linear fractional functionals. *Naval Research Logistics Quarterly, 9*, 181–185.

Charnes, A., Cooper, W. W., & Rhodes, E. (1978). Measuring the efficiency of decision making units. *European Journal of Operational Research, 2*, 429–444.

Chen, Y., Cook, W. D., Li, N., & Zhu, J. (2009). Additive efficiency decomposition in two-stage DEA. *European Journal of Operational Research, 196*, 1170–1176.

Cook, W. D., Zhu, J., Yang, F., & Bi, G.-B. (2010). Network DEA: Additive efficiency decomposition. *European Journal of Operational Research, 207*(2), 1122–1129.

Färe, R., & Grosskopf, S. (1996). Productivity and intermediate products: A frontier approach. *Economics Letters, 50*, 65–70.

Kao, C., & Hwang, S. N. (2008). Efficiency decomposition in two-stage data envelopment analysis: an application to non-life insurance companies in Taiwan. *European Journal of Operational Research, 185*(1), 418–429.

Liang, L., Yang, F., Cook, W. D., & Zhu, J. (2006). DEA models for supply chain efficiency evaluation. *Annals of Operations Research, 145*(1), 35–49.

Liang, L., Cook, W. D., & Zhu, J. (2008). DEA Models for two-stage processes: game approach and efficiency decomposition. *Naval Research Logistics, 55*, 643–653.

Tone, K., & Tsutsui, M. (2009). Network DEA: A slacks-based measure approach. *European Journal of Operational Research, 197*(1), 243–252.

Chapter 6
Scale Efficiency Measurement in Two-Stage Production Systems

Chiang Kao and Shiuh-Nan Hwang

Abstract One important objective in measuring efficiency is to find the factors that cause inefficiencies so that its performance can be improved. The conventional data envelopment analysis approach is able to decompose the overall efficiency of a system into the product of the technical and scale efficiencies when the internal structure is ignored. For two-stage systems, where the inputs are supplied to the first process to produce intermediate products for the second process to produce the final outputs, the system efficiency can be decomposed into process efficiencies. This paper further decomposes each process efficiency into the product of the technical and scale efficiencies via an input-oriented model for the first process and an output-oriented one for the second. The decomposition also reveals that the overall efficiency of the two-stage system, when the operations of the two processes are considered, is still the product of the technical and scale efficiencies. The concept is illustrated using an example of 24 non-life insurance companies in Taiwan.

Keywords Data envelopment analysis • Two-stage system • Scale efficiency • Technical efficiency

---

C. Kao (✉)
Department of Industrial and Information Management, National Cheng Kung University, Tainan, Taiwan
e-mail: ckao@mail.ncku.edu.tw

S.-N. Hwang
Department of Business Administration, Ming Chuan University, Taipei, Taiwan
e-mail: snhwang@mail.mcu.edu.tw

W.D. Cook and J. Zhu (eds.), *Data Envelopment Analysis*,
International Series in Operations Research & Management Science 208,
DOI 10.1007/978-1-4899-8068-7_6,

6.1 Introduction

Charnes et al. (1978) developed a data envelopment analysis (DEA) model, conventionally referred to as the CCR model, to measure the relative efficiency of a set of decision making units (DMUs) that use multiple inputs to produce multiple outputs under the assumption of constant returns to scale (CRS). This model was later extended by Banker et al. (1984), conventionally referred to as the BCC model, to measure efficiency under the assumption of variable returns to scale (VRS). The CCR model is able to detect inefficiency due to the aggregate effect of insufficient technology and improper scale, while the BCC model is only used to detect insufficient technology. By comparing the efficiency scores calculated from these two models, the effects due to improper scale can be identified. The identified sources of inefficiency can then enable decision makers to design suitable alternatives to improve the performance of a system.

Conventional DEA models treat the system as a whole unit, and thus only the inputs supplied to the system and the outputs produced from it are considered in measuring efficiency. However, in many situations a system is composed of several interrelated processes, with the outputs of one process being used by some others for production, and ignoring the operations of the internal processes will produce misleading results. In response to this, Färe and Grosskopf (2000) proposed a network DEA model to take the operations of the internal processes into account.

The simplest structure of network systems is a two-stage system, where all the inputs are supplied to the first process to produce intermediate products for the second process to produce the final outputs. Several models have been proposed to measure the efficiency of this type of system (see the review of Cook et al. 2010), and Kao (2009) classified these into independent, connected, and relational. The independent model is typified by that presented in Seiford and Zhu (1999), which treats the two stages as two independent DMUs, and their efficiencies are calculated separately. Therefore, the scale efficiency of each process can be calculated by applying the CCR and BCC models. However, it is still not possible to measure the scale efficiency of the system when considering the interrelation of the two processes.

Färe and Grosskopf (2000) remains the most representative work with regard to a connected model, in that the technologies of the two processes are considered in measuring the overall efficiency of the system. Although the efficiency can be measured under both CRS and VRS, how to calculate the scale efficiency is still a problem, because the relationship between the overall, technical, and scale efficiencies in the two-stage system is not known.

For relational models, the system and process efficiencies can be calculated at the same time; moreover, there exist mathematical relationships between them. For example, the model in Kao and Hwang (2008) shows a multiplicative relationship between the system and process efficiencies, while that in Chen et al. (2009) shows an additive one. The slacks-based measures model also exhibits an additive relationship (Tone and Tsutsui 2009; Kao 2013). For models with an additive relationship, the efficiencies of both the system and processes can be calculated under both CRS and VRS. However, similar to the case of the connected model, the scale efficiency cannot be obtained from the other two, because the relationship between the overall, technical, and scale efficiencies is not known. It thus seems that only the multiplicative form of the relational model can be applied to measure scale efficiencies, and Kao and Hwang (2011) proposed an approach to accomplish this task.

The major problem in measuring scale efficiency from the overall and technical efficiencies for the two-stage system is that the outputs of the first process are the inputs of the second. If one wishes to improve the efficiency of the first process by increasing its outputs, the efficiency of the second process will then be reduced due to the increased amount of inputs. Similarly, if one tries to raise the efficiency score of the second process by reducing the amount of its input, then the efficiency of the first process will be decreased due to producing output. To resolve this conflict, Kao and Hwang (2011) used an input-oriented model to measure the efficiency of the first process by fixing the amount of the output and an output-oriented model to measure the efficiency of the second process. In this way, the efficiency of the first process can be improved by reducing the amount of the input and the second process by increasing that of the output. The CCR and BCC models are then applied to measure scale efficiencies for the two processes, which then represent the scale efficiency of the system as a whole.

In the following sections, the input- and output-oriented DEA models are first briefly reviewed. The measurement of the scale efficiencies is then illustrated graphically using a simple example. After this, the models for measuring scale efficiencies for general cases are developed, and the technical and scale efficiencies of the system and two processes of non-life insurance companies in Taiwan are calculated. Finally, some conclusions are drawn from the discussion of these results.

6.2 Input- and Output-Oriented Models

Both the CCR and BCC models can be formulated from the input and output sides. The objective of the former is to examine how much of the input can be reduced while producing the same amount of output, while that of the latter is to examine how much of the output can be increased by using the same amount of input for production. In the following discussions, we use X_{ij}, $i = 1, \ldots, m$ and Y_{rj}, $r = 1, \ldots, s$ to denote the ith input and rth output of the jth DMU, $j = 1, \ldots, n$, respectively, with v_i and u_r being the virtual multipliers associated with X_{ij} and Y_{rj}.

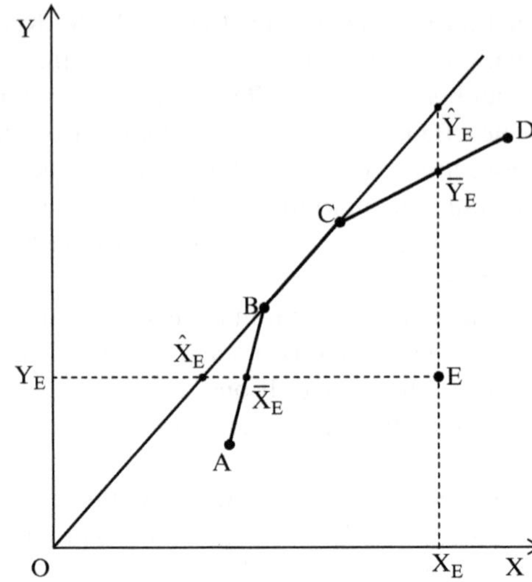

Fig. 6.1 Input- and output-oriented efficiency measures

6.2.1 Input-Oriented Model

Banker et al. (1984) developed the following model to measure the technical efficiency of the *k*th DMU from the input side under VRS:

$$
\begin{aligned}
E_k = \max.&\sum_{r=1}^{s} u_r Y_{rk} - u_0 \\
\text{s.t.} \quad &\sum_{i=1}^{m} v_i X_{ik} = 1 \\
&\left(\sum_{r=1}^{s} u_r Y_{rj} - u_0\right) - \sum_{i=1}^{m} v_i X_{ij} \le 0, \quad j = 1, \ldots, n \\
&u_r, v_i \ge \varepsilon, \quad r = 1, \ldots, s, \quad i = 1, \ldots, m \\
&u_0 \text{ unrestricted in sign}
\end{aligned} \tag{6.1}
$$

where ε is a small non-Archimedean number imposed to avoid ignoring any factor (Charnes et al. 1979; Charnes and Cooper 1984). When the term u_0 is omitted, Model (6.1) becomes the CCR model (Charnes et al. 1978), and the corresponding efficiency is the overall efficiency. The ratio of the overall efficiency to the technical efficiency is the (input) scale efficiency.

Consider a one-input one-output example with five DMUs, labeled as A, B, C, D, and E, as shown in Fig. 6.1. The straight line *OBC* and the connected line segments *ABCD* are the production frontiers constructed under CRS and VRS, respectively. DMUs A, B, C, and D are technically efficient, among which B and C are also overall efficient. The input-oriented model calculates efficiencies based on the amount of input consumed for production. The (input) overall and technical efficiencies for DMU E are the ratios of $\hat{X}_E$ to X_E and $\overline{X}_E$ to X_E, respectively, and the (input) scale efficiency is the ratio of the overall efficiency to the technical efficiency, which is the ratio of $\hat{X}_E$ to $\overline{X}_E$.

6.2.2 *Output-Oriented Model*

Banker et al. (1984) also developed a model for measuring efficiencies from the output side under VRS, which can be formulated as:

$$
\begin{aligned}
\frac{1}{E_k} = \min. & \sum_{i=1}^{m} v_i X_{ik} + v_0 \\
\text{s.t.} & \sum_{r=1}^{s} u_r Y_{rk} = 1 \\
& \left(\sum_{i=1}^{m} v_i X_{ij} + v_0 \right) - \sum_{r=1}^{s} u_r Y_{rj} \geq 0, \quad j = 1, \ldots, n \\
& u_r, v_i \geq \varepsilon, \quad r = 1, \ldots, s, \quad i = 1, \ldots, m \\
& v_0 \text{ unrestricted in sign}
\end{aligned} \tag{6.2}
$$

where the efficiency is expressed in reciprocal form. The efficiency thus obtained is the output-oriented technical one. Similar to the previous case, when the term v_0 is omitted, Model (6.2) becomes the CCR model, and the resulting measure is the overall efficiency. The ratio of the overall efficiency to the (output) technical efficiency is the (output) scale efficiency.

The output-oriented model measures efficiencies based on the amount of output produced. For DMU E in Fig. 6.1, the output-oriented overall and technical efficiencies are the ratios of Y_E to $\hat{Y}_E$ and Y_E to $\overline{Y}_E$, respectively. The (output) scale efficiency, which is the ratio of the overall efficiency to the (output) technical efficiency, is the ratio of $\overline{Y}_E$ to $\hat{Y}_E$.

Note that the overall efficiencies calculated from the input Model (6.1), with the term u_0 omitted, and output Model (6.2), with the term v_0 omitted, are the same. The technical and scale efficiencies calculated from these two models, however, may not be the same. In the example shown in Fig. 6.1, only the DMUs using the frontier facet BC to calculate efficiencies will result in the same measures.

6.3 Graphical Illustration

The two-stage system is a system composed of two processes connected in series, where all the inputs are supplied to the first process to produce intermediate products, and all of them in turn are used by the second process to produce the final outputs of the system. Figure 6.2 shows a typical two-stage system, where Z_{fj}, $f = 1, \ldots, g$ denote the intermediate products.

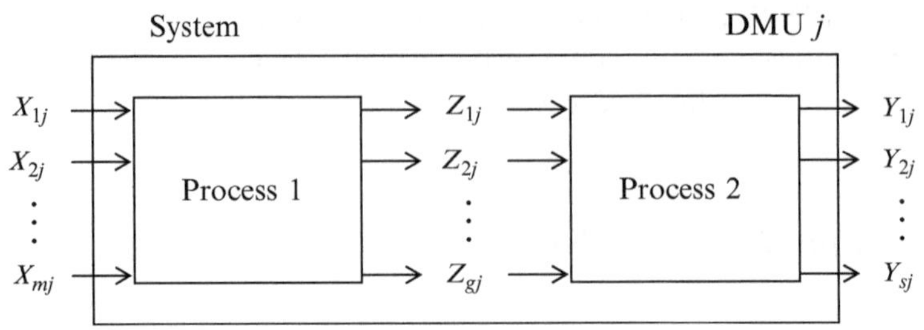

Fig. 6.2 Two-stage system with inputs X, outputs Y, and intermediate products Z

Table 6.1 Data and efficiencies measured from output-oriented Model (6.2) for five DMUs of a two-stage system

DMU	Input X	Intermediate product Z	Output Y	Overall efficiency	Technical efficiency	Scale efficiency
A	4	2	0.5	5/32	1	5/32
B	8	4	1.8	9/32	36/113	113/128
C	6	6	3.3	11/16	1	11/16
D	10	8	8	1	1	1
E	15	12	10.5	7/8	1	7/8

Consider a simple example of five DMUs, A~E, where each applies one input X to produce one intermediate product Z in the first process, and the intermediate product Z is then used in the second process to produce one output Y. The left part of Table 6.1 shows a set of hypothetical data for these five DMUs. The conventional DEA approach ignores the operations of the two processes, assuming that output Y is directly produced by input X. In this case, the production frontiers constructed under CRS and VRS are the dashed straight line OD and connected line segments $ACDE$, respectively, shown in Fig. 6.3. For DMU B, the overall and (output) technical efficiencies are B/B^* and B/B°, respectively, which produce an output scale efficiency of B°/B^*. The right part of Table 6.1 shows the overall, (output) technical, and (output) scale efficiencies of the five DMUs.

The network DEA approach, on the other hand, takes the operations of the two processes into consideration. Figure 6.4 depicts the production process in a counter-clockwise orientation, where the right side shows that Process 1 applies input X to produce intermediate product Z, and the left side shows that Process 2 applies intermediate product Z to produce output Y. The superscripts associated with the DMUs indicate the process. The straight lines OC^1 and OD^2 passing through the origin are the production frontiers under CRS for Processes 1 and 2, respectively. Note that here two production frontiers are constructed for the two processes, which is different from the idea of using one frontier for the two processes, as in Chen et al. (2010).

On the right side, the kinked line $A^1C^1E^1$ is the production frontier for Process 1 under VRS. The three DMUs on the frontier, A, C, and E, are thus technically efficient. DMU C is also overall efficient, because it lies on the frontier constructed

Fig. 6.3 Different types of production frontier on the X-Y plane

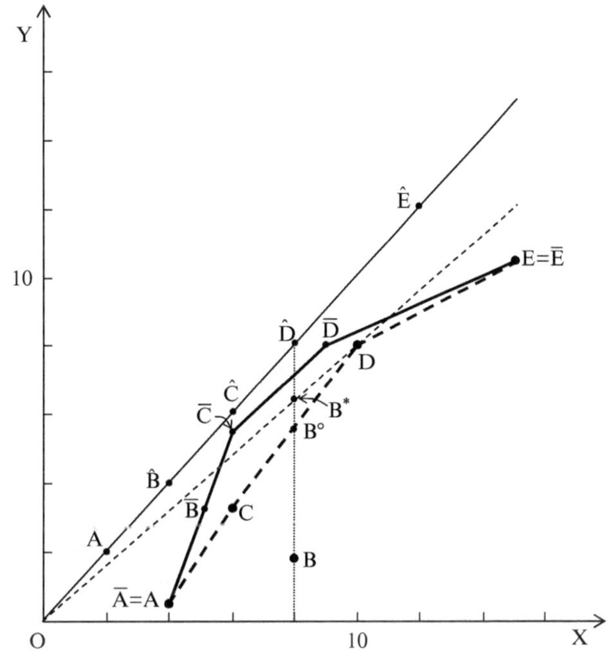

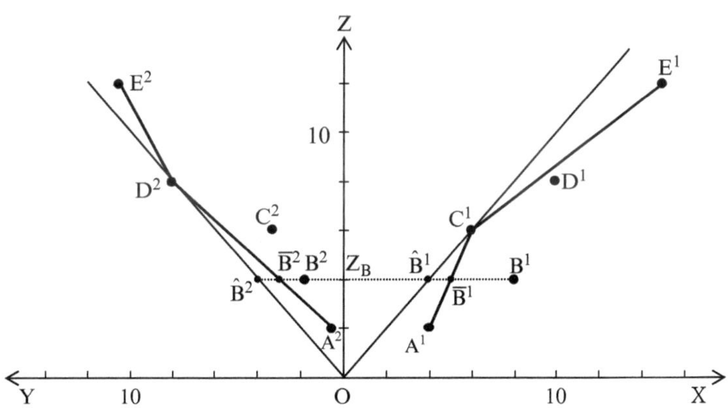

Fig. 6.4 Efficiency measurement for processes in a two-stage system

under CRS. The other two DMUs, B and D, are technically inefficient. Consider DMU B. Its overall and technical efficiencies measured from the input side are $\hat{B}^1 Z_B / B^1 Z_B$ (=1/2) and $\overline{B}^1 Z_B / B^1 Z_B$ (=5/8), respectively, which produce an (input) scale efficiency of $\hat{B}^1 Z_B / \overline{B}^1 Z_B$ (=4/5). These three types of efficiency for the other four DMUs can be calculated similarly, with the results shown in Table 6.2 under the heading of "Process 1."

The left side of Fig. 6.4 shows the production of Process 2. Note that the vertical axis represents the input of this process, the intermediate product Z, and the

Table 6.2 System and process efficiencies for the example, taking into account the operations of the processes

	System			Process 1			Process 2		
DMU	Overall	(Tech.	Scale)	Overall	(Tech.	Scale)	Overall	(Tech.	Scale)
A	1/8	(1	1/8)	1/2	(1	1/2)	1/4	(1	1/4)
B	9/40	(3/8	3/5)	1/2	(5/8	4/5)	9/20	(3/5	3/4)
C	11/20	(3/5	11/12)	1	(1	1)	11/20	(3/5	11/12)
D	4/5	(9/10	8/9)	4/5	(9/10	8/9)	1	(1	1)
E	7/10	(1	7/10)	4/5	(1	4/5)	7/8	(1	7/8)

horizontal axis shows the output Y of this process. The kinked line $A^2D^2E^2$ is the production frontier constructed under VRS, and DMUs A, D, and E are technically efficient. For the two technically inefficient DMUs, B and C, consider B. Its overall and technical efficiencies measured from the output side are $B^2Z_B/\hat{B}^2Z_B$ (=9/20) and $B^2Z_B/\overline{B}^2Z_B$ (=3/5), respectively, which result in an (output) scale efficiency of $\overline{B}^2Z_B/\hat{B}^2Z_B$ (=3/4). The efficiencies of the other four DMUs are measured similarly, with the results shown in Table 6.2, under the heading of "Process 2."

The results from the two processes can be aggregated to form that of the system. Consider DMU B again. This DMU uses 8 units of input X to produce 1.8 units of output Y (via 4 units of intermediate product Z), with a rate of 1.8/8. If it is technically efficient in both processes, then only 5 units of input X (corresponding to $\overline{B}^1$) are needed to produce 3 units of output Y (corresponding to $\overline{B}^2$), with a rate of 3/5. This corresponds to point $\overline{B}$ on the X-Y plane of Fig. 6.3. Comparing the actual rate of 1.8/8 to the technically efficient rate of 3/5, an overall technical efficiency of [(1.8/8)/(3/5)], or 3/8, is obtained for the system. This efficiency is clearly the product of the technical efficiency of the two processes, 5/8 and 3/5. By the same token, if both processes are overall efficient, then DMU B only requires 4 units of input X (corresponding to $\hat{B}^1$) to produce 4 units of output Y (corresponding to $\hat{B}^2$), with a rate of 4/4. This corresponds to point $\hat{B}$ on the X-Y plane of Fig. 6.3. Comparing the actual rate of 1.8/8 to this overall efficient rate of 4/4, an overall efficiency of [(1.8/8)/(4/4)], or 9/40, is obtained for the system. The ratio of the overall efficiency to the technical efficiency of the system, (9/40)/(3/8) = 3/5, is the scale efficiency of the system. From the graphical relationship shown in Fig. 6.4, this value is clearly the product of the scale efficiency of the two processes, 4/5 and 3/4. The efficiencies corresponding to the other four DMUs can be calculated similarly, with the results shown in Table 6.2 under the heading of "System."

From the discussion regarding the target points, we conclude that the straight line $O\hat{A}\,\hat{B}\,\hat{C}\,\hat{D}\,\hat{E}$ in Fig. 6.3 is the system frontier under CRS and the connected line segment $\overline{A}\,\overline{C}\,\overline{D}\,\overline{E}$ is that under VRS. These two frontiers lie above their counterparts OD and $ACDE$, respectively, constructed from the conventional DEA approach, indicating that ignoring the operations of the processes will overstate the measured efficiencies.

In this example, we find, first, the overall efficiency of the system is the product of those of the two processes, second, the overall efficiency of each process is the product of their technical and scale efficiencies, third, the technical efficiency of the system is the product of those of the two processes, and fourth, the scale efficiency of the system is the product of those of the two processes.

6.4 Measurement Models for General Cases

To measure the efficiency of the two-stage system for DMU k, Kao and Hwang (2008) proposed the following model:

$$\begin{aligned}
E_k^S = \max. & \sum_{r=1}^{s} u_r Y_{rk} \\
\text{s.t.} \quad & \sum_{i=1}^{m} v_i X_{ik} = 1 \\
& \sum_{r=1}^{s} u_r Y_{rj} - \sum_{i=1}^{m} v_i X_{ij} \le 0, \quad j = 1, \ldots, n \\
& \sum_{f=1}^{g} w_f Z_{fj} - \sum_{i=1}^{m} v_i X_{ij} \le 0, \quad j = 1, \ldots, n \\
& \sum_{r=1}^{s} u_r Y_{rj} - \sum_{f=1}^{g} w_f Z_{fj} \le 0, \quad j = 1, \ldots, n \\
& u_r, v_i, w_f \ge \varepsilon, \quad r = 1, \ldots, s, \quad i = 1, \ldots, m, \quad f = 1, \ldots, g
\end{aligned} \tag{6.3}$$

where the three sets of constraints correspond to the system, Process 1, and Process 2, respectively. Since the constraints corresponding to the system are the sum of those corresponding to the two processes, they are redundant, and can thus be deleted.

At optimality, the system and process efficiencies, based on Model (6.3), are calculated as:

$$\begin{aligned}
E_k^S &= \sum_{r=1}^{s} u_r Y_{rk} \Big/ \sum_{i=1}^{m} v_i X_{ik} \\
E_k^{(1)} &= \sum_{f=1}^{g} w_f Z_{fk} \Big/ \sum_{i=1}^{m} v_i X_{ik} \\
E_k^{(2)} &= \sum_{r=1}^{s} u_r Y_{rk} \Big/ \sum_{f=1}^{g} w_f Z_{fk}
\end{aligned}$$

Clearly, the system efficiency is the product of the two process efficiencies; that is, $E_k^S = E_k^{(1)} \times E_k^{(2)}$.

Model (6.3) may produce multiple solutions for the two process efficiencies. When this happens, the two process efficiencies do not have common bases for comparison. To make $E_j^{(1)}$ (and $E_j^{(2)}$) of different DMUs comparable, Kao and Hwang (2008) suggested using the maximum value of $E_k^{(1)}$ or $E_k^{(2)}$ for comparison, depending on which process is considered more important. Suppose the first process is of major concern, and the maximum value of $E_k^{(1)}$ is sought. The objective function of Model (6.3) is replaced by the formula of Process 1 efficiency, with the system efficiency maintained at the level of E_k^S obtained from Model (6.3). In symbols, it is:

$$
\begin{aligned}
E_k^{(1)} = \max.&\sum_{f=1}^{g} w_f Z_{fk} \\
\text{s.t.}\ &\sum_{i=1}^{m} v_i X_{ik} = 1 \\
&\sum_{r=1}^{s} u_r Y_{rk} = E_k^S \sum_{i=1}^{m} v_i X_{ik} \\
&\sum_{f=1}^{g} w_f Z_{fj} - \sum_{i=1}^{m} v_i X_{ij} \le 0, \quad j = 1, \ldots, n \\
&\sum_{r=1}^{s} u_r Y_{rj} - \sum_{f=1}^{g} w_f Z_{fj} \le 0, \quad j = 1, \ldots, n \\
&u_r, v_i, w_f \ge \varepsilon, \quad r = 1, \ldots, s, \quad i = 1, \ldots, m, \quad f = 1, \ldots, g
\end{aligned} \tag{6.4}
$$

Note that in this formulation the constraints corresponding to the system have been deleted for simplicity.

To calculate the (input) technical efficiency of Process 1, one simply replaces the part related to the CCR model in Model (6.4) by that of BCC Model (6.1), and the model is thus:

$$
\begin{aligned}
T_k^{(1)} = \max.&\sum_{f=1}^{g} \widetilde{w}_f Z_{fk} - \widetilde{w}_0 \\
\text{s.t.}\ &\sum_{i=1}^{m} \widetilde{v}_i X_{ik} = 1 \\
&\sum_{r=1}^{s} u_r Y_{rk} = E_k^S \sum_{i=1}^{m} v_i X_{ik} \\
&\sum_{f=1}^{g} w_f Z_{fj} - \sum_{i=1}^{m} v_i X_{ij} \le 0, \quad j = 1, \ldots, n \\
&\sum_{r=1}^{s} u_r Y_{rj} - \sum_{f=1}^{g} w_f Z_{fj} \le 0, \quad j = 1, \ldots, n \\
&\left(\sum_{f=1}^{g} \widetilde{w}_f Z_{fj} - \widetilde{w}_0\right) - \sum_{i=1}^{m} \widetilde{v}_i X_{ij} \le 0, \quad j = 1, \ldots, n \\
&u_r, v_i, w_f, \widetilde{w}_f \ge \varepsilon, \ r = 1, \ldots, s, \ i = 1, \ldots, m, \ f = 1, \ldots, g \\
&\widetilde{w}_0 \text{ unrestricted in sign}
\end{aligned} \tag{6.5}
$$

Models (6.4) and (6.5) can be combined to calculate the overall and (input) technical efficiencies of Process 1 at the same time:

$$
\begin{aligned}
\text{max.} \quad & \sum_{f=1}^{g} w_f Z_{fk} + \left(\sum_{f=1}^{g} \widetilde{w}_f Z_{fk} - \widetilde{w}_0 \right) \\
\text{s.t.} \quad & \sum_{i=1}^{m} v_i X_{ik} = 1 \\
& \sum_{i=1}^{m} \widetilde{v}_i X_{ik} = 1 \\
& \sum_{r=1}^{s} u_r Y_{rk} = E_k^S \sum_{i=1}^{m} v_i X_{ik} \\
& \sum_{f=1}^{g} w_f Z_{fj} - \sum_{i=1}^{m} v_i X_{ij} \leq 0, \quad j = 1, \ldots, n \\
& \sum_{r=1}^{s} u_r Y_{rj} - \sum_{f=1}^{g} w_f Z_{fj} \leq 0, \quad j = 1, \ldots, n \\
& \left(\sum_{f=1}^{g} \widetilde{w}_f Z_{fj} - \widetilde{w}_0 \right) - \sum_{i=1}^{m} \widetilde{v}_i X_{ij} \leq 0, \quad j = 1, \ldots, n \\
& u_r, v_i, w_f, \widetilde{w}_f \geq \varepsilon, \quad r = 1, \ldots, s, \quad i = 1, \ldots, m, \quad f = 1, \ldots, g \\
& \widetilde{w}_0 \text{ unrestricted in sign}
\end{aligned}
\tag{6.6}
$$

At optimality, the overall and (input) technical efficiencies of Profess 1 are calculated as:

$$
\begin{aligned}
E_k^{(1)} &= \sum_{f=1}^{g} w_f Z_{fk} \\
T_k^{(1)} &= \sum_{f=1}^{g} \widetilde{w}_f Z_{fk} - \widetilde{w}_0
\end{aligned}
$$

Consequently, the input scale efficiency is calculated as their ratio:

$$
S_k^{(1)} = E_k^{(1)} / T_k^{(1)}
$$

Similarly, the (output) technical efficiency of Process 2 is calculated by replacing the part related to the CCR model in Model (6.4) by that of output

BCC Model (6.2), while maintaining the system efficiency at E_k^S and Process 1 efficiency at $E_k^{(1)}$:

$$
\begin{aligned}
\frac{1}{T_k^{(2)}} = \min. & \sum_{f=1}^{g} \hat{w}_f Z_{fk} + \hat{w}_0 \\
\text{s.t.} \ & \sum_{r=1}^{s} \hat{u}_r Y_{rk} = 1 \\
& \sum_{r=1}^{s} u_r Y_{rk} = E_k^S \sum_{i=1}^{m} v_i X_{ik} \\
& \sum_{f=1}^{g} w_f Z_{fk} = E_k^{(1)} \sum_{i=1}^{m} v_i X_{ik} \\
& \sum_{f=1}^{g} w_f Z_{fj} - \sum_{i=1}^{m} v_i X_{ij} \le 0, \qquad j = 1, \ldots, n \\
& \sum_{r=1}^{s} u_r Y_{rj} - \sum_{f=1}^{g} w_f Z_{fj} \le 0, \qquad j = 1, \ldots, n \\
& \left(\sum_{f=1}^{g} \hat{w}_f Z_{fj} + \hat{w}_0 \right) - \sum_{r=1}^{s} \hat{u}_r Y_{rj} \ge 0, \ j = 1, \ldots, n \\
& u_r, v_i, w_f, \hat{w}_f \ge \varepsilon, \quad r = 1, \ldots, s, \quad i = 1, \ldots, m, \quad f = 1, \ldots, g \\
& \hat{w}_0 \text{ unrestricted in sign}
\end{aligned}
\tag{6.7}
$$

At optimality, the (output) technical efficiency of Process 2 is:

$$T_k^{(2)} = 1 \Big/ \left(\sum_{f=1}^{g} \hat{w}_f Z_{fk} + \hat{w}_0 \right)$$

and the corresponding (output) scale efficiency is:

$$S_k^{(2)} = E_k^{(2)} / T_k^{(2)}$$

If Process 2 is considered more important, then $E_k^{(2)}$ and the associated $T_k^{(2)}$ are calculated first. The overall efficiency of Process 1 is calculated as $E_k^{(1)} = E_k^S / E_k^{(2)}$, and the technical and scale efficiencies are calculated by a similar procedure.

Combining the above discussions, one obtains the following properties:

$$
\begin{aligned}
E_k^S &= E_k^{(1)} \times E_k^{(2)} = \left(T_k^{(1)} \times S_k^{(1)} \right) \times \left(T_k^{(2)} \times S_k^{(2)} \right) \\
E_k^S &= T_k \times S_k = \left(T_k^{(1)} \times T_k^{(2)} \right) \times \left(S_k^{(1)} \times S_k^{(2)} \right)
\end{aligned}
$$

where T_k and S_k are the technical and scale efficiencies of the system, respectively.

6.5 Non-life Insurance Companies in Taiwan

Kao and Hwang (2008) developed a relational model to calculate the system and process efficiencies for two-stage systems, and used an example of 24 non-life insurance companies in Taiwan to illustrate it. In order to have a common basis for comparison, the same data set is used in this paper to calculate the technical and scale efficiencies of the system and processes.

The operations of a non-life insurance company can be separated into two processes, premium acquisition and profit generation. In the first process, clients are attracted to pay direct written premiums, and reinsurance premiums are received from other insurance companies. In the second process, premiums are loaned and invested to earn profit. The inputs are classified into two categories:

Operating expenses (X_1): salaries of the employees and various costs incurred in daily operations.
Insurance expenses (X_2): expenses paid to agencies, brokers, and solicitors, and other expenses associated with marketing insurance.

The intermediate products considered are:

Direct written premiums (Z_1): premiums received from insured clients.
Reinsurance premiums (Z_2): premiums received from ceding companies.

The outputs include two types of profit:

Underwriting profit (Y_1): profit earned from the insurance business.
Investment profit (Y_2): profit earned from the investment portfolio.

Table 6.3 shows the original data.

By applying Model (6.3), the system efficiency, E_k^S, is calculated for each company, as shown in the second column of Table 6.4. Model (6.6) is then applied to calculate the overall and (input) technical efficiencies, $E_k^{(1)}$ and $T_k^{(1)}$, for the first process. The ratio of $E_k^{(1)}$ to $T_k^{(1)}$ is the (input) scale efficiency, $S_k^{(1)}$, and that of E_k^S to $E_k^{(1)}$ is the overall efficiency of the second process, $E_k^{(2)}$. The results are shown in the central part of Table 6.4 under the heading "Process 1". Finally, Model (6.7) is used to calculate the (output) technical efficiencies, $T_k^{(2)}$, for the second process. Similar to Process 1, the ratio of $E_k^{(2)}$ to $T_k^{(2)}$ is the (output) scale efficiency, $S_k^{(2)}$, of Process 2. The results are shown on the right side of Table 6.4 under the heading "Process 2". The products of the two process technical efficiencies and two process scale efficiencies are the system technical and system scale efficiencies, respectively, as shown on the left side of Table 6.4 under the heading "System".

As indicated by the scores shown in the second column of Table 6.4, none of the 24 companies is efficient for the whole system. This is simply because none of the companies is efficient for both Processes 1 and 2, although four companies have an efficient Process 1 and two have an efficient Process 2. In order to get a general idea of the performance of both processes, and to preserve the relationship of

Table 6.3 Data of 24 non-life insurance companies in Taiwan

Company	Operating expenses (X_1)	Insurance expenses (X_2)	Direct written premiums (Z_1)	Reinsurance premiums (Z_2)	Underwriting profit (Y_1)	Investment profit (Y_2)
1. Taiwan Fire	1,178,744	673,512	7,451,757	856,735	984,143	681,687
2. Chung Kuo	1,381,822	1,352,755	10,020,274	1,812,894	1,228,502	834,754
3. Tai Ping	1,177,494	592,790	4,776,548	560,244	293,613	658,428
4. China Mariners	601,320	594,259	3,174,851	371,863	248,709	177,331
5. Fubon	6,699,063	3,531,614	37,392,862	1,753,794	7,851,229	3,925,272
6. Zurich	2,627,707	668,363	9,747,908	952,326	1,713,598	415,058
7. Taian	1,942,833	1,443,100	10,685,457	643,412	2,239,593	439,039
8. Ming Tai	3,789,001	1,873,530	17,267,266	1,134,600	3,899,530	622,868
9. Central	1,567,746	950,432	11,473,162	546,337	1,043,778	264,098
10. The First	1,303,249	1,298,470	8,210,389	504,528	1,697,941	554,806
11. Kuo Hua	1,962,448	672,414	7,222,378	643,178	1,486,014	18,259
12. Union	2,592,790	650,952	9,434,406	1,118,489	1,574,191	909,295
13. Shingkong	2,609,941	1,368,802	13,921,464	811,343	3,609,236	223,047
14. South China	1,396,002	988,888	7,396,396	465,509	1,401,200	332,283
15. Cathay Century	2,184,944	651,063	10,422,297	749,893	3,355,197	555,482
16. Allianz President	1,211,716	415,071	5,606,013	402,881	854,054	197,947
17. Newa	1,453,797	1,085,019	7,695,461	342,489	3,144,484	371,984
18. AIU	757,515	547,997	3,631,484	995,620	692,731	163,927
19. North America	159,422	182,338	1,141,950	483,291	519,121	46,857
20. Federal	145,442	53,518	316,829	131,920	355,624	26,537
21. Royal & Sunalliance	84,171	26,224	225,888	40,542	51,950	6,491
22. Asia	15,993	10,502	52,063	14,574	82,141	4,181
23. AXA	54,693	28,408	245,910	49,864	0.1	18,980
24. Mitsui Sumitomo	163,297	235,094	476,419	644,816	142,370	16,976
Mean	1,544,215	828,963	7,832,893	667,964	1,602,873	477,733

multiplicity, geometric means for the system efficiency, E_k^S, Process 1 efficiency, $E_k^{(1)}$, and Process 2 efficiency, $E_k^{(2)}$, are calculated, as shown in the last row of Table 6.4. As expected, the system mean, 0.3705, is the product of the two process means, 0.7831 and 0.4731. The values also indicate that Process 1 is more efficient than Process 2. This is reasonable, because the task of premium acquisition is relatively straightforward, and neither significant mistakes nor breakthroughs can be made. Profit generation, in contrast, involves high risk, which may produce large differences among companies. Therefore, the efficiencies are high for the former

Table 6.4 Various efficiencies measured from the two-stage model for 24 non-life insurance companies in Taiwan

	System			Process 1			Process 2		
Co.	Overall	(Tech.	Scale)	Overall	(Tech.	Scale)	Overall	(Tech.	Scale)
1	0.6992	(0.7845	0.8914)	0.9926	(1.0000	0.9926)	0.7045	(0.7845	0.8980)
2	0.6246	(0.7240	0.8626)	0.9982	(1.0000	0.9982)	0.6256	(0.7240	0.8642)
3	0.6900	(0.6903	0.9996)	0.6900	(0.6903	0.9996)	1.0000	(1.0000	1.0000)
4	0.3042	(0.3284	0.9264)	0.7242	(0.7258	0.9979)	0.4200	(0.4524	0.9283)
5	0.7670	(1.0000	0.7670)	0.8307	(1.0000	0.8307)	0.9233	(1.0000	0.9233)
6	0.3897	(0.5357	0.7273)	0.9606	(0.9636	0.9969)	0.4057	(0.5559	0.7297)
7	0.2766	(0.4654	0.5943)	0.6706	(0.7520	0.8918)	0.4124	(0.6189	0.6664)
8	0.2752	(0.6816	0.4037)	0.6630	(0.8156	0.8130)	0.4150	(0.8358	0.4966)
9	0.2233	(0.2955	0.7557)	1.0000	(1.0000	1.0000)	0.2233	(0.2955	0.7557)
10	0.4658	(0.6403	0.7275)	0.8611	(0.8612	0.9999)	0.5409	(0.7434	0.7276)
11	0.1637	(0.3710	0.4414)	0.6476	(0.7406	0.8744)	0.2528	(0.5009	0.5047)
12	0.7596	(0.8658	0.8773)	1.0000	(1.0000	1.0000)	0.7596	(0.8658	0.8773)
13	0.2078	(0.7552	0.2752)	0.6720	(0.8652	0.7767)	0.3093	(0.8729	0.3543)
14	0.2886	(0.4236	0.6813)	0.6699	(0.7248	0.9243)	0.4309	(0.5845	0.7371)
15	0.6138	(0.9377	0.6546)	1.0000	(1.0000	1.0000)	0.6138	(0.9377	0.6546)
16	0.3202	(0.4153	0.7709)	0.8856	(0.9107	0.9724)	0.3615	(0.4560	0.7928)
17	0.3600	(0.7239	0.4974)	0.6276	(0.7239	0.8670)	0.5736	(1.0000	0.5736)
18	0.2588	(0.5502	0.4705)	0.7935	(1.0000	0.7935)	0.3262	(0.5502	0.5929)
19	0.4112	(0.7775	0.5288)	1.0000	(1.0000	1.0000)	0.4112	(0.7775	0.5288)
20	0.5465	(0.9990	0.5471)	0.9331	(0.9990	0.9340)	0.5857	(1.0000	0.5857)
21	0.2008	(0.2955	0.6795)	0.7321	(0.9131	0.8018)	0.2743	(0.3236	0.8476)
22	0.5895	(1.0000	0.5895)	0.5895	(1.0000	0.5895)	1.0000	(1.0000	1.0000)
23	0.4203	(0.6042	0.6957)	0.8426	(0.9877	0.8530)	0.4989	(0.6117	0.8155)
24	0.1348	(0.3769	0.3577)	0.4288	(1.0000	0.4288)	0.3144	(0.3769	0.8343)
Mean	0.3705	(0.5921	0.6257)	0.7831	(0.8953	0.8747)	0.4731	(0.6614	0.7153)

and low for the latter. An effective way to increase the efficiency of a company is thus to improve the performance of Process 2.

The overall efficiency of both processes can be decomposed into the product of technical and scale efficiencies. For Process 1, ten companies are technically efficient and four have perfect scale efficiency. Their averages, as shown in the last row of Table 6.4, are 0.8953 and 0.8747. The product of the technical and scale efficiencies is equal to the overall efficiency of the process (0.7831). For Process 2, five companies have perfect technical efficiency and two have perfect scale efficiency. The geometric means are 0.6614 and 0.7153, respectively, whose product is exactly the overall efficiency, 0.4731.

The products of the technical efficiencies and scale efficiencies of the processes are the technical efficiency and scale efficiency of the system, respectively. The last row of Table 6.4, under the heading "System", shows that the average technical and scale efficiencies of the system are 0.5921 and 0.6257, respectively, which are exactly the products of those of the two processes, 0.8953 $\times$ 0.6614 and

0.8747×0.7153. If every company is operating efficiently from the technical point of view, then the system efficiency can be improved from 0.3705 to 0.6257. This improvement is accomplished by reducing the amount of input of Process 1 by 10.47 % (=1−0.8953), and increasing the output of Process 2 by 33.86 % (=1 − 0.6614). Each company can thus identify sources of inefficiency and make appropriate amendments to improve its overall efficiency.

6.6 Conclusion

The measurement of scale efficiency is quite straightforward in conventional DEA when only the aggregate operation of the system is considered. However, when the operations of the individual processes of the system are also considered, the measurement becomes a little complicated. This is primarily because the relationship between the efficiencies calculated under CRS and VRS are not known. In this paper we investigate the simplest case, the two-stage system.

The problem in decomposing the technical efficiency of the system into those of the two processes is that the outputs of the first process are the inputs of the second, such that to improve the efficiency of the first by increasing its outputs will affect that of the second, and to improve the efficiency of the second by reducing its inputs will affect that of the first. To resolve this conflict, this paper fixes the amounts of the intermediate products, which are the outputs of the first process and the inputs of the second, and uses an input-oriented model to measure the technical efficiency of the first process and an output-oriented one to measure that of the second. Based on the relational model of Kao and Hwang (2008), where the overall system and process efficiencies are calculated first and the technical efficiencies of the processes are calculated second, the scale efficiency of each process is calculated as the ratio of their respective overall efficiency to technical efficiency. The product of the technical efficiency of the two processes is that of the system. Similarly, the product of the scale efficiency of the two processes is that of the system. Moreover, the overall efficiency of the system is the product of its technical and scale efficiencies.

Decomposing the system efficiency into the product of the two process efficiencies, and each process efficiency into the product of their respective technical and scale efficiencies, enables decision makers to identify the sources of inefficiency and to find effective alternatives for making improvements to the system.

The efficiency measures used in this paper are radial, but other measures are also discussed in the literature. How to calculate scale efficiency based on these other measures, so that the performance of the system can be improved more effectively, is one direction for future research. Finally, the network system discussed in this paper is the simplest one; and calculating the scale efficiency for more complicated systems will be a more challenging task.

References

Banker, R. D., Charnes, A., & Cooper, W. W. (1984). Some models for estimating technical and scale efficiencies in data envelopment analysis. *Management Science, 30*, 1078–1092.

Charnes, A., & Cooper, W. W. (1984). The non-Archimedean CCR ratio for efficiency analysis: A rejoinder to Boyd and Färe. *European Journal of Operational Research, 15*, 333–334.

Charnes, A., Cooper, W. W., & Rhodes, E. (1978). Measuring the efficiency of decision making units. *European Journal of Operational Research, 2*, 429–444.

Charnes, A., Cooper, W. W., & Rhodes, E. (1979). Short communication: Measuring the efficiency of decision making units. *European Journal of Operational Research, 3*, 339.

Chen, Y., Cook, W. D., Ning, L., & Zhu, J. (2009). Additive efficiency decomposition in two-stage DEA. *European Journal of Operational Research, 196*, 1170–1176.

Chen, Y., Cook, W. D., & Zhu, J. (2010). Deriving the DEA frontier for two-stage processes. *European Journal of Operational Research, 202*, 138–142.

Cook, W. D., Liang, L., & Zhu, J. (2010). Measuring performance of two-stage network structure by DEA: A review. *Omega, International Journal of Management Science, 38*, 423–430.

Färe, R., & Grosskopf, S. (2000). Network DEA. *Socio-Economic Planning Sciences, 34*, 35–49.

Kao, C. (2009). Efficiency decomposition in network data envelopment analysis: A relational model. *European Journal of Operational Research, 192*, 949–962.

Kao, C. (2013). Efficiency decomposition in network data envelopment analysis with slacks-based measures, Omega (2013), http://dx.doi.org/10.1016/j.omega.2013.12.002

Kao, C., & Hwang, S. N. (2008). Efficiency decomposition in two-stage data envelopment analysis: An application to non-life insurance companies in Taiwan. *European Journal of Operational Research, 185*, 418–429.

Kao, C., & Hwang, S. N. (2011). Decomposition of technical and scale efficiencies in two-stage production systems. *European Journal of Operational Research, 211*, 515–519.

Seiford, L. M., & Zhu, J. (1999). Profitability and marketability of the top 55 US commercial banks. *Management Science, 45*, 1270–1288.

Tone, K., & Tsutsui, M. (2009). Network DEA: A slacks-based measure approach. *European Journal of Operational Research, 197*, 243–252.

Chapter 7
Decomposing Efficiency and Returns to Scale in Two-Stage Network Systems

Biresh K. Sahoo, Joe Zhu, and Kaoru Tone

Abstract Most of real-life production technologies are multi-stage in nature. Characterization of such technologies via concept like *network returns to scale* is considered important to firm managers for the stage-specific analysis of their business decisions concerning expansion or contraction so as to improve their firms' overall performance. Similarly, depicting such multi-stage technologies via *network efficiency* is important in identifying the sources of network inefficiency. It is, therefore, imperative to estimate both efficiency and returns to scale of a firm not only for the network technology but also for the sub-technologies so as to locate the sources of efficiency and scale economies. The primary purpose of constructing a network technology is to address allocative efficiency that is associated with the choice of how much of intermediate products to produce and consume, in addition to the economic use of primary inputs and the maximal production of final outputs. Therefore, it is necessary that not only the intermediate products are explicitly modeled, but also their optimal values are considered in the construction of sub-technologies' frontiers so that the issue of allocative efficiency, if exists, can be addressed. Based on the premise concerning whether a network technology considers allocative inefficiency, two approaches are suggested for the estimation of network technology.

Part of this chapter is based upon Sahoo, B.K., Zhu, J., Tone, K. & Klemen. B.M. (2014), "Decomposing scale elasticity in two-stage network DEA", *European Journal of Operational Research, 233*(3), 584–594, with permission from Elsevier.

B.K. Sahoo (✉)
Xavier Institute of Management, Bhubaneswar 751 013, India
e-mail: biresh@ximb.ac.in

J. Zhu
School of Business, Worcester Polytechnic Institute, Worcester, MA 01609, USA
e-mail: jzhu@wpi.edu

K. Tone
National Graduate Institute for Policy Studies, 7-22-1 Roppongi, Minato-ku, Tokyo 106-8677, Japan
e-mail: tone@grips.ac.jp

W.D. Cook and J. Zhu (eds.), *Data Envelopment Analysis*, International Series in Operations Research & Management Science 208, DOI 10.1007/978-1-4899-8068-7_7,

The first approach makes use of a single network technology for two *interdependent* sub-technologies. The second approach, however, assumes complete allocative efficiency by considering two *independent* sub-technology frontiers, one for each sub-technology. These two approaches are, however, necessary, in modeling the output loss of a network firm suffering from allocative inefficiency, which arises due to any possible sub-optimal decision as to how much of intermediate products to produce and consume in the world of changing prices.

Keywords Data envelopment analysis • Network DEA • Returns to scale decomposition • Efficiency decomposition • Modeling output loss due to allocative inefficiency

7.1 Introduction

Most of real-life production technologies are multi-stage in nature. Characterization of such technologies via concept like *returns to scale* (RTS) or *scale elasticity* (SE) is considered important to firm managers for the stage-specific analysis of their business decisions concerning expansion or contraction. Therefore, it is imperative to estimate the SE of a firm not only for the network technology but also for its sub-technologies so as to locate the sources of scale economies. This chapter presents the idea of Sahoo et al. (2014), and develop new approaches in non-parametric data envelopment analysis (DEA) for the decomposition of efficiency and RTS of a network firm into its stage-specific efficiencies and RTS, which are of practical use to firm managers in improving the overall performance of their firms.

Data envelopment analysis (DEA), a linear programming (LP) based technique, has been widely accepted as a competent methodology to estimate the structure of production technology in both *primal* (production) and *dual* (cost) environments. See Scarf (1990) for a discussion on the analogy between economic institutions and algorithms for solving the LP problems where the *simplex method* is interpreted as a search for market prices that equilibrate demand for factors of production with their supply. Much of DEA literature that considers the evaluation of SE treats production technology as a black-box (see, e.g., Sahoo et al. 1999; Fukuyama 2003; Tone and Sahoo 2003; Banker et al. 2004; Sahoo and Tone 2013; Zelenyuk 2013; among others), thus completely ignoring the literature on production control problems dealing with multi-stage production technologies (see, e.g., Aburzzi 1965; Bakshi and Arora 1969; among others).

The DEA literature that considers modeling of multi-stage technology by linking its sub-technologies is fairly recent. To the best of our knowledge, the network structure that links sub-technologies with intermediate products in the DEA framework was first introduced by Färe (1991); was, subsequently, extended in Färe and Grosskopf (2000), and Tone and Tsutsui (2009, 2010); and was, finally, applied in Tone and Sahoo (2003), Prieto and Zofio (2007), Yu and Lin (2008), and Lewis et al. (2013), among others.

A special variant of Färe and Grosskopf's multi-stage technology, i.e., a two-stage technology was developed in a different way by several scholars under multiplier DEA models (see, e.g., Chen and Zhu 2004; Chen et al. 2006, 2009a, b, 2010, 2013; Liang et al. 2006, 2008; Kao and Hwang 2008, 2011; Kao 2009, 2013; Cook et al. 2010; among others). In this set up, sub-technology I consumes *input* resources to produce *intermediate* products, which are all, in turn, used as *inputs* to sub-technology II to produce *final* outputs. A further restricted variant of this two-stage structure is developed by Seiford and Zhu (1999) and Zhu (2000) where sub-technologies are treated *independent*, and network as well as its sub-technologies' efficiencies are estimated independently.

The two-stage DEA literature (Kao and Hwang 2008, 2011, 2014; Liang et al. 2008; Kao 2009, 2013; Chen et al. 2009a, b, 2010, 2013) that addresses the evaluation of the decomposition of network efficiency into the sub-technology specific efficiencies is fairly recent. This decomposition is done under the assumption of *constant returns to scale* (CRS). What seems to be more intriguing but has completely been overlooked is whether this decomposition can be made under the assumption of *variable returns to scale* (VRS). And, if the answer to this question is yes, but at a cost, then it is worth investigating what this cost amounts to, i.e., allocative inefficiency due to any sub-optimal decision by the sub-technology managers as to how much of intermediate products to produce in the world of changing prices. The first objective of this chapter is to address the aforementioned issue.

Another important issue related to the first one, which has also not been addressed in the two-stage DEA literature, is the decomposition of network SE into the sub-technology specific SEs. This issue is related because the SE estimation can be done only under VRS. This decomposition will help a firm manager to not only determine the scale economies of network technology but also locate their sources, which lie in the sub-technologies. To our best knowledge, Kao and Hwang (2011) are the first to propose a scheme to determine only the scale efficiency of independent sub-technologies under the two-stage setting. Therefore, the second objective of this chapter is to propose a scheme to analytically show the SE of network technology as the product of those of its two sub-technologies.

For network SE estimation, two approaches may be considered based on the premise concerning whether the VRS-based network technology construct considers allocative inefficiency. In economics, the primary purpose of constructing a technology is to address allocative efficiency associated with the economic choice of how much of intermediate products to produce and consume, in addition to the economic use of primary inputs and the maximal production of final outputs. Therefore, it is necessary that not only the intermediate products are explicitly modeled, but also their optimal values are considered in the construction of sub-technologies' frontiers so that the issue of allocative efficiency, if exists, can be addressed.

Under the first approach (Approach I), which is ours, one network frontier is constructed for the two *interdependent* sub-technology frontiers, which are linked through optimal values of intermediate products. The dual pricing interpretation of the constraint that the intermediate products are freely determined in our envelopment-based network technology is that the weights for intermediate products as inputs and outputs in our multiplier-based network technology are the same. We maintain that our multiplier-based network technology is *additive*.

The construct of our proposed additive network technology holds under two conditions: (1) weights for intermediate products as inputs and outputs are the same, and (2) intercept multiplier of network technology is the sum of those of the two sub-technologies. The first condition holds due to our constraint that the intermediate products are freely determined in our envelopment-based network technology. The second condition holds under the assumption that the additive network technology can inherit the properties of its sub-technologies, i.e., if the sub-technologies satisfy the properties such as no free lunch, free disposability in inputs and outputs, compactness, convexity, and returns to scale, then so does the additive network technology. The proof of this is made in the spirit of the proof of Proposition 2.3.2 in Färe and Grosskopf (1996, p. 23, pp. 44–45).

The network technical efficiency (TE) decomposition based on Approach I reveals that allocative inefficiency arises only under the VRS specification, but disappears under the CRS specification. It can, therefore, be argued that interpreting the 'same weights' assumption for the intermediate products as outputs and inputs as a perfect coordination between the two sub-technologies, as in Liang et al. (2008), is not sufficient to rule out allocative inefficiency in the VRS environment. Allocative inefficiency is a broader concept that includes inefficiencies arising from possible sub-optimal decisions as to how much of intermediate products to produce and consume in the world of changing prices. Our additive network technology can be used in identifying such inefficiency when optimal values of intermediate products are less than their observed values. Our network TE decomposition reveals that a network firm is fully efficient only when it is efficient in both of its sub-technologies.

The second approach (Approach II), which is due to Kao and Hwang (2011), requires the two sub-technologies to be *independent* for the construction of network frontier. To keep the sub-technologies independent, the input-orientation in the sub-technology I and the output-orientation in the sub-technology II are maintained to keep the level of intermediate products unaltered. This way of modeling network technology assumes the current uses of intermediate products as optimal, thereby effectively rules out allocative inefficiency arising from their possible sub-optimal uses. However, allocative inefficiency of this kind, if exists, may question the very TE estimates estimated against the two assumed independent sub-technology frontiers.

Note that the choice of a particular approach adopted implies whether assuming allocative inefficiency in the underlying technology construct, and hence, yields a distinct set of TE estimates. The distinction between the two approaches is important from a policy point of view as the factors attributing to the network's inefficiency in each approach are distinct. For example, a lower network TE may be due to allocative inefficiency in Approach I as against the same due to lower sub-technologies' efficiencies in Approach II. In this case, policies to remove allocative inefficiency may be more effective in improving the network efficiency in Approach I than the policies directed at improving the sub-technology specific TEs. However, a comparison between the two approaches can be worth revealing in modeling the output loss of a network firm suffering from allocative inefficiency that may arise due to sub-optimal decision as to how much of intermediate products to produce and consume by the sub-technology managers in the world of changing prices. This is the third objective of this chapter.

The remainder of the chapter proceeds as follows. Section 7.2 deals with a discussion on the development of variants of two-stage network DEA models to estimate the TE and SE of firms in the network technology as well as sub-technologies. Section 7.3 provides an illustrative empirical application, showing how the TE and SE estimates of a firm yielded from the two approaches are different due to allocative inefficiency. Section 7.4 provides some concluding remarks.

7.2 Model Development

7.2.1 Two-Stage Network Technology

Consider a two-stage technology in which sub-technologies are connected in a network to form a network technology (T^N) (see Fig. 7.1). Further, assume that there are n firms, and each firm ($h = 1, 2, \ldots, n$) in the first sub-technology (T^I) uses *inputs* x_i ($i = 1, 2, \ldots, m$) to produce *intermediate* outputs z_d ($d = 1, 2, \ldots, p$) and the same firm in the second sub-technology (T^{II}) uses these *intermediate* outputs as *inputs* to produce *final* outputs y_r ($r = 1, 2, \ldots, s$). These z_d are called intermediate measures by Chen and Zhu (2004) and Liang et al. (2008).

7.2.2 TE Estimation

We now discuss the *TE* evaluation using Approach I.

7.2.2.1 TE Estimation Using Approach I

One can evaluate the TE of a network firm either in input-oriented manner or in output-oriented manner or in non-oriented manner. In this study we, however, concentrate on TE evaluation in input-oriented manner. We set up the following

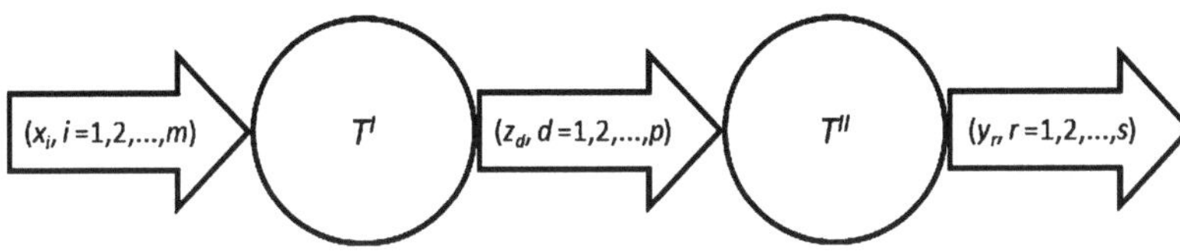

Fig. 7.1 Two-stage network technology

input-oriented VRS-based network DEA model for estimating the input *TE* of firm h ($TE_{ih}^{N(I)}$) in *envelopment* form as

$$
\begin{aligned}
TE_{ih}^{N(I)} &= \min_{\beta,\lambda,\tilde{z}} \beta_h \\
\text{s.t.} \quad & \sum_{j=1}^{n} x_{ij}\lambda_j \leq \beta_h x_{ih} (\forall i),\ \sum_{j=1}^{n} z_{dj}\lambda_j - \tilde{z}_{dh} \geq 0\ (\forall d),\ \sum_{j=1}^{n} \lambda_j = 1,\ \text{(sub-technology I)} \\
& \sum_{j=1}^{n} z_{dj}\mu_j - \tilde{z}_{dh} \leq 0\ (\forall d),\ \sum_{j=1}^{n} y_{rj}\mu_j \geq y_{rh} (\forall r),\ \sum_{j=1}^{n} \mu_j = 1,\ \text{(sub-technology II)} \\
& \beta_h \leq 1, \lambda_j, \mu_j \geq 0\ (\forall j), \tilde{z}_{dh} : \text{free}\ (\forall d)
\end{aligned}
\tag{7.1}
$$

Let $(\beta^*, \lambda^*, \tilde{z}^*)$ be optimal solution vector of model (7.1), which is based on the following VRS-based network technology set ($T_{VRS}^{N(I)}$) defined as

$$
T_{VRS}^{N(I)} = \left\{ (x, y, z) \left| \begin{array}{l} \sum_{j=1}^{n} x_{ij}\lambda_j \leq x_i (\forall i),\ \sum_{j=1}^{n} z_{dj}\lambda_j - z_d \geq 0\ (\forall d),\ \sum_{j=1}^{n} \lambda_j = 1, \lambda_j \geq 0\ (\forall j) \\ \sum_{j=1}^{n} z_{dj}\mu_j - z_d \leq 0\ (\forall d),\ \sum_{j=1}^{n} y_{rj}\mu_j \geq y_r (\forall r), \sum_{j=1}^{n} \mu_j = 1, \mu_j \geq 0\ (\forall j) \end{array} \right. \right\}
\tag{7.2}
$$

$T_{VRS}^{N(I)}$ uses λ and μ as intensity weights to form a linear combinations of n observed firms. Since both $T_{VRS}^{I(I)}$ and $T_{VRS}^{II(I)}$ satisfy VRS (i.e., $\sum_{j=1}^{n} \lambda_j = 1$ and $\sum_{j=1}^{n} \mu_j = 1$), $T_{VRS}^{N(I)}$ satisfies VRS. Similarly, $T_{VRS}^{I(I)}$ and $T_{VRS}^{II(I)}$ satisfy the assumption of *strong* (*free*) *disposability* of inputs and outputs by the use of inequality constraints, and so is the case with $T_{VRS}^{N(I)}$. The most distinguishing feature of $T_{VRS}^{N(I)}$ is that the intermediate products are explicitly modeled to be *freely* determined so as to make the sub-technologies *interdependent*. Chen and Zhu (2004), Liang et al. (2008), and Chen et al. (2010) also used this feature to reveal the frontier points of the two-stage technology.

β_h^* can be regarded as representing the *minimum* input proportion possible in $T_{VRS}^{N(I)}$ to produce y_h. Firm h is technically efficient, i.e., $TE_{ih}^{N(I)} = 1$ if and only if $(\beta_h^* x_h, \tilde{z}_h^*, y_h) \in \partial T_{VRS}^{N(I)}(\cdot)$ where $\partial T_{VRS}^{N(I)}(\cdot)$ represents the boundary of $T_{VRS}^{N(I)}(\cdot)$, and $(\beta_h^* x_h, z_h, y_h) \notin \partial T_{VRS}^{N(I)}(\cdot)$ when $z_h \neq \tilde{z}_h^*$.

One can also set up the *input-oriented* VRS-based network DEA model for estimating the input *TE* of firm h ($TE_{ih}^{N(I)}$) in *multiplier* form as

$$
TE_{ih}^{N(I)} = \max \sum_{r=1}^{s} u_r y_{rh} - \omega_I - \omega_{II}
\tag{7.3}
$$

$$\text{s.t.}\quad \sum_{i=1}^{m} v_i x_{ih} = 1, \tag{7.3.1}$$

$$\sum_{d=1}^{p} w_d z_{dj} - \sum_{i=1}^{m} v_i x_{ij} - \omega_I \le 0\ (\forall j), \tag{7.3.2}$$

$$\sum_{r=1}^{s} u_r y_{rj} - \sum_{d=1}^{p} w_d z_{dj} - \omega_{II} \le 0\ (\forall j), \tag{7.3.3}$$

$$v_i, u_r, w_d \ge 0\ (\forall i, r, d); \omega_I, \omega_{II} : \text{free} \tag{7.3.4}$$

where v_i, w_{dI} and ω_I are the dual decision variables to the respective constraints of sub-technology I, and w_{dII}, u_r and ω_{II} are the dual decision variables to the respective constraints of sub-technology II, in (7.1). Here $w_{dI} = w_{dII}(=w_d)$, which is due to the constraint that $\tilde{z}_d^*$ are free in (7.1). Otherwise, w_{dI} would have been no less than $w_{dII}(w_{dI} \ge w_{dII})$, had $\tilde{z}_d^*$ been non-negative. Note that Liang et al. (2008) model the 'same weights' assumption on z_d as a perfect coordination between the two sub-technologies under the CRS specification.

Constraints (7.3.2) and (7.3.3) correspond to the sub-technologies $T_{VRS}^{I(I)}$ and $T_{VRS}^{II(I)}$, respectively whose respective intercept multipliers are ω_I and ω_{II}. The construct of our network technology is such that the network technology constraint is the sum of the two sub-technology constraints, i.e., $T_{VRS}^{N(I)}$ is *additive*. This proposed additive structure holds under two conditions: (1) weights for the intermediate measures (products as inputs and outputs) are the same, and (2) intercept multiplier of $T_{VRS}^{N(I)}$ is the sum of those of its two sub-technologies. The first condition is satisfied due to the fact that $\tilde{z}_d^*$ are free in (7.1). The second condition holds under the assumption that the additive $T_{VRS}^{N(I)}$ can inherit the properties of its sub-technologies, i.e., if the sub-technologies satisfy the properties such as no free lunch, free disposability in inputs and outputs, compactness, convexity, and returns to scale, then so does the additive network technology. The proof of this is made in the spirit of the proof of Proposition 2.3.2 in Färe and Grosskopf (1996, p. 23, pp. 44–45).

Using optimal multipliers from (7.3), one can obtain the input-oriented TE of firm h in $T_{VRS}^{N(I)}$ ($TE_{ih}^{N(I)}$) and the sub-technologies (($TE_{ih}^{I(I)}$) and $TE_{ih}^{II(I)}$) as:

$$TE_{ih}^{N(I)} = \frac{\sum_{r=1}^{s} u_r^* y_{rh} - \omega_I^* - \omega_{II}^*}{\sum_{i=1}^{m} v_i^* x_{ih}}, TE_{ih}^{I(I)} = \frac{\sum_{d=1}^{p} w_d^* z_{dh} - \omega_I^*}{\sum_{i=1}^{m} v_i^* x_{ih}} \text{ and } TE_{ih}^{II(I)} = \frac{\sum_{r=1}^{s} u_r^* y_{rh} - \omega_{II}^*}{\sum_{d=1}^{p} w_d^* z_{dh}} \tag{7.4}$$

One can express $TE_{ih}^{N(I)}$ as the product of three terms: the first two terms representing the TEs in the sub-technologies – $TE_{ih}^{I(I)}$ and $TE_{ih}^{II(I)}$, respectively, and the third term representing an index (I_h^{VRS}) indicating whether the decision concerning the use of *observed* intermediate products (z) as intermediate measures (outputs and inputs) is optimal, i.e., whether z_h equals $\widetilde{z}_h$. The proposed TE decomposition is given below.

$$
\begin{aligned}
TE_{ih}^{N(I)} &= \left(\sum_{r=1}^{s} u_r^* y_{rh} - \omega_I^* - \omega_{II}^*\right) \Big/ \sum_{i=1}^{m} v_i^* x_{ih} \\
&= \frac{\sum_{d=1}^{p} w_d^* z_{dh} - \omega_I^*}{\sum_{i=1}^{m} v_i^* x_{ih}} \cdot \frac{\sum_{r=1}^{s} u_r^* y_{rh} - \omega_{II}^*}{\sum_{d=1}^{p} w_d^* z_{ih}} \cdot \frac{\left(\sum_{r=1}^{s} u_r^* y_{rh} - \omega_I^* - \omega_{II}^* \Big/ \sum_{r=1}^{s} u_r^* y_{rh} - \omega_{II}^*\right)}{\left(\sum_{r=1}^{s} w_d^* z_{dh} - \omega_I^* \Big/ \sum_{r=1}^{s} w_d^* z_{dh}\right)} \\
&= TE_{ih}^{I(I)} \cdot TE_{ih}^{II(I)} \cdot \frac{1 - \frac{1}{TE_{ih}^{II}}\left(\omega_I^* \Big/ \sum_{r=1}^{s} w_d^* z_{dh}\right)}{1 - \left(\omega_I^* \Big/ \sum_{r=1}^{s} w_d^* z_{dh}\right)} = TE_{ih}^{I(I)} \cdot TE_{ih}^{II(I)} \cdot I_h^{VRS}
\end{aligned}
\tag{7.5}
$$

Assuming unique optimal solutions in (7.3), we have three remarks based on the TE decomposition in (7.5).

Remark 1 I_h^{VRS} represents a proxy for the indication of allocative inefficiency, in which case $I_h^{VRS} > (<)\ 1$. Allocative inefficiency arises under the VRS specification but disappears under the CRS specification. One can therefore infer that maintaining the 'same weight' assumption on z as outputs and inputs under the VRS specification is not sufficient to rule out allocative inefficiency. Allocative inefficiency is a broader concept that includes inefficiencies arising from any possible sub-optimal decision as to how much z to produce and consume in the light of changing prices, i.e., $\widetilde{z}_h < z_h$, in which case $I_h^{VRS} \neq 1$. Our proposed additive $T_{VRS}^{N(I)}$ is helpful in identifying such inefficiency when the optimal intermediate products ($\widetilde{z}^*$) is less than its observed counterparts (z), i.e., $\widetilde{z}_h < z_h$ when $T_{VRS}^{II(I)}$ turns inefficient.

Remark 2 $I_h^{VRS} = 1$ when $TE_{ih}^{II(I)} = 1$, implying the decision concerning the use of *observed* intermediate products (z_h) as outputs and inputs as optimal, i.e., $z_h = \widetilde{z}_h$. This means that there is no allocative inefficiency in the use of observed z_h. In this case, $TE_{ih}^{N(I)} = TE_{ih}^{I(I)}$. Therefore, the TE decomposition under the additive network structure reveals that a network firm is fully efficient only when it is efficient in both of its sub-technologies.

Remark 3 When $TE_{ih}^{II(I)} < 1$, $I_h^{VRS} > (<)\ 1$. (A) $I_h^{VRS} > 1$ when (1) $\omega_I^* < 0$ and (2) $TE_{ih}^{I(I)} > |\omega_I^*|$ in which case firm *h* exhibits *increasing returns to scale* (IRS) in $T_{VRS}^{I(I)}$. (B) $I_h^{VRS} < 1$ when $\omega_I^* > 0$ in which case firm *h* exhibits *decreasing returns to scale* (DRS) in $T_{VRS}^{I(I)}$.

To prove the statement (A) in Remark 3, let us redefine $I_h^{VRS} = 1 - \frac{1}{TE_{ih}^{II}}\left(\omega_I^*/\sum_{r=1}^{s} w_d^* z_{dh}\right)/1 - \left(\omega_I^*/\sum_{r=1}^{s} w_d^* z_{dh}\right)$ as $\left(1 - \frac{1}{TE_{ih}^{II}} \cdot \frac{\omega_I^*}{\omega_I^* + TE_{ih}^{I(I)}}\right)/\left(1 - \frac{\omega_I^*}{\omega_I^* + TE_{ih}^{I(I)}}\right)(= I^1/I^2,\ \text{say})$. $I_h^{VRS} > 1$ implies that $I^1 - I^2 > 0$. This means that $\frac{\omega_I^*}{\omega_I^* + TE_{ih}^{I(I)}} \cdot \frac{TE_{ih}^{II(I)} - 1}{TE_{ih}^{II(I)}} > 0$. One can see that for this strict inequality to hold, two conditions need to hold: (1) $\omega_I^* < 0$ and (2) $TE_{ih}^{I(I)} > |\omega_I^*|$ since $TE_{ih}^{II(I)} < 1$; and firm *h* exhibits IRS since $\omega_I^* < 0$. Similarly, one can prove the statement (B) by examining the value of I_h^{VRS} when it is less than 1. $I_h^{VRS} < 1$ when $I^1 - I^2 < 0$, i.e., $\frac{\omega_I^*}{\omega_I^* + TE_{ih}^{I(I)}} \cdot \frac{TE_{ih}^{II(I)} - 1}{TE_{ih}^{II(I)}} < 0$. This inequality holds only when $\omega_I^* > 0$ irrespective of the values of $TE_{ih}^{I(I)}$ since $TE_{ih}^{II(I)} < 1$; and firm *h* exhibits DRS since $\omega_I^* > 0$. Note that the issue of determination of returns to scale will be dealt with in Sect. 7.2.3.

Note that optimal multipliers obtained from (7.3) may not be unique, implying that $TE_{ih}^{I(I)}$ and $TE_{ih}^{II(I)}$ are not unique. Therefore, in the spirit of Kao and Hwang (2008), assuming $T_{VRS}^{I(I)}$ to be more important, we first determine the maximum value of $TE_{ih}^{I(I)}$ via

$$TE_{ih}^{I} = \max \sum_{d=1}^{p} w_d z_{dh} - \omega_I \tag{7.6}$$

$$\text{s.t.} \sum_{r=1}^{s} u_r y_{rh} - \omega_I - \omega_{II} = TE_{ih}^{N(I)}, \sum_{i=1}^{m} v_i x_{ih} = 1, \sum_{d=1}^{p} w_d z_{dj} - \sum_{i=1}^{m} v_i x_{ij} - \omega_I \le 0\ (\forall j),$$
$$\sum_{r=1}^{s} u_r y_{rj} - \sum_{d=1}^{p} w_d z_{dj} - \omega_{II} \le 0\ (\forall j), v_i, u_r, w_d \ge 0\ (\forall i, r, d); \omega_I, \omega_{II} : \text{free}$$

One can then compute the minimum of $TE_{ih}^{II(I)}$ by using optimal multipliers obtained from model (7.6). However, if $T_{VRS}^{II(I)}$ is considered more important, we first determine the maximum value of $TE_{ih}^{II(I)}$, and then the minimum value of $TE_{ih}^{I(I)}$ in an analogous manner.

We now illustrate how to measure TE using Approach II.

7.2.2.2 TE Estimation Using Approach II

As shown in Chen et al. (2013), since the sub-technology specific TEs can be computed independently of the overall efficiencies, we set up the network technology set ($T_{VRS}^{N(II)}$):

$$T_{VRS}^{N(II)} = \left\{(x, z, y)|T_{VRS}^{I(II)} \cup T_{VRS}^{II(II)}\right\} \tag{7.7}$$

where

$$T_{VRS}^{I(II)} = \left\{(x, z)\middle|\sum_{j=1}^{n} x_{ij}\alpha_j \le x_i(\forall i),\ \sum_{j=1}^{n} z_{dj}\alpha_j \ge z_d(\forall d),\ \sum_{j=1}^{n} \alpha_j = 1,\ \alpha_j \ge 0\ (\forall j)\right\} \tag{7.7.1}$$

$$T_{VRS}^{II(II)} = \left\{(z, y)\middle|\sum_{j=1}^{n} z_{dj}\beta_j \le z_d(\forall d),\ \sum_{j=1}^{n} y_{rj}\beta_j \ge y_r(\forall r),\ \sum_{j=1}^{n} \beta_j = 1,\ \beta_j \ge 0\ (\forall j)\right\} \tag{7.7.2}$$

For the construction of $T_{VRS}^{N(II)}$, Kao and Hwang (2011) maintains input-orientation in $T_{VRS}^{I(II)}$ and output-orientation in $T_{VRS}^{II(II)}$. The input-oriented TE of firm h in $T_{VRS}^{I(II)}(TE_{ih}^{I(II)})$ can be computed by setting up the following linear problem:

$$TE_{ih}^{I(II)} = \min_{\delta, \alpha}\left\{\delta_h : (\delta_h x_h, z_h) \in T_{VRS}^{I(II)}\right\} \tag{7.8}$$

Similarly, the output-oriented TE of firm h in $T_{VRS}^{II(II)}(TE_{oh}^{II(II)})$ can be obtained from the following linear problem:

$$\left(TE_{oh}^{II(II)}\right)^{-1} = \max_{\mu, \beta}\left\{\mu_h : (z_h, \mu_h y_h) \in T_{VRS}^{II(II)}\right\} \tag{7.9}$$

Kao and Hwang (2011) have shown that the network TE of firm h ($TE_h^{N(II)}$) is the product of $TE_{ih}^{I(II)}$ and $TE_{oh}^{II(II)}$, i.e.,

$$TE_h^{N(II)} = TE_{ih}^{I(II)} \times TE_{oh}^{II(II)} \tag{7.10}$$

We now discuss the evaluation of SE.

7.2.3 SE Evaluation

7.2.3.1 Estimating SE Using Approach I

To compute the input-oriented SE of network firm h, we first need to compute its TE using the model (7.1). Let its optimal solution vector be $\left(\beta_h^*, \lambda^*, \mu^*, \tilde{z}^*\right)$. Firm h is (input-oriented) technically efficient if $\beta_h^* = 1$, $z_h = \tilde{z}_h^*$ and input and output slacks are all zero. If it is not, then it needs to be projected onto the network frontier by applying the following formulae:

$$x_h^* \leftarrow \beta_h^* x_h - s^-, \quad \tilde{z}_h^* \leftarrow \tilde{z}_h^* \text{ and } y_h^* \leftarrow y_h + s^+ \tag{7.11}$$

where s^- and s^+ are respectively vectors of input and output slacks under (7.1).

Due to duality theory, the following transformation function $F^{N(I)}\left(x_h^*, y_h^*; \tilde{z}_h^*\right) = 0$ holds:

$$F^{N(I)}\left(x_h^*, y_h^*; \tilde{z}_h^*\right) \equiv \sum_{r=1}^{s} u_r^*\left(y_{rh} + s_r^+\right) - \sum_{i=1}^{m} v_i^*\left(\beta_h^* x_{ih} - s_i^-\right) - \omega_I^* - \omega_{II}^* = 0 \tag{7.12}$$

where u_r^*, v_i^*, w_d^*, ω_I^* and ω_{II}^* are assumed to be the *unique* optimal multipliers obtained from (7.3); otherwise $F^{N(I)}(\cdot)$ is not differentiable at extreme points.

To define the SE in $T_{VRS}^{N(I)}$, $T_{VRS}^{I(I)}$ and $T_{VRS}^{II(I)}$, we consider, respectively, the following input–output vectors from (7.11): (x_h^*, y_h^*), $\left(x_h^*, \tilde{z}_h^*\right)$ and $\left(\tilde{z}_h^*, y_h^*\right)$. Following Baumol et al. (1982), we define the input-oriented (local) SE of firm h in $T_{VRS}^{N(I)}$, $\varepsilon_{ih}^{N(I)}\left(x_h^*, y_h^*; \tilde{z}_h^*\right)$ as:

$$\begin{aligned}\varepsilon_{ih}^{N(I)}\left(x_h^*, y_h^*; \tilde{z}_h^*\right) &\equiv -\sum_{i=1}^{m} x_{ih}\frac{\partial F^{N(I)}(\cdot)}{\partial x_{ih}} \Big/ \sum_{r=1}^{s} y_{rh}\frac{\partial F^{N(I)}(\cdot)}{\partial y_{rh}} \\ &= \frac{\beta_h^* \sum_{i=1}^{m} v_i^* x_{ih}}{\sum_{r=1}^{s} u_r^* y_{rh}} = \frac{\beta_h^*}{\beta_h^* + \omega_I^* + \omega_{II}^*} = \frac{\beta_h^*}{\beta_h^* + \omega_I^* + \omega_{II}^*}\end{aligned} \tag{7.13}$$

Note that in (7.13), $\sum_{i=1}^{m} v_i^* x_{ih} = 1$ due to (7.3.1); and $\sum_{r=1}^{s} u_r y_{rh} = \beta_h^* + \omega_I^* + \omega_{II}^*$, due to duality, the objective function values of (7.1) and (7.3) are the same, i.e., $\beta_h^* = \sum_{r=1}^{s} u_r y_{rh} - \omega_I^* - \omega_{II}^*$.

Based on (7.13), we have now the following proposition.

Proposition 1 *The input-oriented returns to scale are increasing* (*IRS*) (*i.e.*, $\varepsilon_{ih}^{N(I)}(\cdot) > 1$) *if* $\omega_I^* + \omega_{II}^* < 0$ *in all optimal solutions, constant* (*CRS*) (*i.e.*, $\varepsilon_{ih}^{N(I)}(\cdot) = 1$) *if* $\omega_I^* + \omega_{II}^* = 0$ *in an optimal solution, and decreasing* (*DRS*) (*i.e.*, $\varepsilon_{ih}^{N(I)}(\cdot) < 1$) *if* $\omega_I^* + \omega_{II}^* > 0$ *in all optimal solutions.*

Proof The proof is similar to that of determining the RTS underlying black-box DEA model. See Banker and Thrall (1992) and Banker et al. (2004). □

We now discuss the analytical SE evaluation of a fully network efficient firm h in its sub-technologies for which the constraints (7.3.2) and (7.3.3) are of special interest. Note that the network technology constraint is the sum of its two sub-technology constraints – (7.3.2) and (7.3.3), i.e.,

$$\sum_{r=1}^{s} u_r^* \left(y_{rh} + s_r^+\right) - \sum_{i=1}^{m} v_i^* \left(\beta_h^* x_{ih} - s_i^-\right) - \omega_I^* - \omega_{II}^* = \left(\sum_{d=1}^{p} w_d^* \tilde{z}_{dh}^* - \sum_{i=1}^{m} v_i^* \left(\beta_h^* x_{ih} - s_i^-\right) - \omega_I^*\right) + \left(\sum_{r=1}^{s} u_r^* \left(y_{rh} + s_r^+\right) - \sum_{d=1}^{p} w_d^* \tilde{z}_{dh}^* - \omega_{II}^*\right) \tag{7.14}$$

Since $\sum_{r=1}^{s} u_r^* y_{rh} - \sum_{i=1}^{m} v_i^* \left(\beta_o^* x_{ih} - s_i^-\right) - \omega_I^* - \omega_{II}^* = 0$ for the technically efficient firm h in $T_{VRS}^{N(I)}$, h will also be efficient in $T_{VRS}^{I(I)}$ and $T_{VRS}^{II(I)}$, in which case the respective transformation functions are:

$$F^{I(I)}\left(x_h^*, \tilde{z}_h^*\right) \equiv \sum_{d=1}^{p} w_d^* \tilde{z}_{dh} - \sum_{i=1}^{m} v_i^* \left(\beta_h^* x_{ih} - s_i^-\right) - \omega_I^* = 0 \tag{7.15}$$

$$F^{II(I)}\left(\tilde{z}_h^*, y_h^*\right) \equiv \sum_{r=1}^{s} u_r^* \left(y_{rh} + s_r^+\right) - \sum_{d=1}^{p} w_d^* \tilde{z}_{dh} - \omega_{II}^* = 0 \tag{7.16}$$

Using (7.13), one can obtain the respective sub-technology specific input-oriented SEs as:

$$\varepsilon_{ih}^{I(I)}\left(x_h^*, \tilde{z}_h^*\right) \equiv -\sum_{i=1}^{m} x_{ih} \frac{\partial F^{I(I)}(\cdot)}{\partial x_{ih}} \Big/ \sum_{d=1}^{p} \tilde{z}_{dh} \frac{\partial F^{I(I)}(\cdot)}{\partial \tilde{z}_{dh}} = \frac{\beta_h^* \sum_{i=1}^{m} v_i^* x_{ih}}{\sum_{d=1}^{p} w_d^* \tilde{z}_{dh}} = \frac{\beta_h^*}{\beta_h^* + \omega_I^*} \tag{7.17}$$

$$\varepsilon_{ih}^{II(I)}\left(\tilde{z}_h^*, y_h^*\right) \equiv -\sum_{d=1}^{p} \tilde{z}_{ih} \frac{\partial F^{II(I)}(\cdot)}{\partial \tilde{z}_{dh}} \Big/ \sum_{r=1}^{s} y_{rh} \frac{\partial F^{II(I)}(\cdot)}{\partial y_{ro}} = \frac{\sum_{d=1}^{p} w_d^* \tilde{z}_{dh}}{\sum_{r=1}^{s} u_r^* y_{rh}} = \frac{\beta_h^* + \omega_I^*}{\beta_h^* + \omega_I^* + \omega_{II}^*} \tag{7.18}$$

Note that in (7.17), $\sum_{d=1}^{p} w_d^* \widetilde{z}_{dh} = \beta_h^* + \omega_I^*$. This is because $\sum_{i=1}^{m} v_i^* s_i^- = 0$ in (7.15) due to complementary slackness condition. In (7.18), $\sum_{r=1}^{s} u_r^* y_{rh} = \beta_h^* + \omega_I^* + \omega_{II}^*$. This is due to $\sum_{r=1}^{s} u_r^* s_r^+ = 0$ in (7.16) due to complementary slackness condition, and $\sum_{d=1}^{p} w_d^* \widetilde{z}_{dh} = \beta_h^* + \omega_I^*$ in (7.15).

One can now show that the SE of network firm h in $T_{VRS}^{N(I)}$ is the product of those of the two sub-technologies – $T_{VRS}^{I(I)}$ and $T_{VRS}^{II(I)}$, i.e.,

$$\begin{aligned} \varepsilon_{ih}^{I(I)}\left(x_o^*, \widetilde{z}_h^*\right) * \varepsilon_{ih}^{II(I)}\left(\widetilde{z}_h^*, y_h^*\right) &= \frac{\beta_h^*}{\beta_h^* + \omega_I^*} \cdot \frac{\beta_h^* + \omega_I^*}{\beta_h^* + \omega_I^* + \omega_{II}^*} \\ &= \frac{\beta_h^*}{\beta_h^* + \omega_I^* + \omega_{II}^*} = \varepsilon_{ih}^{N(I)}\left(x_h^*, y_h^*; \widetilde{z}_h^*\right) \end{aligned} \tag{7.19}$$

Note that DEA technologies are not differentiable at extreme efficient points due to multiple optimal solutions for $(\omega_I + \omega_{II})$. We, therefore, set up the following linear problem to find out the input-oriented right-hand SE ($\varepsilon_{ih-}^{N(I)}(\,)$) for firm h in $T_{VRS}^{N(I)}$ as.

$$\left[\frac{\beta_h^*}{\varepsilon_{ih-}^{N(I)}(\cdot)} - \beta_h^*\right] = \max \omega_I + \omega_{II} \tag{7.20}$$

$$\begin{aligned} \text{s.t.} \quad & \sum_{i=1}^{m} v_i\left(\beta_h^* x_{ih} - s_i^-\right) = 1, \sum_{d=1}^{p} w_d \widetilde{z}_{dh}^* - \sum_{i=1}^{m} v_i\left(\beta_h^* x_{ih} - s_i^-\right) - \omega_I = 0, \\ & \sum_{r=1}^{s} u_r\left(y_{rh} + s_r^+\right) - \sum_{d=1}^{p} w_d \widetilde{z}_{dh}^* - \omega_{II} = 0, \\ & \sum_{d=1}^{p} w_d z_{dj} - \sum_{i=1}^{m} v_i x_{ij} - \omega_I \le 0\ (\forall j \ne h), \sum_{r=1}^{s} u_r y_{rj} - \sum_{d=1}^{p} w_d z_{dj} - \omega_{II} \le 0\ (\forall j \ne h), \\ & v_i, u_r, w_d \ge \varepsilon\ (\forall i, r, d), \omega_I, \omega_{II} : \text{free} \end{aligned}$$

Similarly, the input-oriented left-hand SE ($\varepsilon_{ih+}^{N(I)}(\cdot)$) can be obtained by replacing the "max" with "min" in objective of model (7.20).

Let the max of $(\omega_I + \omega_{II})$ in (7.20) be $(\omega_I + \omega_{II})^+$ in which the values of ω_I and ω_{II} are $\overline{\omega}_I$ and $\overline{\omega}_{II}$; and let the min of $(\omega_I + \omega_{II})$ in (7.20) be $(\omega_I + \omega_{II})^-$ in which the values of ω_I and ω_{II} are $\underline{\omega}_I$ and $\underline{\omega}_{II}$. Banker and Thrall (1992) used the upper and lower bounds of the intercept multiplier of the (black-box) BCC DEA model to define the left-hand (lower bound) and right-hand (upper-bound) SEs. Following Banker and Thrall (1992), we now use the SE expression (7.17) to determine the left-and right-hand SEs of firm h in $T_{VRS}^{I(I)}$ as

$$\varepsilon_{ih+}^{I(I)}(\cdot) = \frac{\beta_h^*}{\beta_h^* + \overline{\omega}_I} \quad \text{and} \quad \varepsilon_{ih-}^{I(I)}(\cdot) = \frac{\beta_h^*}{\beta_h^* + \underline{\omega}_I} \tag{7.21}$$

Similarly, one can use the SE expression (7.18) to determine the left- and right-hand SEs of firm h in $T_{VRS}^{II(I)}$ as

$$\varepsilon_{ih+}^{II(I)}(\cdot) = \frac{\beta_h^* + \overline{\omega}_I}{\beta_h^* + (\omega_I + \omega_{II})^+} \text{ and } \varepsilon_{ih-}^{II(I)}(\cdot) = \frac{\beta_h^* + \underline{\omega}_I}{\beta_h^* + (\omega_I + \omega_{II})^-} \tag{7.22}$$

While defining these sub-technology specific SEs, we have followed Banker and Thrall (1992) to consider the upper and lower bounds of ($\omega_I + \omega_{II}$) in the program (7.20), i.e., $(\overline{\omega}_I, \overline{\omega}_{II})$ and $(\underline{\omega}_I, \underline{\omega}_{II})$ to determine the left- and right-hand SEs. However, if one considers the individual max (min) values of ω_I and ω_{II} (i.e., $\omega_I^+(\omega_I^-)$ and $\omega_{II}^+(\omega_{II}^-)$), which can be obtained by replacing max (min) ($\omega_I + \omega_{II}$) in the objective of (7.20) with max (min)ω_I and max (min)ω_{II} respectively, then our SE expressions in (7.21) and (7.22) may produce the incorrect values of left- and right-hand SEs. This is possible only when $\overline{\omega}_I(\underline{\omega}_I) \neq \omega_I^+(\omega_I^-)$ and $\overline{\omega}_{II}(\underline{\omega}_{II}) \neq \omega_{II}^+(\omega_{II}^-)$.

We have now our proposition 2.

Proposition 2

(2.1) *Assuming alternate optima in* ($\omega_I + \omega_{II}$), $T_{VRS}^{N(I)}$ *exhibits IRS* ($\varepsilon_{ih-}^{N(I)}(\cdot) > 1$) *if* $(\omega_I + \omega_{II})^- < 0$, *CRS* ($\varepsilon_{ih-}^{N(I)}(\cdot) \leq 1 \leq \varepsilon_{ih+}^{N(I)}(\cdot)$) *if* $(\omega_I + \omega_{II})^- \geq 0 \geq (\omega_I + \omega_{II})^+$ *and DRS* ($\varepsilon_{ih+}^{N(I)}(\cdot) < 1$) *if* $(\omega_I + \omega_{II})^+ > 0$.

(2.2) *Assuming alternate optima in* ω_I, $T_{VRS}^{I(I)}$ *exhibits IRS* ($\varepsilon_{ih-}^{I(I)}(\cdot) > 1$) *if* $\underline{\omega}_I < 0$, *CRS* ($\varepsilon_{ih-}^{I(I)}(\cdot) \leq 1 \leq \varepsilon_{ih+}^{I(I)}(\cdot)$) *if* $\underline{\omega}_I \geq 0 \geq \overline{\omega}_I$, *and DRS* ($\varepsilon_{ih+}^{I}(\cdot) < 1$) *if* $\overline{\omega}_I > 0$.

(2.3) *Assuming alternate optima in* ($\omega_I + \omega_{II}$), $T_{VRS}^{II(I)}$ *exhibits IRS* ($\varepsilon_{ih-}^{II(I)}(\cdot) > 1$) *if* $\underline{\omega}_{II} < 0$, *CRS* ($\varepsilon_{ih-}^{II(I)}(\cdot) \leq 1 \leq \varepsilon_{ih+}^{II(I)}(\cdot)$) *if* $\underline{\omega}_{II} \geq 0 \geq \overline{\omega}_{II}$, *and DRS* ($\varepsilon_{ih+}^{II(I)}(\cdot) < 1$) *if* $\overline{\omega}_{II} > 0$.

Proof The proof is similar to that of determining the RTS underlying black-box DEA model. See Banker and Thrall (1992) and Banker et al. (2004).

Banker et al. (1984) are the first to show that the intercept ω in the multiplier form of the (black-box) BCC DEA model can be used to estimate RTS. Several contributions exist, at the extreme points, on the evaluation of right-hand (upper bound) and left-hand SE (lower bound) measures in the black-box models. See, e.g., among others, Banker and Thrall (1992), Førsund (1996), Tone and Sahoo (2004), Tone and Sahoo (2005), Tone and Sahoo (2006), Hadjicostas and Soteriou (2006), Podinovski et al. (2009), and Sahoo et al. (2012).

7.2.3.2 Estimating SE Using Approach II

To compute the input-oriented SE of firm h in $T_{VRS}^{I(II)}$, we first set up the dual of model (7.8) as

$$TE_{ih}^{I(II)} = \max \sum_{d=1}^{p} w_d' z_{dh} - \omega_I' \tag{7.23}$$

$$\text{s.t.} \quad \sum_{i=1}^{m} v_i x_{ih} = 1, \tag{7.23.1}$$

$$\sum_{d=1}^{p} w_d' z_{dj} - \sum_{i=1}^{m} v_i x_{ij} - \omega_I' \le 0 \ (\forall j), \tag{7.23.2}$$

$$v_i, w_d' \ge 0 \ (\forall i, d); \omega_I' : \text{free} \tag{7.23.3}$$

Assume that unique optimal solutions in (7.23) exist. The duality theory suggests that the following transformation function for firm h, $F^{I(II)}(z_h, x_h) = 0$ holds, i.e.,

$$F^{I(II)}(z_h, x_h) \equiv \sum_{d=1}^{p} w_d'^{*}\left(z_{dh} + s_d^{+}\right) - \sum_{i=1}^{m} v_i^{*}\left(\delta_h^{*} x_{ih} - s_i^{-}\right) - \omega_I'^{*} = 0 \tag{7.24}$$

where s_i^- and s_d^+ are respectively the ith input and d^{th} output slacks in model (7.8). Using the SE formula (7.13), one can obtain the input-oriented SE of firm h in $T_{VRS}^{I(II)}$ as

$$\varepsilon_{ih}^{I(II)}(z_h, x_h) \equiv -\sum_{i=1}^{m} x_{ih} \frac{\partial F^{I(II)}(\cdot)}{\partial x_{ih}} \Big/ \sum_{d=1}^{p} z_{dh} \frac{\partial F^{I(II)}(\cdot)}{\partial z_{dh}} = \frac{\delta_h^{*} \sum_{i=1}^{m} v_i^{*} x_{ih}}{\sum_{d=1}^{p} w_d'^{*} z_{dh}} = \frac{\delta_h^{*}}{\delta_h^{*} + \omega_I'^{*}} \tag{7.25}$$

Notice that in (7.25), $\sum_{i=1}^{m} v_i^{*} x_{ih} = 1$ due to (7.23.1); and $\sum_{d=1}^{p} w_d'^{*} z_{dh} = \delta_h^{*} + \omega_I'^{*}$, which is because, by duality, the objective function values of (7.8) and (7.23) are the same, i.e., $\delta_h^{*} = \sum_{d=1}^{p} w_d'^{*} z_{dh} - \omega_I'^{*}$.

One can compute the output-oriented SE of firm h in $T_{VRS}^{II(II)}$ by setting up the dual of (7.9) as

$$\left(TE_{oh}^{II(II)}\right)^{-1} = \min \sum_{d=1}^{p} w_d' z_{dh} + \omega_{II}' \tag{7.26}$$

$$\text{s.t.} \quad \sum_{r=1}^{s} u_r y_{rh} = 1, \tag{7.26.1}$$

$$\sum_{d=1}^{p} w'_d z_{dj} - \sum_{r=1}^{s} u_r y_{rj} + \omega'_{II} \geq 0 \ (\forall j), \tag{7.26.2}$$

$$u_r, w'_d \geq 0 \ (\forall r, d); \omega'_{II} : \text{free} \tag{7.26.3}$$

Assume that unique optimal solutions in the model (7.26) exist. Due to duality theory, the following transformation function for firm h, $F^{II(II)}(z_h, y_h) = 0$ holds, i.e.,

$$F^{II(II)}(z_h, y_h) \equiv \sum_{i=1}^{m} u_r^* \left(\mu_h^* y_{rh} + s_r^+\right) - \sum_{d=1}^{p} w'^*_d \left(z_{dh} - s_d^-\right) - \omega'^*_{II} = 0 \tag{7.27}$$

where s_d^- and s_r^+ are respectively the dth input slack and r^{th} output slacks of the model (7.9). Using the SE formula (7.13), one can obtain the output-oriented SE of firm h in $T_{VRS}^{II(II)}$ as

$$\begin{aligned} \varepsilon_{oh}^{II(II)}(z_h, y_h) &\equiv -\sum_{d=1}^{p} z_{dh} \frac{\partial F^{II(II)}(\cdot)}{\partial z_{dh}} \Big/ \sum_{r=1}^{s} y_{rh} \frac{\partial F^{II(II)}(\cdot)}{\partial y_{rh}} \\ &= \frac{\sum_{d=1}^{p} w'^*_d z_{dh}}{\mu_h^* \sum_{r=1}^{s} u_r^* y_{rh}} = \frac{\mu_h^* - \omega'^*_{II}}{\mu_h^*} = 1 - \frac{\omega'^*_{II}}{\mu_h^*} \end{aligned} \tag{7.28}$$

Notice that in (7.28), $\sum_{r=1}^{s} u_r^* y_{rh} = 1$ due to (7.26.1); and $\sum_{d=1}^{p} w'^*_d z_{dh} = \mu_h^* - \omega'^*_{II}$, which is because, by duality, the objective function values of (7.9) and (7.26) are the same, i.e., $\mu_h^* = \sum_{d=1}^{p} w'^*_d z_{dh} + \omega'^*_{II}$.

Since in many cases ω'^*_I and ω'^*_{II} are not uniquely determined in (7.23) and (7.26) respectively, the SE estimates are not unique. There is thus a need to find out both right- and left-hand SEs.

We set up the following model to compute the input-oriented right-hand SE of firm h, $\varepsilon_{ih-}^{I(II)}(\cdot)$ in $T_{VRS}^{I(II)}$ as

$$\frac{\delta_h^*}{\varepsilon_{ih-}^{I(II)}(\cdot)} - \delta_h^* = \max \ \omega'_I \tag{7.29}$$

$$\text{s.t.} \quad \sum_{i=1}^{m} v_i \left(\delta_h^* x_{ih} - s_i^-\right) = 1, \sum_{d=1}^{p} w_d^{'} \left(z_{dh} + s_d^+\right) - \sum_{i=1}^{m} v_i \left(\delta_h^* x_{ih} - s_i^-\right) - \omega_I^{'} = 0,$$
$$\sum_{d=1}^{p} w_d^{'} z_{dj} - \sum_{i=1}^{m} v_i x_{ij} - \omega_I^{'} \leq 0 \quad (\forall j \neq h),$$
$$v_i, w_d^{'} \geq 0 \ \forall i, d; \text{and} \ \omega_I^{'} : \text{free}.$$

Denote optimal solution of $\omega_I^{'}$ be ω'^{-}_I. $\varepsilon_{ih-}^{I(II)}(\cdot)$ can be computed as:

$$\varepsilon_{ih-}^{I(II)}(\cdot) = \frac{\delta_h^*}{\delta_h^* + \omega'^{-}_I} \tag{7.30}$$

Similarly, the input-oriented left-hand SE of firm h, $\varepsilon_{ih+}^{I(II)}(\cdot)$ in $T_{VRS}^{I(II)}$ can be computed by running the model (7.29) with 'min' instead of 'max'.

The output-oriented right-hand *SE* of firm h in $T_{VRS}^{II(II)}$ can be computed by setting up the following linear problem:

$$\mu_h^* \left(1 - \varepsilon_{oh-}^{II(II)}(\cdot)\right) = \max \omega_{II}^{'} \tag{7.31}$$
$$\text{s.t.} \quad \sum_{r=1}^{s} u_r \left(\mu_h^* y_{rh} + s_r^+\right) = 1, \ \sum_{d=1}^{p} w_d^{'} \left(z_{dh} - s_d^-\right) - \sum_{r=1}^{s} u_r \left(\mu_h^* y_{rh} + s_r^+\right) + \omega_{II}^{'} = 0,$$
$$\sum_{d=1}^{p} w_d^{'} z_{dj} - \sum_{r=1}^{s} u_r y_{rj} + \omega_{II}^{'} \geq 0 \quad (\forall j \neq h), \ \ u_r, w_d^{'} \geq 0 \ \forall r, d; \ \text{and} \ \omega_{II}^{'} : \text{free}.$$

Denote optimal solution of $\omega_I^{'}$ in (7.31) as ω'^{-}_{II}. The output-oriented right-hand SE of firm h in $T_{VRS}^{II(II)}$ can be computed as

$$\varepsilon_{oh-}^{II(II)}(\cdot) = 1 - \frac{\omega'^{-}_{II}}{\mu_h^*} \tag{7.32}$$

Similarly, the output-oriented left-hand SE of firm $h\varepsilon_{oh+}^{II(II)}(\cdot)$ in $T_{VRS}^{II(II)}$ can be computed by running (7.31) with 'min' instead of 'max'.

Note that unlike in Approach I, it is not possible in Approach II to decompose the network technology SE into its sub-technology specific SEs. We, however, note that Kao and Hwang (2011) develop an *ad hoc* approach to obtain network scale efficiency as the product of the sub-technology specific scale efficiencies.

7.2.4 Modeling Efficiency Against Different Efficiency Frontiers

Real-world firms suffer from profit loss due to allocative inefficiency arising from sub-optimal decision concerning the production and consumption of intermediate products connecting the sub-technologies. This profit loss has implications for revenue growth and cost control exercises. Production managers have every incentive to choose right input- and output-mix; otherwise, the opportunity cost of doing so is surprisingly high. Therefore, it is imperative to know the extent of output loss of firms suffering from such allocative inefficiencies.

In order to compute the loss in final output, one needs to model the TE of a firm against the three frontiers: the two network frontiers revealed from Approach I and Approach II, and the black-box (BB) frontier that ignore intermediate products connecting the sub-technologies. For this purpose, we specifically consider describing the network frontier under both approaches comprising of only inputs (x) and final outputs (y).

Using the model (7.1) under Approach I, we first project all the firms onto the network efficiency frontier. Let their projected input and final output vectors be $\left(\overline{x}_j, \overline{y}_j\right)$ where $\overline{x}_j = \beta_h^* x_j - s^-$ and $\overline{y}_j = y_j + s^+$ for all $j = 1, 2, \ldots, n$. We define the frontier of network technology set ($T_{VRS}^{N(I)}$), $\partial T_{VRS}^{N(I)}$, comprising of all of these projected input and final output vector as

$$\partial T_{VRS}^{N(I)} = \left\{ \left(\widehat{x}, \widehat{y}\right) \middle| \sum_{j=1}^{n} \overline{x}_j \lambda_j \le \widehat{x}, \sum_{j=1}^{n} \overline{y}_j \lambda_j \ge \widehat{y}, \sum_{j=1}^{n} \lambda_j = 1, \lambda_j \ge 0 \right\} \tag{7.33}$$

Similarly, the frontier of network technology set under the Approach II ($T_{VRS}^{N(II)}$), $\partial T_{VRS}^{N(II)}$, comprising of all the projected input and final output vectors $\left(\widetilde{x}_j, \widetilde{y}_j\right)$ can be set up as

$$\partial T_{VRS}^{N(II)} = \left\{ \left(\breve{x}, \breve{y}\right) \middle| \sum_{j=1}^{n} \widetilde{x}_j \lambda_j \le \breve{x}, \sum_{j=1}^{n} \widetilde{y}_j \lambda_j \ge \breve{y}, \sum_{j=1}^{n} \lambda_j = 1, \lambda_j \ge 0 \right\} \tag{7.34}$$

Here $\left(\widetilde{x}_j, \widetilde{y}_j\right) = \left(\delta_j^* x_j - s^-, \mu_j^* y_j + s^+\right)$ for $j = 1, 2, \ldots, n$ and δ_j^* and μ_j^* are obtained from the model (7.8) and model (7.9) respectively.

We then define the BB-based technology set ($T_{VRS}^{BB)}$) comprising of all the observed input and final output vectors as

$$T_{VRS}^{BB}(x, y) = \left\{ (x, y) \middle| \sum_{j=1}^{n} x_j \lambda_j \le x, \sum_{j=1}^{n} y_j \lambda_j \ge y, \sum_{j=1}^{n} \lambda_j = 1, \lambda_j \ge 0 \right\} \tag{7.35}$$

Now consider evaluating the output TE of firm h (TE_{oh}) against $\partial T_{VRS}^{N(I)}$, $\partial T_{VRS}^{N(II)}$ and T_{VRS}^{BB} whose actual input and final output vector is (x_h, y_h). The respective output-oriented *TE*s can be obtained by setting up the following LP programs:

$$\left(TE_{oh}^{N(I)}\right)^{-1} = \max\left\{\theta_I : (x_h, \theta_I y_h) \in \partial T_{VRS}^{N(I)}\right\} \tag{7.36}$$

$$\left(TE_{oh}^{N(II)}\right)^{-1} = \max\left\{\theta_{II} : (x_h, \theta_{II} y_h) \in \partial T_{VRS}^{N(II)}\right\} \tag{7.37}$$

$$\left(TE_{oh}^{BB}\right)^{-1} = \max\left\{\theta_{BB} : (x_h, \theta_{BB} y_h) \in T_{VRS}^{BB}\right\} \tag{7.38}$$

Since no allocative inefficiency is assumed in the construction of network technology under Approach II, one could *a priori* expect, for any given level of input, an output level in $\partial T_{VRS}^{N(II)}$ that is no less than that in $\partial T_{VRS}^{N(I)}$, which allows for inefficiencies. One could, therefore, interpret $(\theta_{II}^* - \theta_I^*)y_h$ as the output loss due to allocative inefficiency. Since the BB technology does not regard efficiencies concerning the internal operations of firm, one could expect, with any given level of input, the least output in this technology, i.e., $\theta_{BB}^* y_h \leq \theta_I^* y_h \leq \theta_{II}^* y_h$.

7.3 An Illustrative Example

Consider a simple hypothetical data set exhibited in Table 7.1. There are nine firms labeled as A, B, C, D, E, F, G, H and I. Each firm in T^I uses one input (x) to produce an intermediate product (measure) (z), which is then taken as input to T^{II} by the same firm to produce one final output (y).

7.3.1 On TE Estimates

Based on Table 7.1, Fig. 7.2 exhibits the two *independent* sub-technology frontiers in a counterclockwise orientation under the VRS specification. These frontiers are drawn by keeping z unaltered. Figure 7.3 exhibits the BB frontier involving observed x and y under an appropriate RTS specification (identified with lines: A-D-H-C), and the network production frontiers revealed from both approaches (model (7.1) under Approach I and models (7.8) and (7.9) under Approach II).

Table 7.1 Example data set

Firms	x	z	y
A	1.5	1	1
B	4	6	4
C	6	7	7
D	2	2.5	3
E	5	4	6
F	4	3.5	2
G	5.5	5	5
H	4.5	6	6
I	7	6.5	6.5

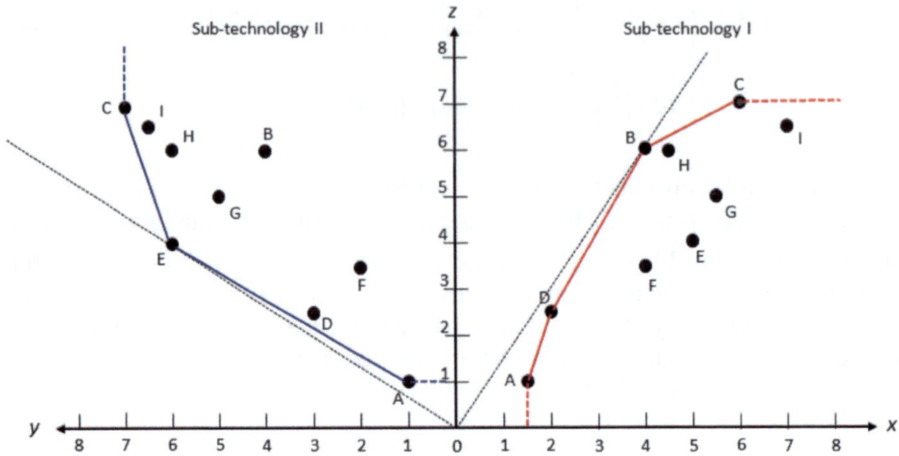

Fig. 7.2 Independent sub-technologies (Approach II)

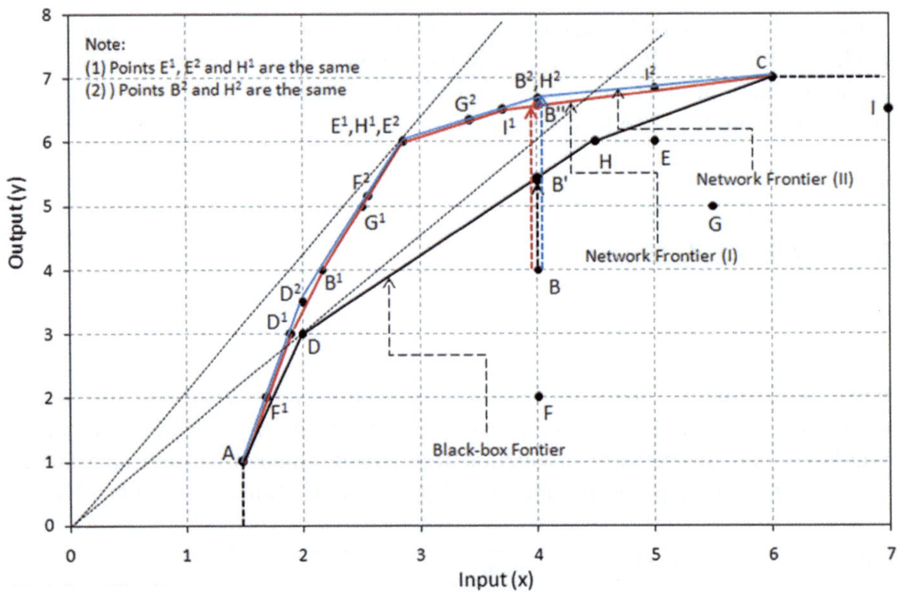

Fig. 7.3 Black-box technology vs. network technologies

Since the BB technology considers only the relation between *inputs* and *final outputs*, and makes no assumptions regarding the internal operations of firm, it provides no insights regarding the locations of inefficiency and scale economies. For example, firms – D and H that appear efficient in the BB technology turn out to be

Table 7.2 TE decomposition results

	Firms	TE_h^N	TE_h^I	TE_h^{II}	I_h^{VRS}
Approach I	A	1	1	1	1
	B	0.543	1	0.467	1.163
	C	1	1	1	1
	D	0.95	1	0.880	1.080
	E	0.571	0.571	1	1
	F	0.425	0.583	0.457	1.594
	G	0.457	0.623	0.680	1.078
	H	0.635	0.889	0.667	1.071
	I	0.531	0.612	0.846	1.024
Approach II	A	1	1	1	Not applicable
	B	0.6	1	0.6	
	C	1	1	1	
	D	0.857	1	0.857	
	E	0.571	0.571	1	
	F	0.249	0.643	0.387	
	G	0.492	0.623	0.789	
	H	0.800	0.889	0.900	
	I	0.679	0.714	0.951	

inefficient in the network technologies (identified with broken lines: A-F^1-D^1-B^1-G^1-E^1(H^1)-I^1-C under Approach I and A-D^2-F^2-E^2-G^2-B^2(H^2)-I^2-C under Approach II). Note that superscripts – 1 and 2 indicate, respectively, the projected points of the corresponding inefficient firms in both approaches. Points such as E^1, E^2 and H^1 are the same projected point for firm E (under both approaches) and for H (under Approach I). Similarly, points – B^2 and H^2 – are the same projected point for firms – B and H under Approach II. As regards the RTS, D that appears exhibiting CRS in the BB technology exhibits IRS (if projected in an input-oriented manner) in the network technologies.

We report in Table 7.2 the TE decomposition results obtained from Approach I (top part) and Approach II (bottom part), which will facilitate managerial insights regarding specific area of improvement for the network inefficient firms. The upshot of these results is summarized below.

1. Both approaches are in complete agreement in identifying the network efficient firms. The examples of such firms are A, C and E.
2. As expected, I_h^{VRS} is greater than 1 for those firms (B, D, F, G, H and I) that are technically inefficient in $T_{VRS}^{II(I)}$. Technical inefficiency arises only when the intermediate products consumed by these firms are not minimal implying that there is an overproduction of these outputs in $T_{VRS}^{I(I)}$. The results of our model (7.1) reveal that the optimal quantities of these products ($\tilde{z}^*$) are 2.8 (6), 2.2 (2.5), 1.6 (3.5), 3.4 (5), 4 (6) and 5.5 (6.5) for B, D, F, G, H and I respectively (the terms in brackets are their respective actual quantities). This is why the estimated sub-technology frontiers in Figs. 7.4 and 7.5 are different from those in Fig. 7.2.

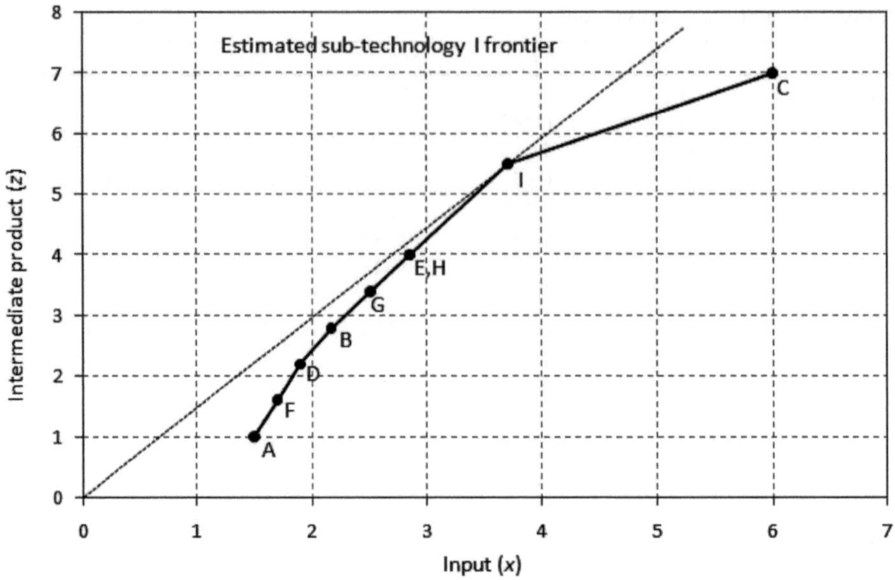

Fig. 7.4 Estimated sub-technology frontier I (Approach I)

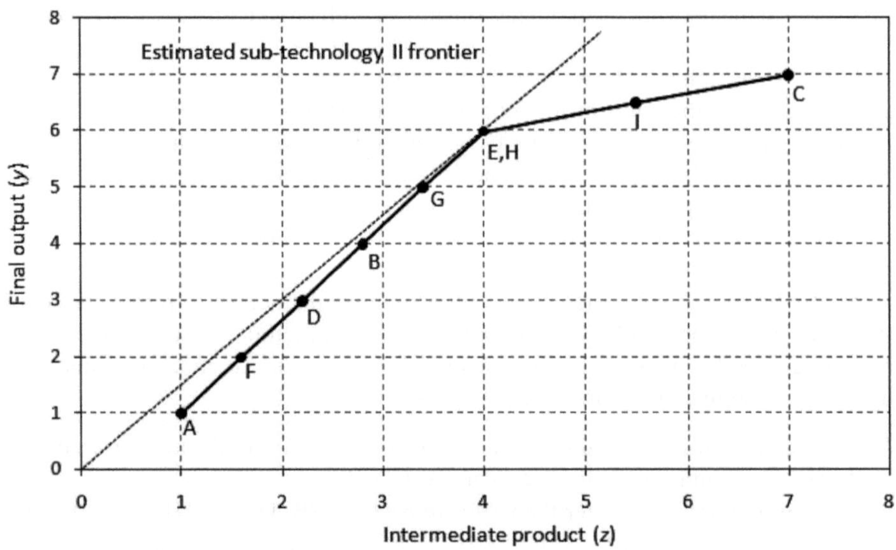

Fig. 7.5 Estimated sub-technology frontier II (Approach I)

3. The finding that the two approaches yield differential TE decomposition results for network inefficient firms is not at all strange. As expected, the decision to allow allocative inefficiency into the system in Approach I yields a frontier different from the one yielded from Approach II with no allocative inefficiency.

7.3.2 On Modeling the Output Losses

From our empirical application, one can observe that $T_{VRS}^{N(II)} \supseteq T_{VRS}^{N(I)} \supseteq T_{VRS}^{BB}$. As a result, one expect the following relationship: $TE_{oh}^{N(II)} \leq TE_{oh}^{N(I)} \leq TE_{oh}^{BB}$. Therefore, modeling of the TE of a firm against three different frontiers yields valuable information concerning whether the output loss is due to either missing of intermediate products connecting the sub-technologies, or the allocative inefficiency arising from any sub-optimal decision as to how much of intermediate products to produce and consume in the world of changing prices.

Let us consider, e.g., the TE evaluation of firm B. If B's output TE is evaluated against ∂T_{VRS}^{BB}, the projection is made on the point B′ (4, 5.4) where $TE_{oB}^{BB} = 0.741$ (=4/5.4). If it is evaluated against $\partial T_{VRS}^{N(I)}$, the projection is made on the point B″ (4, 6.563) where $TE_{oB}^{N(I)} = 0.610(=4/6.563)$; and if it is against the $\partial T_{VRS}^{N(II)}$, then the projection is made onto the point B^2 (4, 6.667) where $TE_{B}^{N(II)} = 0.600(=4/6.667)$. Since the potential output of 5.4 identified against ∂T_{VRS}^{BB} is the least as compared to 6.563 and 6.667 against $\partial T_{VRS}^{N(I)}$ and $\partial T_{VRS}^{N(II)}$, respectively, TE_{oB}^{BB} is highest. The output loss of 1.163 (= 6.563 − 5.4) against $\partial T_{VRS}^{N(I)}$ (with allocative inefficiency) is due to not accounting for the intermediate products connecting the sub-technologies in T_{VRS}^{BB}. This loss is lower as compared to the output loss of 1.267 (= 6.667 − 5.4) against $\partial T_{VRS}^{N(II)}$ with no allocative inefficiency. Therefore, the output loss of 0.104 (= 6.667 − 6.563) can be purely attributed to the allocative inefficiency associated with the possible sub-optimal decisions con-cerning the production and consumption of intermediate products connecting the sub-technologies.

We now discuss in the immediately following section the sources of input-oriented scale effects.

7.3.3 On SE Estimates

7.3.3.1 SE Estimates Using Approach I

Using Approach I we run both max and min forms of model (7.20) to compute the input-oriented left- and right-hand SEs of firms not only for the network technology but also for the sub-technologies (using formulas (7.21) and (7.22)). The SE results are reported in Table 7.3 (top part). The results reveal that $T_{VRS}^{N(I)}$ finds five firms – A, B, D, F and G operating under IRS, two firms – E and H under CRS and two firms – C and I under DRS. While the sources of increasing returns of firms in $T_{VRS}^{N(I)}$ are all located in both of the sub-technologies, the same is not the case for firms exhibiting decreasing and/or constant returns. For example, $T_{VRS}^{N(I)}$ exhibiting DRS for firm I is

Table 7.3 Upper and lower bounds of *SE* estimates

		Network			Sub-technology I			Sub-technology II		
	Firms	$\varepsilon_-^N(\cdot)$	$\varepsilon_+^N(\cdot)$	RTS	$\varepsilon_-^I(\cdot)$	$\varepsilon_+^I(\cdot)$	RTS	$\varepsilon_-^{II}(\cdot)$	$\varepsilon_+^{II}(\cdot)$	RTS
Approach I	A	7.500	∞	IRS	4.500	∞	IRS	1.667	∞	IRS
	B	1.583	2.000	IRS	1.357	1.714	IRS	1.167	1.167	IRS
	C	0	0.187	DRS	0	0.562	DRS	0	0.333	DRS
	D	2.333	3.167	IRS	1.909	2.591	IRS	1.222	1.222	IRS
	E	0.278	1.389	CRS	1.250	1.250	IRS	0.222	1.111	CRS
	F	4.250	4.250	IRS	3.188	3.188	IRS	1.333	1.333	IRS
	G	1.467	1.467	IRS	1.294	1.294	IRS	1.133	1.133	IRS
	H	0.278	1.389	CRS	1.250	1.250	IRS	0.222	1.111	CRS
	I	0.125	0.333	DRS	0.443	1.182	CRS	0.282	0.282	DRS
Approach II	A	7.500	∞	IRS	4.500	∞	IRS	1.667	∞	IRS
	B	1.583	1.583	IRS	0.333	1.167	CRS	0.300	0.300	DRS
	C	0	0.143	DRS	0	0.429	DRS	0	0.333	DRS
	D	3.167	3.167	IRS	1.400	2.400	IRS	1.191	1.191	IRS
	E	0.278	1.389	CRS	1.250	1.250	IRS	0.222	1.111	CRS
	F	4.250	4.250	IRS	1.286	1.286	IRS	1.129	1.129	IRS
	G	1.467	1.467	IRS	1.200	1.200	IRS	0.263	0.263	DRS
	H	0.278	1.389	CRS	0.333	1.167	CRS	0.300	0.300	DRS
	I	0.333	0.333	DRS	0.385	0.385	DRS	0.317	0.317	DRS

due to DRS in $T_{VRS}^{II(I)}$ even though CRS prevails in $T_{VRS}^{I(I)}$. Similarly, $T_{VRS}^{N(I)}$ exhibiting CRS for firms – E and H is precisely due to CRS in $T_{VRS}^{II(I)}$ even though IRS prevails in $T_{VRS}^{I(I)}$.

7.3.3.2 SE Estimates Using Approach II

Using Approach II we run both max and min forms of the model (7.29), which is based on the optimal input TE values of the model (7.8), to compute the input-oriented left- and right-hand SEs of the firms in $T_{VRS}^{I(II)}$. Similarly, we run both max and min forms of the model (7.31), which is based on the optimal output *TE* values of the model (7.9), to compute the output-oriented left- and right-hand SEs of the firms in $T_{VRS}^{II(II)}$. However, under this approach, it is not possible to compute the input-oriented network SEs of firms using (7.29) and (7.31). Therefore, in order to compute the input-oriented left- and right-hand network SEs, we use firms' projected input–output vectors, $(\tilde{x}, \tilde{y}) = (\delta^* x - s^-, \mu^* y + s^+)$ obtained from (7.8) and (7.9), in model (7.29). The input-oriented network SE estimates are reported in Table 7.3 (bottom part). We find five firms – A, B, D, F and G operating under IRS, two firms – E and H under CRS and the remaining two firms – C and I under DRS (which call can be visualized in Fig. 7.3).

Note that since it is not possible in this approach to decompose network SE into its sub-technology specific SEs, the scale economies/diseconomies revealed from sub-technologies [(7.29) and (7.31)] may not attribute to the network scale economy/diseconomy obtained from the use of projected data of network firms $(\widetilde{x}, \widetilde{y})$ in (7.29). For example, consider firm B whose sub-technologies exhibit CRS and DRS (CRS in $T_{VRS}^{I(II)}$ and DRS in $T_{VRS}^{II(II)}$), but its network technology, $T_{VRS}^{N(II)}$ exhibits IRS. It is therefore quite improbable to argue that the sources of increasing returns in the network technology are due to CRS and DRS in the sub-technologies. Note that the very purpose of computing the input-oriented network SE of firms under Approach II is just to compare these SE estimates with those obtained under Approach I.

Notice that though the network technologies revealed from both approaches look similar (Fig. 7.3), and the (input-oriented) RTS possibilities of network firms are the same; the degrees of underlying SE estimates of some network firms are different due to differential nature (flatness/steepness) of some production facets. For example, $T_{VRS}^{N(I)}$ finds B exhibiting IRS whose value ranges from 1.583 to 2 since its SE is estimated against the vertex point B^1 connecting two facets – D^1B^1 and B^1G^1. $T_{VRS}^{I(II)}$ also finds this firm operating under the same IRS but its SE value is exact at 1.583 since it is estimated against a point on the facet D^2F^2. So are the cases with firms – D and I.

On comparison between the two approaches with regard to the sources of scale economies of firms, we find some divergent information on their RTS possibilities. Though both approaches maintain input-orientation in T_{VRS}^{I}, they yield contrasting RTS possibilities for some firms. For example, while $T_{VRS}^{I(I)}$ finds both B and H operating under IRS, and I under CRS, $T_{VRS}^{I(II)}$ finds B and H under CRS, and I under DRS. These contrasting *RTS* information are because the estimated T_{VRS}^{I} revealed from both approaches are different (see Figs. 7.2 (right) and 7.4). However, there are contrasting information on the RTS possibilities in T_{VRS}^{II} even though the estimated sub-technology II frontiers are exactly the same in both approaches (see Figs. 7.2 (left) and 7.5). This is simply due to the different orientations maintained in T_{VRS}^{II} for the measurement of efficiency and scale elasticity (i.e., the input orientation in $T_{VRS}^{II(I)}$ and the output orientation in $T_{VRS}^{II(II)}$). Note that the finding that the estimated sub-technology frontiers in T_{VRS}^{II} are the same in both approaches is just a coincidence.

Finally, the finding that firms – E and H exhibit CRS in the network technology and IRS and CRS in the sub-technology I and sub-technology II respectively reminds one that the CRS assumption maintained in the neoclassical theory for justifying the black-box structure of production technology does not necessarily allow one to infer that there are no scale benefits available in the sub-technologies. One can, therefore, argue that it is crucial for the firm's ownership to locate the sources of scale effects in their sub-technologies, which will enable the firm management to improve productivity.

However, the modeling of a firm technology by considering only the inputs consumed and the final outputs produced often yields the imprecise estimates of production function; and as a result, yields erroneous inferences concerning the RTS behavior of firms (see, e.g., D and H in Fig. 7.3). This is because the black-box characterization obscures important relations by ignoring the interdependencies that exist between the sub-technologies.

7.4 Concluding Remarks

To reveal the sources of efficiency and scale economies, two approaches are suggested, based on the premise as to whether the two-stage network technology considered in each approach allows allocative inefficiency. The first approach is developed by making use of a single network technology for the two interdependent sub-technologies. This approach allows for allocative inefficiency that may arise due to any sub-optimal decision as to how much of intermediate products to produce and consume by the sub-technology managers in the world of changing prices. In the second approach, however, the technology structure is determined by assuming its sub-technologies to be independent, implying that there is no allocative inefficiency.

Instances of real-life firms suffering from profit loss due to allocative inefficiency are not usually uncommon as they most often face uncertainty in forecasting prices in their production decisions. Therefore, production managers are given incentives to choose right output-mix and right input-mix in the world of changing market prices in order to improve upon profit. And, even if managers are not held responsible for the changing prices, management would still like to know the opportunity cost of using the sub-optimal input and output mixes. Therefore, the network production system is modeled by our two approaches to know the extent of output loss of a firm suffering from allocative inefficiencies.

The current study is limited to the estimation of TE and SE only. The potential future research subject could be the one where one could interpret a two-stage network firm as a multi-product firm producing both intermediate products and final outputs, and then, measure *economies of scope* by linking it with I_h^{VRS}, a proxy for the indication of allocative inefficiency.

References

Aburzzi, A. (1965). The production process: Operation characteristics. *Management Science, 11*, B98–B118.

Bakshi, M. S., & Arora, S. R. (1969). The sequencing problem. *Management Science, 16*, B247–B263.

Banker, R. D., & Thrall, R. M. (1992). Estimation of returns to scale using data envelopment analysis. *European Journal of Operational Research, 62*, 74–84.

Banker, R. D., Charnes, A., & Cooper, W. W. (1984). Some models for estimating technical and scale inefficiencies in data envelopment analysis. *Management Science, 30*, 1078–1092.

Banker, R. D., Cooper, W. W., Seiford, L. M., Thrall, R. M., & Zhu, J. (2004). Returns to scale in different DEA models. *European Journal of Operational Research, 154*, 345–362.

Baumol, W. J., Panzar, J. C., & Willig, R. D. (1982). *Contestable markets and the theory of industry structure*. New York: Harcourt Brace Jovanovich.

Chen, Y., & Zhu, J. (2004). Measuring information technology's indirect impact on firm performance. *Information Technology and Management, 5*, 9–22.

Chen, Y., Liang, L., Yang, F., & Zhu, J. (2006). Evaluation of information technology investment: A data envelopment analysis approach. *Computers and Operations Research, 33*, 1368–1379.

Chen, Y., Cook, W. D., Li, N., & Zhu, J. (2009a). Additive efficiency decomposition in two-stage DEA. *European Journal of Operational Research, 196*, 1170–1176.

Chen, Y., Liang, L., & Zhu, J. (2009b). Equivalence in two-stage DEA approaches. *European Journal of Operational Research, 193*, 600–604.

Chen, Y., Cook, W. D., & Zhu, J. (2010). Deriving the DEA frontier for two-stage processes. *European Journal of Operational Research, 202*, 138–142.

Chen, Y., Cook, W. D., Kao, C., & Zhu, J. (2013). Network DEA pitfalls: Divisional efficiency and frontier projection under general network structures. *European Journal of Operational Research, 226*, 507–515.

Cook, W. D., Liang, L., & Zhu, J. (2010). Measuring performance of two-stage network structure by DEA: A review. *Omega, 38*, 423–430.

Färe, R. (1991). Measuring Farrell efficiency for a firm with intermediate inputs. *Academia Economic Papers, 19*, 329–340.

Färe, R., & Grosskopf, S. (1996). *Intertemporal production frontiers: With dynamic DEA*. Boston: Kluwer Academic.

Färe, R., & Grosskopf, S. (2000). Network DEA. *Socio-Economic Planning Sciences, 34*, 35–49.

Førsund, F. R. (1996). On the calculation of scale elasticity in DEA models. *Journal of Productivity Analysis, 7*, 283–302.

Fukuyama, H. (2003). Scale characterizations in a DEA directional technology distance function framework. *European Journal of Operational Research, 144*, 108–127.

Hadjicostas, P., & Soteriou, A. C. (2006). One-sided elasticities and technical efficiency in multi-output production: A theoretical framework. *European Journal of Operational Research, 168*, 425–449.

Kao, C. (2009). Efficiency decomposition in network data envelopment analysis: A relational model. *European Journal of Operational Research, 192*, 949–962.

Kao, C. (2013). Dynamic data envelopment analysis: A relational analysis. *European Journal of Operational Research.* http://dx.doi.org/10.1016/j.ejor.2012.12.012

Kao, C., & Hwang, S.-N. (2008). Efficiency decomposition in two-stage data envelopment analysis: An application to non-life insurance companies in Taiwan. *European Journal of Operational Research, 185*, 418–429.

Kao, C., & Hwang, S.-N. (2011). Decomposition of technical and scale efficiencies in two-stage production systems. *European Journal of Operational Research, 211*, 515–519.

Kao, C., & Hwang, S.-N. (2014). Multi-period efficiency and Malmquist productivity index in two-stage production systems. *European Journal of Operational Research, 232*, 512–521.

Lewis, H. F., Mallikarjun, S., & Sexton, T. R. (2013). Unoriented two-stage DEA: The case of oscillating intermediate products. *European Journal of Operational Research, 229*, 529–539.

Liang, L., Yang, F., Cook, W. D., & Zhu, J. (2006). DEA models for supply chain efficiency evaluation. *Annals of Operations Research, 145*, 35–49.

Liang, L., Cook, W. D., & Zhu, J. (2008). DEA models for two-stage processes: Game approach and efficiency decomposition. *Naval Research Logistics, 55*, 643–653.

Podinovski, V. V., Førsund, F. R., & Krivonozhko, V. E. (2009). A simple derivation of scale elasticity in data envelopment analysis. *European Journal of Operational Research, 197*, 149–153.

Prieto, A. M., & Zofio, J. L. (2007). Network DEA efficiency in input–output models: With an application to OECD countries. *European Journal of Operational Research, 178*, 292–304.

Sahoo, B. K., & Tone, K. (2013). Non-parametric measurement of economies of scale and scope in non-competitive environment with price uncertainty. *Omega, 41*, 97–111.

Sahoo, B. K., Mohapatra, P. K. J., & Trivedi, M. L. (1999). A comparative application of data envelopment analysis and frontier translog production function for estimating returns to scale and efficiencies. *International Journal of Systems Science, 30*, 379–394.

Sahoo, B. K., Kerstens, K., & Tone, K. (2012). Returns to growth in a non-parametric DEA approach. *International Transactions in Operational Research, 19*, 463–486.

Sahoo, B. K., Zhu, J., Tone, K., & Klemen, B.M. (2014). Decomposing technical efficiency and scale elasticity in two-stage network DEA. *European Journal of Operational Research, 233*, 584–594.

Scarf, H. E. (1990). Mathematical programming and economic theory. *Operations Research, 38*, 377–385.

Seiford, L. M., & Zhu, J. (1999). Profitability and marketability of the top 55 US commercial banks. *Management Science, 45*, 1270–1288.

Tone, K., & Sahoo, B. K. (2003). Scale, indivisibilities and production function in data envelopment analysis. *International Journal of Production Economics, 84*, 165–192.

Tone, K., & Sahoo, B. K. (2004). Degree of scale economies and congestion: A unified DEA approach. *European Journal of Operational Research, 158*, 755–772.

Tone, K., & Sahoo, B. K. (2005). Cost efficiency and returns to scale in Life Insurance Corporation of India using data envelopment analysis. *Socio-Economic Planning Sciences, 39*, 261–285.

Tone, K., & Sahoo, B. K. (2006). Re-examining scale elasticity in DEA. *Annals of Operations Research, 145*, 169–187.

Tone, K., & Tsutsui, M. (2009). Network DEA: A slacks-based measure approach. *European Journal of Operational Research, 197*, 243–252.

Tone, K., & Tsutsui, M. (2010). Dynamic DEA: A slacks-based measure approach. *Omega, 38*, 145–156.

Yu, M.-M., & Lin, E. T. J. (2008). Efficiency and effectiveness in railway performance using a multi-activity network DEA model. *Omega, 36*, 1005–1017.

Zelenyuk, V. (2013). A scale elasticity measures for directional distance function and its dual: Theory and DEA estimation. *European Journal of Operational Research, 228*, 592–600.

Zhu, J. (2000). Multi-factor performance measure model with an application to Fortune 500 companies. *European Journal of Operational Research, 123*, 105–124.

Chapter 8
Evaluating Two-Stage Network Structures: Bargaining Game Approach

Juan Du, Yao Chen, Wade D. Cook, Liang Liang, and Joe Zhu

Abstract This chapter presents a Nash bargaining game model to measure the performance of two-stage decision making units (DMUs) in data envelopment analysis (DEA). The two stages are viewed as players to bargain for a better payoff, which is represented by DEA ratio efficiency score. The efficiency model is developed as a cooperative game model. It is shown that when only one intermediate measure exists between the two stages, the newly-developed bargaining approach yields the same results as applying the standard DEA approach to each stage separately.

Part of this chapter is based upon J. Du, L. Liang, Y. Chen, W.D. Cook, and J. Zhu (2011), "A bargaining game model for measuring performance of two-stage network structures", *European Journal of Operational Research* 210, 390–397, with permissions from Elsevier Science.

J. Du
School of Economics and Management, Tongji University, 1239 Siping Road, Shanghai 200092, P.R. China
e-mail: dujuan@tongji.edu.cn

Y. Chen (✉)
Manning School of Business, University of Massachusetts at Lowell, Lowell, MA 01845, USA
e-mail: Yao_Chen@uml.edu

W.D. Cook
Schulich School of Business, York University, Toronto, ON M3J 1P3, Canada
e-mail: wcook@schulich.yorku.ca

L. Liang
School of Management, University of Science and Technology of China, 96 Jinzhai Road, Hefei, Anhui 230026, P.R. China
e-mail: lliang@ustc.edu.cn

J. Zhu
School of Business, Worcester Polytechnic Institute, Worcester, MA 01609, USA
e-mail: jzhu@wpi.edu

W.D. Cook and J. Zhu (eds.), *Data Envelopment Analysis*, International Series in Operations Research & Management Science 208, DOI 10.1007/978-1-4899-8068-7_8, © Springer Science+Business Media New York 2014

Keywords Data envelopment analysis • Efficiency • Nash bargaining game • Two-stage structure

8.1 Introduction

In order to address the potential conflict caused by the dual role of intermediate measures, quite a number of scholars propose their own versions of solutions. For example, Kao and Hwang (2008) combine the efficiency scores of the two stages in a multiplicative (geometric) manner, while Chen et al. (2009) use a weighted additive model. Liang et al. (2008) develop a number of DEA models using game theory concept. Specifically, they develop a leader-follower model borrowed from the notion of Stackelberg games, and a centralized or cooperative game model where the combined stage is of interest.

This chapter presents the study of Du et al. (2011) which applies directly the Nash bargaining game theory to the efficiency of DMUs that have the afore-mentioned two-stage processes. The two stages are regarded as two individuals bargaining with each other for a better payoff, which is characterized by the DEA ratio efficiency of each individual stage. In general, the resulting bargaining game model is highly non-linear, given the nature of ratio forms of DEA efficiency. This chapter shows that this non-linear bargaining model can be converted equivalently into a parametric linear programming problem with one parameter, whose lower and upper bounds can be determined. As a result, a global optimal solution can be found using a heuristic search on the single parameter.

In the bargaining model, the breakdown or status quo point is determined via the standard DEA model. The bargaining efficiency scores of the two stages may depend on the selection of the breakdown point. Thus in applications, a sensitivity analysis is carried out to study the stability of the bargaining DEA efficiency scores with respect to different status quo points. Also, it is shown that when only one intermediate measure exists between the two stages, the Nash bargaining game model in this study yields the same results as applying the standard DEA model to each stage separately.

8.2 Background

Consider a two-stage process shown in Fig. 8.1. Suppose there are n DMUs and each DMU_j ($j = 1, 2, \ldots, n$) has m inputs to the first stage, denoted by $x_{ij}\, (i = 1, 2, \ldots, m)$, and D outputs from this stage, denoted by $z_{dj}\, (d = 1, 2, \ldots, D)$. These D outputs then become the inputs to the second stage, which are referred to as intermediate measures. The s outputs from the second stage are denoted by $y_{rj}\, (r = 1, 2, \ldots, s)$.

Based upon the constant returns to scale (CRS) model (Charnes et al. 1978), the (CRS) efficiency scores for each DMU_j ($j = 1, 2, \ldots, n$) in the first and second stages can be defined by e_j^1 and e_j^2, respectively,

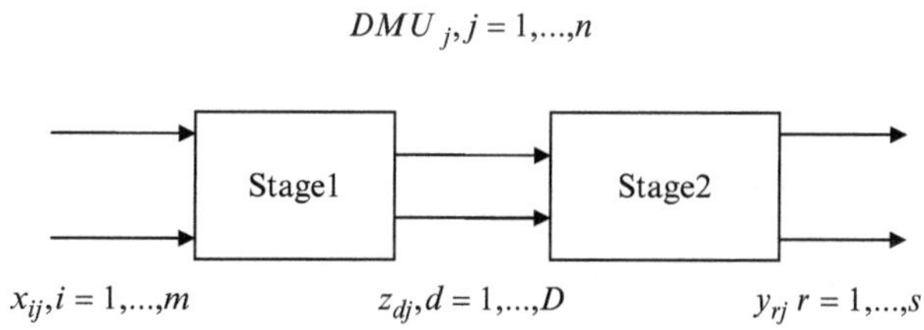

Fig. 8.1 A two-stage process

$$e_j^1 = \frac{\sum_{d=1}^{D} w_d^1 z_{dj}}{\sum_{i=1}^{m} v_i x_{ij}} \leq 1 \quad \text{and} \quad e_j^2 = \frac{\sum_{r=1}^{s} u_r y_{rj}}{\sum_{d=1}^{D} w_d^2 z_{dj}} \leq 1 \tag{8.1}$$

where v_i, w_d^1, w_d^2 and u_r are unknown non-negative weights. These ratios are then optimized in a linear fractional programming problem which can be converted into a linear CRS DEA model (Charnes et al. 1978).

As noted both in Kao and Hwang (2008) and in Liang et al. (2008), it is reasonable to set w_d^1 equal to w_d^2, since the "worth" or value assigned to the intermediate measures should be the same regardless of whether they are viewed as outputs from the first stage or inputs to the second stage. Then in this case, given the individual efficiency scores e_j^1 and e_j^2, it is reasonable to define the overall efficiency of the entire two-stage process for DMU_j ($j = 1, \ldots, n$) as $e_j = e_j^1 \cdot e_j^2$ since

$$e_j = \frac{\sum_{r=1}^{s} u_r y_{rj}}{\sum_{i=1}^{m} v_i x_{ij}} = \frac{\sum_{d=1}^{D} w_d z_{dj}}{\sum_{i=1}^{m} v_i x_{ij}} \cdot \frac{\sum_{r=1}^{s} u_r y_{rj}}{\sum_{d=1}^{D} w_d z_{dj}} = e_j^1 \cdot e_j^2 \tag{8.2}$$

The above overall efficiency definition ensures that $e_j \leq 1$ from $e_j^1 \leq 1$ and $e_j^2 \leq 1$, and the overall process is efficient if and only if $e_j^1 = e_j^2 = 1$.

Clearly, separate DEA analysis can be applied to each individual stage as in Seiford and Zhu (1999). However, as pointed out by Liang et al. (2008), such an approach could cause inherent conflict between these two separate analyses.

The efficiency-evaluation problem can be approached from two game theory perspectives. One is to view the two-stage process as a non-cooperative game model, in which one stage is assumed to be a leader and solved for its CRS efficiency first, and the other stage a follower, whose efficiency is computed without changing the leader's efficiency score. The other approach is to regard the process as a centralized model, where the overall efficiency given in (8.2) is maximized, and a decomposition of the overall efficiency is obtained by finding a set of multipliers producing the largest first (or second) stage efficiency score while maintaining the overall efficiency score.

Note that in fact, the two stages can be regarded as two players in Nash bargaining game. Therefore, the efficiency evaluation of two-stage processes can be approached by using Nash bargaining game theory directly. But before that, we first briefly introduce the Nash bargaining game approach.

Denote the set of two individuals participating in the bargaining by $N = \{1, 2\}$, and a payoff vector is an element of the payoff space R^2, which is the 2-dimensional Euclidean space. A feasible set S is a subset of the payoff space, and a breakdown or status quo point $\vec{b}$ is an element of the payoff space. A bargaining problem can then be specified as the triple $\left(N,\ S,\ \vec{b}\right)$ consisting of participating individuals, feasible set, and breakdown point. Nash (1950) requires that the feasible set be compact, convex, and contain some payoff vector such that each individual's payoff is at least as large as the individual's breakdown payoff. The solution is a function F that is associated with each bargaining problem $\left(N,\ S,\ \vec{b}\right)$, expressed as $F\left(N,\ S,\ \vec{b}\right)$. Nash (1950, 1953) argue that a reasonable solution should satisfy the four properties: (i) Pareto efficiency (*PE*), (ii) invariance with respect to affine transformation (*IAT*), (iii) independence of irrelevant alternatives (*IIA*), and (iv) symmetry (*SYM*). Due to extensive discussion about these properties in the literature, no detailed explanation will be provided here. For the traditional bargaining problem, Nash (1950, 1953) has shown that there exists a unique solution, called the Nash solution, which satisfies the above-mentioned four properties, and can be obtained by solving the following maximization problem

$$\underset{\vec{u}\in S,\, \vec{u}\geq \vec{b}}{Max} \prod_{i=1}^{2} (u_i - b_i) \tag{8.3}$$

where $\vec{u}$ is the payment vector for the individuals, and u_i, b_i is the ith element of vector $\vec{u}, \vec{b}$, respectively.

8.3 Nash Bargaining Game Model for Two-Stage Structures

In the current case, we view the two individual stages as two players in the bargaining procedure, the efficiency ratios as the payoffs, and weights chosen for efficiency scores as strategies. To proceed, one needs to find a breakdown point for stages 1 and 2 which is the starting point for bargaining. Note that the breakdown point or status quo represents possible payoff pairs obtained if one decides not to bargain with the other player. As mentioned in Binmore et al. (1986), the choice of the breakdown point is a matter of modeling judgment. A number of elements in the underlying situation can be natural candidates for this role. For example, Lundberg and Pollak (1993) use a non-cooperative equilibrium as the breakdown point in their bargaining model. In application section, we will use different breakdown points, including the ones based upon the leader-follower (non-cooperative) model of Liang et al. (2008), to perform sensitivity analysis to study the stability of our bargaining DEA efficiency scores with respect to different breakdown points.

We here first construct the least ideal DMU and use its DEA efficiency scores as the breakdown point. By doing that, we assume that if the two stages do not negotiate, their efficiency scores will be the worst. Note that such a DMU may not exist, however, its inputs and outputs are observed. Let $x_i^{\max} = \max_j \{x_{ij}\}$, $y_r^{\min} = \min_j \{y_{rj}\}$, $z_d^{\min} = \min_j \{z_{dj}\}$ and $z_d^{\max} = \max_j \{z_{dj}\}$. Then $(x_i^{\max}, z_d^{\min})$ $(i = 1, \ldots, m,\ d = 1, \ldots, D)$ represents the least ideal DMU in the first stage, which consumes the maximum amount of input values, while producing the least amount of intermediate measures. Similarly, we denote $(z_d^{\max},\ y_r^{\min})$ $(d = 1, \ldots, D,$ $r = 1, \ldots, s)$ the least ideal DMU in the second stage, which consumes the maximum amount of intermediate measures while producing the least amount of output values.

The CRS efficiency for the above two least ideal DMUs is the worst among the existing DMUs. We denote the (CRS) efficiency scores of the two least ideal DMUs in the first and second stage as $\theta^1_{\min}$ and $\theta^2_{\min}$, respectively, and use $\theta^1_{\min}$ and $\theta^2_{\min}$ as the breakdown point. The (input-oriented) DEA bargaining model for a specific DMU_o with respect to (8.3) can be expressed as

$$
\begin{aligned}
&Max \left(\frac{\sum_{d=1}^{D} w_d z_{do}}{\sum_{i=1}^{m} v_i x_{io}} - \theta^1_{\min} \right) \cdot \left(\frac{\sum_{r=1}^{s} u_r y_{ro}}{\sum_{d=1}^{D} w_d z_{do}} - \theta^2_{\min} \right) \\
&s.t. \ \frac{\sum_{d=1}^{D} w_d z_{do}}{\sum_{i=1}^{m} v_i x_{io}} \geq \theta^1_{\min} \\
&\frac{\sum_{r=1}^{s} u_r y_{ro}}{\sum_{d=1}^{D} w_d z_{do}} \geq \theta^2_{\min} \\
&\frac{\sum_{d=1}^{D} w_d z_{dj}}{\sum_{i=1}^{m} v_i x_{ij}} \leq 1, \quad j = 1, \ldots, n \\
&\frac{\sum_{r=1}^{s} u_r y_{rj}}{\sum_{d=1}^{D} w_d z_{dj}} \leq 1, \quad j = 1, \ldots, n \\
&v_i,\ u_r,\ w_d > 0, \quad i = 1, \ldots, m,\ r = 1, \ldots, s,\ d = 1, \ldots, D
\end{aligned}
\tag{8.4}
$$

Denote all the constraints in model (8.4) by S, which represents the feasible set for this bargaining problem. Then the bargaining problem here can be specified as the triple $(\{1, 2\}, S, \{\theta^1_{\min}, \theta^2_{\min}\})$. Next we will prove that the feasible set S is both compact and convex.

Lemma 1 *The feasible set* S *is compact and convex.*

Proof Since the feasible set S is bounded and closed in Euclidean space, then S is compact. Next we will prove that S is also convex.

Suppose $(v'_1, \ldots, v'_m, u'_1, \ldots, u'_s, w'_1, \ldots, w'_D) \in S$ and $(v''_1, \ldots, v''_m, u''_1, \ldots, u''_s, w''_1, \ldots, w''_D) \in S$. For any $\lambda \in [0, 1]$ we have $\lambda v'_i + (1-\lambda) v''_i > 0, i = 1, \ldots, m$, $\lambda u'_r + (1-\lambda) u''_r > 0$, $r = 1, \ldots, s$ and $\lambda w'_d + (1-\lambda) w''_d > 0$, $d = 1, \ldots, D$. Since $\sum_{i=1}^{m} v_i x_{ij} > 0$ and $\sum_{d=1}^{D} w_d z_{dj} > 0$ for all $j = 1, \ldots, n$, the constraints in S, $\frac{\sum_{d=1}^{D} w_d z_{dj}}{\sum_{i=1}^{m} v_i x_{ij}} \le 1$ and $\frac{\sum_{r=1}^{s} u_r y_{rj}}{\sum_{d=1}^{D} w_d z_{dj}} \le 1$ are equivalent to $\sum_{d=1}^{D} w_d z_{dj} \le \sum_{i=1}^{m} v_i x_{ij}$ and $\sum_{r=1}^{s} u_r y_{rj} \le \sum_{d=1}^{D} w_d z_{dj}$, respectively, for all $j = 1, \ldots, n$, and the constraints $\frac{\sum_{d=1}^{D} w_d z_{do}}{\sum_{i=1}^{m} v_i x_{io}} \ge \theta^1_{\min}$ and $\frac{\sum_{r=1}^{s} u_r y_{ro}}{\sum_{d=1}^{D} w_d z_{do}} \ge \theta^2_{\min}$ are equivalent to $\sum_{d=1}^{D} w_d z_{do} \ge \theta^1_{\min} \sum_{i=1}^{m} v_i x_{ij}$ and $\sum_{r=1}^{s} u_r y_{rj} \ge \theta^2_{\min} \sum_{d=1}^{D} w_d z_{do}$, respectively. Then we have

$$
\begin{aligned}
\sum_{d=1}^{D} \left[\lambda w'_d + (1-\lambda) w''_d\right] z_{dj} &= \lambda \sum_{d=1}^{D} w'_d z_{dj} + (1-\lambda) \sum_{d=1}^{D} w''_d z_{dj} \\
&\le \lambda \sum_{i=1}^{m} v'_i x_{ij} + (1-\lambda) \sum_{i=1}^{m} v''_i x_{ij} \\
&= \sum_{i=1}^{m} \left[\lambda v'_i + (1-\lambda) v''_i\right] x_{ij}
\end{aligned}
$$

and

$$\begin{aligned}\sum_{r=1}^{s}\left[\lambda u'_r+(1-\lambda)\,u''_r\right]y_{rj} &= \lambda\sum_{r=1}^{s}u'_r y_{rj}+(1-\lambda)\sum_{r=1}^{s}u''_r y_{rj}\\ &\le \lambda\sum_{d=1}^{D}w'_d z_{dj}+(1-\lambda)\sum_{d=1}^{D}w''_d z_{dj}\\ &= \sum_{d=1}^{D}\left[\lambda w'_d+(1-\lambda)\,w''_d\right]z_{dj}\end{aligned}$$

Similarly, we have $\sum_{d=1}^{D}\left[\lambda w'_d+(1-\lambda)\,w''_d\right]z_{do}\ge\theta^1_{\min}\sum_{i=1}^{m}\left[\lambda v'_i+(1-\lambda)\,v''_i\right]x_{ij}$ and $\sum_{r=1}^{s}\left[\lambda u'_r+(1-\lambda)\,u''_r\right]y_{rj}\ge\theta^2_{\min}\sum_{d=1}^{D}\left[\lambda w'_d+(1-\lambda)\,w''_d\right]z_{do}$.

Therefore $(\lambda v'_i + (1-\lambda)\,v''_i,\ \lambda u'_r + (1-\lambda)\,u''_r,\ \lambda w'_d + (1-\lambda)\,w''_d) \in S$, where $i = 1, \ldots, m$, $r = 1, \ldots, s$, $d = 1, \ldots, D$, or equivalently, $\lambda\,(v'_1, \ldots, v'_m, u'_1, \ldots, u'_s, w'_1, \ldots, w'_D) + (1-\lambda)(v''_1, \ldots, v''_m, u''_1, \ldots, u''_s, \ldots, w''_1, \ldots, w''_D) \in S$. Consequently S is a convex set.

Q.E.D.

Let $t_1=\left(\sum_{i=1}^{m}v_i x_{io}\right)^{-1}$, $t_2=\left(\sum_{d=1}^{D}w_d z_{do}\right)^{-1}$, $\gamma_i=t_1\,v_i$, $\omega_d=t_1\,w_d$, $\mu_{r1}=t_1\,u_r$, $\mu_{r2}=t_2\,u_r$. Note that $\mu_{r1}=t_1\,u_r$ and $\mu_{r2}=t_2\,u_r$ imply a linear relationship of $\mu_{r1}=\frac{t_1}{t_2}\,\mu_{r2}$ between μ_{r1} and μ_{r2}. Therefore, we denote $\frac{t_1}{t_2}$ by α (>0) and have $\mu_{r1}=\alpha\mu_{r2}$ for all $r=1,\ldots,s$. Then model (8.5) is converted into model (8.5).

$$\begin{aligned}
&Max\ \sum_{r=1}^{s}\mu_{r1}y_{ro}-\theta^1_{\min}\sum_{r=1}^{s}\mu_{r2}y_{ro}-\theta^2_{\min}\sum_{d=1}^{D}\omega_d z_{do}+\theta^1_{\min}\cdot\theta^2_{\min}\\
&s.t.\ \sum_{d=1}^{D}\omega_d z_{do}\ge\theta^1_{\min}\\
&\qquad\sum_{r=1}^{s}\mu_{r2}y_{ro}\ge\theta^2_{\min}\\
&\qquad\sum_{i=1}^{m}\gamma_i x_{io}=1\\
&\qquad\sum_{d=1}^{D}\omega_d z_{do}=\alpha\\
&\qquad\sum_{d=1}^{D}\omega_d z_{dj}-\sum_{i=1}^{m}\gamma_i x_{ij}\le 0,\quad j=1,\ldots,n\\
&\qquad\sum_{r=1}^{s}\mu_{r1}y_{rj}-\sum_{d=1}^{D}\omega_d z_{dj}\le 0,\quad j=1,\ldots,n\\
&\qquad\mu_{r1}=\alpha\mu_{r2},\quad r=1,\ldots s\\
&\qquad\alpha>0,\ \gamma_i,\ \omega_d,\ \mu_{r1},\ \mu_{r2}>0,\quad i=1,\ldots,m,\ r=1,\ldots,s,\ d=1,\ldots,D
\end{aligned}\tag{8.5}$$

Model (8.5) is equivalent to the following nonlinear model (8.6).

$$
\begin{aligned}
& Max\ \alpha \times \sum_{r=1}^{s} \mu_{r2} y_{ro} - \theta_{\min}^{1} \sum_{r=1}^{s} \mu_{r2} y_{ro} - \theta_{\min}^{2} \sum_{d=1}^{D} \omega_d z_{do} + \theta_{\min}^{1} \cdot \theta_{\min}^{2} \\
& s.t.\ \sum_{d=1}^{D} \omega_d z_{do} \geq \theta_{\min}^{1} \\
& \sum_{r=1}^{s} \mu_{r2} y_{ro} \geq \theta_{\min}^{2} \\
& \sum_{i=1}^{m} \gamma_i x_{io} = 1 \\
& \sum_{d=1}^{D} \omega_d z_{do} = \alpha \\
& \sum_{d=1}^{D} \omega_d z_{dj} - \sum_{i=1}^{m} \gamma_i x_{ij} \leq 0, \quad j = \quad 1, \ldots, n \\
& \alpha \times \sum_{r=1}^{s} \mu_{r2} y_{rj} - \sum_{d=1}^{D} \omega_d z_{dj} \leq 0, \quad j = 1, \ldots, n \\
& \alpha > 0,\ \gamma_i,\ \omega_d,\ \mu_{r2} > 0, \quad i = 1, \ldots, m,\ r = 1, \ldots, s,\ d = 1, \ldots, D
\end{aligned}
\tag{8.6}
$$

Note the constraints in model (8.6) that $\sum_{i=1}^{m} \gamma_i x_{io} = 1$, $\sum_{d=1}^{D} \omega_d z_{do} \geq \theta_{\min}^{1}$, $\sum_{d=1}^{D} \omega_d z_{do} = \alpha$, and for any $j = 1, \ldots, n$, $\sum_{d=1}^{D} \omega_d z_{dj} - \sum_{i=1}^{m} \gamma_i x_{ij} \leq 0$. Then we have $\theta_{\min}^{1} \leq \alpha = \sum_{d=1}^{D} \omega_d z_{do} \leq \sum_{i=1}^{m} \gamma_i x_{io} = 1$, which provides both upper and lower bounds on α, and indicates that the optimal value of α represents the first-stage efficiency score for each DMU.

Thus α can be treated as a parameter within $[\theta_{\min}^{1}, 1]$. As a result, model (8.6) can be solved as a parametric linear program via searching over the possible α values within $[\theta_{\min}^{1}, 1]$. In computation, we set the initial value for α as the upper bound one, and solve the corresponding linear program. Then we begin to decrease α by a very small positive number ε (=0.0001 for example) for each step t, namely, $\alpha_t = 1 - \varepsilon \times t$, $t = 1, 2, \ldots,$ until the lower bound $\theta_{\min}^{1}$ is reached, and solve each linear program of model (8.6) corresponding to α_t and denote the corresponding optimal objective value by Ω_t. Note that not all values taken by α within $[\theta_{\min}^{1}, 1]$ lead to feasible solutions to program (8.6). Let $\Omega^* = \max_t \Omega_t$ and denote the specific α_t associated with Ω^* as α^*. Note that it is likely that Ω^* is associated with several α^* values.

Then Ω^* associated with α^* is solution to model (8.6). Denote $e_o^{1*} = \alpha^*\left(= \sum_{d=1}^{D} \omega_d^* z_{do}\right)$, $e_o^{2*} = \sum_{r=1}^{s} \mu_{r2}^* y_{ro}$ and $e_o^* = e_o^{1*} \cdot e_o^{2*}$ as DMU_o's bargaining efficiency scores for the first and second stages and the overall process, respectively.

With respect to the four properties associated with a bargaining solution, we have (i) Pareto efficiency (*PE*) indicates that for the bargaining efficiency scores, there is no possibility to improve one stage's individual efficiency score without decreasing the other individual efficiency score; (ii) invariance with respect to affine transformation (*IAT*) reveals that if both the feasible region of bargaining model (8.6) and the breakdown point are subjected to an affine transformation on the payoff space R^2, then the bargaining efficiency scores satisfy the same affine transformation; (iii) independence of irrelevant alternatives (*IIA*) shows that the bargaining efficiency scores will not change when the feasible region of bargaining model (8.6) is decreased but still includes the bargaining solution; and (iv) symmetry (*SYM*) demonstrates that if $\left(S, \vec{b}\right)$ is symmetric, where S is the feasible region of bargaining model (8.6) and $\vec{b}$ is breakdown point, then the bargaining efficiency scores of both individual stages are equal to each other.

8.4 Mathematical Relationship

We finally look at the relationship between the bargaining efficiency scores obtained from model (8.6) and the standard CRS efficiency scores. Let θ_o^1 and θ_o^2 represent the standard (CRS) efficiency scores for the first and second stages, respectively. It will be shown that when there is only one intermediate measure linking the two stages, $e_o^{1*} = \theta_o^1$ and $e_o^{2*} = \theta_o^2$.

Theorem 1 *For any specific* $\mathrm{DMU_o}$, $\Omega_o \leq (\theta_o^1 - \theta_{\min}^1) \cdot (\theta_o^2 - \theta_{\min}^2)$, *where* Ω_o *is the (maximum) optimal value to model (8.6) (or model (8.4)).*

Proof θ_o^1 and θ_o^2 can be obtained by solving the following two regular DEA models (8.7) and (8.8), respectively.

$$
\begin{aligned}
\theta_o^1 = Max\ & \frac{\sum_{d=1}^{D} \hat{w}_d^1 z_{do}}{\sum_{i=1}^{m} \hat{v}_i x_{io}} \\
s.t.\ & \frac{\sum_{d=1}^{D} \hat{w}_d^1 z_{dj}}{\sum_{i=1}^{m} \hat{v}_i x_{ij}} \leq 1,\ j = 1, \ldots, n \\
& \hat{v}_i, \hat{w}_d^1 > 0,\ i = 1, \ldots, m, d = 1, \ldots, D
\end{aligned}
\tag{8.7}
$$

$$
\begin{aligned}
&\theta_o^2 = Max\ \frac{\sum_{r=1}^{s} \hat{u}_r y_{ro}}{\sum_{d=1}^{D} \hat{w}_d^2 z_{do}} \\
&s.t.\ \frac{\sum_{r=1}^{s} \hat{u}_r y_{rj}}{\sum_{d=1}^{D} \hat{w}_d^2 z_{dj}} \le 1,\ j = 1, \ldots, n \\
&\hat{u}_r, \hat{w}_d^2 > 0,\ r = 1, \ldots, s, d = 1, \ldots, D
\end{aligned}
\tag{8.8}
$$

Let v_i^*, w_d^* and u_r^* be an optimal solution to the bargaining model (8.4). By comparing the constraints in models (8.4), (8.7) and (8.8), we note that the feasible regions of model (8.7) and (8.8) both contain the feasible region of model (8.4). Thus, v_i^* and w_d^* are a feasible solution to model (8.7), and w_d^* and u_r^* are a feasible solution to model (8.8). Therefore, we have $\frac{\sum_{d=1}^{D} w_d^* z_{do}}{\sum_{i=1}^{m} v_i^* x_{io}} \le \theta_o^1$ and $\frac{\sum_{r=1}^{s} u_r^* y_{ro}}{\sum_{d=1}^{D} w_d^* z_{do}} \le \theta_o^2$, and furthermore $\Omega_o \le (\theta_o^1 - \theta_{\min}^1) \cdot (\theta_o^2 - \theta_{\min}^2)$.

Q.E.D.

Based upon Theorem 1, under the special case of one intermediate measure, we have

Theorem 2 *If there is only one intermediate measure, then* $\Omega_o = (\theta_o^1 - \theta_{\min}^1) \cdot (\theta_o^2 - \theta_{\min}^2)$.

Proof Under the situation of one intermediate measure where $D = 1$, let $\hat{v}_i^*$ and $\hat{w}_1^{1*}$ be an optimal solution to model (8.7), and $\hat{u}_r^*$ and $\hat{w}_1^{2*}$ be an optimal solution to model (8.8), then we have $\theta_o^1 = \frac{\hat{w}_1^{1*} z_{1o}}{\sum_{i=1}^{m} \hat{v}_i^* x_{io}}$, $\theta_o^2 = \frac{\sum_{r=1}^{s} \hat{u}_r^* y_{ro}}{\hat{w}_1^{2*} z_{1o}}$, and $\left(\frac{\hat{w}_1^{1*}}{\hat{w}_1^{2*}}\right) \hat{u}_r^*$ and $\hat{w}_1^{1*}$ can be another optimal solution to model (8.8). By the definition of $\theta_{\min}^1$ and $\theta_{\min}^2$, we know that $\theta_o^1 \ge \theta_{\min}^1$, and $\theta_o^2 \ge \theta_{\min}^2$. Therefore $\hat{v}_i^*, \hat{w}_1^{1*}$ and $\left(\frac{\hat{w}_1^{1*}}{\hat{w}_1^{2*}}\right) \hat{u}_r^*$ satisfy all the constraints in our bargaining game model (8.4), indicating that $\hat{v}_i^*$, $\hat{w}_1^{1*}$ and $\left(\frac{\hat{w}_1^{1*}}{\hat{w}_1^{2*}}\right) \hat{u}_r^*$ are a feasible solution to model (8.4). Thus, we have $\Omega_o \ge (\theta_o^1 - \theta_{\min}^1) \cdot (\theta_o^2 - \theta_{\min}^2)$.

From Theorem 1, we have $\Omega_o \le (\theta_o^1 - \theta_{\min}^1) \cdot (\theta_o^2 - \theta_{\min}^2)$. Therefore, $\Omega_o = (\theta_o^1 - \theta_{\min}^1) \cdot (\theta_o^2 - \theta_{\min}^2)$.

Q.E.D.

Since the regular (CRS) efficiency scores for the first stage θ_o^1 and for the second stage θ_o^2 are the maximum achievable efficiency scores for individual stages, based upon Theorems 1 and 2, we have

Theorem 3 *If there is only one intermediate measure, then* $e_o^{1*} = \theta_o^1$ *and* $e_o^{2*} = \theta_o^2$*, where* e_o^{1*} *and* e_o^{2*} *represent the bargaining efficiency scores to the first and second stage of any specific* $\mathrm{DMU_o}$ *obtained via model (8.6), respectively.*

Proof In the case of one intermediate measure where $D = 1$, let v_i^*, w_1^* and u_r^* be an optimal solution to the bargaining model (8.4), and then $e_o^{1*} = \frac{w_1^* z_{1o}}{\sum_{i=1}^{m} v_i^* x_{io}} \geq \theta_{\min}^1$ and $e_o^{2*} = \frac{\sum_{r=1}^{s} u_r^* y_{ro}}{w_1^* z_{1o}} \geq \theta_{\min}^2$. From the proof of Theorem 1, we have $e_o^{1*} \leq \theta_o^1$ and $e_o^{2*} \leq \theta_o^2$, and based on Theorem 2, we have $(e_o^{1*} - \theta_{\min}^1) \cdot (e_o^{2*} - \theta_{\min}^2) = \Omega_o$ $=(\theta_o^1 - \theta_{\min}^1) \cdot (\theta_o^2 - \theta_{\min}^2)$. Therefore $e_o^{1*} = \theta_o^1$ and $e_o^{2*} = \theta_o^2$ must be true.

Q.E.D.

Theorem 3 also indicates that a unique pair of (bargaining) efficiency scores for both stages are obtained for each DMU, which is $(\theta_o^1, \ \theta_o^2)$, regardless the choice of breakdown point. i.e., with one intermediate measure, model (8.4) is independent of the breakdown point. However, such independence can no longer hold when multiple intermediate measures are considered, which will be discussed later. Liang et al. (2008) prove the same conclusion with respect to their leader-follower and centralized models. We note, however, that Liang et al. (2008) models are fundamentally different from this bargaining model. To further explain, we present the centralized model in Liang et al. (2008) as follows.

$$
\begin{aligned}
e_o^{centralized} &= Max \sum_{r=1}^{s} \mu_r y_{ro} \\
s.t. \quad & \sum_{r=1}^{s} \mu_r y_{rj} - \sum_{d=1}^{D} \omega_d z_{dj} \leq 0, \quad j = 1, \ldots, n \\
& \sum_{d=1}^{D} \omega_d z_{dj} - \sum_{i=1}^{m} \gamma_i x_{ij} \leq 0, \quad j = 1, \ldots, n \\
& \sum_{i=1}^{m} \gamma_i x_{io} = 1 \\
& \mu_r, \gamma_i, \omega_d \geq \varepsilon, \quad r = 1, \ldots, s, \quad i = 1, \ldots, m, \quad d = 1, \ldots, D
\end{aligned}
\tag{8.9}
$$

It can be seen that the Nash bargaining game model reduces to the centralized model of Liang et al. (2008) when the breakdown point is set equal to (0, 0). Or, when their centralized efficiency scores are used as breakdown points, the Nash bargaining game model cannot improve the breakdown point, namely, the centralized model of Liang et al. (2008) provides a set of "best" overall bargaining efficiency scores. However, this does not necessary imply that the results from the centralized model should be used. The centralized model solution may not be acceptable to the two stages, or ideal with respect to improving the two stages' operations. The bargaining model is not about finding the best overall efficiency score, or the best solution, but rather is about finding the best achievable efficiency through negotiation. A breakdown point (0, 0) only leads to the best overall efficiency score, but not necessarily the best achievable efficiency for Stage 1 or 2. A breakdown point of (0, 0) simply implies that the two stages will get an efficiency score of zero if they do not negotiate. This may further imply that (0, 0) is not a good candidate for a breakdown point in bargaining model.

8.5 Output-Oriented Bargaining Model

The above DEA bargaining model (8.4) is input-oriented. If an output-orientation is taken into account, the bargaining model becomes

$$\begin{aligned}
&Max \left(\frac{\sum_{i=1}^{m} v_i x_{io}}{\sum_{d=1}^{D} w_d z_{do}} - h_{\min}^1 \right) \cdot \left(\frac{\sum_{d=1}^{D} w_d z_{do}}{\sum_{r=1}^{s} u_r y_{ro}} - h_{\min}^2 \right) \\
&s.t. \quad \frac{\sum_{i=1}^{m} v_i x_{io}}{\sum_{d=1}^{D} w_d z_{do}} \le h_{\min}^1 \\
&\qquad \frac{\sum_{d=1}^{D} w_d z_{do}}{\sum_{r=1}^{s} u_r y_{ro}} \le h_{\min}^2 \\
&\qquad \frac{\sum_{i=1}^{m} v_i x_{ij}}{\sum_{d=1}^{D} w_d z_{dj}} \ge 1, \quad j = 1, \ldots, n \\
&\qquad \frac{\sum_{d=1}^{D} w_d z_{dj}}{\sum_{r=1}^{s} u_r y_{rj}} \ge 1, \quad j = 1, \ldots, n \\
&\qquad v_i,\ u_r,\ w_d > 0, \quad i = 1, \ldots, m, \quad r = 1, \ldots, s, \quad d = 1, \ldots, D
\end{aligned} \tag{8.10}$$

where $h^1_{\min}$ and $h^2_{\min}$ represent the output-oriented CRS efficiency scores of the two least ideal DMUs for the first and second stage, respectively. Since $h^1_{\min} = \frac{1}{\theta^1_{\min}}$ and $h^2_{\min} = \frac{1}{\theta^2_{\min}}$, model (8.4) and (8.10) are structurally the same except for the different objective functions.

Let $t_1 = \left(\sum_{d=1}^{D} w_d z_{do}\right)^{-1}$, $t_2 = \left(\sum_{r=1}^{s} u_r y_{ro}\right)^{-1}$, $\gamma_{i1} = t_1\ v_i$, $\gamma_{i2} = t_2\ v_i$, $\omega_d = t_2\ w_d$, $\mu_r = t_2\ u_r$. Note that $\gamma_{i1} = t_1\ v_i$ and $\gamma_{i2} = t_2\ v_i$ imply a linear relationship of $\gamma_{i2} = \frac{t_2}{t_1}\gamma_{i1}$ between γ_{i1} and γ_{i2}. Therefore, we denote $\frac{t_2}{t_1}$ by β (> 0) and have $\gamma_{i2} = \beta\gamma_{i1}$ for all $i = 1, \ldots, m$. Then model (8.10) can be equivalently converted into model (8.11) with parameter β.

$$
\begin{aligned}
&Max\ \beta\sum_{i=1}^{m}\gamma_{i1}x_{io} - h^2_{\min}\sum_{i=1}^{m}\gamma_{i1}x_{io} - h^1_{\min}\sum_{d=1}^{D}\omega_d z_{do} + h^1_{\min}\cdot\ h^2_{\min}\\
&s.t.\quad \sum_{d=1}^{D}\omega_d z_{do} \leq h^2_{\min}\\
&\qquad \sum_{i=1}^{m}\gamma_{i1}x_{io} \leq h^1_{\min}\\
&\qquad \sum_{r=1}^{s}\mu_r y_{ro} = 1\\
&\qquad \sum_{d=1}^{D}\omega_d z_{do} = \beta\\
&\qquad \sum_{d=1}^{D}\omega_d z_{dj} - \sum_{r=1}^{s}\mu_r y_{rj} \geq 0, \quad j = 1, \ldots, n\\
&\qquad \beta\sum_{i=1}^{m}\gamma_{i1}x_{ij} - \sum_{d=1}^{D}\omega_d z_{dj} \geq 0, \quad j = 1, \ldots, n\\
&\qquad \beta > 0,\ \gamma_{i1},\ \gamma_{i2},\ \omega_d,\ \mu_r > 0, \quad i = 1, \ldots, m, \quad r = 1, \ldots, s, \quad d = 1, \ldots, D
\end{aligned}
\tag{8.11}
$$

We have $h^2_{\min} \geq \beta = \sum_{d=1}^{D}\omega_d z_{do} \geq \sum_{r=1}^{s}\mu_r y_{ro} = 1$, which provides both upper and lower bounds on β, and indicates that the optimal value of β represents the second-stage efficiency score for each DMU. Therefore model (8.11) can be solved as a parametric linear program via searching over the possible β values within $[1, h^2_{\min}]$.

We should point out that the DEA bargaining model presented in this chapter is not suitable for situations when one stage is input-oriented and the other is output-oriented. It is because the resulting bargaining model cannot be transformed into a parametric linear program like model (8.6) or (8.11).

8.6 Applications

The Nash bargaining game approach is applied to two real world data sets. The first one consists of 30 top US commercial banks with two intermediate measures, which was used in Seiford and Zhu (1999) first, and then in Liang et al. (2008). The second data set, which was previously studied both in Kao and Hwang (2008) and in Chen et al. (2009), also has two intermediate measures and consists of 24 Taiwanese non-life insurance companies.

8.6.1 Top US Commercial Banks

The data set consisting of 30 top US commercial banks is presented in Table 8.1. The inputs to the first stage are number of employees, assets ($ million) and equity ($ million). The intermediate measures connecting two stages are revenue ($ million) and profit ($ million). The outputs from the second stage are market value ($ million), earning per share ($) and returns to the investors (%). See Seiford and Zhu (1999) for discussion on the above measures.

The CRS efficiency scores for the least ideal DMUs in the first and second stages are calculated as $\theta^1_{\min} = 0.0775$ and $\theta^2_{\min} = 0.0515$, respectively. We next begin with the initial value for α in model (8.6) as one, then decrease α by a small positive number $\varepsilon = 0.0001$ for each step t, namely, $\alpha_t = 1 - 0.0001 \times t,\ t = 1,2, \ldots$, until the lower bound $\theta^1_{\min} = 0.0775$ is reached. Solving the linear program of model (8.6) for each step t corresponding to α_t, we obtain a best heuristic search solution to the bargaining efficiency scores of both individual stages and the overall process, which are reported in columns 2 through 4 in Table 8.2. Column 5 shows the corresponding value of the parameter α when the best heuristic search solution is obtained. In this case, the value for α associated with the optimal solution is unique for each DMU, indicating we have a unique pair of efficiency scores for both individual stages.

For comparison, columns 6 through 8 display the corresponding results from Liang et al. (2008) via the centralized model, which, as indicated above, could be viewed as a special case of Nash bargaining model with breakdown point (0, 0). Note that the efficiency scores of both individual stages and the overall process, obtained through the bargaining game approach, are almost the same with those obtained from Liang et al. (2008)'s centralized model, except for DMU 10. This indicates that bargaining results are very similar to those obtained from the centralized model for this particular data set.

The centralized scores obtained from Liang et al. (2008) represent efficiency pairs under the cooperative game structure that lead to the best overall efficiency scores. Thus, if the centralized efficiency scores are used as breakdown point, model (8.6) cannot further improve the bargaining efficiency scores for the two stages and model (8.6) must yield scores identical to the centralized scores.

Table 8.1 US commercial bank data

Bank	Employees	Assets	Equity	Revenue	Profit	Market value	Earning	Returns
1. Citicorp	85,300	256,853	19,581	31,690	3,464	33,221.7	7.21	66.1
2. BankAmerica Corp.	95,288	232,446	20,222	20,386	2,664	27,148.6	6.49	69.4
3. NationsBank Corp.	58,322	187,298	12,801	16,298	1,950	20,295.9	7.13	59.7
4. Chemical Banking Corp.	39,078	182,926	11,912	14,884	1,805	16,971.3	6.73	70.5
5. J.P. Morgan & Co.	15,600	184,879	10,451	13,838	1,296	15,003.5	6.42	49.4
6. Chase Manhattan Corp.	33,365	121,173	9,134	11,336	1,165	12,616.4	5.76	82.4
7. First Chicago NBD	35,328	122,002	8,450	10,681	1,150	12,351.1	3.45	50
8. First Union Corp.	44,536	131,879.9	9,043.1	10,582.9	1,430.2	16,815	5.04	39.9
9. Banc One Corp.	46,900	90,454	8,197.5	8,970.9	1,277.9	14,807.4	2.91	54.9
10. Bankers Trust New York	14,000	104,000	5,000	8,600	215	5,252.4	2.03	28.3
11. Fleet Financial	30,800	84,432.2	6,364.8	7,919.4	610	10,428.7	1.57	31.8
12. Norwest Corp.	45,404	72,134.4	5,312.1	7,582.3	956	12,268.6	2.76	45.5
13. PNC Bank Corp.	26,757	73,404	5,768	6,389.5	408.1	9,938.2	1.19	61.4
14. KeyCorp	28,905	66,339.1	5,152.5	6,054	825	8,671.2	3.45	51.6
15. Bank of Boston	17,881	47,397	3,751	5,410.6	541	5,310.1	4.55	84.7
16. Wells Fargo & Co.	19,700	50,316	4,055	5,409	1,032	11,342.5	20.37	52.8
17. Bank of New York	15,850	53,685	5,223	5,327	914	10,101.5	4.57	69.9
18. First Interstate Bancorp	27,200	58,071	4,154	4,827.5	885.1	12,138	11.02	108.5
19. Mellon Bank	24,300	40,129	4,106	4,514	691	7,476.7	4.5	83.8
20. Wachovia Corp.	15,996	44,981.3	3,773.8	3,755.4	602.5	7,623.6	3.5	46.9
21. SunTrust Banks	19,415	46,471.5	4,269.6	3,740.3	565.5	7,922.5	4.94	46.9
22. Barnett Banks	20,175	41,553.5	3,272.2	3,680	533.3	5,774.9	5.3	59
23. National City	20,767	36,199	2,921	3,449.9	465.1	4,912.2	3.03	33.9
24. First Bank System	13,231	33,874	2,725	3,328.3	568.1	8,304	4.19	54.3
25. Comerica	13,500	35,469.9	2,607.7	3,112.6	413.4	4,537	3.54	71.7
26. Boatmen's Bancshares	17,023	33,703.8	2,928.1	2,996.1	418.8	4,997	3.25	57.3
27. U.S. Bancorp	14,081	31,794.3	2,617	2,897.3	329	4,865.1	2.09	66.8
28. CoreStates Financial	13,598	29,620.6	2,379.4	2,868	452.2	5,788	3.22	52
29. Republic New York	4,900	43,881.6	3,007.8	2,859.6	288.6	3,218	4.66	41.1
30. MBNA	11,171	13,228.9	1,265.1	2,565.4	353.1	6,543.3	1.54	60.7

Table 8.2 Results for US commercial banks with breakdown point $\{\theta^1_{\min}, \theta^2_{\min}\}$

	Bargaining efficiency scores				Centralized		
Bank	e_o^{1*}	e_o^{2*}	$e_o^{1*} \cdot e_o^{2*}$	α	$e_o^{1,Centralized}$	$e_o^{2,Centralized}$	$e_o^{Centralized}$
1	1.0000	0.4487	0.4487	1.0000	1.0000	0.4487	0.4487
2	0.6821	0.5327	0.3634	0.6821	0.6821	0.5327	0.3634
3	0.7946	0.5305	0.4215	0.7946	0.7946	0.5305	0.4216
4	0.8463	0.5050	0.4274	0.8463	0.8463	0.5050	0.4274
5	1.0000	0.6061	0.6061	1.0000	1.0000	0.6061	0.6061
6	0.8179	0.5111	0.4180	0.8179	0.8180	0.5110	0.4180
7	0.7816	0.5042	0.3941	0.7816	0.7816	0.5042	0.3940
8	0.7451	0.6371	0.4747	0.7451	0.7451	0.6371	0.4747
9	0.7021	0.6389	0.4486	0.7021	0.7021	0.6389	0.4486
10	0.5868	0.5735	0.3365	0.5868	0.4884	0.6946	0.3393
11	0.6619	0.6281	0.4157	0.6619	0.6619	0.6282	0.4158
12	0.6906	0.6576	0.4541	0.6906	0.6906	0.6576	0.4541
13	0.5843	0.7640	0.4464	0.5843	0.5843	0.7641	0.4464
14	0.7131	0.5852	0.4173	0.7131	0.7131	0.5852	0.4173
15	0.8469	0.7582	0.6421	0.8469	0.8469	0.7582	0.6421
16	1.0000	1.0000	1.0000	1.0000	1.0000	1.0000	1.0000
17	1.0000	0.7144	0.7144	1.0000	1.0000	0.7144	0.7144
18	0.7974	1.0000	0.7974	0.7974	0.7974	1.0000	0.7974
19	0.7477	0.7811	0.5840	0.7477	0.7478	0.7811	0.5841
20	0.7541	0.7844	0.5915	0.7541	0.7542	0.7844	0.5916
21	0.6550	0.8660	0.5672	0.6550	0.6550	0.8661	0.5673
22	0.6491	0.8005	0.5196	0.6491	0.6491	0.8005	0.5196
23	0.6280	0.6330	0.3975	0.6280	0.6280	0.6330	0.3975
24	0.8711	0.9478	0.8256	0.8711	0.8711	0.9478	0.8257
25	0.7403	1.0000	0.7403	0.7403	0.7403	1.0000	0.7403
26	0.6344	0.8363	0.5305	0.6344	0.6345	0.8363	0.5306
27	0.6549	1.0000	0.6549	0.6549	0.6549	1.0000	0.6549
28	0.7735	0.8012	0.6197	0.7735	0.7736	0.8011	0.6198
29	0.8092	1.0000	0.8092	0.8092	0.8093	1.0000	0.8093
30	1.0000	1.0000	1.0000	1.0000	1.0000	1.0000	1.0000

Note: The optimal value of parameter α represents the first-stage bargaining efficiency score for the corresponding DMU

In this case, we assume that such a payoff pair or breakdown point is acceptable to the two stages if they do not bargain. Also, any breakdown point other than the centralized efficiency scores will yield a smaller efficiency score for the overall process.

The choice of the breakdown point cannot be arbitrary. For example, it is likely that model (8.4) is infeasible if we use the minimum CRS efficiency score for each stage as the breakdown point. Also, if both breakdown points are greater than the corresponding centralized efficiency scores, model (8.4) will become infeasible. This infeasibility is mainly caused by the fact that some breakdown points will violate the constraints for individual efficiency scores in model (8.4).

Recall that Liang et al. (2008) also develop a non-cooperative leader-follower model where one of the two stages is treated as the leader and is given pre-emptive priority to maximize its efficiency. That is, for example, when the first stage is treated as the leader, the efficiency score for the first stage is calculated CRS score, θ_o^1, because this θ_o^1 is the best efficiency score DMU_o can achieve. Then the efficiency score for the second stage, e_o^2, is maximized given that the first stage's efficiency is fixed at θ_o^1.

Then the breakdown point of ($\min_j \left\{ \theta_j^1 \right\} = 0.6345$, $\min_j \left\{ e_j^2 \right\} = 0.3094$) based upon the leader-follower model of Liang et al. (2008) will ensure that model (8.4) is feasible. Here, 0.6345 is the smallest (CRS) efficiency score for the first stage, and 0.3094 is the smallest leader-follower score for the second stage.

Similarly, based upon the case when the second stage is treated as the leader, another breakdown point of ($\min_j \left\{ e_j^1 \right\} = 0.3056$, $\min_j \left\{ \theta_j^2 \right\} = 0.4859$) can be obtained.

Table 8.3 reports in columns 2 through 4 the bargaining efficiency scores for both individual stages and the overall process corresponding to breakdown point ($\min_j \left\{ \theta_j^1 \right\} = 0.6345$, $\min_j \left\{ e_j^2 \right\} = 0.3094$); columns 5 through 7 report the results corresponding to ($\min_j \left\{ e_j^1 \right\} = 0.3056$, $\min_j \left\{ \theta_j^2 \right\} = 0.4859$).

We note that for DMUs 5, 8, 9, 16, 17, 18, 20, 21, 24, 25, 26, 29, 30, their bargaining efficiency scores remain the same under the three different breakdown points, which also are the centralized efficiency scores. Also bargaining efficiency scores for DMU 26 under the breakdown point ($\min_j \left\{ \theta_j^1 \right\} = 0.6345$, $\min_j \left\{ e_j^2 \right\} = 0.3094$), and scores for DMU 1 under the breakdown point ($\min_j \left\{ e_j^1 \right\} = 0.3056$, $\min_j \left\{ \theta_j^2 \right\} = 0.4859$) are equal to their respective leader-follower (noncooperative) efficiency results. This indicates that under the bargaining model, DMU26 achieves its CRS efficiency score for the first stage, and DMU1 achieves its CRS efficiency score for the second stage.

Model (8.11) is also applied to the banking industry in Table 8.1. $h_{\min}^1$ and $h_{\min}^2$ are calculated as $h_{\min}^1 = \frac{1}{\theta_{\min}^1} = 12.9032$ and $h_{\min}^2 = \frac{1}{\theta_{\min}^2} = 19.4175$. Table 8.4 reports the results from (8.11). To make it comparable with the input-oriented bargaining results, we list the reciprocal of each output-oriented efficiency score, and the input-oriented results are listed in columns 2–4.

Seven DMUs, namely, DMUs 4, 10, 12, 13, 14, 22, and 23, have different efficiency decompositions under input- and output-orientations. This indicates that output-orientation can lead to different bargaining efficiency results from the input-oriented ones.

Table 8.3 Bargaining efficiency scores with breakdown points based upon the leader-follower model of Liang et al. (2008)

	Breakdown point {0.6345, 0.3094}			Breakdown point {0.3056, 0.4859}		
Bank	e_o^{1*}	e_o^{2*}	$e_o^{1*} \cdot e_o^{2*}$	e_o^{1*}	e_o^{2*}	$e_o^{1*} \cdot e_o^{2*}$
1	1.0000	0.4487	0.4487	0.8381	0.4859	0.4072
2	0.6823	0.5324	0.3633	0.6793	0.5331	0.3621
3	0.7946	0.5305	0.4215	0.6858	0.5669	0.3888
4	0.8721	0.4882	0.4258	0.8171	0.5221	0.4266
5	1.0000	0.6061	0.6061	1.0000	0.6061	0.6061
6	0.8180	0.5110	0.4180	0.6898	0.5881	0.4057
7	0.7842	0.5021	0.3937	0.6624	0.5546	0.3674
8	0.7451	0.6371	0.4747	0.7451	0.6371	0.4747
9	0.7022	0.6388	0.4486	0.7021	0.6389	0.4486
10	0.8110	0.4058	0.3291	0.4884	0.6946	0.3392
11	0.7413	0.5164	0.3828	0.5659	0.6955	0.3936
12	0.7089	0.6344	0.4497	0.6684	0.6756	0.4516
13	0.6809	0.6098	0.4152	0.5702	0.7807	0.4452
14	0.7139	0.5843	0.4171	0.6831	0.5938	0.4056
15	0.8478	0.7565	0.6414	0.8469	0.7582	0.6421
16	1.0000	1.0000	1.0000	1.0000	1.0000	1.0000
17	1.0000	0.7144	0.7144	1.0000	0.7144	0.7144
18	0.7974	1.0000	0.7974	0.7974	1.0000	0.7974
19	0.7484	0.7795	0.5834	0.7477	0.7811	0.5840
20	0.7541	0.7844	0.5915	0.7541	0.7844	0.5915
21	0.6550	0.8661	0.5673	0.6550	0.8661	0.5673
22	0.6732	0.7673	0.5165	0.6489	0.8007	0.5196
23	0.6429	0.6130	0.3941	0.6115	0.6479	0.3962
24	0.8711	0.9478	0.8256	0.8711	0.9478	0.8256
25	0.7403	1.0000	0.7403	0.7403	1.0000	0.7403
26	0.6345	0.8363	0.5306	0.6344	0.8363	0.5305
27	0.6573	0.9787	0.6433	0.6549	1.0000	0.6549
28	0.7736	0.8010	0.6197	0.7735	0.8012	0.6197
29	0.8092	1.0000	0.8092	0.8092	1.0000	0.8092
30	1.0000	1.0000	1.0000	1.0000	1.0000	1.0000

Note: The optimal value of parameter α represents the first-stage bargaining efficiency score (e_o^{1*}) for the corresponding DMU, and therefore we do not report α values in this table

8.6.2 Taiwanese Non-life Insurance Companies

Kao and Hwang (2008) describe a two-stage process where 24 non-life insurance companies use operational and insurance expenses to generate premiums in the first stage, and then underwriting and investment profits in the second stage. The inputs to the first stage are operational expenses and insurance expenses, and the outputs

Table 8.4 Output-oriented bargaining results for US commercial banks

	Input-oriented			Output-oriented		
Bank	e_o^{1*}	e_o^{2*}	$e_o^{1*} \cdot e_o^{2*}$	e_o^{1*}	e_o^{2*}	$e_o^{1*} \cdot e_o^{2*}$
1	1.0000	0.4487	0.4487	1.0000	0.4487	0.4487
2	0.6821	0.5327	0.3634	0.6821	0.5327	0.3634
3	0.7946	0.5305	0.4215	0.7946	0.5305	0.4216
4	0.8463	0.5050	0.4274	0.8172	0.5216	0.4262
5	1.0000	0.6061	0.6061	1.0000	0.6061	0.6061
6	0.8179	0.5111	0.4180	0.8180	0.5110	0.4180
7	0.7816	0.5042	0.3941	0.7816	0.5042	0.3940
8	0.7451	0.6371	0.4747	0.7451	0.6371	0.4747
9	0.7021	0.6389	0.4486	0.7022	0.6387	0.4485
10	0.5868	0.5735	0.3365	0.6909	0.4828	0.3336
11	0.6619	0.6281	0.4157	0.6619	0.6282	0.4158
12	0.6906	0.6576	0.4541	0.6996	0.6482	0.4535
13	0.5843	0.7640	0.4464	0.6617	0.6369	0.4214
14	0.7131	0.5852	0.4173	0.7139	0.5843	0.4172
15	0.8469	0.7582	0.6421	0.8469	0.7582	0.6421
16	1.0000	1.0000	1.0000	1.0000	1.0000	1.0000
17	1.0000	0.7144	0.7144	1.0000	0.7144	0.7144
18	0.7974	1.0000	0.7974	0.7974	1.0000	0.7974
19	0.7477	0.7811	0.5840	0.7477	0.7810	0.5840
20	0.7541	0.7844	0.5915	0.7542	0.7844	0.5916
21	0.6550	0.8660	0.5672	0.6550	0.8661	0.5673
22	0.6491	0.8005	0.5196	0.6732	0.7673	0.5166
23	0.6280	0.6330	0.3975	0.6430	0.6130	0.3941
24	0.8711	0.9478	0.8256	0.8711	0.9478	0.8257
25	0.7403	1.0000	0.7403	0.7403	1.0000	0.7403
26	0.6344	0.8363	0.5305	0.6345	0.8363	0.5306
27	0.6549	1.0000	0.6549	0.6549	1.0000	0.6549
28	0.7735	0.8012	0.6197	0.7736	0.8011	0.6198
29	0.8092	1.0000	0.8092	0.8093	1.0000	0.8093
30	1.0000	1.0000	1.0000	1.0000	1.0000	1.0000

from the second stage are underwriting profit and investment profit. Direct written premiums and reinsurance premiums act as the intermediate measures connecting the two stages.

Table 8.5 shows the original data, and Table 8.6 reports the efficiency results obtained from both noncooperative (leader-follower) model and centralized model developed by Liang et al. (2008).

The same three breakdown points are considered in the bargaining game approach as in the previous bank application. First of all, from models (8.7) and (8.8), we get the CRS efficiency scores for stage 1's and stage 2's least ideal DMU as $\theta^1_{\min} = 0.001725$ and $\theta^1_{\min} = 0.001058$, respectively. Also the smallest leader-follower efficiency scores when either stage acts as the leader are calculated

Table 8.5 Taiwanese non-life insurance company data

Company	Operation expenses	Insurance expenses	Direct written premiums	Reinsurance premiums	Underwriting profit	Investment profit
1. Taiwan Fire	1,178,744	673,512	7,451,757	856,735	984,143	681,687
2. Chung Kuo	1,381,822	1,352,755	10,020,274	1,812,894	1,228,502	834,754
3. Tai Ping	1,177,494	592,790	4,776,548	560,244	293,613	658,428
4. China Mariners	601,320	594,259	3,174,851	371,863	248,709	177,331
5. Fubon	6,699,063	3,531,614	37,392,862	1,753,794	7,851,229	3,925,272
6. Zurich	2,627,707	668,363	9,747,908	952,326	1,713,598	415,058
7. Taian	1,942,833	1,443,100	10,685,457	643,412	2,239,593	439,039
8. Ming Tai	3,789,001	1,873,530	17,267,266	1,134,600	3,899,530	622,868
9. Central	1,567,746	950,432	11,473,162	546,337	1,043,778	264,098
10. The First	1,303,249	1,298,470	8,210,389	504,528	1,697,941	554,806
11. Kuo Hua	1,962,448	672,414	7,222,378	643,178	1,486,014	18,259
12. Union	2,592,790	650,952	9,434,406	1,118,489	1,574,191	909,295
13. Shingkong	2,609,941	1,368,802	13,921,464	811,343	3,609,236	223,047
14. South China	1,396,002	988,888	7,396,396	465,509	1,401,200	332,283
15. Cathay Century	2,184,944	651,063	10,422,297	749,893	3,355,197	555,482
16. Allianz President	1,211,716	415,071	5,606,013	402,881	854,054	197,947
17. Newa	1,453,797	1,085,019	7,695,461	342,489	3,144,484	371,984
18. AIU	757,515	547,997	3,631,484	995,620	692,731	163,927
19. North America	159,422	182,338	1,141,950	483,291	519,121	46,857
20. Federal	145,442	53,518	316,829	131,920	355,624	26,537
21. Royal & Sunalliance	84,171	26,224	225,888	40,542	51,950	6,491
22. Asia	15,993	10,502	52,063	14,574	82,141	4,181
23. AXA	54,693	28,408	245,910	49,864	0.1	18,980
24. Mitsui Sumitomo	163,297	235,094	476,419	644,816	142,370	16,976

according to the results from Table 8.6, which are ($\min_j\left\{\theta_j^1\right\} = 0.5895$, $\min_j\left\{e_j^2\right\} = 0.0870$) and ($\min_j\left\{e_j^1\right\} = 0.2507$, $\min_j\left\{\theta_j^2\right\} = 0.2795$).

Table 8.7 reports the bargaining results for both individual stages and the overall process associated with breakdown point $\{\theta_{\min}^1, \theta_{\min}^2\}$, ($\min_j\left\{\theta_j^1\right\} = 0.5895$, $\min_j\left\{e_j^2\right\} = 0.0870$) and ($\min_j\left\{e_j^1\right\} = 0.2507$, $\min_j\left\{\theta_j^2\right\} = 0.2795$) in columns 2 through 4, columns 5 through 7, and columns 8 through 10, respectively.

Table 8.6 Efficiency results for Taiwanese non-life insurance companies

	Stage 1 as the leader			Stage 2 as the leader			Centralized		
DMU	θ_j^1	e_j^2	$\theta_j^1 \cdot e_j^2$	e_j^1	θ_j^2	$e_j^1 \cdot \theta_j^2$	$e_j^{1,Centralized}$	$e_j^{2,Centralized}$	$e_j^{Centralized}$
1	0.9926	0.7045	0.6993	0.9260	0.7134	0.6606	0.9926	0.7045	0.6993
2	0.9985	0.6257	0.6248	0.9908	0.6275	0.6217	0.9985	0.6257	0.6248
3	0.6900	1.0000	0.6900	0.6900	1.0000	0.6900	0.6900	1.0000	0.6900
4	0.7243	0.4200	0.3042	0.4981	0.4323	0.2153	0.7243	0.4200	0.3042
5	0.8375	0.8060	0.6750	0.7376	1.0000	0.7376	0.8306	0.9234	0.7670
6	0.9637	0.4010	0.3864	0.9606	0.4057	0.3897	0.9606	0.4057	0.3897
7	0.7521	0.3522	0.2649	0.3000	0.5378	0.1613	0.6706	0.4124	0.2766
8	0.7256	0.3780	0.2743	0.3898	0.5114	0.1993	0.6631	0.4150	0.2752
9	1.0000	0.2233	0.2233	0.4388	0.2920	0.1281	1.0000	0.2233	0.2233
10	0.8615	0.5409	0.4660	0.2587	0.6736	0.1743	0.8615	0.5409	0.4660
11	0.7405	0.1677	0.1242	0.4718	0.3267	0.1541	0.6468	0.2534	0.1639
12	1.0000	0.7596	0.7596	1.0000	0.7596	0.7596	1.0000	0.7596	0.7596
13	0.8107	0.2431	0.1970	0.3384	0.5435	0.1839	0.6719	0.3093	0.2078
14	0.7246	0.3740	0.2710	0.3097	0.5178	0.1604	0.6699	0.4309	0.2887
15	1.0000	0.6138	0.6138	0.7102	0.7047	0.5005	1.0000	0.6138	0.6138
16	0.9072	0.3356	0.3045	0.5980	0.3848	0.2301	0.8856	0.3615	0.3201
17	0.7233	0.4557	0.3296	0.2507	1.0000	0.2507	0.6276	0.5736	0.3600
18	0.7935	0.3262	0.2588	0.6549	0.3737	0.2447	0.7935	0.3262	0.2588
19	1.0000	0.4112	0.4112	0.9787	0.4158	0.4069	1.0000	0.4112	0.4112
20	0.9332	0.5857	0.5466	0.4073	0.9014	0.3671	0.9332	0.5857	0.5466
21	0.7505	0.2623	0.1969	0.6918	0.2795	0.1934	0.7321	0.2743	0.2008
22	0.5895	1.0000	0.5895	0.5895	1.0000	0.5895	0.5895	1.0000	0.5895
23	0.8501	0.4509	0.3833	0.6812	0.5599	0.3814	0.8425	0.4989	0.4203
24	1.0000	0.0870	0.0870	0.3987	0.3351	0.1336	0.4287	0.3145	0.1348

Note that as with the previous bank data, in this application, the value for parameter α associated with the optimal solution is unique for each DMU throughout the entire searching range, which also leads to a unique pair of efficiency scores for both individual stages.

It can be seen from Table 8.7 that with breakdown point $\{\theta_{\min}^1, \theta_{\min}^2\}$, the bargaining efficiency results are exactly the same as those obtained from Liang et al. (2008)'s centralized model. Also from Table 8.7, note that for some DMUs, such as DMUs 1, 2, 3, 4, 5, 6, 10, 12, 15, 22, 23, their respective bargaining efficiency results remain unchanged under all three breakdown points, while for the rest DMUs, such as DMUs 7, 8, 9, 11, 13, 14, 16, 17, 18, 19, 20, 21, 24, their respective bargaining efficiency scores are varied according to different breakdown points.

Table 8.7 Bargaining efficiency scores with three breakdown points

	$\{\theta^1_{\min}, \theta^2_{\min}\}$			{0.5895, 0.0870}			{0.2507, 0.2795}		
DMU	e_o^{1*}	e_o^{2*}	$e_o^{1*} \cdot e_o^{2*}$	e_o^{1*}	e_o^{2*}	$e_o^{1*} \cdot e_o^{2*}$	e_o^{1*}	e_o^{2*}	$e_o^{1*} \cdot e_o^{2*}$
1	0.9926	0.7045	0.6993	0.9926	0.7045	0.6993	0.9926	0.7045	0.6993
2	0.9985	0.6257	0.6248	0.9985	0.6257	0.6248	0.9985	0.6257	0.6248
3	0.6900	1.0000	0.6900	0.6900	1.0000	0.6900	0.6900	1.0000	0.6900
4	0.7243	0.4200	0.3042	0.7243	0.4200	0.3042	0.7243	0.4200	0.3042
5	0.8306	0.9234	0.7670	0.8306	0.9234	0.7670	0.8306	0.9234	0.7670
6	0.9606	0.4057	0.3897	0.9606	0.4057	0.3897	0.9606	0.4057	0.3897
7	0.6706	0.4124	0.2766	0.7521	0.3522	0.2649	0.6200	0.4317	0.2677
8	0.6631	0.4150	0.2752	0.7256	0.3780	0.2743	0.6630	0.4150	0.2751
9	1.0000	0.2233	0.2233	1.0000	0.2233	0.2233	0.4390	0.2920	0.1282
10	0.8615	0.5409	0.4660	0.8615	0.5409	0.4660	0.8615	0.5409	0.4660
11	0.6468	0.2534	0.1639	0.7292	0.2066	0.1507	0.4718	0.3267	0.1541
12	1.0000	0.7596	0.7596	1.0000	0.7596	0.7596	1.0000	0.7596	0.7596
13	0.6719	0.3093	0.2078	0.8107	0.2431	0.1971	0.4600	0.4344	0.1998
14	0.6699	0.4309	0.2887	0.7246	0.3740	0.2710	0.6699	0.4309	0.2887
15	1.0000	0.6138	0.6138	1.0000	0.6138	0.6138	1.0000	0.6138	0.6138
16	0.8856	0.3615	0.3201	0.8856	0.3615	0.3201	0.8687	0.3651	0.3172
17	0.6276	0.5736	0.3600	0.7231	0.4598	0.3325	0.6276	0.5736	0.3600
18	0.7935	0.3262	0.2588	0.7935	0.3262	0.2589	0.6551	0.3737	0.2448
19	1.0000	0.4112	0.4112	1.0000	0.4112	0.4112	0.9788	0.4158	0.4070
20	0.9332	0.5857	0.5466	0.9332	0.5857	0.5466	0.8159	0.6561	0.5353
21	0.7321	0.2743	0.2008	0.7505	0.2623	0.1969	0.6918	0.2795	0.1934
22	0.5895	1.0000	0.5895	0.5895	1.0000	0.5895	0.5895	1.0000	0.5895
23	0.8425	0.4989	0.4203	0.8425	0.4989	0.4203	0.8425	0.4989	0.4203
24	0.4287	0.3145	0.1348	0.7752	0.1390	0.1078	0.3987	0.3351	0.1336

Note: The optimal value of parameter α represents the first-stage bargaining efficiency score (e_o^{1*}) for the corresponding DMU, and therefore we do not report α values in this table

8.7 Conclusions

This chapter introduces the Nash bargaining game model as a way of addressing the conflict arising from intermediate measures, and presents an alternative approach to evaluate the efficiency scores for both stages and the overall process. Furthermore, it is proved that in the case of only one intermediate measure, the bargaining game approach yields the same efficiency results as obtained from the separately-applied standard DEA approach, and also with the non-cooperative and centralized approaches in Liang et al. (2008).

Different breakdown points can be used to calculate the bargaining efficiency scores. As a matter of fact, each DMU can use a specific breakdown point. For example, based upon the leader-follower model of Liang et al. (2008), both $\left(\min_j\left\{\theta_j^1\right\}, \min_j\left\{e_j^2\right\}\right)$ and $\left(\min_j\left\{e_j^1\right\}, \min_j\left\{\theta_j^2\right\}\right)$ can be used as breakdown points, where θ_j^1 and e_j^2 respectively represent the efficiency scores for stages 1 and 2 of DMU_j when Stage 1 is treated as the leader, whereas e_j^1 and θ_j^2 respectively represent the efficiency scores for stage 1 and 2 when Stage 2 takes the leader's role.

References

Binmore, K., Rubinstein, A., & Wolinsky, A. (1986). The Nash bargaining solution in economic modeling. *The RAND Journal of Economics, 17*(2), 176–188.

Charnes, A., Cooper, W. W., & Rhodes, E. (1978). Measuring the efficiency of decision making units. *European Journal of Operational Research, 2*, 429–444.

Chen, Y., Cook, W. D., Li, N., & Zhu, J. (2009). Additive efficiency decomposition in two-stage DEA. *European Journal of Operational Research, 196*, 1170–1176.

Du, J., Liang, L., Chen, Y., Cook, W. D., & Zhu, J. (2011). A bargaining game model for measuring performance of two-stage network structures. *European Journal of Operational Research, 210*, 390–397.

Kao, C., & Hwang, S. N. (2008). Efficiency decomposition in two-stage data envelopment analysis: An application to non-life insurance companies in Taiwan. *European Journal of Operational Research, 185*(1), 418–429.

Liang, L., Cook, W. D., & Zhu, J. (2008). DEA models for two-stage processes: Game approach and efficiency decomposition. *Naval Research Logistics, 55*, 643–653.

Lundberg, S., & Pollak, R. A. (1993). Separate spheres bargaining and the marriage market. *Journal of Political Economy, 101*(6), 988–1010.

Nash, J. F. (1950). The bargaining problem. *Econometrica, 18*(2), 155–162.

Nash, J. F. (1953). Two-person cooperative games. *Econometrica, 21*, 128–140.

Seiford, L. M., & Zhu, J. (1999). Profitability and marketability of the top 55 US commercial banks. *Management Science, 45*(9), 1270–1288.

Chapter 9
Shared Resources and Efficiency Decomposition in Two-Stage Networks

Yao Chen, Juan Du, H. David Sherman, and Joe Zhu

Abstract In many real world scenarios, decision making units (DMUs) may have a two-stage structure with input resources shared by both stages of operations. The distinguishing characteristic is that some of the inputs to the first stage are also consumed by the second stage, and some of the shared inputs cannot be conveniently split up and allocated to operations of the two stages. Recognizing this distinction is critical for these types of DEA applications because measuring the efficiency of the production for first-stage outputs can be misleading and understate the efficiency if DEA fails to consider that some of the inputs generate other second-stage outputs. This chapter presents DEA models for measuring the performance of two-stage network processes with non-splittable shared inputs.

Part of this chapter is based upon Y. Chen, J. Du, H.D. Sherman, and J. Zhu, 2010, "DEA model with shared resources and efficiency decomposition", *European Journal of Operational Research*, Vol. 207, 339–349, with permissions from Elsevier Science.

Y. Chen
Manning School of Business, University of Massachusetts at Lowell,
Lowell, MA 01845, USA
e-mail: Yao_Chen@uml.edu

J. Du (✉)
School of Economics and Management, Tongji University, 1239 Siping Road,
Shanghai 200092, People's Republic of China
e-mail: dujuan@tongji.edu.cn

H.D. Sherman
D'Amore-McKim School of Business, Northeastern University,
422 Hayden Hall, Boston, MA 02115-5000, USA
e-mail: h.sherman@neu.edu

J. Zhu
School of Business, Worcester Polytechnic Institute, Worcester, MA 01609, USA
e-mail: jzhu@wpi.edu

W.D. Cook and J. Zhu (eds.), *Data Envelopment Analysis*,
International Series in Operations Research & Management Science 208,
DOI 10.1007/978-1-4899-8068-7_9, © Springer Science+Business Media New York 2014

Keywords Efficiency • Intermediate measures • Shared resources • Two-stage network

9.1 Introduction

Recently, a number of studies have looked at DMUs that have a two-stage structure where besides inputs and outputs, there are a set of intermediate measures that exist in-between the two stages. These intermediate measures are the outputs from the first stage that become the only inputs to the second stage. In many real world settings, inputs to the first stage are actually shared by both stages, and some of the shared inputs cannot be split up and allocated to the two stage operations. For example, consider the profitability and marketability stages in Seiford and Zhu (1999). The first stage of profitability is measured using labor and assets as inputs and profits and revenue as outputs. In the second stage of marketability, the profits and revenue are then used as inputs, while market value, returns and earnings per share are used as outputs. Note, however, that in this example, labor and assets are actually shared inputs for both stages, i.e., both stages will use the labor and assets of the bank, and many of these inputs cannot be separated into the elements that are directly associated with generating profits and revenues versus the resources that augment the investor estimate of the bank market value.

Hospitals represent another clear example of this phenomenon as is suggested by Fig. 9.1. A common set of inputs including plant, equipment, administrative staff, information technology, is used to provide stage 1 outputs such as medical records, laundry, housekeeping, lab tests, and radiology treatments. Some hospitals have more than 14 of these types of stage 1 activities, and these draw on a common group of resource inputs. For example, medical records generates patient care tracking and billing information using resources such as information technology, plant, equipment, and administrative oversight. Laundry outputs include providing clean linens and towels for patients as well as lab coats and operating room uniforms and use hospital resources including plant, equipment, administrative and utilities.

These stage 1 activities provide services to patients in tracking medical records to manage care as well as accumulate costs for billing. Housekeeping maintains the rooms for patients along with other parts of the hospital. For patient care, in addition to inputs from the stage one activities, other resources are used directly such as the plant and equipment in patient rooms, administrative oversight of the inpatient activities, and management time overseeing the nursing staff. These are just a few examples of the similar activities in a hospital that generate stage 1 and stage 2 outputs. Analyzing the efficiency of stage 1 activities such as laundry without recognizing that many of the personnel, plant and other resources are also used to provide stage 2 outputs would not correctly characterize the efficiency of the laundry.

Shared inputs have been studied in DEA literature. For example, Cook and Hababou (2001), and Cook et al. (2000) develop models that accommodate both dedicated and shared inputs. Note that their models do not have to specify the sharing proportions of shared inputs. This is a desirable feature because in

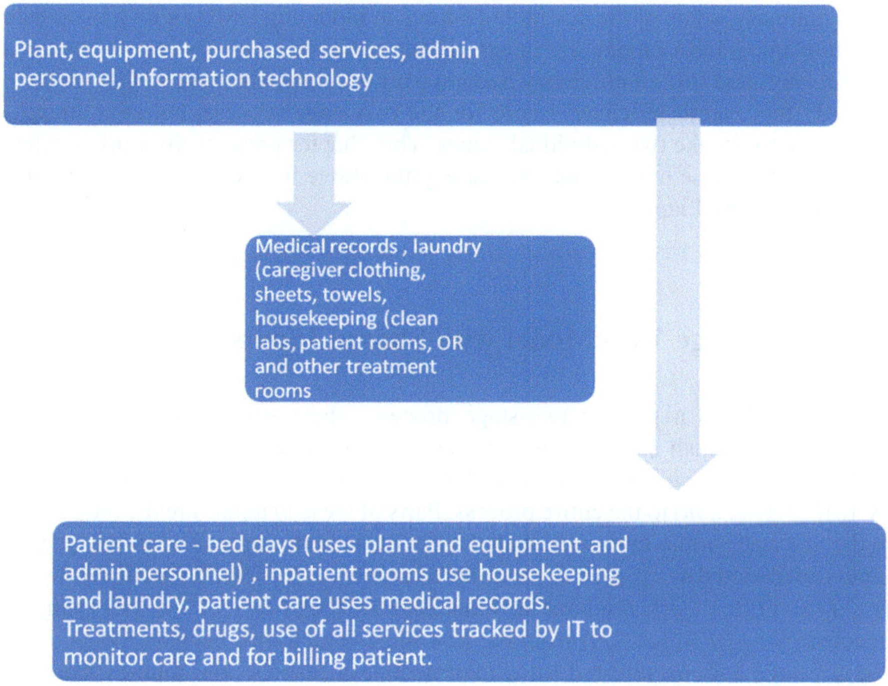

Fig. 9.1 Hospital example of shared resources

reality those shared inputs cannot be conveniently split up and allocated to the outputs. However, their DMUs do not have the afore-mentioned two-stage structure.

Although the models developed in Seiford and Zhu (1999), Zhu (2000), and Sexton and Lewis (2003) consider the two-stage structures, these studies utilize the standard DEA models without modeling the issue of shared inputs. As demonstrated in Chen and Zhu (2004), standard DEA models are not a good choice for measuring the performance of two-stage processes, because they do not address the potential conflict arising from the dual role of intermediate measures. For example, the second stage may have to reduce its inputs (intermediate measures) in order to achieve an efficient status. Such an action would, however, imply a reduction in the first-stage outputs, thereby reducing the efficiency of that stage. Some studies such as Kao and Hwang (2008) and Liang et al. (2008) correctly address this conflict, and explicitly provide an efficiency decomposition of the overall efficiency into both individual stages. However, shared inputs are not modeled. If we model the shared inputs as in Cook and Hababou (2001) or Cook et al. (2000), the models proposed by Kao and Hwang (2008) and Liang et al. (2008) will become highly non-linear and a global optimal solution cannot be guaranteed.

In summary, some of the desirable features in modeling DMUs with a two-stage structure are (i) DEA models are linear; (ii) intermediate measures are modeled in a correct way; and (iii) an efficiency decomposition can be obtained so that we have not only the overall efficiency score for the two-stage process, but also the efficiency scores for the two individual stages. This chapter presents the work in Chen et al. (2010), whose models not only satisfy the above important features, but also address the shared inputs.

9.2 Two-Stage VRS Model with Shared Inputs

Figure 9.2 shows a generic two-stage process where some inputs are directly associated with both stages. Suppose that there are a set of n DMUs denoted by DMU_j ($j = 1, \ldots, n$) and that each DMU_j ($j = 1, \ldots, n$) has m inputs denoted by x_{ij} ($i = 1, \ldots, m$) to the entire process. Parts of these m inputs are the only inputs to the first stage while others are shared as inputs in both stages. These two types of inputs are denoted as x_{i_1j} $(i_1 \in I_1)$ and shared inputs x_{i_2j} $(i_2 \in I_2)$, respectively, where $I_1 \cup I_2 = \{1,2, \ldots,m\}$. Suppose also that each DMU_j ($j = 1, \ldots, n$) has t outputs denoted by z_{dj} ($d = 1, \ldots, t$) from the first stage, which then become inputs to the second stage and are referred to as intermediate measures. The outputs from the second stage are denoted by y_{rj} ($r = 1, \ldots s$).

Since inputs $i_2 \in I_2$ are shared by both stages, we assume that all x_{i_2j} $(i_2 \in I_2)$ are divided into $\alpha_{i_2j}x_{i_2j}$ and $(1 - \alpha_{i_2j})x_{i_2j}$ $(0 \le \alpha_{i_2j} \le 1)$, corresponding to the portions of shared inputs used by the first and second stage, respectively. Similar to the constraints in Cook and Hababou (2001), all α_{i_2j} $(i_2 \in I_2,\ j = 1, \ldots, n)$ will be required to be within certain intervals, namely $L^1_{i_2j} \le \alpha_{i_2j} \le L^2_{i_2j}$.

Based upon the variable returns to scale (VRS) model of Banker et al. (1984), the VRS efficiency scores for DMU_o in the first and second stages are calculated respectively by

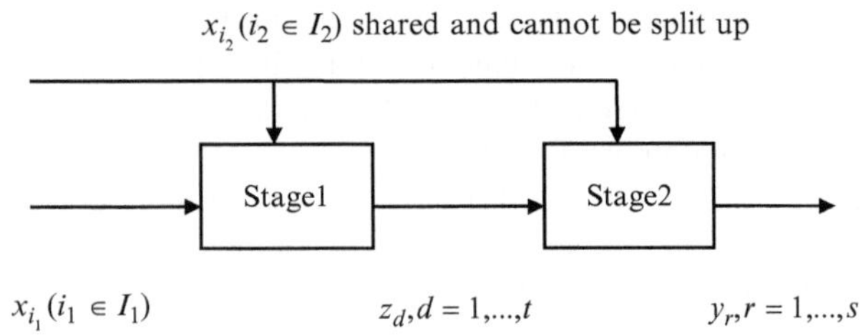

Fig. 9.2 Two-stage process with shared non-separable inputs

$$\begin{aligned}
&Max \quad \frac{\sum_{d=1}^{t} \eta_d^1 z_{do} + u^A}{\sum_{i_1 \in I_1} v_{i_1} x_{i_1 o} + \sum_{i_2 \in I_2} v_{i_2}^1 \alpha_{i_2 o} x_{i_2 o}} \\
&s.t. \quad \frac{\sum_{d=1}^{t} \eta_d^1 z_{dj} + u^A}{\sum_{i_1 \in I_1} v_{i_1} x_{i_1 j} + \sum_{i_2 \in I_2} v_{i_2}^1 \alpha_{i_2 j} x_{i_2 j}} \le 1, \quad j = 1, \ldots, n \\
&\qquad L_{i_2 j}^1 \le \alpha_{i_2 j} \le L_{i_2 j}^2, \quad i_2 \in I_2, \quad j = 1, \ldots, n \\
&\qquad \eta_d^1, \; v_{i_1}, v_{i_2}^1 \ge \varepsilon, \quad d = 1, \ldots, t, \quad i_1 \in I_1, \quad i_2 \in I_2; \quad u^A free
\end{aligned} \tag{9.1}$$

$$\begin{aligned}
&Max \quad \frac{\sum_{r=1}^{s} u_r y_{ro} + u^B}{\sum_{i_2 \in I_2} v_{i_2}^2 (1 - \alpha_{i_2 o}) x_{i_2 o} + \sum_{d=1}^{t} \eta_d^2 z_{do}} \\
&s.t. \quad \frac{\sum_{r=1}^{s} u_r y_{rj} + u^B}{\sum_{i_2 \in I_2} v_{i_2}^2 \left(1 - \alpha_{i_2 j}\right) x_{i_2 j} + \sum_{d=1}^{t} \eta_d^2 z_{dj}} \le 1, \quad j = 1, \ldots, n \\
&\qquad L_{i_2 j}^1 \le \alpha_{i_2 j} \le L_{i_2 j}^2, \quad i_2 \in I_2, \quad j = 1, \ldots, n \\
&\qquad u_r, \eta_d^2, v_{i_2}^2 \ge \varepsilon, \quad r = 1, \ldots, s, \quad d = 1, \ldots, t, \quad i_2 \in I_2; \quad u^B free
\end{aligned} \tag{9.2}$$

As pointed out in a number of studies (e.g., Chen and Zhu (2004), applying models (9.1) and (9.2) separately does not correctly model the intermediate measures z_{dj} ($d = 1, \ldots, t$). Because model (9.2) tries to reduce z_{dj} ($d = 1, \ldots, t$), which is assumed to be kept at its current level in model (9.1). An alternative approach to measuring the efficiency of the two-stage process is to view them from a centralized perspective, and determine a set of optimal weights on the intermediate measures that maximize the aggregate or global efficiency score, as would be true where the manufacturer and retailer jointly determine the price, order quantity, etc. to achieve maximum profit (Huang and Li 2001).

Therefore, similar to Kao and Hwang's (2008) assumption and the centralized model in Liang et al. (2008), we assume that $\eta_d^1 = \eta_d^2 = \eta_d$ for all $d = 1, \ldots, t$ in models (9.1) and (9.2). We also assume that $v_{i_2}^1 = v_{i_2}^2 = v_{i_2}$ for all $i_2 \in I_2$ because these are the same types of inputs.

We propose to combine the two stages in a weighted average of efficiency scores of stages 1 and 2 as follows

$$w_1 \cdot \frac{\sum_{d=1}^{t} \eta_d z_{do} + u^A}{\sum_{i_1 \in I_1} v_{i_1} x_{i_1 o} + \sum_{i_2 \in I_2} v_{i_2} \alpha_{i_2 o} x_{i_2 o}} + w_2 \cdot \frac{\sum_{r=1}^{s} u_r y_{ro} + u^B}{\sum_{i_2 \in I_2} v_{i_2} (1 - \alpha_{i_2 o}) x_{i_2 o} + \sum_{d=1}^{t} \eta_d z_{do}} \tag{9.3}$$

where w_1 and w_2 are user-specified weights such that $w_1 + w_2 = 1$.

Since w_1 and w_2 are intended to represent the relative importance or contribution of the performances of the first and second stage, respectively, to the overall performance of the DMU in the whole process, one reasonable choice of each weight is the proportion of total resources devoted to each stage, reflecting the relative size of a stage. To be more specifically, we define

$$w_1 = \frac{\sum_{i_1 \in I_1} v_{i_1} x_{i_1 o} + \sum_{i_2 \in I_2} v_{i_2} \alpha_{i_2 o} x_{i_2 o}}{\sum_{i_1 \in I_1} v_{i_1} x_{i_1 o} + \sum_{i_2 \in I_2} v_{i_2} x_{i_2 o} + \sum_{d=1}^{t} \eta_d z_{do}} \quad \text{and}$$
$$w_2 = \frac{\sum_{i_2 \in I_2} v_{i_2} (1 - \alpha_{i_2 o}) x_{i_2 o} + \sum_{d=1}^{t} \eta_d z_{do}}{\sum_{i_1 \in I_1} v_{i_1} x_{i_1 o} + \sum_{i_2 \in I_2} v_{i_2} x_{i_2 o} + \sum_{d=1}^{t} \eta_d z_{do}} \tag{9.4}$$

where $\sum_{i_1 \in I_1} v_{i_1} x_{i_1 o} + \sum_{i_2 \in I_2} v_{i_2} x_{i_2 o} + \sum_{d=1}^{t} \eta_d z_{do}$ represents the total size of or total amount of resources consumed by the whole two-stage process, while $\sum_{i_1 \in I_1} v_{i_1} x_{i_1 o} + \sum_{i_2 \in I_2} v_{i_2} \alpha_{i_2 o} x_{i_2 o}$ and $\sum_{i_2 \in I_2} v_{i_2} (1 - \alpha_{i_2 o}) x_{i_2 o} + \sum_{d=1}^{t} \eta_d z_{do}$ represent the sizes of the first and second stages, respectively. These weights themselves are not optimization variables, but rather are functions of the optimization variables.

Thus, under VRS, the overall efficiency score of the two-stage process for DMU_o can be evaluated by solving the fractional program (9.5).

$$
\begin{aligned}
\theta_o^* = Max \quad & \frac{\sum_{d=1}^{t} \eta_d z_{do} + u^A + \sum_{r=1}^{s} u_r y_{ro} + u^B}{\sum_{i_1 \in I_1} v_{i_1} x_{i_1 o} + \sum_{i_2 \in I_2} v_{i_2} x_{i_2 o} + \sum_{d=1}^{t} \eta_d z_{do}} \\
s.t. \quad & \frac{\sum_{d=1}^{t} \eta_d z_{dj} + u^A}{\sum_{i_1 \in I_1} v_{i_1} x_{i_1 j} + \sum_{i_2 \in I_2} v_{i_2} \alpha_{i_2 j} x_{i_2 j}} \le 1, \quad j = 1, \ldots, n \\
& \frac{\sum_{r=1}^{s} u_r y_{rj} + u^B}{\sum_{i_2 \in I_2} v_{i_2} \left(1 - \alpha_{i_2 j}\right) x_{i_2 j} + \sum_{d=1}^{t} \eta_d z_{dj}} \le 1, \quad j = 1, \ldots, n \\
& L^1_{i_2 j} \le \alpha_{i_2 j} \le L^2_{i_2 j}, \quad i_2 \in I_2, \ j = 1, \ldots, n \\
& u_r, \eta_d, v_{i_1}, v_{i_2} \ge \varepsilon, \quad r = 1, \ldots, s, \ d = 1, \ldots, t, \ i_1 \in I_1, i_2 \in I_2 \\
& u^A, u^B \ free
\end{aligned}
\tag{9.5}
$$

By applying the Charnes-Cooper transformation, fractional program (9.5) can be converted to model (9.6).

$$
\begin{aligned}
\theta_o^* = Max \ & \sum_{d=1}^{t} \pi_d z_{do} + \sum_{r=1}^{s} \mu_r y_{ro} + u^1 + u^2 \\
s.t. \ & \sum_{d=1}^{t} \pi_d z_{dj} + u^1 - \left(\sum_{i_1 \in I_1} \omega_{i_1} x_{i_1 j} + \sum_{i_2 \in I_2} \omega_{i_2} \alpha_{i_2 j} x_{i_2 j} \right) \le 0, \quad j = 1, \ldots, n \\
& \sum_{r=1}^{s} \mu_r y_{rj} + u^2 - \left[\sum_{i_2 \in I_2} \omega_{i_2} \left(1 - \alpha_{i_2 j}\right) x_{i_2 j} + \sum_{d=1}^{t} \pi_d z_{dj} \right] \le 0, \quad j = 1, \ldots, n \\
& \sum_{i_1 \in I_1} \omega_{i_1} x_{i_1 o} + \sum_{i_2 \in I_2} \omega_{i_2} x_{i_2 o} + \sum_{d=1}^{t} \pi_d z_{do} = 1 \\
& L^1_{i_2 j} \le \alpha_{i_2 j} \le L^2_{i_2 j}, \quad i_2 \in I_2, \ j = 1, \ldots, n \\
& \mu_r, \pi_d, \omega_{i_1}, \ \omega_{i_2} \ge \varepsilon, \quad r = 1, \ldots, s, \ d = 1, \ldots, t, \ i_1 \in I_1, \ i_2 \in I_2 \\
& u^1, u^2 \ free
\end{aligned}
\tag{9.6}
$$

Model (9.6) is still non-linear since there exist the non-linear item $\sum_{i_2 \in I_2} \omega_{i_2} \alpha_{i_2 j} x_{i_2 j}$ in some constraints. By letting $\beta_{i_2 j} = \omega_{i_2} \alpha_{i_2 j}$ $(j = 1, \ldots, n)$, model (9.6) is linearized as

$$
\begin{aligned}
\theta_o^* = Max & \sum_{d=1}^{t} \pi_d z_{do} + \sum_{r=1}^{s} \mu_r y_{ro} + u^1 + u^2 \\
s.t. & \sum_{d=1}^{t} \pi_d z_{dj} + u^1 - \left(\sum_{i_1 \in I_1} \omega_{i_1} x_{i_1 j} + \sum_{i_2 \in I_2} \beta_{i_2 j} x_{i_2 j} \right) \leq 0, \quad j = 1, \ldots, n \\
& \sum_{r=1}^{s} \mu_r y_{rj} + u^2 - \left[\sum_{i_2 \in I_2} \left(\omega_{i_2} - \beta_{i_2 j} \right) x_{i_2 j} + \sum_{d=1}^{t} \pi_d z_{dj} \right] \leq 0, \quad j = 1, \ldots, n \\
& \sum_{i_1 \in I_1} \omega_{i_1} x_{i_1 o} + \sum_{i_2 \in I_2} \omega_{i_2} x_{i_2 o} + \sum_{d=1}^{t} \pi_d z_{do} = 1 \\
& L_{i_2 j}^1 \omega_{i_2} \leq \beta_{i_2 j} \leq L_{i_2 j}^2 \omega_{i_2}, \quad i_2 \in I_2, \ j = 1, \ldots, n \\
& \mu_r, \pi_d, \omega_{i_1}, \omega_{i_2} \geq \varepsilon, \quad r = 1, \ldots, s, \ d = 1, \ldots, t, \ i_1 \in I_1, \ i_2 \in I_2 \\
& u^1, u^2 \, free
\end{aligned}
\tag{9.7}
$$

9.3 Efficiency Decomposition

Once we obtain an optimal solution to linear program (9.7), the efficiency scores for both individual stages can be calculated accordingly. However, since model (9.7) can have alternative optimal solutions, the decomposition of the overall efficiency defined in (9.3) may not be unique. Therefore we follow Kao and Hwang's (2008) approach to find a set of multipliers which produce the highest first or second stage efficiency score while maintaining the overall efficiency score of the entire process.

Let w_1^* and w_2^* represent optimal weights based upon model (9.7). Note that in model (9.7), we have $\sum_{i_1 \in I_1} \omega_{i_1}^* x_{i_1 o} + \sum_{i_2 \in I_2} \omega_{i_2}^* x_{i_2 o} + \sum_{d=1}^{t} \pi_d^* z_{do} = 1$. Thus, we have $w_1^* = \sum_{i_1 \in I_1} \omega_{i_1}^* x_{i_1 o} + \sum_{i_2 \in I_2} \beta_{i_2 o}^* x_{i_2 o}$ and $w_2^* = 1 - w_1^*$, where $\omega_{i_1}^*, \omega_{i_2}^*, \beta_{i_2 o}^*, \pi_d^*$ $(i_1 \in I_1, \ i_2 \in I_2, \ d = 1, \ldots, t)$ represent optimal values for $\omega_{i_1}, \omega_{i_2}, \beta_{i_2 o}, \pi_d$ in model (9.7).

Denote the overall efficiency score for DMU_o obtained from (9.7) as θ_o^*. Suppose that the first-stage efficiency is maximized first while maintaining the overall performance. We have

$$
\begin{aligned}
\theta_o^{1*} = Max \quad & \frac{\sum_{d=1}^{t} \eta_d z_{do} + u^A}{\sum_{i_1 \in I_1} v_{i_1} x_{i_1 o} + \sum_{i_2 \in I_2} v_{i_2} \alpha_{i_2 o} x_{i_2 o}} \\
s.t. \quad & \frac{\sum_{d=1}^{t} \eta_d z_{dj} + u^A}{\sum_{i_1 \in I_1} v_{i_1} x_{i_1 j} + \sum_{i_2 \in I_2} v_{i_2} \alpha_{i_2 j} x_{i_2 j}} \le 1, \quad j = 1, \ldots, n \\
& \frac{\sum_{r=1}^{s} u_r y_{rj} + u^B}{\sum_{i_2 \in I_2} v_{i_2} \left(1 - \alpha_{i_2 j}\right) x_{i_2 j} + \sum_{d=1}^{t} \eta_d z_{dj}} \le 1, \quad j = 1, \ldots, n \\
& \frac{\sum_{d=1}^{t} \eta_d z_{do} + u^A + \sum_{r=1}^{s} u_r y_{ro} + u^B}{\sum_{i_1 \in I_1} v_{i_1} x_{i_1 o} + \sum_{i_2 \in I_2} v_{i_2} x_{i_2 o} + \sum_{d=1}^{t} \eta_d z_{do}} = \theta_o^* \\
& w_1^* \cdot \frac{\sum_{d=1}^{t} \eta_d z_{do} + u^A}{\sum_{i_1 \in I_1} v_{i_1} x_{i_1 o} + \sum_{i_2 \in I_2} v_{i_2} \alpha_{i_2 o} x_{i_2 o}} \le \theta_o^* \\
& L_{i_2 j}^1 \le \alpha_{i_2 j} \le L_{i_2 j}^2, \quad i_2 \in I_2, \; j = 1, \ldots, n \\
& u_r, \eta_d, v_{i_1}, v_{i_2} \ge \varepsilon, \quad r = 1, \ldots, s, \; d = 1, \ldots, t, \; i_1 \in I_1, \; i_2 \in I_2; \; u^A, u^B \, free
\end{aligned}
\tag{9.8}
$$

Model (9.8) can be converted into linear program (9.9).

$$
\begin{aligned}
\theta_o^{1*} = Max \quad & \sum_{d=1}^{t} \pi_d z_{do} + u^1 \\
s.t. \quad & \sum_{d=1}^{t} \pi_d z_{dj} + u^1 - \left(\sum_{i_1 \in I_1} \omega_{i_1} x_{i_1 j} + \sum_{i_2 \in I_2} \beta_{i_2 j} x_{i_2 j} \right) \le 0, \quad j = 1, \ldots, n \\
& \sum_{r=1}^{s} \mu_r y_{rj} + u^2 - \left[\sum_{i_2 \in I_2} \left(\omega_{i_2} - \beta_{i_2 j} \right) x_{i_2 j} + \sum_{d=1}^{t} \pi_d z_{dj} \right] \le 0, \quad j = 1, \ldots, n \\
& \left(1 - \theta_o^*\right) \sum_{d=1}^{t} \pi_d z_{do} + \sum_{r=1}^{s} \mu_r y_{ro} + u^1 + u^2 - \theta_0^* \left(\sum_{i_1 \in I_1} \omega_{i_1} x_{i_1 o} + \sum_{i_2 \in I_2} \omega_{i_2} x_{i_2 o} \right) = 0 \\
& w_1^* \left(\sum_{d=1}^{t} \pi_d z_{do} + u^1 \right) \le \theta_o^* \\
& \sum_{i_1 \in I_1} \omega_{i_1} x_{i_1 o} + \sum_{i_2 \in I_2} \beta_{i_2 o} x_{i_2 o} = 1 \\
& L_{i_2 j}^1 \omega_{i_2} \le \beta_{i_2 j} \le L_{i_2 j}^2 \omega_{i_2}, \quad i_2 \in I_2, \; j = 1, \ldots, n \\
& \mu_r, \pi_d, \omega_{i_1}, \omega_{i_2} \ge \varepsilon, \quad r = 1, \ldots, s, \; d = 1, \ldots, t, \; i_1 \in I_1, \; i_2 \in I_2; \; u^1, u^2 \, free
\end{aligned}
\tag{9.9}
$$

Then the second-stage efficiency score for DMU_o can be calculated as $\theta_o^2 = \frac{\theta_o^* - w_1^* \theta_o^{1*}}{w_2^*}$. Note that (*) is used in θ_o^{1*} to indicate that the first-stage efficiency score is optimized first. In this case, the resulting efficiency score for the second stage is denoted as θ_o^2 without (*).

In a similar way, the second stage can be optimized first without hurting the overall efficiency θ_o^*.

$$
\begin{aligned}
\theta_o^{2*} = Max & \sum_{r=1}^{s} \mu_r y_{ro} + u^2 \\
s.t. & \sum_{d=1}^{t} \pi_d z_{dj} + u^1 - \left(\sum_{i_1 \in I_1} \omega_{i_1} x_{i_1 j} + \sum_{i_2 \in I_2} \beta_{i_2 j} x_{i_2 j} \right) \le 0, \quad j = 1, \ldots, n \\
& \sum_{r=1}^{s} \mu_r y_{rj} + u^2 - \left[\sum_{i_2 \in I_2} \left(\omega_{i_2} - \beta_{i_2 j} \right) x_{i_2 j} + \sum_{d=1}^{t} \pi_d z_{dj} \right] \le 0, \quad j = 1, \ldots, n \\
& \left(1 - \theta_o^*\right) \sum_{d=1}^{t} \pi_d z_{do} + \sum_{r=1}^{s} \mu_r y_{ro} + u^1 + u^2 - \theta_0^* \left(\sum_{i_1 \in I_1} \omega_{i_1} x_{i_1 o} + \sum_{i_2 \in I_2} \omega_{i_2} x_{i_2 o} \right) = 0 \\
& w_2^* \left(\sum_{r=1}^{s} \mu_r y_{ro} + u^2 \right) \le \theta_o^* \\
& \sum_{i_2 \in I_2} \left(\omega_{i_2} - \beta_{i_2 o} \right) x_{i_2 o} + \sum_{d=1}^{t} \pi_d z_{do} = 1 \\
& L_{i_2 j}^1 \omega_{i_2} \le \beta_{i_2 j} \le L_{i_2 j}^2 \omega_{i_2}, \; i_2 \in I_2, j = 1, \ldots, n \\
& \mu_r, \pi_d, \omega_{i_1}, \omega_{i_2} \ge \varepsilon, \; r = 1, \ldots, s, d = 1, \ldots, t, i_1 \in I_1, i_2 \in I_2; u^1, u^2 \, free
\end{aligned}
\tag{9.10}
$$

Then the first-stage efficiency score for DMU_o can be calculated as $\theta_o^1 = \frac{\theta_o^* - w_2^* \theta_o^{2*}}{w_1^*}$.

If the results satisfy $\theta_o^1 = \theta_o^{1*}$ or $\theta_o^2 = \theta_o^{2*}$, then the conclusion can be reached that an unique efficiency decomposition is obtained.

9.4 Two-Stage CRS Model with Shared Inputs

9.4.1 Overall Efficiency

Note that if we assume $u^A = u^B = 0$, the above models become the constant returns to scale (CRS) models of Charnes et al. (1978). Thus all previous discussions are open to be applied to the CRS situation.

The following models (9.11) and (9.12) will assess the efficiency of the first and second stages of DMU_o, respectively.

$$
\begin{aligned}
&Max \quad \frac{\sum_{d=1}^{t} \eta_d^1 z_{do}}{\sum_{i_1 \in I_1} v_{i_1} x_{i_1 o} + \sum_{i_2 \in I_2} v_{i_2}^1 \alpha_{i_2 o} x_{i_2 o}} \\
&s.t. \quad \frac{\sum_{d=1}^{t} \eta_d^1 z_{dj}}{\sum_{i_1 \in I_1} v_{i_1} x_{i_1 j} + \sum_{i_2 \in I_2} v_{i_2}^1 \alpha_{i_2 j} x_{i_2 j}} \leq 1, \quad j = 1, \ldots, n \\
&\qquad L_{i_2 j}^1 \leq \alpha_{i_2 j} \leq L_{i_2 j}^2, \ i_2 \in I_2, j = 1, \ldots, n \\
&\qquad \eta_d^1, \ v_{i_1}, v_{i_2}^1 \geq \varepsilon, \ d = 1, \ldots, t, i_1 \in I_1, i_2 \in I_2
\end{aligned}
\tag{9.11}
$$

$$
\begin{aligned}
&Max \quad \frac{\sum_{r=1}^{s} u_r y_{ro}}{\sum_{i_2 \in I_2} v_{i_2}^2 (1 - \alpha_{i_2 o}) x_{i_2 o} + \sum_{d=1}^{t} \eta_d^2 z_{do}} \\
&s.t. \quad \frac{\sum_{r=1}^{s} u_r y_{rj}}{\sum_{i_2 \in I_2} v_{i_2}^2 (1 - \alpha_{i_2 j}) x_{i_2 j} + \sum_{d=1}^{t} \eta_d^2 z_{dj}} \leq 1, \quad j = 1, \ldots, n \\
&\qquad L_{i_2 j}^1 \leq \alpha_{i_2 j} \leq L_{i_2 j}^2, \ i_2 \in I_2, j = 1, \ldots, n \\
&\qquad u_r, \eta_d^2, v_{i_2}^2 \geq \varepsilon, \ r = 1, \ldots, s, d = 1, \ldots, t, i_2 \in I_2
\end{aligned}
\tag{9.12}
$$

Under the assumption of $\eta_d^1 = \eta_d^2 = \eta_d$ for all $d = 1, \ldots, t$ and $v_{i_2}^1 = v_{i_2}^2 = v_{i_2}$ for all $i_2 \in I_2$, the weighted average efficiency of both stages is presented as

$$
w_1 \cdot \frac{\sum_{d=1}^{t} \eta_d z_{do}}{\sum_{i_1 \in I_1} v_{i_1} x_{i_1 o} + \sum_{i_2 \in I_2} v_{i_2} \alpha_{i_2 o} x_{i_2 o}} + w_2 \cdot \frac{\sum_{r=1}^{s} u_r y_{ro}}{\sum_{i_2 \in I_2} v_{i_2} (1 - \alpha_{i_2 o}) x_{i_2 o} + \sum_{d=1}^{t} \eta_d z_{do}}
\tag{9.13}
$$

where w_1 and w_2 are user-specified weights such that $w_1 + w_2 = 1$. The same choice on w_1 and w_2 is made with (9.4) to reflect the relative importance of either stage. The overall efficiency is evaluated by fractional program (9.14), which is then linearized as model (9.15).

$$\theta_o^* = Max \frac{\sum_{d=1}^{t} \eta_d z_{do} + \sum_{r=1}^{s} u_r y_{ro}}{\sum_{i_1 \in I_1} v_{i_1} x_{i_1 o} + \sum_{i_2 \in I_2} v_{i_2} x_{i_2 o} + \sum_{d=1}^{t} \eta_d z_{do}}$$

$$s.t. \quad \frac{\sum_{d=1}^{t} \eta_d z_{dj}}{\sum_{i_1 \in I_1} v_{i_1} x_{i_1 j} + \sum_{i_2 \in I_2} v_{i_2} \alpha_{i_2 j} x_{i_2 j}} \le 1, \quad j = 1, \ldots, n \tag{9.14}$$

$$\frac{\sum_{r=1}^{s} u_r y_{rj}}{\sum_{i_2 \in I_2} v_{i_2} \left(1 - \alpha_{i_2 j}\right) x_{i_2 j} + \sum_{d=1}^{t} \eta_d z_{dj}} \le 1, \quad j = 1, \ldots, n$$

$$L_{i_2 j}^1 \le \alpha_{i_2 j} \le L_{i_2 j}^2, \ i_2 \in I_2, j = 1, \ldots, n$$

$$u_r, \eta_d, v_{i_1}, v_{i_2} \ge \varepsilon, \ r = 1, \ldots, s, d = 1, \ldots, t, i_1 \in I_1, i_2 \in I_2$$

$$\theta_o^* = Max \sum_{d=1}^{t} \pi_d z_{do} + \sum_{r=1}^{s} \mu_r y_{ro}$$

$$s.t. \quad \sum_{d=1}^{t} \pi_d z_{dj} - \left(\sum_{i_1 \in I_1} \omega_{i_1} x_{i_1 j} + \sum_{i_2 \in I_2} \beta_{i_2 j} x_{i_2 j} \right) \le 0, \ j = 1, \ldots, n$$

$$\sum_{r=1}^{s} \mu_r y_{rj} - \left[\sum_{i_2 \in I_2} \left(\omega_{i_2} - \beta_{i_2 j} \right) x_{i_2 j} + \sum_{d=1}^{t} \pi_d z_{dj} \right] \le 0, \ j = 1, \ldots, n$$

$$\sum_{i_1 \in I_1} \omega_{i_1} x_{i_1 o} + \sum_{i_2 \in I_2} \omega_{i_2} x_{i_2 o} + \sum_{d=1}^{t} \pi_d z_{do} = 1$$

$$L_{i_2 j}^1 \omega_{i_2} \le \beta_{i_2 j} \le L_{i_2 j}^2 \omega_{i_2}, \ i_2 \in I_2, j = 1, \ldots, n$$

$$\mu_r, \pi_d, \omega_{i_1}, \omega_{i_2} \ge \varepsilon, \ r = 1, \ldots, s, d = 1, \ldots, t, i_1 \in I_1, i_2 \in I_2 \tag{9.15}$$

9.4.2 *Efficiency Decomposition*

Let w_1^* and w_2^* represent optimal weights based upon model (9.15), which are calculated as $w_1^* = \sum_{i_1 \in I_1} \omega_{i_1}^* x_{i_1 o} + \sum_{i_2 \in I_2} \beta_{i_2 o}^* x_{i_2 o}$ and $w_2^* = 1 - w_1^*$, where $\omega_{i_1}^*, \omega_{i_2}^*, \beta_{i_2 o}^*, \pi_d^* \ (i_1 \in I_1, \ i_2 \in I_2, \ d = 1, \ldots, t)$ represent optimal values obtained from model (9.15).

If Stage 1 is given priority, the decomposition model maintaining the overall performance θ_o^* is developed as

$$
\begin{aligned}
\theta_o^{1*} = Max & \sum_{d=1}^{t} \pi_d z_{do} \\
s.t. & \sum_{d=1}^{t} \pi_d z_{dj} - \left(\sum_{i_1 \in I_1} \omega_{i_1} x_{i_1 j} + \sum_{i_2 \in I_2} \beta_{i_2 j} x_{i_2 j} \right) \leq 0, \quad j = 1, \ldots, n \\
& \sum_{r=1}^{s} \mu_r y_{rj} - \left[\sum_{i_2 \in I_2} \left(\omega_{i_2} - \beta_{i_2 j} \right) x_{i_2 j} + \sum_{d=1}^{t} \pi_d z_{dj} \right] \leq 0, \quad j = 1, \ldots, n \\
& \left(1 - \theta_o^* \right) \sum_{d=1}^{t} \pi_d z_{do} + \sum_{r=1}^{s} \mu_r y_{ro} - \theta_0^* \left(\sum_{i_1 \in I_1} \omega_{i_1} x_{i_1 o} + \sum_{i_2 \in I_2} \omega_{i_2} x_{i_2 o} \right) = 0 \\
& w_1^* \left(\sum_{d=1}^{t} \pi_d z_{do} \right) \leq \theta_o^* \\
& \sum_{i_1 \in I_1} \omega_{i_1} x_{i_1 o} + \sum_{i_2 \in I_2} \beta_{i_2 o} x_{i_2 o} = 1 \\
& L_{i_2 j}^1 \omega_{i_2} \leq \beta_{i_2 j} \leq L_{i_2 j}^2 \omega_{i_2}, \ i_2 \in I_2, j = 1, \ldots, n \\
& \mu_r, \pi_d, \omega_{i_1}, \omega_{i_2} \geq \varepsilon, \ r = 1, \ldots, s, d = 1, \ldots, t, i_1 \in I_1, i_2 \in I_2
\end{aligned}
\tag{9.16}
$$

Then the second-stage efficiency score can be calculated as $\theta_o^2 = \frac{\theta_o^* - w_1^* \theta_o^{1*}}{w_2^*}$. Similarly, to optimize Stage 2 first, we propose

$$
\begin{aligned}
\theta_o^{2*} = Max & \sum_{r=1}^{s} \mu_r y_{ro} \\
s.t. & \sum_{d=1}^{t} \pi_d z_{dj} - \left(\sum_{i_1 \in I_1} \omega_{i_1} x_{i_1 j} + \sum_{i_2 \in I_2} \beta_{i_2 j} x_{i_2 j} \right) \leq 0, \quad j = 1, \ldots, n \\
& \sum_{r=1}^{s} \mu_r y_{rj} - \left[\sum_{i_2 \in I_2} \left(\omega_{i_2} - \beta_{i_2 j} \right) x_{i_2 j} + \sum_{d=1}^{t} \pi_d z_{dj} \right] \leq 0, \quad j = 1, \ldots, n \\
& \left(1 - \theta_o^* \right) \sum_{d=1}^{t} \pi_d z_{do} + \sum_{r=1}^{s} \mu_r y_{ro} - \theta_0^* \left(\sum_{i_1 \in I_1} \omega_{i_1} x_{i_1 o} + \sum_{i_2 \in I_2} \omega_{i_2} x_{i_2 o} \right) = 0 \\
& w_2^* \left(\sum_{r=1}^{s} \mu_r y_{ro} \right) \leq \theta_o^* \\
& \sum_{i_2 \in I_2} \left(\omega_{i_2} - \beta_{i_2 o} \right) x_{i_2 o} + \sum_{d=1}^{t} \pi_d z_{do} = 1 \\
& L_{i_2 j}^1 \omega_{i_2} \leq \beta_{i_2 j} \leq L_{i_2 j}^2 \omega_{i_2}, i_2 \in I_2, j = 1, \ldots, n \\
& \mu_r, \pi_d, \omega_{i_1}, \omega_{i_2} \geq \varepsilon, r = 1, \ldots, s, d = 1, \ldots, t, i_1 \in I_1, i_2 \in I_2
\end{aligned}
\tag{9.17}
$$

The efficiency score for Stage 1 is $\theta_o^1 = \frac{\theta_o^* - w_2^* \theta_o^{2*}}{w_1^*}$. A unique efficiency decomposition is obtained when $\theta_o^1 = \theta_o^{1*}$ or $\theta_o^2 = \theta_o^{2*}$.

9.5 Illustrative Application

Wang et al. (1997) utilize DEA to study the marginal benefits of information technology (IT) with respect to a two-stage process in firm-level banking industry. In their first stage, the banks use fixed assets, IT investment, and personnel as inputs to generate deposit dollars as output (intermediate measure). During the second stage, banks use the deposit dollars as a source of funds to invest in securities and to provide loans. Profit and fraction of loans recovered are regarded as two outputs from the second stage. Obviously, the IT budget, personnel support and fixed assets needed in stage 2 are ignored. In other words, fixed assets, IT budget and employees are directly associated with both stages and they should be treated as shared inputs. We therefore use model (9.7) to re-visit their data set as shown in Table 9.1.

Table 9.1 Data

Bank	Fixed assets ($ billion)	IT budget ($ billion)	Employees (thousand)	Deposits ($ billion)	Profit ($ billion)	Fraction of loans recovered
1	0.713	0.15	13.3	14.478	0.232	0.986
2	1.071	0.17	16.9	19.502	0.34	0.986
3	1.224	0.235	24	20.952	0.363	0.986
4	0.363	0.211	15.6	13.902	0.211	0.982
5	0.409	0.133	18.485	15.206	0.237	0.984
6	5.846	0.497	56.42	81.186	1.103	0.955
7	0.918	0.06	56.42	81.186	1.103	0.986
8	1.235	0.071	12	11.441	0.199	0.985
9	18.12	1.5	89.51	124.072	1.858	0.972
10	1.821	0.12	19.8	17.425	0.274	0.983
11	1.915	0.12	19.8	17.425	0.274	0.983
12	0.874	0.05	13.1	14.342	0.177	0.985
13	6.918	0.37	12.5	32.491	0.648	0.945
14	4.432	0.44	41.9	47.653	0.639	0.979
15	4.504	0.431	41.1	52.63	0.741	0.981
16	1.241	0.11	14.4	17.493	0.243	0.988
17	0.45	0.053	7.6	9.512	0.067	0.98
18	5.892	0.345	15.5	42.469	1.002	0.948
19	0.973	0.128	12.6	18.987	0.243	0.985
20	0.444	0.055	5.9	7.546	0.153	0.987
21	0.508	0.057	5.7	7.595	0.123	0.987
22	0.37	0.098	14.1	16.906	0.233	0.981
23	0.395	0.104	14.6	17.264	0.263	0.983
24	2.68	0.206	19.6	36.43	0.601	0.982
25	0.781	0.067	10.5	11.581	0.12	0.987
26	0.872	0.1	12.1	22.207	0.248	0.972
27	1.757	0.0106	12.7	20.67	0.253	0.988

In this case, model (9.7) becomes

$$
\begin{aligned}
\theta_o^* = &\ Max\ \pi D_o + \mu_P P_o + \mu_R R_o + u^1 + u^2 \\
&s.t.\ \ \pi D_j + u^1 - \left(\beta_{Fj}F_j + \beta_{Ij}I_j + \beta_{Ej}E_j\right) \le 0, \quad j = 1, \ldots, n \\
&\left(\mu_P P_j + \mu_R R_j\right) + u^2 - \left[\left(\omega_F - \beta_{Fj}\right)F_j + \left(\omega_I - \beta_{Ij}\right)I_j + \left(\omega_E - \beta_{Ej}\right)E_j + \pi D_j\right] \\
&\quad \le 0, \quad j = 1, \ldots, n \\
&\omega_F F_o + \omega_I I_o + \omega_E E_o + \pi D_o = 1 \\
&0.4\,\omega_F \le \beta_{Fj} \le 0.6\,\omega_F, \quad j = 1, \ldots, n \\
&0.25\,\omega_I \le \beta_{Ij} \le 0.75\,\omega_I, \quad j = 1, \ldots, n \\
&0.55\,\omega_E \le \beta_{Ej} \le 0.75\omega_E, \quad j = 1, \ldots, n \\
&\mu_P, \mu_R, \pi, \omega_F, \omega_I, \omega_E \ge \varepsilon; u^1,\ \ u^2 free
\end{aligned}
\tag{9.18}
$$

where fixed assets (F), IT budget (I), and employees (E) are treated as shared inputs as $\alpha_F F$ and $(1 - \alpha_F)\,F$, $\alpha_I I$ and $(1 - \alpha_I)\,I$, as well as $\alpha_E E$ and $(1 - \alpha_E)\,E$ between two stages. The lower and upper bounds are specified as $0.40 \le \alpha_F \le 0.60$, $0.25 \le \alpha_I \le 0.75$ and $0.55 \le \alpha_E \le 0.75$. These constraints are converted into the last three constraints in model (9.18).

Tables 9.2 and 9.3 report the results of the overall efficiency and efficiency decomposition when the first-stage efficiency score is maximized first under VRS and CRS, respectively. Tables 9.4 and 9.5 report overall efficiency and efficiency

Table 9.2 VRS results (stage 1 takes priority)

Bank	Overall efficiency θ_o^*	Deposit efficiency θ_o^{1*}	Loan efficiency θ_o^2	w_1^*	w_2^*	α_F^*	α_I^*	α_E^*
1	0.817416	0.902544	0.735635	0.48997	0.51003	0.4000	0.5000	0.5500
2	0.884392	0.894746	0.876417	0.43506	0.56494	0.4000	0.5000	0.5500
3	0.745634	0.697699	0.779187	0.41176	0.58824	0.4000	0.5000	0.5500
4	1.000000	1.000000	1.000000	0.47677	0.52323	0.6000	0.5000	0.7500
5	0.998006	1.000000	0.996396	0.44665	0.55335	0.5231	0.5000	0.4639
6	0.810887	1.000000	0.718858	0.32734	0.67266	0.4582	0.5000	0.5500
7	1.000000	1.000000	1.000000	0.41767	0.58233	0.6000	0.5000	0.7500
8	0.758876	0.769498	0.748291	0.49909	0.50091	0.4000	0.2500	0.5500
9	1.000000	1.000000	1.000000	0.37193	0.62807	0.4000	0.5000	0.5500
10	0.641277	0.700543	0.563920	0.56621	0.43379	0.4000	0.2500	0.5500
11	0.638457	0.696925	0.562096	0.56635	0.43365	0.4000	0.2500	0.5500
12	0.891486	0.970490	0.843485	0.37794	0.62206	0.4000	0.2500	0.4917
13	0.950591	1.000000	0.908177	0.46190	0.53810	0.6000	0.5000	0.7385
14	0.680918	0.850776	0.534268	0.46333	0.53667	0.4000	0.5000	0.5500
15	0.804134	0.869254	0.761576	0.39523	0.60477	0.4000	0.5000	0.5500
16	1.000000	1.000000	1.000000	0.34468	0.65532	0.4000	0.2500	0.5500
17	1.000000	1.000000	0.999998	0.34806	0.65194	0.6000	0.7500	0.5985
18	1.000000	1.000000	1.000000	0.30843	0.69157	0.4000	0.5000	0.5500

(continued)

Table 9.2 (continued)

Bank	Overall efficiency θ_o^*	Deposit efficiency θ_o^{1*}	Loan efficiency θ_o^2	w_1^*	w_2^*	α_F^*	α_I^*	α_E^*
19	0.822764	1.000000	0.677141	0.45104	0.54896	0.5997	0.5000	0.5500
20	1.000000	1.000000	1.000000	0.45904	0.54096	0.6000	0.5000	0.7500
21	1.000000	1.000000	1.000000	0.56425	0.43575	0.6000	0.5000	0.6066
22	1.000000	1.000000	1.000000	0.47247	0.52753	0.6000	0.5000	0.7500
23	1.000000	1.000000	1.000000	0.45146	0.54854	0.6000	0.5000	0.6539
24	1.000000	1.000000	1.000000	0.36579	0.63421	0.6000	0.5000	0.5724
25	0.767069	0.890467	0.647451	0.49222	0.50778	0.4000	0.2500	0.5500
26	0.872988	1.000000	0.782691	0.41552	0.58448	0.6000	0.5000	0.7500
27	1.000000	1.000000	1.000000	0.52843	0.47157	0.6000	0.7500	0.7500

Table 9.3 CRS results (stage 1 takes priority)

Bank	Overall efficiency θ_o^*	Deposit efficiency θ_o^{1*}	Loan efficiency θ_o^2	w_1^*	w_2^*	α_F^*	α_I^*	α_E^*
1	0.809051	0.877548	0.771997	0.35105	0.64895	0.4000	0.5000	0.5500
2	0.855189	0.893773	0.824813	0.44049	0.55951	0.4000	0.5000	0.5500
3	0.730706	0.711527	0.748419	0.48015	0.51985	0.4000	0.5000	0.5500
4	0.904087	0.713145	0.999999	0.33436	0.66564	0.4000	0.5000	0.5500
5	0.857950	0.641374	0.974511	0.34989	0.65011	0.4000	0.2500	0.5500
6	0.750373	0.984808	0.634402	0.32982	0.67081	0.4000	0.2500	0.5500
7	1.000000	1.000000	1.000000	0.41563	0.58437	0.6000	0.5000	0.7500
8	0.749277	0.736279	0.762046	0.49554	0.50446	0.4000	0.2500	0.5500
9	0.611823	0.723234	0.533634	0.41239	0.58761	0.4000	0.5000	0.5500
10	0.635599	0.680372	0.602233	0.42700	0.57300	0.4000	0.2500	0.5500
11	0.633172	0.677297	0.600303	0.42690	0.57310	0.4000	0.2500	0.5500
12	0.773044	0.622073	0.840380	0.30845	0.69155	0.4000	0.2500	0.5500
13	0.927622	1.000000	0.876124	0.41572	0.58428	0.6000	0.5000	0.7115
14	0.629430	0.773005	0.516998	0.43916	0.56084	0.4000	0.2500	0.5500
15	0.696454	0.859753	0.543272	0.48402	0.51598	0.4000	0.2500	0.5500
16	0.721130	0.884557	0.596487	0.43266	0.56734	0.4000	0.2500	0.5500
17	0.897636	1.000000	0.830094	0.39752	0.60248	0.6000	0.2500	0.5563
18	1.000000	1.000000	1.000000	0.41058	0.58942	0.4000	0.5000	0.6024
19	0.815065	1.000000	0.689946	0.40354	0.59646	0.5868	0.5000	0.5500
20	1.000000	1.000000	1.000000	0.32872	0.67128	0.4000	0.2500	0.5500
21	1.000000	1.000000	1.000000	0.33484	0.66516	0.4402	0.2500	0.5500
22	0.949163	0.899387	0.972158	0.31600	0.68400	0.4000	0.5000	0.5500
23	0.958367	0.869308	0.999999	0.31855	0.68145	0.4000	0.5000	0.5500
24	0.922196	1.000000	0.879230	0.35576	0.64424	0.5525	0.7500	0.5500
25	0.734708	0.910991	0.603834	0.42608	0.57392	0.4000	0.2500	0.5500
26	0.866254	1.000000	0.772482	0.41215	0.58785	0.6000	0.5000	0.7500
27	1.000000	1.000000	1.000000	0.40300	0.59700	0.6000	0.7500	0.7500

Table 9.4 VRS results (stage 2 takes priority)

Bank	Overall efficiency θ_o^*	Deposit efficiency θ_o^1	Loan efficiency θ_o^{2*}	w_1^*	w_2^*	α_F^*	α_I^*	α_E^*
1	0.817416	0.647684	0.980472	0.48997	0.51003	0.6000	0.5000	0.7500
2	0.884392	0.734271	1.000000	0.43506	0.56494	0.4000	0.5000	0.6933
3	0.745634	0.382241	1.000000	0.41176	0.58824	0.5251	0.5000	0.7500
4	1.000000	1.000000	1.000000	0.47677	0.52323	0.5356	0.5000	0.7500
5	0.998006	0.995536	1.000000	0.44665	0.55335	0.5254	0.5000	0.7313
6	0.810887	0.700251	0.864725	0.32734	0.67266	0.6000	0.7500	0.7500
7	1.000000	1.000000	1.000000	0.41767	0.58233	0.6000	0.5000	0.5500
8	0.758876	0.516874	1.000000	0.49909	0.50091	0.4000	0.7500	0.6595
9	1.000000	1.000000	1.000000	0.37193	0.62807	0.6000	0.5000	0.3670
10	0.641277	0.420902	0.928921	0.56621	0.43379	0.6000	0.7500	0.7500
11	0.638457	0.418934	0.925161	0.56635	0.43365	0.6000	0.7500	0.7500
12	0.891486	0.712880	1.000000	0.37794	0.62206	0.6000	0.3060	0.8089
13	0.950591	0.984595	0.921401	0.46190	0.53810	0.3073	0.5000	0.7500
14	0.680918	0.614391	0.738354	0.46333	0.53667	0.6000	0.5000	0.7500
15	0.804134	0.631571	0.916907	0.39523	0.60477	0.6000	0.5000	0.7500
16	1.000000	1.000000	1.000000	0.34468	0.65532	0.4000	0.2500	0.5500
17	1.000000	1.000000	1.000000	0.34806	0.65194	0.4322	0.7500	0.7500
18	1.000000	1.000000	1.000000	0.30843	0.69157	0.4000	0.5000	0.5500
19	0.822764	0.808300	0.834647	0.45104	0.54896	0.6000	0.5000	0.7500
20	1.000000	1.000000	1.000000	0.45904	0.54096	0.4000	0.5000	0.5500
21	1.000000	1.000000	1.000000	0.56425	0.43575	0.4000	0.5000	0.5681
22	1.000000	1.000000	1.000000	0.47247	0.52753	0.6000	0.5000	0.5500
23	1.000000	1.000000	1.000000	0.45146	0.54854	0.4711	0.5000	0.5500
24	1.000000	1.000000	1.000000	0.36579	0.63421	0.6000	0.5000	0.5724
25	0.767069	0.526775	1.000000	0.49222	0.50778	0.4000	0.6722	0.5500
26	0.872988	0.999988	0.782699	0.41552	0.58448	0.6000	0.5000	0.7500
27	1.000000	1.000000	1.000000	0.52843	0.47157	0.6000	0.7500	0.7500

decomposition when stage 2 is given priority under VRS and CRS, respectively. Each table demonstrates the overall efficiency for the whole process in column 2, the efficiency scores for the first (deposit) and second (loan) stages in columns 3 and 4, the optimal weights attached to either stage in columns 5 and 6, and the values of α_F^*, α_I^*, α_E^* from columns 7 to 9. α_F^*, α_I^*, α_E^* indicate the optimal proportions of three inputs distributed between two stages.

From the efficiency decomposition results in Tables 9.2, 9.3, 9.4 and 9.5, we notice that under VRS, DMU 4, 7, 9, 16, 18, 20, 21, 22, 23, 24, 27 have a unique efficiency decomposition while DMU 17 and 26 have a unique efficiency decomposition in an approximate sense when ignoring trivial errors. DMU 7, 18, 20, 21, 27 have a unique efficiency decomposition while DMU 4 and 23 have an approximately unique efficiency decomposition when ignoring trivial errors under CRS.

Table 9.5 CRS results (stage 2 takes priority)

Bank	Overall efficiency θ_o^*	Deposit efficiency θ_o^1	Loan efficiency θ_o^{2*}	w_1^*	w_2^*	α_F^*	α_I^*	α_E^*
1	0.809051	0.490733	0.981248	0.35105	0.64895	0.6000	0.5000	0.7500
2	0.855189	0.671250	1.000000	0.44049	0.55951	0.4637	0.5000	0.7500
3	0.730706	0.472306	0.969374	0.48015	0.51985	0.6000	0.5000	0.7500
4	0.904087	0.713143	1.000000	0.33436	0.66564	0.4000	0.5000	0.5500
5	0.857950	0.594015	1.000000	0.34989	0.65011	0.4136	0.2500	0.7500
6	0.750373	0.657693	0.795236	0.32982	0.67081	0.6000	0.5000	0.7500
7	1.000000	1.000000	1.000000	0.41563	0.58437	0.4000	0.5000	0.7500
8	0.749277	0.494043	1.000000	0.49554	0.50446	0.6000	0.5006	0.7500
9	0.611823	0.462001	0.716969	0.41239	0.58761	0.6000	0.7500	0.7500
10	0.635599	0.232047	0.936330	0.42700	0.57300	0.6000	0.7500	0.7500
11	0.633172	0.230624	0.933029	0.42690	0.57310	0.6000	0.7500	0.7500
12	0.773044	0.264194	1.000000	0.30845	0.69155	0.6000	0.3718	0.7500
13	0.927622	0.950684	0.911212	0.41572	0.58428	0.7945	0.5000	0.7500
14	0.629430	0.525840	0.710535	0.43916	0.56084	0.6000	0.5000	0.7500
15	0.696454	0.609251	0.778255	0.48402	0.51598	0.6000	0.5000	0.7500
16	0.721130	0.606705	0.808384	0.43266	0.56734	0.6000	0.7500	0.7500
17	0.897636	0.742494	1.000000	0.39752	0.60248	0.6000	0.3241	0.7500
18	1.000000	1.000000	1.000000	0.41058	0.58942	0.4000	0.5000	0.5500
19	0.815065	0.803035	0.823204	0.40354	0.59646	0.6000	0.5000	0.7500
20	1.000000	1.000000	1.000000	0.32872	0.67128	0.4000	0.2500	0.5500
21	1.000000	1.000000	1.000000	0.33484	0.66516	0.4000	0.2500	0.5591
22	0.949163	0.839120	1.000000	0.31600	0.68400	0.4000	0.5000	0.6463
23	0.958367	0.869306	1.000000	0.31855	0.68145	0.4000	0.5000	0.5500
24	0.922196	0.781304	1.000000	0.35576	0.64424	0.5917	0.2500	0.7500
25	0.734708	0.512093	0.899980	0.42608	0.57392	0.6000	0.7500	0.7500
26	0.866254	0.999984	0.772493	0.41215	0.58785	0.6000	0.5000	0.7500
27	1.000000	1.000000	1.000000	0.40300	0.59700	0.4000	0.2500	0.5500

9.6 Conclusions

This chapter presents models for a two-stage process with shared input resources between both stages. In reality, many DMUs actually have this kind of structures. Beyond the hospital and banking examples referenced earlier, many non-profit and government organizations also have two-stage processes. Other examples can be found in corporations, which are ultimately accountable to shareholders. Increasing shareholder value is one of their primary objectives or outputs. Increasing value requires generating a return on funds invested by shareholders – return on equity. This process is a two-stage process. A set of resources is used by the corporation including plant and equipment, information technology and senior management. Such set of corporate resources are inputs used to generate operating results measured by revenues and profits, the stage 1 outputs. However, these outputs alone are not sufficient to generate increased shareholder value. While sales growth and profits are a key measure of performance, the ability to attract financing through debt at favorable interest rates to invest in future product and market development directly influences the estimated future value of the business. Arranging favorable financing through debt or equity to support investments in new products and markets or acquisitions draws on the corporate resources such as treasury and strategic planning and uses the plant and equipment and information technology – all inputs – share resources – also are used in stage 1. In addition, a critical element is that the stage 1 operating results are also a key input to obtaining loans at attractive interest rates and equity funds to grow through investment or acquisitions to increase shareholder value. Hence, measuring the efficiency of the business in using it resources to generate profits and revenues understates the efficiency because many of those resources are also used to generate increased shareholder value along with the stage 1 operating outputs. Here again, the example is just an illustration of the two stage aspects of the corporation. Individual industries will have differing elements of the two stage process aimed at increasing shareholder value.

This chapter extends the work of Kao and Hwang (2008), and Liang et al. (2008) under the assumption that DMUs have control over both stages, and provides a mechanism for developing network DEA models with shared inputs/outputs. In situations where one stage's operation possesses the priority to be optimized, one can adopt the leader-follower model of Liang et al. (2008) with one stage of the operations as the leader and the other as the follower. This is a future research topic.

References

Banker, R. D., Charnes, A., & Cooper, W. W. (1984). Some models for estimating technical and scale inefficiencies in data envelopment analysis. *Management Science, 30*, 1078–1092.

Charnes, A., Cooper, W. W., & Rhodes, E. (1978). Measuring the efficiency of decision making units. *European Journal of Operational Research, 2*, 429–444.

Chen, Y., & Zhu, J. (2004). Measuring information technology's indirect impact on firm performance. *Information Technology & Management Journal, 5*(1–2), 9–22.

Chen, Y., Du, J., Sherman, H. D., & Zhu, J. (2010). DEA model with shared resources and efficiency decomposition. *European Journal of Operational Research, 207*, 339–349.

Cook, W. D., & Hababou, M. (2001). Sales performance measurement in bank branches. *OMEGA, 29*, 299–307.

Cook, W. D., Hababou, M., & Tuenter, H. (2000). Multicomponent efficiency measurement and shared inputs in data envelopment analysis: An application to sales and service performance in bank branches. *Journal of Productivity Analysis, 14*(3), 209–224.

Huang, Z. M., & Li, S. X. (2001). Co-op advertising models in a manufacturing-retailing supply chain: A game theory approach. *European Journal of Operational Research, 135*, 527–544.

Kao, C., & Hwang, S. N. (2008). Efficiency decomposition in two-stage data envelopment analysis: An application to non-life insurance companies in Taiwan. *European Journal of Operational Research, 185*(1), 418–429.

Liang, L., Cook, W. D., & Zhu, J. (2008). DEA models for two-stage processes: Game approach and efficiency decomposition. *Naval Research Logistics, 55*, 643–653.

Seiford, L. M., & Zhu, J. (1999). Profitability and marketability of the top 55 US commercial banks. *Management Science, 45*(9), 1270–1288.

Sexton, T. R., & Lewis, H. F. (2003). Two-stage DEA: An application to major league baseball. *Journal of Productivity Analysis, 19*(2–3), 227–249.

Wang, C. H., Gopal, R., & Zionts, S. (1997). Use of data envelopment analysis in assessing information technology impact on firm performance. *Annals of Operations Research, 73*, 191–213.

Zhu, J. (2000). Multi-factor performance measure model with an application to Fortune 500 companies. *European Journal of Operational Research, 123*(1), 105–124.

Chapter 10
A Network-DEA Model with Internal Dynamic Effects

Chien-Ming Chen

Abstract Modern production networks are comprised of a large collection of interrelated value-adding processes. Adding to this, the flow time of such a complex network can easily go from weeks to months, making time an influential factor in constructing efficient frontier for the network technology. Both the network structure and dynamics in the production environment stand in sharp contrast with the standard DEA model, which assumes that the internal process is a "black box" and that inputs and outputs are independent across time periods. In light of the above limitations, this chapter introduces an approach to computing the technical efficiency scores for a dynamic production network and its sub-processes.

Keywords Network • Productive efficiency • Dynamic effects

10.1 Introduction

A production network can be viewed as a generalization of a value chain and consists of a set of inter-related sub-processes with directed links. These sub-processes are inter-related in a sense that one sub-process's outputs may be used by another sub-process as inputs. As a production network may span across multiple companies in different geographical regions or countries, the productive activities engaged by different sub-processes may be dependent on each other

---

This article is prepared based on Chen (2009). The author thanks Elsevier for granting reproducing the content of that paper.

C.-M. Chen (✉)
Nanyang Business School, Nanyang Technological University, 50 Nanyang Avenue, Singapore 639798, Singapore
e-mail: cmchen@ntu.edu.sg

W.D. Cook and J. Zhu (eds.), *Data Envelopment Analysis*,
International Series in Operations Research & Management Science 208,
DOI 10.1007/978-1-4899-8068-7_10,

across time. Hence in the analysis of technical efficiencies of a network, this dynamic interdependence must be taken into account. The network model considered in this article is one in which a sub-process's outputs in period t may influence other sub-processes in both t and a finite number of subsequent periods ($t + 1, t + 2, \ldots, t + n$). For example, in a multi-echelon production network, the stock level of a supplier in a certain month may influence the service level of a retailer in the following months.

The model described in this chapter is therefore distinct from the traditional DEA in two major ways. The first distinction is that our unit of analysis is a network of multiple processes, instead of a single process or decision-making unit (DMU). This type of model is usually referred to network-DEA models in the literature (e.g.,Yang et al. 2000; Castelli et al. 2001, 2004; Lewis and Sexton 2004). In these studies, the interdependence between different sub-decision-making units (SDMUs) is represented by an intra-connected production network, in which SDMUs may consume inputs (which can be either exogenous inputs or intermediate outputs produced by other SDMUs) to yield outputs (which again can be intermediate outputs or final outputs of the DMU). These underlying interrelationships among SDMUs are hidden in the traditional DEA methodology.

The studies on network-DEA models in the literature, however, assume that the input-output correspondence is complete and close in each evaluation period. This assumption can be interpreted from two different angles. On the DMU level, this means that inputs used by the DMU as a whole in one period will not affect its output in any subsequent periods. Down to the SDMU level, this assumption means that inputs used by one SDMU will not affect the output production of any downstream SDMUs that are connected to it. In practice, however, one can easily find examples in which such an assumption would be violated. One straightforward example is the use of inventory. Other common examples include (1) capital accumulation, (2) the use of fertilizer or pesticide in agriculture and the cross-period impact of pollution in the environmental context, and (3) various managerial activities used to improve organizational performance, such as the investment in advertising (Clarke 1976) and the implementation of a new human-resource strategy (Huselid and Becker 1996). In some situations, the intermediate output can even have a negative short-term influence on production (see, e.g., De Meyer and Ferdows 1990; Cooper et al. 2004).

Färe and Grosskopf (1996) introduce the formulation of storable inputs to allow asynchronism between the appearance of inputs and the use of inputs in the dynamic production model. While their approach considers dynamics of production, Färe and Grosskopf, instead of adopting a broader network perspective, confine their analysis to the dynamics of a single production process linked over multiple time periods. In their study, the intertemporal effect is limited to inputs only, and the perishableness of storable inputs is not considered. Moreover, the emphases of Färe and Grosskopf (1996) and other recent studies concerning quasi-fixed inputs within the dynamic framework (e.g., Nemoto and Goto 2003; Ouellette and Vierstraete 2004) center primarily on the efficient allocation or adjustment of inputs over time. The literature does not provide a clear guideline as for how to incorporate dynamic effects in production networks into efficiency measurement—these effects have been largely neglected or assumed to be nonexistent, say, by imposing balancing constraints on the network production model (e.g., Castelli et al. 2004). Therefore the impact of dynamic effects on efficiencies remains unclear and still requires formal and systematic treatment.

This chapter discusses a unified framework to analyze the performance of a dynamic production network. I first provide a analytical definition of the structure of production networks. The intent is to develop a systematic view on the structure of production networks, so as to facilitate legitimate comparisons among the production units in the sub-network. I then introduce a new efficiency measure to assess the performance of different hierarchical levels in the dynamic production system. The proposed efficiency measure can be decomposed in a way similar to the approach used to analyze the structure of the network. The output of the analysis can provide specific recommendations to managers at different organizational levels. In addition, I show that the new measure is closely related to the conventional DEA efficiency indexes in the literature. Finally, I investigate the relationship between the returns-to-scale properties of DMUs and those of its constituting SDMUs. This result is crucial to determining the minimum input requirement in the general network production model. Revealing this linkage also sheds new light on how a DMU can improve its scale performance from within.

10.2 Network DEA Models

To lay the groundwork for subsequent discussions, I next briefly introduce the conventional DEA. In the conventional DEA model, the production processes within a DMU are treated as a black box (i.e., only the input/output quantities at the DMU level are considered). I then formalize the concept of dynamic production networks.

10.2.1 Conventional DEA-Efficiency

Consider a set of DMUs indexed by $\mathbf{K}$, operating at a particular time period t_m within the observation window indexed by T. For all $k \in \mathbf{K}$, DMU$_k$ uses inputs $x_k^{t_m} = [x_{pk}^{t_m}]_{p=1}^{|P|} \in \mathfrak{R}_+^{|P|}$ to produce outputs $z_k^{t_m} = [z_{uk}^{t_m}]_{u=1}^{|U|} \in \mathfrak{R}_+^{|U|}$, where P and U are respectively the index sets for inputs and outputs, and $\mathfrak{R}_+^*$ represents the $*$-dimensional semipositive real space. For the time being, all inputs used are assumed to have a *contemporaneous correspondence* to outputs, meaning that inputs contribute only to the production in the same time period and vice versa.

The input-oriented technical efficiency of DMU$_0$ can be measured by the CCR model below (Charnes et al. 1978):

$$
\begin{aligned}
& f_{CCR}\left((x_0^{t_m}, z_0^{t_m}) | (x_k^{t_m}, z_k^{t_m}) \forall k \in \mathbf{K}\right) = \left(\tilde{\vartheta}_0^{t_m}, \tilde{\lambda}_0, \tilde{s}_{P0}^{t_m -}, \tilde{s}_{U0}^{t_m +}\right) \\
& = \operatorname{argmin}\left\{ \vartheta_0^{t_m} - \varepsilon \left(\sum_{p \in P} s_p^{t_m -} + \sum_{u \in U} s_u^{t_m -} \right) \right| \\
& \sum_{k \in K} \lambda_k x_{pk}^{t_m} + s_{p0}^{t_m -} = \vartheta_0^{t_m} x_{p0}^{t_m} \ \forall p \in P, \\
& \sum_{k \in \mathbf{K}} \lambda_k z_{uk}^{t_m} - s_{u0}^{t_m +} = z_{u0}^{t_m} \ \forall u \in U, \qquad (10.1) \\
& \left. \lambda_k, s_{p0}^{t_m +}, s_{u0}^{t_m +} \text{are nonnegative real numbers} \right\}, \\
& \text{where } \tilde{\lambda}_0 = [\lambda_k]_{k=1}^{|K|}, \tilde{s}_{P0}^{t_m -} = [\tilde{s}_{p0}^{t_m -}]_{p=1}^{|P|}, \tilde{s}_{U0}^{t_m +} = [\tilde{s}_{u0}^{t_m +}]_{u=1}^{|U|}, \\
& \varepsilon \text{ is a non-Archimedean infinitesimal.}
\end{aligned}
$$

The first argument of the optimal set mapping f_{CCR} is the input-output ordered pair of DMU$_0$. The second argument, namely $(x_k^{t_m}, z_k^{t_m}) \ \forall k \in \mathbf{K}$, represents the collection of all input/output data used to construct the referenced technology with which DMU$_0$ is compared. The objective of the LP in (10.1) is to proportionally minimize the input vector of DMU$_0$ and simultaneously maximize possible input and output slacks, provided that the output vector is feasible in the program. In particular, denoting the optimal solution of LP (10.1) by $(\tilde{\vartheta}_0^{t_m}, \tilde{\lambda}_0, \tilde{s}_{P0}^{t_m -}, \tilde{s}_{U0}^{t_m +})$, it can be shown that $\tilde{\vartheta}_0^{t_m} \in (0, 1]$, and $\tilde{\vartheta}_0^{t_m} x_0^{t_m} - \tilde{s}_{P0}^{t_m -}$ represents the minimal input consumption while $z_0^{t_m} + \tilde{s}_{U0}^{t_m +}$ is still producible. Then DMU$_0$ is called *weakly efficient* if $\tilde{\vartheta}_0^{t_m} = 1$, and *CCR-efficient* if it is weakly efficient and the slack vectors $\tilde{s}_{P0}^{t_m -}, \tilde{s}_{U0}^{t_m +}$ are componentwise zero. Note that *CCR-efficiency* is by definition a stronger property than *weakly efficiency*. Assumption on variable returns-to-scale (VRS) technology can be implemented by appending one additional constraint such that λ_k's sum up to one (see Banker and Thrall (1992) for discussions on returns-to-scale in DEA). To make a clear distinction, the result obtained from LP (10.1) is henceforth referred to as "DEA-efficiencies".

10.2.2 Motivational Example

I next use an example to exemplify that conventional DEA and network DEA models in the literature can break down in the presence of dynamic effects. In this example, we want to measure the efficiency of three supply chains (DMUs), each consisting of a manufacturing plant (Mfg) and a distribution center (DC) to perform production activities (see Fig. 10.1). The input, output, inventory quantities (inv.),

Fig. 10.1 Illustration of the supply chain example

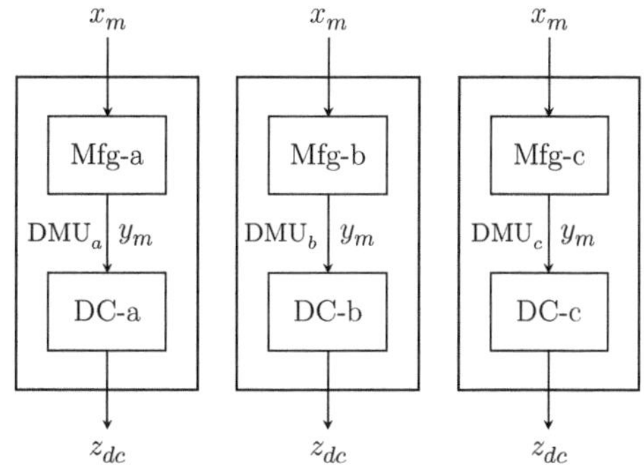

Table 10.1 Data and evaluation results of the supply chain example

DMU	x_m	y_m	eff_m	y_m	inv.	z_{dc}	eff_{dc}^{a}	eff_{DMU}^{b}	eff_{DMU}^{c}
$(t_0)a$	8	5	1.00(1)	5(*3*)	2	6	1.00	0.75(2)	0.60(2)
b	10	5	0.80(2)	5(*5*)	0	10	1.00	1.00(1)	0.80(1)
c	12	5	0.67(3)	5(*4*)	1	8	1.00	0.67(3)	0.53(3)
$(t_1)a$	12	5	0.67(3)	5(*7*)	0	14	1.00	0.78(2)	0.67(2)
b	10	5	0.80(2)	5(*5*)	0	10	1.00	0.67(3)	0.57(3)
c	8	5	1.00(1)	5(*6*)	0	12	1.00	1.00(1)	0.85(1)

[a]DC's real efficiency
[b]DMU's efficiency scores given by the CCR DEA model
[c]DMU's efficiency scores given by the index from Lewis and Sexton (2004)

and efficiencies scores are shown in Table 10.1. Particularly, in Column 4 the italic figures in the parentheses indicate the level of intermediate outputs that in effect contributes to the concurrent final outputs, and Column 5 displays the inventory level of DCs. For example, the Mfg-a uses 8 units of input to produce 5 units of intermediate product in time period t_0; the corresponding DC receives the intermediate outputs and processes it with its labor input to produce 6 units of final output. For the moment let us assume all DCs use labor in proportion to the level of intermediate outputs used. So its effect can be neglected, and we can focus on the dynamic effect of intermediate outputs. More specifically, in t_0 DC-a and DC-c actually used 3 units and 4 units of the intermediate outputs, so in t_0 2 units of intermediate outputs were inventoried in DC-a, 1 unit for DC-c, and none for DC-b. These inventoried intermediate outputs are then used to produce DC's outputs in t_1, and for the moment the quality of the inventory are assumed to remain constant over time (i.e., one unit of input stored in this period can be used equivalently as one unit of input in the next period). Thus only DCs are influenced by the dynamic effect.

The Mfgs (DCs) of these three supply chains are benchmarked with Mfgs (DCs) of the same period. The figures in the parentheses of Columns 3, 8 and 9 give the ranks of the efficiency scores. When dynamic effects are considered in the analysis, all DCs are CCR-efficient in time period t_0 and t_1 (Column 7). Given the fact that all DCs are actually CCR-efficient, the operation of Mfg should be the only source of inefficiency and therefore the supply chains' efficiency ranking should follow those of Mfgs (Column 3). However, the rankings obtained from the CCR model

Fig. 10.2 Production networks (no dynamic effect)

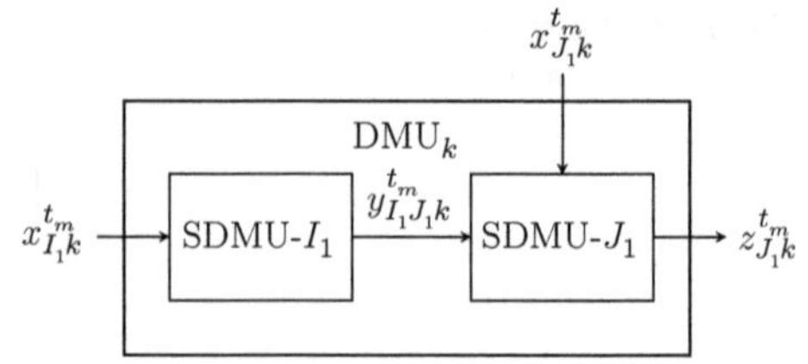

(Column 8) and the model of Lewis and Sexton (2004) (Column 9) deviate from the anticipated results in both t_0 and t_1, indicating that these two models may produce misleading results in the presence of dynamic effects.

10.2.3 Production Networks

Next I formally introduce the analysis of production networks. Consider the case where two SDMUs I_1 and J_1 are found in DMU$_k$ $\forall k \in \mathbf{K}$ (see Fig. 10.2). Then, in the case of no dynamic effect, the input/output vectors related to DMU$_k$ can be recast as $x_k^{t_m} = [x_{I_1k}^{t_m}\ x_{J_1k}^{t_m}]$ and $z_k^{t_m} = [z_{J_1k}^{t_m}]$, where $x_{I_1k}^{t_m} \in \mathfrak{R}_+^{|P_I|}$ and $x_{J_1k}^{t_m} \in \mathfrak{R}_+^{|P_J|}$ are the external inputs used by I_1 and J_1 in time period t_m, respectively. Clearly $|P_I| + |P_J| = |P|$. In particular, SDMU I_1 uses $x_{I_1k}^{t_m}$ to produce intermediate outputs $y_{I_1J_1k}^{t_m} \in \mathfrak{R}_+^{|Q|}$, where the subscripts specify the origin I_1 and the destination J_1 of the intermediate outputs indexed by Q. The intermediate outputs can be alternatively expressed as a vector $[y_{qI_1J_1k}^{t_m}]_{q=1}^{|Q|}$. SDMU J_1 employs both $x_{J_1k}^{t_m}$ and $y_{I_1J_1k}^{t_m}$ produced by I_1 to yield the final outputs $z_{J_1k}^{t_m}$. Thus homogeneity of SDMUs can be defined as:

Definition 1 (Homogeneous SDMUs) Two SDMUs I_1 and I_2 are homogeneous if and only if they employ the same inputs to produce the same outputs. Two SDMUs belong to the same layer if they are homogeneous.

By Definition 1, the membership of a layer is actually defined in terms of the homogeneity of SDMUs. Consequently, homogenous SDMUs can measure their relative efficiencies by making use of LP (10.1) with all other SDMUs in the same layer being the reference group. Moreover I_1 and J_1 can only be compared with their counterparts in their own layers according to this definition.

10.2.3.1 Layers in a DMU

We can further extend the concept to the case where individual layers within one DMU consist of multiple SDMUs. In this extended framework, the environment described in the preceding subsection becomes a special case. However, by treating individual layers in the multi-SDMU model as black boxes, the model will reduce to the single-SDMU case. Now consider the case where there are two layers I and J in each DMU (i.e., two distinct groups of homogeneous SDMUs). SDMUs in these two layers in DMU_k can be represented by the non-empty layer sets $\mathcal{L}_I^k$ and $\mathcal{L}_J^k$, respectively. Further, denote two universal layer sets by $\mathcal{L}_I$ and $\mathcal{L}_J$, where $\mathcal{L}_I = \bigcup_{k \in \mathbf{K}} \mathcal{L}_I^k$ and $\mathcal{L}_J = \bigcup_{k \in \mathbf{K}} \mathcal{L}_J^k$. These notations allows is to analytically describe each DMU_k in terms of its constituent SDMUs: $(\mathcal{L}_I^k, \mathcal{L}_J^k, \mathcal{A}^k)$, where $\mathcal{L}_I^k \subseteq \mathcal{L}_I$ and $\mathcal{L}_J^k \subseteq \mathcal{L}_J$ and $\mathcal{A}^k$ denotes the arc set of DMU_k. The arc set represents the connectivities between SDMUs in one layer to those in the other. More generally, the production process of an arbitrary DMU_k comprising a total of l layers, can be described as an ordered $(l + 1)$-tuples $(\mathcal{L}_1^k, \mathcal{L}_2^k, \ldots, \mathcal{L}_l^k, \mathcal{A}^k)$.

Furthermore, let $\mathcal{S}^k$ denote the collection of all SDMUs in DMU_k, then $\bigcup_{i=1}^{l} \mathcal{L}_i^k = \mathcal{S}^k$, and $\mathcal{L}_{l_1}^k \bigcap \mathcal{L}_{l_2}^k = \varnothing$ for any l_1, $l_2 \in l$, $l_1 \neq l_2$. These expressions imply that each SDMU can belong to one layer only, and each DMU has at least one SDMU in each layer. Here it is assumed that SDMUs do not consume intermediate outputs from SDMUs in the same layer, and therefore SDMUs within each layer are not interconnected. Now we are ready to define the structural homogeneity of DMUs that have multiple layers of SDMUs:

Definition 2 (Structurally homogeneous DMUs) DMU-k_1 and DMU-k_2 are structurally homogeneous if, and only if $\mathcal{A}^{k_1} = \mathcal{A}^{k_2}$ and $n(\mathcal{L}_l^{k_1}) = n(\mathcal{L}_l^{k_2})$ for all l, where $n(\cdot)$ is equal to the cardinality of the set.

Definition 2 defines the homogeneity of DMUs in terms of their layer structure. Similar to the homogeneity notion in conventional DEA, Definition 2 determines which DMUs are amenable to the analysis of our model. Also, it is clear that structural homogeneity implies the homogeneity in the conventional DEA models but the reverse does not necessarily hold true. In conclusion, I show that production networks can be characterized by the structural relationship between different hierarchical production units: a DMU's operation comprises the flows from and between the exterior and its internal layers, and the layers' production activities are fulfilled by their subordinate SDMUs.

10.2.4 Dynamic Effects in the Duo-Layer Network

I now introduce the analysis to the dynamic effect in a duo-layer production network. To facilitate the presentation, later discussions will be limited to the single-SDMU model, which is shown in Figs. 10.3a and 10.4. Nevertheless, the concept used to construct the single-SDMU model (hereafter "the model") is equally applicable in the extended multi-SDMU model (see Fig. 10.3b). In the model, dynamic effects prevail only in the layer $\mathcal{L}_J$ and are represented by the shaded triangle in Figs. 10.3 and 10.4. The notations used here follow those

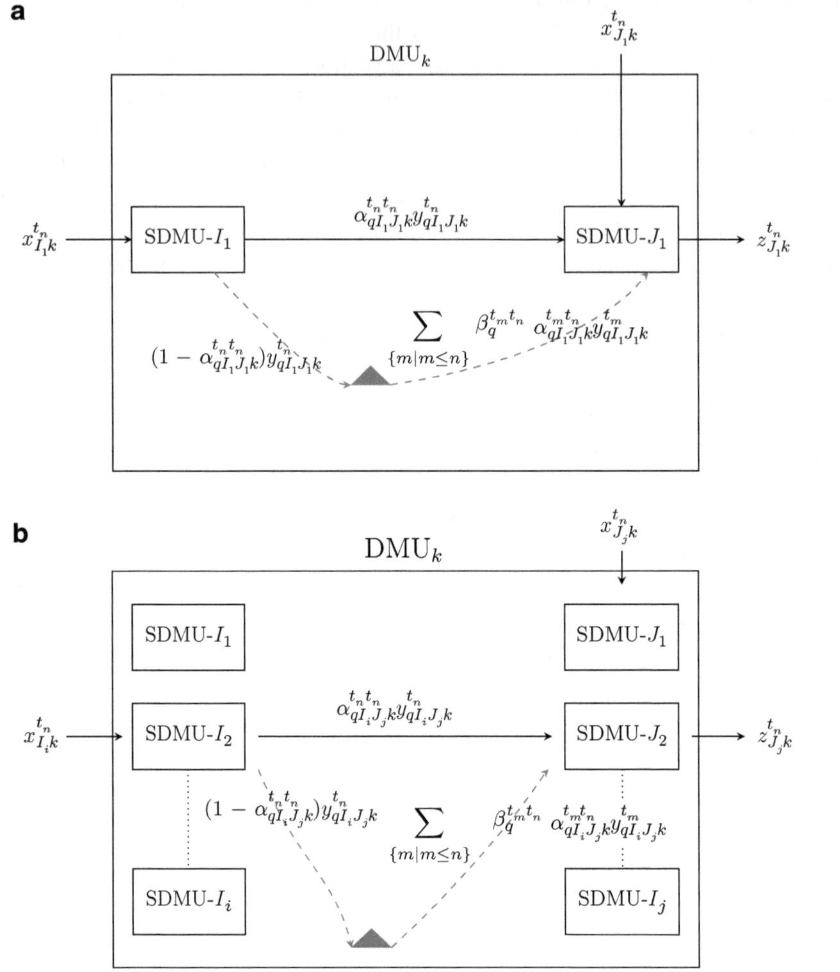

Fig. 10.3 Single-SDMU and multi-SDMU model (cross-section view). (**a**) Single-SDMU model. (**b**) Multi-SDMU model

introduced in the preceding subsections. So in $t_m \in T$, DMU$_k$ uses $x_k^{t_m} = [x_{I_1k}^{t_m}\ x_{J_1k}^{t_m}]$ to produce $z_k^{t_m} = [z_{J_1k}^{t_m}]$. Specifically, if we denote the i-th SDMU in layer I of DMU$_k$ by $s(i, I, k)$, then $s(1, I, k)$ consumes $x_{I_1k}^{t_m}$ to produce the intermediate output $y_{I_1J_1k}^{t_m}$, and $s(1, J, k)$ employs both $y_{I_1J_1k}^{t_m}$ and $x_{J_1k}^{t_m}$ to yield $z_{J_1k}^{t_m}$.

As for the dynamic factors, let us define $\alpha_{I_1J_1k}^{t_mt_n} = [\alpha_{qI_1J_1k}^{t_mt_n}]_{q=1}^{|Q|}$ where $\alpha_{qI_1J_1k}^{t_mt_n} \in [0, 1]\ \forall q \in Q$. For DMU$_k$, each component in this vector specifies the proportion of an intermediate output q that was produced by $s(1, I, k)$ in t_m, received by

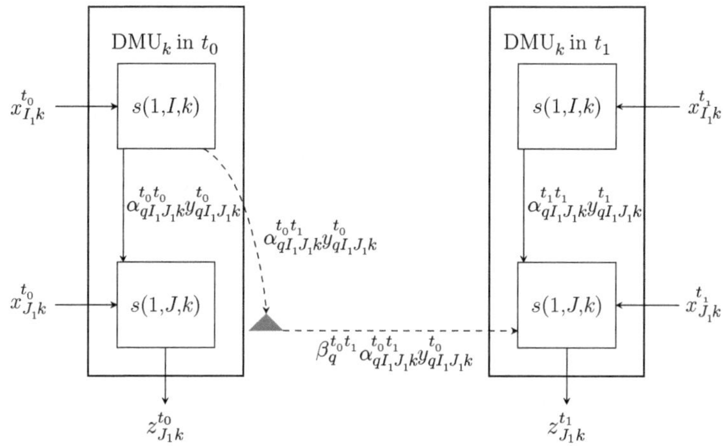

Fig. 10.4 Dynamic structure of the single-SDMU model

$s(1, J, k)$ and takes effect in t_n. The effectiveness of the *unconsumed* intermediate output is contingent on the factor $\beta^{t_mt_n} = [\beta_q^{t_mt_n}]_{q=1}^{|Q|}$, where $\beta_q^{t_mt_n} \geq 0 \quad \forall q \in Q$. The value of this factor will depend on the operational environment. This factor can readily express the degree of perishableness of intermediate outputs when the value is strictly less than one. We can further assume that (i) the dynamic effects influence the target periods only, and these effects will be fully exploited in the target periods (i.e, no compound effect exists), and (ii) the system does not produce residual dynamic effect affecting the production in all subsequent periods beyond the observation window T (see, e.g., Fig. 10.4). Then by definition it follows that:

$$\sum_{\{n|n\geq m;t_m,\,t_n\in T\}} \alpha^{t_mt_n}_{qI_1J_1k} = 1 \ \forall t_m \in T, q\in Q, k\in \mathbf{K} \tag{10.2}$$

Note that zero is not a permissible value for $\alpha^{t_mt_n}_{qI_1J_1k}$ when $t_m = t_n = t_0$ (i.e., the first period of production), since in this case the model would violate the production axiom—null inputs produce non-zero outputs in the first period t_0 of production (see, e.g., Färe and Grosskopf 1996, p. 12). Similarly it also follows that $\beta^{t_mt_n}$ is a non-zero vector. Then the effective intermediate outputs used by $s(1, J, k)$ in t_n for production are (see also Fig. 10.3):

$$\alpha^{t_nt_n}_{qI_1J_1k}y^{t_n}_{qI_1J_1k} + \sum_{\{m|m\leq n,\,t_m,\,t_n\in T\}} \beta_q^{t_mt_n}\alpha^{t_mt_n}_{qI_1J_1k}y^{t_m}_{qI_1J_1k} \ \forall q\in Q \tag{10.3}$$

To succinctly demonstrate the core ideas of the proposed model, in the remainder I focus on the efficiency measuring in the two-period case, i.e., $T = \{t_0, t_1\}$. In this setting, the dynamic effects only influence the adjacent time period. Formally, this means $\alpha_{I_1J_1k}^{t_mt_n} = 0$ for all $\{t_m, t_n \in T \mid t_n - t_m > 1\}$ and $k \in \mathbf{K}$. As can be seen in Fig. 10.4, only $\alpha_{I_1J_1k}^{t_0t_0}$ of the intermediate outputs of $s(1, I, k) \in \mathcal{L}_I$ is contemporaneous, contributing their effects to the production of $s(1, J, k) \in \mathcal{L}_J$ concurrently, and the rest will dynamically influence $s(1, J, k)$'s production in the next time period. Thus in time t_1, $s(1, J, k)$ uses not only part of the intermediate outputs produced by $s(1, I, k)$ in time t_1, but also those produced in the previous time period due to the dynamic effect. Finally, note that the dynamic effect can be either physically observable (e.g., inventory) or only conceptual (e.g., improvement in human resources) in nature, so the shaded triangles in the figures only symbolize the effect.

10.3 Efficiency Measurement

The mathematical formulations of the proposed network-DEA model and efficiency measures will be introduced in this section. Following the formulation of LP (10.1) shown earlier, I limit the discussion to the input-oriented measure only, and the technology is assumed to exhibit constant returns-to-scale (CRS). Output-oriented measures can be developed and implemented analogously. Our model describes the situation where each DMU consists of two SDMUs and the first SDMU provides its intermediate outputs to the other SDMU. More specifically, the model depicts a two-SDMU production network, in which the production of the second SDMU is affected by the dynamic effect. The DEA-efficiencies of SDMUs are derived and interpreted as in LP (10.1), and the results are subsequently used to compute our new efficiency measure. I use the new measure to evaluate DMUs and their SDMUs according to their ability to use minimal inputs to produce a given level of outputs in the dynamic production network.

10.3.1 DEA Efficiencies: SDMUs

Consider a set of homogeneous DMUs operating over time periods T. Each DMU has two layers of SDMUs I, J, and each DMU has one SDMU within each layer. Referring to Fig. 10.3a, we can see that $s(1, I, k) = \mathcal{L}_I^k$ and $s(1, J, k) = \mathcal{L}_J^k$. Using the notations defined earlier, we can identify the following properties:

$$n(\mathcal{L}_I) = n(\mathcal{L}_J) = k, n(\mathcal{L}_I^k) = n(\mathcal{L}_J^k) = 1$$
$$n(\mathcal{S}^k) = n(\mathcal{L}_I^k \bigcup \mathcal{L}_J^k) = n(\mathcal{L}_I^k) + n(\mathcal{L}_J^k) = 2, \mathcal{A}^{k_1} = \mathcal{A}^{k_2} \text{ for all } k, k_1, k_2 \in \mathbf{K}.$$

Measuring SDMU's DEA-efficiency is straightforward. Each SDMU is benchmarked with other SDMUs in the same layer set operating in the same time period. Formally, the efficiency of SDMUs can be measured by invoking LP (10.1):

Efficiency of $s(1, I, 0)$ in time t_0:

$$(\tilde{\vartheta}^{t_0}_{I_1 0}, \tilde{\lambda}_0, \tilde{s}^{t_0 -}_{P_I I_1 0}, \tilde{s}^{t_0 +}_{Q I_1 0}) = f_{CCR}\left((x^{t_0}_{I_1 0}, y^{t_0}_{I_1 0}) \middle| (x^{t_0}_{I_1 k}, y^{t_0}_{I_1 k})\ \forall k \in \boldsymbol{K} \right) \tag{10.4}$$

Efficiency of $s(1, I, 0)$ in time t_1:

$$(\tilde{\vartheta}^{t_1}_{I_1 0}, \tilde{\lambda}_0, \tilde{s}^{t_1 -}_{P_I I_1 0}, \tilde{s}^{t_1 +}_{Q I_1 0}) = f_{CCR}\left((x^{t_1}_{I_1 0}, y^{t_1}_{I_1 0}) \middle| (x^{t_1}_{I_1 k}, y^{t_1}_{I_1 k})\ \forall k \in K \right) \tag{10.5}$$

Efficiency of $s(1, J, 0)$ in time t_0:

$$\begin{aligned}(\tilde{\vartheta}^{t_0}_{J_1 0}, \tilde{\lambda}_0, \tilde{s}^{t_0 -}_{P_J J_1 0}, \tilde{s}^{t_0 -}_{Q J_1 0}, \tilde{s}^{t_0 +}_{U J_1 0}) &= f_{CCR}\left((x^{t_0}_{J_1 0}, z^{t_0}_{J_1 0}) \middle| (x^{t_0}_{J_1 k}, z^{t_0}_{J_1 k})\ \forall k \in K \right), \\ \text{where } x^{t_0}_{J_1 k} &= \left[x^{t_0}_{J_1 k}\ \left(\alpha^{t_0 t_0}_{I_1 J_1 k} \cdot y^{t_0}_{I_1 J_1 k} \right) \right]\end{aligned} \tag{10.6}$$

Efficiency of $s(1, J, 0)$ in time t_1:

$$\begin{aligned}(\tilde{\vartheta}^{t_1}_{J_1 0}, \tilde{\lambda}_0, s^{t_1}_{P_J J_1 0}, s^{t_1}_{Q J_1 0}, \tilde{s}^{t_1 +}_{U J_1 0}) &= f_{CCR}\left((x^{t_1}_{J_1 0}, z^{t_1}_{J_1 0}) \middle| (x^{t_1}_{J_1 k}, z^{t_1}_{J_1 k})\ \forall k \in K \right), \\ \text{where } x^{t_1}_{J_1 k} &= \left[x^{t_1}_{J_1 k}\ \left(\alpha^{t_1 t_1}_{I_1 J_1 k} \cdot y^{t_1}_{I_1 J_1 k} + \beta^{t_0 t_1} \cdot \alpha^{t_0 t_1}_{I_1 J_1 k} \cdot y^{t_0}_{I_1 J_1 k} \right) \right]\end{aligned} \tag{10.7}$$

where "·" in (10.6) and (10.7) denotes the componentwise multiplication of two vectors. Observe that in (10.4) and (10.5) y is the production output, while in (10.6) and (10.7) it is treated as an input. Since t_1 is the final period, it holds that $\alpha^{t_0 t_0}_{I_1 J_1 k} + \alpha^{t_0 t_1}_{I_1 J_1 k} = \alpha^{t_1 t_1}_{I_1 J_1 k} = i_{|Q|}$ where $i_{|Q|}$ is an $|Q|$ vector with all components equal to one. Consequently, model (10.6) and (10.7) will reduce to the conventional DEA model without dynamic effects if $\alpha^{t_0 t_0}_{I_1 J_1 k} = i_{|Q|}$. Before introducing the new efficiency measure in the dynamic environment, let us first prove the following theorem, which shows the relationship between DEA-efficiencies of SDMUs and that of their parent DMU.

Theorem 1 *If $s(1, I, k)$ and $s(1, J, k)$ are both CCR-efficient in some period, then DMU_k is CCR-efficient in that period.*

Proof Suppose that DMU_k is not CCR-efficient. Then there must exist some vectors $\tilde{s}^{-}_{P_I I_1 k}$, $\tilde{s}^{-}_{P_J J_1 k}$ and $\tilde{s}^{+}_{U J_1 k}$ and at least one of them is semipositive, such that $(x_{I_1 k} - \tilde{s}^{-}_{P_I I_1 k}, y_{I_1 J_1 k})$ and $([y_{I_1 J_1 k}\ x_{J_1 k} - \tilde{s}^{-}_{P_J J_1 k}], z_{J_1 k} + \tilde{s}^{+}_{U J_1 k})$ are both feasible in $s(1, I, k)$'s and $s(1, J, k)$'s respective DEA models LP (10.1). This contradicts the assumption that $s(1, I, k)$ and $s(1, J, k)$ are both CCR-efficient. Thus the result follows. □

Note that the converse of the theorem is not necessarily true. This can be observed from the fact that CCR-efficiencies of a DMU do not give further information about the levels of intermediate outputs in the DMU. Thus CCR-efficiencies of SDMUs are not assured. To illustrate, consider a simple two-SDMU example consisting of DMU_a and DMU_b as in Fig. 10.2. Let (2, 3, 1, 1) and (1, 1, 1, 1) denote (*input of* $s(1, I, \cdot)$, *intermediate output, input of* $s(1, J, \cdot)$, *final output*) of these two DMUs. Note that DMU_b is CCR-efficient but $s(1, I, b)$ is not.

10.3.2 Ψ-Efficiencies of SDMUs and DMU

According to the layer structure developed earlier, we can see that the operation of a production network system depends on the collaborative production of its internal multiple layers, whose tasks are further carried out by the SDMUs belonged to the corresponding layer. Likewise, we can measure the efficiency of production networks by first estimating the efficiencies of SDMUs, then of layers, then finally of the entire DMU. Our new efficiency measure considers both the network structure and the dynamic effect in production. The underlying concept of our approach still adheres to the classical notion of productivity in production economics, namely "*consuming less inputs to produce equivalent outputs*." In our network production model, the minimal input requirements of layers are computed by applying backward-induction-like techniques to the second layer then to the first, assuming that all SDMU are technically efficient.

Here, in the case of two-period, duo-layer with one SDMU in each layer, I define two input-oriented efficiency indices (10.8) and (10.9) with respect to $s(1, J, k)$ and $s(1, I, k)$, i.e., Ψ_{J_1k} and Ψ_{I_1k}, respectively.

$$\Psi_{J_1k} := \max_{p\in P_J, q\in Q}\left\{\frac{\sum_{t\in T} x^{*t}_{pJ_1k}}{\sum_{t\in T} x^{t}_{pJ_1k}}, \frac{\sum_{t\in T} y^{*t}_{qI_1J_1k}}{\sum_{t\in T} y^{t}_{qI_1J_1k}}\right\}$$

$$= \max_{p\in P_J, q\in Q}\left\{\frac{\sum_{t\in\{t_0,t_1\}}\left(\tilde{\vartheta}^{t}_{J_1k}x^{t}_{pJ_1k}-\tilde{s}^{t-}_{pJ_1k}\right)}{\sum_{t\in\{t_0,t_1\}} x^{t}_{pJ_1k}}, \frac{\sum_{t\in\{t_0,t_1\}}\left(\tilde{\vartheta}^{t}_{J_1k}y^{t}_{qI_1J_1k}-\tilde{s}^{t-}_{qJ_1k}\right)}{\sum_{t\in\{t_0,t_1\}} y^{t}_{qI_1J_1k}}\right\}, \tag{10.8}$$

where x^* and y^* represent the possible minimized input use.

$$\Psi_{I_1k} := \max_{p\in P_I}\left\{\frac{\sum_{t\in T} x^{*t}_{pI_1k}}{\sum_{t\in T} x^{t}_{pI_1k}}\right\}$$

$$= \max_{p\in P_I}\left\{\frac{\max_{q\in Q}\{\xi_q\}\cdot\left(\tilde{\vartheta}^{t_0}_{I_1k}x^{t_0}_{pI_1k}-\tilde{s}^{t_0-}_{pI_1k}\right)+\tilde{\vartheta}^{t_1}_{J_1k}\left(\tilde{\vartheta}^{t_1}_{I_1k}x^{t_1}_{pI_1k}-\tilde{s}^{t_1-}_{pI_1k}\right)}{x^{t_0}_{pI_1k}+x^{t_1}_{pI_1k}}\right\}$$

$$\text{where } \xi_q = \frac{\left(\tilde{\vartheta}^{t_0}_{J_1k}\alpha^{t_0t_0}_{qI_1J_1}y^{t_0}_{qI_1J_1}-\tilde{s}^{t_0-}_{qI_1J_1k}\right)+\left(\tilde{\vartheta}^{t_1}_{J_1k}\alpha^{t_0t_1}_{qI_1J_1}\beta^{t_0t_1}_{q}y^{t_0}_{qI_1J_1}-\tilde{s}^{t_1-}_{qI_1J_1k}\right)}{y^{t_0}_{qI_1J_1}}$$

$$= \tilde{\vartheta}^{t_0}_{J_1k}\alpha^{t_0t_0}_{qI_1J_1}+\tilde{\vartheta}^{t_1}_{J_1k}\left(1-\alpha^{t_0t_0}_{qI_1J_1}\right)\beta^{t_0t_1}_{q}-\frac{\tilde{s}^{t_0-}_{qI_1J_1k}+\tilde{s}^{t_1-}_{qI_1J_1k}}{y^{t_0}_{qI_1J_1}},$$

x^* represents the possible minimized input use.

(10.9)

Derivations of (10.8) and (10.9) are recounted as follows. The numerator of (10.8) and (10.9) represents the minimal aggregate input requirement with respect to the aggregate final output in these two periods. The denominators consist of the aggregate inputs used by the SDMU. To compute the numerator of Ψ_{J_1k}, the efficiencies and input slacks of $s(1, J, k)$ in t_0 and t_1 are first derived from (10.6) and (10.7). The maximum operator, instead of the minimum, is applied to the ratio because otherwise one cannot ensure that the final output vectors are still producible after implementing this reduction ratio. Ψ_{I_1k} bears a relatively complex structure, because $s(1, I, k)$ is entangled in the dynamic effect that it imposes on $s(1, J, k)$.

Similar to the SDMU in $\mathcal{L}_J$, $s(1, I, k)$ will invoke (10.4) and (10.5) to obtain the required entries of the efficiency index (10.9). Subsequently, we can derive the numerator of (10.9) in two steps corresponding to the production in t_0 and t_1. In t_1, $s(1, I, k)$ first has to reduce its input vector to $\tilde{\vartheta}^{t_1}_{I_1k}x^{t_1}_{I_1k} - \tilde{s}^{t_1-}_{P_I I_1k}$ to render itself technically efficient. Secondly, $s(1, I, k)$ has to further reduce its outputs, and thereby its inputs, by a ratio $\tilde{\vartheta}^{t_1}_{J_1k}$ in order to accommodate itself to the input reduction from $s(1, J, k)$ in t_1. So in t_1, $s(1, I, k)$ can reduce its input to $\tilde{\vartheta}^{t_1}_{J_1k}(\tilde{\vartheta}^{t_1}_{I_1k} x^{t_1}_{I_1k} - \tilde{s}^{t_1-}_{P_I I_1k})$ and $s(1, J, k)$ can still produce $z^{t_1}_{J_1k}$. Similarly, in t_0, $s(1, I, k)$ can first reduce its input levels to $\tilde{\vartheta}^{t_o}_{I_1k}x^{t_0}_{I_1k} - \tilde{s}^{t_0-}_{P_I I_1k}$. In the second step, we need to consider the input reduction from $s(1, J, k)$ in both t_0 and t_1 due to the dynamic effect (see Fig. 10.4 for an illustration). This reduction factor, denoted by ξ_q in (10.9), is the ratio of the minimally required level of intermediate outputs to the observed intermediate outputs produced at time period t_0. In particular, the terms within the first pair of parentheses in the numerator of ξ_q correspond to the minimal requirement of intermediate output q for the production of $s(1, J, k)$ in t_0 (results derived from (10.6)); the terms within the second pair of parentheses have a similar meaning except for the term t_1 and the additional decay factor in the formulation (results derived from (10.7)). So (10.9) indicates that the input requirement of $s(1, I, k)$ in t_0 also relates to the performance of $s(1, J, k)$ in both t_0 and t_1 due to the intra-connected network structure and the dynamic effect. The influence of the performance of $s(1, J, k)$ in t_1 on the index depends on the intensity of dynamic effects that $s(1, I, k)$ contributes to $s(1, J, k)$. This dynamic interrelation will be further discussed in the next subsection.

A SDMU $s(i, I, k)$ is called input-oriented Ψ-efficient if, and only if $\Psi_{I_ik} = 1$. Based on (10.8) and (10.9), we can observe several properties of these two Ψ-efficiency indexes:

Property 1 $\Psi_{J_1k} \in (0, 1]$, *and* $s(1, J, k)$ *is* Ψ*-efficient*

(a) *If and only if* $s(1, J, k)$ *is weakly efficient in both* t_0 *and* t_1, *and either* $s^{t_0-}_{pJ_1k} = s^{t_1-}_{pJ_1k} = 0$ *for at least one* $p \in P_J$ *or* $s^{t_0-}_{qJ_1k} = s^{t_1-}_{qJ_1k} = 0$ *for at least one* $q \in Q$.

(b) *If* $s(1, J, k)$ *is CCR-efficient in both* t_0 *and* t_1.

Proof From LP (10.1) we know $x^{t*}_{pJ_1k} \geq \tilde{\vartheta}^{t*}_{J_1k} x^{t*}_{pJ_1k} - \tilde{s}^{t*-}_{pJ_1k}$, and $y^{t*}_{qJ_1k} \geq \tilde{\vartheta}^{t*}_{J_1k} y^{t*}_{qJ_1k}$ $-\tilde{s}^{t*-}_{qJ_1k}$ for $* = 1, 2$ and all $p \in P_J$, $q \in Q$. Hence $\Psi_{J_1k} \in (0, 1]$ because input vectors are semipositive (see also Theorem 3.3 in Cooper et al. 2006). Given $\Psi_{J_1k} = 1$, there must exist at least one $p \in P_J$ (or one $q \in Q$), such that equalities hold in the above two inequalities. Thus $\tilde{\vartheta}^{t_0}_{J_1k} = \tilde{\vartheta}^{t_1}_{J_1k} = 1$, and either $\tilde{s}^{t_0-}_{pJ_1} = \tilde{s}^{t_1-}_{pJ_1}$ $= 0$ for some $p \in P_J$ or $\tilde{s}^{t_0-}_{qJ_1} = \tilde{s}^{t_1-}_{qJ_1} = 0$ for some $q \in Q$. Then the sufficiency of (a) is proved. The necessity of (a) can be shown by simple algebraic substitutions. The proof of (b) follows immediately from (a). □

Property 2 $\Psi_{I_1k} \in (0, 1]$, *and* $s(1, I, k)$ *is* Ψ*-efficient*

(a) *If and only if the following three conditions are all met: (1)* $s(1, I, k)$ *and* $s(1, J, k)$ *are weakly efficient in both* t_0 *and* t_1, *(2)* $s^{t_0-}_{pI_1k} = s^{t_1-}_{pI_1k} = 0$ *for at least one* $p \in P_I$, *and (3)* $s^{t_0-}_{qI_1J_1k} = s^{t_1-}_{qI_1J_1k} = 0$, *and either* $\beta^{t_0t_1}_q = 1$ *or* $\alpha^{t_0t_0}_{qI_1J_1k}$ $= 1$ *for some* $q \in Q$.

(b) *If* $s(1, I, k)$ *and* $s(1, J, k)$ *are CCR-efficient in both* t_0 *and* t_1 *and either* $\beta^{t_0t_1}_q = 1$ *or* $\alpha^{t_0t_0}_{qI_1J_1k} = 1$ *for some* $q \in Q$. □

Proof Similar to the proof of Property 1.

The Ψ-efficiency of a DMU is defined as in Definition 3. This definition is also applicable in the general case where multiple SDMUs exist in each layer.

Definition 3 (Ψ-efficiencies of DMUs) DMU$_k$'s input-oriented Ψ-efficiency is defined by

$$\Psi_k = \left(\prod_{i \in \mathcal{L}^k_I} \Psi_{I_ik} \right)^{1/n(\mathcal{L}^k_I)} \left(\prod_{j \in \mathcal{L}^k_J} \Psi_{J_jk} \right)^{1/n(\mathcal{L}^k_J)}$$

and DMU$_k$ is called input-oriented Ψ-efficient if, and only if $\Psi_k = 1$.

By Definition 3, two properties follow immediately from Properties 1 and 2.

Property 3 $\Psi_k \in (0, 1]$ *and DMU$_k$ is* Ψ*-efficient if, and only if* $s(1, I, k)$ *and* $s(1, J, k)$ *are both* Ψ*-efficient.*

Property 4 *If DMU$_k$* is Ψ*-efficient, then DMU$_k$ is at least weakly efficient in* t_0 *and* t_1.

10.3.3 Discussions on Ψ-Efficiencies

The first remark is that the Ψ-efficiency of $s(1, I, k)$ depends not only on its own performance, but also on the performance of $s(1, J, k)$ to an extent moderated by the dynamic parameters. Observe that $s(1, I, k)$ and $s(1, J, k)$ will become

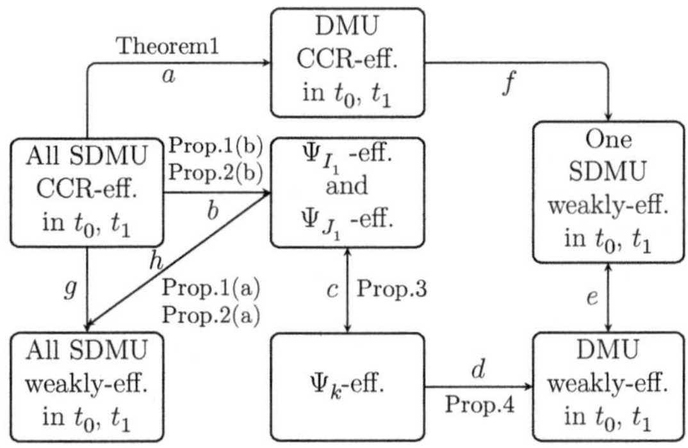

Fig. 10.5 Relationships among Ψ-efficiencies and DEA-efficiencies

independent of each other in production, if there does not exist any connection between these two layers. In this case, the former intermediate output changes into the final output, and $s(1, I, k)$ is no longer associated with $s(1, J, k)$ in terms of the DMU's final outputs. Therefore its Ψ-efficiency can be measured in a way similar to (10.8).

Property 1 indicates that CCR-efficiency is a sufficient condition for $s(1, J, k)$ to be Ψ-efficient, while weakly efficiency is not. Properties 1 and 2 together imply that by Definition 3 there can be no Ψ-efficient DMU at all, either because of the internal inefficiency or the production externality due to the dynamic effect. However, the non-existence of efficient DMUs can be considered a relative merit of our approach, as compared to the conventional DEA, because the network-DEA model is more sensitive in detecting inefficiencies. If we assume that either intermediate outputs are of equivalent effect in t_1 as in t_0 (i.e., $\beta_q^{t_0t_1} = 1$, no decay effect), or they are contemporaneous (i.e., $\alpha_{qI_1J_1}^{t_0t_0} = 1$, so the decay factor becomes irrelevant), Property 2 is actually quite similar to Property 1, except for those conditions related to the efficiencies of $s(1, J, k)$. However, Property 2 does not hold when $\beta^{t_0t_1}$ and $\alpha_{I_1J_1k}^{t_0t_0}$ are strictly less than one. In fact, $s(1, I, k)$ is then Ψ-inefficient by default in this two-period model. This is because the utility of a proportion of the intermediate outputs will inevitably be nullified by the decay factor. Definition 3 specifies the Ψ-efficiency of a DMU as the product of two geometric means, which individually can be interpreted as the average Ψ-efficiency of a layer in the DMU. In other words, a DMU's Ψ-efficiencies depend on the Ψ-efficiencies of its layers, which will further rely on the performance of those SDMUs within the corresponding layer. Property 3 shows that a DMU is Ψ-efficient if and only if all SDMUs within the DMU are also Ψ-efficient. Therefore, under some appropriate assumptions on the parameters of dynamic effects, SDMU's CCR-efficiencies can imply DMU's Ψ-efficiencies, which also suggests that the DMU is weakly efficient in both periods (Property 4).

Finally, we can schematically summarize the relationships between the Ψ-efficiency and the DEA-efficiency of different production units (see Fig. 10.5). This figure shows a new perspective on the connections between static and dynamic efficiencies, and it also provides a road map for decision-makers to reexamine and improve their performance. Links in the figure (i.e., *a*, *b*, *c* and *d*) are substantiated by the corresponding Theorem or Property annotated beside the link. Link *f* and *g* come directly from the definition via *e*. Link *h* is affirmed by the above discussion, from which we know that Ψ-efficiencies of SDMUs imply their own weakly efficiencies. Yang et al. (2000) proved that a DMU is CCR-efficient if and only if all SDMUs in the DMU are also CCR-efficient. This finding is not entirely compatible to our model (cf. link *a* in Fig. 10.5) due to the discrepant internal structures: in our terminology, the production network in Yang et al. can be described as a single-SDMU, multi-layer production network, in which no linkage exists between layers. Link *e* was proved in Castelli et al. (2004).

10.4 RTS of Production Networks

By exploring the internal structure of a DMU, we can identify the relationship between the RTS property of a DMU and that of its constituting layers of SDMUs. Disclosing this interrelationship helps signify new insights that the decision-maker at the DMU level can employ to improve the scale efficiency from within.

Before investigating the RTS properties, let us first check whether introducing network structures will interfere with the propensity of the RTS of a DMU. Let us first look at the invariance property of production networks. Consider now for every DMU_k there are two layers $\mathcal{L}_I^k$ and $\mathcal{L}_J^k$ inside, and each layer can include single or multiple SDMUs. I use the backward-induction technique introduced in the previous section to calculate the Ψ-efficiency index. Then, if $\mathcal{L}_J^k$ needs to reduce the aggregate use of intermediate outputs produced by $\mathcal{L}_I^k$ by, say, 20 %, will the subsequent input reduction of $\mathcal{L}_I^k$ vary if output reductions are not allocated evenly to each SDMU in $\mathcal{L}_I^k$? One can also relate this situation to the scenario where the demand for a certain product is declining in the market but its impact to different suppliers is asymmetrical. This is an important issue, because otherwise the RTS property of a DMU will lose tractability in the network environment. Define the degree of RTS as the ratio between the proportional increase in inputs and the corresponding proportional increase in outputs. Then the *back-attributively invariant property* means that, if the SDMUs that are being back-attributed have the same degree of RTS,[1] the total amount of input saved in this layer is invariant even under different back-attribution schemes, provided that the mix of aggregate intermediate outputs of the downstream layer remains constant after reductions.

[1] This is not to be confused with the CRS assumption in the conventional DEA model, as the latter is a special case of the former.

Specifically, for some $\theta \in (0, 1]$, the constant-mix condition between two layers can be mathematically represented as:

$$\sum_{i\in\mathcal{L}_I^k}\sum_{j\in\mathcal{L}_J^k}\tilde{y}_{qI_iJ_jk} = \theta\sum_{i\in\mathcal{L}_I^k}\sum_{j\in\mathcal{L}_J^k} y_{qI_iJ_jk} \ \ \forall q\in Q \tag{10.10}$$

where $\tilde{y}_{qI_iJ_jk}$ denotes the reduced level of intermediate output q corresponding to the aggregate input reduction of $\mathcal{L}_J^k$. So the mix among different intermediate outputs is maintained after reduction. Additionally, denoting the degree of RTS of $s(i, I, k)$ by ζ_{I_i}, the same degree of RTS assumption is equivalent to the following condition:

$$\zeta_{I_i} = \zeta_I \ \text{ for all } \ i\in\mathcal{L}_I^k. \tag{10.11}$$

The above statements can be formally summarized in the following proposition.

Proposition 1 *Suppose $\mathcal{L}_I^k$ precedes $\mathcal{L}_J^k$ in terms of flows, then the total amount of inputs reduced in $\mathcal{L}_I^k$ by performing backward-induction is back-attributively invariant if and only if all SDMUs in $\mathcal{L}_I^k$ have the same degree of RTS, provided that the intermediate output mix remains constant.*

Proof of Proposition 1 Let $\tilde{\vartheta}_{J_jk}$ be the DEA-efficiency score of $s(j, J, k) \in \mathcal{L}_J^k$. By this score we can identify $s(j, J, k)$'s efficient target, which uses intermediate outputs $\tilde{\vartheta}_{J_jk}y_{qI_iJ_j} - \tilde{s}^-_{qJ_jk}\forall q\in Q$ and input $\tilde{\vartheta}_{J_jk}x_{pJ_j} - \tilde{s}^-_{pJ_jk}\forall p\in P_J$ to produce the given level of final outputs. By the constant-mix assumption (10.10), the following equation holds for the aggregate intermediate output reduction of $\mathcal{L}_J^k$:

$$\sum_{j\in\mathcal{L}_J^k}\left((1-\tilde{\vartheta}_{J_jk})\sum_{i\in\mathcal{L}_I^k} y_{qI_iJ_jk} + \tilde{s}^-_{qJ_jk}\right) = \theta\mathrm{Y}_q \ \forall q\in Q$$

$$\text{where } \theta\in(0,1] \text{ is some constant and } \mathrm{Y}_q = \sum_{i\in\mathcal{L}_I^k}\sum_{j\in\mathcal{L}_J^k} y_{qI_iJ_jk} \tag{10.12}$$

Thus we can omit the subscript q without the risk of confusion. Let $\phi_{I_i}\in[0, 1]$ be the ratio of the total reductions in the intermediate outputs allocated to $s(i, I, k)$, and it thereby has to reduce its outputs by the amount equal to (10.13).

$$\phi_{I_ik}\theta\mathrm{Y}, \text{ where } \sum_{i\in\mathcal{L}_I^k}\phi_{I_i} = 1 \tag{10.13}$$

The reduction in inputs consumed by $s(i, I, k)$ is proportional to its output reductions, because of the assumption that $\mathcal{L}_I^k$ exhibits the same degree of RTS as defined in (10.11). Thus $\zeta_{I_i} = \zeta_I$ holds for all $i \in \mathcal{L}_I^k$. Then the total input reductions can be expressed via:

$$\sum_{i \in \mathcal{L}_I^k} (\zeta_{I_i} \phi_{I_i} \theta \mathrm{Y}) = \zeta_I \theta \mathrm{Y} \sum_{i \in \mathcal{L}_I^k} \phi_{I_i} = \zeta_I \theta \mathrm{Y} \quad (10.14)$$

Equation (10.14) indicates that the total input reduction is invariant with different values of ϕ_{I_i}. This shows the necessity of the condition. To prove the sufficiency, the total reduction in inputs used by $\mathcal{L}_I^k$ can be written as:

$$\sum_{i \in \mathcal{L}_I^k} \zeta_{I_i} \phi_{I_i} \theta \mathrm{Y} = \theta \mathrm{Y} \sum_{i \in \mathcal{L}_I^k} \phi_{I_i} \zeta_{I_i} \quad (10.15)$$

Given that total amount of input saved in $\mathcal{L}_I^k$ is attributively invariant, the summation term on the right side of (10.15) has to be a constant for different $\phi_{I_i} \in [0, 1]$. Thus $\zeta_{I_i} = \zeta_I$ must hold for all $i \in \mathcal{L}_I^k$. This also implies that all $s(i, I, k) \in \mathcal{L}_I^k$ exhibit the same degree of RTS ζ_I in production. □

As opposed to the assumption of the same degree of RTS in Proposition 1, if the technology of $\mathcal{L}_I^k$ exhibits variable degrees of RTS (i.e., for different $i \in I$, $s(i, I, k)$ can produce at different degrees of RTS), then the problem will become complicated, since now different allocation schemes can result in different total input reductions of the precedent layer. Moreover, output reductions itself can lead to changes in RTS of $\mathcal{L}_I^k$. Thus here in this study constant RTS (CRS) is assumed in all layers.

Following Proposition 1, two corollaries can be derived (cf. Färe and Grosskopf 1996, p. 163).

Corollary 1 *If $\mathcal{L}_I^k$ and $\mathcal{L}_J^k$ exhibit CRS, then DMU_k also exhibits CRS.*

Corollary 2 *Suppose DMU_k is Ψ_k-efficient and denote ζ_k as the degree of RTS of DMU_k, ζ_I as that of $s(i, I, k) \forall i$, ζ_J as that of $s(j, J, k) \forall j$. Then $\zeta_k = \zeta_I \zeta_J$.*

Corollary 1 suggests that the DMU can achieve CRS if all of its constituent SDMUs also operate on CRS. However the reverse is not necessarily true, since one layer with increasing RTS and the other with decreasing RTS can also result in CRS at the DMU level. Färe and Grosskopf (1996) proved the sufficient condition for a network consisting of a series of technologies to exhibit CRS. Corollary 1 is in line with their finding. However the production network in their study corresponds to a multi-layer version of the single-SDMU model without dynamic effects, which is different from our model. Corollary 2 reveals the causal relationship of the RTS properties between a DMU and its SDMUs. Thus decision-makers have to seek the balance between two layers to optimize the scale performance. For example, a DMU can increase the overall scale of the layer exhibiting increasing RTS, and diminish the overall scale of the layer exhibiting decreasing RTS. Additionally, the DMU must simultaneously maintain the efficiencies of all its SDMUs and promote a well-connected coordination between two layers (i.e., incorporating the influence of dynamic effects into decision-makings). Consequently, one can also consider increasing or decreasing the number of SDMUs that possess a specific RTS, so as to improve scale performance and balance supply and demand between layers (e.g., downsize, merge, enlarge or acquire new SDMUs).

10.5 Numerical Example

The proposed efficiency measure is applied to a numerical example consisting of six DMUs and two observation periods t_0 and t_1 (see Fig. 10.3a and Table 10.2). Using the notation defined earlier, we now have $s(1, I, k)$ and $s(1, J, k)$ for $k = 1$ to 6, and $T = \{t_0, t_1\}$. The control factor $\beta_q^{t_0 t_1}$ is assumed to be unity. $\alpha_{qI_1J_1k}^{t_0 t_0}$ is designated to be 0.7 for all k. The data and the efficiency scores are tabulated in Table 10.2. It can be seen that no SDMU in $\mathcal{L}_I$ is Ψ-efficient, whereas one SDMU in layer $\mathcal{L}_J$ is Ψ-efficient. Moreover, the mean Ψ-efficiency score of the SDMUs in $\mathcal{L}_I$ (≈ 0.72) is lower than that of $\mathcal{L}_J$ (≈ 0.81). This result indicates $\mathcal{L}_J$ outperforms $\mathcal{L}_I$ as a whole. Specifically, only $s(1, J, 3)$ is Ψ-efficient because it is CCR-efficient in t_0 and t_1 (Columns 14 and 15). This result is clear from Properties 1 and 2. Unlike in DEA models, none of the six DMUs achieves Ψ-efficiency (Column 13), because they did not use minimal inputs vector x_{I_1} and x_{J_1} to produce the given level of final outputs. Insights and directions for improvements can be discovered by decomposing the DMU's Ψ-efficiency score into SDMU's Ψ-efficiencies (Columns 11 and 12). The feature of efficiency decomposition is a clear advantage over the ordinary DEA analysis, which comparatively reveals deficient information for improving the internal production processes.

10.6 Discussion and Conclusion

This study shows that conventional DEA approaches may lead to biased results due to the dynamic effect in production networks, and proposes a systematic approach to incorporate the effect into efficiency measurement. The proposed Ψ-efficiency measure is built upon the hierarchical and interrelated production structure within a DMU. Using this approach, decision-makers can inerrably delve into the production network and pinpoint areas for improvement. Various connections between the DEA efficiency and the Ψ-efficiency have been established. This chapter also shows, in the network production, the RTS properties of DMUs can be characterized by those of its constitutive SDMUs. In all, decision-makers can substantially benefit from our approach to methodically analyzing and seeking for performance improvements in the dynamic production network.

This study has several additional implications. This study adverts to an important, yet much-ignored issue in efficiency measuring caused by dynamic effects in a production network. So decision-makers should carefully consider dynamic effects

Table 10.2 A numerical example and the efficiency

	$I_1(t_0)$		$J_1(t_0)$			$I_1(t_1)$		$J_1(t_1)$							
	Input	Output	Input	Int.[a]	Output	Input	Output	Input	Int.[a]	Output	Ψ_{I_1}	Ψ_{J_1}	Ψ_k	$\vartheta_k^{t_0}$[b]	$\vartheta_k^{t_1}$[b]
DMU_1	87	87	164	60.9	178	82	93	195	119.1	184	0.69	0.73	0.51	0.63	0.73
DMU_2	79	98	195	60.8	184	94	75	165	104.4	147	0.57	0.62	0.36	0.72	0.51
DMU_3	95	77	213	53.9	293	75	96	215	119.1	232	0.90	1.00	0.90	1.00	1.00
DMU_4	75	79	193	55.3	156	97	96	192	119.7	180	0.63	0.67	0.42	0.64	0.60
DMU_5	92	82	155	57.4	192	70	72	161	96.6	192	0.75	0.84	0.63	0.69	0.89
DMU_6	78	76	279	53.2	216	98	77	292	99.8	237	0.79	0.95	0.75	0.85	0.78

[a] Effective intermediate outputs to $s(1, J, k)$

[b] DMU's efficiency scores given by the CCR DEA model

when assessing organizational performance, especially for those production units with identifiable internal structures. Similarly, decision-makers should also attend to external dynamic effects in production. Therefore the management should pay equivalent attention to the dynamic interactions among DMUs within a larger body of production. Secondly, DMU's efficiencies relate closely to SDMUs' efficiencies, but the former in general do not imply the latter. So exploring the internal structure of a DMU should help detect additional areas for improvement. In practice, however, decision-makers may need to consider the trade-off between the cost of obtaining detailed information about internal activities, and the ensuing economic benefit from additional knowledge of inefficiencies. As for the scale performance in production networks, the finding here indicates that decision-makers should improve DMU's scale efficiency by first adjusting the scales of all SDMUs inside the DMU to an optimal size. Finally, as the proposed model is conditional on the values of dynamic parameters, one interesting follow-up direction is to incorporate econometric methodologies, stochastic modeling techniques or resort to expert opinion to appropriately determine the empirical values of the dynamic parameters in the proposed model.

As noted earlier, a concomitant downside of parameterization of the dynamic effect is that analysts need to find a reliable way to estimate this effect. The estimation issue is examined in Chen and Van Dalen (2010), in which an econometric model for panel data is used to estimate the dynamic effect. In retrospect, the above methodology may not be completely satisfactory because inefficiency is not considered during the estimation of the dynamic parameters. One interesting future direction is to develop a single-stage efficiency model in which the dynamic effect and inefficiencies can be estimated simultaneously; see, for example, the single-stage stochastic frontier model by Battese and Coelli (1995).

References

Banker, R., & Thrall, R. (1992). Estimation of returns to scale using data envelopment analysis. *European Journal of Operational Research, 62*, 74–84.

Battese, G., & Coelli, T. (1995). A model for technical inefficiency effects in a stochastic frontier production function for panel data. *Empirical Economics, 20*(2), 325–332.

Castelli, L., Pesenti, R., & Ukovich, W. (2001). DEA-like models for efficiency evaluations of specialized and interdependent units. *European Journal of Operational Research, 132*, 274–286.

Castelli, L., Pesenti, R., & Ukovich, W. (2004). DEA-like models for the efficiency evaluation of hierarchically structured units. *European Journal of Operational Research, 154*, 465–476.

Charnes, A., Cooper, W., & Rhodes, E. (1978). Measuring the efficiency of decision making units. *European Journal of Operational Research, 2*, 429–444.

Chen, C. (2009). A network-dea model with new efficiency measures to incorporate the dynamic effect in production networks. *European Journal of Operational Research, 194*(3), 687–699.

Chen, C., & Van Dalen, J. (2010). Measuring dynamic efficiency: Theories and an integrated methodology. *European Journal of Operational Research, 203*(3), 749–760.

Clarke, D. (1976). Econometric measurement of the duration of advertising effect on sales. *Journal of Marketing Research, 13*, 345–357.

Cooper, W., Deng, H., Seiford, L., & Zhu, J. (2004). Congestion: Its identification and management with DEA. In W. Cooper, L. Seiford, & J. Zhu (Eds.), *Handbook on data envelopment analysis* (pp. 177–202). New York: Kluwer Academic.

Cooper, W., Tone, K., & Seiford, L. (2006). Introduction to data envelopment analysis and its uses: With DEA-solver software and references. New York: Springer.

De Meyer, A., & Ferdows, K. (1990). Influence of manufacturing improvement programmes on performance. *International Journal of Operations & Production Management, 10*, 120–131.

Färe, R., & Grosskopf, S. (1996). *Intertemporal production frontiers: With dynamic DEA*. Boston: Kluwer Academic.

Huselid, M., & Becker, B. (1996). Methodological issues in cross-sectional and panel estimates of the human resource-firm performance link. *Industrial Relations, 35*, 400–422.

Lewis, H., & Sexton, T. (2004). Network DEA: Efficiency analysis of organizations with complex internal structure. *Computers & Operations Research, 31*, 1365–1410.

Nemoto, J., & Goto, M. (2003). Measurement of dynamic efficiency in production: An application of data envelopment analysis to Japanese electric utilities. *Journal of Productivity Analysis, 19*, 191–210.

Ouellette, P., & Vierstraete, V. (2004). Technological change and efficiency in the presence of quasi-fixed inputs: A DEA application to the hospital sector. *European Journal of Operational Research, 154*, 755–763.

Yang, Y., Ma, B., & Koike, M. (2000). Efficiency-measuring DEA model for production system with k independent subsystems. *Journal of the Operations Research Society of Japan, 43*, 343–354.

Chapter 11
Slacks-Based Network DEA

Kaoru Tone and Miki Tsutsui

Abstract Traditional DEA models deal with measurements of relative efficiency of DMUs regarding multiple-inputs versus multiple-outputs. One of the drawbacks of these models is the neglect of intermediate products or linking activities. After pointing out needs for inclusion of them to DEA models, we propose a slacks-based network DEA model that can deal with intermediate products formally. Using this model we can evaluate divisional efficiencies along with the overall efficiency of decision making units (DMUs).

Keywords Data envelopment analysis • Network DEA • SBM • WSBM • Divisional efficiency • Overall efficiency

11.1 Introduction

Traditional DEA models deal with measurements of relative efficiency of DMUs regarding multiple-inputs versus multiple-outputs. One of the drawbacks of these models is the neglect of internal or linking activities. For example, many companies are comprised of several divisions that are linked as illustrated in Fig. 11.1.

Part of the material in this chapter is adapted from *European Journal of Operational Research*, Vol. 197, Tone K., & Tsutsui M., Network DEA: A slacks-based measure approach, 243–252, 2009, with permission from Elsevier Science

K. Tone (✉)
National Graduate Institute for Policy Studies, 7-22-1 Roppongi, Minato-ku,
Tokyo 106-8677, Japan
e-mail: tone@grips.ac.jp

M. Tsutsui
Central Research Institute of Electric Power Industry, 1-6-1 Otemachi,
Chiyoda-ku, Tokyo 100-8126, Japan

W.D. Cook and J. Zhu (eds.), *Data Envelopment Analysis*,
International Series in Operations Research & Management Science 208,
DOI 10.1007/978-1-4899-8068-7_11,

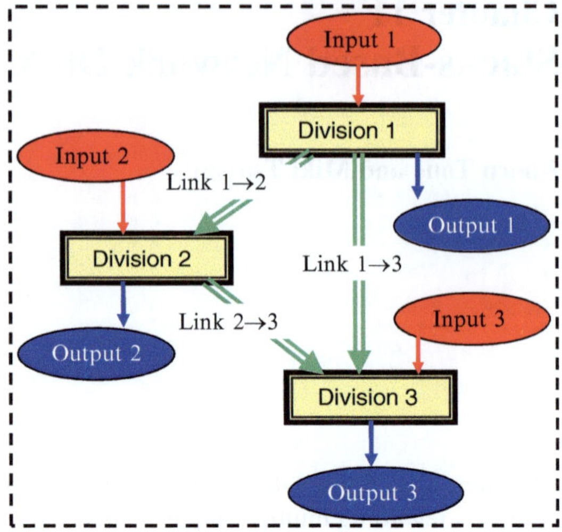

Fig. 11.1 Company with three linked divisions

In the example, the company has three divisions. Each division utilizes its own input resources for producing its own outputs. However, there are linking activities (or intermediate products) as shown by Link 1 → 2, Link 1 → 3 and Link 2 → 3. Link 1 → 2 indicates that parts of the outputs from Division 1 are utilized as inputs to Division 2. In traditional DEA models, every activity should belong to either input or output but not to both. So usually they employ multiple steps for evaluation, using intermediate products as outputs in one step and as inputs in another step. Thus, these models cannot deal with intermediate products directly in a single step.

Although there may be many variants of this process flow, the existence of linking activities is an indispensable part of Network DEA models.

Within traditional DEA models there are at least two approaches for evaluating the efficiency of multi-division organizations.

11.1.1 Aggregation (Black Box)

A simple approach is to aggregate these divisions into a single company which utilizes Inputs 1, 2 and 3, and produces Outputs 1, 2 and 3 (Fig. 11.2). However, using this approach we neglect internal linking activities, and thus, we cannot evaluate the impact of division-specific inefficiencies on the overall efficiency of the company as a whole. Furthermore, this model might choose an inappropriate pair of input vs. output for evaluation and assign an unreasonable score to the concerned DMU, since DEA selects the most favorable pair for the DMU in the sense of maximizing the ratio scale (see Cooper et al. 2007, p. 25). In other words, the analysis does not fully access the underlying diagnostic value potentially available to management. This model often rouses a problem involving degree of freedom in that the number of input and output items increases relative to the

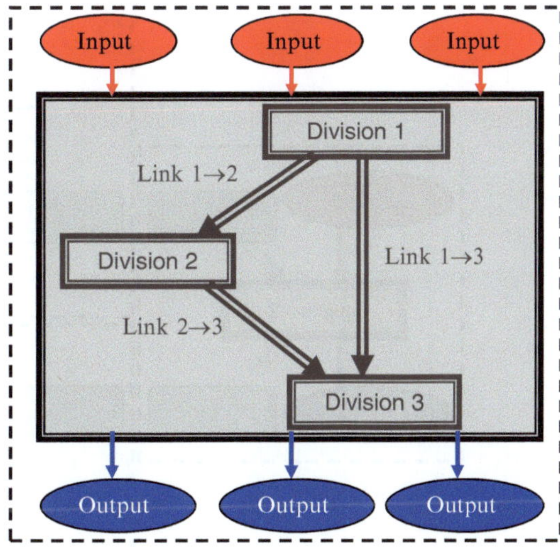

Fig. 11.2 Black box

number of DMUs. As a rule of thumb, DEA demands that the number of DMUs should be at least three times larger than the sum of numbers of inputs outputs as otherwise DEA is apt to lose discriminating power (see Cooper et al. 2007, p. 284). We will point to this in Sects. 11.5 and 11.6.

11.1.2 Separation

The second approach is to evaluate divisional efficiency individually (Fig. 11.3). In this case, we evaluate the efficiency of Division 1 of each company among the set of DMUs using Input 1 as input, and Output 1, Link 1 $\rightarrow$ 2 and Link 1 $\rightarrow$ 3 as outputs. Similarly we evaluate the efficiency of Division 2 of each company among the set of DMUs using Link 1 $\rightarrow$ 2 and Input 2 as inputs, and Link 2 $\rightarrow$ 3 and Output 2 as outputs. In this way, we can evaluate efficiency of each division of a company among the set of DMUs, and hence can find benchmarks for each division. However, this approach does not account for the continuity of links between divisions.

11.1.3 Needs for Network DEA

The above observations lead us to consider a DEA model called "Network DEA model" that accounts for divisional efficiencies as well as the overall efficiency in a unified framework. This means that we evaluate the total efficiency of DMUs as the main objective which involves divisional efficiencies as its components. Network DEA models were introduced in the innovative book (Färe and Grosskopf 1996) by

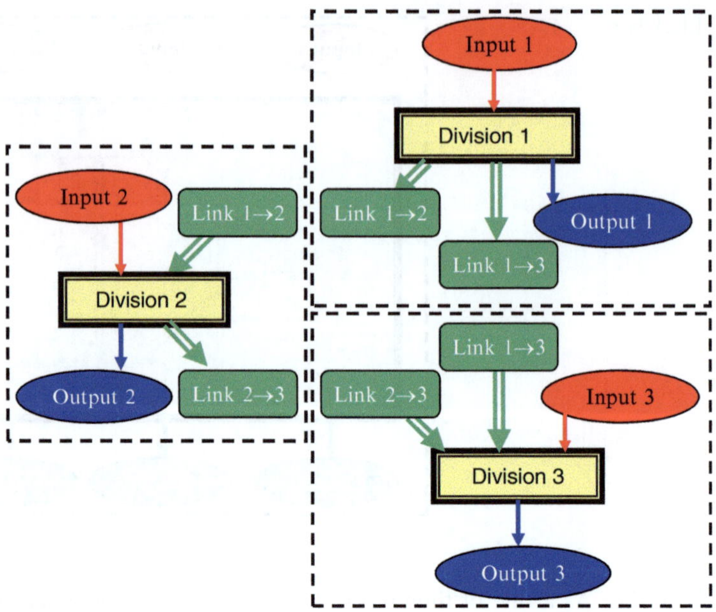

Fig. 11.3 Separation

Färe and Grosskopf (see also Färe 1991; Färe and Grosskopf 2000). They investigated the so-called "black box" for the first time. Their models were extended by several authors.

The network DEA model (Lewis and Sexton 2004) proposed by Lewis and Sexton has a multi-stage structure as an extension of the two-stage DEA model proposed in Sexton and Lewis (2003). This study solves a DEA model for each node independently. For an output-oriented model, firstly a general DEA model is solved for the upstream node at the 1st stage to obtain the optimal solution of outputs. At the next stage, a part of (or all of) optimal outputs obtained at the upstream node are applied as intermediate inputs to the next node. After solving DEA models for all nodes in turn, a final optimal output is obtained at the last node. The firm-level efficiency score is measured as the final optimal output divided by an observed output.

Prieto and Zofio (2007) applied network efficiency analysis within an input–output model initiated by Koopmans (1951). They optimized primary input allocations, intermediate products and final demand products by way of Network DEA techniques and succeeded in applying their models to input–output database of OECD countries.

Löthgren and Tambour (1999) applied Network DEA model to a sample of Swedish pharmacies with organizational objectives that necessitates a monitoring of efficiency and productivity as well as customer satisfaction. They compared the results of Network DEA models with those of traditional DEA models (see also Chen 2009; Kao 2009).

The above Network DEA models utilize the radial measure of efficiency, e.g. the CCR (Charnes et al. 1978) or the BCC (Banker et al. 1984) models as the basic DEA methodology and the production possibility set. The radial models stand on the assumption that inputs or outputs undergo proportional changes. However, this assumption needs care. For example, if we employ labor, materials and capital as inputs, some of them are substitutional and do not change proportionally.

This chapter introduces a network DEA model, that uses the slacks-based measure (SBM: Tone 2001; Pastor et al. 1999) approach for evaluating efficiency. The SBM is a non-radial method and is suitable for measuring efficiencies when inputs and outputs may change non-proportionally. This model can decompose the overall efficiency into divisional ones. Furthermore, we employ the weighted SBM model (Cooper et al. 2007; Tsutsui and Goto 2009) in order to incorporate the importance of divisions. These weights are set exogenously. We also investigate several properties of Network DEA models and show that, under the variable returns-to-scale assumption, every division has at least one efficient DMU (decision making unit) for the division, whereas under the constant returns-to-scale assumption it is possible that some division has no efficient DMUs for the division.

The remainder of this chapter unfolds as follows. In the next section, we introduce several network structures in actual business situations. Then in Sect. 11.3, we propose Network DEA (NDEA) models based on the weighted slacks-based measure (WSBM) approach. We discuss the characteristics of the divisional efficiencies in Sect. 11.4. Illustrative examples are introduced in Sect. 11.5. We extend our models in Sect. 11.6. We summarize the results and conclude the chapter in the last section. This chapter is written based on Tone and Tsutsui (2009).

11.2 Several Examples of Network Structures

We introduce network structures from actual businesses which motivated this study.

11.2.1 Electric Power Companies

Figure 11.4 exhibits typical vertically integrated electric utility companies consisting of generation, transmission and distribution divisions.

The generation division (Division 1) uses several inputs such as capital, labor and fuel (Input 1) and produces electric power. Then it becomes an intermediate input for the transmission division (Link 1–2). In the transmission division (Division 2), companies utilize capital and labor inputs (Input 2) as well as the intermediate inputs from generation division (Link 1–2). Electricity through transmission lines is sent to distribution division as intermediate output (Link 2–3) or sales to

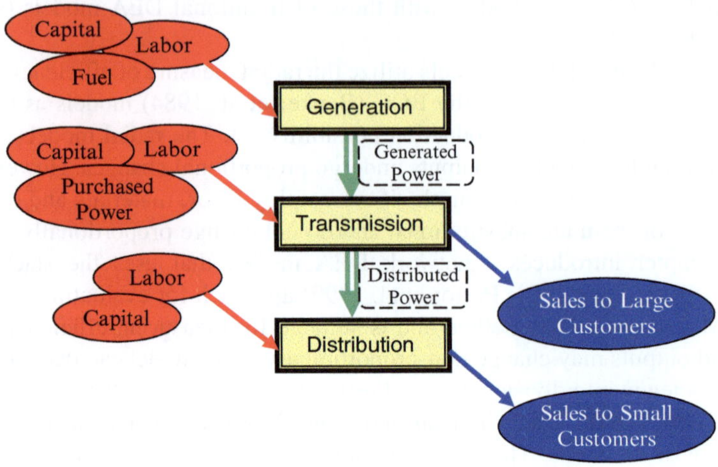

Fig. 11.4 Vertically integrated electric power companies

large customers (Output 2) that do not utilize distribution line. The distribution division (Division 3) uses capital and labor inputs (Input 3) and the intermediate input from the transmission division (Link 2–3) and provides electricity to small customers (Output 3).

11.2.2 Hospitals

Kaihara et al. (2007) private communication report the standard structure of Japanese general hospitals as depicted in Fig. 11.5. A general hospital consists of divisions, such as medical department, clinical laboratory, radiology, pharmacy and dietetic department. Each division has its own inputs; labor, materials and capital, and outputs; incomes. These divisions are connected by internal links. For example, a part of patients checked up at medical department is sent to radiology department. In order to evaluate the efficiency of general hospitals we need to account these divisions as a whole including linking activities. Thus, a network DEA model is appropriate for this purpose.

11.2.3 Broadcasting Companies

Broadcasting companies consist of two divisions; production and transmission. Using labor, materials and capital, the production division produces programs. A part of these products can be marketed to other media, while they are intermediate products to the transmission division. This division utilizes its own labor, materials and capital to send the programs to audiences. Figure 11.6 displays this

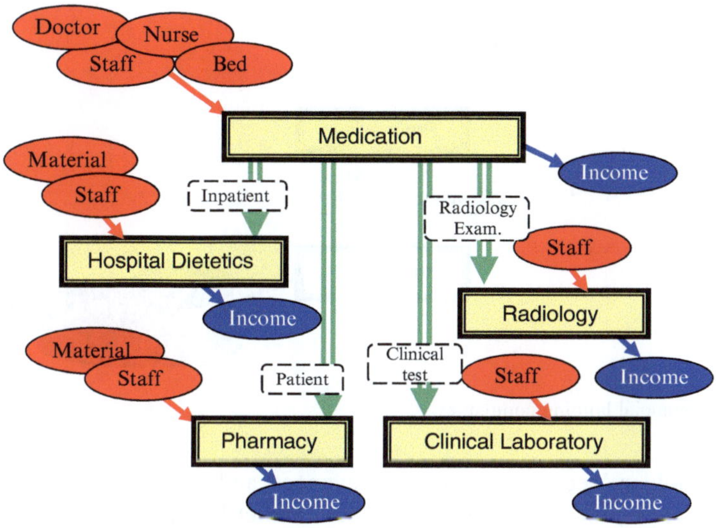

Fig. 11.5 General hospital

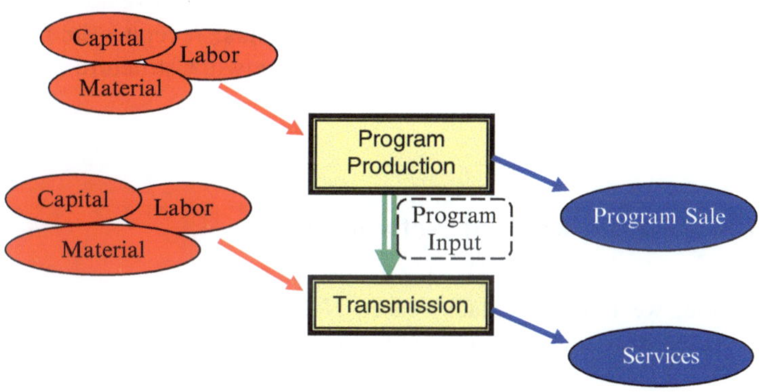

Fig. 11.6 Broadcasting companies

network structure. Product of the production division is the link (intermediate product) to the transmission division. This network structure is reported by (Asai (2007), private communication).

11.2.4 Financial Holding Companies

Seiford and Zhu (1999) pointed out that financial holding companies have two stages; profit generation and market value creation as exhibited in Fig. 11.7. Usually this process is studied in the two stage approaches; profitability and marketability. In the first stage, the profitability sector utilizes employees, assets

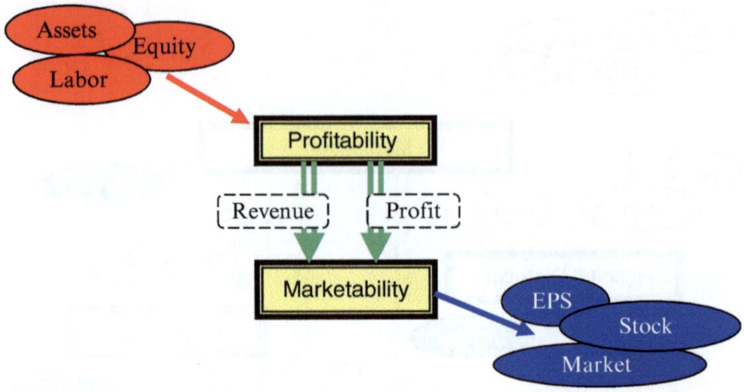

Fig. 11.7 Financial holding companies

and stockholders' equity to produce revenues and profits. The second stage measures (stock) marketability in the stock market by the revenue and profits it generates. It can be seen that revenues and profits serve as intermediate factors in the sense that they are outputs from the first stage and inputs to the second stage. The market sector produces market values, total returns to investors and earnings per share as outputs (Seiford and Zhu 1999). Thus, revenues and profits are linking activities between the two sectors and a network structure is recognized in this field.

11.3 Basic Framework of Network DEA

In this section, we introduce slacks-based Network DEA model referring to its production possibility set, efficiency and projection.

11.3.1 Notation and Production Possibility Set

We deal with n DMUs ($j = 1,\ldots, n$) consisting of K divisions ($k = 1,\ldots, K$). Let m_k and r_k be the numbers of inputs and outputs to Division k, respectively. We denote the link leading from Division k to Division h by (k,h) and the set of links by L. The observed data are $\left\{\mathbf{x}_j^k \in R_+^{m_k}\right\}(j = 1, \ldots, n;\ \ k = 1, \ldots, K)$ (input resources to DMU$_j$ at Division k), $\left\{\mathbf{y}_j^k \in R_+^{r_k}\right\}(j = 1, \ldots, n;\ \ k = 1, \ldots, K)$ (output products from DMU$_j$ at Division k) and $\left\{\mathbf{z}_j^{(k,h)} \in R_+^{t_{(k,h)}}\right\}(j = 1, \ldots, n;\ \ (k,h) \in L)$ (linking intermediate products from Division k to Division h) where $t_{(k,h)}$ is the number of items in Link (k,h).

The production possibility set $\{(\mathbf{x}^k, \mathbf{y}^k, \mathbf{z}^{(k,h)})\}$ is defined by

$$
\begin{aligned}
\mathbf{x}^k &\geq \sum_{j=1}^{n} \mathbf{x}_j^k \lambda_j^k \quad (k = 1, \ldots, K) \\
\mathbf{y}^k &\leq \sum_{j=1}^{n} \mathbf{y}_j^k \lambda_j^k \quad (k = 1, \ldots, K) \\
\mathbf{z}^{(k,h)} &= \sum_{j=1}^{n} \mathbf{z}_j^{(k,h)} \lambda_j^k (\forall (k,h)) \quad (\text{as outputs from } k) \\
\mathbf{z}^{(k,h)} &= \sum_{j=1}^{n} \mathbf{z}_j^{(k,h)} \lambda_j^h (\forall (k,h)) \quad (\text{as inputs to } h) \\
\sum_{j=1}^{n} \lambda_j^k &= 1 \ (\forall k), \ \lambda_j^k \geq 0 \ (\forall j, k)
\end{aligned}
\tag{11.1}
$$

where $\boldsymbol{\lambda}^k \in R_+^n$ is the intensity vector corresponding to Division k ($k = 1, \ldots, K$).

We notice that the above model assumes the variable returns-to-scale (VRS) for production. That is, the production frontiers are spanned by the convex hull of the existing DMUs. However, if we neglect the last constraint $\sum j{=}1^n \lambda_j^k = 1(\forall\ k)$, we can deal with the constant returns-to-scale (CRS) case as well.

DMU $_o(o = 1, \ldots, n)$ can be represented by

$$
\begin{aligned}
\mathbf{x}_o^k &= \mathbf{X}^k \boldsymbol{\lambda}^k + \mathbf{s}_o^{k-} \quad (k = 1, \ldots, K) \\
\mathbf{y}_o^k &= \mathbf{Y}^k \boldsymbol{\lambda}^k - \mathbf{s}_o^{k+} \quad (k = 1, \ldots, K) \\
\mathbf{e}\boldsymbol{\lambda}^k &= 1 \quad (k = 1, \ldots, K) \\
\boldsymbol{\lambda}^k &\geq \mathbf{0}, \ \mathbf{s}_o^{k-} \geq \mathbf{0}, \ \mathbf{s}_o^{k+} \geq \mathbf{0}, \quad (\forall k)
\end{aligned}
\tag{11.2}
$$

where

$$
\begin{aligned}
\mathbf{X}^k &= \left(\mathbf{x}_1^k, \ldots, \mathbf{x}_n^k\right) \in R^{m_k \times n} \\
\mathbf{Y}^k &= \left(\mathbf{y}_1^k, \ldots, \mathbf{y}_n^k\right) \in R^{r_k \times n}.
\end{aligned}
\tag{11.3}
$$

and $\mathbf{s}_o^{k-} (\mathbf{s}_o^{k+})$ are the input (output) slack vectors.

As regard to the linking constraints, we have several options of which we present two possible cases.

(a) The "free" link value case.
The linking activities are freely determined (discretionary) while keeping continuity between input and output:

$$
\mathbf{Z}^{(k,h)} \boldsymbol{\lambda}^h = \mathbf{Z}^{(k,h)} \boldsymbol{\lambda}^k. \quad (\forall (k,h)) \tag{11.4a}
$$

where

$$
\mathbf{Z}^{(k,h)} = \left(\mathbf{z}_1^{(k,h)}, \ldots, \mathbf{z}_n^{(k,h)}\right) \in R^{t_{(k,h)} \times n}. \tag{11.5}
$$

This case can serve to see if the current link flow is appropriate or not in the light of other DMUs', i.e. the link flow may increase or decrease in the optimal solution of the linear programs which we will introduce in the next section.

(b) The "fixed" link value case.
The linking activities are kept unchanged (non-discretionary):

$$\begin{aligned} \mathbf{z}_o^{(k,h)} &= \mathbf{Z}^{(k,h)}\boldsymbol{\lambda}^h \quad (\forall (k,h)) \\ \mathbf{z}_o^{(k,h)} &= \mathbf{Z}^{(k,h)}\boldsymbol{\lambda}^k. \quad (\forall (k,h)) \end{aligned} \tag{11.4b}$$

This case corresponds to the situation where the intermediate products are beyond the control of DMUs. However, if all link values are fixed, this case reduces structurally to the separation model described in Sect. 11.1.2 with non-discretionary inputs and outputs.

Throughout this chapter, we assume that all data are positive, since basically we employ the slacks-based measure (SBM) that demands positive data.

11.3.2 Efficiency

For each DMU$_o$, we define several efficiency scores depending on the selected orientation, input, output or non-oriented, as follows.

11.3.2.1 Input–Oriented Efficiency θ_o^*

We evaluate the input-oriented efficiency of DMU$_o$ by solving the following linear program:

$$\theta_o^* = \min_{\lambda^k, s_o^{k-}} \sum_{k=1}^K w^k \left[1 - \frac{1}{m_k} \left(\sum_{i=1}^{m_k} \frac{s_{io}^{k-}}{x_{io}^k} \right) \right] \tag{11.6}$$
$$\text{subject to (11.2), and (11.4a) or (11.4b)}$$

where $\sum_{k=1}^K w^k = 1,\ w^k \geq 0\ (\forall\ k)$ and w^k is the relative weight of Division k which is determined corresponding to its importance, e.g. cost share and supplied exogenously.

This model is called the weighted SBM model (WSBM), an extension of the SBM. See Cooper et al. (2007) for details.

Definition 1 (Input-oriented overall efficiency)
We call θ_o^* the *overall input-efficiency* of DMU$_o$. If we have $\theta_o^* = 1$, the DMU$_o$ is called *overall input-efficient*.

Definition 2 (Input-oriented divisional efficiency)
Using the optimal input slacks $\mathbf{s}_o^{k-*}$ of (11.6), we define the input-oriented divisional efficiency by

$$\theta_o^k = 1 - \frac{1}{m_k}\left(\sum\nolimits_{i=1}^{m_k} \frac{s_{io}^{k-*}}{x_{io}^k}\right) \quad (k = 1, \ldots, K). \tag{11.7}$$

θ_o^k is the divisional efficiency index which optimizes the overall efficiency θ_o^*. If we have $\theta_o^k = 1$, then the DMU_o is called *input-efficient* for the division k.

We notice that the above divisional efficiency score is not always uniquely determined,[1] although the overall efficiency is uniquely obtained as the linear program optimum. In Sect. 11.6.1, we present a scheme for deciding divisional efficiency scores uniquely.

The overall input-oriented efficiency score is the weighted arithmetic mean of the divisional scores

$$\theta_o^* = \sum\nolimits_{k=1}^{K} w^k \theta_o^k. \tag{11.8}$$

This measure is useful for comparing the total productivity of DMU_o among the concerned DMUs. It will serve not only managers but also regulatory agencies to compare DMUs in the firm-level view point.

11.3.2.2 Output-Oriented Efficiency τ_o^*

We evaluate the output-oriented efficiency of DMU_o by solving the following linear program:

$$\begin{aligned} 1/\tau_o^* = \max_{\lambda^k, s_o^{k+}} \sum\nolimits_{k=1}^{K} w^k \left[1 + \frac{1}{r_k}\left(\sum\nolimits_{r=1}^{r_k} \frac{s_{ro}^{k+}}{y_{ro}^k}\right)\right] \\ \text{subject to (2), and (4a) or (4b)} \end{aligned} \tag{11.9}$$

where $\sum_{k=1}^{K} w^k = 1$, $w^k \geq 0\ (\forall\ k)$, and w^k is the relative weight of Division k which is determined corresponding to its importance.

Definition 3 (Output-oriented overall efficiency)
We call τ_o^* the *overall output-efficiency* of DMU_o. If we have $\tau_o^* = 1$, the DMU_o is called overall output-efficient.

[1] In order to see the range in which a divisional efficiency may vary, we can solve the maximum and the minimum of θ_o^k subject to (11.2), (11.4a) or (11.4b) while keeping the overall efficiency at the optimal value θ_o^*.

Definition 4 (Output-oriented divisional efficiency)
In order to confine the score into the range [0, 1], we define the output-oriented divisional efficiency score by

$$\tau_o^k = \frac{1}{1 + \frac{1}{r_k}\left(\sum_{r=1}^{r_k} \frac{s_{ro}^{k+*}}{y_{ro}^k}\right)} \quad (k = 1, \ldots, K). \tag{11.10}$$

where $\mathbf{s}_o^{k+*}$ is the optimal output-slacks for (11.9).

The output-oriented overall efficiency score is the weighted harmonic mean of the divisional scores

$$\frac{1}{\tau_o^*} = \sum_{k=1}^{K} \frac{w^k}{\tau_o^k}. \tag{11.11}$$

11.3.2.3 Non-oriented Efficiency ρ_o^*

Accounting for both input and output slacks, we can evaluate the non-oriented efficiency of DMU$_o$ as follows:

$$\rho_o^* = \min_{\lambda^k, s_o^{k-}, s_o^{k+}} \frac{\sum_{k=1}^{K} w^k \left[1 - \frac{1}{m_k}\left(\sum_{i=1}^{m_k} \frac{s_{io}^{k-}}{x_{io}^k}\right)\right]}{\sum_{k=1}^{K} w^k \left[1 + \frac{1}{r_k}\left(\sum_{r=1}^{r_k} \frac{s_{ro}^{k+}}{y_{ro}^k}\right)\right]} \tag{11.12}$$
$$\text{subject to } (2), \text{and } (4\text{a}) \text{ or } (4\text{b}).$$

where $\sum_{k=1}^{K} w^k = 1, \;\; w^k \geq 0 \; (\forall\, k)$, and w^k is the relative weight of Division k which is determined corresponding to its importance.[2] We can solve this problem by transforming into a linear program using Charnes and Cooper transformation (see Tone 2001).

Definition 5 (Non-oriented overall efficiency)
We call ρ_o^* the *non-oriented overall efficiency* of DMU$_o$. If we have $\rho_o^* = 1$, the DMU$_o$ is called *overall efficient*.

[2] Although other forms of the objective function might be possible, we chose (11.12) for alignment with the non-oriented SBM model proposed in Tone (2001). This form serves to interpret its dual linear programming problem as the virtual profit efficiency model (see (Tone 2001)).

Definition 6 (Non-oriented divisional efficiency)
We define the non-oriented divisional efficiency score by

$$\rho_o^k = \frac{1 - \frac{1}{m_k}\left(\sum_{i=1}^{m_k} \frac{s_{io}^{k-*}}{x_{io}^k}\right)}{1 + \frac{1}{r_k}\left(\sum_{r=1}^{r_k} \frac{s_{ro}^{k+*}}{y_{ro}^k}\right)} \quad (k = 1, \ldots, K). \tag{11.13}$$

where $\mathbf{s}_o^{k-*}$ and $\mathbf{s}_o^{k+*}$ are optimal input- and output-slacks for (11.12).

The overall non-oriented efficiency score is a weighted mean of the divisional efficiency scores but is neither their arithmetic nor their harmonic mean.

We notice that the above divisional and overall efficiencies are units-invariant, i.e. they are independent of the units in which the inputs, outputs and links are measured.

Since the number of inputs and outputs may differ division by division and DEA scores are affected by the number, i.e. large number tends to give a high average score, care is needed in comparing divisional scores mutually.

Comparing the results by (11.4a) and (11.4b), we can see how the linking activities exert influence over the efficiency of each division.

11.3.3 Projection

Let an optimal solution to (11.6), (11.9) or (11.12) be ($\boldsymbol{\lambda}^{*k}$,$\mathbf{s}_o^{k-*}$,$\mathbf{s}_o^{k+*}$). Then we have the projection onto the frontier as follows:

$$\begin{aligned} \mathbf{x}_o^{k*} &\leftarrow \mathbf{x}_o^k - \mathbf{s}_o^{k-*} \ (k = 1, \ldots, K) \\ \mathbf{y}_o^{k*} &\leftarrow \mathbf{y}_o^k + \mathbf{s}_o^{k+*}. \ (k = 1, \ldots, K) \end{aligned} \tag{11.14}$$

If we employ the constraints (11.4b) for links, then the link values are unchanged (fixed). If we utilize the constraints (11.4a) (free link case), then we have the projection as follows:

$$\mathbf{z}_o^{(k,h)*} \leftarrow \mathbf{Z}^{(k,h)}\boldsymbol{\lambda}^{k*}. \ (\forall (k,h)) \tag{11.15}$$

11.3.4 Reference Set

Using the optimal intensity vector $\boldsymbol{\lambda}^{*k}$ we have:

Definition 7 (Reference set)
We define the reference set of the division k for DMU$_o$ by

$$\mathrm{R}_o^k = \left\{ j \middle| \lambda_j^{k*} > 0 \right\} \ (j \in \{1, \ldots, n\}). \tag{11.16}$$

Using this notation we can express $\mathbf{x}_o^k$ and $\mathbf{y}_o^k$ as

$$\mathbf{x}_o^k = \sum_{j \in R_o^k} \mathbf{x}_j^k \lambda_j^{k*} + \mathbf{s}_o^{k-*}, \quad \mathbf{y}_o^k = \sum_{j \in R_o^k} \mathbf{y}_j^k \lambda_j^{k*} - \mathbf{s}_o^{k+*}. \tag{11.17}$$

11.4 Several Properties of Slacks-Based Network DEA Models

In this section we discuss several properties of the slacks-based NDEA models.

11.4.1 Overall Versus Divisional Efficiencies

We have defined the overall efficiencies corresponding to input, output and non-oriented orientations by (11.6), (11.9) and (11.12), and then the divisional efficiencies corresponding to these models are defined respectively by (11.7), (11.10) and (11.13).

Between the overall and divisional efficiencies we have:

Theorem 1 *A DMU is overall efficient if and only if it is efficient for all divisions.*

We notice that it can happen that there exists no overall efficient unit, in contrast to the traditional DEA models (see examples in Sect. 11.5.3), and furthermore that in a certain NDEA model some division may have no divisional efficient DMUs (see an example in Sect. 11.5.4).

11.4.2 Divisional Efficiency

Let us denote the sets of inputs, outputs, incoming links and outgoing links for Division k, respectively by $\mathbf{X}^k = \{\mathbf{x}_j^k\}$, $\mathbf{Y}^k = \{\mathbf{y}_j^k\}$, $\mathbf{Z}^{(pk)} = \{\mathbf{z}_j^{(pk)}\}$ and $\mathbf{Z}^{(kq)} = \{\mathbf{z}_j^{(kq)}\}$ where $j = 1, \ldots, n$. We notice that some of inputs and outputs may be vacant. However, we assume that all divisions in the model are at least indirectly connected by links.

In this section, we demonstrate that under the variable returns-to-scale (VRS) assumption every division has at least one divisionally efficient DMU. However, the constant returns-to-scale (CRS) cases are mixed. For the fixed link case under CRS, every division has at least one divisionally efficient DMU whereas for the free link case under CRS it is possible that some division has no divisionally efficient DMU.

11.4.2.1 The Variable Returns-to-Scale (VRS) Case

Under the VRS assumption, we have the following theorem:

Theorem 2 *Under the variable returns-to-scale assumption, every division has at least one divisionally efficient DMU.*

Proof We sort the n DMUs in the division k in ascending order in input values using Input i as the ith key. We further sort the resultant in descending order in output values using Output r as the $m_k + r$ th key. Then the lexicographical minimum (top) DMU has $\mathbf{s}_o^{k-} = \mathbf{0}$ and $\mathbf{s}_o^{k+} = \mathbf{0}$ for every feasible $\boldsymbol{\lambda}^k$ under the VRS assumption, even if there are tied DMUs. Thus, the division has at least one efficient DMU regardless of the orientation.

Q.E.D

11.4.2.2 The Constant Returns-to-Scale (CRS) Case

For the CRS assumption, we have two options; the free link case (11.4a) and the fixed link case (11.4b). For the later we have:

Theorem 3 *Under the constant returns-to-scale assumption with the fixed link case, every division has at least one divisionally efficient DMU.*

Proof As we noticed earlier in Sect. 11.3.1, the fixed link case reduces to the separation model with non-discretionary inputs and outputs corresponding to the fixed links. Hence we can solve this case separately division by division. Therefore, every division has at least one efficient DMU in the division.

Q.E.D.

As a consequence of the separation model, we have:

Corollary 1 *For the fixed link case, DMUs in the reference set are divisionally efficient.*

So far, we have demonstrated the existence of the divisionally efficient ($\theta_o^k = 1$) DMU for slacks-based NDEA models under the VRS assumption as well as for the CRS with the fixed link case. The remaining is the case of the CRS with the free link. In Sect. 11.5.4, we show a counter example that has no divisionally efficient DMU in this case.

11.4.3 Efficiency of the Projected DMU

We defined the projection of DMU_o by (11.14) and (11.15) (free link case).

Theorem 4 *The projected DMU is overall efficient.*

Proof We prove the theorem in the input-oriented case.

We evaluate the efficiency of the projected DMU ($\mathbf{x}_o^{k*}$,$\mathbf{y}_o^{k*}$,$\mathbf{z}_o^{k*}$) ($k = 1, \ldots, K$). Let an optimal solution be $\left(\overline{\boldsymbol{\lambda}}^{k*}, \overline{\mathbf{s}}_o^{k-*}, \overline{\mathbf{s}}_o^{k+*}\right)(k = 1, \ldots, K)$. Then we have:

$$\mathbf{x}_o^{k*} = \mathbf{X}^k\overline{\boldsymbol{\lambda}}^{k*} + \overline{\mathbf{s}}_o^{k-*}, \mathbf{y}_o^{k*} = \mathbf{Y}^k\overline{\boldsymbol{\lambda}}^{k*} - \overline{\mathbf{s}}_o^{k+*}, \mathbf{z}_o^{(k,h)*} = \mathbf{Z}^k\overline{\boldsymbol{\lambda}}^{k*}. \tag{11.18}$$

Replacing ($\mathbf{x}_o^{k*}$,$\mathbf{y}_o^{k*}$) by (11.14), we have

$$\mathbf{x}_o^{k} = \mathbf{X}^k\overline{\boldsymbol{\lambda}}^{k*} + \overline{\mathbf{s}}_o^{k-*} + \mathbf{s}_o^{k-*} \text{ and } \mathbf{y}_o^{k} = \mathbf{Y}^k\overline{\boldsymbol{\lambda}}^{k*} - \overline{\mathbf{s}}_o^{k+*} - \mathbf{s}_o^{k+*}. \tag{11.19}$$

Corresponding to this expression we have the overall efficiency:

$$\overline{\rho}_o = \min\sum\nolimits_{k=1}^{K} w^k\left[1 - \frac{1}{m_k}\left(\sum\nolimits_{i=1}^{m_k} \frac{s_{io}^{k-*} + \overline{s}_{io}^{k-*}}{x_{io}^k}\right)\right]. \tag{11.20}$$

If any member of $\{\overline{\mathbf{s}}_o^{k-*}\}(k = 1, \ldots, K)$ is positive, then it holds that

$$\overline{\rho}_o < \rho_o^*. \tag{11.21}$$

This contradicts the optimality of ρ_o^*. Thus, we have $\overline{\mathbf{s}}_o^{k-*} = \mathbf{0}$ $(k = 1, \ldots, K)$. Hence, the projected DMU is overall efficient.

Similarly we can prove the theorem in the output-oriented and non-oriented models.

Q.E.D.

11.5 Illustrative Examples

We present an illustrative example of electric power companies for describing slacks-based Network DEA and compare the results with traditional approaches. Also we demonstrate an example with the free link case that has no divisionally efficient DMUs.

11.5.1 Data

As introduced in Sect. 11.2 (Fig. 11.4), the vertically integrated electric power companies consist of several divisions such as generation, transmission and distribution. For illustrative purpose, we choose ten vertically integrated power companies in the U.S. in 1994 obtained from "Form No.1" published by the Federal Energy Regulatory Commission (FERC). The inputs, outputs and links are as follows:

Table 11.1 Sample data

	Div1	Div2		Div3		Link	
DMU	Input1	Input2	Output2	Input3	Output3	Link12	Link23
A	0.838	0.277	0.879	0.962	0.337	0.894	0.362
B	1.233	0.132	0.538	0.443	0.18	0.678	0.188
C	0.321	0.045	0.911	0.482	0.198	0.836	0.207
D	1.483	0.111	0.57	0.467	0.491	0.869	0.516
E	1.592	0.208	1.086	1.073	0.372	0.693	0.407
F	0.79	0.139	0.722	0.545	0.253	0.966	0.269
G	0.451	0.075	0.509	0.366	0.241	0.647	0.257
H	0.408	0.074	0.619	0.229	0.097	0.756	0.103
I	1.864	0.061	1.023	0.691	0.38	1.191	0.402
J	1.222	0.149	0.769	0.337	0.178	0.792	0.187
Average	1.020	0.127	0.763	0.560	0.273	0.832	0.290

Generation (Div 1):

Input 1 = Labor input (number of employees)

Transmission (Div 2):

Input 2 = Labor input (number of employees)
Output 2 = Electric power sold to large customers

Distribution (Div 3):

Input 3 = Labor input (number of employees)
Output 3 = Electric power sold to small customers

Link (1–2) = Electric power generated

(Output from Generation Division and Input to Transmission Division)

Link (2–3) = Electric power distributed

(Output from Transmission Division and Input to Distribution Division)

Table 11.1 exhibits data for inputs, outputs and links of the ten DMUs; A to J. Numbers in each column of the table are obtained from the source data by dividing some standard of the column. So we do not denote the units. This has no effect on the efficiency scores, since all DEA models employed are units-invariant.

11.5.2 Results of Black Box and Separation Models

First, we solved the aggregated (black box) model explained in Sect. 11.1.1, using Inputs 1, 2 and 3, and Outputs 2 and 3 where Links were neglected (see Fig. 11.2). Throughout this section, we utilized the input-oriented SBM (slacks-based measure) under the variable returns-to-scale (VRS) assumption for evaluating

Table 11.2 SBM scores for black box and separation models

		Separation			
			Divisional score		
DMU	Aggregation (Black box)	Overall score[a]	Div1	Div2	Div3
A	1.000	0.659	0.633	0.662	0.684
B	0.531	0.657	0.260	0.763	1.000
C	1.000	0.984	1.000	1.000	0.959
D	1.000	0.719	0.297	1.000	1.000
E	1.000	0.547	0.202	1.000	0.665
F	0.681	0.844	1.000	0.635	0.792
G	1.000	0.855	0.712	1.000	0.926
H	1.000	0.893	0.787	0.890	1.000
I	1.000	0.915	1.000	1.000	0.786
J	1.000	0.640	0.263	0.672	1.000
Average	0.921	0.771	0.615	0.862	0.881

[a]Overall score indicates 0.4*Div1 + 0.2*Div2 + 0.4* Div3

efficiency (see (Cooper et al. 2007)). The column "Aggregation (Black box)" in Table 11.2 reports the results.

Next, we solved the separation model explained in Sect. 11.1.2.[3] Table 11.2 reports the results where "Overall score" indicates the weighted average 0.4 × Div1 + 0.2 × Div2 + 0.4 × Div3. The numbers 0.4, 0.2 and 0.4 are weights to Div 1, Div 2 and Div 3, respectively, which are utilized in the following Network DEA model too. This weight selection is just for illustrative purpose. No significant correlation is observed between the two efficiencies; Aggregation and Overall. This is quite natural, since we neglected the internal linking activities in the former.

The scores of the black box model tend to be higher than those of the separation model (Fig. 11.8). Actually, these two models cannot be fairly comparable, because the number of inputs is different between the two models. However, this figure clearly explains that the discriminate power of the black box model is inferior to that of the separation model. In addition, it shows that the ranks of the scores of the two models are not always corresponding, e.g. F is scored worse in the black box model, while better in the separation model.

11.5.3 Results of Slacks-Based Network DEA

We now return to the Network DEA model taking account the links inside the black box. We minimize the objective function (11.6) subject to the constraints (11.2),

[3] In solving the separation model, links were treated as ordinary (discretionary) inputs or outputs as explained in Sect. 1.2, and hence the continuity of link values between divisions were not assured. Also, the separation model takes into account the inefficiency associated with the link variables, whereas the NSBM does not.

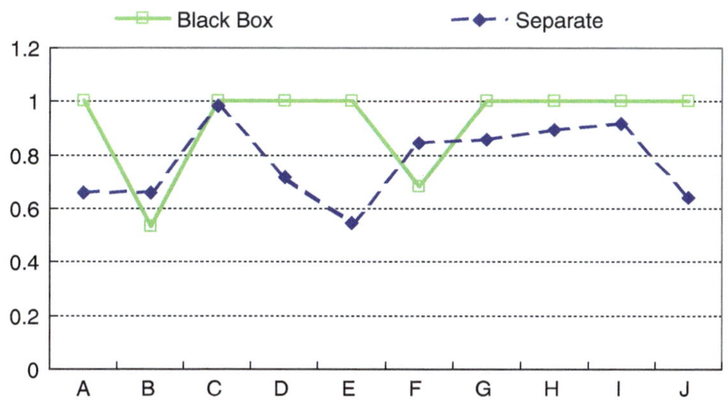

Fig. 11.8 Comparisons of scores between black box and separation models

Table 11.3 Slacks-based network DEA: fixed link case

		Divisional score			Reference			Link	
DMU	Overall score	Div1 (0.4)	Div2 (0.2)	Div3 (0.4)	Div1	Div2	Div3	Link12	Link23
A	0.478	0.633	0.339	0.393	C1,F1	C2,D2,E2,I2	D3,H3	0.894	0.362
B	0.739	0.349	1.000	1.000	C1,G1	B2	B3	0.678	0.188
C	0.968	1.000	1.000	0.919	C1	C2	B3,D3,J3	0.836	0.207
D	0.719	0.297	1.000	1.000	C1,F1	D2	D3,H3	0.869	0.516
E	0.456	0.263	1.000	0.377	C1,G1	E2	D3,H3	0.693	0.407
F	0.719	1.000	0.403	0.596	F1	C2,H2,I2	D3,H3	0.966	0.269
G	0.947	1.000	1.000	0.868	G1	G2	D3,H3	0.647	0.257
H	0.969	0.922	1.000	1.000	C1,G1	H2	H3	0.756	0.103
I	0.832	1.000	1.000	0.581	I1	I2	D3,H3	1.191	0.402
J	0.590	0.288	0.377	1.000	C1,G1	C2,G2,H2	J3	0.792	0.187
Average	0.742	0.675	0.812	0.773				0.832	0.29

and (11.4a) or (11.4b), i.e. the input-oriented network model under VRS assumption. As weights to objective function, we employ $w^1 = 0.4$ (Division 1), $w^2 = 0.2$ (Division 2) and $w^3 = 0.4$ (Division 3). This set of weights conforms to the above weights in Sect. 11.5.2. The results of the fixed link case (11.4b) are displayed in Table 11.3 while the free link case (11.4a) is exhibited in Table 11.4 where the overall efficiency (θ_o^*) together with divisional efficiencies is displayed.[4] The divisional efficiency means the individual term (11.7) in the objective function. In the "Reference" column, A1 indicates DMU A in the Division 1. This means $\lambda_A^1 > 0$ in the optimal solution. Since the constraint (11.4a) is tighter than (11.4b),

[4] We checked the uniqueness of the divisional efficiency scores as described in Footnote 1 and found no alternate optima.

Table 11.4 Slacks-based network DEA: free link case

		Divisional score			Reference			Projected link			
DMU	Overall score	Div1(0.4)	Div2(0.2)	Div3(0.4)	Div1	Div2	Div3	Link12	Link12/Data	Link23	Link23/Data
A	0.385	0.383	0.383	0.389	C1	C2,D2,E2,I2	D3,H3	0.836	0.935	0.355	0.979
B	0.433	0.260	0.341	0.652	C1	C2	D3,H3	0.836	1.233	0.207	1.101
C	0.968	1.000	1.000	0.919	C1	C2	B3,D3,J3	0.836	1.000	0.207	1.000
D	0.719	0.297	1.000	1.000	C1,F1	D2	D3,H3	0.869	1.000	0.516	1.000
E	0.456	0.263	1.000	0.377	C1,G1	E2	D3,H3	0.693	1.000	0.407	1.000
F	0.484	0.406	0.420	0.593	C1	C2,D2,G2	D3,H3	0.836	0.865	0.267	0.991
G	0.778	0.712	0.740	0.863	C1	C2,D2,G2	D3,H3	0.836	1.292	0.254	0.988
H	0.969	0.922	1.000	1.000	C1,G1	H2	H3	0.756	1.000	0.103	1.000
I	0.832	1.000	1.000	0.581	I(1)	I2	D3,H3	1.191	1.000	0.402	1.000
J	0.506	0.271	0.338	0.825	C1,G1	C2,H2	D3,H3	0.821	1.037	0.188	1.005
Average	0.653	0.551	0.722	0.720				0.851	1.036	0.291	1.006

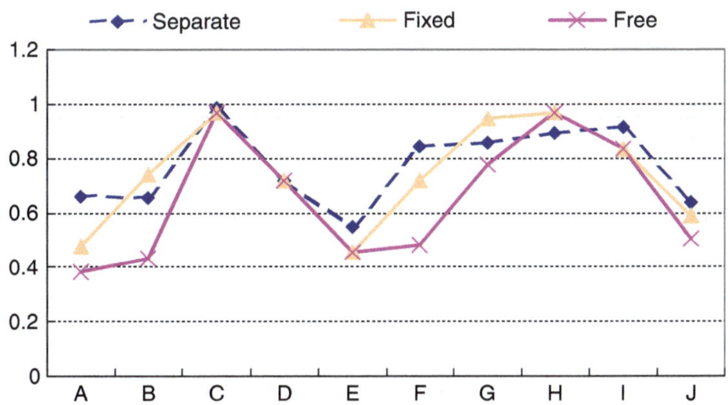

Fig. 11.9 Comparisons of scores among separate and two network models

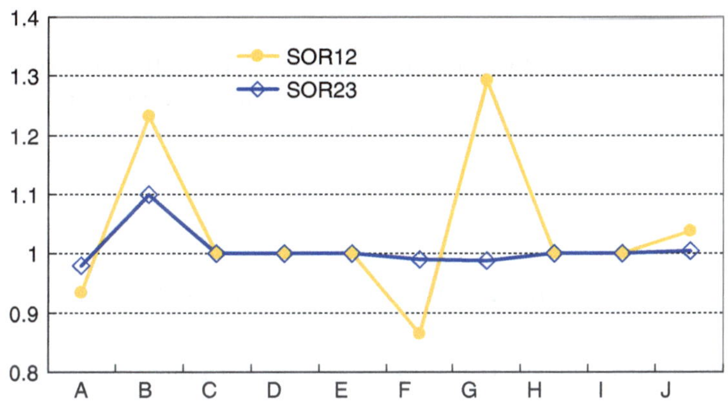

Fig. 11.10 Link effect ratio

the overall score of the former (fixed link case) is larger than that of the latter (free link case) for every DMU.

Figure 11.9 compares scores of the separate model and network models (fixed and free link cases). The trends of three models are roughly similar but exhibit sharp contrast to that of the black box model explained in Fig. 11.8. However, we can find gaps among three models, which must be caused by the difference of assumption on the links among divisions. As we mentioned, the separation model does not take account of the links, and therefore, the gap between the separation and network models implies the "linking effects". The separation model cannot catch the full story in the case when the linking effects inside DMUs actually exist.

Concerning two network models, the scores of the fixed link case exceed or equal to those of the free case. The gap of two models explains "link effects". Figure 11.10 shows the "Link effect ratio (LER)" of links measured as projected

Table 11.5 Data for four DMUs

	Div1	Div2		Div3		Link	
DMU	Input1	Input2	Output2	Input3	Output3	Link12	Link23
K	3	10	2	5	2	8	2
L	14	1	1	5	5	9	5
M	16	2	2	11	4	7	4
N	19	0.5	2	7	4	11	4

Table 11.6 Results of the input-oriented free link CRS model

DMU	Overall Score	Div1(0.4)	Div2(0.2)	Div3(0.4)
K	0.71	0.875	0.2	0.8
L	0.6723	0.3683	0.625	1
M	0.2986	0.2578	0.25	0.3636
N	0.5154	0.2171	1	0.5714

links in free case divided by actual links (see Table 11.4). If there exists the gap between the two network models in Fig. 11.10, the link effect ratio is not equal to unity, and if the ratio is larger than unity, the DMU should increase the link value, and vice versa.

11.5.4 Example with No Divisionally Efficient DMUs

As we have demonstrated in Theorems 2 and 3, the VRS models and the fixed link CRS model have at least one efficient DMU within every division. However, as to the free link CRS model, the proposition is not always effected. In this section, we exhibit a counter example which has no efficient DMU within a certain division. We observe 4 DMUs with the same network structure as the previous example. Table 11.5 exhibits the data. We solved this problem using the input-oriented free link NDEA model under the CRS assumption and obtained the results exhibited in Table 11.6. We found no efficient DMU in Division 1, while other divisions have an efficient DMUs; N for Division 2 and L for Division 3. This indicates that all DMUs in Division 1 need improvement. Table 11.7 reports the projection of inputs, outputs and links onto the efficient frontiers by the formulas (11.14) and (11.15). Actually, all inputs to Division 1 are reduced proportionally to their scores of Division 1. On the other hand, other divisions and links have benchmarks that remain unchanged in the projection. This occurrence of vacancy of divisionally efficient DMUs in some division is one of characteristics of this model which cannot be expected by traditional DEA models.

Table 11.7 Projection onto efficient frontiers

	Div1		Div2				Div3				Link			
DMU	Input1	Prj[a]	Input2	Prj	Output2	Prj	Input3	Prj	Output3	Prj	Link12	Prj	Link23	Prj
K	3	2.625	10	2	2	2	5	4	2	4	8	7	2	4
L	14	5.156	1	0.625	1	2.5	5	5	5	5	9	13.75	5	5
M	16	4.125	2	0.5	2	2	11	4	4	4	7	11	4	4
N	19	4.125	0.5	0.5	2	2	7	4	4	4	11	11	4	4

[a]Prj indicates projection onto efficient frontiers

11.6 Extensions

In this section, we introduce several extensions of the NDEA model.

11.6.1 Uniqueness Issue of Divisional Efficiencies

Although the overall efficiency is uniquely determined by the program (11.6) in the input-oriented model, slacks are not necessarily unique. Hence, the divisional efficiency in (11.7) may suffer from non-uniqueness issue.

In the model (11.6), we use w^k as the relative weight of Division k, which reflects importance of each division. Based on w^k, we can prioritize divisions. Under this priority principle, we propose the following scheme for overcoming this non-uniqueness problem. If any other priority rule exists, we can cope with it in the similar way.

For convenience sake, here we define the last division K has the top priority and those of $K-1$, $K-2$,. . ., 1 decrease in this order.

11.6.1.1 Divisional Efficiency in K

First, we solve the program (11.6) and obtain the overall efficiency θ_o^*. Then we minimize divisional efficiency in K while keeping the overall efficiency at θ_o^*.

Let us denote the divisional efficiency in K thus obtained by θ_o^{K*}.

$$\theta_o^{K*} = \min 1 - \frac{1}{m_K}\left(\sum\nolimits_{i=1}^{m_K} \frac{s_{io}^{K-*}}{x_{io}^K}\right) \tag{11.22}$$

subject to

$$\sum\nolimits_{k=1}^{K} w^k \left[1 - \frac{1}{m_k}\left(\sum\nolimits_{i=1}^{m_k} \frac{s_{io}^{k-}}{x_{io}^k}\right)\right] = \theta_o^* \tag{11.23}$$

and (11.2), (11.4a) or (11.4b).

11.6.1.2 Divisional Efficiency in k

We repeat this process in the descending order of priority until $k = 2$. Thus, divisional efficiency in k (θ_o^{k*}) is measured by the following program.

$$\theta_o^{k*} = \min 1 - \frac{1}{m_k}\left(\sum\nolimits_{i=1}^{m_K} \frac{s_{io}^{k-*}}{x_{io}^k}\right) \tag{11.24}$$

subject to

$$1-\frac{1}{m_K}\left(\sum_{i=1}^{m_K}\frac{s_{io}^{K-}}{x_{io}^{K}}\right)=\theta_o^{K*}$$
$$\vdots \tag{11.25}$$
$$1-\frac{1}{m_k}\left(\sum_{i=1}^{m_k}\frac{s_{io}^{k-}}{x_{io}^{k}}\right)=\theta_o^{k+1*}$$

and (11.2), (11.4a) or (11.4b), and (11.23).

Divisional efficiency in the division 1 can be obtained from θ_o^*, θ_o^{K*}, ..., θ_o^{2*}.

Through this scheme, we can obtain unique divisional efficiency scores θ_o^{k*} ($\forall k$) for the input-oriented model. As for the output-oriented and non-oriented models, we can develop similar processes for uniqueness issues.

11.6.2 Incorporation of Link Flows in Efficiency Measurements

In the above cases, link flows do not directly concern with the objective function. They are related with efficiency scores only through link constraints (11.4a) or (11.4b). However, if we want to account their excesses (in the input-oriented case) or shortfalls (in the output-oriented case) into the objective function, we can modify the model as follows.

(i) In the input-oriented case, we consider the slacks of the link (*f*,*k*) as input to Division *k* and set link constraints as

$$\begin{aligned}\mathbf{z}_o^{(f,k)} &= \mathbf{Z}^{(f,k)}\boldsymbol{\lambda}^k+\mathbf{s}_o^{(f,k)-}\\ \mathbf{Z}^{(f,k)}\boldsymbol{\lambda}^f &= \mathbf{Z}^{(f,k)}\boldsymbol{\lambda}^k\\ \mathbf{s}_o^{(f,k)-} &\geq \mathbf{0}\end{aligned} \tag{11.4c}$$

The objective function is modified as:

$$\theta_o^*=\min\sum_{k=1}^{K}w^{k-}\left[1-\frac{1}{m_k+\sum_{f\in P_k}t_{(f,k)}}\left(\sum_{i=1}^{m_k}\frac{s_{io}^{k-}}{x_{io}^{k}}+\sum_{f\in P_k}\frac{s_{fo}^{(f,k)-}}{z_{fo}^{(f,k)}}\right)\right]$$
$$\text{subject to (2) and (4c),} \tag{11.26}$$

where $\sum_{k=1}^{K}w^{k-}=1$, $w^{k-}\geq 0$ ($\forall$ k) and P_k is the set of divisions having the link $(f,k)\in L$ (antecessor of Division *k*) and $t_{(f,k)}$ is the number of items in the link. We optimize (11.26) in terms of $\{\boldsymbol{\lambda}^k\}$, $\{\mathbf{s}_o^{k-}\}$ and $\{\mathbf{s}_o^{(f,k)-}\}$.

(ii) In the output-oriented case, we consider the slacks of the link (*k*,*h*) as output from Division *k* and set link constraints as;

$$\begin{aligned} \mathbf{z}_o^{(k,h)} &= \mathbf{Z}^{(k,h)}\boldsymbol{\lambda}^k - \mathbf{s}_o^{(k,h)+} \\ \mathbf{Z}^{(k,h)}\boldsymbol{\lambda}^h &= \mathbf{Z}^{(k,h)}\boldsymbol{\lambda}^k \\ \mathbf{s}_o^{(k,h)+} &\geq \mathbf{0} \end{aligned} \tag{11.4d}$$

The objective function is modified to

$$1/\tau_o^* = \max \sum_{k=1}^{K} w^k \left[1 + \frac{1}{r_k + \sum_{h \in F_k} t_{(k,h)}} \left(\sum_{r=1}^{r_k} \frac{s_{ro}^{k+}}{y_{ro}^k} + \sum_{h \in F_k} \frac{s_{ho}^{(k,h)+}}{z_{ho}^{(k,h)}} \right) \right]$$

subject to (2) and (4d).

(11.27)

where $\sum_{k=1}^{K} w^k = 1$, $w^k \geq 0 \ (\forall\ k)$ and F_k is the set of divisions having the link $(k,h) \in L$ (successor of Division *k*). We optimize (11.27) in terms of $\{\boldsymbol{\lambda}^k\}$, $\{\mathbf{s}_o^{k+}\}$ and $\{\mathbf{s}_o^{(k,h)+}\}$.

(iii) In the case that links are categorized into either input type (the less the better) or output type (the more the better), we can unify the above (i) and (ii) models into the non-oriented case in the similar way as the non-oriented model described in Sect. 11.3.2.3.

11.6.3 The Role of Intensity Vector λ

One of the characteristics of the NDEA is that it has an intensity vector $\boldsymbol{\lambda}^k = (\lambda_1^k, \ldots, \lambda_n^k)^T \in R^n \ (\boldsymbol{\lambda}^k \geq \mathbf{0})$ specific to each Division *k* ($k = 1, \ldots, K$). We observe the role of this vector in this section.

11.6.3.1 The Identical Intensity Vector Case

In this case we assume a common intensity vector $\boldsymbol{\lambda} = \boldsymbol{\lambda}^k$ for every division k ($k = 1, \ldots, K$). Thus, DMU_o can be expressed as follows:

$$\begin{aligned} \mathbf{x}_o^k &= \mathbf{X}^k\boldsymbol{\lambda} + \mathbf{s}_o^{k-} \\ \mathbf{y}_o^k &= \mathbf{Y}^k\boldsymbol{\lambda} - \mathbf{s}_o^{k+} \\ \mathbf{z}_o^{(k,h)} &= \mathbf{Z}^{(k,h)}\boldsymbol{\lambda} \\ \mathbf{e}\boldsymbol{\lambda} &= 1 \\ \boldsymbol{\lambda} \geq \mathbf{0},\ & \mathbf{s}_o^{k-} \geq \mathbf{0},\ \mathbf{s}_o^{k+} \geq \mathbf{0}. \end{aligned} \tag{11.28}$$

Now let us define matrices **X**, **Y** and **Z** as follows:

$$\mathbf{X} = \begin{pmatrix} \mathbf{X}^1 \\ \mathbf{X}^2 \\ \vdots \\ \mathbf{X}^K \end{pmatrix} \in R_+^{(m_1+\cdots+m_K)\times n}, \mathbf{Y} = \begin{pmatrix} \mathbf{Y}^1 \\ \mathbf{Y}^2 \\ \vdots \\ \mathbf{Y}^K \end{pmatrix} \in R_+^{(r_1+\cdots+r_K)\times n}, \tag{11.29}$$

$$\mathbf{Z} = \left(\mathbf{Z}^{(k,h)} | (k,h) \in L\right) \in R_+^{\left(\sum_{(k,h)\in L} t_{(k,h)}\right)\times n}. \tag{11.30}$$

Using these notations, DMU_o can be expressed as,

$$\begin{aligned} \mathbf{x}_o &= \mathbf{X}\boldsymbol{\lambda} + \mathbf{s}_o^- \\ \mathbf{y}_o &= \mathbf{Y}\boldsymbol{\lambda} - \mathbf{s}_o^+ \\ \mathbf{z}_o &= \mathbf{Z}\boldsymbol{\lambda} \end{aligned} \tag{11.31}$$

where $\mathbf{s}_o^- = \left(s_o^{1-}, \ldots, s_o^{K-}\right)^T \in R^{m_1+\cdots+m_K}$ and $\mathbf{s}_o^+ = \left(s_o^{1+}, \ldots, s_o^{K+}\right)^T \in R^{r_1+\cdots+r_K}$.

Thus this case reduces to a traditional DEA model added by the last linking constraint. This model has $(m_1 + \cdots + m_K)$ inputs, $(s_1 + \cdots + s_K)$ outputs and $\sum_{(k,h) \in L} t_{(k,h)}$ linking constraints. In the case the sum of these numbers grows up to n (the number of DMUs), this model might lose discriminating power. As a rule of thumb, DEA demands that the number of DMUs should be at least three times larger than the sum of the number of inputs and outputs. The equality condition for the linking constraints will further narrow the feasible region and many DMUs may be judged as efficient in consequence.

11.6.3.2 Connectivity Among Divisions

In the preceding section, we have observed a special case regarding the decision variable $\boldsymbol{\lambda}$; identical. In this case, all divisions of DMU_o are evaluated by an identical set of referent DMUs, i.e. all divisions have the same benchmarks. In the NDEA models, however, benchmarks can vary division by division. Such diversified benchmarks among divisions might embrace supervisors in choosing peers to follow as a company.

These two extreme cases can be unified via the following connectivity index $\delta^{(h,k)}(\geq 0)$ $((h,k) \in L)$ as

$$\left|\lambda_j^h - \lambda_j^k\right| \leq \delta^{(h,k)} \quad (j = 1, \ldots n; (h,k) \in L) \tag{11.32}$$

The case $\delta^{(h,k)} = 0$ $(\forall\ (h,k))$ corresponds to the identical $\boldsymbol{\lambda}$, while the case $\delta^{(h,k)} = \infty$ $(\forall\ (h,k))$ corresponds to the independent $\boldsymbol{\lambda}$ setting, i.e. slacks-based NDEA models.

In our experiments with 56 Japanese (11.9) and the US (47) electric power companies, the identical **λ** case, i.e. $\delta^{(h,k)} = 0$ (∀ (h,k)), almost lost discriminating power and many companies were judged efficient, whereas $\delta^{(h,k)} = 0.01$ (∀ (h,k)) case demonstrated discrimination of efficiency and reasonable connectivity among divisions, i.e. compelling benchmarks. Appropriate setting of the connectivity index is an experimental issue. See Tsutsui (2007) for details.

11.7 Concluding Remarks

In this chapter, we have proposed a network DEA model based on the weighted SBM (WSBM) approach which accounts for the importance of each division. Thus, we can evaluate multi-divisional efficiencies and the overall efficiency in a unified framework.

The following subjects are discussed.

1. We have developed the slacks-based NDEA model under the fixed (non-discretionary) link and the free (discretionary) link assumptions. In the latter case, the optimal link values may increase or decrease from the observed ones. Comparisons of both results (fixed and free) give suggestions for improvements in the intermediate production policy. Thus, we can analyze economy and diseconomy of internal links by comparing fixed-link and free-link models.
2. We have proved that, under the VRS assumption, every division has at least one divisionally efficient DMU. This also holds for the case the fixed link under the CRS.
3. For the CRS and free link case, we have demonstrated a counter example in which a division has no divisionally efficient DMU. This may suggest improvements of the division as a whole. Also, it may reflect an unstable or unbalanced network structure in the problem of concern.

As an extension model of slacks-based NDEA to intertemporal analysis, Tone and Tsutsui (2010) proposed slacks-based dynamic DEA model, which takes account carry-over activities of DMUs such as retained earnings and facilities. These are incorporated into the model as input from the previous period and output to the next period.

Network and dynamic model is also proposed in Tone and Tsutsui (2014). Vertically, this model deals with multiple divisions connected by links of network structure within each period and, horizontally, it combines the network structure by means of carry-over activities between two succeeding periods.

Finally, we hope that these studies serve as a basis for extending theory and applications of DEA models which have been growing rapidly worldwide.

References

Banker, R., Charnes, A., & Cooper, W. W. (1984). Some models for estimating technical and scale inefficiencies in data envelopment analysis. *Management Science, 30*, 1078–1092.

Charnes, A., Cooper, W. W., & Rhodes, E. (1978). Measuring the efficiency of decision making units. *European Journal of Operational Research, 2*, 429–444.

Chen, C. M. (2009). A network-DEA model with new efficiency measures to incorporate the dynamic effect in production networks. *European Journal of Operational Research, 194*, 687–699.

Cooper, W. W., Seiford, L. M., & Tone, K. (2007). *Data envelopment analysis: A comprehensive text with models, applications, references and DEA-solver software* (2nd ed.). New York: Springer.

Färe, R. (1991). Measuring Farrell efficiency for a firm with intermediate inputs. *Academia Economic Papers, 19*(2), 329–340.

Färe, R., & Grosskopf, S. (1996). *Intertemporal production frontiers: With dynamic DEA*. Boston: Kluwer Academic.

Färe, R., & Grosskopf, S. (2000). Network DEA. *Socio-Economic Planning Sciences, 34*, 35–49.

Kao, C. (2009). Efficiency decomposition in network data envelopment analysis: A relational model. *European Journal of Operational Research, 192*, 949–962.

Koopmans, T. (1951). An analysis of production as an efficient combination of activities. In activity analysis of production and allocation. In T. Koopmans (Ed.), *Cowles commission for research in economics, monograph* (Vol. 13). New York: Wiley.

Lewis, H. F., & Sexton, T. R. (2004). Network DEA: Efficiency analysis of organisations with complex internal structure. *Computers & Operations Research, 31*, 1365–1410.

Löthgren, M., & Tambour, M. (1999). Productivity and customer satisfaction in Swedish pharmacies: A DEA network model. *European Journal of Operational Research, 115*, 449–458.

Pastor, J. T., Ruiz, J. L., & Sirvent, I. (1999). An enhanced DEA Russell graph efficiency measure. *European Journal of Operational Research, 115*, 596–607.

Prieto, A. M., & Zofio, J. L. (2007). Network DEA efficiency in input–output models: With an application to OECD countries. *European Journal of Operational Research, 178*, 292–304.

Seiford, L. M., & Zhu, J. (1999). Profitability and marketability of the top 55 U.S. commercial banks. *Management Science, 45*(9), 1270–1288.

Sexton, T. R., & Lewis, H. F. (2003). Two-stage DEA: An application to major league baseball. *Journal of Productivity Analysis, 19*, 227–249.

Tone, K. (2001). A slacks-based measure of efficiency in data envelopment analysis. *European Journal of Operational Research, 130*, 498–509.

Tone, K., & Tsutsui, M. (2009). Network DEA: A slacks-based measure approach. *European Journal of Operational Research, 197*, 243–252.

Tone, K., & Tsutsui, M. (2010). Dynamic DEA: A slacks-based measure approach. *Omega, 38*, 145–156.

Tone, K., & Tsutsui, M. (2014). Dynamic DEA with network structure: A slacks-based measure approach. *Omega, 42*, 124–131.

Tsutsui, M. (2007). *Measuring the management efficiency of vertically integrated electric utilities*, Doctoral dissertation, National Graduate Institute for Policy Studies.

Tsutsui, M., & Goto, M. (2009). A multi-division efficiency evaluation of U.S. electric power companies using a weighted slacks-based measure. *Socio-Economic Planning Sciences, 43*, 201–208.

Chapter 12
DEA Models for Extended Two-Stage Network Structures

Yongjun Li, Yao Chen, Liang Liang, and Jianhui Xie

Abstract This chapter discusses DEA modeling technique for a two-stage network process where the inputs of the second stage include both the outputs from the first stage and additional inputs to the second stage. Two models are proposed to evaluate the performance of this type two-stage network structures. One is a non-linear centralized model whose global optimal solutions can be estimated using a heuristic search procedure. The other is a non-cooperative model, in which one of the stages is regarded as the leader and the other is the follower. The newly developed models are illustrated with a case of regional R&D of China.

Keywords Data envelopment analysis (DEA) • Two-stage • Game

12.1 Introduction

Data envelopment analysis (DEA), developed by Charnes et al. (1978), is a mathematical programming approach for analyzing the relative performance of peer decision making units (DMUs) which have multiple inputs and multiple outputs. Cooper et al. (2004) has shown that DEA can be applied in various of settings, such as performance evaluation in education, bank and bank branch, sports, retailing, engineering and health care. In conventional DEA models, DMUs are seen as black-boxes and the internal structure of DMUs is ignored. In recent years, a number of studies have looked at DMUs with network structures (see, e.g., Färe and

Y. Li (✉) • L. Liang • J. Xie
School of Business, University of Science and Technology of China, He Fei, An Hui Province 230026, P.R. China
e-mail: lionli2012@gmail.com

Y. Chen
Manning School of Business, University of Massachusetts at Lowell, Lowell, MA 01845, USA
e-mail: Yao_Chen@uml.edu

W.D. Cook and J. Zhu (eds.), *Data Envelopment Analysis*,
International Series in Operations Research & Management Science 208,
DOI 10.1007/978-1-4899-8068-7_12,

Grosskopf 2000; Liang et al. 2006; Tone and Tsutsui 2009; Castelli et al. 2004; Kao 2009a, b; Kao and Hwang 2008; Liang et al. 2008). In a recent survey by Cook et al. (2010a), the authors point out several approaches in modeling DMUs with a two-stage network structure. Typically, models are developed based upon additive or geometric mean efficiency decompositions.

For example, Cook et al. (2010b) develop models for DMUs with network structures based upon additive efficiency decomposition. Their approach can be viewed as a centralized model of Liang et al. (2008). The centralized model of Liang et al. (2008) assumes the overall efficiency is a product or sum of divisional efficiencies. For example, consider the approach of Kao and Hwang (2008) where a set of insurance companies are assumed to have a two-stage operations of premium acquisition and profit generation. The overall efficiency is then a product of premium acquisition efficiency and profit generation efficiency. Liang et al. (2008) classify this type of modeling technique or efficiency decomposition as cooperative or centralized game approach, as the efficiency scores of all sub-DMUs or stages are simultaneously optimized.

Liang et al. (2008) further introduce modeling two-stage network DMUs from the perspective of the non-cooperative game. The non-cooperative approach is characterized by the leader-follower, or Stackelberg game. For example, if we assume that the first stage of premium acquisition is the leader, then the first stage performance is more important, and the efficiency or performance of the second stage of profit generation is computed subject to the requirement that the efficiency of the first stage is to stay fixed. In a similar manner, we can also assume the second stage is the leader and the first stage is the follower.

Note that while the centralized model approach of Liang et al. (2008) can be applied to DMUs with any network structures by assuming the overall efficiency is a weighted average of individual stage (or divisional) efficiencies, the leader-follower cannot be easily applied. Note also that the approach of Liang et al. (2008) or Kao and Hwang (2008) is developed under the assumption that the outputs from the first stage all become the only inputs to the second stage. The current chapter extends Liang et al. (2008) and Kao and Hwang (2008) by assuming that the second stage has its own inputs in addition to outputs from the first stage.

In fact, Liang et al. (2008) studied this type of two-stage network structure in analyzing the performance of a set of hypothetical supply chains. Other examples can be found in manufacturing with two sub-processes, one is production and the other is distribution. The inputs of first stage are manufacturing facilities, raw materials and components, laborers and operating fees of manufacturing department, the outputs of first stage are finished goods which are also part of the inputs to the second stage. Another part of inputs to the second stage are advertisement fee, employees of market department.

Due to the existence of additional inputs to the second stage, the approach of Liang et al. (2008) results a non-linear program that cannot be converted into linear programming problems if we assume that the overall efficiency is a geometric mean of two stages' efficiency. The current chapter develops procedures to convert the resulting non-linear programs into parametric linear programs so that the global optimal solution can be found if one adopts the centralized and leader-follower approaches of Liang et al. (2008). Therefore, the current chapter extends the approach of Liang et al. (2008) to more general two-stage network structures.

The remainder of the chapter is organized as follows. In the next section we extend the models of Liang et al. (2008) to evaluate performance of the two-stage network structure with additional inputs to the second stage. Relations between the two approaches are established. The two approaches are then illustrated with an example about regional R&D process in China. We demonstrate how to estimate the global optimal solution from our converted non-linear program. Conclusions are given in the last section.

12.2 DEA Models

Figures 12.1 and 12.2 graphically illustrate two types of two-stage network structure. Figure 12.1 studied by Liang et al. (2008) or Kao and Hwang (2008) assumes that the outputs from the first stage all become the only inputs to the second stage. These measures in-between the two stages are called intermediate measures. Figure 12.2 relaxes the above assumption by introducing inputs to the second stage in addition to the intermediate measures.

We assume that each DMU_j (j = 1, 2, . . ., n) has m inputs to the first stage, x_{ij}, (i = 1, 2, . . ., m) and D outputs (intermediate measures) from the first stage, z_{dj}, (d = 1, 2, . . ., D). These D outputs then become part of the inputs to the second stage, Another part of inputs are x_{hj}^2 (h = 1, 2, . . ., H). The outputs from the second stage are y_{rj}, (r = 1, 2, . . ., s).

We next develop models based upon the approaches of Liang et al. (2008) to analyze the performance of extended two-stage network structure as depicted in Fig. 12.2.

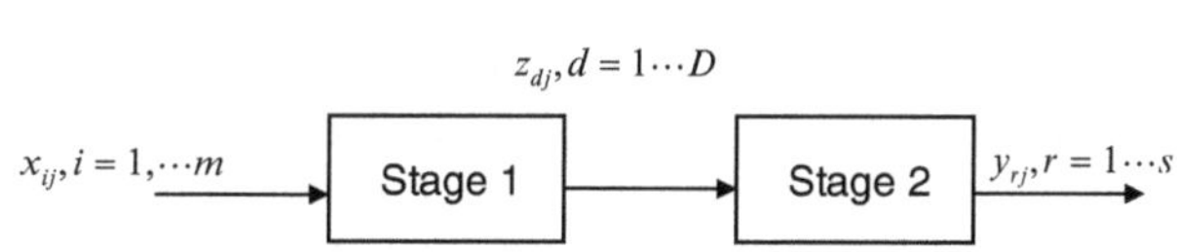

Fig. 12.1 Two-stage process of DMU_j

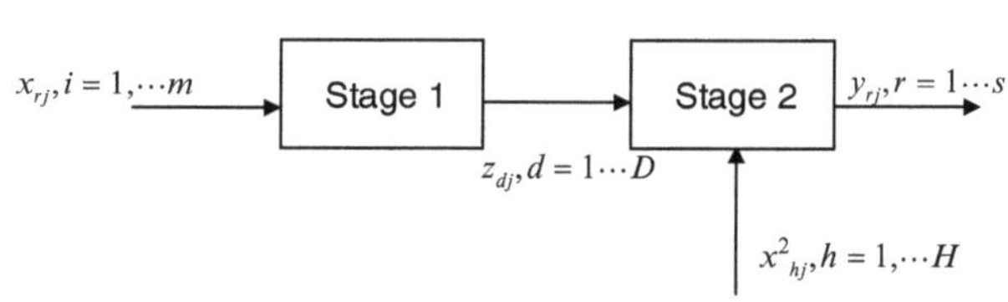

Fig. 12.2 Two-stage process with additional inputs to the second stage

12.2.1 Centralized Model

There are many cases that each sub-DMU works together to reach the optimal performance of the overall DMU. For example, marketing and production departments would cooperate to maximize company's profit. Liang et al. (2008) developed a centralized approach to analyze the performance of two-stage network structure described in Fig. 12.1. In their model, overall efficiency of the two-stage process is defined as the product of two stages' efficiencies. In a similar manner, based upon the ratio efficiency of the CCR model (Charnes et al. 1978), we can establish have the following model for Fig. 12.2.

$$\theta^{cen} = \max\theta_1^o * \theta_2^o = \max \frac{\sum_{d=1}^{D} w_d z_{do}}{\sum_{i=1}^{m} v_i x_{io}} * \frac{\sum_{r=1}^{s} u_r y_{ro}}{\sum_{d=1}^{D} w_d z_{do} + \sum_{h=1}^{H} Q_h x_{ho}^2}$$

$$s.t. \frac{\sum_{d=1}^{D} w_d z_{dj}}{\sum_{i=1}^{m} v_i x_{ij}} \leq 1 \quad \forall j \tag{12.1}$$

$$\frac{\sum_{r=1}^{s} u_r y_{rj}}{\sum_{d=1}^{D} w_d z_{dj} + \sum_{h=1}^{H} Q_h x_{hj}^2} \leq 1 \quad \forall j$$

$$v_i, w_d, Q_h, u_r \geq 0, \forall i, d, h, r$$

Where θ_1^o and θ_2^o represent the ratio efficiencies for stages 1 and 2, respectively. As in Liang et al. (2008), it is assumed that a same set of weights (w_d) is applied to the intermediate measures (z_{dj}) for both stages. For example, the manufacturer and retailer jointly determine the price, order quantity, etc. to achieve maximum profit (Huang and Li 2001). Herein, as in Liang et al. (2008), we also assume that the "worth" or value accorded to the intermediate variables is the same regardless of whether they are being viewed as inputs or outputs.

Due to the additional inputs to the second stage $\left(\sum_{h=1}^{H} Q_h x_{ho}^2\right)$, model (12.1) cannot be converted into a linear program. We here introduce a heuristic method to solve this problem.

Consider the following model

$$
\begin{aligned}
\theta_1^{o^{\max}} &= \max \frac{\sum_{d=1}^{D} w_d z_{do}}{\sum_{i=1}^{m} v_i x_{io}} \\
s.t. & \frac{\sum_{d=1}^{D} w_d z_{dj}}{\sum_{i=1}^{m} v_i x_{ij}} \leq 1 \quad \forall j \\
& \frac{\sum_{r=1}^{s} u_r y_{rj}}{\sum_{d=1}^{D} w_d z_{dj} + \sum_{h=1}^{H} Q_h x_{hj}^2} \leq 1 \quad \forall j \\
& v_i, w_d, Q_h, u_r \geq 0, \forall i, d, h, r
\end{aligned}
\tag{12.2}
$$

In model (12.2), the two sets of constraints are the same to the ones in model (12.1), which ensure the efficiencies for the first and the second stage do not exceed one, respectively. Therefore, model (12.2) can be used to estimate the best possible efficiency for stage 1. Denote the optimal value to model (12.2) as $\theta_1^{o^{\max}}$, then the efficiency for the first stage θ_1^o must satisfy $\theta_1^o \in \left[0, \theta_1^{o^{\max}}\right]$.

Model (12.2) is a non-linear model, but can be converted into a linear program through the Charnes-Cooper transformation as follows:

$$
\begin{aligned}
\theta_1^{o^{\max}} &= \max \sum_{d=1}^{D} w_d z_{do} \\
s.t. & \sum_{d=1}^{D} w_d z_{dj} - \sum_{i=1}^{m} v_i x_{ij} \leq 0 \quad \forall j \\
& \sum_{r=1}^{s} u_r y_{rj} - \sum_{h=1}^{H} Q_h x_{hj}^2 - \sum_{d=1}^{D} w_d z_{dj} \leq 0 \quad \forall j \\
& \sum_{i=1}^{m} v_i x_{io} = 1 \\
& v_i, w_d, Q_h, u_r \geq 0, \forall i, d, h, r
\end{aligned}
\tag{12.3}
$$

Therefore, the efficiency of the first stage θ_1^o can be treated as a variable $\theta_1^o \in \left[0, \theta_1^{o^{\max}}\right]$ and the overall efficiency denoted as $\theta^{cen,1,*}$ can be considered as a function of θ_1^o as follows: (or model (12.1) can be written as)

$$\theta^{cen,1,*} = \max \theta_1^o * \frac{\sum_{r=1}^{s} u_r y_{ro}}{\sum_{d=1}^{D} w_d z_{do} + \sum_{h=1}^{H} Q_h x_{ho}^2}$$

$$\begin{aligned} s.t. & \frac{\sum_{d=1}^{D} w_d z_{dj}}{\sum_{i=1}^{m} v_i x_{ij}} \leq 1 \quad \forall j \\ & \frac{\sum_{r=1}^{s} u_r y_{rj}}{\sum_{d=1}^{D} w_d z_{dj} + \sum_{h=1}^{H} Q_h x_{hj}^2} \leq 1 \quad \forall j \\ & \frac{\sum_{d=1}^{D} w_d z_{do}}{\sum_{i=1}^{m} v_i x_{io}} = \theta_1^o \\ & \theta_1^o \in \left[0, \theta_1^{o^{\max}}\right] \\ & v_i, w_d, Q_h, u_r \geq 0, \forall i, d, h, r \end{aligned} \tag{12.4}$$

Model (12.4) now can be transformed via the Charnes-Cooper transformation as follows:

$$\begin{aligned} \theta^{cen,1,*} = & \max \theta_1^o * \sum_{r=1}^{s} u_r y_{ro} \\ s.t. & \sum_{d=1}^{D} w_d z_{dj} - \sum_{i=1}^{m} v_i x_{ij} \leq 0 \quad \forall j \\ & \sum_{r=1}^{s} u_r y_{rj} - \sum_{h=1}^{H} Q_h x_{hj}^2 - \sum_{d=1}^{D} w_d z_{dj} \leq 0 \quad \forall j \\ & \sum_{h=1}^{H} Q_h x_{ho}^2 + \sum_{d=1}^{D} w_d z_{do} = 1 \\ & \sum_{d=1}^{D} w_d z_{do} - \theta_1^o \sum_{i=1}^{m} v_i x_{io} = 0 \\ & \theta_1^o \in \left[0, \theta_1^{o^{\max}}\right] \\ & v_i, w_d, Q_h, u_r \geq 0, \forall i, d, h, r \end{aligned} \tag{12.5}$$

Let $\theta_1^o = \theta_1^{o^{\max}} - k\Delta\varepsilon$, Here $\Delta\varepsilon$ is a step size,[1] $k = 0, 1, 2 \ldots [k^{\max}] + 1$, where $[k^{\max}]$ is the maximal integer which is smaller than or equal to $\theta_1^{o^{\max}}/\Delta\varepsilon$. Now, given each θ_1^o, model (12.5) can be solved as a linear program.

In solving model (12.5), we set the initial k value as the lower bound, $k = 0$. Then we increase k at each step. We solve each linear program of model (12.5) corresponding to each k and denote the optimal value to model (12.5) as $\theta^{cen,1}(k)$. Therefore, the global optimal efficiency of the system under evaluation can be estimated as $\hat{\theta}^{cen,1,*} = \max_k \theta^{cen,1}(k)$.

Note, when the efficiency of the entire two-stage system under evaluation is $\hat{\theta}^{cen,1,*}$, the maximal efficiency for its first stage is $\hat{\theta}_1^{o+} = \theta_1^o(k^*)$, where $k^* = \min\{k|\hat{\theta}^{cen,1,*} = \theta^{cen,1}(k)\}$. Therefore, the minimal efficiency for its second stage is $\hat{\theta}_2^{o-} = \frac{\hat{\theta}^{cen,1,*}}{\hat{\theta}_1^{o+}}$.

Similarly, we can also treat stage 2 as a variable. The optimal efficiency of the second stage $\theta_2^{o^{\max}}$ can be calculated using a model similar to model (12.2). Then, according to the above-mentioned algorithm, we can get the global efficiency $\hat{\theta}^{cen,2,*}$ and its corresponding maximal efficiency for the second stage $\hat{\theta}_2^{o+}$, respectively. And then the minimal efficiency for the first stage is $\hat{\theta}_1^{o-} = \frac{\hat{\theta}^{cen,2,*}}{\hat{\theta}_2^{o+}}$.

The best possible efficiency for stage 2 can be obtained via the following model

$$
\begin{aligned}
\theta_2^{o^{\max}} &= \frac{\sum_{r=1}^{s} u_r y_{ro}}{\sum_{d=1}^{D} w_d z_{do} + \sum_{h=1}^{H} Q_h x_{ho}^2} \\
s.t.\ & \frac{\sum_{d=1}^{D} w_d z_{dj}}{\sum_{i=1}^{m} v_i x_{ij}} \le 1 \quad \forall j \\
& \frac{\sum_{r=1}^{s} u_r y_{rj}}{\sum_{d=1}^{D} w_d z_{dj} + \sum_{h=1}^{H} Q_h x_{hj}^2} \le 1 \quad \forall j \\
& v_i, w_d, Q_h, u_r \ge 0, \forall i, d, h, r
\end{aligned}
\tag{12.6}
$$

Denote the optimal value to model (12.6) as $\theta_2^{o^{\max}}$, then the efficiency for the second stage θ_2^o must satisfy $\theta_2^o \in [0, \theta_2^{o^{\max}}]$.

[1] The smaller the $\Delta\varepsilon$ value we select, the more precise results we obtain.

Model (12.6) is a non-linear model, and can be converted into a linear program through the Charnes-Cooper transformation as follows:

$$
\begin{aligned}
\theta_2^{o^{\max}} &= \max\sum_{r=1}^{s} u_r y_{rj_0} \\
s.t.&\sum_{d=1}^{D} w_d z_{dj} - \sum_{i=1}^{m} v_i x_{ij} \le 0 \quad \forall j \\
&\sum_{r=1}^{s} u_r y_{rj} - \sum_{h=1}^{H} Q_h x_{hj}^2 - \sum_{d=1}^{D} w_d z_{dj} \le 0 \quad \forall j \\
&\sum_{h=1}^{H} Q_h x_{ho}^2 + \sum_{d=1}^{D} w_d z_{do} = 1 \\
&v_i, w_d, Q_h, u_r \ge 0, \forall i, d, h, r
\end{aligned}
\tag{12.7}
$$

The efficiency of the second stage θ_2^o can be treated as a variable $\theta_2^o \in \left[0, \theta_2^{o^{\max}}\right]$ and the overall efficiency $\theta^{cen,2,*}$ can be considered as a function of θ_2^o as follows: (or model (12.1) can be written as)

$$
\begin{aligned}
\theta^{cen,2,*} &= \max\theta_2^o * \frac{\sum_{d=1}^{D} w_d z_{do}}{\sum_{i=1}^{m} v_i x_{io}} \\
s.t.&\frac{\sum_{d=1}^{D} w_d z_{dj}}{\sum_{i=1}^{m} v_i x_{ij}} \le 1 \quad \forall j \\
&\frac{\sum_{r=1}^{s} u_r y_{rj}}{\sum_{d=1}^{D} w_d z_{dj} + \sum_{h=1}^{H} Q_h x_{hj}^2} \le 1 \quad \forall j \\
&\frac{\sum_{r=1}^{s} u_r y_{ro}}{\sum_{d=1}^{D} w_d z_{do} + \sum_{h=1}^{H} Q_h x_{ho}^2} = \theta_2^o \\
&\theta_2^o \in \left[0, \theta_2^{o^{\max}}\right] \\
&v_i, w_d, Q_h, u_r \ge 0, \forall i, d, h, r
\end{aligned}
\tag{12.8}
$$

Model (12.8) now can be transformed via the Charnes-Cooper transformation as follows:

$$
\begin{aligned}
\theta^{cen,2,*} &= \max \theta_2^o * \sum_{d=1}^{D} w_d z_{do} \\
s.t. &\sum_{d=1}^{D} w_d z_{dj} - \sum_{i=1}^{m} v_i x_{ij} \le 0 \quad \forall j \\
&\sum_{r=1}^{s} u_r y_{rj} - \sum_{h=1}^{H} Q_h x_{hj}^2 - \sum_{d=1}^{D} w_d z_{dj} \le 0 \quad \forall j \\
&\sum_{i=1}^{m} v_i x_{io} = 1 \\
&\sum_{r=1}^{s} u_r y_{ro} - \theta_2^o \left(\sum_{h=1}^{H} Q_h x_{ho}^2 + \sum_{d-1}^{D} w_d z_{do} \right) = 0 \\
&\theta_2^o \in \left[0, \theta_2^{o^{\max}}\right] \\
&v_i, w_d, Q_h, u_r \ge 0, \forall i, d, h, r
\end{aligned}
\tag{12.9}
$$

Let $\theta_2^o = \theta_2^{o^{\max}} - \rho\Delta\varepsilon$, $\rho = 0, 1, 2 \ldots [\rho^{\max}] + 1$, where $[\rho^{\max}]$ is the maximal integer which is smaller than or equal to $\theta_2^{o^{\max}}/\Delta\varepsilon$. Given each θ_2^o, model (12.9) can be solved as a linear program.

By solving model (12.9), the global optimal efficiency of the system under evaluation can be estimated as $\hat{\theta}^{cen,2,*} = \max_{\rho} \theta^{cen,2}(\rho)$. Then, when the efficiency of the entire two-stage system under evaluation is $\hat{\theta}^{cen,2,*}$, the maximal efficiency for its second stage is $\hat{\theta}_2^{o+} = \theta_2^o(\rho^*)$, where $\rho^* = \min\{\rho | \hat{\theta}^{cen,2,*} = \theta^{cen,2}(\rho)\}$. Finally, the minimal efficiency for its first stage is $\hat{\theta}_1^{o-} = \frac{\hat{\theta}^{cen,2,*}}{\hat{\theta}_2^{o+}}$.

Note that, no matter which stage's efficiency is assumed as a variable in deriving the efficiency for the entire two-stage system, the same optimal global optimal efficiency should be obtained, i.e., $\theta^{cen,1,*} = \theta^{cen,2,*}$. And the efficiency decomposition is unique if $\theta_1^{o+} = \theta_1^{o-}$ and $\theta_2^{o+} = \theta_2^{o-}$.

12.2.2 Non-cooperative Model

The models presented in former section for analyzing the extended two-stage network structure with additional inputs to the second stage are under centralized decision-making environment. In this section, a non-cooperative approach is introduced to analyze this network structure. We first treat stage 1 as the leader (this sub-process is assumed to be more important) and stage 2 as the follower.

The efficiency of the first stage (the leader) for a specific DMUo is calculated by using the CCR model (Charnes et al. 1978) as follows:

$$
\begin{aligned}
e_1^{o*} &= \max \sum_{d=1}^{D} w_d z_{do} \\
s.t. &\sum_{d=1}^{D} w_d z_{dj} - \sum_{i=1}^{m} v_i x_{ij} \le 0 \quad \forall j \\
&\sum_{i=1}^{m} v_i x_{io} = 1 \\
&v_i, w_d, Q_h, u_r \ge 0, \forall i, d, h, r
\end{aligned}
\tag{12.10}
$$

Getting the efficiency of stage 1 e_1^{o*} by model (12.10), the optimal weights v_i^*, $i = 1 \cdots m$, w_d^*, $d = 1 \cdots D$ are also given. Since the two sub-processes are connected with each other by intermediate measures, v_i^*, w_d^* need to be introduced to the next model for calculating the efficiency of stage 2. However, the weights v_i^*, w_d^* may not be unique. Doyle and Green (1994) develop second goal models to solve a similar problem about cross-efficiency in DEA. Follow this idea, we develop a model that maximize the efficiency of stage 2 as the objective function while remain the efficiency of stage 1 as a constraint. The model is as follows:

$$
\begin{aligned}
e_2^{o*} &= \max \frac{\sum_{r=1}^{s} u_r y_{ro}}{\sum_{d=1}^{D} w_d z_{do} + \sum_{h=1}^{H} Q_h x_{ho}^2} \\
s.t. &\frac{\sum_{d=1}^{D} w_d z_{dj}}{\sum_{i=1}^{m} v_i x_{ij}} \le 1 \quad \forall j \\
&\frac{\sum_{r=1}^{s} u_r y_{rj}}{\sum_{d=1}^{D} w_d z_{dj} + \sum_{h=1}^{H} Q_h x_{hj}^2} \le 1 \quad \forall j \\
&\frac{\sum_{d=1}^{D} w_d z_{do}}{\sum_{i=1}^{m} v_i x_{io}} = e_1^{o*} \\
&v_i, w_d, Q_h, u_r \ge 0, \forall i, d, h, r
\end{aligned}
\tag{12.11}
$$

In model (12.11), the efficiency for the second stage of DMUo is optimized based upon that the efficiency of the first stage e_1^{o*} remains unchanged. Model (12.11) can be transformed as:

$$
\begin{aligned}
e_2^{o*} = \max \sum_{r=1}^{s} u_r y_{rjo} \\
s.t. \sum_{d=1}^{D} w_d z_{dj} - \sum_{i=1}^{m} v_i x_{ij} \le 0 \quad \forall j \\
\sum_{r=1}^{s} u_r y_{rj} - \sum_{h=1}^{H} Q_h x_{hj}^2 - \sum_{d=1}^{D} w_d z_{dj} \le 0 \quad \forall j \\
\sum_{h=1}^{H} Q_h x_{ho}^2 + \sum_{d=1}^{D} w_d z_{do} = 1 \\
\sum_{d=1}^{D} w_d z_{do} - e_1^{o*} \sum_{i=1}^{m} v_i x_{io} = 0 \\
v_i, w_d, Q_h, u_r \ge 0, \forall i, d, h, r
\end{aligned}
\tag{12.12}
$$

Denote the optimal value to model (12.12) as e_2^{o*}, then the efficiency for the entire two-stage system or DMUo is $e^{non,1} = e_1^{o*} * e_2^{o*}$.

In a similar manner, if we assume the second stage is the leader, the regular CCR efficiency π_2^{o*} for that stage can be calculated via using the standard CCR model with inputs (z_{dj} and x_{hj}^2) and outputs (y_{rj}). Then, the efficiency score (π_1^{o*}) for the first stage (follower) can be obtained by solving a model with the restriction that the second stage score π_2^{o*} remains unchanged. The overall efficiency of the entire system in this situation is $\pi^{non,2} = \pi_1^{o*} * \pi_2^{o*}$.

We first calculate the efficiency of the second stage (the leader) for a specific DMUo by using the standard CCR model (Charnes et al. 1978) as follows:

$$
\begin{aligned}
\pi_2^{o*} = \max \sum_{r=1}^{s} u_r y_{rjo} \\
s.t. \sum_{r=1}^{s} u_r y_{rj} - \sum_{h=1}^{H} Q_h x_{hj}^2 - \sum_{d=1}^{D} w_d z_{dj} \le 0 \quad \forall j \\
\sum_{h=1}^{H} Q_h x_{ho}^2 + \sum_{d=1}^{D} w_d z_{do} = 1 \\
w_d, Q_h, u_r \ge 0, \forall d, h, r
\end{aligned}
\tag{12.13}
$$

The model for calculating the efficiency of follower (the first stage) is as follows:

$$\begin{aligned}
\pi_1^{o*} &= \max \frac{\sum_{d=1}^{D} w_d z_{do}}{\sum_{i=1}^{m} v_i x_{io}} \\
s.t.\ & \frac{\sum_{d=1}^{D} w_d z_{dj}}{\sum_{i=1}^{m} v_i x_{ij}} \leq 1 \quad \forall j \\
& \frac{\sum_{r=1}^{s} u_r y_{rj}}{\sum_{d=1}^{D} w_d z_{dj} + \sum_{h=1}^{H} Q_h x_{hj}^2} \leq 1 \quad \forall j \\
& \frac{\sum_{r=1}^{s} u_r y_{ro}}{\sum_{d=1}^{D} w_d z_{do} + \sum_{h=1}^{H} Q_h x_{ho}^2} = \pi_2^{o*} \\
& v_i, w_d, Q_h, u_r \geq 0, \forall i, d, h, r
\end{aligned} \tag{12.14}$$

In model (12.14), the efficiency for the first stage of DMUo is optimized based upon that the efficiency of the second stage π_2^{o*} remains unchanged. Model (12.14) can be transformed as:

$$\begin{aligned}
\pi_1^{o*} &= \max \sum_{d=1}^{D} w_d z_{do} \\
s.t.\ & \sum_{d=1}^{D} w_d z_{dj} - \sum_{i=1}^{m} v_i x_{ij} \leq 0 \quad \forall j \\
& \sum_{r=1}^{s} u_r y_{rj} - \sum_{h=1}^{H} Q_h x_{hj}^2 - \sum_{d=1}^{D} w_d z_{dj} \leq 0 \quad \forall j \\
& \sum_{i=1}^{m} v_i x_{io} = 1 \\
& \sum_{r=1}^{s} u_r y_{ro} - \pi_2^{o*} \left(\sum_{h=1}^{H} Q_h x_{ho}^2 + \sum_{d=1}^{D} w_d z_{do} \right) = 0 \\
& v_i, w_d, Q_h, u_r \geq 0, \forall i, d, h, r
\end{aligned} \tag{12.15}$$

In model (12.15), denote π_1^{o*} as the optimal efficiency for first stage. The overall efficiency of the entire system in this situation is $\pi^{non,2} = \pi_1^{o*} * \pi_2^{o*}$.

12.2.3 Relations Between the Two Models

This section gives three theorems to illustrate the relations between the centralized model and the non-cooperative model.

Theorem 1 *$e_1^{o*} \geq \pi_1^{o*}$, $e_2^{o*} \leq \pi_2^{o*}$, where e_1^{o*} and e_2^{o*} are the efficiencies for the first stage and the second stage, respectively, when stage 1 is assumed as leader. π_1^{o*} and π_2^{o*} are the efficiencies for the first stage and the second stage, respectively, when stage 2 is assumed as leader.*

Proof Denote an optimal solution to model (12.8) as ($v_i^{non,1,*}$, $w_d^{non,1,*}$, $Q_h^{non,1,*}$, $u_r^{non,1,*}$, $\forall$ i, d, h, r) and the optimal efficiency for the stage 2 as e_2^{o*}.

Let $\varsigma = (w_d^{non,1,*}, Q_h^{non,1,*}, u_r^{non,1,*}, \forall\ d, h, r)$, then, ς is also a feasible solution to model (A.5). Note that, model (A.5) calculates the optimal efficiency for stage 2 when stage 2 is treated as leader. Therefore, its optimal efficiency is π_2^{o*}. Thus, the efficiency for stage 2 based upon ς is not bigger than π_2^{o*}. So we have $e_2^{o*} \leq \pi_2^{o*}$.

Similarly, we can get the result $e_1^{o*} \geq \pi_1^{o*}$. Q.E.D.

Theorem 2 *(1) To each DMU, $\theta^{cen,1,*} = \theta^{cen,2,*}$ where $\theta^{cen,1,*}$ and $\theta^{cen,2,*}$ are the optimal efficiencies for the system based upon the centralized model when stage 1 and stage 2 are treated as variables, respectively; (2) $\theta^{cen} \geq e^{non,1,*}$, $\theta^{cen} \geq \pi^{non,2,*}$, where $\theta^{cen} = \theta^{cen,1,*} = \theta^{cen,2,*}$, and $e^{non,1,*}$ and $\pi^{non,2,*}$ are the optimal efficiencies for the system when stage 1 or stage 2 is assumed leader, respectively.*

Proof

1. Either stage 1 or stage 2 is treated as a variable, the maximal efficiency for the system is unique. So $\theta^{cen,1,*} = \theta^{cen,2,*}$.
2. Denote an optimal solution to model (12.7) as ($v_i^{non,1,*}$, $w_d^{non,1,*}$, $Q_h^{non,1,*}$, $u_r^{non,1,*}$, $\forall$ i, d, h, r), accordingly, the optimal efficiency for the system as $e^{non,1} = e_1^{o*} * e_2^{o*}$.

Let $\xi = (w_d^{non,1,*}, Q_h^{non,1,*}, u_r^{non,1,*}, \forall\ d, h, r)$, then ξ is also a feasible solution to model (12.1), therefore, the optimal efficiency based upon model (12.1) is bigger than or equal to the efficiency based upon the feasible solution ξ. So we get:

$$\theta^{cen} \geq \frac{\sum_{d=1}^{D} w_d^{non,1,*} z_{do}}{\sum_{i=1}^{m} v_i^{non,1,*} x_{io}} * \frac{\sum_{r=1}^{s} u_r^{non,1,*} y_{ro}}{\sum_{d=1}^{D} w_d^{non,1,*} z_{do} + \sum_{h=1}^{H} Q_h^{non,1,*} x_{ho}^2} = e^{non,1,*}.$$

Similarly, we can get $\theta^{cen} \geq \pi^{non,2,*}$. Q.E.D.

Theorem 3 *If there is only one intermediate measure, then, the optimal efficiency for the system is unique either based upon the centralized model or based upon non-cooperative model, such that $\theta^{cen} = \theta^{cen,1,*} = \theta^{cen,2,*} = e^{non,1,*} = \pi^{non,2,*}$. And the efficiency decomposition is also unique such that $e_1^{o*} = \pi_1^{o*} = \theta_1^{CCR}$, $e_2^{o*} = \pi_2^{o*} = \theta_2^{CCR}$, where θ_1^{CCR} and θ_2^{CCR} are the efficiencies for the first and second stage as applying the standard CCR model.*

Proof Liang et al. (2008) has proven that $e_1^{o*} = \pi_1^{o*} = \theta_1^{CCR}$ and $e_2^{o*} = \pi_2^{o*} = \theta_2^{CCR}$. Based upon Theorem 2, we have $\theta^{cen} \geq e^{non,1,*}$. Thus $\theta^{cen} = \theta^{cen,1,*} \geq \theta_1^{CCR} * \theta_2^{CCR}$. On the other hand, the efficiency of the entire two-stage system based upon the centralized model is not bigger than the product of two efficiencies of the first and second stages using the standard CCR models. Therefore, $\theta_1^{CCR} * \theta_2^{CCR} \geq \theta^{cen}$. Thus, $\theta^{cen} = \theta_1^{CCR} * \theta_2^{CCR} = \theta^{cen,1,*} = e^{non,1,*}$.

Similarly, we can have $\theta^{cen} = \theta_1^{CCR} * \theta_2^{CCR} = \theta^{cen,2,*} = \pi^{non,2,*}$. Q.E.D.

12.3 An Illustrative Application

This section presents a real example about regional R&D process of 30 Provincial Level Regions in Mainland of China. Figure 12.3 shows a regional R&D process which contains two sub-processes, one is technology development and the other is economic application.

In the technology development process, the inputs are: R&D expenditure (R&DE), R&D personnel (R&DP) and the proportion of regional science and technology funds in regional total financial expenditure (S&TF/TFE), the outputs are: patents and papers. Among them, R&DE and R&DP are two core indexes in science and technology activities (see Zhong et al. 2011), S&TF/TFE is a important index for reflecting government's support. The outputs of the first stage are the number of patents and papers which are also inputs to the second stage, namely, these are intermediate measures. The second stage also has an input of contract value (CV) in technology market. Economic application process transforms technology development into economic benefits. CV presents the function of intermediary services institution which provides services in this process. The final outputs are complex economic indexes which embody the regional economic performance affected by R&D: GDP represents the macro-economy performance, total exports (TE) is important to depict international competency, urban per capita disposable annual income (UPCDAI) depicts people's living level and gross output of high-tech industry (GOHI) is directly to depict the condition of high-tech industry.

Table 12.1 provides the data for the above R&D system of for the 30 Provincial Level Regions in Mainland of China. The data for Tibet Autonomous Region are

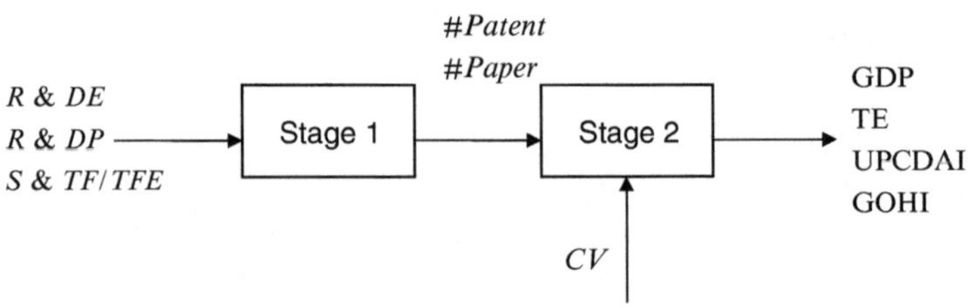

Fig. 12.3 A two-stage regional R&D system

Table 12.1 Inputs and outputs of R&D for 30 provincial level regions in Mainland China

Region type	DMU	Region	R&DP	R&DE	S&TF/ TFE	Patent	Paper	CV	GDP	TE	UPCDAI	GOHI
Municipality	1	Beijing	10.34786	668.6351	5.445765	9,157	65,951	1,236.245	12,153.03	483.7932	26,738.48	2,757.14
	2	Chongqing	2.00665	79.45994	1.203814	834	13,737	38.31581	6,530.01	42.80071	15,748.67	352.84
	3	Shanghai	6.46163	423.3774	7.201873	5,997	32,733	435.4108	15,046.45	1,417.96027	28,837.78	5,557.45
	4	Tianjin	2.8783	178.4661	3.023746	1,889	12,472	105.4611	7,521.85	298.92719	21,402.01	1,901.07
Province	5	Anhui	3.01654	135.9535	1.702644	795	13,699	35.61736	10,062.82	88.86487	14,085.74	460.31
	6	Fujian	2.27886	135.3819	1.975486	824	9,075	23.25944	12,236.53	533.1911	19,576.83	1,972.01
	7	Gansu	1.27445	37.26124	0.817095	227	7,856	35.62869	3,387.56	7.35512	11,929.78	67.39
	8	Guangdong	12.97681	652.982	3.887568	11,355	35,773	170.985	39,482.56	3,589.54893	21,574.72	17,161.94
	9	Guizhou	0.77328	26.41343	1.040003	322	4,946	1.780611	3,912.68	13.56612	12,862.53	293.64
	10	Hainan	0.17583	5.7806	1.249058	84	2,726	0.555627	1,654.21	13.08632	13,750.85	54.75
	11	Hebei	3.8808	134.8446	1.125763	691	17,970	17.21118	17,235.48	156.88902	14,718.25	629.17
	12	Heilongjiang	3.70197	109.1704	1.062859	1,142	14,553	48.855	8,587	100.82127	12,565.98	311.4
	13	Henan	4.79963	174.7599	1.222261	1,129	21,188	26.30461	19,480.46	73.45376	14,371.56	953.23
	14	Hubei	5.12124	213.449	1.211472	1,478	25,268	77.03287	12,961.1	99.78796	14,367.48	1,039.52
	15	Hunan	3.49591	153.4995	1.339839	1,752	21,042	44.04324	13,059.69	54.92034	15,084.31	648.75
	16	Jiangsu	10.67826	701.9529	2.912858	5,322	47,441	108.2184	34,457.3	1,991.9919	20,551.72	13,015.35
	17	Jiangxi	1.83522	75.8936	0.857803	386	6,811	9.78927	7,655.18	73.68488	14,021.54	755.65
	18	Jilin	2.60875	81.36019	1.28305	719	8,987	19.75983	7,278.75	31.24935	14,006.27	537.66
	19	Liaoning	5.43947	232.3687	2.143081	1,993	20,801	119.7095	15,212.49	334.14928	15,761.38	1,313.84
	20	Qinghai	0.30013	7.59379	0.982114	35	1,240	8.496721	1,081.27	2.51876	12,691.85	19.22
	21	Shandong	8.33303	519.592	1.924254	2,865	26,941	71.9391	33,896.65	794.90706	17,811.04	4,555.71

(continued)

Table 12.1 (continued)

Region type	DMU	Region	R&DP	R&DE	S&TF/TFE	Patent	Paper	CV	GDP	TE	UPCDAI	GOHI
	22	Shanxi	2.52624	80.85633	1.127415	603	6,757	16.20675	7,358.31	28.37455	13,996.55	196.47
	23	Shanxi	4.23465	189.5063	1.131443	1,342	26,403	69.80741	8,169.8	39.88149	14,128.76	717.04
	24	Sichuan	4.87863	214.459	0.79759	1,596	22,568	54.59769	14,151.28	141.69447	13,839.4	1,766.76
	25	Yunnan	1.22051	37.23044	0.972869	476	7,101	10.24687	6,169.75	45.13252	14,423.93	147.17
	26	Zhejiang	5.90844	398.8367	3.74258	4,818	25,638	56.45805	22,990.35	1,330.12954	24,610.81	2,672.09
Autonomous region	27	Guangxi	1.56993	47.20277	1.114432	326	9,982	1.766182	7,759.16	83.7537	15,451.48	273.67
	28	Inner Mongolia	1.27057	52.07259	0.937557	178	3,214	14.76515	9,740.25	23.15476	15,849.19	236.61
	29	Ningxia	0.33954	10.44221	1.018058	52	1,365	0.898229	1,353.31	7.4293	1,4024.7	32.89
	30	Xinjiang	0.82683	21.80426	1.198296	120	5,688	1.207767	4,277.05	109.34563	12,257.52	23.74

Table 12.2 Efficiency for Zhejiang Province (DMU26) corresponding to each k based upon the centralized model

k	$\theta_1^o(k) = \theta_1^{o^{\max}} - k * 0.01$	$\theta^{cen,1,*}(k)$
0–10	0.9111–0.8111	0.4712–0.6231
11–18	0.8011–0.7311	06339–0.6682
19	0.7211	0.6688 (Global optimal efficiency)
20–30	0.7111–0.6111	0.6685–0.5926
31–40	0.6011–0.5111	0.5830–0.4971
41–50	0.5011–0.4111	0.4876–0.4017
51–60	0.4011–0.3111	0.3922–0.3063
61–70	0.3011–0.2111	0.2968–0.2109
71–80	0.2011–0.1111	0.2014–0.1155
81–90	0.1011–0.0111	0.1060–0.0201
91–92	0.0011–0.0000	0.0106–0.0000

incomplete and are not included in the current study. The data are derived from "China statistical yearbook, 2009" and "China science and technology statistical yearbook, 2009".

We now illustrate the proposed computation procedure in estimating the global optimal efficiency for each Provincial Level Region. Consider Zhejiang Province (DMU 26). The maximal score for its first stage is $\theta_1^{o^{\max}} = 0.9111$ based upon model (12.3). Now, let $\theta_1^o = 0.9111 - k\Delta\varepsilon, k = 0, 1, 2 \ldots [k^{\max}] + 1$, and set the step size as $\Delta\varepsilon = 0.01$. Therefore, $[k^{\max}] = [\theta_1^{o^{\max}}/\Delta\varepsilon] = [0.9111/0.01] = 91$, i.e., $k = 0, 1, 2...92$. Table 12.2 shows the results from model (12.5) for Zhejiang Province (DMU 26) corresponding to each k from 0 to 92. For example, when we set $k = 10$ and $k = 41$, the optimal efficiency for the Zhejiang Province (DMU 26) is 0.6231 and 0.3922, respectively.

Figure 12.4 shows the change of the optimal value to model (12.5) as k increases from 1 to 92. It can be seen that its efficiency increases until $k = 19$. When k exceeds 19, the optimal efficiency for Zhejiang province (DMU 26) starts to decrease. Thus, the global optimal efficiency for Zhejiang Province (DMU 26) is $\hat{\theta}^{cen,1,*} = 0.6688$ when $k = 19$.

Table 12.3 reports the results based upon the proposed approaches in This chapter. The results based upon centralized models ($\Delta\varepsilon = 0.01$) are shown in columns 4–9, where columns 4, 5, 6 are the efficiencies when stage 1 is assumed as a variable and columns 7, 8, 9 present the efficiencies when stage 2 is assumed as a variable. The results based upon non-cooperative models are shown in the last six columns, in which columns 10, 11, 12 shows the results when stage 1 is assumed as the leader, and the last three columns gives the efficiencies when stage 2 is assumed as the leader.

First, the results in Table 12.3 verify our Theorem 1. For example, to each Provincial Level Region, its efficiency for the first stage in column 10 is always bigger than or equal to the one in column 13. Similarly, the efficiency for the second stage in column 11 is always less than or equal to the one in column 14. So the Theorem 1 can be verified such that $e_1^{o*} \geq \pi_1^{o*}$, $e_2^{o*} \leq \pi_2^{o*}$.

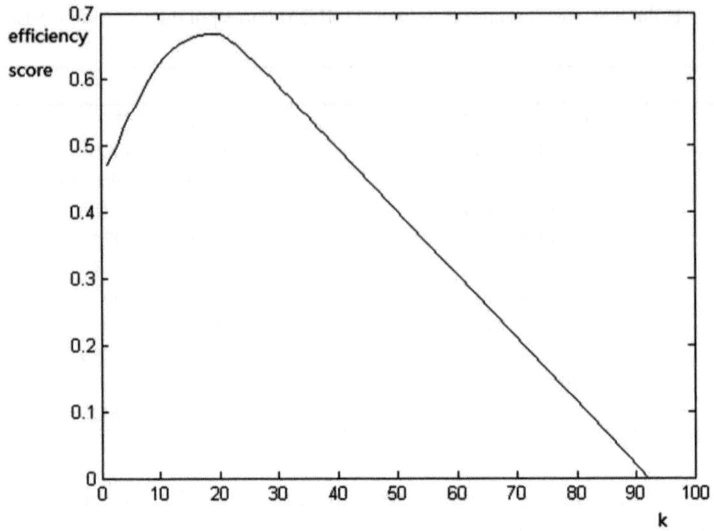

Fig. 12.4 The efficiency changes of Zhejiang Province (DMU 26) corresponding to each k

However, some results in Table 12.3 are not consistent with our Theorem 2. First, for some Provincial Level Regions, their efficiencies are not equal based upon the centralized model when stage 1 and 2 are treated as variables, respectively. For example, for Shanxi Province (DMU 22), $\hat{\theta}^{cen,1,*} = 0.2943$, but $\hat{\theta}^{cen,2,*} = 0.2942$. Therefore, the result does not support the first part of Theorem 2 such that $\theta^{cen,1,*} = \theta^{cen,2,*}$. The similar situation occurs to Shanghai (DMU 3), Gansu (DMU 7), Heilongjiang (DMU 12), Henan (DMU 13), Hubei (DMU 14), Hunan (DMU 15), Jiangxi (DMU 17), Jilin (DMU 18), Liaoning (DMU 19), Shandong (DMU 21), Shanxi (DMU 22), Sichuan (DMU 24) and Yunnan Province (DMU 25). The results for the other 16 Provincial Level Regions satisfy the first part of Theorem 2.

Furthermore, some results in Table 12.3 are not consistent with the second part of our Theorem 2. For example, for Jilin Province (DMU 18) $\hat{\theta}^{cen,2,*} = 0.3518$, while $e^{non,1,*} = 0.3533$. So the result is inconsistent with the second part of our Theorem 2 such that $\theta^{cen} \geq e^{non,1,*}$, $\theta^{cen} \geq \pi^{non,2,*}$. The similar situation occurs to Heilongjiang (DMU 12), Gansu (DMU 7), Henan (DMU 13), Hubei (DMU 14), Hunan (DMU 15), Shandong (DMU 21), Shanxi (DMU 22), Sichuan (DMU 24) and Yunnan Province (DMU 25). The results for the other 20 Provincial Level Regions satisfy the second part of Theorem 2.

The reason for the above inconsistency is due to the fact that the step size $\Delta\varepsilon$ we use is not small enough. If the $\Delta\varepsilon$ is adequately small, we can get the results in consistent with the theorems.

Table 12.4 reports the results based upon centralized model with $\Delta\varepsilon = 0.0001$ and $\Delta\varepsilon = 0.00001$, respectively. It shows there are only three Provincial Level

Table 12.3 Results based upon the centralized model with $\Delta\varepsilon$ =0.01 and the non-cooperative model

			Centralized model ($\Delta\varepsilon$ =0.01)						Non-cooperative model					
			Stage 1 as a variable			Stage 2 as a variable			Stage 1 as leader			Stage 2 as leader		
Region type	DMU	Region	$\hat{\theta}_1^{o+}$	$\hat{\theta}_2^{o-}$	$\hat{\theta}^{cen,1,*}$	$\hat{\theta}_1^{o-}$	$\hat{\theta}_2^{o+}$	$\hat{\theta}^{cen,2,*}$	e_1^{o*}	e_2^{o*}	$e^{non,1,*}$	π_1^{o*}	π_2^{o*}	$\pi^{non,2,*}$
Municipality	1	Beijing	1.0000	0.1598	0.1598	1.0000	0.1598	0.1598	1.0000	0.1598	0.1598	1.0000	0.1598	0.1598
	2	Chongqing	1.0000	0.2489	0.2489	1.0000	0.2489	0.2489	1.0000	0.2489	0.2489	1.0000	0.2489	0.2489
	3	Shanghai	0.8900	0.5394	0.4801	0.8842	0.5428	0.4799	1.0000	0.4433	0.4433	0.7942	0.5728	0.4549
	4	Tianjin	0.6774	0.5704	0.3864	0.6726	0.5704	0.3864	0.7426	0.4657	0.3459	0.6774	0.5704	0.3864
Province	5	Anhui	0.6697	0.3895	0.2609	0.6697	0.3895	0.2609	0.6697	0.3895	0.2609	0.6697	0.3895	0.2609
	6	Fujian	0.5668	1.0000	0.5668	0.5668	1.0000	0.5668	0.5668	1.0000	0.5668	0.5668	1.0000	0.5668
	7	Gansu	1.0000	0.2207	0.2207	0.9869	0.2221	0.2192	1.0000	0.2207	0.2207	0.3504	0.3121	0.1094
	8	Guangdong	1.0000	1.0000	1.0000	1.0000	1.0000	1.0000	1.0000	1.0000	1.0000	1.0000	1.0000	1.0000
	9	Guizhou	0.9398	1.0000	0.9398	0.9398	1.0000	0.9398	0.9398	1.0000	0.9398	0.9398	1.0000	0.9398
	10	Hainan	1.0000	1.0000	1.0000	1.0000	1.0000	1.0000	1.0000	1.0000	1.0000	1.0000	1.0000	1.0000
	11	Hebei	0.8885	0.8351	0.7420	0.8885	0.8351	0.7420	0.8885	0.8351	0.7420	0.8885	0.8351	0.7420
	12	Heilongjiang	0.9328	0.2648	0.2470	0.9328	0.2603	0.2428	0.9328	0.2648	0.2470	0.8536	0.2703	0.2308
	13	Henan	0.8504	0.7329	0.6233	0.8493	0.7373	0.6262	0.8504	0.7329	0.6233	0.8493	0.7373	0.6262
	14	Hubei	0.9060	0.2816	0.2551	0.9060	0.2760	0.2501	0.9060	0.2816	0.2551	0.4177	0.3360	0.1404
	15	Hunan	1.0000	0.3685	0.3685	1.0000	0.3680	0.3630	1.0000	0.3685	0.3685	0.9336	0.3780	0.3529
	16	Jiangsu	0.9225	1.0000	0.9225	0.9225	1.0000	0.9225	0.9225	1.0000	0.9225	0.9225	1.0000	0.9225
	17	Jiangxi	0.5647	0.9877	0.5577	0.5646	0.9900	0.5539	0.5647	0.9877	0.5577	0.4812	1.0000	0.4812
	18	Jilin	0.7158	0.4936	0.3533	0.7058	0.4984	0.3518	0.7158	0.4936	0.3533	0.6030	0.5184	0.3126
	19	Liaoning	0.6669	0.3669	0.2447	0.6713	0.3642	0.2445	0.6969	0.3488	0.2431	0.6335	0.3742	0.2371
	20	Qinghai	0.4573	1.0000	0.4573	0.4573	1.0000	0.4573	0.4573	1.0000	0.4573	0.4573	1.0000	0.4573

(continued)

Table 12.3 (continued)

			Centralized model ($\Delta\varepsilon$ =0.01)						Non-cooperative model					
			Stage 1 as a variable			Stage 2 as a variable			Stage 1 as leader			Stage 2 as leader		
Region type	DMU	Region	$\hat{\theta}_1^{o+}$	$\hat{\theta}_2^{o-}$	$\hat{\theta}^{cen,1,*}$	$\hat{\theta}_1^{o-}$	$\hat{\theta}_2^{o+}$	$\hat{\theta}^{cen,2,*}$	e_1^{o*}	e_2^{o*}	$e^{non,1,*}$	π_1^{o*}	π_2^{o*}	$\pi^{non,2,*}$
	21	Shandong	0.7101	0.8176	0.5806	0.7068	0.8198	0.5795	0.7101	0.8176	0.5806	0.5700	0.8498	0.4844
	22	Shanxi	0.5759	0.5109	0.2943	0.5664	0.5195	0.2942	0.5864	0.4996	0.2930	0.4774	0.5709	0.2725
	23	Shanxi	1.0000	0.1941	0.1941	1.0000	0.1909	0.1909	1.0000	0.1941	0.1941	0.4073	0.2509	0.1022
	24	Sichuan	1.0000	0.4566	0.4566	1.0000	0.4517	0.4517	1.0000	0.4566	0.4566	0.6851	0.4817	0.3300
	25	Yunnan	1.0000	0.5846	0.5846	0.9955	0.5859	0.5832	1.0000	0.5846	0.5846	0.8938	0.6159	0.5504
	26	Zhejiang	0.7311	0.9148	0.6688	0.7316	0.9141	0.6688	0.9111	0.5172	0.4712	0.6298	0.9541	0.6009
Autonomous region	27	Guangxi	1.0000	1.0000	1.0000	1.0000	1.0000	1.0000	1.0000	1.0000	1.0000	1.0000	1.0000	1.0000
	28	Inner Mongolia	0.3599	1.0000	0.3599	0.3599	1.0000	0.3599	0.3599	1.0000	0.3599	0.3599	1.0000	0.3599
	29	Ningxia	0.4300	1.0000	0.4300	0.4300	1.0000	0.4300	0.4300	1.0000	0.4300	0.4300	1.0000	0.4300
	30	Xinjiang	1.0000	1.0000	1.0000	1.0000	1.0000	1.0000	1.0000	1.0000	1.0000	1.0000	1.0000	1.0000

Table 12.4 Results based upon the centralized model when $\Delta\varepsilon = 0.0001$ and 0.00001, respectively

			$\Delta\varepsilon = 0.0001$						$\Delta\varepsilon = 0.00001$					
			Stage 1 as a variable			Stage 2 as a variable			Stage 1 as a variable			Stage 2 as a variable		
Region type	DMU	Region	$\hat{\theta}_1^{o+}$	$\hat{\theta}_2^{o-}$	$\hat{\theta}^{cen,1,*}$	$\hat{\theta}_1^{o-}$	$\hat{\theta}_2^{o+}$	$\hat{\theta}^{cen,2,*}$	$\hat{\theta}_1^{o+}$	$\hat{\theta}_2^{o-}$	$\hat{\theta}^{cen,1,*}$	$\hat{\theta}_1^{o-}$	$\hat{\theta}_2^{o+}$	$\hat{\theta}^{cen,2,*}$
Municipality	1	Beijing	1.0000	0.1598	0.1598	1.0000	0.1598	0.1598	1.0000	0.1598	0.1598	1.0000	0.1598	0.1598
	2	Chongqing	1.0000	0.2489	0.2489	1.0000	0.2489	0.2489	1.0000	0.2489	0.2489	1.0000	0.2489	0.2489
	3	Shanghai	0.8950	0.5365	0.4802	0.8949	0.5366	0.4802	0.8950	0.5365	0.4802	0.8950	0.5365	0.4802
	4	Tianjin	0.6774	0.5704	0.3864	0.6774	0.5704	0.3864	0.6774	0.5704	0.3864	0.6774	0.5704	0.3864
Province	5	Anhui	0.6697	0.3895	0.2609	0.6697	0.3895	0.2609	0.6697	0.3895	0.2609	0.6697	0.3895	0.2609
	6	Fujian	0.5668	1.0000	0.5668	0.5668	1.0000	0.5668	0.5668	1.0000	0.5668	0.5668	1.0000	0.5668
	7	Gansu	1.0000	0.2207	0.2207	1.0000	0.2207	0.2207	1.0000	0.2207	0.2207	1.0000	0.2207	0.2207
	8	Guangdong	1.0000	1.0000	1.0000	1.0000	1.0000	1.0000	1.0000	1.0000	1.0000	1.0000	1.0000	1.0000
	9	Guizhou	0.9398	1.0000	0.9398	0.9398	1.0000	0.9398	0.9398	1.0000	0.9398	0.9398	1.0000	0.9398
	10	Hainan	1.0000	1.0000	1.0000	1.0000	1.0000	1.0000	1.0000	1.0000	1.0000	1.0000	1.0000	1.0000
	11	Hebei	0.8885	0.8351	0.7420	0.8885	0.8351	0.7420	0.8885	0.8351	0.7420	0.8885	0.8351	0.7420
	12	Heilongjiang	0.9328	0.2648	0.2470	0.9326	0.2648	0.2470	0.9328	0.2648	0.2470	0.9328	0.2648	0.2470
	13	Henan	0.8493	0.7373	0.6262	0.8493	0.7373	0.6262	0.8493	0.7373	0.6262	0.8493	0.7373	0.6262
	14	Hubei	0.9060	0.2816	0.2551	0.9060	0.2815	0.2551	0.9060	0.2816	0.2551	0.9060	0.2816	0.2551
	15	Hunan	1.0000	0.3685	0.3685	1.0000	0.3684	0.3684	1.0000	0.3685	0.3685	1.0000	0.3685	0.3685
	16	Jiangsu	0.9225	1.0000	0.9225	0.9225	1.0000	0.9225	0.9225	1.0000	0.9225	0.9225	1.0000	0.9225
	17	Jiangxi	0.5644	0.9914	0.5595	0.5644	0.9913	0.5595	0.5644	0.9914	0.5595	0.5644	0.9914	0.5595
	18	Jilin	0.7152	0.4947	0.3538	0.7151	0.4947	0.3538	0.7152	0.4947	0.3538	0.7152	0.4947	0.3538
	19	Liaoning	0.6671	0.3668	0.2447	0.6671	0.3668	0.2447	0.6671	0.3668	0.2447	0.6671	0.3668	0.2447
	20	Qinghai	0.4573	1.0000	0.4573	0.4573	1.0000	0.4573	0.4573	1.0000	0.4573	0.4573	1.0000	0.4573

(continued)

Table 12.4 (continued)

Region type	DMU	Region	$\Delta\varepsilon$ =0.0001						$\Delta\varepsilon$ =0.00001					
			Stage 1 as a variable			Stage 2 as a variable			Stage 1 as a variable			Stage 2 as a variable		
			$\hat{\theta}_1^{o+}$	$\hat{\theta}_2^{o-}$	$\hat{\theta}^{cen,1,*}$	$\hat{\theta}_1^{o-}$	$\hat{\theta}_2^{o+}$	$\hat{\theta}^{cen,2,*}$	$\hat{\theta}_1^{o+}$	$\hat{\theta}_2^{o-}$	$\hat{\theta}^{cen,1,*}$	$\hat{\theta}_1^{o-}$	$\hat{\theta}_2^{o+}$	$\hat{\theta}^{cen,2,*}$
	21	Shandong	0.7101	0.8176	0.5806	0.7099	0.8177	0.5805	0.7101	0.8176	0.5806	0.7101	0.8176	0.5806
	22	Shanxi	0.5708	0.5156	0.2943	0.5708	0.5156	0.2943	0.5708	0.5156	0.2943	0.5708	0.5156	0.2943
	23	Shanxi	1.0000	0.1941	0.1941	1.0000	0.1940	0.1940	1.0000	0.1941	0.1941	1.0000	0.1941	0.1941
	24	Sichuan	1.0000	0.4566	0.4566	1.0000	0.4566	0.4566	1.0000	0.4566	0.4566	1.0000	0.4566	0.4566
	25	Yunnan	1.0000	0.5846	0.5846	1.0000	0.5847	0.5846	1.0000	0.5846	0.5846	1.0000	0.5846	0.5846
	26	Zhejiang	0.7293	0.9171	0.6688	0.7293	0.9171	0.6688	0.7293	0.9171	0.6688	0.7293	0.9171	0.6688
Autonomous region	27	Guangxi	1.0000	1.0000	1.0000	1.0000	1.0000	1.0000	1.0000	1.0000	1.0000	1.0000	1.0000	1.0000
	28	Inner Mongolia	0.3599	1.0000	0.3598	0.3599	1.0000	0.3599	0.3599	1.0000	0.3599	0.3599	1.0000	0.3599
	29	Ningxia	0.4300	1.0000	0.4300	0.4300	1.0000	0.4300	0.4300	1.0000	0.4300	0.4300	1.0000	0.4300
	30	Xinjiang	1.0000	1.0000	1.0000	1.0000	1.0000	1.0000	1.0000	1.0000	1.0000	1.0000	1.0000	1.0000

Regions (Shandong (DMU 21), Hunan (DMU 15) and Shanxi (DMU 22)) whose two efficiencies are not equal which is inconsistent with Theorem 2 when $\Delta\varepsilon = 0.0001$. When $\Delta\varepsilon = 0.00001$, the results for all the 30 Provincial Level Regions verify Theorem 2. This indicates that the choice of $\Delta\varepsilon$ is important and we should always use a very small $\Delta\varepsilon$ in order to reach the global optimal solution.

The results in the last six columns of Table 12.4 with $\Delta\varepsilon$ =0.00001 shows that the efficiency decomposition is unique for all Provincial Level Regions. For example, for Zhejiang Province (DMU 26), $\hat{\theta}_1^{o+} = \hat{\theta}_1^{o-} = 0.7293$ and $\hat{\theta}_2^{o+} = \hat{\theta}_2^{o-} = 0.9171$.

Finally, note also that the two efficiencies based upon the centralized model and the non-cooperative model with stage 1 as leader are the same for the majority of Provincial Level Regions. This may indicate that the first stage or the technology development stage is more important.

12.4 Conclusions

The current chapter extends the approach of Liang et al. (2008) to analyze the efficiency of two-stage network structures where the second stage has its own inputs in addition to the outputs from the first stage. In the current chapter, a centralized model and a non-cooperative model are proposed to evaluate the efficiency of such a two-stage process and to further decompose the overall efficiency as a product of efficiency scores of the two individual stages as in Kao and Hwang (2008).

Unlike the models in Liang et al. (2008) or Kao and Hwang (2008), the centralized model cannot be transformed to a linear program due to the existence of additional inputs to the second stage. The current chapter proposes a heuristic method to estimate the global optimal efficiency. The proposed approaches are illustrated with a data set for measuring the R&D performance of 30 Provincial Level Regions in Mainland of China. As demonstrated in the application, the developed relations between the centralized and non-cooperative approaches can help test for whether a global optimal solution is found.

Although the current chapter assumes that all the outputs from the first stage become inputs to the second stage, similar development can be made for cases when only portion of the outputs from the first stage become inputs to the second stage. That is, we can provide similar models for a more general two-stage network structure where each stage has its own inputs and outputs.

Finally, our models can only give solutions for general two-stage network structure, it is desirable to improve these approaches to decompose efficiency for complex network structure (not restrict to two-stage) in future research. And, the current models are under the assumption of CRS (constant return to scale), how to modify these models to decompose efficiency for general network structure by VRS (variable return to scale) model is also a direction for future research.

Acknowledgements This research is supported by National Natural Science Foundation of China under Grants (No. 70901070, 71271196 and 70821001), the Foundation for Innovative Research Groups of the National Natural Science Foundation of China (No. 71121061), the National Science Foundation of China for Distinguished Youth Scholars (No. 71225002), the Chinese Universities Scientific Fund (WK2040150005) and the Fund for International Cooperation and Exchange of the National Natural Science Foundation of China (No. 71110107024).

Part of the materials are based upon Li et al. (2012) with permission from Elsevier.

References

Castelli, L., Pesenti, R., & Ukovich, W. (2004). DEA-like models for the efficiency evaluation of hierarchically structured units. *European Journal of Operational Research, 154*, 465–476.

Charnes, A., Cooper, W. W., & Rhodes, E. (1978). Measuring the efficiency of decision making units. *European Journal of Operational Research, 2*, 429–444.

Cook, W. D., Liang, L., & Zhu, J. (2010a). Measuring performance of two-stage network structures by DEA: a review and future perspective. *Omega, 38*, 423–430.

Cook, W. D., Zhu, J., Yang, F., & Bi, G. B. (2010b). Network DEA: Additive efficiency decomposition. *European Journal of Operational Research, 207*, 1122–1129.

Cooper, W. W., Seiford, L. M., & Zhu, J. (2004). *Handbook on data envelopment analysis*. Boston: Springer.

Doyle, J., & Green, R. (1994). Efficiency and cross-efficiency in DEA: Derivations, meanings and uses. *Journal of the Operational Research Society, 5*, 567–578.

Färe, R., & Grosskopf, S. (2000). Network DEA. *Socio-Economic Planning Sciences, 34*, 35–49.

Huang, Z. M., & Li, S. X. (2001). Co-op advertising models in a manufacturing-retailing supply chain: A game theory approach. *European Journal of Operational Research, 135*, 527–544.

Kao, C. (2009a). Efficiency decomposition in network data envelopment analysis: A relational model. *European Journal of Operational Research, 192*, 949–962.

Kao, C. (2009b). Efficiency measurement for parallel production systems. *European Journal of Operational Research, 196*, 1107–1112.

Kao, C., & Hwang, S. N. (2008). Efficiency decomposition in two-stage data envelopment analysis: An application to non-life insurance companies in Taiwan. *European Journal of Operational Research, 185*, 418–429.

Li, Y., Chen, Y., Liang, L., & Xie, J. (2012). DEA models for extended two-stage network structures. *Omega, 40*(5), 611–618.

Liang, L., Yang, F., Cook, W. D., & Zhu, J. (2006). DEA models for supply chain efficiency evaluation. *Annals of Operations Research, 145*, 35–49.

Liang, L., Cook, W. D., & Zhu, J. (2008). DEA models for two-stage processes: Game approach and efficiency decomposition. *Naval Research Logistics, 55*, 643–653.

Tone, K., & Tsutsui, M. (2009). Network DEA: A slacks-based measure approach. *European Journal of Operational Research, 197*, 243–252.

Zhong, W., Yuan, W., Li, S. X., & Huang, Z. M. (2011). The performance evaluation of regional R&D investments in China: An application of DEA based on the first official China economic census data. *Omega, 39*, 447–455.

Chapter 13
An Efficiency Measurement Framework for Multi-stage Production Systems

Boaz Golany, Steven T. Hackman, and Ury Passy

Abstract We develop an efficiency measurement framework for systems composed of two subsystems arranged in series that simultaneously computes the efficiency of the aggregate system and each subsystem. Our approach expands the technology sets of each subsystem by allowing each to acquire resources from the other in exchange for delivery of the appropriate (intermediate or final) product, and to form composites from both subsystems. Managers of each subsystem will not agree to "vertical integration" initiatives unless each subsystem will be more efficient than what each can achieve by separately applying conventional efficiency analysis. A Pareto Efficient frontier characterizes the acceptable set of efficiencies of each subsystem from which the managers will negotiate to select the final outcome. Three proposals for the choice for the Pareto efficient point are discussed: the one that achieves the largest equiproportionate reduction in the classical efficiencies; the one that achieves the largest equal reduction in efficiency; and the one that maximizes the radial contraction in the aggregate consumption of resources originally employed before integration. We show how each choice for the Pareto efficient point determines a derived measure of aggregate efficiency. An extensive numerical example is used to illustrate exactly how the two subsystems can significantly improve their operational efficiencies via integration beyond what would be predicted by conventional analysis.

Keywords Multi-stage production systems • Productivity and efficiency measurement • Data envelopment analysis

---

B. Golany (✉) • U. Passy
Faculty of Industrial Engineering and Management, Technion—Israel Institute of Technology, Haifa 32000, Israel
e-mail: golany@ie.technion.ac.il

S.T. Hackman
School of Industrial and Systems Engineering, Georgia Institute of Technology, Atlanta, GA 30332, USA

W.D. Cook and J. Zhu (eds.), *Data Envelopment Analysis*, International Series in Operations Research & Management Science 208,
DOI 10.1007/978-1-4899-8068-7_13,

13.1 Introduction

Data Envelopment Analysis (DEA) was developed to measure the relative efficiency of operational units known in the literature as "Decision Making Units" — DMUs (see Charnes et al. 1994). Given a pair of observed input-output vectors (X_0, Y_0), DEA assesses its efficiency by comparing it to other choices in the technology set $\mathcal{T} = \{(X, Y)\}$, which characterizes the collection of all input vectors X that can produce the output vector Y. For example, the classical DEA radial input measure of efficiency (Charnes et al. 1978) is calculated as

$$\text{Min}\ \{\theta :\ (\theta X_0, Y_0) \in \mathcal{T}\}. \tag{13.1}$$

The technology set $\mathcal{T}$ is extrapolated from the observed data on input-output pairs (X_j, Y_j), $j = 1, 2, \ldots, N$, for N DMUs, and is typically defined via a set of linear inequalities, which turns (13.1) into a linear program. The optimal input-output pair (X^*, Y^*) of the linear program to which (X_0, Y_0) is compared is a linear combination of existing DMUs, and is commonly referred to as a "composite" unit or system.

A significant portion of DEA research to date has focused on defining the rules for constructing $\mathcal{T}$, and defining corresponding measures of efficiency. Regardless of the definition of efficiency, most DEA models treat each DMU as a non-separable entity without attempting to probe the internal mechanisms of how each DMU converts its inputs into outputs. With today's information systems it is now much easier to collect data on how capital and labor are used to transform raw materials through various stages to produce final products. The availability of such data presents an opportunity to explore efficiency measurement of the stages within complex, multi-stage DMUs.

In particular, we are interested in DMUs that consist of several stages arranged in series where succeeding stages (or subsystems) are fed by a mixture of external inputs and intermediate factors which are outputs of preceding stages. The focus of the present paper is on how to assess the efficiency of each stage within the aggregate system and how to explore possible tradeoffs of these efficiencies. As a starting point, of course, one can treat each subsystem as a system in its own right. In this manner, the technology for each subsystem is constructed using the relevant input-output data from its own peers, and the technology of the aggregate system is constructed on the basis of aggregated inputs and outputs and without regard to intermediate input-output factors that link the various stages. As we subsequently demonstrate, this approach exhibits the phenomena in which it is possible for the aggregate system to be rated very inefficient, while each subsystem is rated efficient, and for the aggregate system to be rated near efficient, while each subsystem is rated highly inefficient.

In this paper, we propose an expansion of the ordinary technology sets which were used till now in DEA and develop a corresponding efficiency measurement framework that simultaneously computes the efficiency of each subsystem and the aggregate system. The measurement framework has the following properties.

First, if each subsystem is rated efficient, then so must the aggregate system. Second, each subsystem's efficiency and the aggregate efficiency cannot exceed the efficiency obtained using the classical DEA approach, and may be expected to be far lower. Third, the methodology described herein provides the recipe on how to obtain the operational improvements at *all* levels of the hierarchy, as it explicitly integrates the computation of the various efficiencies.

With the assumptions of our expanded technology set, we show how it may be possible to obtain significant improvements in operations when the subsystems agree to "vertically integrate". The key to our approach is that we allow each subsystem to construct composite units from both subsystems as in the classical approach, and we allow each subsystem to *acquire* resources from the other stage in return for delivering that stage's output. We insist that acquisitions by one subsystem from another have to make economic sense in that both parties have to benefit and consent. A Pareto Efficient frontier of acceptable efficiencies for each subsystem will be introduced. While in principle any choice along the Pareto Efficient frontier could be selected, depending on which subsystem has the greater market power, we discuss several reasonable choices. We show how each choice for the Pareto efficient point by the subsystems determines a derived measure of aggregate efficiency. Our choice for the measure of aggregate efficiency corresponds to the Pareto efficient point that achieves the largest radial contraction in the aggregate amount of capital and labor used before integration.

To achieve the benefits listed above, we limit our application to two stages in series, and we make the following simplifying assumptions:

1. *Technology.* Each subsystem 1 uses capital and labor to produce an intermediate product used by subsystem 2 to produce final product. Subsystem 2 also requires capital and labor, which are assumed to be completely transferrable resources between stages. All technologies described herein exhibit constant returns-to-scale.
2. *Market.* Each subsystem 2 has a unique supplier given by its subsystem 1. There is a market for the intermediate product, and the transfer price subsystem 1 charges subsystem 2 is the prevailing market price. Competitive markets exist so that either subsystem is a *price-taker* in the input and output markets. That is, it may expand or contract its output without affecting its price or cost of inputs.
3. *Organization.* Monitoring and information costs may make it difficult (and perhaps unwise) for senior management to dictate sweeping changes to the allocation of capital and labor between stages. Organizationally, each subsystem is viewed as a profit center, and each manager is given decision-making authority. Although we do not explicitly model the incentive scheme for the managers, which is beyond the scope of this paper, we assume each manager is highly motivated to improve his own system's efficiency. For our purposes, efficiency may be thought of as the proxy for performance, which is why each manager will not consent to an acquisition of resources unless he directly benefits from it. From an organizational perspective, we view our modeling approach described herein as a natural starting point for efficiency improvement.

Attempts to model DMUs that exhibit known internal structures started in the mid-1980s. Färe and Primont (1984) constructed multi-plant efficiency measures and illustrated their models by analyzing utility firms each of whom operated several electric generation plants. Their structure may be characterized as *horizontal integration* since the plants they model operate in parallel and there is no flow of intermediate inputs or outputs between them. Färe et al. (1992) further expand this modeling approach to describe firm and industry performance where, in some cases, reallocation of resources among firms is allowed to improve the industry performance. Cook et al. (1998) define various hierarchies and groupings of DMUs in DEA, and apply DEA formulations for the different groupings.

Another research thrust explored *vertical integration* structures in which a series of stages is connected through intermediate input-output factors. Charnes et al. (1986a, see also Charnes et al. 1994, p. 432) developed a two-stage model in the context of the US Army Recruiting Command. The first stage used advertising to generate awareness and propensity to enlist. The two outputs generated by the first stage were joined by other external inputs (e.g. recruiters) to produce the second stage outputs, which were the actual recruitment contracts. Färe and Whittaker (1995) developed a linear programming model that focused on the role played by the intermediate factors and demonstrated its potential through an application to dairy farms. Hoopes et al. (2000) developed a goal-programming DEA formulation that models serial manufacturing processes and applied it to data on circuit board manufacturing. Chen and Zhu (2004) explored two-stage systems in the banking industry (where the first stage produces deposits which are then used to produce loans) and developed a model that identifies the efficiency frontier that characterizes such systems.

Färe and Grosskopf (1996, 2000) deserve special mention, as they pioneered a line of research, coined *network DEA*, aimed at developing a general multi-stage model with intermediate inputs-outputs. Their representation of the flow of product is consistent with the industrial engineering and operations research literature on multi-stage systems (e.g., Graves et al. 1993; Hackman and Leachman 1989; Johnson and Montgomery 1974; Troutt et al. 2001). Each internal stage's technology is modeled using a single-stage DEA model. The conventional radial-based measure of aggregate efficiency would still be determined as in (13.1) using their more extensive description of technology. For an application of the Färe and Grosskopf framework, see Löthgren and Tambour (1999), who applied their modeling approach to evaluate the performance of Swedish pharmacies.

It is important to point out here that our two-stage model of the *flow of material* is a special case of Färe and Grosskopf's multi-stage framework; however, our proposed aggregate efficiency measure is fundamentally different. In particular, in cases when our proposed aggregate efficiency is higher, it necessarily follows that it would not be possible to disaggregate the Fare-Grosskopf aggregate efficiency measure into separate efficiency measures for which each submanager would consent. If there is an aggregate manager who may unilaterally reallocate resources without consent of the submanagers, then assessing aggregate efficiency using the Färe-Grosskopf framework may lead to superior results for the whole system.

However, as we have noted, due to the linkage of inputs and outputs between the stages, in such a context one subsystem's efficiency may be vastly improved at the expense of potential improvement in the other subsystem, which may render meaningless the assessment of subsystem efficiency.

The outline of the paper is as follows. Section 13.2 introduces the specific network structure and data used throughout the paper, and discusses a motivating example to illustrate how it is possible to achieve significant improvements in efficiency via acquisition. Section 13.3 presents the formal descriptions of the newly expanded models of technology that are used to assess the efficiencies of each stage, and discusses their application to our dataset. Section 13.4 introduces the aforementioned Pareto Efficient frontier. Section 13.5 develops the aggregate measure of efficiency and compares the numerical results obtained from the dataset to both the classical approach and the Färe and Grosskopf approach. Section 13.6 presents the concept of "consistent pricing" which characterizes both our proposed models and that of Färe and Grosskopf. Section 13.7 discusses extensions to the basic modeling approach and analyzes, as well as suggestions for further research. Section 13.8 closes this paper with a few concluding remarks.

13.2 Preliminaries

13.2.1 A Representative Multi-stage System

To ease notational burdens and to make concrete the conceptual discussions to follow, we shall analyze multi-stage systems such as the one depicted in Fig. 13.1. Each DMU_j ($j = 1,\ldots,N$) consists of two subsystems in series. Subsystem $1j$ (hereafter abbreviated S_{1j}) uses capital K_{1j} and labor L_{1j} to produce intermediate product I_j. Subsystem $2j$ (hereafter abbreviated by S_{2j}) uses capital K_{2j} and labor L_{2j} together with I_j to produce final output F_j. Constant returns-to-scale (CRS) will be assumed throughout.

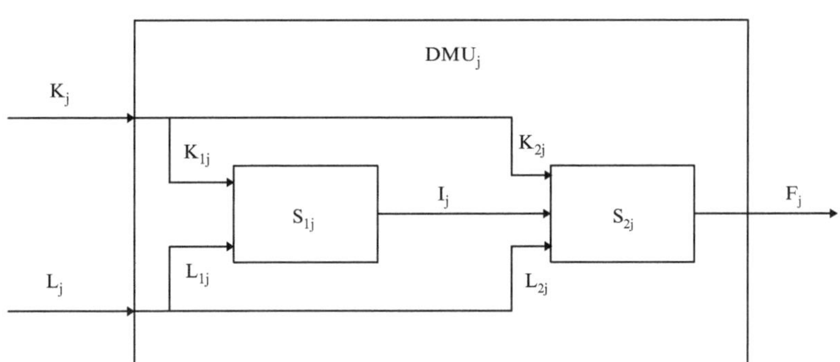

Fig. 13.1 Aggregate DMU with two stages in tandem

Table 13.1 Data for the numerical example

DMU	K_1	K_2	K	L_1	L_2	L	I	F
1	32	81	113	67	83	150	46.928	64.941
2	96	28	124	40	81	121	47.431	46.492
3	79	51	130	89	79	168	79.694	67.388
4	41	80	121	35	26	61	32.978	31.124
5	99	8	107	33	74	107	45.921	35.018
6	72	29	101	15	36	51	24.861	29.146
7	21	88	109	64	23	87	32.250	34.049
8	60	39	99	71	49	120	64.659	45.176
9	7	86	93	80	16	96	21.531	21.062
10	10	40	50	33	11	44	12.519	10.189

The models we develop will be illustrated with a 10 DMU numerical example constructed as follows. We modeled each subsystem's technology via a Cobb-Douglas production function so that the observed output for S_{1j} was $K_{1j}^{\alpha}L_{1j}^{1-\alpha} - 10\,\mu_1$, and the observed output for S_{2j} was $K_{2j}^{\beta}L_{2j}^{\gamma}I^{1-\beta-\gamma} - 10\,\mu_2$. The capital and labor inputs were randomly generated in the range [1,100], and we set the other parameters as $\alpha = 0.\,4$, $\beta = 0.\,25$, $\gamma = 0.\,45$ and $\mu_1, \mu_2 \sim$ Uniform[0,1]. The resulting data on inputs and outputs for S_{1j}, (K_{1j}, L_{1j}, I_j), and S_{2j}, $(K_{2j}, L_{2j}, I_j, F_j)$, are given in Table 13.1.

13.2.2 Classical Models of Technology

Following Shephard (1970), Charnes et al. (1978), Banker et al. (1984), and Färe and Grosskopf (1996), a model of technology $\mathcal{T} \in R_+^n \times R_+^m$ characterizes the collection of all input vectors $X \in R_+^n$ that can be used to produce the output vector $Y \in R_+^m$. For a given dataset of input-output pairs (X_j, Y_j), $j = 1, 2, \ldots, N$, the classical DEA approach models the technology as

$$\mathcal{T} \equiv \left\{ (X,Y) : \sum_j \lambda_j X_j \leq X,\ \sum_j \lambda_j Y_j \geq Y \right\}, \tag{13.2}$$

where we now and hereafter suppress the nonnegativity constraints imposed on the intensity variables (the λ_j's). Given the technology $\mathcal{T}$ and using the subscript "0" to denote a DMU in our dataset which is to be analyzed (i.e., $(X_0, Y_0) \in \mathcal{T}$), the classical (CL) radial measure of input efficiency is defined as

$$\theta_0^{CL} \equiv \text{Min}\ \{\theta_0 :\ (\theta_0 X_0, Y_0\} \in \mathcal{T}\}. \tag{13.3}$$

(In principle, any measure of input efficiency would suffice for the developments to follow, since our focus is to expand the classical model of technology when detailed information about its structure is available.) For the multi-stage systems depicted in Fig. 13.1, the classical descriptions of technology for each subsystem are

Table 13.2 Classical efficiency evaluation for S_{1j} and S_{j2}

	S_{1j}		S_{2j}	
DMU_j	θ_{1j}^{CL}	Benchmarks	θ_{2j}^{CL}	Benchmarks
1	1.00	1	1.00	1
2	0.943	5,8	1.00	2
3	0.973	5,8	1.00	3
4	0.958	5,8	0.904	3,6
[5]	1.00	5	1.00	5
6	1.00	6	1.00	6
7	0.941	1,9	1.00	7
8	1.00	8	1.00	8
9	1.00	9	0.918	1,7
10	0.748	1,9	0.735	1,7

$$\mathcal{T}_1 \equiv \left\{((K,L),\, I) : \sum_j \lambda_{1j} K_{1j} \le K,\ \sum_j \lambda_{1j} L_{1j} \le L,\ \sum_j \lambda_{1j} I_j \ge I \right\}, \tag{13.4}$$

$$\mathcal{T}_2 \equiv \left\{((K,L,I),F): \sum_j \lambda_{2j} K_{2j} \le K,\ \sum_j \lambda_{2j} L_{2j} \le L,\ \sum_j \lambda_{2j} I_j \le I,\ \sum_j \lambda_{2j} F_j \ge F \right\}. \tag{13.5}$$

For each DMU_0 in our dataset the classical measures of input efficiency for each stage are computed as follows:

$$\theta_{10}^{CL} \equiv \text{Min}\{\theta_{10} : ((\theta_{10} K_{10}, \theta_{10} L_{10}),\ I_0) \in \mathcal{T}_1\}, \tag{13.6}$$

$$\theta_{20}^{CL} \equiv \text{Min}\{\theta_{20} : ((\theta_{20} K_{20}, \theta_{20} L_{20}, \theta_2 I_0),\ F_0) \in \mathcal{T}_2\}. \tag{13.7}$$

Computational results are reported in Table 13.2.

13.2.3 An Expanded Model of Technology: A Motivating Example

A numerical example using one of the DMUs in our dataset will be used to explain how to expand the technology to provide better opportunities for each subsystem to improve its efficiency. The model used to generate this example is formally described in the next section. In what follows, we have made the following assumptions: (1) Each stage is managed as a profit center; (2) S_{1j} may sell its intermediate product on the open market for the same price it charges S_{2j}; and (3) Both S_{1j} and S_{2j} may sell any amount of their respective outputs on the open market without affecting input cost or price.

The observed S_{25} of DMU_5 uses 8 units of capital, 74 units of labor and 45.92 units of intermediate product to produce 35.02 units of final product. The manager of S_{25} (hereafter named 'M_{25}'), while always looking to improve efficiency, is content for now as his system is rated efficient by classical efficiency analysis. Now suppose the manager of S_{15} (hereafter named 'M_{15}') comes to M_{25} with the following proposal: "I can show you how to increase your output by 6.3 %, while *simultaneously* reducing your cost of inputs by 11.75 %. Interested?" That is, M_{15} is proposing a way for M_{25} to use $(K, L, I) = (7.06, 65.29, 40.52)$ to produce $F = 37.23$ instead of M_{25}'s current production plan that uses $(K, L, I) = (8, 74, 45.92)$ to produce $F = 35.02$. To expand his output by 6.3 %, M_{25} would normally expect (under CRS) to have to increase his inputs by 6.3 %, and so, in effect, M_{15} is offering M_{25} to consume only $100(1 - 0.1175)/1.063 = 83$ % of his input to achieve the same output level. While M_{25} is obviously intrigued by M_{15}'s proposal, M_{25} demands an explanation as to how M_{15} proposes to accomplish this seemingly impossible task, as M_{25} knows that *both* S_{15} and S_{25} were rated input efficient by classical analysis. M_{15} obliges with the following explanation.

Using classical descriptions of technology for each subsystem, M_{15} found a composite subsystem 1 process that uses (34.48, 40.80) units of capital and labor to produce 37.16 units of intermediate product, and a composite subsystem 2 process that uses (37.04, 45.98, 31.76) units of capital, labor and intermediate product to produce the 37.23 units of final product, which M_{15} promised to deliver to M_{25}. The total amounts of capital and labor required by these two composite processes are 71.52 and 86.78, respectively. With the 7.06 units of capital and 65.29 units of labor acquired from M_{25}, M_{15} still needs 64.46 units of capital and 21.49 units of labor, which he possesses as these totals represent only 65.1 % of his current capacity of (99, 33) units of capital and labor. With respect to the intermediate product, M_{25} notes that while he is now purchasing $45.92 - 40.52 = 5.40$ *less* units of intermediate product from M_{15}, the difference between what the subsystem 2 composite requires and what the subsystem 1 composite currently produces of intermediate product is also $5.40 = 37.16 - 31.76$ units, which M_{15} will sell on the open market to compensate him for the loss in revenue from M_{25}. M_{25} is satisfied that M_{15}'s proposal is conceptually sound.

M_{25} now understands why M_{15} is so eager to offer this proposal to M_{25}: under the proposal, M_{15} will be able to free up 34.9 % of *his* inputs, a considerable savings, while still producing his same level of output. Since M_{15} cannot achieve this savings without M_{25}'s consent, M_{25} realizes he must understand exactly how M_{15} was able to devise this seemingly ingenious plan, so that he will be in position to negotiate with M_{15} a better deal for himself.

13.3 The Expanded Technology Sets for S_{1j} and S_{2j}

For every DMU$_j$, the manager of S_{1j} now realizes that the classical efficiency analysis constructed the efficient frontier using *only* subsystem 1 processes. It does not consider the possibility that M_{1j} may have the options of adopting an alternative subsystem 2 production process *and* acquiring resources from M_{2j}

(as long as M_{2j} would agree). With these options, the technology set for S_{1j}, which defines the collection of input pairs (K, L) that can produce at least I_j, has been expanded.

Under the CRS assumption, M_{10} knows that $\omega_{20}((K_{20}, L_{20}, I_0),\ F_0) \in \mathcal{T}_2$ for all $\omega_{20} \geq 0$. In order to entice M_{20} to agree, M_{10} selects a value $\theta_{20} < 1$, and offers M_{20} the opportunity to achieve the input-output point of $\omega_{20}((\theta_{20}K_{20},\ \theta_{20}L_{20},\ \theta_{20}I_0), F_0)$. In order for M_{10} to meet his obligation to M_{20} and his objective, namely to produce I_0 with resources (K, L), he must find two composite processes $((\hat{K}_1, \hat{L}_1),\ \hat{I}_1) \in \mathcal{T}_1$ and $((\hat{K}_2, \hat{L}_2, \hat{I}_2),\ \hat{F}) \in \mathcal{T}_2$ for which the following four *inventory balance equations* must hold:

[E1] *Capital*. The "supply" of capital from M_{10} and M_{20}, $K + \omega_{20}(\theta_{20}K_{20})$, must be no smaller than the "demand" for capital by both composite subsystems, $\hat{K}_1 + \hat{K}_2$;

[E2] *Labor*. The "supply" of labor from M_{10} and M_{20}, $L + \omega_{20}(\theta_{20}L_{20})$, must be no smaller than the "demand" for labor by both composite subsystems, $\hat{L}_1 + \hat{L}_2$;

[E3] *Intermediate Product*. The "supply" of intermediate product from M_{20} and the composite Stage 1 process, $\hat{I}_1 + \omega_{20}(\theta_{20}I_0)$, must be no smaller than the "demand" for intermediate product by M_{10} and the composite Stage 2 process, $I_0 + \hat{I}_2$; and

[E4] *Final Product*. The "supply" of final product from the composite Stage 2 process, $\hat{F}$, must be no smaller than the "demand" for final product by M_{20}, $\omega_{20}F_0$.

Let $\mathcal{T}_1^E(\theta_{20})$ denote the collection of input-output pairs ((K, L), I_0) that satisfy the inventory balance equations [E1–E4] listed above for DMU_0. Given θ_{20}, it would make sense for M_{10} to find the least amount of capital and labor to satisfy his own output requirement of I_0. Accordingly, he should solve the following linear programming model, which we shall denote as the *Acquisition* (AQ) model:

$$\theta_{10}^{AQ}(\theta_{20}) \equiv \min \theta_{10} \tag{13.8}$$

$$\sum_j \lambda_{1j}K_{1j} + \sum_j \lambda_{2j}K_{2j} \leq \theta_{10}K_{10} + \omega_{20}[\theta_{20}K_{20}]$$

$$\sum_j \lambda_{1j}L_{1j} + \sum_j \lambda_{2j}L_{2j} \leq \theta_{10}L_{10} + \omega_{20}[\theta_{20}L_{20}]$$

$$\sum \lambda_{1j}I_j + \omega_{20}[\theta_{20}I_0] \geq I_0 + \sum \lambda_{2j}I_j$$

$$\sum \lambda_{2j}F_j \geq \omega_{20}F_0$$

In the proposal of M_{15} to M_{25} that was described in Sect. 13.2.3, M_{15} selected $\theta_{25} = 0.83$, and solved the *AQ* model, whose solution was $\omega_{25} = 1.063$ with $\theta_{15}^{AQ}(\theta_{25}) = 0.651$.

Now that M_{20} understands how M_{10} was able to achieve *his* objective, M_{20} realizes he can play the same game. Let $\mathcal{T}_2^E(\theta_{10})$ denote the collection of input-output pairs ((K, L, I), F_0) that satisfy analogous four inventory balance requirements as described above. Given θ_{10} it would make sense for M_{20} to find the least amount of capital and labor to satisfy his own output requirement of F_0. Accordingly, he would solve his own *Acquisition* (AQ) model, namely, the following linear programming model:

$$\theta_{20}^{AQ}(\theta_{10}) \equiv \min \theta_{20} \tag{13.9}$$

$$\sum_j \lambda_{1j}K_{1j} + \sum_j \lambda_{2j}K_{2j} \leq \theta_{20}K_{20} + \omega_{10}[\theta_{10}K_{10}]$$

$$\sum_j \lambda_{1j}L_{1j} + \sum_j \lambda_{2j}L_{2j} \leq \theta_{20}L_{20} + \omega_{10}[\theta_{10}L_{10}]$$

$$\sum \lambda_{1j}I_j + \theta_{20}I_0 \geq \sum \lambda_{2j}I_j + \omega_{10}I_0$$

$$\sum \lambda_{2j}F_j \geq F_0$$

For example, suppose M_{25} selects $\theta_{15} = 0.9$. Solution of his *AQ* model gives $\omega_{15} = 0.686$ and $\theta_{25}^{AQ}(\theta_{15}) = 0.697$. Note how much better off M_{25} is and worse off M_{15} is as compared to M_{15}'s original proposal. Both managers will agree that either proposal will outperform the classical analysis.

We close this section by emphasizing the following point about describing the subsystem technologies. Since we allow the possibility of one subsystem manager to acquire resources from the other, as long as they can agree, the potential acquisition of resources consistent with the "θ_{10}–θ_{20}" agreement must now be embedded in the respective descriptions of technology given by $\mathcal{T}_1^E(\theta_{20})$ and $\mathcal{T}_2^E(\theta_{10})$ to reflect the set of all production possibilities.

13.4 Pareto Efficient Frontiers

It should be intuitively clear that for the serial system we discuss here a gain by one manager is a loss by the other manager. Regardless of the final choice for how the two subsystems shall vertically integrate, the agreed-upon choice for θ_{10} and θ_{20} should minimally result in a Pareto efficient outcome; that is, $(\theta_{10}, \theta_{20}) = (\theta_{10}^{AQ}(\theta_{20}), \theta_{20}^{AQ}(\theta_{10}))$. Otherwise, neither manager, M_{10} nor M_{20}, would agree to the vertical integration.

The efficient frontier corresponding to DMU_5 in our example is depicted in Fig. 13.2. This frontier was constructed using a recently developed algorithm by Hackman and Passy (2002). When θ_{15} is set to 1.00, θ_{25} is at its lowest value 0.66. On the other hand, when θ_{25} is set to 1.00, θ_{15} is assigned its lowest value 0.43. We remark that the frontier does not always span an interval as depicted in Fig. 13.2. In extreme cases the frontier may consist of a single point. Such cases are similar, in a sense, to the class of "weakly efficient" DMUs that were discussed in Charnes et al. (1986b).

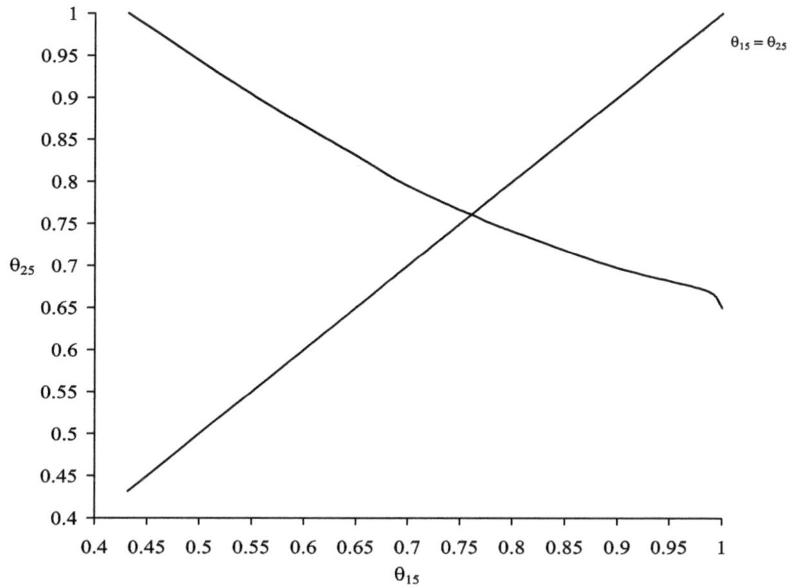

Fig. 13.2 Efficient frontier for DMU_5, θ_{15} *vs.* θ_{25}

Two remarks concerning the Acquisition Models (13.8) and (13.9) that determine the Pareto Efficient frontier are in order. First, it is not necessary to solve both Acquisition Models, as there is a one-to-one correspondence between the solutions for each Acquisition Model: the solution to Model (13.9) may be obtained from the solution to Model (13.8) by dividing λ_{1j}^* and λ_{2j}^* by ω_{20}^*, and setting $\omega_{10}^* = (\omega_{20}^*)^{-1}$. Second, the solutions to either Model (13.8) or Model (13.9) must necessarily lie below their respective classical efficiency counterparts: the linear program to compute θ_{10}^{CL} is a special case of Model (13.8) in which $\omega_{20} = 0$ and $\lambda_{2j} = 0$, and the linear program to compute θ_{20}^{CL} is a special case of Model (13.9) in which $\omega_{10} = 0$ and $\lambda_{1j} = 0$. From an economic perspective, M_{20} would never agree to a proposal from M_{10} if the proposed θ_{20} exceeds what he could achieve on his own, and M_{10} would never offer a proposal to M_{20} in which he receives an efficiency θ_{10} that exceeds what he could achieve on his own, too.

In principle, any point on the Pareto efficient frontier is a candidate. One natural choice is to select the point that achieves the largest *equiproportionate* reduction in the respective classical single-stage efficiencies θ_{10}^{CL} and θ_{20}^{CL}. For example, the 45° line in Fig. 13.2 intersects the frontier at the equiproportional point $\theta_{15} = \theta_{25} = 0.762$, a point which might be considered as "fair" for both subsystems. A second choice is to select the point that achieves the largest equal reduction in efficiency. (When $\theta_{10}^{CL} = \theta_{20}^{CL} = 1$, as is the case for DMU_5, these two choices will obviously coincide.) A third choice is to select the point that achieves for the vertically integrated unit the largest radial contraction in the aggregate amounts of capital and labor originally employed, which we more fully discuss in the next section.

13.5 Aggregate Efficiency

13.5.1 Measures of Aggregate Input Efficiency

From the perspective of an aggregate DMU, the classical model of technology is given by

$$\mathcal{T}_A = \left\{ ((K,L),\ F) :\ \sum_j \lambda_j (K_{1j} + K_{2j}) \leq K, \right. \tag{13.10}$$

$$\left. \sum_j \lambda_j (L_{1j} + L_{2j}) \leq L,\ \sum_j \lambda_j F_j \geq \mathrm{F} \right\}.$$

For the aggregate DMU_0 (denoted hereafter as $A0$) the classical model ignores the intermediate product I_0 as it represents internal production. The corresponding classical input efficiency measure would be computed as:

$$\theta_{A0}^{CL} = \mathrm{Min}\{\theta_{A0} :\ ((\theta_{A0}(K_{10} + K_{20}), \theta_{A0}(L_{10} + L_{20})),\ F_0) \in \mathcal{T}_A\}. \tag{13.11}$$

Färe and Grosskopf (1996) provide an in-depth development of models of technology for general multi-stage systems. One of their basic models (Färe and Grosskopf 1996, pp. 20–23) allows *complete transferability* (CT) of capital and labor flows between the stages. Applied to the two-stage systems we analyze in this paper, their model is formulated as:

$$\mathcal{T}_A^{CT} = \left\{ ((K,L),\ F) :\ \sum_j \lambda_{1j} K_{1j} + \sum_j \lambda_{2j} K_{2j} \leq K, \right. \tag{13.12}$$

$$\sum_j \lambda_{1j} L_{1j} + \sum_j \lambda_{2j} L_{2j} \leq L, \tag{13.13}$$

$$\sum_j \lambda_{1j} I_j - \sum_j \lambda_{2j} I_j \geq 0 \tag{13.14}$$

$$\left. \sum_j \lambda_{2j} F_j \geq F \right\}. \tag{13.15}$$

The third constraint above represents *inventory balance* of intermediate product to ensure that the supply of I produced by the composite S_1 will be sufficient to satisfy the demand for I by the composite S_2. Assuming complete transferability of resources, the Färe-Grosskopf measure of input efficiency would be computed as:

$$\theta_{A0}^{CT} = \mathrm{Min}\{\theta_{A0} : ((\theta_{A0}(K_{10} + K_{20}), \theta_{A0}(L_{10} + L_{20})),\ F_0) \in \mathcal{T}_A^{CT}\}. \tag{13.16}$$

13.5.2 A Derived Measure of Aggregate Efficiency

For each Pareto efficient point (θ_{10}, θ_{20}), let $K(\theta_{10}, \theta_{20})$ and $L(\theta_{10}, \theta_{20})$ denote, respectively, the aggregate amounts of capital and labor which the vertically integrated unit would use to produce *both* F_0 and I_0. A natural choice for a derived measure of aggregate input efficiency is

$$\theta_{A0}^{D}(\theta_{10},\theta_{20}) \equiv \text{Max}\left\{\frac{K(\theta_{10},\theta_{20})}{K_{10}+K_{20}},\ \frac{L(\theta_{10},\theta_{20})}{L_{10}+L_{20}}\right\}. \tag{13.17}$$

We now show how to compute $K(\theta_{10}, \theta_{20})$ using both Acquisition Models (13.8) and (13.9). (The derivation for $L(\theta_{10}, \theta_{20})$ is analogous.) First suppose that $\omega_{20} \leq 1$. Examine the right-hand side of the first constraint in (13.8). In return for delivering $\omega_{20}F_0$ units of final product to S_{20} and meeting its own requirements of producing I_0, S_{10} uses $\omega_{20}[\theta_{20}K_{20}]$ units of capital it acquires from S_{20} and $\theta_{10}K_{10}$ for its own production needs. For the vertically integrated unit to produce a total of F_0, S_{20} will have to produce the remaining amount $(1 - \omega_{20})F_0$ by its own production process and thereby consume $(1-\omega_{20})K_{20}$ units of capital. In this case $K(\theta_{10},\theta_{20}) = \theta_{10}K_{10} + \omega_{20}[\theta_{20}K_{20}] + (1-\omega_{20})K_{20}$. Now suppose $\omega_{20} \geq 1$. Since $\omega_{10} = \omega_{20}^{-1} \leq 1$, we shall examine the right-hand side of the first constraint in (13.9). Here, in return for delivering $\omega_{10}I_0$ units of intermediate output to S_{10}, S_{20} uses $\omega_{10}[\theta_{10}K_{10}]$ of capital it acquires from S_{10}, and $\theta_{20}K_{20}$ units for its own production needs. For S_{10} to produce a total of I_0, it will need $(1 - \omega_{10})K_{10}$ units of capital to produce the remaining amount $(1 - \omega_{10})I_0$ using its current production process. In this case $K(\theta_{10},\theta_{20}) = \omega_{10}[\theta_{10}K_{10}] + \theta_{20}K_{20} + (1-\omega_{10})K_{10}$. To summarize, we have

$$K(\theta_{10},\theta_{20}) \equiv \begin{cases} \theta_{10}K_{10} + \omega_{20}[\theta_{20}K_{20}] + (1-\omega_{20})K_{20}, & \omega_{20} \leq 1 \\ \omega_{10}[\theta_{10}K_{10}] + \theta_{20}K_{20} + (1-\omega_{10})K_{10}, & \omega_{10} \leq 1 \end{cases} \tag{13.18}$$

$$L(\theta_{10},\theta_{20}) \equiv \begin{cases} \theta_{10}L_{10} + \omega_{20}[\theta_{20}L_{20}] + (1-\omega_{20})L_{20}, & \omega_{20} \leq 1 \\ \omega_{10}[\theta_{10}L_{10}] + \theta_{20}L_{20} + (1-\omega_{10})L_{10}, & \omega_{10} \leq 1 \end{cases} \tag{13.19}$$

A numerical example will help explain our proposed derived measure of aggregate efficiency. We solved Model (13.8) for DMU_3 with $\theta_{23} = 0.9$. The result is $\theta_{13} = 0.9667$ and $\omega_{23} = 0.1439$. The two composite subsystems constructed by the linear program are, respectively, $((\hat{K}_1,\hat{L}_1),\hat{I})_1 = ((70.88, 83.87),\ 76.38)$ and $((\hat{K}_2,\hat{L}_2,\hat{I}_2),\ \hat{F}_2) = ((12.09, 12.39, 7.01),\ 9.696)$. Now consider the four inventory balance equations [E1–E4] associated with this Pareto efficient point ($\theta_{13} = 0.9667$, $\theta_{23} = 0.9$):

$$70.88 + 12.09 \leq (0.9667)[79] + 0.1439[0.90 \cdot 51] = 82.97 \tag{13.20}$$

$$83.87 + 12.39 \leq (0.9667)[89] + 0.1439[0.90 \cdot 79] = 96.26 \tag{13.21}$$

Table 13.3 Aggregate efficiency measures

	Classical efficiency		Complete transferability		Expanded technology	
DMU_j	θ_{Aj}^{CL}	Benchmarks	θ_{Aj}^{CT}	Benchmarks	θ_{Aj}^{P}	Benchmarks
1	1.00	1	0.983	[1,8], [6]	0.988	[8], [6]
2	0.829	1,6	0.777	[5,8] [1]	0.842	[8] [1]
3	0.922	1,6	0.899	[1,8], [6]	0.953	[1], [6]
4	0.893	6	0.853	[5], [1,7]	0.922	[5], [1]
5	0.710	1,6	0.666	[5,8], [1]	0.781	[8], [6]
6	1.00	6	0.956	[5], [1,7]	0.981	[5], [7]
7	0.799	1,6	0.751	[5,8], [1]	0.867	[8], [6]
8	0.854	1,6	0.816	[8], [1,6]	0.868	[–], [6]
9	0.480	1,6	0.449	[5,8], [1]	0.621	[8], [6]
10	0.486	1,6	0.456	[5,8], [1]	0.712	[8], [6]

$$76.38 + 0.1439[0.90 \cdot 79.694] \geq [79.694] + 7.01 \tag{13.22}$$

$$9.696 \geq 0.1439(67.39) \tag{13.23}$$

Note how S_{13} is only promising to deliver 14.39 % of final output; the remaining 85.61 % must be produced by S_{23} using its own production process. The derived capital in this case is $(0.9667)[79] + 0.1439[0.90 \cdot 51] + (1 - 0.1439) \cdot 51 = 126.63$, and the derived labor is $(0.9667)[89] + 0.1439[0.90 \cdot 79] + (1 - 0.1439) \cdot 79 = 163.89$. When (126. 63, 163. 89) is compared to the original values of (130, 168), we obtain $\theta_{A3}^{D} = 0.9755$.

The derived aggregate measure of efficiency is measured along the Pareto efficient frontier corresponding to Models (13.8) and (13.9). It can never be larger than 1.0. Conceptually, any point on the Pareto efficient frontier could be used to define the aggregate efficiency. As discussed at the end of the last section there are two obvious choices: the equiproportional solution, $(\theta_{10} = \rho\theta_{10}^{CL}, \theta_{20} = \rho\theta_{20}^{CL})$, where $\rho \leq 1$, and the equal contraction solution in which $\theta_{10} = \theta_{20}$. We propose a third alternative: Minimize θ_{A0}^{D} on the Pareto efficient frontier, which we shall denote by θ_{A0}^{P}. To compute θ_{A0}^{P}, a bi-level programming problem, we iteratively solve Model (13.9) (resp. Model (13.8)) for different θ_{10} (resp. θ_{20}) values.

13.5.3 *Computational Results*

Table 13.3 reports the computational results for each measure of aggregate efficiency. First, we compare θ_{Aj}^{CT} to θ_{Aj}^{CL} for $j = 1, \ldots, 10$. In stark contrast to the relative efficiency nature of DEA, when additional flexibility of transferring resources between stages is available, the CT model is able to use this flexibility to identify potential improvement opportunities for *all* DMUs. Indeed, the relevant benchmarks, which report the reference sets for each of the evaluated DMUs, contain *both* S_{1j} and S_{2j} stages (first and second rectangular brackets, respectively, in column 5 of Table 13.3). Of course, from a measurement perspective it will always be the case that $\theta_{Aj}^{CT} \leq \theta_{Aj}^{CL}$.

When comparing θ_{Aj}^{CT} to θ_{Aj}^{P} in Table 13.3, we see that $\theta_{Aj}^{P} > \theta_{Aj}^{CT}$ always holds. (We have been unable to establish any definitive relationship between θ_{Aj}^{CL} and θ_{Aj}^{P}.) Thus, for our numerical example, the CT model indeed finds the maximal possible contraction from the point of view of the aggregate DMU. From an organizational perspective, it may not be possible to implement this solution. We know it is impossible to achieve a better result than θ_{Aj}^{P} *along the Pareto Efficient frontier*. Consequently, to implement the solution proposed by θ_{Aj}^{CT} when it is smaller than θ_{Aj}^{P} will require either M_{1j} or M_{2j} to consent to a restructuring that would make him *worse* off than he can achieve by negotiating directly with the manager of the other subsystem. When consent is required, it would make more sense for an aggregate manager to select a Pareto efficient point that both managers will accept. Table 13.4 records the Pareto efficient points for the two stages and their corresponding θ_{Aj}^{D} values. For every DMU$_j$, the minimal value for the derived aggregate efficiency (θ_{Aj}^{P}) is given in a box and the equiproportional choice is highlighted in boldface. Observe the wide disparity in efficiencies for each stage corresponding to the Pareto aggregate efficiency. Since the equiproportionate choice seems to sacrifice little in the way of aggregate efficiency, it may be a practical alternative that is easier for the managers to agree on.

13.6 The Consistent Pricing Principle

The dual linear fractional program (known in the DEA literature as the *multiplier* formulation) to each manager's Acquisition Model provides an alternative means to understand the tradeoff inherent in the Pareto Efficient frontier for managers M_{10} and M_{20}. For M_{10} we have

$$\theta_{10}^{AQ} = \max \frac{\pi_I I_0}{\pi_K K_{10} + \pi_L L_{10}} \tag{13.24}$$

$$\left.\begin{aligned} \frac{\pi_I I_j}{\pi_K K_{1j} + \pi_L L_{1j}} &\leq 1 \\ \frac{\pi_F F_j}{\pi_K K_{2j} + \pi_L L_{2j} + \pi_I I_j} &\leq 1 \end{aligned}\right\} j = 1, \cdots, n$$

$$\frac{\pi_F F_0}{\pi_K K_{20} + \pi_L L_{20} + \pi_I I_0} \geq \theta_{20},$$

Table 13.4 Derived aggregate efficiency along the Pareto frontier of the two stages

DMU	Measures	Efficiency values								
	θ_{11}	1.000	0.9999	**0.9865**	0.9703	0.9272				
1	θ_{21}	0.9837	0.9837	**0.9865**	0.9900	1.000				
	θ_{A1}^{D}	0.991	0.990	0.989	0.988	0.990				
	θ_{12}	0.9428	0.9409	0.9344	**0.8358**	0.8037	0.7638	0.7427	0.7014	0.5196
2	θ_{22}	0.7970	0.7980	0.800	**0.8358**	0.8500	0.8600	0.8800	0.9000	1.000
	θ_{A2}^{D}	0.924	0.923	0.919	0.861	0.845	0.842	0.844	0.867	0.892
	θ_{13}	0.9731	0.9718	0.9667	**0.9288**	0.8408	0.766	0.5266		
3	θ_{23}	0.8753	0.8800	0.9000	**0.9288**	0.9500	0.9600	1.000		
	θ_{A3}^{D}	0.986	0.981	0.976	0.953	0.958	0.961	0.965		
	θ_{14}	0.9580	0.9436	0.9292	**0.8936**	0.5912				
4	θ_{24}	0.8600	0.8700	0.8800	**0.8936**	0.900				
	θ_{A4}^{D}	0.986	0.947	0.922	0.952	0.952				
	θ_{15}	1.000	0.8924	**0.7608**	0.6922	0.6511	0.5547	0.4318		
5	θ_{25}	0.6600	0.7000	**0.7608**	0.8000	0.8300	0.9000	1.000		
	θ_{A5}^{D}	1.000	0.908	0.819	0.781	0.781	0.843	0.892		
	θ_{16}	1.000	0.9827	**0.9612**	0.9523	0.7469	0.6419			
6	θ_{26}	0.9000	0.9300	**0.9612**	0.9750	0.9900	1.000			
	θ_{A6}^{D}	1.000	0.981	0.981	0.981	0.982	0.988			
	θ_{17}	0.9416	0.8728	**0.7666**	0.7367	0.6962	0.5793			
7	θ_{27}	0.4917	0.6250	**0.7666**	0.800	0.85	1.000			
	θ_{A7}^{D}	0.988	0.930	0.867	0.879	0.888	0.985			
	θ_{18}	0.9976	0.9055	**0.8401**	0.6560	0.4822	0.1625			
8	θ_{28}	0.8170	0.8300	**0.8401**	0.8740	0.9155	1.000			
	θ_{A8}^{D}	0.927	0.919	0.913	0.892	0.868	0.907			
	θ_{19}	1.000	0.8086	0.6005	**0.5424**	0.4719	0.3966	0.1618		
9	θ_{29}	0.3402	0.400	0.500	**0.5424**	0.600	0.700	0.9065		
	θ_{A9}^{D}	0.825	0.771	0.687	0.621	0.723	0.813	0.913		
	$\theta_{1,10}$	0.7476	0.7000	**0.5589**	0.5226	0.4896	0.4386			
10	$\theta_{2,10}$	0.3351	0.4000	**0.5589**	0.65	0.6670	0.6967			
	$\theta_{A,10}^{D}$	0.949	0.821	0.712	0.726	0.763	0.809			

and for M_{20} we have

$$\theta_{20}^{AQ} = \max \frac{\pi_F F_0}{\pi_K K_{20} + \pi_L L_{20} + \pi_I I_0} \tag{13.25}$$

$$\left.\begin{aligned} \frac{\pi_I I_j}{\pi_K K_{1j} + \pi_L L_{1j}} &\leq 1 \\ \frac{\pi_F F_j}{\pi_K K_{2j} + \pi_L L_{2j} + \pi_I I_j} &\leq 1 \end{aligned}\right\} j = 1, \cdots, n$$

$$\frac{\pi_I I_0}{\pi_K K_{10} + \pi_L L_{10}} \geq \theta_{10}.$$

The last constraint in each model ensures a lower bound on the efficiency of the counterpart subsystem. As the lower bound parameter varies it changes the efficiency in the obvious way: for example, raising θ_{20} lowers S_{10}'s efficiency in (13.24), and raising θ_{10} lowers S_{20}'s efficiency in (13.25).

Observe that there is a single multiplier π_K for both K_1 and K_2, a single multiplier π_L for both L_1 and L_2, and a single multiplier π_I that is used to weigh the intermediate factor both when it is an output (of the first stage) and when it is an input (to the second stage). Since the capital, labor and intermediate product are freely transferable between stages, their respective weights in the multiplier formulation should be the same. We shall call this the *Consistent Pricing Principle*. Consistent pricing holds for the Färe-Grosskopf model as well. There, the linear fractional programming dual is given by:

$$\theta_{20}^{CT} = \max \frac{\pi_F F_0}{\pi_K(K_{10}+K_{20}) \; + \; \pi_L(L_{10}+L_{20}) + \pi_I I_0} \tag{13.26}$$

$$\left.\begin{array}{r} \dfrac{\pi_I I_j}{\pi_K K_{1j} + \pi_L L_{1j}} \leq 1 \\ \dfrac{\pi_F F_j}{\pi_K K_{2j} + \pi_L L_{2j} + \pi_I I_j} \leq 1 \end{array}\right\} j = 1, \cdots, n$$

In an ordinary application of DEA, M_{10} would prefer a larger value of the multiplier π_I, whereas M_{20} would prefer a smaller value of π_I. When both output-input ratios appear in the same optimization, necessarily there will be a tradeoff between the measurement of efficiency of both stages. The consistent pricing principle leads to a natural conflict between the efficiency measures of the two stages. A weighting scheme that might make M_{10} efficient might very well make M_{20} look inefficient, and vice-versa. Thus, there will be a need to coordinate the choice for this multiplier. Regardless of the weighting scheme ultimately agreed upon, it should not be possible to select an alternative set of weights that would make both stages at least as efficient while making one of them more efficient. That is, it should minimally result in a Pareto efficient outcome; otherwise, neither manager M_{10} nor M_{20} would agree to the vertical integration.

13.7 Extensions and Directions for Further Research

Our aim in this paper was to present a novel approach that opens new ways to evaluate the efficiency of DMUs that are composed of subsystems arranged in series. Due to the complexity of the various relationships we model, we left many possible extensions to the basic model and to the analyzes for further research. These extensions are outlined below.

13.7.1 Extensions to the Basic Model

Technology: Our basic model is based on the CCR model that generates the standard radial measure of efficiency. It may be easily extended to the BCC model (Banker et al. 1984) or, more generally, to other technologies used to generate "Russell-type" measures of efficiency that eliminate the slacks in resource use (Russell 1985).

Structure: Our basic model contains just two stages. An obvious extension is to increase the number of serial stages in the model. Such an extension is possible as the notion of Pareto efficient frontier and the definition of Pareto aggregate efficiency easily generalize. The main (and significant) difficulty here would be the time it would take to compute the various measures.

Choice of variables: The model may also be extended to allow more flexibility in the definition of the inputs and outputs. First, it is straightforward to incorporate additional input and output factors. Second, the model can be extended to allow the presence of inputs and outputs that are not completely transferrable in some subsystems. For example, some inputs may be specific to a particular subsystem and can not be shared. Then, for the purpose of modeling and computing a subsystem's efficiency, it may be easier to work with the multiplier formulation using the consistent pricing principle.

Transaction costs: The basic model assumes no transaction costs when resources are moved between the two stages. In real-world cases, such movements are often accompanied by some transaction costs. Formulating these costs can be done either by adding some terms to the balance equation of the capital or by applying a certain "depreciation" term on each amount that is transferred.

13.7.2 Extensions to the Analysis

Principal-agent issues: We have assumed that subsystem managers have full decision-making power on how to reallocate capital and labor to improve their respective operations. An agreed-upon Pareto efficient point of subsystem efficiencies may or may not lead to the best improvement in aggregate efficiency, which can only be achieved if there is a single "aggregate" manager who has the full authority to unilaterally make decisions. A potentially important and fruitful line of research would be to explicitly define the relationship between the aggregate and subsystem managers using a principal-agent framework (see Jehle and Reny (2001) for a discussion). That is, do compensation schemes exist that will provide the necessary incentives to subsystem managers to choose the Pareto efficient point desired by the aggregate manager? Typically, such investigations also assume that the principle (the aggregate manager) does not observe all of the actions taken by the agent (the subsystem managers).

Cost analysis: When market prices are known, one can extend the analysis to explore the cost (or profit) efficiency of the various subsystems using models such as those proposed by Färe et al. (1994). Such an analysis may reveal interesting situations where the market prices of the input and output commodities may differ from the shadow prices obtained from the optimal solution of the models we propose here.

Reallocation: We have shown how it is possible for two subsystems to significantly improve their operational efficiencies via reallocation beyond what would be predicted by conventional analysis. However, such improvements may not be realized for certain DMU's, and it may be fruitful to obtain an understanding of when the reallocations proposed herein do not improve operational efficiencies.

Inter-DMUs trading: Our framework allows the manager of a subsystem (say, M_{11}) to trade resources only with the relevant counterpart in the same DMU (here, M_{21}). It might be useful to explore what would happen if resources may be traded among *different* DMUs. (For example, M_{11} acquires resources from M_{24} in return for delivering the amount of intermediate factor (I_4) to which the latter is committed). This might be a prelude to considering mergers or cooperation among independent entities in certain settings.

13.8 Concluding Remarks

DEA is a methodology aimed at evaluating the relative efficiency of DMUs. The construction of composite units that serve as benchmarks against which the performance of observed units are compared lies at the core of DEA. Each DEA model is characterized by a set of assumptions that are translated into a specific mathematical formulation that defines the possible configurations for the composite units, namely, the technology.

We have presented an approach to simultaneously measure the efficiency of aggregate DMUs with two subsystems in series, which goes beyond simply applying standard DEA analysis to each subsystem separately. The main novelty in our proposed approach lies in the more flexible manner in which we model the technology sets for each subsystem by allowing each subsystem to acquire resources from the other and to construct composites from both subsystems.

Our approach is potentially useful in various settings, both at the manufacturing and the service industries. In manufacturing, it can be implemented in the petrochemical or refinery industries which are characterized by sequential processes where the output of a given stage (e.g., oil refined to a certain Octan level) enters the next stage as an input. Similar processes can be found in food manufacturing plants and other sequential industries. Another area of implementation that bridges the manufacturing and service industries is the warehouse and distribution industry. In many cases, a warehouse may be modelled as a two-stage system in which the first stage uses labor, equipment and space to unload and store inventory, while the second stage uses labor, equipment, space and the inventory to prepare the packages shipped to end-users (see Hackman et al. 2001). An example of a possible implementation in the service industry was pointed out by Chen and Zhu (2004) in the banking context where the first stage uses labor, capital and space to generate deposits and the second stage uses labor, capital and deposits to generate loans.

Recently, there were some attempts to link DEA with current issues in the field of Supply Chain Management (SCM) (see Zhu 2002). We believe that our approach may further contribute in this context. For example, the implicit competition between S_{1j} and S_{2j} on the weight of the intermediate factor I_j that we describe in Sect. 13.6 resembles the *double marginalization* effect that is described in Chap. 5 of Tayur et al. (1999) where the competition revolves around the pricing of intermediate products. When the chain is owned and operated by a single party (i.e., when vertical integration has been affected) the proposed models could become useful in evaluating the performance of nodes in the chain and making appropriate managerial decisions. For example, the strongest could be allocated additional resources while the activities performed by the weakest nodes could be outsourced.

Acknowledgements The authors wish to express their gratitude to the editor and two anonymous reviewers for their helpful comments.

References

Banker, R. D., Charnes, A., & Cooper, W. W. (1984). Some models for estimating technical and scale inefficiencies in data envelopment analysis. *Management Science, 30*, 1078–1092.

Charnes, A., Cooper, W. W., & Rhodes, E. (1978). Measuring the efficiency of decision making units. *European Journal of Operational Research, 2*, 429–444.

Charnes, A., Cooper, W. W., Golany, B., Halek, R., Klopp, G., Schmitz, E., & Thomas, D. (1986a). *Two phase data envelopment analysis approaches to policy evaluation and management of army recruiting activities: Tradeoffs between joint services and army advertising.* (Research Report CCS no. 532). Center for Cybernetic Studies, The University of Texas at Austin.

Charnes, A., Cooper, W. W., & Thrall, R. (1986b). Identifying and classifying scale and technical efficiencies in data envelopment analysis. *Operations Research Letters, 5*, 105–116.

Charnes, A., Cooper, W. W., Lewin, A. R., & Seiford, L. M. (1994). *Data envelopment analysis: Theory, methodology and applications*. Boston: Kluwer Academic.

Chen, Y., & Zhu, J. (2004). Measuring information technology's indirect impact on firm performance. *Information Technology & Management Journal, 5*, 9–22.

Cook, W., Chai, D., Doyle, J., & Green, R. (1998). Hierarchies and groups in DEA. *Journal of Productivity Analysis, 10*, 177–198.

Färe, R., & Grosskopf, S. (1996). *Intertemporal production frontiers: With dynamic DEA*. Boston: Kluwer Academic.

Färe, R., & Grosskopf, S. (2000). Network DEA. *Socio-Economic Planning Sciences, 34*, 35–49.

Färe, R., & Primont, D. (1984). Efficiency measures for multiplant firms. *Operations Research Letters, 3*, 257–260.

Färe, R., & Whittaker, G. (1995). An intermediate input model of dairy production using complex survey data. *Journal of Agricultural Economics, 46*, 201–213.

Färe, R., Grosskopf, S., & Li, S. K. (1992). Linear programming models for firm and industry performance. *The Scandinavian Journal of Economics, 94*, 599–608.

Färe, R., Grosskopf, S., & Lovell, C. A. K. (1994). *Production frontiers*. New York: Cambridge University Press.

Graves, S. C., Rinnoy Kan, A. H. G., & Zipkin, P. H. (1993). *Logistics of production and inventory*. Amsterdam: North-Holland.

Hackman, S. T., & Leachman, R. C. (1989). A general framework for modeling production. *Management Science, 35*, 478–495.

Hackman, S. T., & Passy, U. (2002). Maximizing a linear fractional function on a pareto efficient frontier. *Journal of Optimization Theory and Applications, 113*, 83–103.

Hackman, S. T., Frazelle, E. H., Griffin, P. M., Griffin, S. O., & Vlatsa, D. A. (2001). Benchmarking warehousing and distribution operations: An input-output approach. *Journal of Productivity Analysis, 16*, 79–100.

Hoopes, B., Triantis, K. P., & Partangel, N. (2000). The relationship between process and manufaturing plant performance: A goal programming approach. *International Journal on Operations and Quantitative Management, 6*, 287–310.

Jehle, G. A., & Reny, P. J. (2001). *Advanced microeconomic theory*, (2nd ed.). New York: Addison-Wesley.

Johnson, L. A., & Montgomery, D. C. (1974). *Operations research in production planning, scheduling and inventory control*. New York: Wiley.

Löthgren, M., & Tambour, M. (1999). Productivity and customer satisfaction in Swedish pharmacies: A DEA network model. *European Journal of Operational Research, 115*, 449–458.

Russell, R. (1985). Measures for technical efficiency. *Journal of Economic Theory, 35*, 109–126.

Shephard, R.W. (1970). *Theory of cost and production functions*. Princeton: Princeton University Press.

Tayur, S., Ganeshan, R., & Magazine, M. (1999). *Quantitative models for supply chain management*. Boston: Kluwer.

Troutt, M.D., Ambrose, P., & Chan, C.K. (2001). Optimal throughput for multistage input – output processes. *International Journal of Operations and Production Management, 21*, 148–158.

Zhu, J. (2002). *Quantitative models for performance evaluation and benchmarking: Data envelopment analysis with spreadsheets*. Boston: Kluwer Academic.

Chapter 14
Network DEA II

Rolf Färe, Shawna Grosskopf, and Gerald Whittaker

Abstract The original DEA model by Charnes et al. (Eur. J. Oper. Res. 2:429–444, 1978) is set to analyze production as a black box, i.e., there is no information about the processes inside. Network DEA was proposed for analysis of the contents of the black box. This theory allows the researcher to model processes within the black box by formulating sub-technology DEA models. The interaction of sub-technology DEA models preserves the DEA structure, and the network model can therefore be solved using linear programming. This chapter discusses network DEA models, both static and dynamic. The discussion also explores various useful objective functions that can be applied to the models to find the optimal allocation of resources for processes within the black box that are normally invisible to DEA.

Keywords Data Envelopment Analysis (DEA) • Network • Intermediate products • Dynamic production

---

R. Färe
Economics Unit and Department of Applied Economics, Oregon State University, Corvallis, OR 97331, USA

S. Grosskopf
Department of Economics, Oregon State University, Corvallis, OR 97331, USA

CERE, Umea, Sweden
e-mail: shawna.grosskopf@orst.edu

G. Whittaker (✉)
National Forage Seed Production Research Center, Agricultural Research Service, USDA, Corvallis, OR 97331, USA
e-mail: Jerry.Whittaker@ars.usda.gov

W.D. Cook and J. Zhu (eds.), *Data Envelopment Analysis*,
International Series in Operations Research & Management Science 208,
DOI 10.1007/978-1-4899-8068-7_14,

14.1 Introduction

It is commonly observed that the DEA model proposed by Charnes et al. (1978) is a "black box" that receives inputs and produces outputs, but the transformation process by which this occurs is opaque to the analyst. As Tone and Tsutsui (2014) remark, "One of the drawbacks of these models is the omission of the internal structure of the DMUs." Färe et al. (2007a) built on Shephard and Färe (1975) with a sequence of models where the interior ("black box") of the CCR model could be analyzed. The primary device for achieving this was the use of a network. The insight they had was that when multiple DEA models are connected in a network, the network itself is a DEA model, and can be calculated using linear programming.

In this chapter we extend and update our paper (Färe et al. 2007a) with additional discussion of DEA sub-technologies, objective functions, and static and dynamic DEA models. We start with a discussion of the CCR model from an axiomatic perspective. Then we turn to objective functions, which can be either price dependent or not. In the case where objective functions are price independent, we discuss both distance functions and slack-based functions. The static network model is introduced next, starting with a generic model and extending it to three common cases. We end the chapter with dynamic DEA. The models discussed in this chapter are relatively simple, but provide the tools for construction of models describing arbitrarily complex processes.

14.2 The Black Box and Sub-technologies

Production models are frequently modeled as a black box, where inputs $x = (x_1, \ldots, x_N) \in \mathbb{R}^N_+$ are transformed into outputs $y = (y_1, \ldots, y_M) \in \mathbb{R}^M_+$ (Fig. 14.1).

In this chapter we go inside the black box and define sub-technologies as its smallest building blocks. First, we establish an axiomatic structure for the models, then discuss the connection of sub-technologies through a directed network.

A technology or sub-technology may modeled as

$$T = \{(x, y) : x \textit{ can produce } y\}$$

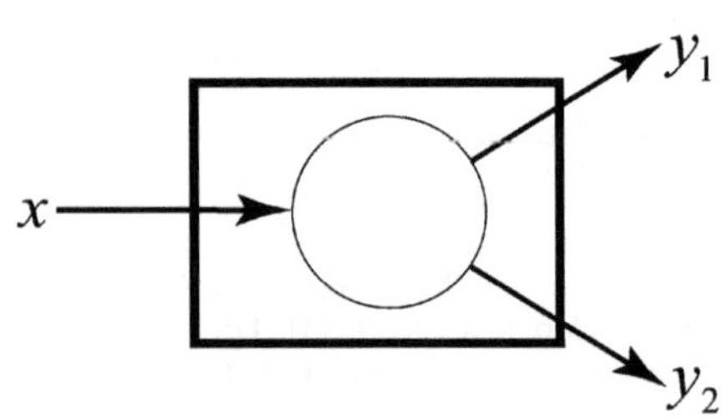

Fig. 14.1 The black box

or by its output sets,

$$P(x) = \{y : (x, y) \in T\} \ x \in \mathbb{R}_{+}^{N}$$

or by its input sets,

$$L(y) = \{x : (x, y) \in T\},$$

and it holds that

$$y \in P(x) \Leftrightarrow (x, y) \in T \Leftrightarrow x \in L(y).$$

Suppose that we have $k = k, \ldots, K$ observations of inputs and good outputs (x^k, y^k), then we can formulate the activity analysis or data envelopment analysis model as

$$T = \Bigg\{ (x, y) : \sum_{k=1}^{K} z_k x_{kn} \leq x_n, \ n = 1, \ldots, N, \\ \sum_{k=1}^{K} z_n y_{kn} \geq y_m, \ m = 1, \ldots, M, \\ z_k \geq 0, \ k = 1, \ldots, K \Bigg\}$$

or equivalently,

$$P(x) = \Bigg\{ y : \sum_{k=1}^{K} z_n x_{kn} \leq x_n, \ n = 1, \ldots, N, \\ \sum_{k=1}^{K} z_m y_{km} \geq y_m, \ m = 1, \ldots, M, \\ z_k \geq 0, \ k = 1, \ldots, K \Bigg\}$$

i.e.,

$$L(y) = \Bigg\{ x : \sum_{k=1}^{K} z_k x_{kn} \leq x_n, \ n = 1, \ldots, N, \\ \sum_{k=1}^{K} z_k y_{km} \geq y_m, \ m = 1, \ldots, M, \\ z_k \geq 0, \ k = 1, \ldots, K \Bigg\}.$$

If the data meet the Kemeny et al. (1956) conditions, then the technology satisfies the following conditions;

$$(i)\ P(0) = \{0\};\ \text{no free lunch,}$$
$$(ii)\ P(x)\ \text{is bounded for each x; scarcity,}$$
$$(iii)\ T\ \text{is closed.}$$

Additional properties, from the definitions are;

$$(iv)\ T\ \text{is convex}\ (\Rightarrow L(y)\ \text{and}\ P(x)\ \text{are convex}),$$
$$(v)\ x' \geq x \in L(y) \Rightarrow x' \in L(y);\ \text{free disposability of inputs,}$$
$$(vi)\ y' \leq y \in P(x) \Rightarrow y' \in P(x);\ \text{free disposability of outputs,}$$
$$(vii)\ T\ \text{is a cone; constant returns to scale (CRS).}$$

The last condition holds, since the intensity variables z_i are only non-negative. If, in addition to non-negativity, $\sum_{k=1}^{K} z_k = 1$, variable returns to scale are modeled. Conditions (v) and (vi) follow from the inequalities of the inputs and outputs expressions.

$$(a) \qquad \sum_{k=1}^{K} y_{km} > 0 \qquad m = 1, \ldots, M$$
$$(b) \qquad \sum_{m=1}^{M} y_{km} > 0 \qquad k = 1, \ldots, K$$
$$(c) \qquad \sum_{k=1}^{K} x_{kn} > 0 \qquad n = 1, \ldots, N$$
$$(d) \qquad \sum_{n=1}^{N} x_{kn} > 0 \qquad k = 1, \ldots, K$$

Output condition (a) states that each output is produced by some k (DMU), and (b) requires that each activity produce some output. The input conditions (c) and (d) say that each input is used by some k and that each k uses at least one input. If $z_k \geq 0$, $k = 1, \ldots, K$ and $\sum_{k=1}^{K} z_k \leq 1$, the technology exhibits non-increasing returns to scale.

Before we link technologies (and/or sub-technologies) together to model processes within the black box, we study optimization problems on a technology. These problems include profit and revenue maximization, cost minimization and distance function measures.

14.3 Objective Functions

Network models are frequently applied in performance measurement, including optimal resource allocations. An objective function that specifies the evaluation of measurement or allocation is required for optimization. Two types of objective functions are used; (i) those that require prices, and (ii) those that require only measures of inputs and outputs (quantity functions). These quantity functions can be either slack based or distance functions.

The distance functions have their duals among the objective functions involving prices. The most common are; the profit function dual to the directional technology function, the revenue function dual to Shephard's output distance function, and the cost function dual to Shephard's input distance function. The slack based objective functions do not have natural duals (Färe et al. 2007b).

Let $g = (g_x, g_y) \in \mathbb{R}_+^{N+M}$ be a directional vector for inputs (g_x) and outputs (g_y). This vector provides the direction in which a given input/output observation is projected onto the frontier of T. The optimization problem that defines the directional technology distance function is

$$\vec{D}_T(x, y; g) = \max\left\{\beta : \left(x - \beta g_x, y + \beta g_y\right) \in T\right\}$$

Inputs are contracted while outputs are expanded.[1] Under "g-disposability" this function characterizes T, i.e.,

$$\vec{D}_T(x, y; g) \geq 0 \Leftrightarrow (x, y) \in T,$$

and it can be used as a measure of technical efficiency.

If we choose the directional vector to equal $g = (0, g_y)$, then we have a directional output distance function

$$\vec{D}_0\left(x, y; g_y\right) = \max\left\{\beta : \left(y + \beta g_y\right) \in P(x)\right\},$$

which may also be a measure of technical efficiency, now output based.

It is interesting to note how $\vec{D}_0(x, y; g_y)$ is related to Shephard's output distance function. The latter is defined as

$$D_i(x, y) = \min\{\lambda : y/\lambda \in P(x)\}.$$

Shephard's output distance function or its reciprocal is the Farrell output oriented measure of technical efficiency.

[1] This distance function was introduced by Luenberger as shortage function (see for example Luenberger 1995).

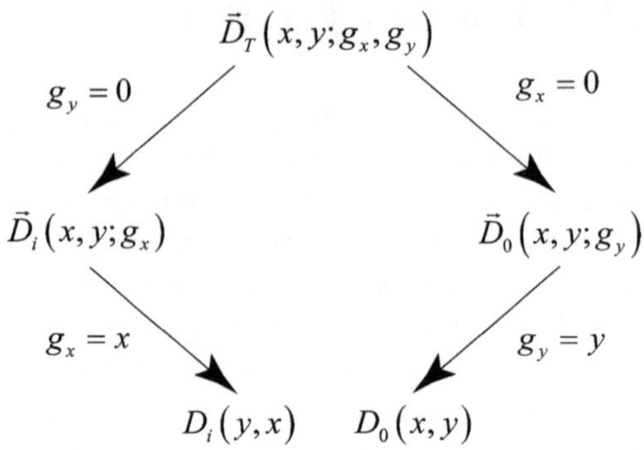

Fig. 14.2 Relation of directional distance functions

To relate the two output distance function to each other, choose $g_y = y$, then

$$\vec{D}_O(x, y; y) = 1 - 1/D_O(x, y)$$

Thus, we have shown how the directional output distance function generalizes Shephard's (radial) output distance function. Similarly, we may choose $g = (g_x, 0)$ to obtain a directional input distance function

$$\vec{D}_i(x, y; g_x) = \max\{\beta : (x - \beta g_x) \in L(y)\},$$

and relate it to Shephard's input distance function defined as

$$D_i(y, x) = \max\{\lambda : x/\lambda \in L(y)\}.$$

Choose $g_x = x$ and we have

$$\vec{D}_i(y, x; x) = -1 + 1/D_i(y, x).$$

Our derivations above show that the directional output distance function is the origin of the other four distance functions. This can be illustrated as (Fig. 14.2).

Finally we have (Färe and Lovell 1978)

$$D_O(x, y) = 1/D_i(y, x) \Leftrightarrow \text{CRS}.$$

The above distance functions are associated with measurement of technical efficiency. Each projects the observation in question onto a corresponding isoquant.[2]

[2] Note that an isoquant may contain the efficient subset as a proper subset and does not reflect Pareto/Koopmans efficiency.

Turning to the duals, assume input prices $w \in \mathbb{R}_+^N$ and output prices $p \in \mathbb{R}_+^M$ are known. Then we can define the profit function as

$$\begin{aligned}\Pi(p,w) &= \max\{py - wx : (x,y) \in T\} \\ &= \max\{py - wx : \vec{D}_T(x,y;g) \geq 0\}\end{aligned}$$

The last inequality holds, since the directional technology distance function characterizes T. From this it follows that (Färe and Grosskopf 2004)

$$\frac{\Pi(p,w) - (py - wx)}{pg_y + wg_x} \geq \vec{D}_T(x,y;g)$$

where the LHS is the Nerlovian profit indicator, which is the normalized difference between maximal profit $\Pi(p,w)$ and observed profit $(py - wx)$. This indicator is larger than the corresponding directional distance function (a measure of technical efficiency). By adding a measure of allocative efficiency, the inequality becomes an equality, and the Nerlovian indicator is decomposed into technical and allocative efficiency (Chambers et al. 1998).

In order to derive results similar to the Nerlovian indicator, we first define the revenue and cost functions

$$\begin{aligned}R(x,p) &= \max\{py : y \in P(x)\} \\ &= \max\left\{py : \vec{D}_0\left(x,y;g_y\right) \geq 0\right\}\end{aligned}$$

and

$$\begin{aligned}C(y,w) &= \min\{wx : x \in L(y)\} \\ &= \min\{wx : \vec{D}_i(x,y;g_x) \geq 0\}.\end{aligned}$$

From these expressions we get

$$\frac{R(x,p) - py}{pg_y} \geq \vec{D}_0\left(x,y;g_y\right)$$

and

$$\frac{wx - C(y,w)}{wg_x} \geq \vec{D}_i(x,y;g_x)$$

respectively.

Each of these inequalities can be closed by adding, as above, an allocative inefficiency component. Hence arriving at a revenue and a cost indicator with their corresponding decompositions.

Prior to a discussion of the radial distance function (Shephard 1953, 1970), we provide activity analysis/DEA models for calculating profit and the directional technology distance function. Since, under constant returns to scale, profit is zero, we choose the variable returns to scale formulation allowing for losses and profit. Maximal profit is then estimated as

$$\begin{aligned}
&\Pi(p, w) = \max_{(x,y,z)}(py - wx)\\
&s.t. \sum_{k=1}^{K} z_k x_{kn} \leq x_n, \ n = 1, \ldots, N\\
&\sum_{k=1}^{K} z_k y_{km} \geq y_m, \ m = 1, \ldots, M\\
&\sum_{k=1}^{K} z_k = 1, \ z_k \geq 0, \ k = 1, \ldots, K
\end{aligned}$$

One may, of course, allow the price vectors p and w to vary with the observation.

To estimate the technology distance function, the researcher may choose the directional vectors or endogenize them.[3] Here we consider the case of $g = (g_x, g_y)$ without any specific choice. Thus, for observation k', we have

$$\begin{aligned}
&\vec{D}_T\left(x^{k'}, y^{k'}; g_x, g_y\right) = \max\beta\\
&s.t. \sum_{k=1}^{K} z_k x_{kn} \leq x_{k'n} - \beta g_{x_n}, \ n = 1, \ldots, N\\
&\sum_{k=1}^{K} z_k y_{km} \geq y_{k'm} + \beta g_{y_m}, \ m = 1, \ldots, M\\
&\sum_{k=1}^{K} z_k = 1, \ z_k \geq 0, \ k = 1, \ldots, K.
\end{aligned}$$

From our two calculations and the data on $\left(x^{k'}, y^{k'}\right), (p, w)$ we may calculate the Nerlovian profit indicator. Similar problems can be formulated for the revenue and cost indicators.[4]

In the discussion above, profit, revenue and cost indicators took an additive structure. The traditional Farrell (1957) decomposition of cost and revenue efficiency is multiplicative, not additive, and is based on Shephard's (1953, 1970) distance function.

[3] For endogenous directions, see (Färe et al. 2013)

[4] For an example of aggregation of these indicators, see (Färe and Grosskopf 2004).

To derive the revenue measure, note that

$$\begin{aligned} R(x,p) &= \max\{py : y \in P(x)\} \\ &= \max\{py : D_O(x,y) \leq 1\}, \end{aligned}$$

where the last equality holds because

$$D_O(x,y) \leq 1 \Leftrightarrow y \in P(x).$$

From this definition of revenue, it follows (Färe and Grosskopf 2004) that

$$\frac{R(x,p)}{py} \geq 1/D_O(x,y),$$

i.e., the ratio of maximal revenue to observed revenue is larger than the reciprocal of the output distance function.[5] By multiplying the RHS with an allocative efficiency component, the Farrell decomposition of revenue efficiency is obtained. It is the product, not sum, of technical and allocative efficiencies.[6]

To set the stage for the slack-based measure of technical efficiency consider the Leontief production function

$$y = \min\{x_1, x_2\}$$

The isoquant for $y = 1$, is

$$\{(x_1, x_2) : \min\{x_1, x_2\} = 1\}$$

and the Koopmans efficient input set is

$$\{x_1, x_2\} = \{1, 1\}$$

Thus the isoquant and the efficient subset do not coincide. The following figure illustrates the relation between the isoquant and efficient subset (Fig. 14.3).

The isoquant consists of the points along II while E is the only efficient point.

The input distance functions introduced above have the property that they project an input vector onto the isoquant, and not necessarily onto the efficient point(s). Measures that project an input vector onto efficient points are the multiplicative Russell measure (Färe and Lovell 1978) and the additive slack-based measures (Färe and Grosskopf 2010; Tone 2001).

[5] This expression is referred to as the Mahler inequality.

[6] For the decomposition of Farrell's cost efficiency measure, see (Färe and Grosskopf 2004).

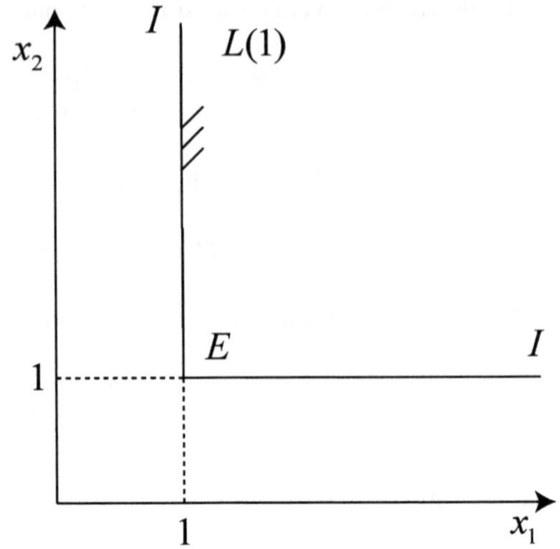

Fig. 14.3 Leontief isoquant (I,I) and efficient set (E)

The input oriented Russell measure (RM) has the following DEA formulation

$$RM_i\left(x^{k'}, y^{k'}\right) = \frac{1}{N}\min\sum_{n=1}^{N}\lambda_n$$

$$\begin{aligned} s.t. \quad & \sum_{k=1}^{K} z_k x_{kn} \leq \lambda_n x_{k'n}, \quad n = 1, \ldots, N \\ & \sum_{k=1}^{K} z_m y_{km} \geq y_{k'm}, \quad m = 1, \ldots, M \\ & z_k \geq 0, \quad k = 1, \ldots, K\ . \end{aligned}$$

If an input $x_{k'n}$ is zero, we modify the measure and set $\lambda_n = 1$. This measure is one if and only if $x^{k'}$ belongs to the efficient subset of $L\left(y^{k'}\right)$, where

$$Eff\,L(y) = \{x : x \in L(y), x' \leq x \Rightarrow x' \notin L(y)\}$$

The input oriented slack-based measure (SB) by Färe and Grosskopf (2010) is based on the input oriented directional distance function, and has the following DEA formulation

$$SB_i\left(x^{k'}, y^{k'}\right) = \max\sum_{n=1}^{N}\beta_n$$

$$\begin{aligned} & \sum_{k=1}^{K} z_n y_{kn} \leq x_{k'n} - \beta_n \bullet 1_n, \quad n = 1, \ldots, N \\ & \sum_{k=1}^{K} z_m y_{km} \geq y_{k'm}, \quad m = 1, \ldots, M \\ & z_k \geq 0, \quad k = 1, \ldots, K. \end{aligned}$$

This measure is zero if and only if $x^{k'}$ belongs to the efficient subset of $L(y^{k'})$.[7] Note that β_n is independent of the unit of measurement, since the direction $g_n = 1_n$ is in the same units as x_n, thus $\beta_{n_i} \forall i$ can be added.

14.4 Static Network Models

In this section we go inside the black box and model it as a network of sub-technologies. This approach has its origin in Shephard and Färe (1975), who wrote that "many production systems (technologies) may be conceptualized as the joint interaction of a finite number of production sub-technologies called activities." The static model is useful for analyzing the allocation of intermediate products and also provides the basic structure of dynamic DEA models.

We restrict our presentation to three sub-technologies P^1, P^2, and P^3. These three sub-technologies are connected by the directed network shown in Fig. 14.4.

To complete a network of these three sub-technologies, we add a distribution process and a sink, or collection of final outputs.[8] Inputs are denoted by $x = (x_1, \ldots, x_N) \in \Re_+^N$, the network exogenous vector, i.e., total availability is attached to the three sub-technologies, ${}_0^i x$, $i = 1, 2, 3 \ldots$ where *0* denotes source and *i* denotes use (Fig. 14.5). For example, ${}_0^1 x$ is the input vector from source *0* used

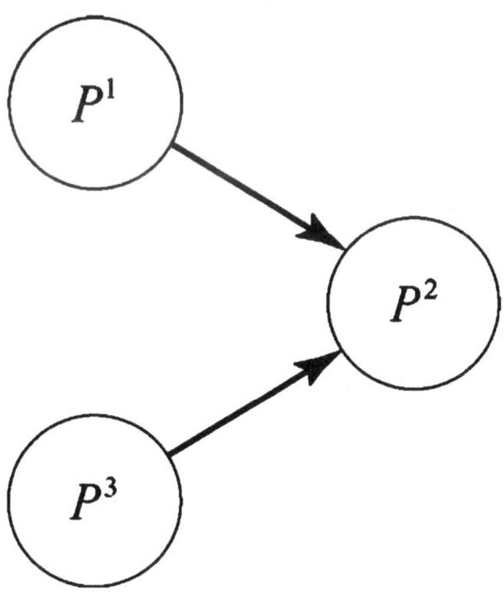

Fig. 14.4 Representation of sub-technologies

[7] We note that the input oriented Tone (2001) measure requires all inputs to be positive. See Färe and Grosskopf (2010) for a discussion inputs to this measure.

[8] This model is adopted from Färe and Grosskopf (1996a).

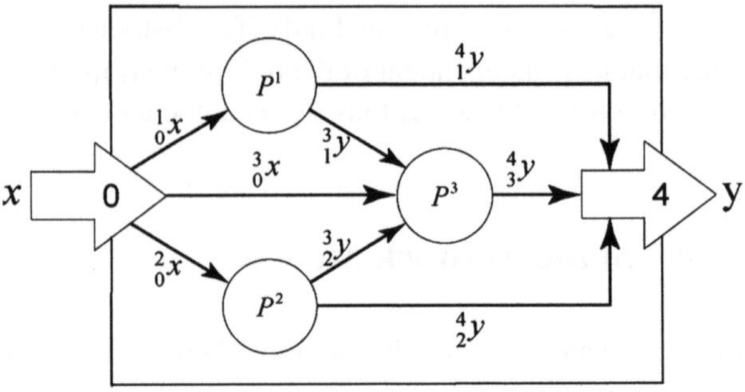

Fig. 14.5 A network technology

in activity *1*. The total amounts used in the three activities cannot exceed the total amount available, i.e.,

$$x \geq \sum_{i=1}^{3} {}_0^i x \ .$$

In Fig. 14.5 activity P^1 uses ${}_0^1 x$ as an exogenous input and produces

$$ {}_1^3 y + {}_1^4 y $$

as outputs. The ${}_1^3 y$ is an input into activity 3, while ${}_1^4 y$ is the final output from P^1. Activity P^3 uses ${}_0^3 x$ as an exogenous input and ${}_1^3 y$, ${}_2^3 y$ as intermediate inputs. The final product from activity P^3 is output vector ${}_3^4 y$. The total network output is the sum of the final output of the three activities,

$$ {}_1^4 y + {}_2^4 y + {}_3^4 y . $$

Note that even if an activity does not produce one of the listed outputs, that output is set at zero and the network structure remains the same. It is also noteworthy that some outputs may be both final and intermediate outputs, e.g. spare parts.

Based on the description above, a generic network model takes the form

$$\begin{aligned} P(x) = \big\{ \big({}_1^4 y + {}_2^4 y + {}_3^4 y \big) : \\ \big({}_1^4 y + {}_1^3 y \big) \in P^1 \big({}_0^1 x \big) \\ \big({}_2^4 y + {}_2^3 y \big) \in P^2 \big({}_0^2 x \big) \\ {}_3^4 y \in P^3 \big({}_0^3 x, {}_1^3 y + {}_2^3 y \big) \\ {}_0^1 x + {}_0^2 x + {}_0^3 x \leq x \big\} \ , \end{aligned}$$

where $P^i(\bullet)$ $i = 1, 2, 3$ are output sets, i.e., $P(x) = \{y : x \text{ can produce } y\}$. Thus the network model $P(x)$ is formed by the individual sub-technologies.

Next we formulate the network technology as a DEA model, i.e.,

$$
\begin{aligned}
&P(x) = \left\{y = \left({}_{1}^{4}y + {}_{2}^{4}y + {}_{3}^{4}y\right) : \right. \\
&\text{(a)}\ \ {}_{3}^{4}y_m \le \sum_{k=1}^{K} z_k^3\, {}_{3}^{4}y_{km},\ \ m = 1, \ldots, M, \\
&\text{(b)}\ \ \sum_{k=1}^{K} z_k^3\, {}_{0}^{3}x_{kn} \le {}_{0}^{3}x_n,\ \ n = 1, \ldots, N, \\
&\text{(c)}\ \ \sum_{k=1}^{K} z_k^3\, {}_{1}^{3}y_{km} \le {}_{1}^{3}y_m, m = 1, \ldots, M, \\
&\text{(d)}\ \ \sum_{k=1}^{K} z_k^3\, {}_{2}^{3}y_{km} \le {}_{2}^{3}y_m, m = 1, \ldots, M, \\
&\text{(e)}\ \ z_k^3 \ge 0, k = 1, \ldots, K, \\
&\text{(f)}\ \ \left({}_{1}^{3}y_m + {}_{1}^{4}y_m\right) \le \sum_{k=1}^{K} z_k^1 \left({}_{1}^{3}y_{km} + {}_{1}^{4}y_{km}\right), m = 1, \ldots, M, \\
&\text{(g)}\ \ \sum_{k=1}^{K} z_k^1\, {}_{0}^{1}x_{kn} \le {}_{0}^{1}x_n, n = 1, \ldots, N, \\
&\text{(h)}\ \ z_k^1 \ge 0, k = 1, \ldots, K, \\
&\text{(i)}\ \ \left({}_{2}^{3}y_m + {}_{2}^{4}y_m\right) \le \sum_{k=1}^{K} z_k^2 \left({}_{2}^{3}y_{km} + {}_{2}^{4}y_{km}\right), m = 1, \ldots, M, \\
&\text{(j)}\ \ \sum_{k=1}^{K} z_k^2\, {}_{0}^{2}x_{kn} \le {}_{0}^{2}x_n, n = 1, \ldots, N, \\
&\text{(k)}\ \ z_k^2 \ge 0, k = 1, \ldots, K, \\
&\text{(l)}\ \ {}_{0}^{1}x_n + {}_{0}^{2}x_n + {}_{0}^{3}x_n \le x_n,\ \ n = 1, \ldots, N \left.\right\}.
\end{aligned}
$$

where ${}_{0}^{i}y_{km}$ and ${}_{0}^{i}x_{kn}$, $k = 1, \ldots, K$, $m = 1, \ldots, M$ and $n = 1, \ldots, N$ are observed data. In the network DEA model the first sub-technology is specified by (f)–(h), the second by (i)–(k) and the third by (a)–(e). Note that each sub-technology has a vector of intensity variables (e), (h) and (k). Here they all satisfy constant returns to scale.

Färe and Grosskopf (1996b) have shown that if each sub-technology satisfies standard axioms such as free disposability of inputs and outputs, compactness of the output set and other axioms, then the network technology has the same properties as the sub-technologies from which it is constructed.

If we wish to compute, for example, output efficiency of the network DEA model, we may use the standard Farrell (1957) output measure using linear programming;

$$
F_O\left(x^{k'}, y^{k'}\right) = \max\left\{\theta : \theta y^{h'} \in P\left(x^{k'}\right)\right\},
$$

where $P(x^{k'})$ is the network for observation k'. Note that in this model the ratio of outputs from P^1 and P^2 to observed output may vary. If this effect is not appropriate, we may modify the LHS of (f) and (i) to read

$$(\mathrm{f}')\ \theta\left({}^3_1y_{k'm} + {}^4_1y_{k'm}\right) \leq$$

and

$$(\mathrm{i}')\ \mu\left({}^3_2y_{k'm} + {}^4_2y_{k'm}\right) \leq$$

where k' is the observed data and θ, μ are non-negative scalars. When $F_0\left(x^{k'}, y^{k'}\right)$ is computed under these more restrictive conditions, the optimal θ and μ are the sub-vector Farrell efficiency scores.

From standard DEA modeling practice (Cooper et al. 2004), one knows that the primal envelopment model has a dual multiplier formulation. Hence, when we use the "fixed" mix formulation, i.e., when (f') and (i') are used, we may write the network in its dual form, as in Kao (2009). To see how the network model may be specified as a traditional neoclassic production model, suppose for simplicity that ${}^4_1y = {}^4_2y = 0$ and that ${}^3_1y$, ${}^3_2y$, and ${}^4_3y \in \mathbb{R}_+$, so that each sub-technology produces a single output and no final products are produced by P^1 and P^2. If we define a production function by

$$F(x) = \max\{y : y \in P(x)\},$$

then the network model can be written as follows[9];

$$\begin{aligned} {}^4_1y &= F^3\left(x^3_0, F^1\left({}^1_0x\right), F^2\left({}^2_0x\right)\right) \\ &= F^3\left(x - {}^1_0x - {}^2_0x, F^1\left({}^1_0x\right), F^2\left({}^2_0x\right)\right), \end{aligned}$$

i.e., as function of functions. This example illustrates the creation of "parametric" neoclassical models of network technologies.

Next, we formulate three commonly used network models; (i) the Johansen (1972) model, (ii) the two stage model and (iii) the externality model. These three models are set up here generically, and like the network models above can be extended to any finite set of sub-technologies. We start with the Johansen model as formulated by Färe et al. (1992). Here we express it and the other models by means of diagrams (Fig. 14.6).

We have two technologies P^1 and P^2 producing two vectors of outputs y^1 and y^2. They use fixed inputs (non-allocable) x^1_f and x^2_f respectively, and variable allocable

[9] For this to exist $P(x)$ must be nonempty and compact.

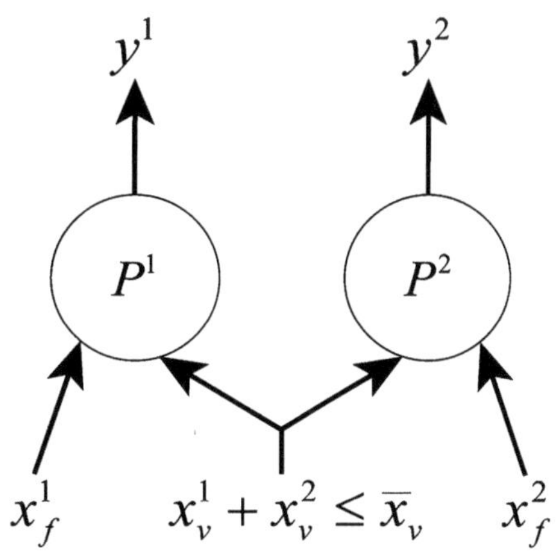

Fig. 14.6 The Johansen model

inputs x_v^1 and x_v^2. The sum of these input vectors cannot exceed a given vector $\bar{x}_v$. Thus the total can be allocated between the two technologies P^1 and P^2. Such an allocation can be implemented by choosing one of the objective functions discussed in Sect. 14.2. For example, if output prices are known, one may maximize total revenue for the network, viz,

$$\begin{aligned}&\max p^1 y^1 + p^2 y^2\\ &s.t.\ \ y^1 \in P^1, y^2 \in P^2\end{aligned}$$

where p^1 and p^2 are two output price vectors. The solution to this problem yields the optimal output vector for each technology and the optimal allocation of the variable inputs.[10] This model may be used in the discussion of merger and coalition formation (Bogetoft and Wang 2005; Färe et al. 2011).

While the Johansen model organizes the technologies in parallel, our second model organizes the technologies in a sequence (Chen et al. 2009) (Fig. 14.7).[11]

Again we have two technologies P^1 and P^2, now ordered sequentially with the output from P^1 being an intermediate input into P^2 to produce the final output y^2. Each technology has its own input x^1, x^2. It is not hard to adapt the concept of allocation from the Johansen model to introduce savings and borrowing into the model (Färe and Grosskopf 1996a).

Suppose the externality model consists of one polluter and one receptor, with the technologies P^1 and P^2, respectively. Each technology uses an input, x^1 or x^2, to produce desirable outputs y^1 and y^2, and the polluter also produces an undesirable output u. The undesirable output u is an input into the receptor technology, as illustrated below (Fig. 14.8).

[10] One may, of course, solve this problem using DEA.

[11] For a review see (Cook et al. 2010)

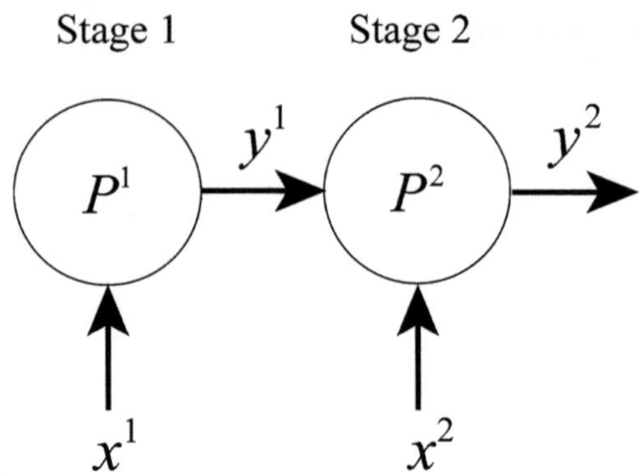

Fig. 14.7 A two stage model of production

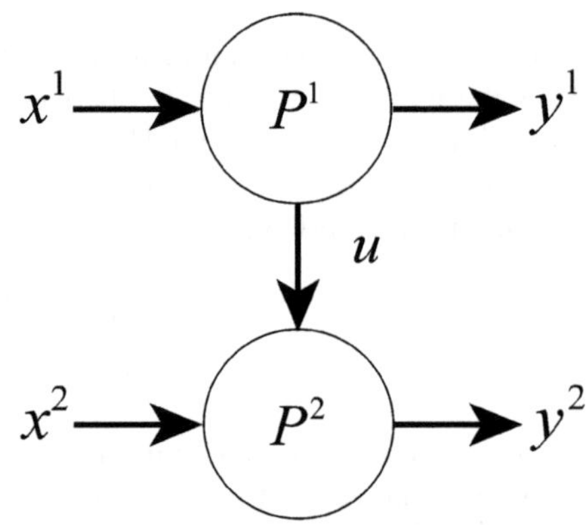

Fig. 14.8 The network model with externalities

The upstream firm (technology P^1) may be, for example, a paper mill, and the downstream firm P^2 a fishery. The upstream firm produces y^1 and u jointly and in the terminology of Shephard and Färe (1974) we say that y^1 is nulljoint with u if

$$(y^1, u) \in P^1(x^1) \text{ and } u = 0 \Rightarrow y^1 = 0,$$

i.e., if no undesirable outputs are produced, no desirable outputs can be produced. The downstream technology takes u as an additional input with negative implications, i.e.,

$$u^1 \geq u \Rightarrow P^2(x^2, u^1) \leq P^2(x^2, u).$$

This means that more desirable input does not increase production, but may decrease it. This type of network models have also been applied in the studies of property rights (see Färe and Grosskopf 2004, for a DEA formulation of such a problem).

14.5 Dynamic Network Models[12]

"Late in 1974, Dr Thomas Varley, Office of Naval Research, asked whether I could formulate a production function for shipbuilding. In thinking about this matter it became apparent that the usualsteady state (static) production function could at best provide a faint model of this production technology. Imagine that you visit a shipyard. Day by day a tremendous amount of production activity of great variety is carried on, yet no ships are turned out. This goes on for a long time. Eventually a ship emerges. What was being produced day by day all during this time? It is clear that the daily, weekly, monthly outputs of the system were intermediate products. The shipbuilding production system, like construction, is a dynamically evolving process."(Shephard and Färe 1980, page V)

With this description of a production process as a motivation for the dynamics, let us start by formulating a dynamic model as a network. Assume there are three time periods $t-1, t, t+1$, and that each has its production technology P^{τ}, $\tau = t-1,\ t, t+1$. A dynamic model has the property that a decision in one time period impacts on later time periods. For example, if I save now, then my possible consumption may increase later. Therefore we introduce intermediate products, i.e., those products that are held over between time periods, ${}^{\tau+1}_{\tau}y \in \Re^{M}_{+}$. If $\tau = 1$, then ${}^{t+1}_{t}y$ is the intermediate vector of outputs produced at time t and entering the production process at $t+1$, i.e., an input at P^{t+1}. Figure 14.9 illustrates this setup.

Each of the production sub-technologies P^{τ} uses exogenous inputs x^{τ} to produce the final output ${}^{\tau}_{\tau-1}y$ and intermediate inputs ${}^{i}_{\tau-1}y$. To complete the network model (see the figure below) we add initial conditions (distribution process in the static model) and transversality conditions (sink in the static model).

The initial condition is given by ${}^{i}\bar{y}$ and may be thought of as the stocks of "capital" initially available. The transversality condition could include the number of periods, say $t+1 = T$, the state of the system at T, ${}^{i}_{t+1}\bar{y}$, and the final output vector

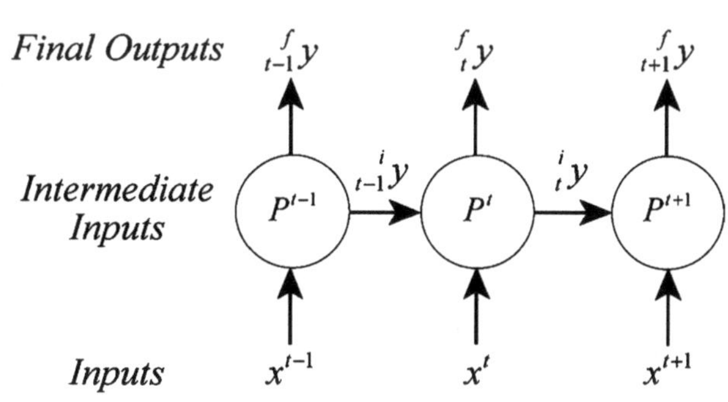

Fig. 14.9 Dynamic sub-technologies

[12] This section is to a large extent adapted from Färe, Grosskopf and Whittaker (2007) For a recent survey of nonparametric dynamic efficiency see (Fallah-Fini et al. 2013).

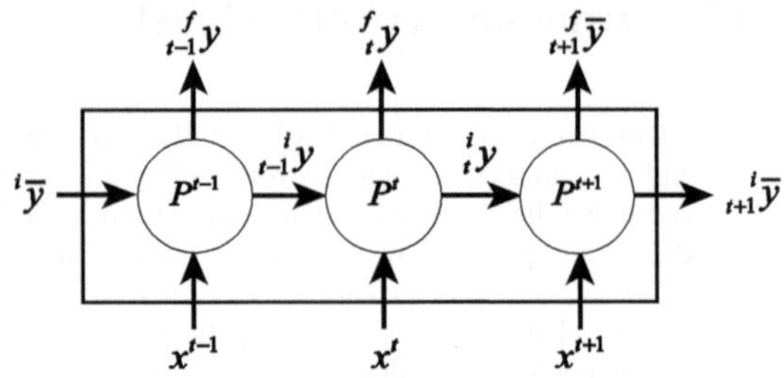

Fig. 14.10 A dynamic network model

from the last period, ${}^{f}_{t+1}\bar{y}$. The chosen conditions are specific to the research problem to be analyzed.

Thus far we have not introduced discounting. However, if outputs are given in value terms, a discount factor of δ^{τ}, $0 \leq \delta^{\tau} \leq 1$, may be introduced to account for the lesser of future income compared to present income. For example, $\delta^{tf}_{t}y$ are the discounted values of the final output in period t (Fig. 14.10).

The dynamic network DEA model consists of the interaction of a finite number of static models. Therefore let us start by studying the sub-technology P^t from above. This technology uses inputs x^t and intermediate inputs ${}^{i}_{t-1}y$ to produce outputs ${}^{f}_{t}y$ + ${}^{i}_{t}y$, where ${}^{f}_{t}y$ is the final output and ${}^{i}_{t}y$ is the intermediate output that is used as an input in the next period. Thus we may write

$$
\begin{aligned}
P^t\left(x^t, {}_{t-1}^{\;t}y\right) = \Big\{\left({}^{f}_{t}y + {}^{i}_{t}y\right) : \\
\left({}^{f}_{t}y_m + {}^{i}_{t}y_m\right) \leq \sum_{k=1}^{K^t} z_k^t\left({}^{f}_{t}y_{km} + {}^{i}_{t}y_{km}\right) \qquad & m = 1, \ldots, M, \\
\sum_{k=1}^{K^t} z_k^{tt-1}{}_{i}y_{km} \leq {}^{t-1}{}_{i}y_{k'm}, \qquad & m = 1, \ldots, M, \quad , \\
\sum_{k=1}^{K^t} z_k^t x_{km}^t \leq x_{k'n}^t, \qquad & n = 1, \ldots, N, \\
z_k^t \geq 0, \qquad & k = 1, \ldots, K^t \Big\}
\end{aligned}
$$

where ${}^{f}_{t}y_{km}$, ${}^{i}_{t}y_{km}$, ${}_{t-1}^{\;i}y_{km}$, and x_{kn} are observed inputs and outputs. We allow the number of observations to differ between periods, hence the notation K^t. The output vector $({}^{f}_{t}y + {}^{i}_{t}y)$ is specified so that one can include M^t for all t, i.e., $M = M^{t-1} + M^t + \ldots$ by including appropriate zeroes.

Recall from the static model, that if the sub-technologies have the properties *(i)–(vii)*, then the whole network model also has those properties. This observation also applies to the dynamic model below:

$$
\begin{aligned}
P^t\left(x^{t-1}, x^t, x^{t+1}, {}^{i}\bar{y}\right) = \Big\{ \Big({}_{t-1}^{\,f}y\, {}_{t}^{f}y, \left({}_{t+1}^{\,f}y + {}_{t+1}^{\,i}y \right) \Big) : \\
\left({}_{t-1}^{\,f}y_m + {}_{t-1}^{\,i}y_m \right) \leq \sum_{k=1}^{K^{t-1}} z_k^{t-1} \left({}_{t-1}^{\,f}y_{km} + {}_{t-1}^{\,i}y_{km} \right) \quad & m = 1, \ldots, M, \\
\sum_{k=1}^{K^{t-1}} z_k^{t-1}\, {}^{i}y_{km} \leq {}^{i}\bar{y}_{km}, \quad & m = 1, \ldots, M, \\
\sum_{k=1}^{K^{t-1}} z_k^{t-1} x_{kn}^{t-1} \leq x_n^{t-1}, \quad & n = 1, \ldots, N, \\
z_k^{t-1} \geq 0, \quad & k = 1, \ldots, K^{t-1}, \\
\left({}_{t}^{f}y_m + {}_{t}^{i}y_m \right) \leq \sum_{k=1}^{K^{t}} z_k^{t} \left({}_{t}^{f}y_{km} + {}_{t}^{i}y_{km} \right) \quad & m = 1, \ldots, M, \\
\sum_{k=1}^{K^{t}} z_k^{t}\, {}_{t-1}^{\,i}y_{km} \leq {}_{t-1}^{\,i}\bar{y}_{m}, \quad & m = 1, \ldots, M, \\
\sum_{k=1}^{K^{t}} z_k^{t} x_{kn}^{t} \leq x_n^{t}, \quad & n = 1, \ldots, N, \\
z_k^{t} \geq 0, \quad & k = 1, \ldots, K^{t}, \\
\left({}_{t+1}^{\,f}y_m + {}_{t+1}^{\,i}y_m \right) \leq \sum_{k=1}^{K^{t+1}} z_k^{t+1} \left({}_{t+1}^{\,f}y_{km} + {}_{t+1}^{\,i}y_{km} \right) \quad & m = 1, \ldots, M, \\
\sum_{k=1}^{K^{t+1}} z_k^{t+1}\, {}_{t}^{i}y_{km} \leq {}_{t}^{i}\bar{y}_{km}, \quad & m = 1, \ldots, M, \\
\sum_{k=1}^{K^{t+1}} z_k^{t+1} x_{kn}^{t+1} \leq x_n^{t+1}, \quad & n = 1, \ldots, N, \\
z_k^{t+1} \geq 0, \quad & k = 1, \ldots, K^{t+1} \Big\},
\end{aligned}
$$

Note that each sub-technology has its own intensity vector z^{τ}, $\tau = t - 1$, t, $t + 1$, and that the interaction between time periods comes through the intermediate outputs.

Färe and Grosskopf (1997) use this model to study the inefficiency of APEC countries due to dynamic misallocation of resources. They used the sum of Shephard (1970) sub-technology distance functions as their optimization criterion. Nemota and Goto (2003) applied the dynamic network model to study Japanese electricity production over time. They used cost minimization for the optimization criterion. Jaenicke (2000) applied the dynamic model in the analysis of the yield effects of crop rotation. Nemota and Goto (1999) applied the dual linear programming problem formulation to the cost minimization and derived the fundamental equation (Hamilton-Jacobi-Bellman) of dynamic programming. We also refer the reader to the work of J.K. Sengupta. A search for "J.K. Sengupta dynamic models" using any one of the major internet search engines produces a large number of dynamic non-parametric models that he has analyzed.

References

Bogetoft, P., & Wang, D. (2005). Estimating the potential gains from mergers. *Journal of Productivity Analysis, 23*(2), 145–171.

Chambers, R. G., Chung, Y., & Färe, R. (1998). Profit, directional distance functions, and Nerlovian efficiency. *Journal of Optimization Theory and Applications, 98*(2), 351–364.

Charnes, A., Copper, W. W., & Rhodes, E. (1978). Measuring the efficiency of decision making units. *European Journal of Operational Research, 2*, 429–444.

Chen, Y., Cook, W. D., Li, N., & Zhu, J. (2009). Additive efficiency decomposition in two-stage DEA. *European Journal of Operational Research, 196*(3), 1170–1176.

Cook, W. D., Liang, L., & Zhu, J. (2010). Measuring performance of two-stage network structures by DEA: A review and future perspective. *Omega, 38*(6), 423–430.

Cooper, W. W., Seiford, L., & Zhu, J. (2004). *Handbook on data envelopment analysis*. Boston: Kluwer Academic.

Fallah-Fini, S., Triantis, K., & Johnson, A. L. (2013). Reviewing the literature on non-parametric dynamic efficiency measurement: state-of-the-art. *Journal of Productivity Analysis*, in print.

Färe, R., Grosskopf, S., & Brännlund R. et al. (1996a). *Intertemporal production frontiers: With dynamic DEA*. Boston/London/Dordrecht: Kluwer Academic.

Färe, R., & Grosskopf, S. (1996). Productivity and intermediate products: A frontier approach. *Economics Letters, 50*(1), 65–70.

Färe, R., & Grosskopf, S. (1997). Efficiency and productivity in rich and poor countries. In B. S. Jensen & K. Wong (Eds.), *Dynamics, economic growth, and international trade*, (pp. 243–263). Ann Arbor: University of Michigan Press, Studies in International Economics.

Färe, R., & Grosskopf, S. (2004). *New directions: Efficiency and productivity*. Boston: Kluwer Academic.

Färe, R., & Grosskopf, S. (2010). Directional distance functions and slacks-based measures of efficiency. *European Journal Of Operational Research, 200*(1), 320–322.

Färe, R., & Lovell, C. A. K. (1978). Measuring the technical efficiency of production. *Journal of Economic Theory, 19*(1), 150.

Färe, R., Grosskopf, S., & Li, S.-K. (1992). Linear programming models for firm and industry performance. *Scandinavian Journal of Economics, 94*(4), 599–608.

Färe, R., Grosskopf, S., & Whittaker, G. (2007a). Network DEA. In J. Zhu & W. Cook (Eds.), *Modeling data irregularities and structural complexities in data envelopment analysis* (pp. 209–240). New York: Springer.

Färe, R., Grosskopf, S., Zelenyuk, V., Färe, R., Grosskopf, S., & Zelenyuk, V. (2007b). Finding common ground: Efficiency indices. In R. Färe, S. Grosskopf, D. Primont, R. Färe, S. Grosskopf, & D. Primont (Eds.), *Aggregation, efficiency, and measurement* (pp. 83–95). New York: Springer.

Färe, R., Grosskopf, S., & Margaritis, D. (2011). Coalition formation and data envelopment analysis. *Journal of Centrum Cathedra, 4*, 216–223.

Färe, R., Grosskopf, S., & Whittaker, G. (2013). Directional output distance functions: endogenous direction based on exogenous normalization constraints. *Journal of Productivity Analysis, 40*(3), 267–269.

Farrell, M. J. (1957). The measurement of productive efficiency. *Journal of the Royal Statistical Society, 120*(3), 253–281.

Jaenicke, E. C. (2000). Testing for intermediate outputs in dynamic DEA models: Accounting for soil capital in rotational crop production and productivity measures. *Journal of Productivity Analysis, 14*(3), 247–266.

Johansen, L. (1972). *An Integration of micro and macro short run and long run aspects*. Amsterdam: North-Holland Publishing.

Kao, C. (2009). Efficiency decomposition in network data envelopment analysis: A relational model. *European Journal of Operational Research, 192*(3), 949–962.

Kemeny, J. G., Morgenstern, O., & Thompson, G. L. (1956). A generalization of the von Neumann model of an expanding economy. *Econometrica, 24*(2), 115–135.

Luenberger, D. G. (1995). *Microeconomic theory*. New York: McGraw-Hill.

Nemota, J., & Goto, M. (1999). Dynamic data envelopment analysis: Modeling intertemporal behavior of a firm in the presence of productive inefficiencies. *Economic Letters, 64*, 51–56.

Nemota, J., & Goto, M. (2003). Measuring dynamic efficiency in production: An application of data envelopment analysis to Japanese electric utilities. *Journal of Productivity Analysis, 19*, 191–210.

Shephard, R. W. (1953). *Cost and production functions*. Princeton: Princeton University Press.

Shephard, R. W. (1970). *Theory of cost and production functions*. Princeton: Princeton University Press.

Shephard, R. W., & Färe, R. (1974). The law of diminishing returns. *Zeitschrift für Nationalökonomie Journal of Economics, 34*(1–2), 69–90.

Shephard, R. W., & Färe, R. (1975). *A dynamic theory of production correspondences*. ORC 75–13, Berkeley: Operations Research Center, University of California.

Shephard, R. W., & Färe, R. (1980). *Dynamic theory of production correspondences*. Königstein: Verlag Anton Hain.

Tone, K. (2001). Slacks-based measure of efficiency in data envelopment analysis. *European Journal of Operational Research, 130*(3), 498–509.

Tone, K., & Tsutsui, M. (2014). Dynamic DEA with network structure: A slacks-based measure approach. *Omega, 42*(1), 124–131.

Chapter 15
Network, Shared Flow and Multi-level DEA Models: A Critical Review

Lorenzo Castelli and Raffaele Pesenti

Abstract In the last two decades, complex and detailed DEA models that consider the internal structure of DMUs have been proposed by several authors. This chapter describes the mathematical formulations, along with their main variants, extensions and applications, of three large and popular model families: network (with special emphasis on multi-stage), shared flow (also known as multi-component or multi-activity), and multi-level models. Each family is a different generalization of the same elementary internal structure. This review extends and updates the classification presented in Castelli et al. (Ann Oper Res 173(1):207–235, 2010).

Keywords Data envelopment analysis • Network-DEA • Shared-flows • Multi-level • Multi-stage • Multi-component • Survey

15.1 Introduction

Data Envelopment Analysis (DEA) has been a standard tool for evaluating the relative efficiencies of Decision Making Units (DMUs) since the paper of Charnes et al. (1978) based on the seminal work of Farrell (1957). Some underlying assumptions are common to standard DEA models. The *efficiency* of a DMU is defined as the weighted ratio of the outputs (products or outcomes) yielded by the DMU over the inputs (resources used or consumed). DMUs are *homogeneous*, i.e., they all have the same types of inputs and outputs, and *independent*, i.e., no constraint binds input

L. Castelli (✉)
Dipartimento di Ingegneria e Architettura, Università degli Studi di Trieste,
Via A.Valerio 10, 34127 Trieste, Italy
e-mail: castelli@units.it

R. Pesenti
Dipartimento di Management, Università Ca' Foscari di Venezia,
San Giobbe – Cannaregio 837, 30121 Venezia, Italy

W.D. Cook and J. Zhu (eds.), *Data Envelopment Analysis*,
International Series in Operations Research & Management Science 208,
DOI 10.1007/978-1-4899-8068-7_15,

and output levels of a DMU with the inputs and outputs of other DMUs. Furthermore, DMUs are seen as *black boxes*, i.e., their internal structures are not considered. As a consequence, generally, there is no clear evidence of the transformations to which the inputs are subject to within the DMUs.

In the last two decades, several authors have explored the possibility of abandoning the black box perspective and of considering the internal structures of the DMUs (see, e.g., Sect. 3 of the paper by Cook and Seiford (2009) devoted to the major research thrusts over 30 years since the work of Charnes et al. (1978), or the dedicated chapter in the book by Cook et al. (2007) or, finally, the specific subsections in the citation-based DEA literature survey by Liu et al. (2013b)). These authors justify their approach by observing that, in some particular contexts, the knowledge of the internal structure of DMUs can give further insights for the DMU performance evaluation.

The aim of this chapter is to survey the models that consider internal structures of DMUs. The main rationale of the classification is driven by identifying three families of models as different generalizations of the same elementary formulation.

In particular, we analyze a specific model by comparing a set of homogeneous and independent DMUs, each composed of a set of Decision Making SubUnits (DMSUs). In the literature, subunit, component, activity, division, (sub)structure and (sub)process are synonyms of DMSU and are reported as such in this review. Each subunit is allowed to perform a unique function or activity. Only to keep the notation simple, we also assume that all the DMUs under comparison have the same internal structure.

All the models that we consider can be derived from an *elementary* one that assumes that each DMU internal structure complies with the following assumptions:

Assumptions

1. No intermediate flows among DMSUs exist. In other words, the output of a DMSU cannot be the input of another DMSU (and also cannot re-enter the same DMSU).
2. All the subunits of the DMU do not have shared inputs and shared outputs, i.e., the DMU does not have the opportunity to decide how to allocate its inputs or outputs among its subunits in order to maximize its efficiency (Cook et al. 2000).
3. Any input (output) of the DMU is also an input (output) of one of its subunits.

Here note that Assumption 2 implies that the components of an elementary DMU do not compete for the same resource and do not synergically yield the same product. It follows that the combined presence of Assumptions 1 and 2 makes all the DMSUs of an elementary DMU independent (Fig. 15.1).

By dropping one of the three above assumptions at a time, we obtain different families of DEA models. Specifically:

- We refer to *network* DEA models when Assumption 1 is neglected. Here DMUs have at least one output of a DMSU which is an input of a different DMSU (see Fig. 15.2). These models are of interest because they also allow to describe

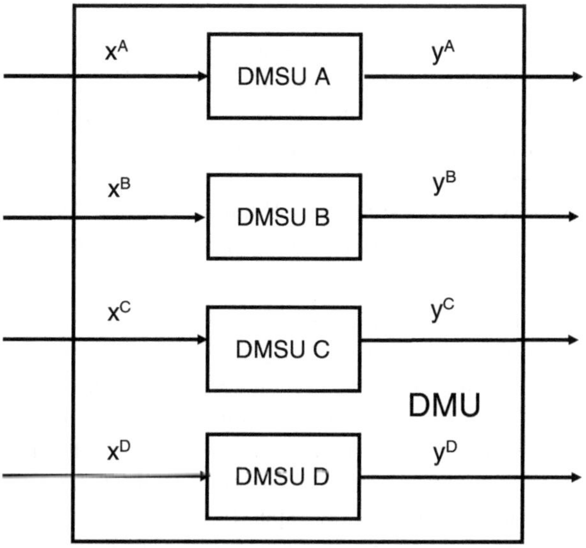

Fig. 15.1 Elementary DMU

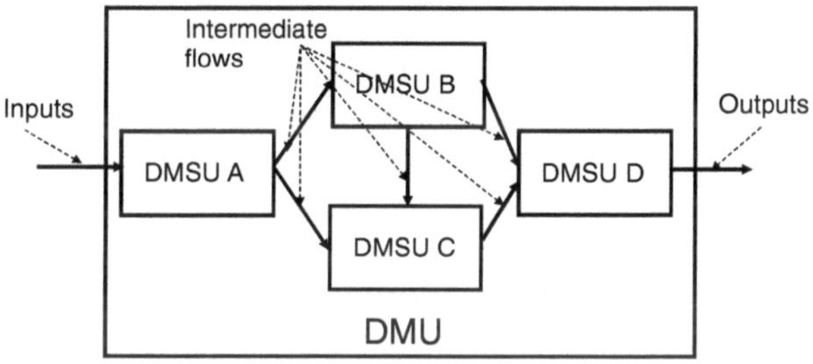

Fig. 15.2 Network DMU

systems where the DMUs are organized in networks so that the outputs of some of them become inputs for other ones. This framework may encompass manufacturing production systems, and in general supply chains, in which some DMUs yield intermediate products that feed other DMUs. The same framework may also include the dynamic DEA models in which some outputs of the DMUs at period t become their inputs in the next period, $t + 1$ (Färe and Grosskopf 2000). Finally, this framework may also possibly cover a further line of research (not discussed in this chapter) that is in fact not specifically devoted to just assessing the efficiency of DMUs but to considering DMUs as components of a greater structure which is interested in maximizing its future efficiency by either re–allocating resources or fixing targets to DMUs (see, e.g., Sect. 5 of the previous version of this survey (Castelli et al. 2010)).

- We refer to *shared flow (or multi-activity or multi-component)* DEA models when Assumption 2 no longer holds (see Fig. 15.4). As an example, this situation may occur when DMUs are divided into different components that require common resources (e.g., money) or produce goods or services obtained through the synergy and collaboration among them (e.g., the quality of service provided to customers). Then, a DMU may maximize its efficiency also by choosing the most appropriate allocation of the common flows among the subunits and not only by optimizing the weight associated to each flow as it happens in standard DEA.

- We refer to *multi-level* DEA models when Assumption 3 is dropped, i.e., when DMU inputs (outputs) are not necessarily inputs (outputs) of its subunits (see Fig. 15.8).

In the following sections, we first describe the formulation to maximize the relative efficiency of an elementary DMU (Sect. 15.2). Then we introduce the basic reference models (typically with constant returns to scale) for network (Sect. 15.3), shared flow (Sect. 15.5), and multi-level (Sect. 15.6) DEA models. Section 15.4 is specifically devoted to the evaluation of multi-stage processes, a special class of network DEA models. We provide interpretations and applications proposed by different authors, and specify the possible variations from the basic model. In Sect. 15.7 conclusions are drawn. Throughout the paper we assume that the reader is familiar with at least the seminal works on DEA (see, e.g., Banker et al. 1984; Charnes et al. 1978), as we will not define or justify basic concepts such as, e.g., positive non-Archimedean value ε, slack variables, production set, virtual inputs and outputs, Constant or Variable Returns to Scale (CRS or VRS, respectively), allocative and technical efficiencies.

15.2 Elementary Model

In this section, we introduce a DEA model for assessing the efficiency of elementary DMUs (i.e., whose internal structure follows Assumptions 1–3). To this end, for each elementary DMU k, let us define

- i, j, r: the indexes of the generic input, output, and DMSU, respectively,
- $X_k^r = \{x_{ik}^r\}$: the vector of the inputs of DMSU r,
- $Y_k^r = \{y_{jk}^r\}$: the vector of the outputs of DMSU r,
- $\nu^r = \{\nu_i^r\}$: the vector of weights of the inputs of DMSU r
- $\mu^r = \{\mu_j^r\}$: the vector of weights of the outputs of DMSU r.

For an elementary DMU 0 belonging to a set of N homogeneous and independent DMUs with the same internal structure, the CRS input-oriented version of the envelopment-based DEA model can be written as:

$$\theta_0^* = \min \theta_0 - \varepsilon\left(\sum_r\left(\sum_i s_i^{r-} + \sum_j s_j^{r+}\right)\right) \tag{15.1a}$$

$$\sum_k \lambda_k^r x_{ik}^r = \theta_0 x_{i0}^r - s_i^{r-} \quad \forall i, r \tag{15.1b}$$

$$\sum_k \lambda_k^r y_{jk}^r = y_{j0}^r + s_j^{r+} \quad \forall j, r \tag{15.1c}$$

$$\lambda_k^r, s_i^{r-}, s_j^{r+} \geq 0 \quad \forall i, j, k, r \tag{15.1d}$$

where λ_k^r is the multiplier of DMSU r belonging to DMU k, and s_i^{r-}, s_j^{r+} are the slack variables.

The dual formulation of (15.1) is the following multiplier-based DEA model:

$$e_0^* = \max \sum_{j,r} \mu_j^r y_{j0}^r \tag{15.2a}$$

$$\sum_{i,r} \nu_i^r x_{i0}^r = 1 \tag{15.2b}$$

$$\sum_j \mu_j^r y_{jk}^r \leq \sum_i \nu_i^r x_{ik}^r \quad \forall k, r \tag{15.2c}$$

$$\nu_i^r, \mu_j^r \geq \varepsilon \quad \forall i, j, r. \tag{15.2d}$$

In Model (15.2) the maximum relative efficiency e_0^* is assessed by comparing DMU 0 with all the existing subunits. Then, as shown in Yang et al. (2000), Castelli et al. (2004), and Kao (2009b), e_0^* is equal to the maximum relative efficiency of its subunits, and DMU 0 is:

- Weakly efficient if and only if there exists at least one of its subunits which is weakly efficient relative to the corresponding subunits of other DMUs;
- CRS-efficient if and only if each of its subunits is CRS-efficient relative to the corresponding subunits of other DMUs.

A multiple input and single output elementary configuration is also proposed by Färe and Primont (1984). Specifically, the authors, relying on the Farrell (1957) output-based efficiency measure, construct a reference technology for DMUs using their subunit data. Next, they compare this efficient technology against the reference frontier of the subunits, i.e., as if the subunits were independent DMUs and not part of a larger DMU. Kao (2000) generalizes this model for cases of multiple outputs and multiple inputs.

15.3 Network DEA Models

In this section, we describe DEA models for DMUs that present intermediate flows between subunits. In this case, the subunits are neither independent nor homogeneous. They are interdependent in the sense that part of the output produced by some of them may be used as an input by other ones. In addition, their interdependency leads to their non-homogeneity as they may present different inputs and/or different outputs.

The basic network DEA models have been introduced by Färe (1991), Färe and Whittaker (1995) and Färe and Grosskopf (1996b). These models represent DMUs composed of two consecutive subunits with one intermediate flow: the output from the first subunit is used as input in the second one. Then, Färe and Grosskopf (2000) extend these models to consider DMUs made of more subunits (see also the book by Färe et al. 2007).

Since the above seminal papers, many different models, both envelopment- and multiplier-based, have appeared in the literature. Here, as an illustrative example, we provide a CRS envelopment-based (input oriented) model under the assumption that all DMUs have exactly the same internal structure in terms of DMSUs. Specifically, we assess the relative efficiency θ_0^* of the whole DMU 0 using the following notation: for each DMU k, r indicates a generic DMSU of k, then x_{ik}^r is the amount of the i-th external input of the DMU entering subunit r, y_{jk}^r is the amount of the j-th final output of the DMU produced by subunit r, and z_{lk}^{rt} is the l-th intermediate flow of DMU produced by subunit r and used by subunit t (Fig. 15.2); *pred*(r) represents the set of predecessors of subunit r, i.e., the set of subunits which have at least one output used as input by subunit r, similarly, *succ*(r) is the set of successors of subunit r; finally s_k^r are slack variables.

$$\theta_0^* = \min \theta_0 - \varepsilon \sum_r \sum_i s_i^{r-} \tag{15.3a}$$

$$\sum_k \lambda_k^r x_{ik}^r = \theta_0 x_{i0}^r - s_i^{r-} \quad \forall i, r \tag{15.3b}$$

$$\sum_k \lambda_k^r \sum_{t \in pred(r)} z_{lk}^{tr} = \sum_{t \in pred(r)} z_{l0}^{tr} - s_l^{r-} \quad \forall l, r \tag{15.3c}$$

$$\sum_k \lambda_k^r \sum_{t \in succ(r)} z_{lk}^{rt} = \sum_{t \in succ(r)} z_{l0}^{rt} + s_l^{r+} \quad \forall l, r \tag{15.3d}$$

$$\sum_k \lambda_k^r y_{jk}^r = y_{j0}^r + s_j^{r+} \quad \forall j, r \tag{15.3e}$$

$$s_i^{r-}, s_l^{r-}, s_l^{r+}, s_j^{r+}, \lambda_k^r \geq 0 \quad \forall k, i, j, l, r. \tag{15.3f}$$

As for standard envelopment-based DEA formulations, model (15.3) considers a radial measure of efficiency (as ε is a positive non-Archimedean parameter) and is based upon the definition of the Production Possibility Set (PPS) of DMU 0. Indeed, provided that $\theta_0 = 1$, constraints (15.3b)–(15.3f) describe the PPS of DMU 0 in the following terms. For each subunit r, constraints (15.3b) and (15.3c) indicate that the value of each external input flow i or intermediate input flow l cannot be less than the conic combination of the values of the corresponding input flows of the analogous DMSUs r from all the observed DMUs. Similarly, constraints (15.3d) and (15.3e) indicate that the value of each final output flow j or intermediate output flow l cannot exceed a conic combination of the values of the corresponding output flows of the analogous DMSUs r from all the observed DMUs. Model (15.3) describes a *closed* (network) process since each subunit either receives only external input flows or only intermediate flows and, analogously, it either produces only final output flows or only intermediate flows. Model (15.3) trivially generalizes the model proposed by Färe and Whittaker (1995), where the slack variables are omitted, and implies that the observed DMUs and their DMSUs exhibit constant returns to scale (CRS) and strong disposability of inputs and outputs (see Färe and Grosskopf 1996b).

In the rest of the paper, for both CRS and VRS situations we will introduce envelopment- and multiplier-based DEA models. Differently from the standard DEA models, the multiplier- and envelopment-based network DEA models are not, in general, dual of each others (Chen et al. 2010a, 2013b). They represent two different approaches that may produce different efficiency results. For this reason, Chen et al. (2010a, 2013b) suggest that envelopment based network DEA models should be used for determining the frontier projection for inefficient DMUs. Differently, multiplier-based network DEA models should be used for determining the DMSU (called *division* by the authors) efficiency. In addition, the authors also point out that, contrary to what it is sometimes suggested, it is not sufficient to add convexity constraints to an envelopment-based network DEA model or free variables to a multiplier-based network DEA model to make these models capable of describing VRS network processes.

15.3.1 Non-radial Measures of Efficiency

Leaving aside the radial measure of efficiency considered in model (15.3), some authors propose different non-radial measures of efficiency for network DEA models.

Tone and Tsutsui (2009) introduce a VRS *Slack-Based Measure* (SBM) of efficiency. Following Pastor et al. (1999) for standard DEA models, this efficiency measure is a function of the slack variables and is appropriate when we employ flows, such as labor, materials and capital, that are substitutional and do not change proportionally. Specifically, Tone and Tsutsui (2009) substitute objective (15.3a) with

$$\theta_0^* = \min_{\lambda, s} \sum_r w_r \left(1 - \frac{1}{m_r} \sum_{i=1}^{m_r} \frac{s_i^{r-}}{x_{i0}^r} \right), \tag{15.4}$$

where, for each subunit r, w_r is a constant parameter that weighs the relative importance of the subunit and m_r is the number of its inputs. Even though objective (15.4) is adequate only for an input oriented model, Tone and

Tsutsui (2009) propose analogous measures of efficiencies for output oriented and non-oriented models. They also introduce a *discretionary formulation* that is applied when the DMU 0 under assessment may decide the values of its intermediate flows in the light of other DMUs' intermediate flow values.

In practice, the discretionary formulation requires the substitution of the two sets of constraints (15.3c) and (15.3d) with constraints

$$\sum_k \lambda_k^r \sum_{t \in pred(r)} z_{lk}^{tr} = \sum_k \lambda_k^r \sum_{t \in succ(r)} z_{lk}^{rt} \quad \forall l, r. \tag{15.5}$$

Tone and Tsutsui (2009) finally claim that their approach has the further advantage that it can be trivially modified to also model CRS processes.

Fukuyama and Weber (2010) introduce the network *directional* slack-based measures. In these measures, the values of the slack variables are normalized on the basis of user defined coefficients. For example, the coefficients $x_{i0}{}^r$ in objective (15.4) would be substituted by generic positive coefficients $g_i{}^r$, being the vector $g_x = \{g_i{}^r\}$ the desired direction of input contraction. These efficiency measures are then extended to account also for possible undesirable (or *bad*) outputs.

In a paper addressing sensitivity analysis in network DEA models, Avkiran and McCrystal (2012) introduce a *Range Adjusted Measure* (RAM) of efficiency. This measure builds upon the one by Cooper et al. (1999) for standard DEA models and, again, it is a function of the values of the slack variables. Then, the authors compare the results obtained with the application of sensitivity analysis to envelopment-based RAM network DEA models and to corresponding SBM network DEA models.

15.3.2 Simultaneous Evaluation of DMU and DMSU Efficiencies

Some authors specifically focus their work on developing models aiming at evaluating subunit efficiencies and at studying the influence of such values to the efficiency of the DMU the subunits belong to.

Their research is justified by the following facts. The knowledge of the internal structure of the observed DMUs allows to determine whether better performances could be obtained by a DMU that merged the technologies of the most efficient substructures of the observed DMUs. In addition, the assessment of the efficiency of each subunit might prevent that in a DMU the inefficiency of some of its DMSUs may be compensated by the efficiency of others.

Castelli et al. (2001) introduce a DEA-like model to compare non-homogenous and interdependent subunits belonging to the same DMU. A given subunit r may be evaluated according to three different sets: (a) all the subunits homogeneous to it, (b) all the subunits of the DMU, and (c) with respect to a given output, all the subunits yielding that output. In this last case, the rationale is that these subunits, although not necessarily homogeneous, have a certain degree of commonality because they can be considered as potential substitutes for each other, as far as the production of that output is concerned. Thus the interest in comparing them. As a possible limitation, Lewis and Sexton (2004) point out that this approach may lead to small reference sets. Castelli et al. (2001) also link the subunits' and DMU efficiencies by defining an efficiency value W obtained by maximizing the product of the efficiency of the subunit under evaluation and the efficiency of the DMU it belongs to. In this way, subunits not only maximize their own efficiency, but also positively contribute to the efficiency of the whole system they are part of. Indeed, the authors prove that a subunit seeking to optimize its W efficiency behaves with a benevolent attitude, i.e., being equal to other conditions, it also maximizes a combination of the efficiencies of the other subunits. In addition, the authors show that the whole DMU is efficient if and only if all its subunits are W efficient.

Sexton and Lewis (2003) and Lewis and Sexton (2004) explicitly compute the efficiencies of the subunits using both input and output oriented formulations. Their basic models can be seen as an adaptation of model (15.3), where the efficiency of subunit r belonging to DMU 0 is optimized and constraints (15.3b)–(15.3e) are adequately rewritten. In a simple case of DMUs composed of two subunits S_1 and S_2 in series, the authors show that DMU 0 is efficient when its output values are equal to the output values produced in the case that S_2 is efficient and uses the intermediate product levels that it would have received, had S_1 been efficient. Lewis and Sexton (2004) describe the internal structure of each DMU as an acyclic direct graph. This graph has a node for each subunit plus one origin and one destination node. In this case, the authors show that a necessary (but not sufficient) condition for a whole DMU to be efficient is the existence of a path from origin to destination along with every subunit is efficient. As a consequence, it is possible that, when considering the internal structure, all DMUs under evaluation are inefficient. Lewis et al. (2009) use the model presented in Lewis and Sexton (2004) to assess simultaneously organizational capability, efficiency, and effectiveness in Major League Baseball.

Kao (2009a) proposes a *relational* approach (see also Kao and Hwang 2008, Sect. 15.4.2.1), whose underlying concept is that some relationship exists between the measure of the overall DMU efficiency and the measure of its DMSUs' efficiencies, for example, a simple multiplication, as in Kao and Hwang (2008), or a weighted average, as in Chen et al. (2009a). The authors' assumptions imply that the relational network DEA models, when formulated with a multiplier-based measure of efficiency, are also characterized by the fact that the same flows have associated the same weights no matter which subunits these flows belong to. In other words, an intermediate flow presents the same weight both when is considered as an output flow of a DMSU and when is considered as an input flow of a different DMSU. In the same context, Lozano (2011) introduces an envelopment-based relational network DEA model to asses the technical, scale, cost and allocative efficiency scores of the DMUs. To this end, he proposes an axiomatic approach to define the PPS of a DMU through the composition of the PPS of each of the DMUs' subunits. Then, Lozano et al. (2013) generalize the previous model to take into account processes with undesirable outputs and apply this new model to assess airport performances.

Relying on work by Chen et al. (2010a), Fukuyama and Mirdehghan (2012) propose a two-phase slacks-based network model to assess the efficiency of a set of DMUs and their DMSUs. To this aim, the authors first consider an additive envelopment-based network DEA model that optimizes the slacks of exogenous inputs and final outputs. Then, they use a linear programming model to assess the efficiency status of each DMSU.

Most recently, Kao and Chan (2013) have introduced a multi-objective programming method that computes both the overall efficiencies of the DMUs and the divisional efficiencies of the DMSUs of network DEA models.

15.3.3 DMSU Ownership

Some authors have considered the consequences of having DMSUs of the same DMU controlled by different agents that may have different agendas (see also the game theory approach discussed in Sect. 15.4.2.5).

Chang et al. (2011) discuss the importance of taking into account the ownership of the different DMSUs composing the DMU under assessment. The authors argue that an agent interested in assessing the efficiency of a DMU cannot include in her DEA model the external inputs and final outputs of the DMSUs that she does not own. In fact, these flows are usually unknown to her. Differently, she can assume the knowledge of the internal flows if they are regulated by contracts between the different DMSUs. Accordingly, the authors introduce three ownership-specified (centralized, distributed and hybrid) network DEA models which take into account the different possibility of ownership of the DMSUs. Similar problems are also considered by Chen and Yan (2011). These latter authors are motivated by the necessity of assessing the efficiency of DMUs representing supply chains. As main result, they prove that a supply chain is weakly efficient only if there exists a path from the external inputs to the final outputs along which all DMSUs are weakly efficient. They also show that it is never appropriate to ignore the internal structure of the supply chain. In fact, standard DEA models may lead to overrate the efficiency of a supply chain not only when different agents pursuing their own agenda own the different DMSUs, but also when all the DMSUs are owned by a single agent and the DMSUs are then centralized controlled.

15.3.4 Applications

Not surprisingly, many network DEA models have been proposed to evaluate the performances of different processes, applied in particular to the top-five industries (transportation, banking, agriculture and farm, healthcare, and education) addressed by the standard DEA literature (Liu et al. 2013a).

15.3.4.1 Hospitality and Transportation Industries

Hsieh and Lin (2010) apply a relational network model to the tourist hotels in Taiwan and present a survey on the efficiency assessment in the hospitality industry. In the same context, Zhang and Ma (2011) apply a network DEA model to assess the business efficiency of Chinese hotel and tourism firms.

Yu (2008a) compares the results obtained through standard DEA and network DEA models for assessing the performances of 40 global railways in terms of technical efficiency, service effectiveness, and technical effectiveness. The author suggests to include the environmental factors as non-discretionary inputs and underlines how the network DEA model provides deeper insight regarding the sources of inefficiency. The importance of environmental factors is discussed also in Yu (2010) where a network DEA model is proposed to deal with both production and service efficiency in airports.

Sheth et al. (2007) apply network DEA models to the assessment of bus routes by expanding the Färe and Grosskopf (2000) approach to account for the different perspectives of operators and users, and for multiple goals. Hahn et al. (2011) propose a network DEA model to assess the efficiency of Seoul arterial bus routes. Zhao et al. (2011) assess the efficiency of a transportation system by considering the perspectives of the different stakeholders, such as transportation service providers, users, and the community. The authors propose a model that includes undesirable outputs and where the different perspectives are inter-related through intermediate flows. Finally, Li (2012) uses a network DEA model to assess the China's railway transport industry.

15.3.4.2 Production of Goods or Services

Färe and Whittaker (1995) apply a model similar to model (15.3) to a diary production problem and compare the result obtained with the ones obtained with a standard DEA model. The former model turns out to have greater discrimination power: only 23 % of the DMUs are on the efficiency frontier compared to 65 % when intermediate flows are not explicitly taken into account.

Liu et al. (2012) introduce a network DEA model to assess non-profit farmers associations in Taiwan. Lin and Chiu (2013) and Matthews (2013) propose SBM network DEA models (see Sect. 15.3.1) to improve Taiwan bank performance evaluation and assess Chinese bank income efficiency, respectively. Matthews (2013) uses metrics of risk management practice and risk management organization as intermediate inputs. Vaz et al. (2010) exploit a network DEA model to assess the performances of retail stores, and Lee and Johnson (2012) use a relational network model (see Sect. 15.3.2) to decompose the efficiency of profitability for a general production system.

Löthgren and Tambour (1999) estimate efficiency and productivity for a sample of Swedish pharmacies taking also into account customer satisfaction. The pharmacy technology is represented by a production and a consumption node. The production node yields (final) outputs (e.g., outpatient prescriptions) and also produces nonmarketable characteristics and attributes (e.g., the service level) that are considered as intermediate inputs of the consumption node. Together with external inputs (e.g., customer-service labor hours) the consumption node provides customer quality assessments on pharmacy service.

Chilingerian and Sherman (2004) study network DEA applications to health care systems.

15.3.4.3 Governmental Entities

Prieto and Zofio (2007) employ the network DEA model introduced by Färe and Grosskopf (2000) to assess the economies of a set of countries belonging to the Organisation for Economic Co-operation and Development with the aim of identifying best practices. Each national economy is described in terms of a network where different nodes use primary inputs to produce intermediate input and outputs, and satisfy final demand. Each node represents a basic economic sector, such as agriculture, manufacturing, construction, and services.

Guan and Chen (2010, 2012) and Chen and Guan (2012) apply network DEA to measure the efficiency of China's regional innovation systems, whereas Amatatsu et al. (2012) assess efficiency and returns-to-scale of Japanese local governments.

15.4 Multi-stage Network DEA Models

In this section, we introduce a particular type of network DEA models that thanks to their simple structure and wide applicability have been extensively studied in the past years. Specifically, we consider the two-stage, or in general multi-stage, network DEA models. Such models assume that each DMU is composed of two, or in general more, consecutive stages, each one being a single DMSU or a set of parallel DMSUs (Fig. 15.3). For example, the very first network DEA models by

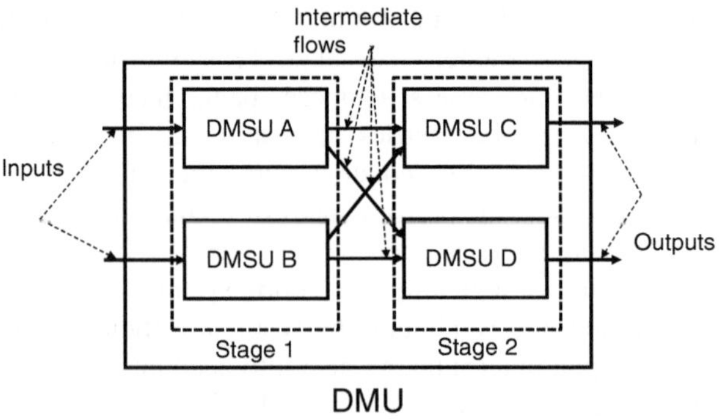

Fig. 15.3 Two-stage DMU

Färe (1991), Färe and Whittaker (1995) and Färe and Grosskopf (1996b) introduced in Sect. 15.3 are two-stage models. Multi-stage models are obviously used for the efficiency evaluation of multi-stage processes but they are also studied because they can model the evolution of processes over time. In this latter case the multi-stage models are referred to as dynamic network models and each stage represents the same DMU at different times (see Sect. 15.4.4).

Seiford and Zhu (1999) are among the first authors to deal with multi-stage processes. They consider each stage and the whole DMU as independent and evaluate the efficiencies of these structures using conventional DEA models. Differently, most of the papers that we analyze in this section take into account some form of interaction between consecutive stages, as it is usually done in network DEA models. In this context, Cook et al. (2010a) and Agrell and Hatami-Marbini (2013) propose two interesting surveys on the multi-stage network literature, respectively considering a game theoretic and a supply chain management perspective. Some alternative DEA models for two-stage process are also surveyed in Wang and Chin (2010).

As an illustrative example, we provide a general model for a two-stage process where each stage is made in turn of parallel DMSUs (see Fig. 15.3). To this aim, we extend the notation used in model (15.3). For each DMU k, x^f_{ik} is the amount of the i-th input of the DMU entering subunit f in the first stage; y^s_{jk} is the amount of the j-th output of the DMU produced by subunit s in the second stage; z^f_{lk} is the amount of the l-th intermediate flow of the DMU output of the subunit f of the first stage; whereas z^f_{lk} is the amount of the same flow input of the subunit s of the second stage; finally, z^{fs}_{lk} is the amount of flow l of subunit f that feeds subunit s, with $\sum_s z^{fs}_{lk} = z^f_{hk}$.

We define the efficiency of DMU 0 as

$$e^*_0 = \frac{\sum_j \sum_s \mu_{js} y^s_{j0}}{\sum_i \sum_f \nu_{if} x^f_{i0}}. \tag{15.6}$$

Then a possible multiplier-based two-stage (input oriented) DEA model that assesses the relative efficiency of DMU 0 is

$$e^*_0 = \max \sum_j \sum_s \mu_{js} y^s_{j0} \tag{15.7a}$$

$$\sum_i \sum_f \nu_{if} x^f_{i0} = 1 \tag{15.7b}$$

$$\sum_l \upsilon_{lf} z^f_{lk} \le \sum_i \nu_{if} x^f_{ik} \quad \forall k,f \tag{15.7c}$$

$$\sum_j \mu_{js} y^s_{jk} \leq \sum_l \eta_{hl} z^s_{lk} \quad \forall k, s \tag{15.7d}$$

$$G(\upsilon_{lf}, z^f_{lk}, \eta_{ls}, z^s_{lk}) \geq 0 \quad \forall k, l, f, s \tag{15.7e}$$

$$\mu_{js}, \nu_{if}, \upsilon_{lf}, \eta_{ls} \geq \varepsilon \quad \forall i, j, l, f, s \tag{15.7f}$$

where conditions (15.7e) represent a set of constraints that link the output flows of the subunits of the first stage with the input flows of the subunits of the second stage. Note that the above formulation, as well as formulation (15.3) for general network structured DMUs, applies to closed processes.

15.4.1 Balancing Intermediate Flows

Castelli et al. (2004) consider two-stage processes where each second stage DMSU receives as input only intermediate flows and discuss two different kinds of constraints (15.7e): virtual *weights balancing constraints* and *flow balancing constraints*.

In the former case, the overall perceived values of the intermediate flows are balanced, i.e., virtual weights of the input flows of the second stage are equal to virtual weights of the feeding flows:

$$\eta_{ls} z^s_{lk} = \sum_f \upsilon_{lf} z^{fs}_{lk} \quad \forall k, l, s.$$

Under this assumption, the authors prove that the relative efficiency of the DMU under evaluation is equal to the product of the maximum relative efficiency of each single stage calculated according to model (15.2).

In the latter case, not only the perceived values but also the flows themselves are balanced:

$$\eta_{ls} = \upsilon_{lf} \quad \forall k, l, f, s \quad \text{and} \quad z^s_{lk} = \sum_f z^{fs}_{lk} \quad \forall k, l, s.$$

Under this second assumption, the authors prove that the relative efficiency of a DMU is assessed by comparing it with each observed DMUs together with all DMUs that could be obtained by composing their two stages with any possible combination of the subunits of the observed DMUs. Castelli et al. (2004) finally point out that the flow balancing constraints model can also be derived as dual of model (15.3) when specialized to a two-stage process, and claim that their results could be generalized to second stage DMSUs with multiple inputs.

15.4.2 Extensions

15.4.2.1 Relational Models

Kao and Hwang (2008) customize model (15.7) to multi-stage processes such that each stage includes a single DMSU, and each DMSU may have multiple inputs. In particular, in Kao and Hwang (2008) the relational approach (see Sect. 15.3.2) is introduced for the first time in the context of two-stage network DEA. In their paper, the authors also discuss possible solutions for dealing with multiple optimal weights. For instance, the authors suggest to choose as optimal the weights that maximize the efficiency of the first stage while maintaining the overall efficiency score of the DMU. Subsequently, Kao (2009a) extends the relational model in Kao and Hwang (2008) to series-parallel networks by utilizing dummy DMSUs such that a DMU structured as a network of DMSUs can be represented by a multi-stage structure where each stage can be composed of a set of parallel DMSUs. Here again, the flow balancing constraints are imposed and the same flow has the same weight all over the network, no matter if it is used as an input or as an output. Kao and Hwang (2010) apply the relational network model to assess information technology on firm performance in a banking industry.

Differently from Kao and Hwang (2008), which assess DMU and DMSU efficiencies of a two-stage DEA model in two separate and consecutive steps, Liu (2011) explains, in a short note, how to assess such efficiencies simultaneously. Liu and Lu (2012) introduce a network-based method for ranking of efficient units in two-stage DEA models. Specifically, each DMU is a node in a network and is linked with its peers. Links are weighted on the basis of the peer importance. Efficient DMUs are then ranked on the basis of their centrality in such a network.

Chen and Zhu (2004) propose a DEA framework that considers a two-stage process as efficient when each stage is efficient. Chen et al. (2009b) prove the equivalence between the CRS version of the Chen and Zhu (2004) model and the Kao and Hwang (2008) model. In this context, the interested reader is also referred to the survey by Agrell and Hatami-Marbini (2013). This paper consider the different two-stage models presented in the literature and points out which of them provides the equivalent results.

15.4.2.2 Variable Returns to Scale and Additive Measures of Efficiency

Chen et al. (2009a) observe that the multi-stage model by Kao and Hwang (2008) is applicable to CRS only. Indeed, it assesses the efficiency of the overall process as the product of the efficiencies of the different stages (i.e., the geometric mean of stage efficiencies). As an example, in the specific case of a two-stage process composed of a single DMSU f in the first stage and a single

DMSU s in the second stage, holding flow balancing constraints, the DMU efficiency (15.6) is:

$$e_0^* = \frac{\sum_j \mu_{js} y_{j0}^s}{\sum_i \nu_{if} x_{i0}^f} = \frac{\sum_j \mu_{js} y_{j0}^s}{\sum_l \upsilon_{ls} z_{l0}^s} \times \frac{\sum_l \upsilon_{ls} z_{l0}^s}{\sum_i \nu_{if} x_{i0}^f}. \tag{15.8}$$

To extend the two-stage models to VRS, Chen et al. (2009a), within the same relational model framework, measure the efficiency of the overall process as a weighted sum of the efficiencies of the two stages:

$$e_0^* = w_s \frac{\sum_j \mu_{js} y_{j0}^s + \omega^s}{\sum_l \upsilon_{ls} z_{l0}^s} + w_f \frac{\sum_l \upsilon_{ls} z_{l0}^s + \omega^f}{\sum_i \nu_{if} x_{i0}^f}, \tag{15.9}$$

where w_s and w_f are user-specified weights such that $w_s + w_f = 1$ and the terms ω^f and ω^s, free variables, express the scale efficiencies of the first and second stage, respectively. As pointed out by Cook et al. (2010b), Eq. (15.9) evaluates the overall performance of the network also in terms of the performances of the individual DMSUs.

Chen et al. (2009a) also show that efficiency measure (15.9) cannot be linearized in the same way efficiency measure (15.6) is turned into Eqs. (15.7a) and (15.7b), unless weights w_s and w_f are chosen to be proportional to the "sizes" of each stage, in terms of total resources devoted to each stage, that is,

$$w_s = \frac{\sum_l \upsilon_{ls} z_{l0}^s}{\sum_i \nu_{if} x_{i0}^f + \sum_l \upsilon_{ls} z_{l0}^s}, \quad w_f = \frac{\sum_i \nu_{if} x_{i0}^f}{\sum_i \nu_{if} x_{i0}^f + \sum_l \upsilon_{ls} z_{l0}^s}. \tag{15.10}$$

In a subsequent study, Chen et al. (2010a) point out that, differently from the standard DEA models, the multiplier and envelopment-based two-stage DEA models are not, in general, dual of each others, but represent two different approaches that provide different information and may produce different efficiency results (see also Chen et al. (2013b) in Sect. 15.3). Specifically, the authors show how some two-stage models in the literature may fail to provide the complete information on how to project inefficient DMUs on to the DEA frontier. Then, they develop two-stage models capable of determining these DEA frontier projections for inefficient DMUs at least in the CRS case. Finally, they indicate that further study is then needed to develop models capable of determining the DEA frontier projections for VRS inefficient DMUs since even their own previous model

(Chen et al. 2009a), which assesses correctly both the overall DMU efficiency and the efficiency of each stage, is not sufficient to yield these projections.

Chiou and Lan (2007) address two-stage VRS models, too. They propose an additive measure of efficiency equal to the one proposed in (15.9) when $w_s = w_f = 1$. They use this measure to asses both efficiency and effectiveness of a transportation system. In Chiou et al. (2010), the same authors discuss in detail the properties of their two-stage VRS model, that they call *integrated* DEA model, and generalize their efficiency measure to obtain exactly (15.9), where w_s and w_f are arbitrarily fixed. Differently form Chen et al. (2009a), these authors do not linearize their model but claim the existence and the uniqueness of optimal weights μ_{js}, v_{ls}, ν_{if}, ω^f and ω^s. Unfortunately, Lim and Zhu (2013) show that such a conclusion is a false statement. They also show how the two-stage DEA model proposed in Chiou and Lan (2007) can be transformed into a parametric linear program.

Finally, Kao and Hwang (2011) propose a multiplier-based relational VRS two-stage model. By solving both an output-oriented and input-oriented model, the authors are able to separate the technical and the scale efficiencies of the DMUs.

15.4.2.3 Open Multi-stage Processes

Cook et al. (2010b) introduce multi-stage DEA models for *open* serial processes, i.e., where some outputs from a given stage may leave the system while new inputs can enter at any stage. As in Chen et al. (2009a), the authors represent the overall efficiency as an additive weighted average of the efficiencies of the DMSUs. These results are also applied to general series-parallel network structures. Open multi-stage processes are considered also by Golany et al. (2006). These latter authors assume that each stage is governed by a different manager that will not agree to "vertical integration" initiatives unless higher efficiency (with respect to separately applying conventional DEA) is achieved. For this reason these authors propose a measure that identifies a Pareto optimal point for the efficiency values of the DMSUs that compose their system. As multiple Pareto optimal point may exist, they discuss the properties of three different possible ways of choosing the Pareto efficient point of interest.

15.4.2.4 Unoriented Models

Holod and Lewis (2011) present a two-stage DEA unoriented model, i.e., a DEA model that seeks to simultaneously decrease input levels and increase output levels (the interested reader is referred to Färe et al. (2002) for standard hyperbolic/unoriented DEA models). The authors use this model to assess bank efficiency and address what they call the DEA literature "deposit dilemma", that is, the lack of agreement on whether deposits should be considered as an input or an output. The authors solve this dilemma by representing deposits as intermediate flows in a two-stage unoriented DEA model. A similar model is also introduced by Lewis et al. (2013), who show how to solve it through an iterative algorithm that alternates between an input-oriented push backward step and an output-oriented push forward step. The same authors are currently working on a general network DEA unoriented model (Mallikarjun et al., 2014)

Yu and Chen (2011) use also a similar measure of efficiency to assess the air routes performance of an airline in Taiwan. In their paper, the authors initially present an interesting discussion on the definition of the performances of airlines in term of production efficiency, service effectiveness and operational effectiveness and a critical analysis of their own previous works. Then, they compare the results obtained through their model with the ones yielded by a corresponding multi-stage DEA model proposed by Chiou and Chen (2006), even though Lin (2008) identifies in this last paper some methodological and terminological inaccuracies.

15.4.2.5 Game Theoretic Perspective

The assessment of two-stage processes has been studied also relying on game theory. In particular, Liang et al. (2006) compare a leader-follower and a cooperative relationship between DMSUs of a supply chain. Liang et al. (2008) show that in a cooperative contest, when different intermediate flows between the two stages are present, then multiple efficiency values for the two stages may emerge. Differently, in a non-cooperative context a two-stage network DEA model just produces the same results as applying a standard DEA model to the two stages consecutively. Li et al. (2012) generalize the result proposed in Liang et al. (2008) by also allowing external inputs to the second stage. Chen et al. (2006) propose a DEA game model in a two-stage supply chain and prove the existence of numerous Nash equilibria efficiency points for the DMSUs.

As already pointed out, recently Cook et al. (2010a) have published an interesting survey that analyzes the DEA models used to assess the efficiency of two-stage processes from a game theoretic perspective. The authors categorize this literature using either Stackelberg (leader-follower) or cooperative game concepts. In this framework, only the multi-stage processes referring to cooperative game or, equivalently, to centralized control concepts have their overall efficiencies assessed through network models like model (15.7) or its variations. Differently, the processes referring to leader-follower concepts have the efficiency of their two stages assessed through two separated non-network DEA models. In this work, Cook et al. (2010a) also point out the equivalence of different two-stage DEA models available in the literature.

Zha et al. (2008) propose a two-stage VRS DEA model where the measure of the overall efficiency is given by the geometric mean of the efficiencies of the two-stages. Specifically, the efficiency of the first stage is evaluated with the input-oriented VRS model and the second stage with the output-oriented VRS model. Then, the overall efficiency is evaluated in a cooperative manner. In the same context, Zha and Liang (2010) introduce a two-stage DEA model with shared inputs to be allocated among the two stages (see also Sect. 15.5). Again, the efficiency measure is in the product-form and the process overall efficiency is assessed assuming that the two stages participate in a cooperative game. Also Wu (2010) considers a two-stage DEA model where stages share some inputs. Here the author assumes that there exists a Stackelberg-game relationship between the two stages and proposes a bilevel programming DEA model, which is solved using a branch and bound algorithm. Wu (2010) provides as case studies the application of his model to a banking chain and a manufacturing supply chain.

15.4.2.6 Processes with Feedback

Liang et al. (2011) consider two-stage processes with feedback, that is, processes in which some of the final outputs of the second stage become inputs of the first stage. In this context, the authors propose two multiplier-based network DEA models. The first (and simpler) one aims at maximizing the average efficiency of the two individual stages. The second model instead ranks the two stages in accordance with their relative importance and is formulated as a bilevel model. In both cases, the authors assume that the weights applied to the intermediate and feedback flows are the same for both stages. In addition, they assume that the weights of the intermediate and feedback flows are fixed when they play the role of outputs of the associated stage. This latter assumption is important in the second model, which maximizes the efficiency of the first stage and let the efficiency of the second stage depend on the first stage's one. In fact, the efficiency of the first stage depends in turn on the value of the weights of the feedback flows, which are fixed when the efficiency of the second stage is assessed. Both models are nonlinear, but their nonlinearity is only due to one or two variables, respectively. Hence, they can be practically solved by iteratively and tentatively assigning values to such few variables.

15.4.3 Applications

Besides some exceptions as in Wei and Chang (2011) who face the problem of *designing* an efficient multi-stage process (the authors propose a DEA approach to support the optimal design of DMU external input, intermediate flow and final output portfolios), in most cases DEA models are used to assess the efficiency of existing processes. This section illustrates several applications of two-stage DEA models.

15.4.3.1 Banking Sector

Avkiran (2009) employs a two-stages DEA model to assess United Arab Emirates (UAE) banks using a slacks-based inefficiency measure. Similarly, Paradi et al. (2011) introduce a SBM two-stage DEA model to study the performance of banks when bad outputs are present. Bad outputs are also considered by Fukuyama and Weber (2010) that introduce a two-stage model to study Japanese banks' performances. This last model accounts for slacks in the input and output constraints defining the technology, and allows inefficiency to be measured with non-radial contractions in inputs and expansions in outputs, even when slack does not exist. This model is also applied by Fukuyama and Matousek (2011) to assess the efficiency of Turkish bank system. Akther et al. (2013) introduce bad outputs while assessing 19 Bangladesh banks and use a slacks-based inefficiency measure within a two-stages DEA model. Huang et al. (2009) assess the efficiency of Chinese banks with a relational two-stage model. Yang and Liu (2012) prove, by integrating a two-stage DEA model and a fuzzy multiobjective model, that in Taiwan mixed ownership banks are more efficient than the fully state-owned ones. Grigoroudis et al. (2013) present a three-stage DEA model to assess banks in terms of satisfaction, employee appraisal, and business performance. Their paper is also a good introduction to the literature that links operating efficiency and quality of service in the bank sector. Wu and Birge (2012) introduce what they call a two-stage serial-chain merger DEA model to evaluate mortgage banking operations. Premachandra et al. (2012) apply a two-stage model to assess the performance of mutual funds.

15.4.3.2 Production Processes and Supply Chains

Liu and Wang (2009) use a two-stage relational DEA model to assess the efficiency of printed circuit board industry in Taiwan. Lee and Johnson (2011) use a multi-stage DEA model to represent the production processes in the semiconductor manufacturing industry. Saranga and Moser (2010) apply what they call classical two-stage Value Chain DEA models to assess the performances of purchasing and supply management activities. Yang et al. (2011) propose a envelopment-based multi-stage DEA model for assessing the performances of supply chains. These authors state the novelty of their model affirming that, even though there is a rich literature on DEA models for supply chains, the exact definition for supply chain production possibility set is still unclear. For this reason, the authors propose two possible types of supply chain production possibility sets that then they prove equivalent. Mirhedayatian et al. (2013) propose a model for assessing "green" supply chains. Chen et al. (2012a) use two-stage DEA model for evaluating sustainable product design performances. They propose both centralized and decentralized models as in Chen and Yan (2011) to analyze the simultaneous, proactive, and reactive approaches adopted by firms for sustainable design.

Bai-Chen et al. (2012) apply a two-stage model to assess both economic benefits and carbon emissions of China's power plants. The author call their model "environmental" network DEA model as it takes into account environmental factors as non-discretionary inputs.

Cao and Yang (2011) measure the performance of Internet companies, whereas Asai (2011) employs a two-stages DEA model to assess Japanese broadcasters.

15.4.3.3 Transportation

Lu et al. (2012) use a two-stage additive DEA model based on the works of Chen et al. (2009a) and of Cook et al. (2010b) to assess the production and marketing efficiency of airline industry. The authors show that low-cost carriers, on average, are more efficient than the full-services ones from a production perspective, but they are less efficient marketers.

Chang and Yu (2012) also deal with low-cost carriers. Specifically, the authors use a SBM two-stage DEA model to assess production and consumption efficiencies. Yu (2010) adopt a SBM efficiency measure to model an open process and assess both production and service efficiency in airports. In this work, the author points out that environmental factors have an important influence in the performances of transportation systems and hence they must be taken into consideration even if that are beyond managerial control. For these reasons, on one side he models these factors as quasi-fixed/non-discretionary inputs; on the other side, he associates no slack variable to them and consequently he does not include them in the SBM efficiency.

Zhu (2011) applies the centralized model by Liang et al. (2008) to asses the efficiency of a set of airlines. Wanke (2013a) (respectively Wanke (2013b)) applies an analogous model to assess the physical infrastructure and flight (respectively shipment and consolidation) efficiency drivers in Brazilian airports (respectively ports). Adler et al. (2013) use a two-stage DEA model for benchmarking airports taking into account of both terminal and airside activities. These last authors point out how previous benchmarking studies based on standard DEA models may arrive to opposing conclusions, whereas a network DEA structure provide more meaningful benchmarks with comparable peer units and target values that are achievable in the medium term. To reach such results, the authors apply a dynamic clustering approach (Golany and Thore 1997) that, for each DMU_0, restricts the set of possible peers to include only DMUs with similar mixes of flows. The rationale of this choice is to set a target for an inefficient DMU_0 which is accessible in the short to medium term.

15.4.3.4 Sports

Moreno and Lozano (2012) introduce an interesting survey of DEA models to analyze sport performances and then compare the results of a standard DEA model with a generalized two-stage one to assess the efficiency of NBA teams. In both models they use SBM efficiency (Tone and Tsutsui 2009). The authors finally conclude that the two-stage DEA model has more discriminating power and provides more insight than the standard one.

15.4.4 Dynamic Networks

Dynamic networks DEA models are multi-stage models that describe the evolution of processes over time. A recent survey of this network DEA literature sub-area can be found in Fallah-Fini et al. (2013) which review all the literature (including non DEA works) on non-parametric dynamic efficiency measurement.

In the basic version of these models, each stage represents the same DMU, as a black box, at different times. Färe and Grosskopf (2000) consider the same production process in two successive periods/DMSUs with period-specific inputs and outputs. Some of the outputs produced in the first period, that is by the first DMSU, are used as inputs in the second period, that is by the second DMSU (see also Färe and Grosskopf 1996a). The authors model these time-intermediate products as intermediate flows of a (dynamic) network DEA model and, hence, they may evaluate the relative efficiency of the involved process using Model (15.3). An illustration of this kind of dynamic network DEA models can be found in Bogetoft et al. (2009). Another basic dynamic network DEA model is introduced by Troutt et al. (2001) who, strangely enough, do not present appropriate bibliography except for two seminal papers on standard DEA.

Nemoto and Goto (1999) use a dynamic network DEA model to describe the intertemporal behavior of a firm. The authors identify the intertemporal efficient cost frontier using an envelopment-based network DEA model. Their model includes both discretionary and quasi-fixed inputs. Discretionary inputs are period-specific, whereas quasi-fixed inputs are the only time-intermediate flows. Both kinds of inputs are assumed variable (instead of, e.g., being considered constant and possibly multiplied by variable scaling factors θ, as it is customary in standard DEA input oriented models) and a linear combination of them is minimized. In a subsequent work, Nemoto and Goto (2003) apply this model to Japanese electric utilities to show how to evaluate the efficiencies of quasi-fixed inputs and describe their adjustment processes. Sueyoshi and Sekitani (2005) propose the VRS formulation of the this model. Later, also Von Geymueller (2009) applies a variation of this model to assess the efficiency of electricity transmission operations.

Tone and Tsutsui (2010) introduce the slacks-based version of the above network dynamic model. In addition, the authors indicate how to deal with both discretionary and non-discretionary intermediate flows, and point out that these flows must be dealt with differently depending on their desirability. Earnings carried forward are a possible example of desirable intermediate flows; on the contrary, losses carried forward are a possible example of undesirable ones.

Kao (2012) proposes a relational approach for dynamic multi-stage processes and underlines that the previous methods described the literature for calculating the efficiency of these processes may produce over-estimated scores if their dynamic nature is disregarded.

In more complex dynamic models, each stage represents again the same DMU at different times, but now this DMU in turn models a multi-stage process. Chen (2009) introduces such a dynamic network DEA model to represent a production network. Let $DMSU_r^k$ be the generic r-th DMSU of the k-th DMU. The author defines a (dynamic) network, the nodes of which are the subunits $DMSU_r{}^k$ at the different times t. Then, he assumes that, at each time t, only a fraction of the intermediate output flow of $DMSU_r^k$ is received immediately as intermediate input flow by the successive $DMSU_{(r+1)}^k$. The complementary fraction of the intermediate output flow is stored and received by $DMSU_{(r+1)}^k$ in successive times, possibly with some losses if this intermediate flow consists of a perishable material. Tone and Tsutsui (2014) extend these kind of dynamic network DEA models to situations in which SBM efficiency is taken into consideration and the DMUs observed over time model general network process.

Finally, other authors (see, e.g., Chen and van Dalen 2010; Emrouznejad and Thanassoulis 2005; Sengupta 1995) consider dynamic DEA models in order to take into account input flows received at a time period t, e.g., capital, that may have a productive effect not only in the same time period t but also over future time periods. These models, however, usually do not consider time-intermediate flows between DMSUs. As an example, Chen and van Dalen (2010) propose an envelopment-based dynamic DEA model assuming that the input received at a time period t may have a productive effect not only in the same time period but also over a given time horizon of, say, length g. On the basis of this observation, for each time period t, they assess the process performances using an efficiency measure that considers the input flow x^t and a value $\tilde{y}^t$ function of the output flows, for $r = 0, \ldots, g$, produced between t and $t + g$.

15.5 Shared Flow DEA Models

In this section, we deal with DEA models for DMUs that include DMSUs that either share some of their inputs or their outputs. These models assume that the total amount of each input (or output) flow entering (or exiting) the whole DMU is known and a-priori fixed, as it is customary in standard DEA models. However, they also assume that the amount of shared flow allocated to each subunits may be considered as a decision variable to be used to maximize the DMU efficiencies (see Fig. 15.4). Even in this case, the subunits of the DMUs cannot be considered independent since they compete for the allocation of the flows that they share. Beasley (1995) introduces one of the first examples of a shared flow DEA model, even if it was not originally referred to as such. The model is applied to departments

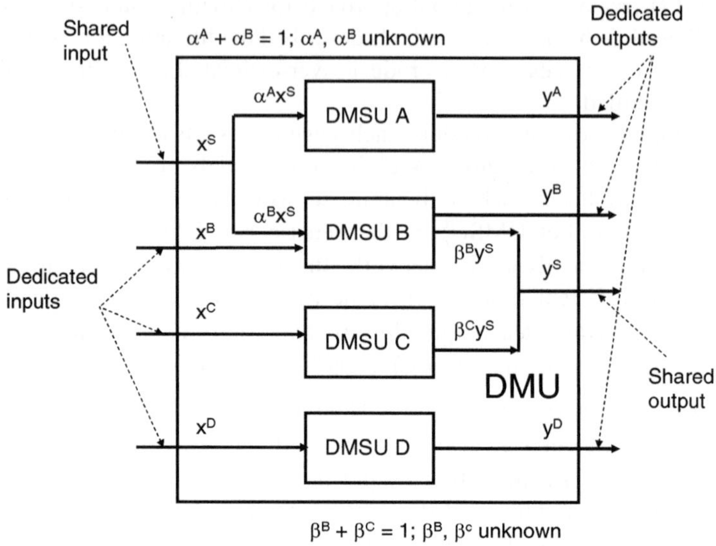

Fig. 15.4 A shared flow DMU: DMSUs A and B are not independent because they compete for the same shared resource. Similarly DMSUs B and C are not independent because of the shared output. DMSU D is independent of DMSUs A, B and C

of different universities devoted to the same disciplines. The departments are homogeneous and independent DMUs. Within each of them, the teaching and research activities clearly define two different separable functions. One of the DMU inputs, research income, is specifically *dedicated* to the research function. The other DMU inputs, general and equipment expenditure, are *shared (joined)* between the two functions. DMU outputs are split, i.e., no shared outputs exist: the number of undergraduates and of taught postgraduates are outputs of the teaching function; the number of research postgraduates, research income, and research rating are outputs of the research function. Kao and Lin (2012) extend this application to the situation in which some input/output data are fuzzy numbers.

15.5.1 Formulation of Shared Flow DEA Models

Referring to r as the generic component of DMU k, now vectors X_k^r, Y_k^r, ν^r, and μ^r introduced in Sect. 15.2 are defined as the vectors of dedicated inputs, dedicated outputs, weights of the dedicated inputs, and weights of dedicated outputs of component r, respectively. In addition, we define

- $X_k^S = \{x_{ik}^S\}$: the vector of shared inputs,
- $Y_k^S = \{y_{jk}^S\}$: the vector of shared outputs,
- $\nu^S = \{\nu_i^S\}$: the vector of weights of shared inputs,

- $\mu^S = \{\mu_j^S\}$: the vector of weights of shared outputs,
- $\alpha^r = \{\alpha_i^r\}$: the vector of proportions of the shared inputs allocated to component r,
- $\beta^r = \{\beta_j^r\}$: the vector of proportions of the shared outputs attributed to component r.

With a little abuse of notation we also define $\alpha^r X_k^S$ as the column vector whose generic entry is $\alpha_i^r x_{ik}^S$. In this context, $\alpha_i^r x_{ik}^S$ is the amount of shared input i allocated to component r by DMU k to maximize its efficiency. When a shared input cannot be clearly divided among functions (e.g., general expenditure), then α_i^r can be seen as the proportion of the (virtual) value of the input i allotted to component r. Similarly, we define $\beta^r Y_k^S$ as the column vector whose generic entry is $\beta_j^r y_{jk}^S$ where β_j^r is always seen as the proportion of the (virtual) value of output j that can be attributed to component r because it is assumed that no component can produce a shared output by itself but needs synergy with other components. As an example, the quality of service level provided by an organization to its customers depends on the degree of collaboration and integration among its subdivisions, each of them sharing with other subunits the responsibility for such output. When outputs common to different components are produced without the need of synergy among them, the literature refers to them as *overlapping* outputs (see Sect. 15.5.2.5 for details).

15.5.1.1 Primal Formulation

Consider, for the sake of simplicity, the case when shared outputs are not present. The efficiency of DMU k is expressed as

$$e_k = \frac{\sum_r \mu^r Y_k^r}{\sum_r \nu^r X_k^r + \sum_r \nu^S (\alpha^r X_k^S)},$$

the *partial* efficiency of the single component r is defined as

$$e_k^r = \frac{\mu^r Y_k^r}{\nu^r X_k^r + \nu^S (\alpha^r X_k^S)},$$

and the *aggregate* efficiency $\hat{e}_k = \sum_r q_k^r e_k^r$ as the weighted combination of the partial efficiencies of its components, where the weight q_k^r of each component r is

$$q_k^r = \frac{\nu^r X_k^r + \nu^S (\alpha^r X_k^S)}{\sum_p \nu^p X_k^p + \sum_p \nu^S (\alpha^p X_k^S)}.$$

Hence q_k^r is the fraction of DMU k total weighted inputs that are consumed by component r: $\sum_r q_k^{\ r} = 1\ \forall k$. Also Yang et al. (2000) introduced the concept of partial efficiency measures but they applied it on an elementary model (see Sect. 15.2). The general model proposed by Beasley (1995) is

$$e_0^* = \max e_0 \tag{15.11a}$$

$$e_k^r \leq 1 \quad \forall k, r \tag{15.11b}$$

$$\sum_r \alpha_i^r = 1 \quad \forall i \tag{15.11c}$$

$$\nu_i^r, \nu_i^S, \alpha_i^r, \mu_j^r \geq \varepsilon \quad \forall i, j, r. \tag{15.11d}$$

Condition (15.11b) imposes that the partial efficiency of each DMU component cannot exceed 1. Beasley (1995) proves that when each DMU is free to allocate the value of the shared inputs among its different components, the aggregate efficiency $\hat{e}_k$ and the efficiency e_k are coincident when maximized.

As for the standard DEA formulations, model (15.11) can be rewritten as follows

$$e_0^* = \max \sum_r \mu^r Y_0^r \tag{15.12a}$$

$$\sum_r \nu^r X_0^r + \sum_r \nu^S (\alpha^r X_0^S) = 1 \tag{15.12b}$$

$$\mu^r Y_k^r \leq \nu^r X_k^r + \nu^S (\alpha^r X_k^S) \quad \forall k, r \tag{15.12c}$$

$$\sum_r \alpha_i^r = 1 \quad \forall i \tag{15.12d}$$

$$\nu_i^r, \nu_i^S, \alpha_i^r, \mu_j^r \geq \varepsilon \quad \forall i, j, r. \tag{15.12e}$$

Model (15.12) is not linear because of inequalities (15.12b) and (15.12c). When no shared inputs exist, model (15.12) easily reduces to the elementary model (15.2) as $X_k^S = 0\ \forall k$. Hence the terms $\sum_r \nu^S (\alpha^r X_0^S)$ in constraint (15.12b) and $\nu^S (\alpha^r X_k^S)$ in constraint (15.12c), and constraint (15.12d) are no longer necessary.

15.5.1.2 Dual Formulation

Mar Molinero (1996) and Mar Molinero and Tsai (1997) propose an approach dual to model (15.11). In addition, the authors include *shared outputs*, i.e., outputs yielded synergically by two or more components. Their output oriented model for what they call a *multi-activity* process is

$$e_0^* = \max \sum_r q_0^r \theta_0^r + \varepsilon \left(\sum_i \left(s_i^{S-} + \sum_r s_i^{r-} \right) + \sum_j \left(s_j^{S+} + \sum_r s_j^{r+} \right) \right) \tag{15.13a}$$

$$\sum_k \lambda_k^r x_{ik}^r = x_{i0}^r - s_i^{r-} \quad \forall i, r \tag{15.13b}$$

$$\sum_k \sum_r \lambda_k^r (\alpha_i^r x_{ik}^S) = x_{i0}^S - s_i^{S-} \quad \forall i \tag{15.13c}$$

$$\sum_k \lambda_k^r y_{jk}^r = \theta_0^r y_{j0}^r + s_j^{r+} \quad \forall j, r \tag{15.13d}$$

$$\sum_k \sum_r \lambda_k^r (\beta_j^r y_{jk}^S) = \sum_r \theta_0^r (\beta_j^r y_{j0}^S) + s_j^{S+} \quad \forall j \tag{15.13e}$$

$$\sum_r \alpha_i^r = 1 \quad \forall i \tag{15.13f}$$

$$\sum_r \beta_j^r = 1 \quad \forall j \tag{15.13g}$$

$$\sum_r q_0^r = 1 \tag{15.13h}$$

$$\lambda_k^r, q_0^r, \alpha_i^r, \beta_j^r, s_i^{r-}, s_i^{S-}, s_j^{r+}, s_j^{S+} \geq 0 \quad \forall i, j, r, k. \tag{15.13i}$$

where q_0^r are positive weights representing the relative importance of each component r for DMU 0, and θ_0^r are measures of the inefficiencies of the components of DMU 0. Actually, θ_0^r are the reciprocals of the *distance functions* defined by Shephard (1970). Note that in the models proposed by Mar Molinero (1996) and Mar Molinero and Tsai (1997) the slack variables s_i^{r-}, s_i^{S-}, s_j^{r+}, s_j^{S+} are not present. Here they are imposed for coherence with the standard DEA dual models (see, e.g., Cooper et al. 2000).

When the values α_i^r, β_j^r, and q_0^r are not decision variables but are fixed, still satisfying conditions (15.13f), (15.13g) and (15.13i), the dual of model (15.13) is

$$e_0^* = \min \sum_r \nu^r \overset{r}{\underset{0}{X}} + \sum_r \nu^S (\alpha^r X_0^S) \tag{15.14a}$$

$$\mu^r Y_0^r + \mu^S (\beta^r Y_0^S) = q_0^r \quad \forall r \tag{15.14b}$$

$$\mu^r Y_k^r + \mu^S (\beta^r Y_k^S) \leq \nu^r X_k^r + \nu^S (\alpha^r X_k^S) \quad \forall k, r \tag{15.14c}$$

$$\nu_i^r, \nu_i^S, \mu_j^r, \mu_j^S \geq \varepsilon \quad \forall i, j, r. \tag{15.14d}$$

The above model parallels the output oriented version of model (15.12) when shared outputs are considered. Besides model (15.14) being linear, the main difference between the two models is the presence of the multiple constraints (15.14b) instead of the single one $\sum_r \mu^r Y_0^r + \sum_r \mu^S (\beta^r Y_0^S) = 1$. This latter constraint is a relaxation of the former ones because $\sum_r q_0^r = 1$. Conditions (15.14b) state a precise relationship between the relative importance attributed to a component and the optimal amount of outputs allocated to it (respectively, the optimal amount of allocated inputs if an input oriented model is considered). Then, conditions (15.14b) justify the choice in Beasley (1995) of expressing the weight q_k^r of the component r in the aggregated efficiency as equal to the fraction of DMU k total weighted inputs that are consumed by component r. Without conditions (15.14b), such a choice might appear arbitrary, although reasonable.

15.5.2 Extensions

Many authors have extended models (15.12) and (15.13). Common features of the different variants are that the aggregate efficiency of a DMU cannot exceed unity, and that a DMU is efficient if and only if it is efficient in all its components. In this section, we describe the peculiarity of each available modeling advance.

15.5.2.1 Weight Restrictions

Beasley (1995) himself does not present model (15.11), but he incorporates the additional constraints

$$(\nu^S, \nu^r, \forall r) \in \Omega_{in} \tag{15.15}$$

$$(\mu^r, \forall r) \in \Omega_{out} \tag{15.16}$$

where the sets Ω_{in} and Ω_{out} are *assurance regions* as defined in Thompson et al. (1990). Constraints (15.15) and (15.16) involve value judgements concerning the proportions α^r and the weights μ^r, and ν^r of the different DMU components. They are not strictly necessary for the definition of a shared flow DEA model, but might prevent the model from yielding unreasonable results. In this context, Beasley (1995) provides an example where, in the absence of constraints (15.15) and (15.16), one research postgraduate was worth about 880,000 undergraduates for a given department.

Assurance regions are also introduced by Yu (2012) to measure the performance of two-division international tourist hotels in Taiwan, which exhibit both shared inputs and shared outputs.

15.5.2.2 Variable Returns to Scale

Mar Molinero and Tsai (1997) prove that the feasible solutions of model (15.13) define a convex set and the objective (15.13a) is a convex function. Tsai and Mar Molinero (2002), considering the problem of assessing the performances of individual specialties of National Health Services Trusts in the UK, introduce and discuss a variable returns to scale version of model (15.13). The efficiency of each component r of DMU k is then defined as

$$e_k^r = \frac{\mu^r Y_k^r + \mu^S(\beta^r Y_k^S)}{\nu^r X_k^r + \nu^S(\alpha^r X_k^S) + \delta_k^r} \tag{15.17}$$

where the variable δ_k^r is unrestricted and its optimal value defines the component's returns to scale status. The aggregate efficiency of DMU k is

$$e^r = \frac{\sum_r \mu^r Y_k^r + \sum_r \mu^S(\beta^r Y_k^S)}{\sum_p \nu^p X_k^p + \sum_p \nu^S(\alpha^p X_k^S) + \sum_p \delta_k^p}. \tag{15.18}$$

Note that the optimal value of $\sum_p \delta_k^p$ may be zero even if some or all elements in the sum are different from zero. In this case, DMU k may appear to be operating under constant returns to scale and technically efficient when analyzed as a black box but, when its individual components are analyzed, it may be found scale inefficient in each of its activities (Tsai and Mar Molinero 2002). It follows that a DMU, that is efficient when considered as a black box, may be inefficient when its different components are taken into account, independently of its returns to scale status.

Variable returns to scale are also considered by Diez-Ticio and Mancebon (2002) to assess the efficiency of Spanish Police Service.

15.5.2.3 Different Weights on Shared Inputs

Cook et al. (2000) allow a same shared input i to be weighted differently by the subunits of the same DMU. The rationale behind such a choice is that different components may disagree on the importance of a same input. Consequently, the shared flow model as in Cook et al. (2000) includes in constraints (15.12b) and (15.12c) a set of vectors ν^{Sr}, one for each component r, instead of a single one. Also, a change of variables is proposed. In particular, let $i = 1, \ldots, s$ be the index of the shared inputs, then $\overline{\nu}_i^{Sr} = \nu_i^{Sr}\alpha_i^r$ for $i = 1, \ldots, s-1$ and $\overline{\nu}_s^{Sr} = \nu_s^{Sr}\left(1 - \sum_{i=1}^{s-1}\alpha_i^r\right)$. Because of these new variables, the authors obtain a linear model. The terms $\nu_r^S(\alpha^r X_k^S)$ in conditions (15.12b) and (15.12c) become $\overline{\nu}^{Sr}X_k^S$ and $\nu_r^S \geq \varepsilon$ in constraint (15.12e) turns $\overline{\nu}_i^{Sr} \geq \varepsilon\alpha_i^r$. Unfortunately, non-linearity may arise again when additional constraints concerning value judgements as constraints (15.15) and (15.16) are necessary. If such judgements are expressed also in terms of ν_r^S, the variable substitution may not lead to a linear model.

15.5.2.4 Additive Objective Function

Cook and Hababou (2001) and Cook and Zhu (2005, Chap. 6) present variables and constraints as in Cook et al. (2000) but differ in the objective function. They formulate an additive objective function representing an aggregate measure of the efficiencies of all the DMU subunits. In the classical additive DEA models (Charnes et al. 1985), a possible measure of the inefficiency of DMU k is given by the difference between the weighted sum of the inputs minus the weighted outputs of DMU k. Here Cook and Hababou (2001) suggest a multiobjective approach where the partial inefficiencies of all components are considered. For each subunit, the weighted sum of its inputs minus the weighted sum of its outputs is considered. In particular, the authors minimize the maximum partial inefficiency in order to give equal importance to each component, i.e., their objective function is

$$\min \max\{\nu^r X_k^r + \nu^{Sr}(\alpha^r X_k^S) - \mu^r Y_k^r : \quad \forall r \text{ subunit of } DMU_0\}. \tag{15.19}$$

Finally, the authors linearize their model with the same variable changes proposed in Cook et al. (2000).

15.5.2.5 Overlapping Outputs

Cook and Green (2004) deal with a manufacturing multi-plant company and point out that some outputs of different components of the same DMU can partially overlap, i.e., some outputs may be common to different components. In particular, each DMSU can yield a given amount of overlapping output j, with no need of synergy with the other components. Hence there is no possibility of attributing the considered amount to the other subunits. From this point of view, the overlapping outputs are different from the shared outputs considered in Mar Molinero and Tsai (1997) and Mar Molinero (1996). In fact, Cook and Green (2004) cannot approach what they call the *overlap problem* by introducing variables β^r as in model (15.13) to determine which proportions of shared outputs are attributed to each component: the efficiency of a single subunit r remains $e_k^r = \frac{\mu^r Y_k^r}{\nu^r X_k^r + \nu^{Sr}(\alpha^r X_k^S)}$ and, consequently, the aggregate efficiency of a whole DMU k is $e_k = \frac{\sum_r \mu^r Y_k^r}{\sum_r \nu^r X_k^r + \sum_r \nu^{Sr}(\alpha^r X_k^S)}$. However, the shared inputs are no longer allocated to the components because such task could hardly be performed without introducing some ambiguities due to the component overlapping. Shared inputs are allocated directly to the outputs. In particular, consider model (15.11) and the extension proposed in Cook et al. (2000). Cook and Green (2004) introduce a new set of variables α_i^j as the proportions of the shared inputs i allocated for outputs j. In addition, they replaced condition (15.11c) with $\sum_j \alpha_i^j = 1$, for all i. Finally, they defined α_i^r as $\alpha_i^r = \sum_{i \in O^r} \alpha_i^j$, where O^r is the set outputs of subunit r. Note that now, in general, $\sum_r \alpha_i^r \geq 1$.

The allocation of shared inputs directly to outputs was originally introduced in Färe et al. (1997). Even though the concept of DMSU is not explicitly mentioned, it can easily be inferred since one input can be allocated among various outputs.

15.5.2.6 Core Business Identification

Cook and Green (2004) and Cook and Zhu (2005, Chap. 11) address the problem of determining in which areas a DMU would perform better. Such areas form the *core business* of a DMU and should be privileged even at the cost of possibly forcing the DMU to abandon the components with less satisfactory performances. To this aim, Cook and Green (2004) modify the objective function of model (15.11) and add assignment constraints (each DMU must have at least one component assigned and each component must be assigned to at least one DMU).

To overcome the insufficiency associated with the black box approach that generally makes DMU inner data not to be available, Bi et al. (2012) consider DMUs with parallel structure and propose to divide the production activities within a DMU into two subsets or units. The first unit is termed as the core business unit (CBU), which includes the main production functions of DMU; the second unit is referred to as the non-core business unit (NCBU). The authors introduce a solution method that assumes that the information related to inner inputs/outputs is available for the DMU under evaluation. For the other DMUs, however, these data are generated by using the Pareto principle: as a rule of thumb, the CBU produces 80 % of total outputs of a DMU, while consumes only 20 % of total inputs. Accordingly, NCBU produces 20 % of the total outputs, while consumes 80 % of all inputs.

15.5.2.7 Resource Allocation

Shared flow models have also been used to allocate input costs among different subunits or activities. da Cruz et al. (2013) propose a model for estimating not only the overall efficiency of water utilities, but also the cost efficiency of drinking water and wastewater services. Using a shared input DEA methodology, the authors are also able to report estimates for the cost shares that correspond to each service. Similarly, Rogge and De Jaeger (2012) evaluate the cost efficiency of Flemish municipalities in the collection and processing of municipal solid waste by considering only one input ("waste cost") that is shared among different collection and treatment activities. In Rogge and De Jaeger (2013) this shared input DEA-model is further developed to make the partial and aggregate cost efficiency scores robust and also corrected for the impact of influences related to the operating environment and long-term policy variables. The same robust shared-input DEA approach has been applied by Broekel et al. (2013) to evaluate for multiple years the innovation efficiency of 150 German market labor regions, using as unique input "R&D employment" figures. Input costs are jointly allocated also by Salerno (2006) to estimate higher education institutions' per-student education costs in The Netherlands.

Barnum et al. (2011) introduce a DEA-based procedure for estimating the overall efficiency of metropolitan public transportation agencies in the United States. Specifically, the authors use a six-step shared flow model to allocate operating expenses among the agencies' organizational subunits that supply transit service. In each of the main step the authors use DEA to asses either the efficiency of the whole system or of each of the transportation modality. Finally, the sixth step, which involves a non-DEA mathematical program, estimates how inputs should be allocated among the target agency's subunits in order to minimize total expenses, while holding output constant.

15.5.2.8 Non-radial Measures of Efficiency

Chen et al. (2013a) describes an empirical study on Taiwan's farmers' cooperatives to offer policy suggestions as to how fixed resources can be effectively reallocated among different departments in a team production environment. The authors adopt Luenberger (1992)'s directional distance function to scale inputs and outputs, but not necessarily along the rays from the input and output origin (Fukuyama 2003). In such a way, the optimal input/output adjustment and the optimal allocation of shared inputs among different activities are taken into consideration simultaneously. Furthermore, the use of a directional distance function allows to easily incorporate an undesirable/bad output as a byproduct of desirable/good production activities. In fact, when we seek a reduction in the bad output and simultaneous increases in the good output, then the directional distance function will be a preferred method because it allows non-proportional adjustments of the good and bad outputs.

Yu and Lee (2009) use instead a *hyperbolic* network DEA model to evaluate the performances of hotels in Taiwan. Specifically, the authors extend the models introduced in Färe and Grosskopf (2000) and Färe and Whittaker (1995) by combining both the input and the output orientation in a non-linear fashion.

15.5.2.9 Two-Stage Networks

In the recent years, the integration between network and shared flow models has been addressed by some authors. Chen et al. (2010b) propose a DEA model to evaluate either the VRS or the CRS efficiency of a two-stage network process where some inputs are directly associated with both stages or shared by the two stages (Fig. 15.5). The DMU efficiency is computed as a convex combination of efficiency scores of the first and second stage, thus ensuring that a DMU is overall efficient if and only if each stage is efficient. In the case of an inefficient DMU, however, the decomposition of the overall DMU efficiency between the two stages may not be unique. Hence, following Kao and Hwang (2008), the authors propose, under both VRS and CRS, an approach to find a set of multipliers that maximize either the first or the second stage efficiency score while maintaining the overall efficiency score.

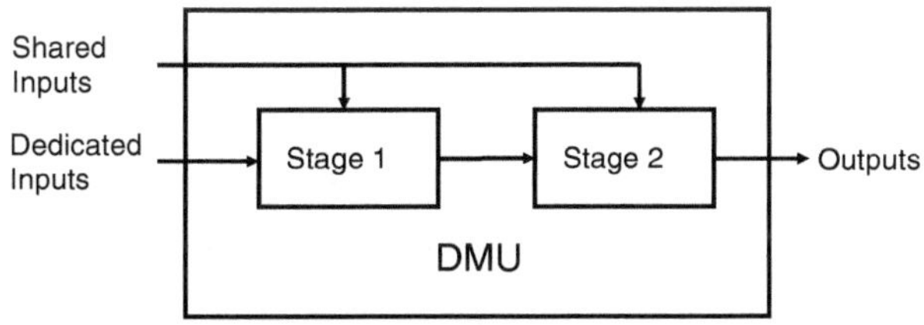

Fig. 15.5 Shared inputs in a two-stage network process

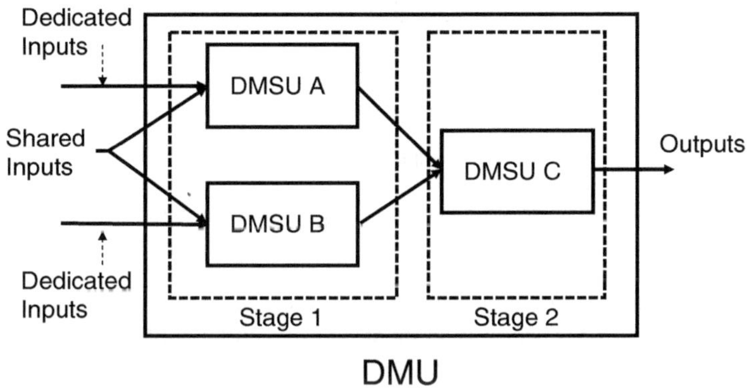

Fig. 15.6 Shared inputs in a open two-stage network process

Zha and Liang (2010) analyze the two-stage network process with shared inputs as in Fig. 15.5. Differently from Chen et al. (2010b), the authors propose to determine the overall DMU efficiency score as the product of the efficiency scores of the two stages, thus optimizing the overall efficiency through cooperation of the different stages, as suggested by Castelli et al. (2004).

To assess the efficiency of multimode bus transit systems, Yu and Fan (2006) introduce a two-stage shared input DEA model that incorporates both desirable and undesirable outputs, and also environmental (non-discretionary) inputs. Following Yu and Fan (2006) and Yu (2008b), a two-stage network with shared inputs between two parallel subunits of the first stage (see Fig. 15.6) has been proposed by Yu and Fan (2009) to simultaneously estimate the production efficiency, service effectiveness and operational effectiveness of Taiwan's bus transit system. Their network model, also called mixed structure network DEA model, extends both the network DEA model introduced by Färe and Grosskopf (2000) (series structure network) and the network DEA model developed by Mar Molinero (1996) (parallel structure network).

Chen et al. (2010b) show that their approach can be easily extended to open two-stage network processes where some inputs from the first stage do not become inputs to the second stage, and the second stage has its own inputs (Fig. 15.7).

Amirteimoori (2013) addresses the same two-stage network process with shared inputs as in Fig. 15.7 using the approach of Chen et al. (2010b), with the only

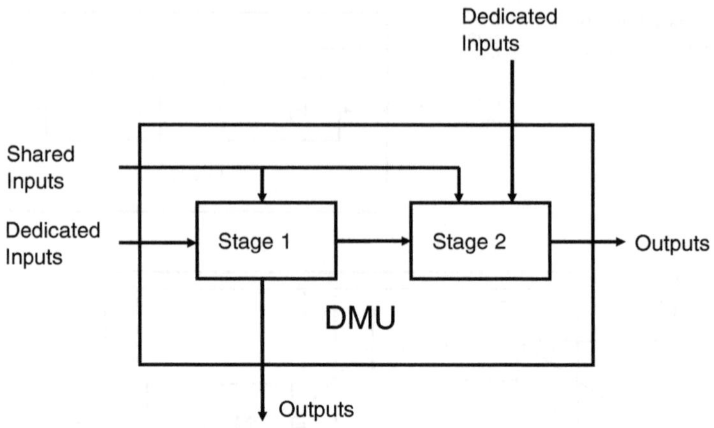

Fig. 15.7 Shared inputs in a two-stage network process with parallel subunits

difference that the intermediate flows are to be considered as undesirable outputs for the first stage. In the same two-stage setting, undesirable intermediate flows were earlier addressed by Yang (2009) who moreover simultaneously considers both DMU desirable and undesirable outputs to measure productive and environmental efficiency in farrow-to-finish pig production in Taiwan. Similarly, also Chen et al. (2012b) evaluate the relative performance of incineration plants in Taiwan by including desirable and undesirable outputs. To allow inputs and outputs to change non-proportionally the directional slacks-based inefficiency measure developed by Fukuyama and Weber (2009) is incorporated into their model (see also Sect. 15.5.2.8).

Two-Stage Network and Non-radial Measures of Efficiency

Sometimes the technology used to measure DMU efficiency has to deal with input excesses and output shortfalls simultaneously. In this case, the graph-oriented DEA model can be applied (Färe et al. 1985). In contrast to input-oriented and output-oriented DEA models, both inputs and outputs are allowed to vary by the same (or different) proportion, but inputs are proportionately decreased while outputs are simultaneously increased by the same (or different) proportion. Graph efficiency measurement has been used by Yu and Lin (2008) who present a multi-activity network DEA model to simultaneously estimate passenger and freight technical efficiency, service effectiveness, and technical effectiveness for 20 selected railways for the year 2002. This model extends the work from Mar Molinero and Tsai (1997). In particular, it generalizes model (15.13) for multi-stage processes and uses an objective function that penalizes all the external input and final output inefficiencies of all the components, with the exception of non-discretionary inputs (while the θ_0^r terms in model (15.13) penalize only the final outputs). For standard DEA models similar measures were proposed in Pastor et al. (1999).

Similarly, a graph-oriented DEA model is proposed by Chao et al. (2010) who apply the multi-activity DEA model to explore the relative efficiency of 12 financial holding companies in Taiwan.

Two-Stage and Dynamic Networks

Chen (2012) proposes a dynamic shared input DEA model to assess the efficiency of the swine production in Taiwan. The model is dynamic in the sense that a same DMU, made of two parallel DMSUs with shared inputs, is observed over time. Hence, some of the outputs of a period become some of the inputs of the following period. Efficiency is not measured radially. Instead, as in Chen et al. (2012b, 2013a), the directional Russell measure of slack-based inefficiency developed by Fukuyama and Weber (2009) is introduced to allow inputs and outputs with non-proportional changes.

15.6 Multi-level DEA Models

In this section, we deal with DEA models for DMUs exhibiting autonomous activities that cannot be associated to any of their subunits. In other words, these DMU present additional inputs/outputs not considered by their DMSUs. For example, in Cook et al. (1998), DMSUs are highway maintenance patrols and DMUs are the districts in which the maintenance patrols are grouped. The subunits have traffic and road conditions as possible inputs, while DMUs may include additional inputs that can be applied only to districts such as the extent of privatization and district engineers' experience. The same authors also introduced possible applications of their model to power plants and hospitals. These models are defined as *multi-level models* (Cook et al. 1998) where the top level, referred to as level n DMU, includes independent and homogeneous subunits, referred to as level $n - 1$ DMUs. Recursively, the level $n - 1$ DMUs include smaller independent and homogeneous subunits, level $n - 2$ DMUs, and so on. Unlike shared flow models, the amount of input and output of each subunit is fixed. In this work, we introduce only two–level structures, and we simply refer to DMU for the level 2 DMU and to DMSU or subunit for level 1 DMUs (see Fig. 15.8).

By denoting i; j; k as the indexes of the generic input, output, and DMU, respectively, the following notation is introduced:

- $R_k = \{r_k\}$: the set of indexes r_k of all DMSUs belonging to DMU k,
- $X_k^r = \{x_{ik}^r\}$: the vector of the inputs of DMSU r_k,
- $X_k = \{x_{ik}\}$: the vector of the additional inputs of DMU k,
- $Y_k^r = \{y_{jk}^r\}$: the vector of the outputs of DMSU r_k,

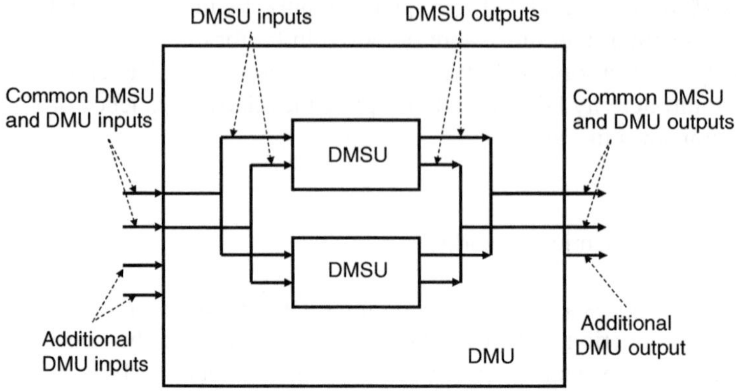

Fig. 15.8 A multi-level DMU: the DMU includes two homogeneous and independent subunits

- $Y_k = \{y_{jk}\}$: the vector of the additional outputs of DMU k,
- $\nu^1 = \{\nu_i^1\}$: the vector of weights of the inputs common to both DMSUs and DMUs,
- $\nu^2 = \{\nu_i^2\}$: the vector of weights of the additional inputs of DMUs,
- $\mu^1 = \{\mu_i^1\}$: the vector of weights of the outputs common to both DMSUs and DMUs,
- $\mu^2 = \{\mu_i^2\}$: the vector of weights of the additional outputs of DMUs.

Accordingly, the efficiency of a DMSU r_k is expressed as

$$e_k^r = \frac{\mu^1 Y_k^r}{\nu^1 X_k^r} \tag{15.20}$$

and the efficiency of a DMU k as

$$e_k = \frac{\mu^1 \sum_{r_k \in R_k} Y_k^r + \mu^2 Y_k}{\nu^1 \sum_{r_k \in R_k} X_k^r + \nu^2 X_k}. \tag{15.21}$$

Cook et al. (1998) present a unifying model for multi-level structures that assesses the efficiency of DMUs of different levels. The authors argue that the efficiency of a DMSU r_k should be evaluated only relative to those other subunits operating under the same conditions, in practice belonging to the same DMU k.

On the other hand, they also assert that the subunits in R_k should be taken into account when evaluating the efficiency of a DMU k. On the basis of these assumptions, Cook et al. (1998) propose that the efficiency of a DMSU 0_0 in DMU 0 is evaluated through the following model

$$e_0^{0*} = \max \mu^1 Y_0^0 \tag{15.22a}$$

$$\nu^1 X_0^0 = 1 \tag{15.22b}$$

$$\mu^1 Y_0^r \leq \nu^1 X_0^r \quad \forall r_0 \in R_0 \tag{15.22c}$$

$$\nu_i^1, \mu_j^1 \geq \varepsilon \quad \forall i, j. \tag{15.22d}$$

This is a standard DEA model that evaluates DMSU 0_0 relative only to subunits included in the same DMU 0. The efficiency of a DMU 0 is evaluated through the following model:

$$e_0^* = \max \mu^1 \sum_{r \in R_0} Y_0^r + \mu^2 Y_0 \tag{15.23a}$$

$$\nu^1 \sum_{r \in R_0} X_0^r + \nu^2 X_0 = 1 \tag{15.23b}$$

$$\mu^1 \sum_{r \in R_k} Y_k^r + \mu^2 Y_k \leq \nu^1 \sum_{r \in R_k} X_k^r + \nu^2 X_k \quad \forall k \tag{15.23c}$$

$$\mu^1 Y_k^r \leq \nu^1 X_k^r \quad \forall k, \forall r \in R_k \tag{15.23d}$$

$$\nu_i^2, \nu_i^1, \mu_j^2, \mu_j^1 \geq \varepsilon \quad \forall i, j. \tag{15.23e}$$

This model compares DMU 0 with all other DMUs. It is different from the linear programming model considering DMUs as black boxes due to the presence of constraints (15.23d). These constraints take into account the DMU internal structure by imposing that their efficiency is related to the efficiencies of their subunits. In particular, constraints (15.23d) force that the optimal values for weights ν_i^1, μ_j^1 are feasible for the DMSUs, i.e., the efficiency of each subunit should not exceed unity. Cook et al. (1998) present a unifying model for multi-level structures that includes both models (15.22) and (15.23). When the DMUs do not have additional inputs/outputs, model (15.23) reduces to the elementary model (15.2). In such case, constraints (15.23c) turn out to be redundant since they are implied by constraints (15.23d). Cook and Green (2005) apply the hierarchical model described in Cook et al. (1998) to the evaluation of power plants. These works are continued by Azadeh et al. (2008, 2011) who use hierarchical models for optimal location of solar plants and wind plants, respectively.

15.6.1 Comparing Subunits Belonging to Different DMUs

In model (15.22) any subunit is compared only against the other subunits belonging to the same DMU. The rationale is that inputs received and decisions taken by each DMU influence the efficiency of its subunits, then comparing subunits belonging to different DMUs would be questionable. In fact, DEA models assess the efficiency of a DMU as a function of its distance from the production frontier defined by the other observed DMUs. In a mathematical programming perspective, DEA models determine the efficiency of a DMU with respect to the other DMUs. In an econometric perspective, the observed DMUs are a sample of a larger population, and DEA is a biased estimator of the efficiency of a DMU with respect to the unknown real production set (Simar and Wilson 2000). In both situations, the larger the sample is, the more likely the DMU under assessment is inefficient. Also, the average efficiency of the DMUs of the sample decreases (Zhang and Bartles 1998). This is why Staat (2001) invites to interpret very carefully possible differences in the efficiencies of subunits belonging to different DMUs when the cardinalities of sets R_k vary. Cook et al. (1998) propose a way of correcting the possible biases by adjusting the efficiency of subunit r_k taking into account the size of the DMU k, the average efficiency of all the subunits in R_k, and the efficiency of DMU k. However, Staat (2002) points out that such a procedure returns different corrections for samples of equal size. He suggests to use bootstrap techniques (see, e.g., Simar and Wilson 2000) to overcome such deficiencies.

In a later paper Cook and Zhu (2007), always dealing with power plants, propose a different model to rectify the weaknesses in the one of Cook et al. (1998). In the new model, the efficiency of each DMSU is now assessed against all the other subunits even if they do not belong to the same DMU. Then, for each DMU, a common set of multipliers applicable to all its DMSUs is determined. Specifically, goal programming is used to identify the multipliers that minimize the maximum discrepancy among the DMSUs' efficiencies from their ideal levels computed in the previous step.

15.6.2 Shared and Multi-level Models

Wu et al. (2008) evaluate the efficiency and performance of the healthcare system in 23 counties and cities in Taiwan for the year 2003. In this paper, the authors propose that each county or city has a budget to produce all the different outputs that can be optimally distributed between such outputs. Hence they propose an input-shared flow model with respect to the available budget. However, Wu et al. (2008) also consider additional inputs (e.g., healthcare manpower and number of facilities) that are not explicitly linked to the different outputs, as in Fig. 15.8. It is then a multi-level model where the amount of input of each subunit is not fixed.

Let us conclude this section underlining that there exists an other DEA literature sub-area called "multi-level". The works in this sub-area deal with the presence of too many input or output flows. Then, they aggregate them in different groups and subgroups (see, e.g., Meng et al. 2008; Kao 2008; Eilat et al. 2008; Rezai and Davoodi 2011). We do not survey these works as they do non assume that DMUs present any internal structure.

15.7 Conclusions

In this work we provide a classification of the main DEA models assessing the efficiency of Decision Making Units when their internal structure is no longer considered as a black box, but insight on their inner processes is available. The interaction in each DMU among the input and output flows and its subunits identifies three broad categories of models. In particular, network DEA models are introduced when intermediate flows among the subunits are taken into account. Shared flow DEA models apply when it is possible to partition a DMU as a collection of components that contend their inputs and/or outputs to other components of the same DMU. Multi-level DEA models are referred to when some of the inputs (or outputs) of a DMU are also inputs (or outputs) of its subunits, and some other inputs (or outputs) are not. We show that these formulations are different generalizations of the same elementary model.

From a theoretical point of view, the knowledge of the internal structure of DMUs should spot the sources of organizational inefficiency by, e.g., preventing compensations among the subunits. In mathematical terms this translates into linking a DMU and its subunits' efficiencies. This relationship may vary across the different models. But, as a general result, a DMU cannot be efficient if none of its subunits are efficient. Furthermore, several applications show that the discrimination power of a DEA model which considers the internal structure of the DMU always increases with respect to the black box approach. As an extreme case, in some situations all DMUs may turn out to be inefficient.

There is large scope of research in the area of this type of DEA models both from a theoretical and application-oriented perspective. In the standard DEA literature, besides the original DEA formulation (Charnes et al. 1978) representing DMUs as black boxes in a CRS environment, many authors have proposed more sophisticated or alternative approaches taking into account, e.g., nonradial measures of efficiency, value judgments, economic measures of efficiency (see Fried et al. 2008, Chap. 3, for a comprehensive survey of such DEA models).

In the recent years, different works have devoted attention to these extensions even when DMUs have an internal structure. However, as pointed out by Chen et al. (2013b), issues still remain and need to be addressed. The presence of an internal structure in fact prevents from generlaizing some of the more obvious properties of the standard DEA models, as an example, the duality relationship between the multiplier-and envelopment-based DEA models. In the authors' opinion, other difficulties may also arise from the level of detail used to describe the internal structure of DMUs. In fact, the greater the detail of the internal structure of a DMU, the greater the discrimination power of a DEA model, but also usually the more difficult to find a sufficient number of homogeneous DMUs to compare. Another promising line of research considers the DEA models from a game theoretic or, in any case, multi-agent perspective. Indeed, from an applicative point of view, these DEA models may find application in the design of more efficient complex processes. In this context, the role of asymmetric information between DMSUs and their DMU (see, e.g., Bogetoft 2000) can be extended to the case when such asymmetry exists among DMSUs which make their decisions by means of a negotiation process.

References

Adler, N., Liebert, V., & Yazhemsky, E. (2013). Benchmarking airports from a managerial perspective. *Omega–International Journal of Management Science, 41*(2), 442–458.

Agrell, P. J., & Hatami-Marbini, A. (2013, to appear). Frontier-based performance analysis models for supply chain management: State of the art and research directions. *Computers & Industrial Engineering*. doi:10.1016/j.cie.2013.02.014.

Akther, S., Fukuyama, H., & Weber, W. L. (2013). Estimating two-stage network Slacks-based inefficiency: An application to Bangladesh banking. *Omega–International Journal of Management Science, 41*(1), 88–96.

Amatatsu, H., Ueda, T., & Amatatsu, Y. (2012). Efficiency and returns-to-scale of local governments. *Journal of the Operational Research Society, 63*(3), 299–305.

Amirteimoori, A. (2013). A DEA two-stage decision processes with shared resources. *Central European Journal of Operational Research, 21*(1), 141–151.

Asai, S. (2011). Efficiency of Japanese local broadcasters. *Journal of Media Economics, 24*(3), 158–173.

Avkiran, N. K. (2009). Opening the black box of efficiency analysis: An illustration with UAE banks. *Omega–International Journal of Management Science, 37*(4), 930–941.

Avkiran, N. K., & McCrystal, A. (2012). Sensitivity analysis of network dea: NSBM versus NRAM. *Applied Mathematics and Computation, 218*(22), 11226–11239.

Azadeh, A., Ghaderi, S., & Maghsoudi, A. (2008). Location optimization of solar plants by an integrated hierarchical DEA PCA approach. *Energy Policy, 36*(10), 3993–4004.

Azadeh, A., Ghaderi, S., & Nasrollahi, M. (2011). Location optimization of wind plants in Iran by an integrated hierarchical data envelopment analysis. *Renewable Energy, 36*, 1621–1631.

Bai-Chen, X., Ying, F., & Qian-Qian, Q. (2012). Does generation form influence environmental efficiency performance? An analysis of China's power system. *Applied Energy, 96*, 261–271.

Banker, R., Charnes, A., & Cooper, W. (1984). Some models for estimating technical and scale inefficiencies in DEA. *Management Science, 30*(9), 1078–1092.

Barnum, D. T., Karlaftis, M. G., & Tandon, S. (2011). Improving the efficiency of metropolitan area transit by joint analysis of its multiple providers. *Transportation Research Part E: Logistics and Transportation Review, 47*(6), 1160–1176.

Beasley, J. (1995). Determining teaching and research efficiencies. *Journal of the Operational Research Society, 46*(4), 441–452.

Bi, G., Feng, C., Ding, J., & Khan, M. R. (2012). Estimating relative efficiency of DMU: Pareto principle and Monte Carlo oriented DEA approach. *INFOR, 50*(1), 44–57.

Bogetoft, P. (2000). DEA and activity planning under asymmetric information. *Journal of Productivity Analysis, 13*(1), 7–48.

Bogetoft, P., Färe, R., Grosskopf, S., Hayes, K., & Taylor, L. (2009). Dynamic network DEA: An illustration. *Journal of the Operations Research Society of Japan, 52*(2), 147–162.

Broekel, T., Rogge, N., & Brenner, T. (2013). *The innovation efficiency of German regions – A shared-input DEA approach*. Technical report, Philipps-Universitët Marburg.

Cao, X., & Yang, F. (2011). Measuring the performance of internet companies using a two-stage data envelopment analysis model. *Enterprise Information Systems, 5*(2), 207–217.

Castelli, L., Pesenti, R., & Ukovich, W. (2001). DEA–like models for efficiency evaluations of specialized and interdependent units. *European Journal of Operational Research, 132*(2), 274–286.

Castelli, L., Pesenti, R., & Ukovich, W. (2004). DEA–like models for the efficiency evaluation of hierarchically structured units. *European Journal of Operational Research, 154*(2), 465–476.

Castelli, L., Pesenti, R., & Ukovich, W. (2010). A classification of DEA models when the internal structure of the Decision Making Units is considered. *Annals of Operations Research, 173*(1), 207–235.

Chang, Y.-C., & Yu, M.-M. (2012, to appear). Measuring production and consumption efficiencies using the slack-based measure network data envelopment analysis approach: The case of low-cost carriers. *Journal of Advanced Transportation*. doi:10.1002/atr.198.

Chang, T.-S., Tone, K., & Wei, Q. (2011, to appear). Ownership-specified network DEA models. *Annals of Operations Research*. doi:10.1007/s10479-011-0949-5.

Chao, C.-M., Yu, M.-M., & Chen, M.-C. (2010). Measuring the performance of financial holding companies. *The Service Industries Journal, 30*(6), 811–829.

Charnes, A., Cooper, W., & Rhodes, E. (1978). Measuring the efficiency of decision making units. *European Journal of Operational Research, 2*(6), 429–444.

Charnes, A., Cooper, W., Golany, B., Seiford, L., & Stutz, J. (1985). Foundation of data envelopment analysis for Pareto-Koopmans efficient empirical production functions. *Journal of Econometrics, 30*(12), 91–107.

Chen, C.-M. (2009). A network-DEA model with new efficiency measures to incorporate the dynamic effect in production networks. *European Journal of Operational Research, 194*(3), 687–699.

Chen, P.-C. (2012). Measurement of technical efficiency in farrow-to-finish swine production using multi-activity network data envelopment analysis: Evidence from Taiwan. *Journal of Productivity Analysis, 38*, 319–331.

Chen, K., & Guan, J. (2012). Measuring the efficiency of China's regional Innovation systems: Application of network data envelopment analysis (DEA). *Regional Studies, 46*(3), 355–377.

Chen, C.-M., & van Dalen, J. (2010). Measuring dynamic efficiency: Theories and an integrated methodology. *European Journal of Operational Research, 203*(3), 749–760.

Chen, C., & Yan, H. (2011). Network DEA model for supply chain performance evaluation. *European Journal of Operational Research, 213*(1), 147–155.

Chen, Y., & Zhu, J. (2004). Measuring information technology's indirect impact on firm performance. *Information Technology and Management, 5*(1–2), 9–22.

Chen, Y., Liang, L., & Yang, F. (2006). A DEA game model approach to supply chain efficiency. *Annals of Operations Research, 145*(1), 5–13.

Chen, Y., Cook, W. D., Ning, L., & Zhu, J. (2009a). Additive efficiency decomposition in two-stage DEA. *European Journal of Operational Research, 196*(3), 1170–1176.

Chen, Y., Liang, L., & Zhu, J. (2009b). Equivalence in two-stage DEA approaches. *European Journal of Operational Research, 193*(2), 600–604.

Chen, Y., Cook, W. D., & Zhu, J. (2010a). Deriving the DEA frontier for two-stage processes. *European Journal of Operational Research, 202*(1), 138–142.

Chen, Y., Du, J., Sherman, H. D., & Zhu, J. (2010b). DEA model with shared resources and efficiency decomposition. *European Journal of Operational Research, 207*(1), 339–349.

Chen, C., Zhu, J., Yu, J.-Y., & Noori, H. (2012a). A new methodology for evaluating sustainable product design performance with two-stage network data envelopment analysis. *European Journal of Operational Research, 221*(2), 348–359.

Chen, P.-C., Chang, C.-C., Yu, M.-M., & Hsu, S.-H. (2012b). Performance measurement for incineration plants using multi-activity network data envelopment analysis: The case of Taiwan. *Journal of Environmental Management, 93*(1), 95–103.

Chen, P.-C., Hsu, S.-H., Chang, C.-C., & Yu, M.-M. (2013a). Efficiency measurements in multi-activity data envelopment analysis with shared inputs: An application to Farmers' Organizations in Taiwan. *China Agricultural Economic Review, 5*(1), 24–42.

Chen, Y., Cook, W. D., Kao, C., & Zhu, J. (2013b). Network DEA pitfalls: Divisional efficiency and frontier projection under general network structures. *European Journal of Operational Research, 226*, 507–515.

Chilingerian, J., & Sherman, H. D. (2004). Health care applications: From hospitals to physician, from productive efficiency to quality frontiers. In W. W. Cooper, L. M. Seiford, & J. Zhu (Eds.), *Handbook on data envelopment analysis*. Boston: Springer.

Chiou, Y.-C., & Chen, Y.-H. (2006). Route-based performance evaluation of Taiwanese domestic airlines using data envelopment analysis. *Transportation Research Part E: Logistics and Transportation Review, 42*(2), 116–127.

Chiou, Y.-C., & Lan, L. W. (2007). Integrated data envelopment analysis models for measuring transport efficiency and effectiveness. *Journal of the Eastern Asia Society for Transportation Studies, 7*, 427–440.

Chiou, Y.-C., Lan, L. W., & Yen, B. T. (2010). A joint measurement of efficiency and effectiveness for non-storable commodities: Integrated data envelopment analysis approaches. *European Journal of Operational Research, 201*(2), 477–489.

Cook, W. D., & Green, R. (2004). Multicomponent efficiency measurement and core business identification in multiplant firms: A DEA model. *European Journal of Operational Research, 157*(3), 540–551.

Cook, W. D., & Green, R. (2005). Evaluating power plant efficiency: A hierarchical model. *Computers and Operations Research, 4*(9), 813–823.

Cook, W. D., & Hababou, M. (2001). Sales performance measurement in bank branches. *Omega–International Journal of Management Science, 29*(4), 299–307.

Cook, W. D., & Seiford, L. M. (2009). Data envelopment analysis (DEA) – Thirty years on. *European Journal of Operational Research, 192*(1), 1–17.

Cook, W. D., & Zhu, J. (2005). *Modeling performance measurement: Applications and implementation issues in DEA*. Berlin: Springer.

Cook, W. D., & Zhu, J. (2007). Within-group common weights in DEA: An analysis of power plant efficiency. *European Journal of Operational Research, 178*(1), 207–216.

Cook, W. D., Chai, D., Doyle, J., & Green, R. (1998). Hierarchies and groups in DEA. *Journal of Productivity Analysis, 10*(2), 177–198.

Cook, W. D., Hababou, M., & Tuenter, H. (2000). Multicomponent efficiency measurement and shared inputs in data envelopment analysis: An application to sales and service performance in bank branches. *Journal of Productivity Analysis, 14*(3), 209–224.

Cook, W. D., Liang, L., Yang, F., & Zhu, J. (2007). DEA models for supply chain or multi-stage structure. In J. Zhu & W. D. Cook (Eds.), *Modeling data irregularities and structural complexities in data envelopment analysis* (pp. 189–208). New York: Springer.

Cook, W. D., Liang, L., & Zhu, J. (2010a). Measuring performance of two-stage network structures by DEA: A review and future perspective. *Omega–International Journal of Management Science, 38*(6), 423–430.

Cook, W. D., Zhu, J., Bi, G., & Yang, F. (2010b). Network DEA: Additive efficiency decomposition. *European Journal of Operational Research, 207*(2), 1122–1129.

Cooper, W., Park, K., & Pastor, J. (1999). RAM: A range adjusted measure of inefficiency for use with additive models, and relations to other models and measures in DEA. *Journal of Productivity Analysis, 11*(1), 5–42.

Cooper, W., Seiford, L., & Tone, K. (2000). *Data envelopment analysis*. Boston: Kluwer.

da Cruz, N. F., Carvalho, P., & Marques, R. C. (2013). Disentangling the cost efficiency of jointly provided water and wastewater services. *Utilities Policy, 24*, 70–77.

Diez-Ticio, A., & Mancebon, M. (2002). The efficiency of the Spanish police service: An application of the multiactivity DEA model. *Applied Economics, 34*(3), 351–362.

Eilat, H., Golany, B., & Shtub, A. (2008). R&D project evaluation: An integrated DEA and balanced scorecard approach. *Omega–International Journal of Management Science, 36*(5), 895–912.

Emrouznejad, A., & Thanassoulis, E. (2005). A mathematical model for dynamic efficiency using data envelopment analysis. *Applied Mathematics and Computation, 160*(2), 363–378.

Fallah-Fini, S., Triantis, K., & Johnson, A. (2013, to appear). Reviewing the literature on non-parametric dynamic efficiency measurement: State-of-the-art. *Journal of Productivity Analysis*. doi:10.1007/s11123-013-0349-8.

Färe, R. (1991). Measuring Farrell efficiency for a firm with intermediate inputs. *Academia Economic Papers, 19*, 329–340.

Färe, R., & Grosskopf, S. (1996a). *Intertemporal production frontier: With dynamic DEA*. Boston: Kluwer.

Färe, R., & Grosskopf, S. (1996b). Productivity and intermediate products: A frontier approach. *Economics Letters, 50*(1), 65–70.

Färe, R., & Grosskopf, S. (2000). Network DEA. *Socio-Economic Planning Sciences, 34*(1), 35–49.

Färe, R., & Primont, D. (1984). Efficiency measures for multiplant firms. *Operations Research Letters, 3*(5), 257–260.

Färe, R., & Whittaker, G. (1995). An intermediate input model of dairy production using complex survey data. *Journal of Agricultural Economics, 46*(2), 201–213.

Färe, R., Grosskopf, S., & Lovell, C. (1985). *The measurement of efficiency of production*. Dordrecht: Kluwer-Nijhoff.

Färe, R., Grabowski, R., Grosskopf, S., & Kraft, S. (1997). Efficiency of a fixed but allocable input: A non-parametric approach. *Economics Letters, 56*(2), 187–193.

Färe, R., Grosskopf, S., & Zaim, O. (2002). Hyperbolic efficiency and return to the dollar. *European Journal of Operational Research, 136*(3), 671–679.

Färe, R., Grosskopf, S., & Whittaker, G. (2007). Network DEA. In J. Zhu & W D. Cook (Eds.), *Modeling data irregularities and structural complexities in data envelopment analysis* (pp. 209–240). New York: Springer.

Farrell, M. (1957). The measurement of productive efficiency. *Journal of the Royal Statistical Society, Series A (General), 120*(3), 253–281.

Fried, H., Knox Lovell, C., & Schmidt, S. (2008). *The measurement of productive efficiency and productive Growth*. New York: Oxford University Press.

Fukuyama, H. (2003). Scale characterizations in a DEA directional technology distance function framework. *European Journal of Operational Research, 144*(1), 108–127.

Fukuyama, H., & Matousek, R. (2011). Efficiency of Turkish banking: Two-stage network system. Variable returns to scale model. *Journal of International Financial Markets, Institutions and Money, 21*(1), 75–91.

Fukuyama, H., & Mirdehghan, S. (2012). Identifying the efficiency status in network DEA. *European Journal of Operational Research, 220*(1), 85–92.

Fukuyama, H., & Weber, W. L. (2009). A directional slack-based measure of technical inefficiency. *Socio-Economic Planning Sciences, 43*(4), 274–287.

Fukuyama, H., & Weber, W. L. (2010). A slacks-based inefficiency measure for a two-stage system with bad outputs. *Omega–International Journal of Management Science, 38*(5), 398–409.

Golany, B., & Thore, S. (1997). Restricted best practice selection in DEA: An overview with a case study evaluating the socio-economic performance of nations. *Annals of Operations Research, 73*, 117–140.

Golany, B., Hackman, S., & Passy, U. (2006). An efficiency measurement framework for multi-stage production systems. *Annals of Operations Research, 145*(1), 51–68.

Grigoroudis, E., Tsitsiridi, E., & Zopounidis, C. (2013). Linking customer satisfaction, employee appraisal, and business performance: An evaluation methodology in the banking sector. *Annals of Operations Research, 205*(1), 5–27.

Guan, J., & Chen, K. (2010). Measuring the innovation production process: A cross-region empirical study of China's high-tech innovations. *Technovation, 30*(5–6), 348–358.

Guan, J., & Chen, K. (2012). Modeling the relative efficiency of national innovation systems. *Research Policy, 41*(1), 102–115.

Hahn, J.-S., Kim, H.-R., & Kho, S.-Y. (2011). Analysis of the efficiency of Seoul Arterial Bus routes and its determinant factors. *KSCE Journal of Civil Engineering, 15*(6), 1115–1123.

Holod, D., & Lewis, H. F. (2011). Resolving the deposit dilemma: A new DEA bank efficiency model. *Journal of Banking and Finance, 35*(11), 2801–2810.

Hsieh, L.-F., & Lin, L.-H. (2010). A performance evaluation model for international tourist hotels in Taiwan-an application of the relational network dea. *International Journal of Hospitality Management, 29*(1), 14–24.

Huang, Y., Ge, H., & Feng, Y.-J. (2009). Relational network DEA model based on series system: Application of 14 commercial banks in China. *Xitong Gongcheng Lilun yu Shijian/System Engineering Theory and Practice, 29*(5), 106–114.

Kao, C. (2000). Short-run and long-run efficiency measures for multiplant firms. *Annals of Operations Research, 97*(1), 379–388.

Kao, C. (2008). A linear formulation of the two-level DEA model. *Omega–International Journal of Management Science, 36*, 958–962.

Kao, C. (2009a). Efficiency decomposition in network data envelopment analysis: A relational model. *European Journal of Operational Research, 192*(3), 949–962.

Kao, C. (2009b). Efficiency measurement for parallel production systems. *European Journal of Operational Research, 196*(3), 1107–1112.

Kao, C. (2012, to appear). Dynamic data envelopment analysis: A relational analysis. *European Journal of Operational Research* (0). doi:10.1016/j.ejor.2012.12.012.

Kao, H.-Y., & Chan, C.-Y. (2013). A discriminative multi-objective programming method for solving network DEA. *Smart Innovation, Systems and Technologies, 20*, 327–335.

Kao, C., & Hwang, S. (2008). Efficiency decomposition in two-stage data envelopment analysis: An application to non-life insurance companies in Taiwan. *European Journal of Operational Research, 185*(1), 418–419.

Kao, C., & Hwang, S.-N. (2010). Efficiency measurement for network systems: IT impact on firm performance. *Decision Support Systems, 48*(3), 437–446.

Kao, C., & Hwang, S.-N. (2011). Decomposition of technical and scale efficiencies in two-stage production systems. *European Journal of Operational Research, 211*(3), 515–519.

Kao, C., & Lin, P.-H. (2012). Efficiency of parallel production systems with fuzzy data. *Fuzzy Sets and Systems, 198*, 83–98.

Lee, C.-Y., & Johnson, A. (2011). A decomposition of productivity change in the semiconductor manufacturing industry. *International Journal of Production Research, 49*(16), 4761–4785.

Lee, C.-Y., & Johnson, A. L. (2012). Two-dimensional efficiency decomposition to measure the demand effect in productivity analysis. *European Journal of Operational Research, 216*(3), 584–593.

Lewis, H., & Sexton, T. (2004). Network DEA: Efficiency analysis of organizations with complex internal structure. *Computers and Operations Research, 31*(9), 1365–1410.

Lewis, H., Lock, K., & Sexton, T. (2009). Organizational capability, efficiency, and effectiveness in Major League Baseball: 1901–2002. *European journal of Operational Research, 197*, 731–740.

Lewis, H. F., Mallikarjun, S., & Sexton, T. R. (2013). Unoriented two-stage DEA: The case of the oscillating intermediate products. *European Journal of Operational Research, 229*(2), 529–539.

Li, W. (2012). Research on the organizational efficiency evaluation of China's railway transport industry with network DEA. *Advances in Information Sciences and Service Sciences, 4*(23), 772–778.

Li, Y., Chen, Y., Liang, L., & Xie, J. (2012). DEA models for extended two-stage network structures. *Omega–International Journal of Management Science, 40*(5), 611–618.

Liang, L., Yang, F., Cook, W., & Zhu, J. (2006). DEA models for supply chain efficiency evaluation. *Annals of Operations Research, 145*(1), 35–49.

Liang, L., Cook, W. D., & Zhu, J. (2008). DEA models for two-stage processes: Game approach and efficiency decomposition. *Naval Research Logistics, 55*(7), 643–653.

Liang, L., Li, Z.-Q., Cook, W., & Zhu, J. (2011). Data envelopment analysis efficiency in two-stage networks with feedback. *IIE Transactions, 43*, 309–322.

Lim, S., & Zhu, J. (2013). Integrated data envelopment analysis: Global vs. local optimum. *European Journal of Operational Research, 229*(1), 276–278.

Lin, E. T. (2008). Route-based performance evaluation of Taiwanese domestic airlines using data envelopment analysis: A comment. *Transportation Research Part E: Logistics and Transportation Review, 44*(5), 894–899.

Lin, T.-Y., & Chiu, S.-H. (2013). Using independent component analysis and network DEA to improve bank performance evaluation. *Economic Modelling, 32*(1), 608–616.

Liu, S.-T. (2011). A note on efficiency decomposition in two-stage data envelopment analysis. *European Journal of Operational Research, 212*(3), 606–608.

Liu, J. S., & Lu, W.-M. (2012). Network-based method for ranking of efficient units in two-stage DEA models. *Journal of the Operational Research Society, 63*, 1153–1164.

Liu, S.-T., & Wang, R.-T. (2009). Efficiency measures of PCB manufacturing firms using relational two-stage data envelopment analysis. *Expert Systems with Applications, 36* (3, Part 1), 4935–4939.

Liu, W.-B., Tso, C.-W., Peng, K.-C., Chiu, C.-C., & Shih, M.-L. (2012). Application of network DEA in the assessment of operating performances of farmers associations in Taiwan. *Journal of Interdisciplinary Mathematics, 15*(1), 65–81.

Liu, J., Lu, L., Lu, W.-M., & Lin, B. (2013a). A survey of DEA applications. *Omega–International Journal of Management Science, 41*(5), 893–902.

Liu, J. S., Lu, L. Y., Lu, W.-M., & Lin, B. J. (2013b). Data envelopment analysis 1978–2010: A citation-based literature survey. *Omega–International Journal of Management Science, 41*(1), 3–15.

Löthgren, M., & Tambour, M. (1999). Productivity and customer satisfaction in Swedish pharmacies. A DEA network model. *European Journal of Operational Research, 115*(3), 449–458.

Lozano, S. (2011). Scale and cost efficiency analysis of networks of processes. *Expert Systems with Applications, 38*(6), 6612–6617.

Lozano, S., Gutiérrez, E., & Moreno, P. (2013). Network DEA approach to airports performance assessment considering undesirable outputs. *Applied Mathematical Modelling, 37*(4), 1665–1676.

Lu, W.-M., Wang, W.-K., Hung, S.-W., & Lu, E.-T. (2012). The effects of corporate governance on airline performance: Production and marketing efficiency perspectives. *Transportation Research Part E: Logistics and Transportation Review, 48*(2), 529–544.

Luenberger, D. G. (1992). Benefit function and duality. *Journal of Mathematical Economics, 26*(5), 461–481.

Mallikarjun, S., Lewis, H., & Sexton, T. (2014). Operational Performance of US Public Rail Transit and Implications for Public Policy Socio-Economic Planning Sciences, Socio-Economic Planning Sciences (in press).

Mar Molinero, C. (1996). On the joint determination of efficiencies in a data envelopment analysis context. *Journal of the Operational Research Society, 47*(10), 1273–1279.

Mar Molinero, C., & Tsai, P. (1997). Some mathematical properties of a DEA model for the joint determination of efficiencies. *Journal of the Operational Research Society, 48*(1), 51–56.

Matthews, K. (2013). Risk management and managerial efficiency in Chinese banks: A network DEA framework. *Omega–International Journal of Management Science, 41*(2), 207–215.

Meng, W., Zhang, D., Qi, L., & Liu, W. (2008). Two-level DEA approaches in research evaluation. *Omega–International Journal of Management Science, 36*, 950–957.

Mirhedayatian, S., Azadi, M., & Farzipoor Saen, R. (2013, to appear). A novel network data envelopment analysis model for evaluating green supply chain management. *International Journal of Production Economics*. doi:10.1016/j.ijpe.2013.02.009.

Moreno, P., & Lozano, S. (2012, to appear). A network DEA assessment of team efficiency in the NBA. *Annals of Operations Research*. doi:10.1007/s10479-012-1074-9.

Nemoto, J., & Goto, M. (1999). Dynamic data envelopment analysis modeling intertemporal behavior of a firm in the presence of productive inefficiencies. *Economics Letters, 64*, 51–56.
Nemoto, J., & Goto, M. (2003). Measuring dynamic efficiency in production: An application of data envelopment analysis to Japanese electric utilities. *Journal of Productivity Analysis, 19*, 191–210.
Paradi, J., Rouatt, S., & Zhu, H. (2011). Two-stage evaluation of bank branch efficiency using data envelopment analysis. *Omega–International Journal of Management Science, 39*(1), 99–109.
Pastor, J., Ruiz, J., & Sirvent, I. (1999). An enhanced DEA Russell graph efficiency measure. *European Journal of Operational Research, 115*(3), 596–607.
Premachandra, I. M., Zhu, J., Watson, J., & Galagedera, D. U. A. (2012). Best-performing US mutual fund families from 1993 to 2008: Evidence from a novel two-stage DEA model for efficiency decomposition. *Journal of Banking & Finance, 36*(12), 3302–3317.
Prieto, A., & Zofio, J. (2007). Network DEA efficiency in input-output models: With an application to OECD countries. *European Journal of Operational Research, 178*(1), 292–304.
Rezai, H. Z., & Davoodi, A. (2011). Some remarks on the two-level DEA model. *Applied Mathematics Letters, 24*(6), 969–974.
Rogge, N., & De Jaeger, S. (2012). Evaluating the efficiency of municipalities in collecting and processing municipal solid waste: A shared input DEA-model. *Waste Management, 32*(10), 1968–1978.
Rogge, N., & De Jaeger, S. (2013). Measuring and explaining the cost efficiency of municipal solid waste collection and processing services. *Omega–International Journal of Management Science, 41*(4), 653–664.
Salerno, C. (2006). Using data envelopment analysis to improve estimates of higher education institution's per-student education costs. *Education Economics, 14*(3), 281–295.
Saranga, H., & Moser, R. (2010). Performance evaluation of purchasing and supply management using value chain DEA approach. *European Journal of Operational Research, 207*(1), 197–205.
Seiford, L., & Zhu, J. (1999). Profitability and marketability of the top 55 U.S. commercial banks. *Management Science, 45*(9), 1270–1288.
Sengupta, J. K. (1995). *Dynamics of data envelopment analysis: Theory of systems efficiency*. Dordrecht/Boston: Kluwer.
Sexton, T., & Lewis, H. (2003). Two–stage DEA: An application to Major League Baseball. *Journal of Productivity Analysis, 19*(2), 227–249.
Shephard, R. (1970). *Theory of cost and production frontier*. Princeton: Princeton University Press.
Sheth, C., Triantis, K., & Teodorović, D. (2007). Performance evaluation of bus routes: A provider and passenger perspective. *Transportation Research Part E: Logistics and Transportation Review, 43*, 453–478.
Simar, L., & Wilson, P. (2000). Statistical inference in non papametic frontier models: The state of the art. *Journal of Productivity Analysis, 13*(1), 49–78.
Staat, M. (2001). The effect of sample size on the mean efficiency in DEA: Comment. *Journal of Productivity Analysis, 15*(2), 129–137.
Staat, M. (2002). Bootstrapped efficiency estimates for a model for groups and hierarchies in DEA. *European Journal of Operational Research, 138*(1), 1–8.
Sueyoshi, T., & Sekitani, K. (2005). Returns to scale in dynamic DEA. *European Journal of Operational Research, 161*, 536–544.
Thompson, R., Langemeier, L., Lee, C., Lee, E., & Thrall, R. (1990). The role of multiplier bounds in efficiency analysis with application to Kansas farming. *Journal of Econometrics, 46*(1–2), 93–108.
Tone, K., & Tsutsui, M. (2009). Network DEA: A slacks-based measure approach. *European Journal of Operational Research, 197*(1), 243–252.
Tone, K., & Tsutsui, M. (2010). Dynamic DEA: A slacks-based measure approach. *Omega–International Journal of Management Science, 38*(3–4), 145–156.

Tone, K., & Tsutsui, M. (2014). Dynamic DEA with network structure: A slacks-based measure approach. *Omega–International Journal of Management Science, 42*(1), 124–131.

Troutt, M., Ambrose, P., & Chan, C. (2001). Optimal throughput for multistage input-output processes. *International Journal of Operations and Production Management, 21*, 148–158.

Tsai, P., & Mar Molinero, C. (2002). A variable returns to scale data envelopment analysis model for the joint determination of efficiencies with an example of the UK health service. *European Journal of Operational Research, 141*(1), 21–38.

Vaz, C. B., Camanho, A. S., & Guimaraes, R. C. (2010). The assessment of retailing efficiency using network data envelopment analysis. *Annals of Operations Research, 173*(1), 5–24.

Von Geymueller, P. (2009). Static versus dynamic DEA in electricity regulation: The case of US transmission system operators. *Central European Journal of Operational Research, 17*, 397–413.

Wang, Y.-M., & Chin, K.-S. (2010). Some alternative DEA models for two-stage process. *Expert Systems with Applications, 37*(12), 8799–8808.

Wanke, P. (2013a). Physical infrastructure and flight consolidation efficiency drivers in Brazilian airports: A two-stage network-DEA approach. *Journal of Air Transport Management, 31*, 1–5.

Wanke, P. F. (2013b). Physical infrastructure and shipment consolidation efficiency drivers in Brazilian ports: A two-stage network-DEA approach. *Transport Policy, 29*, 145–153.

Wei, Q. L., & Chang, T.-S. (2011). Optimal system design series-network DEA models. *Journal of the Operational Research Society, 62*, 1109–1119.

Wu, D. D. (2010). Bilevel programming data envelopment analysis with constrained resource. *European Journal of Operational Research, 207*, 856–864.

Wu, D. D., & Birge, J. R. (2012). Serial chain merger evaluation model and application to mortgage banking. *Decision Sciences, 43*(1), 5–36.

Wu, C.-H., Chang, C.-C., & Kuo, K. K. (2008). Evaluating the resource allocation efficiency of the healthcare system in Taiwan. *International Journal of Public Policy, 3*(5/6), 403–418.

Yang, C.-C. (2009). Productive efficiency, environmental efficiency and their determinants in farrow-to-finish pig farming in Taiwan. *Livestock Science, 126*(1–3), 195–205.

Yang, C., & Liu, H.-M. (2012). Managerial efficiency in Taiwan bank branches: A network DEA. *Economic Modelling, 29*(2), 450–461.

Yang, Y., Ma, B., & Koike, M. (2000). Efficiency–measuring DEA model for production system with k independent subsystems. *Journal of the Operations Research Society of Japan, 43*(3), 343–354.

Yang, F., Wu, D., Liang, L., Bi, G., & Wu, D. (2011). Supply chain DEA: Production possibility set and performance evaluation model. *Annals of Operations Research, 185*(1), 195–211.

Yu, M.-M. (2008a). Assessing the technical efficiency, service effectiveness, and technical effectiveness of the world's railways through NDEA analysis. *Transportation Research Part A: Policy and Practice, 42*(10), 1283–1294.

Yu, M.-M. (2008b). Measuring the efficiency and return to scale status of multi-mode bus transit – Evidence from Taiwan's bus system. *Applied Economics Letters, 15*(8), 647–653.

Yu, M.-M. (2010). Assessment of airport performance using the SBM-NDEA model. *Omega–International Journal of Management Science, 38*(6), 440–452.

Yu, M.-M. (2012). An integration of the multi-component DEA and GAR models to the measurement of hotel performance. *Current Issues in Tourism, 15*(5), 461–476.

Yu, M.-M., & Chen, P.-C. (2011). Measuring air routes performance using a fractional network data envelopment analysis model. *Central European Journal of Operational Research, 19*(1), 81–98.

Yu, M.-M., & Fan, C.-K. (2006). Measuring the cost effectiveness of multimode bus transit in the presence of accident risks. *Transportation Planning and Technology, 29*(5), 383–407.

Yu, M.-M., & Fan, C.-K. (2009). Measuring the performance of multimode bus transit: A mixed structure network DEA model. *Transportation Research Part E: Logistics and Transportation Review, 45*(3), 501–515.

Yu, M.-M., & Lee, B. C. (2009). Efficiency and effectiveness of service business: Evidence from international tourist hotels in Taiwan. *Tourism Management, 30*(4), 571–580.

Yu, M.-M., & Lin, E. (2008). Efficiency and effectiveness in railway performance using a multi-activity network DEA model. *Omega–International Journal of Management Science, 36*, 1005–1017.

Zha, Y., & Liang, L. (2010). Two-stage cooperation model with input freely distributed among the stages. *European Journal of Operational Research, 205*(2), 332–338.

Zha, Y., Liang, L., & Xu, C. y. (2008). Two-stage BCC model for cooperative efficiency evaluation using a geometric mean method. *Systems Engineering – Theory & Practice, 28* (10), 53–58.

Zhang, Y., & Bartles, R. (1998). The effect of sample size on the mean efficiency in DEA with an application to electricity distribution in Australia, Sweden and New Zealand. *Journal of Productivity Analysis, 9*(2), 187–204.

Zhang, Q., & Ma, J. b. (2011). Research on business efficiency of hotel and tourism enterprises based on the influence of innovation factors. *Energy Procedia, 5*, 742–746.

Zhao, Y., Triantis, K., Murray-Tuite, P., & Edara, P. (2011). Performance measurement of a transportation network with a downtown space reservation system: A network-DEA approach. *Transportation Research Part E: Logistics and Transportation Review, 47*(6), 1140–1159.

Zhu, J. (2011). Airlines performance via Two-stage network DEA approach. *Journal of CENTRUM Cathedra: The Business and Economics Research Journal, 4*(2), 260–269.

Chapter 16
Multicomponent Efficiency Measurement in Banking

Wade D. Cook, M. Hababou, and H. Tuenter

Abstract There is a growing need to view performance in organizations in a more disaggregated sense, paying specific attention to different components of the operation. In this chapter we present models for deriving aggregate measures of bank-branch performance, with accompanying component measures that make up that aggregate value. The technical difficulty surrounding the development of an appropriate model is the presence of shared resources on the input side and mechanisms for allocating such resources to the individual components. The chapter presents both a conventional radial model as well as an additive model for handling multiple components in an organization. The models are applied to data for a set of bank branches.

Keywords Multi component • Bank branches • Shared resources • Efficiency

---

This chapter is based upon Cook et al. 2000, with permission from Kluwer Academic Publishers, and Cook and Hababou (2001), with permissions from Elsevier Science

W.D. Cook (✉)
Schulich School of Business, York University, 4700 Keele Street, M3J 1P3 Toronto, ON, Canada
e-mail: wcook@schulich.yorku.ca

M. Hababou
Royal Bank of Scotland, New York, NY, USA
e-mail: Moez.hababou@rbsnb.com

H. Tuenter
Hydro One, 483 Bay Street North Tower, Toronto, Canada
e-mail: htuenter@gmail.com

W.D. Cook and J. Zhu (eds.), *Data Envelopment Analysis*, International Series in Operations Research & Management Science 208,
DOI 10.1007/978-1-4899-8068-7_16,

16.1 Introduction

Banks have evolved over time from their traditional role as reactive monetary intermediaries, and *service* providers, toward a more general and proactive function as universal financial agents with a distinct *sales* culture. This new status has resulted in the introduction of a broad range of financial products to the market place. Under the Canadian Bank Act of 1991, it became legal for an institution to engage in a broad range of financial activities. Technology has contributed as well to the changes that banks are undergoing; a range of convenient customer access points has emerged such as ATMs (Automatic Teller Machines), debit cards, telephone- and PC banking, to name a few.

Banks generate profits from two main sources – (1) interest income, which captures the spread realized on loans and traditional activities, and (2) non-interest income from fees and financial services activities. While historically interest income was the principal source of profits for the bank, the importance of non-interest income has grown significantly over time. It is interesting to note that the profitability ratio, that is the profit as a percentage of assets, has increased dramatically since 1991. Specifically, for the period 1980–1990, the ratio ranged from 0.24 % to 0.79 %, with an average of 0.43 %; the corresponding figures for the period 1991–1995 are 0.59–1.90 % with an average of 1.20 %. This dramatic change has been due in part to the revised regulations in the Bank Act, and partially to improved access to financial services, coupled with a more active sales orientation.

Performance measurement, using tools such as Data Envelopment Analysis (DEA), as proposed by Charnes et al. (1978), has tended to concentrate on achieving a *single measure* for each member of a set of decision making units (DMUs). In most applications, a *single* measure of production or profit efficiency provided by the DEA methodology has been an adequate and useful means of comparing units and identifying best performance. This has been particularly true in the case of banks, where the primary candidates for DMUs are branches, and in their traditional setting, product and prices have tended to be undifferentiated. Numerous studies of bank-branch efficiency using DEA have been conducted over the past 15 years – see, for instance, Charnes et al. (1990), Oral and Yolalan (1990), Schaffnit et al. (1997), Sherman and Gold (1985), and Sherman and Ladino (1995).

There is now a desire to create value-added customer segments by identifying their specific needs. The new challenge is to optimize resource allocation, with most of the industry now allocating 60–80 % of its human capital to customers and markets that represent less than 20 % of its customer base. There is a growing need to view performance in a more dis-aggregated sense, paying specific attention to different components of the operation. These components include different classes of products or sales activities, such as mutual funds and mortgages, and different elements of service. By measuring a branch's performance on each of a set of such components, particular areas of strength and weakness can be identified and addressed, where necessary.

In this chapter we present models for deriving *aggregate* measures of bank-branch performance, with accompanying *component* measures that make up that aggregate value. The technical difficulty surrounding the development of an appropriate model has to do with the presence of *shared resources* on the input side, and mechanisms for allocating such resources to the individual components.

The idea of measuring efficiency relative to certain subprocesses or components of a DMU is not new. Färe and Grosskopf (1996), for example, look at a multistage process wherein intermediate products or outputs at one stage, can be both final products and inputs to later stages of production. Those authors are not explicitly interested in obtaining measures of efficiency at each stage, but rather are concerned with overall efficiency measurement, whereby the network structure of the intermediate activity explicitly enters into the model description. Hence, they are able to provide a better representation of the technology than would a 'black box' input and final output model. Another example is due to Färe and Primont (1984) and involves the evaluation of efficiency of a set of multiplant firms as DMUs, while at the same time measuring the efficiency of plants within firms.

These applications of multicomponent efficiency measurement do not involve shared resources as does the situation examined herein. The work of Beasley (1995) on separating teaching and research, most closely compares to the present application, although we show herein that our treatment of shared resources leads to a linear rather than a nonlinear model. Section 16.2 modifies the conventional radial projection DEA model for bank-branch performance by providing a methodology for splitting shared inputs among the identified components. For development purposes, we concentrate on two specific components, namely service-specific and product-specific sales activities. The model structure used is based on the original CRS model of Charnes et al. (1978). An application is examined in Sect. 16.3. In Sect. 16.4 we present an additive form of the multi-component model. Discussion and conclusions follow in Sect. 16.5.

16.2 A Multicomponent Performance Measurement Model

With the increased emphasis on sales and the differentiation of products and customer segments, there is a need to provide a performance measurement tool with component-based information as part of the aggregate efficiency score.

16.2.1 Multiple Functions and Shared Resources

While one may wish to measure the performance of *several* components of the DMU, we will, for purposes of development in this chapter, assume that transactions can be separated into exactly *two* distinct classes: service and sales. It should be emphasized that this split is not always transparent; the opening of a mortgage loan would generally be classified as a "sales" transaction, although there are "service" activities that must be performed from time to time pertaining to that loan, such as

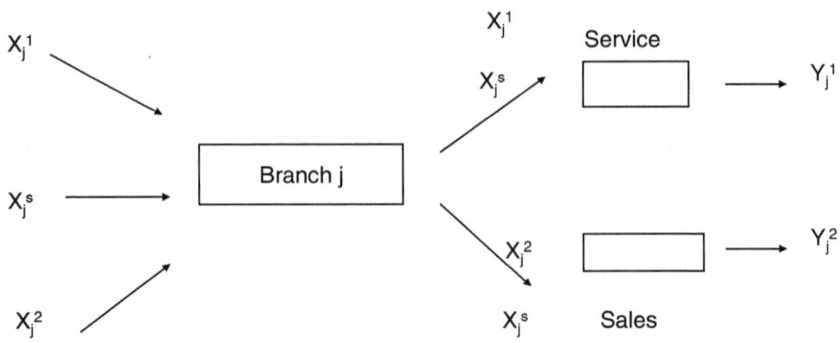

Fig. 16.1 Production process for a DMU*j with Shared Resources*

loan renewal. Thus, a particular transaction may contain both sales and service components. Care should, therefore, be exercised in clearly delineating those activities that belong to each function. Furthermore, one would generally need to separate those sales activities that are volume related (and pertain to specific products), from those that involve the "selling" part of the sales activities. The latter would include reviewing customer portfolios, answering customer requests on various products, and so on. The former would involve the transaction tasks performed after the customer has chosen a particular product. In summary, the selling aspect of sales does not relate to specific sales products while the transaction part of sales is product-specific. In this section we consider only those sales activities that are product or volume specific. We take up the non-volume related activities in a later section.

For notational purposes, let (Y_j^1, Y_j^2) denote the sets of service and sales transactions, respectively, i.e. the two sets of *outputs* are

$$Y_j^1 = \left(y_{j1}^1 \dots, y_{jJ_1}^1\right) \quad \text{and} \quad Y_j^2 = \left(y_{j1}^2 \dots, y_{jJ_2}^2\right).$$

On the *input* side, this split is more complex. Some resources can be designated as *dedicated* service inputs, some as dedicated to sales, and still others are *shared* by the two functions. If, for example, branch staff are classified as Sales, Service, and Support, we can, for illustrative purposes, assume that Support staff are shared by the two functions while the other two classes are dedicated. In some branches this distinction may be less clear than in others. Technology resources may as well be classified as shared.

A schematic of the *production process* for a particular DMU is given in Fig. 16.1.

Here, X_j^1, X_j^2 and X_j^s denote I_1, I_2 and I_s -dimensional vectors of service dedicated-, sales dedicated-, and shared inputs, respectively. Some portion α_i $(0 \leq \alpha_i \leq 1)$ of the shared resource x_{ij}^s is allocated to the service function of DMU j, with the remainder $(1 - \alpha_i)$ being allocated to sales. In the model to be developed herein, α_i is a decision variable to be set by the DMU. At least two difficulties arise in attempting to capture a measure of performance of the DMU on both service and sales functions within

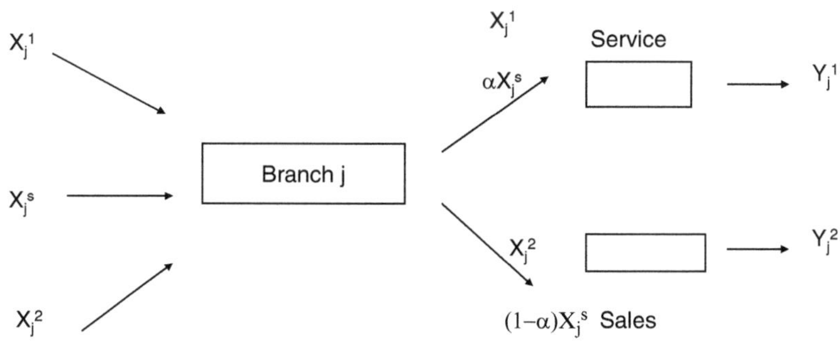

Fig. 16.2 Splitting shared resources

some *overall* efficiency measure. First, if one attempts to derive an overall measure of performance that somehow incorporates sales and service components, the importance of the components of X^s relative to one another, and relative to the dedicated resources X^1 and X^2 (as reflected in the v -vectors v^1, v^2 and v^s), may be different when considering the impact of X^s on Y^1 as compared to its impact on Y^2. For example, consider the simple case of one staff type for each dedicated class (X^1 = no. service staff, X^2 = no. sales staff), and two resources, support staff and available technology, as shared inputs. One may argue that in evaluating service efficiency, technology is more important than support staff. As an example, a constraint such as $v_2^s \geq 2v_1^s$ might be imposed. On the other hand, if technology such as ATMs play a minor role in sales, then a constraint such as $v_2^s \leq 0.3v_1^s$ may be an accurate reflection of the importance of the two shared resources relative to one another. Clearly, these constraints are infeasible if imposed simultaneously. Moreover, even if this issue could be resolved, there would be no clear way of separating the resulting aggregate measure into separate sales and service indicators.

A second difficulty arises if instead of developing an aggregate measure, one attempts to derive separate measures of performance relative to sales and service, with the intention of combining these separate measures into an aggregate score after the fact. The problem here is that the shared resources X^s would need to be *apportioned* to these two functions in some manner consistent with their usage in creating the outputs of the functions. With any shared resources, however, branches do not generally maintain a record of the usage split at the function level. Consequently, a mechanism is needed to split shared resources across functions in some equitable manner. To motivate the development, reconsider Fig. 16.1, but with the shared resources X_j^s allocated to the two functions according to *proportionality variables*, α_i as depicted in Fig. 16.2. The issue of how α_i should be derived is discussed below. Let $\alpha = (\alpha_1, \alpha_2, \ldots \alpha_{I_s})^T$ denote the column vector of proportionality variables, and let $\alpha\ X_j^s$ denote the column vector $(\alpha_1 x_{1j}^s,\ \alpha_2\ x_{2j}^s,\ \ldots\ \alpha_I x_{jI}^s)^T$. Further, we let $(1-\alpha)\,X_j^s$ denote the column vector $((1-\alpha_1)\,x_{1j}^s, (1-\alpha_2)\,x_{2j}^s, \ldots,$ $(1-\alpha_{I_s})\,x_{jI_s}^s)\,^T$.

16.2.2 The Aggregate Performance Measure

From Fig. 16.2 one can argue that since the total bundles of outputs Y_j^1 and Y_j^2 are produced from the inputs X_j^1, X_j^2 and X_j^s, a measure of aggregate performance e_j^a can be represented by:

$$e_j^a = \frac{u^1 Y_j^1 + u^2 Y_j^2}{v^1 X_j^1 + v^{s_1}\left(\alpha X_j^s\right) + v^{s_2}\left((1-\alpha) X_j^s\right) + v^2 X_j^2} \tag{16.1}$$

For this representation, the vectors of multipliers u^ℓ and v^t would be determined in a DEA manner to be discussed below. The rationale for allowing for the possibility of different vectors v^{s_1} and v^{s_2} for the shared service and sales resources, respectively, is that the relative importance of the components of X^s in generating Y^1 may be different than their importance in generating Y^2. This was discussed earlier. In this manner, we avoid the possibility of infeasibilities created by possibly conflicting restrictions on the multipliers v^s. There is yet another rationale for permitting v^{s_1} and v^{s_2} to be different multiplier vectors. It can be argued that normally in a DEA analysis there is no clear connection between subsets of outputs and subsets of inputs. In this event, it is certainly the case that v^{s_1} and v^{s_2} should be the same vectors since they pertain to the same inputs (for example, support staff). When a direct link can be made between such subsets of input and output bundles, however, one might then attempt to impose some form of linking constraints as discussed in earlier literature. We do this in the model discussed below. Such constraints may only be feasible if v^{s_1} and v^{s_2} are, in fact, permitted to be different vectors.

16.2.3 Function-Specific Performance Measures

From e_j^a, performance measures for DMU j that capture service and sales efficiency would appear to be appropriately represented by e_j^1 and e_j^2, respectively, as defined by:

$$e_j^1 = \frac{u^1 Y_j^1}{v^1 X_j^1 + v^{s_1}\left(\alpha X_j^s\right)} \tag{16.2}$$

and

$$e_j^2 = \frac{u^2 Y_j^2}{v^{s_2}\left((1-\alpha) X_j^s\right) + v^2 X_j^2}. \tag{16.3}$$

Property 16.1 *The aggregate performance measure e_j^a is a convex combination of the service and sales measures.*

Specifically $e_j^a = \beta_j e_j^1 + (1 - \beta_j)e_j^2$, where β is the portion of all inputs utilized in e_j^1 (applied to the service component), i.e.

$$\beta_j = \frac{\left[v^1 X_j^1 + v^{s_1}\left(\alpha X_j^s\right)\right]}{\left[v^1 X_j^1 + v^{s_1}\left(\alpha X_j^s\right) + v^{s_2}\left((1-\alpha)X_j^s\right) + v^2 X_j^2\right]}.$$

The aggregate measure is, therefore, a weighted average of the performance across the various functions of the organization, as one would intuitively expect. From this property it is seen that a DMU will be deemed efficient, if and only if it is efficient in both service and sales components. Again we point to the importance of *separate* vectors v^{s_1}, v^{s_2} being permitted in the aggregate measure (16.1). If v^{s_1} and v^{s_2} are forced to be the same in (16.1), yet are permitted to be different in (16.2) and (16.3), then no connection between the aggregate and function-specific measures, as per Property 1, can be made.

16.2.4 Derivation of e_j^a, e_j^1, e_j^2

The defined measures are based upon proportionality variables α which will be treated as DMU-specific variables. Thus, it will be at the discretion of each DMU j to allocate X_j^s across the two functions. Furthermore, the model will make the necessary provisions to ensure that all three measures are appropriately scaled, specifically they will not exceed unity.

Consider the following mathematical programming model:

$$\begin{aligned}
&\max e_o^a \\
&\text{subject to:} \\
&e_j^a \le 1 && \forall j \\
&e_j^1 \le 1, && \forall j \\
&e_j^2 \le 1, && \forall j \\
&0 \le \alpha_i \le 1, && \forall i \\
&(\mu^1, \mu^2) \in \Omega_1 \\
&(v^1, v^2, v^{s_2}, v^2) \in \Omega_2 \\
&u_r^1, u_r^2, v_i^1, v_i^2, v_i^{s_1}, v_i^{s_2} \ge \delta, \quad \forall i, j
\end{aligned} \tag{16.4}$$

In this formulation, the objective is to maximize the aggregate efficiency rating for each DMU "o", while ensuring that the function level ratings (for sales and service) do not exceed 1. We replace ε by δ here to denote the fact that an absolute lower bound δ may be in effect. The sets Ω_1 and Ω_2 are *assurance regions* (see Thompson et al. 1990) defined by any restrictions imposed on the multipliers.

Similar work was done by Beasley and Wong (1990). The set Ω_1 may, for example, contain ratio constraints on the components μ_j^1 and μ_j^2 (the output multipliers), dictated by ranges on transaction processing times. The region Ω_2 would be defined by any restrictions expressing the relative importance of the various inputs pertaining to their impacts on outputs. More will be said regarding such assurance regions later. In general, (16.4) is a constrained version of the original model of Charnes et al. (1978) wherein *linking* constraints that connect output and input bundles are present.

16.2.5 An Alternative Formulation

Model (16.4) can be reduced to a non-ratio format in the usual manner of Charnes and Cooper (1962), yielding:

$$\begin{aligned}
& e_o^a = \max \mu^1 Y_o^1 + \mu^2 Y_o^2 \\
& \text{subject to :} \\
& v^1 X_o^1 + v^{s_1}\left(\alpha X_o^{s_1}\right) + v^{s_2}\left((1-\alpha) X_o^{s_2}\right) + v^2 X_o^2 = 1 \\
& \mu^1 Y_j^1 + \mu^2 Y_j^2 - v^1 X_j^1 - v^{s_1}\left(\alpha X_j^s\right) - v^{s_2}(1-\alpha) X_j^s - v^2 X_j^2 \le 0, \forall j \\
& \mu^1 Y_j^1 - v^1 X_j^1 - v^{s_1}\left(\alpha X_j^2\right) \le 0 \qquad \forall j \\
& \mu^2 Y_j^2 - v^{s2}\left((1-\alpha) X_j^2\right) - v^2 X_j^2 \le 0 \qquad \forall j \\
& 0 \le \alpha_i \le 1, \qquad \forall i \\
& (\mu^1, \mu^2)\, \varepsilon\, \Omega_1, \left(v^1, v^{s_1}, v^{s_2}, v^2\right) \varepsilon\, \Omega_2 \\
& \mu_r, v_i \ge \delta, \qquad \forall i, j
\end{aligned} \tag{16.5}$$

Since α_i is a decision variable, this problem is clearly nonlinear. If we make the change of variables $\overline{v}^{s1} = \alpha v^{s1}$ and $\overline{v}^{s2} = (1-\alpha) v^{s2}$, then problem (16.5) reduces to the following form:

$$\begin{aligned}
& e_o^a = \max \mu^1 Y_o^1 + \mu^2 Y_o^2 \\
& \text{subject to :} \\
& v^1 X_o^1 + \overline{v}^{s_1} X_o^s + \overline{v}^{s_2} X_o^s + v^2 X_o^2 = 1 \\
& \mu^1 Y_j^1 + \mu^2 Y_j^2 - v^1 X_j^1 - \overline{v}^{s1} X_j^s - \overline{v}^{s_2} X_j^s - v^2 X_j^2 \le 0, \quad \forall j \\
& \mu^1 Y_j^1 - v^1 X_j^1 - \overline{v}^{s_1} X_j^2 \le 0 \qquad \forall j \\
& \mu^2 Y_j^2 - \overline{v}^{s_2} X_j^2 - v^2 X_j^2 \le 0, \qquad \forall j \\
& 0 \le \alpha_i \le 1, \qquad \forall i \\
& (\mu^1, \mu^2) \in \Omega_1, \left(v^1, \overline{v}^{s_1}, \overline{v}^{s_2}, v^2\right) \in \overline{\Omega}_2 \\
& \mu_r^1, \mu_r^2, v_i^1, v_i^2 \ge \delta \\
& \overline{v}_i^{s_1} \ge \alpha_i \delta, \overline{v}_i^{s_2} \ge (1-\alpha_i)\, \delta
\end{aligned} \tag{16.6}$$

The form of $\overline{\Omega}_2$ depends upon how Ω_2 is structured. Clearly, if Ω_2 is the full real space, as is the case when no additional restrictions are imposed on the input multipliers, then (16.6) is a linear programming problem whose solution will immediately yield a solution to the nonlinear model (16.5). In the case that Ω_2 is a proper subset of the real space, defined by restrictions on the input multipliers, then (16.6) may or may not be linear. We consider various types of restrictions on the vectors v, and their impact on the linearity of $\overline{\Omega}_2$, hence model formulation (16.6). Again, we point out that this model is similar to that developed by Beasley (1995) for analyzing the efficiency of universities in terms of teaching and research. In that case the same vector v^s was used for both functions (teaching and research), rather than allowing for different multipliers for vectors on the two components. As a result, Beasley's model does not have an LP equivalent.

16.2.6 Types of Constraints in Ω_2

1. Absolute bounds on the components of $(v^1, v^2, v^{s_1}, v^{s_2})$.
   In the case of upper and lower bounds of the form $\delta_1 \leq v_i^e \leq \delta_2$, where $e = 1, 2, s_1, s_2$,, then $\overline{\Omega}_2$ will consist of linear restrictions since, for example, $\delta_1 \leq v_i^{s_1} \leq \delta_2$ becomes $\alpha_i\delta_1 \leq \overline{v}_i^{s_1} \leq \alpha_i\delta_2$.
2. Share of total virtual input occupied by a particular subset of inputs.
   Here, we might have constraints of the form

$$\frac{v^{s_1}(\alpha X^s)}{v^{s_1}\left(\alpha X^s + v^{s_2}((1-\alpha)X^s\right)} \leq c.$$

   Again, such constraints are linear and do not result in nonlinear restrictions in $\overline{\Omega}_2$.
3. Ratio constraints
   Restrictions of the cone-ratio variety, see Charnes et al. (1990), may result in nonlinearities in $\overline{\Omega}_2$, depending upon which components of the v -vectors are compared. Specifically, cone-ratio restrictions that do not involve v^{s_1} or v^{s_2} will result in linear constraints in $\overline{\Omega}_2$, for instance the cone-ratio restriction $v_{i_1}^1/v_{i_2}^2 \geq c$ can be rewritten as the linear constraint $v_{i_1}^1 \geq cv_{i_2}^2$. Ratio constraints on the multipliers of the shared resources *will* render $\overline{\Omega}_2$ nonlinear; for example, restrictions of the form

$$\frac{v_{i_1}^{s_1}}{v_{i_2}^{s_1}} \geq c,$$

are transformed to

$$\frac{\alpha_{i_1} v_{i_1}^{s_1}}{\alpha_{i_2} v_{i_2}^{s_1}} \geq c \frac{\alpha_{i_1}}{\alpha_{i_2}} \text{ or } \frac{\overline{v}_{i_1}^{s_1}}{v_{i_2}^{s_1}} \geq c \frac{\alpha_{i_1}}{\alpha_{i_2}},$$

in order to take account of the sharing of resources between sales and service activities.

16.2.7 Special Cases

The extent to which both shared and dedicated resources exist can vary from one situation to another. There can be special circumstances where, for example, there are no dedicated resources and all resources are shared. This does not change the general structure of the constrained DEA model (16.4), nor the requirement that component measures must fall out of the results. One special case is worth noting, namely, when no shared resources are present, and only resources dedicated to the separate components are involved. In this situation, (16.4) is completely separable in the sense that one can derive the individual component measures e_o^1 and e_o^2 by two separate DEA analyses; one for sales and one for service. The overall aggregate measure e_o^a is then a convex combination of these two measures.

In the following section an application of this multi-component model to a set of bank branches is provided. Due to the presence of ratio constraints of this latter type in the example, the resulting model is nonlinear. In a practical setting with a large number of bank branches to evaluate, solving a quadratic programming problem for each would probably prove to be problematic. A linear relaxation of this nonlinear model is discussed, and outputs from the example are presented. Such a relaxation would prove to be more tractable in the situation where many DMUs are present.

16.3 An Application

The model presented herein evolved from an earlier conventional DEA study of branch efficiency in a major Canadian bank. A total of approximately 1,300 branches was involved, with the aim of the study being to identify benchmark branches for purposes of establishing cost targets. While data on several hundred different transactions is available from bank records, 13 of the major ones (some grouped) account for approximately 80 % of branch workload, and were used as outputs in the analysis. The only inputs considered in that study were

Table 16.1 Input- and output measures used in an application of the model

Inputs		Outputs	
FSE	# service staff	MDP	# counter level deposits
FSA	# sales staff	MTR	# transfers between accounts
FSU	# support staff	RSP	# retirement savings plan openings
FOT	# other staff	MOR	# mortgage accounts opened

personnel counts. Time studies were conducted previously on a small sample of typical branches, and provide ranges on unit processing times for all transactions. These ranges were the basis for the cone-ratio constraints on output multipliers for the DEA runs performed. One result of the aforementioned study was that members of the set of branches identified as being efficient, were those that were primarily *service* oriented units – specifically those with low levels of activity on the sales side while being very efficient in terms of routine counter transactions. The clear desire of the organization was a methodology that could provide a measure of performance on both components as well as an overall efficiency score. In this way one can identify not only those branches that are underperforming, but also the component that is weakest. The model discussed in Sect. 16.2 was applied to a dataset of 20 branches out of the full set of bank branches. These were all chosen from one district. For purposes of illustration only, a subset of transaction types was chosen as outputs, and only personnel counts were used as inputs. The chosen input- and output measures used are summarized in Table 16.1.

The relevant data for a 1 year period is displayed in Table 16.2. To provide for a realistic picture of branch performance, a number of restrictions were imposed:

Type 1: Ratio constraints on multipliers
Ratio constraints of the form $a \leq \mu_{r_1}/\mu_{r_2} \leq b$ on output multipliers were imposed to reflect processing times. Ratio constraints on the shared input multipliers were applied to reflect the relative importance of the two inputs (support and other staff) that are split between sales and service.

Type 2: Limitations on α_i
It is generally the case that some bounds need to be imposed on the fraction α_i of shared resource i being allocated to service activities. For illustrative purposes the range $1/3 \leq \alpha_i \leq 2/3$ was chosen.

Type 3: Constraints on the ratios of total service inputs to total inputs.
Here constraints are imposed to restrict the portion of virtual inputs being allocated to the service component. Recalling the definition of β_j in Property 1, restrictions were imposed on the range over which β_j could vary. For present purposes the limits $1/3 \leq \beta_j \leq 2/3$ were applied. While the same limits were used for all branches j in the example herein, it may be the case that different ranges would apply to different classes of branches. Large urban branches may allocate different mixes of resources to sales than small or mid-size branches.

Table 16.2 Branch data for a selection of 20 bank branches

	Service outputs		Sales outputs		Inputs		Shared inputs	
DMU	MDP	MTR	RSP	MOR	FSE	FSA	FSU	FOT
01	2.873	1.498	03.6	04.2	0.455	0.492	0.17	0.73
02	3.093	1.226	05.9	09.7	0.942	0.661	1.88	1.00
03	1.857	0.865	03.7	04.9	0.510	0.293	0.47	1.01
04	8.532	3.290	04.8	12.2	1.239	0.916	1.13	0.10
05	4.304	1.777	07.9	16.8	1.015	0.724	4.48	0.12
06	4.340	0.110	00.5	00.9	0.883	1.474	3.61	0.33
07	4.640	1.493	08.7	05.2	0.594	0.320	2.86	0.21
08	6.821	3.243	07.4	11.0	0.815	0.669	2.99	0.16
09	4.709	2.599	06.5	06.3	0.862	0.670	0.92	1.21
10	0.015	0.037	00.6	02.9	0.000	0.060	5.45	1.55
11	8.532	4.332	09.7	07.2	0.972	1.216	0.12	0.14
12	5.312	2.718	03.5	03.5	0.035	1.007	0.42	0.31
13	3.643	2.115	08.4	06.4	1.317	0.550	2.59	0.17
14	4.878	3.010	05.9	06.0	0.610	0.939	0.54	0.12
15	4.109	1.993	06.0	06.2	0.511	0.659	1.96	0.01
16	4.950	2.950	05.3	04.7	0.719	0.602	1.17	0.49
17	6.389	2.415	12.3	07.8	1.485	0.689	5.03	0.26
18	2.939	1.377	09.0	04.3	0.528	0.436	0.39	0.13
19	6.184	1.975	02.7	04.3	0.743	0.546	0.83	0.56
20	3.053	0.951	01.0	03.2	0.508	0.395	1.44	1.25

16.3.1 Model Relaxation

The model presented in the previous section is nonlinear in the presence of ratio constraints (Type 1) on shared input multipliers. Specifically, when we impose constraints $a \le v_1^{S_1}/v_2^{S_1} \le b$, these take the form

$$a\,\frac{\alpha_1}{\alpha_2} \le \frac{v_1^{S_1}\alpha_1}{v_2^{S_1}\alpha_2} \le b\,\frac{\alpha_1}{\alpha_2}$$

in the presence of the transformation discussed in Sect. 16.3. To render the model more tractable, various *linear relaxations* are possible. One approach attempted was iterative. Specifically, in the first stage all α_i are assumed to be equal for any given branch (i.e., $\alpha_i = d$, a single variable), and the resulting linear problem was solved to determine a starting solution. This yields an optimal solution ($\mu^*_{(1)}$, $v^*_{(1)}$,$\alpha^*_{(1)}$). Fixing $\mu = \mu^*_{(1)}$ and $v = v^*_{(1)}$, the second stage derives a best set of $\alpha^*_i(2)$ relative to the constants $\mu^*_{(1)}$ and $v^*_{(1)}$. In subsequent stages one alternately fixes either $\alpha^*(n)$ or the pair ($\mu^*_{(n)}$,$v^*_{(n)}$), and optimizes (16.4) on the other. One of the difficulties encountered with this method was that many iterations were required in order to converge to a solution that was reasonably close to the optimum.

An alternative and somewhat more practical method was investigated. This amounted to choosing a grid of points in each α_i range. In the present case, each of the two α_i ranged from 0.25 to 0.75 and the grid of five values 0.25, 0.35, 0.45, 0.65, 0.75 was used. Recall that α_1 is the percent of "support staff" allocated to service transactions and that α_2 represents the split of "other staff". This resulted in $5 \times 5 = 25$ different combinations for (α_1,α_2).

Given the relatively small sample of DMUs in this particular example (20 DMUs), the problem can easily be treated directly in its nonlinear form, and was solved using a standard spreadsheet solver.

16.3.2 Results

A proper evaluation of data such as that in Table 16.2 is complicated by the fact that the sales component is a two-level process as discussed earlier. The ranges for average processing times, as reflected in the cone-ratio constraints imposed upon the output multipliers, pertain only to the second of these two levels, namely the *transaction* part of sales. These average times do not account for the *level of effort* required to transact the sale. This effort would involve activities such as interaction with customers, review of portfolios, etc. To compensate for the *understated* values of the μ_j components, one must either scale up these values, or adjust (downward) the resources (inputs) allotted to the sales component. The latter option becomes problematic in that the portion of sales resources not allocated to the transaction part of sales is left as unassigned inputs (i.e., they appear to not contribute to any of the outputs). In the present situation, the former option of scaling up the sales output multipliers was chosen. The scaling factor γ, defined as the ratio of the "Total Sales effort" to the "Transaction effort" was based on an estimate provided by the organization. The ranges provided for μ_j, namely $a \leq \mu_j \leq b$, were replaced by scaled ranges $\gamma a \leq \mu_j \leq \gamma b$. The resulting aggregate, service and sales efficiency scores are displayed in Table 16.3. It is noted that only one of the branches (#11), is efficient in the aggregate sense, that is in both sales and service. Clearly, branches may be efficient in one component only, such as is the case for branches #12 and #18. The respective α_1 and α_2 values are also shown.

16.4 Measuring Multi Component Efficiency: An Additive Model

16.4.1 Addressing Some Shortcomings

The model described above, when applied within the organization, did help to point to areas where inefficiency existed within branches, and aided in setting targets for improvements. Two suggestions from management for enhancement of performance measurement arose from this application.

Table 16.3 Efficiency scores and optimal split of shared resources

DMU	Aggregate e_k^a	Service e_k^1	Sales e_k^2	α_1	α_2
01	0.47972	0.52172	0.45354	0.72676	0.75000
02	0.40499	0.17158	0.52749	0.75000	0.25000
03	0.41946	0.23162	0.50145	0.75000	0.25000
04	0.74913	0.51905	0.91297	0.75000	0.64929
05	0.54472	0.17250	0.54472	0.75000	0.75000
06	0.14925	0.17663	0.03273	0.75000	0.66891
07	0.47257	0.28014	0.55697	0.75000	0.75000
08	0.58236	0.38787	0.70302	0.36427	0.29968
09	0.41178	0.36773	0.43157	0.25000	0.55019
10	0.07307	0.00570	0.09894	0.26281	0.68108
11	1	1	1	0.75000	0.66891
12	0.57384	1	0.29015	0.75000	0.74959
13	0.40464	0.17685	0.53991	0.53334	0.75000
14	0.70675	0.71811	0.70001	0.53334	0.75000
15	0.49252	0.36720	0.55537	0.75000	0.75000
16	0.44784	0.46087	0.43869	0.25000	0.54547
17	0.36581	0.19350	0.45445	0.25000	0.72687
18	0.85924	0.46010	1	0.25000	0.72687
19	0.49243	0.52389	0.37181	0.72682	0.72188
20	0.21444	0.26296	0.18235	0.72676	0.72224

16.4.1.1 Non-volume Related Activities

The first issue has to do with the characterization of those activities surrounding the sales function. The sales function within the bank environment can be viewed as consisting of two sets of activities. The first set, and those examined in the previous sections, would be classified as *volume-related* activities. These activities consist of those tasks linked directly to sales products, *after* the decision to purchase has been made. These would include the filing of documents, preparation of certificates, etc. Such tasks are characterized by known time estimates, arrived at in the same manner as is the case for service transactions.

The second set, the *non-volume-related* activities, may not be directly linked to any specific product. Such activities would include responding to customer queries, routine tasks such as reproduction of forms, reviewing customer portfolios, carrying out computer searches, and so on. Support costs for print materials, computer expenses, etc. would, as well, fall into this category.

16.4.1.2 Providing a Fair Balance Between Sales and Service Performance Measures

The model of the previous section, because of the form of the objective function, will often produce component measures e_j^1 and e_j^2 that differ from each other in an unreasonable way. Essentially, the model, in setting out to maximize the aggregate score e_j^a will do so by maximizing one of the two component measures *at the expense of the other*. A suggestion raised by management was to attempt to derive measures with the idea of showing both sales and service performance in the best light. To address the above two concerns, an additive form of the DEA model was adopted.

16.4.2 The General Additive Model

In the next subsection we develop a dual-component DEA model for evaluating both sales and transaction functions within bank branches. For purposes of that development, the Pareto-Koopmans, or additive model structure is exploited. While the additive model is seldom the structure of choice in most DEA analyses (one generally utilizes one of the radial models), it is demonstrated that it's structure is, in fact, a general framework containing the radial models as special cases. Specifically, any of the standard models are obtainable by way of constrained versions of the additive model. For development purposes herein, it is convenient to approach the standard models from this angle, rather than in the more conventional way.

It is instructive to examine both dual and primal forms of the additive model:

The Dual

$$\min \sum_i \nu_i x_{i_o} - \sum_r \mu_r y_{r_o} - \mu_o \tag{16.7a}$$

subject to:

$$\sum_i \nu_i x_{ij} - \sum_r \mu_r y_{rj} - \mu_o \geq 0, \forall j \tag{16.7b}$$

$$\mu_r \geq 1/y_{r_o}, \forall r \tag{16.7c}$$

$$\nu_i \geq 1/x_{i_o}, \forall i \tag{16.7d}$$

It is noted that we have chosen lower bounds on the multipliers (16.1c) and (16.7d) that are DMU-specific. This is usually referred to as the *units invariant* form of the model. The "dual" of (16.7a) is the model:

The primal

$$\max \sum_i \left(s_i^1/x_{i_o}\right) + \sum_r s_r^2/y_{r_o} \tag{16.8a}$$

subject to:

$$\sum_j \lambda_i x_{ij} + s_i^1 \le x_{i_o},\ \forall i, \tag{16.8b}$$

$$\sum_j \lambda_j y_{rj} - s_r^2 \ge y_{ro},\quad \forall r, \tag{16.8c}$$

$$\sum_j \lambda_j = 1, \tag{16.8d}$$

$$s_i^1, s_r^2, \lambda_j \ge 0,\ \ \forall i, r, j. \tag{16.8e}$$

If we adopt the notation

$$\theta_i = 1 - s_i^1/x_{i_o}, \phi_r = 1 + s_r^2/y_{r_o} \tag{16.9}$$

and let $\overline{\theta}_i = 1 - \theta_i, \overline{\phi}_r = \phi_r - 1$, model (16.8a, 16.8b, 16.8c, 16.8d and 16.8e) becomes

$$\max \sum_i \overline{\theta}_i + \sum_r \overline{\phi}_r \tag{16.10a}$$

subject to :

$$\sum_j \lambda_i x_{ij} + \overline{\theta}_i x_{i_o} \le x_{i_o}, \forall i \tag{16.10b}$$

$$\sum_j \lambda_j y_{rj} - \overline{\phi}_r y_{ro} \ge y_{r_o}, \forall r \tag{16.10c}$$

$$\sum_j \lambda_j = 1 \tag{16.10d}$$

$$\overline{\theta}_i, \overline{\phi}_r, \lambda_j \ge 0, \forall i, r, j \tag{16.10e}$$

This format is a particularly convenient way to view the additive model, as it exhibits an immediate connection to other models. This form is related to the "Russell Measure" as discussed in Fare and Lovell (1978). There, the objective function takes the form

$$\min R = \left[\sum_i \theta_i + \sum_r (1/\phi_r) \right] / (I + R),$$

where I, R are the numbers of inputs and outputs, respectively. Cooper et al. (1999) discuss several variations on the additive model, as does Thrall (1996).

It is immediately clear that one can adopt a purely *input oriented* variation on the additive model concept, by setting $\overline{\phi}_r = 0$ for all r, and replacing constraints (16.10b) and (16.10c) by

$$\sum_j \lambda_j x_{ij} + \overline{\theta}_i x_{io} \leq x_{io} \tag{16.11a}$$

$$\sum_j \lambda_j y_{rj} \geq y_{ro} \tag{16.11b}$$

This type of structure is discussed in Zieschang (1984). In the section to follow we focus attention on the input oriented model. Furthermore, if we restrict the $\overline{\theta}_i$ further by requiring that they all be equal, then we have a structure equivalent to the standard input oriented *radial* model of Charnes et al. (1978) (or at least Banker et al. (1984)).

In the case that the input oriented approach is to be taken, in which case (16.11a) and (16.11b) replace (16.10b) and (16.10c) in the primal problem (16.10a), the equivalent modification to the dual problem (16.7a) is to replace the lower bound on μ_r (constraint (16.7c) by $\mu_r \geq 0$. As with the Russell measure, an appropriate measure of performance in the input oriented additive model is

$$R_I = \sum_{i=1}^{I} (1 - \overline{\theta}_i)/I = \sum_{i=1}^{I} \theta_i / I. \tag{16.12}$$

It is noted that in the restricted case where $\theta_i = \theta$ for all i (the BCC radial model), $R_I = \theta$. In any event, it will be the case that $0 \leq R_I \leq 1$, with $R_I = 1$ if all $\overline{\theta}_i = 0$; for example, in this case the pair (Y^0, X^0) is on the frontier or an extension.

Stated formally then, the pure input version of (16.10a, 16.10b, 16.10c, 16.10d and 16.10e) is:

$$\begin{aligned}
&\max \sum_i \overline{\theta}_i / I \\
&\text{subject to:} \\
&\sum_j \lambda_j x_{ij} + \overline{\theta}_i x_{i_o} \leq x_{i_o}, \forall i \\
&\sum_j \lambda_j y_{rj} \geq y_{ro}, \ \forall r \\
&\sum_j \lambda_j = 1 \\
&\overline{\theta}_i, \lambda_j \geq 0
\end{aligned} \tag{16.13}$$

Thus, the additive model can be viewed as a flexible mechanism for capturing different aspects of efficiency. Admittedly, restricted versions of the model can fail to be comprehensive in the sense discussed by Cooper et al. (1999). Obviously, it will be true that restricting attention to the input side of the problem, for example, can mean that improper envelopment can occur, as is well known in the radial models.

16.4.3 An Additive Model for Sales and Service Components

The notation of the previous section will be used in the current model, but with the one addition, namely, to use two output multipliers μ^{21} for the per unit processing times for volume-related and μ^{22} for non-volume related portions of the sales outputs Y_j^2. We also chose in this second analysis to use the VRS DEA model, hence defined output variables μ_o^1 and μ_o^2 for service and sales components.

An alternative to optimizing the aggregate efficiency measure as in the previous sections, is to attempt to optimize, in some manner, both the service measure

$$e_{1_o} = \nu^1 X_o^1 + \nu^{s1}\left(\alpha X_o^s\right) - \mu^1 Y_o^1 - \mu_o^1 \tag{16.14}$$

and sales measure[1]

$$e_{2_o} = \nu^2 X_o^2 + \nu^{s2}\left((1-\alpha) X_o^s\right) - \mu^{21} Y_o^2 - \mu^{22} Y_o^2 - \mu_o^2. \tag{16.15}$$

One approach is to minimize the maximum inefficiency, for example, we solve the goal programming problem.

$$\begin{aligned} &\min d \\ &\text{subject to:} \\ &e_{1_o} \le d, e_{2_o} \le d \\ &S_e I_j - S_e O_j \ge 0,\ \ S_a I_j - S_a O_j \ge 0, \forall j. \end{aligned} \tag{16.16}$$

In attempting to reduce the maximum inefficiency (d), the model has the tendency to equalize the sales and service performance measures if feasibility permits. In some respects this could be justified insofar as one can argue that a branch will, or should, give equal importance to all components of its business. It must be pointed out that additional restrictions may be imposed on the multipliers in (16.16) (e.g., assurance regions as per Thompson et al. (1990)). For example, the components of μ^1 would be related to one another through limits arising from branch time studies. For model development purposes in this section, however, we avoid applying specific additional restrictions. This permits us to obtain primal and dual efficiency measurement models, not tied to application-specific situations. The inclusion of these in the models is examined in the next section dealing with the application of the tools in a specific setting.

---

[1]In the context of the VRS structure, we let μ_o^1, μ_o^2 denote service and sales variables, respectively.

Formally, the dual form of the proposed model is given by (16.17).

$$
\begin{aligned}
&\min d \\
&\text{subject to:} \\
&-v^1X_o^1 - v^{s1}\left(\alpha X_o^s\right) + \mu^1 Y_o^1 + \mu_o^1 + d \geq 0 \\
&v^1X_j^1 + v^{s1}\left(\alpha X_j^s\right) - \mu^1 Y_j^1 - \mu_o^1 \geq 0, \forall j, \\
&-v^2X_o^2 - v^{s2}\left((1-\alpha)X_o^s\right) + \mu^{21}Y_o^2 \\
&+\mu^{22}Y_o^2 + \mu_o^2 + d \geq 0 \\
&v^2X_j^2 + v^{s2}\left((1-\alpha)X_j^s\right) - \mu^{21}Y_j^2 \\
&-\mu^{22}Y_j^2 - \mu_o^2 \geq 0, \forall j, \\
&v_i^1 \geq 1/\left(x_{i_o}^1 . |I_1|\right), \forall i \in I_1, \\
&v_i^2 \geq 1/\left(x_{i_o}^2 . |I_2|\right), \forall i \in I_2, \\
&v_i^{s1} \geq 1/\left(x_{i_o}^s . |I_s|\right), \quad \forall i \in I_s, \\
&v_i^{s2} \geq 1/\left(x_{i_o}^s . |I_s|\right), \quad \forall i \in I_s, \\
&\mu_r^1 \geq 0, \quad \forall r \in R, \\
&\mu_r^{21} \geq 0, \quad \forall r \in R, \\
&\mu_r^{22} \geq 0, \quad \forall r \in R,
\end{aligned}
\tag{16.17}
$$

Note that we have introduced the lower bounds $1/x_o^1.\ |I_1|$, etc., to force $0 \leq d \leq 1$. Here, $|I_1|$ denotes the cardinality of the input set I_1.

To deal with the nonlinearity created by the products $\alpha\nu^{s1}$ and $(1-\alpha)\nu^{s2}$, introduce the change of variables $\overline{v}^{s1} = \alpha v^{s1}$, and $\overline{v}^{s2} = (1-\alpha)\, v^{s2}$.

Then, replace the two constraints $\nu_i^{s1} \geq 1/(x_{io}^s.\ |I_s|)$ and $v_j^{s2} \geq 1/(x_{io}^s|I_s|)$ by $\alpha\nu_i^{s1} \geq \alpha\ 1/(x_{io}^s.\ |I_s|)$ and $(1-\alpha)\ v_j^{s2} \geq 1/(x_{io}^s|I_s|) - \alpha 1/(x_{io}^s|I_s|)$.

It is generally the case that constraints will be imposed on the α_i; specifically, the percent of any resource that can be allocated to the service component will be required to be within some interval, namely

$$L_i^1 \le \alpha_i \le L_i^2.$$

Model (16.17) can now be rewritten in the form:

$$\begin{aligned}
e_p = \max\Bigg\{ & \sum_{i\in I_1} s_i^1/\left(x_{io}^s|I_1|\right) + \sum_{i\in I_2} s_i^2/\left(x_{io}^2|I_2|\right) \\
& + \sum_{i\in I_s}\left[s_i^{s_2}/\left(x_{io}^s|I_s|\right) + L_i^1\gamma_i^1 - L_i^2\gamma_i^2\right]\Bigg\}
\end{aligned}$$

$$\begin{aligned}
&\text{subject to:}\\
&\sum_k \lambda_k^1 x_{ik}^1 - \lambda_{n+1}^1 x_{io}^1 + s_i^1 \le 0, \ i\in I_1,\\
&\sum_k \lambda_k^1 x_{ik}^s - \lambda_{n+1}^1 x_{io}^s + s_i^{s1} \le 0, \ i\in I_s,\\
&\sum_k \lambda_k^2 x_{ik}^2 - \lambda_{n+1}^2 x_{io}^2 + s_i^2 \le 0, \ i\in I_2,\\
&\sum_k \lambda_k^2 x_{ik}^s - \lambda_{n+1}^2 x_{io}^s + s_i^{s2} \le 0, \ i\in I_s,\\
&-\sum_k \lambda_k^1 y_{rk}^1 - \lambda_{n+1}^1 y_{ro}^1 \le 0, \ r\in R_1,\\
&-\sum_k \lambda_k^2 y_{rk}^2 + \lambda_{n+1}^2 y_{ro}^2 \le 0, \ r\in R_2,\\
&\lambda_{n+1}^1 + \lambda_{n+1}^2 = 1\\
&-s_i^{s1}/\left(x_{io}^s\cdot|I_s|\right) + s_i^{s2}/\left(x_{io}^s\cdot|I_s|\right)\\
&+\gamma_i^1 - \gamma_i^2 \le 0,\\
&\gamma_i^1, \gamma_i^2, \lambda_k^1, \lambda_k^2, s_i^1, s_i^2 \ge 0.
\end{aligned} \tag{16.18}$$

The dual of this problem is given by

$$e_p = \max\left\{ \sum_{i \in I_1} s_i^1 / \left(x_{io}^s |I_1|\right) + \sum_{i \in I_2} s_i^2 / \left(x_{io}^2 |I_2|\right) \right.$$

$$\left. + \sum_{i \in I_s} \left[s_i^{s_2} / \left(x_{io}^s |I_s|\right) + L_i^1 \gamma_i^1 - L_i^2 \gamma_i^2 \right] \right\}$$

subject to:

$$\begin{aligned}
&\sum_j \lambda_j^1 x_{ij}^1 - \lambda_{n+1}^1 x_{io}^1 + s_i^1 \leq 0, \ i \in I_1, \\
&\sum_j \lambda_j^1 x_{ij}^s - \lambda_{n+1}^1 x_{io}^s + s_i^{s1} \leq 0, \ i \in I_s, \\
&\sum_j \lambda_j^2 x_{ij}^2 \quad \lambda_{n+1}^2 x_{io}^2 + s_i^2 \leq 0, \ i \in I_2, \\
&\sum_j \lambda_j^2 x_{ij}^s - \lambda_{n+1}^2 x_{io}^s + s_i^{s2} \leq 0, \ i \in I_s, \\
&-\sum_j \lambda_j^1 y_{rj}^1 - \lambda_{n+1}^1 y_{ro}^1 \leq 0, \ r \in R_1, \\
&-\sum_j \lambda_j^2 y_{rj}^2 + \lambda_{n+1}^2 y_{ro}^2 \leq 0, \ r \in R_2, \\
&\lambda_{n+1}^1 + \lambda_{n+1}^2 = 1 \\
&-s_i^{s1} / \left(x_{io}^s \cdot |I_s|\right) + s_i^{s2} / \left(x_{io}^s \cdot |I_s|\right) \\
&+\gamma_i^1 - \gamma_i^2 \leq 0, \\
&\gamma_i^1, \gamma_i^2, \lambda_j^1, \lambda_j^2, s_i^1, s_i^2 \geq 0.
\end{aligned} \tag{16.19}$$

Letting $\overline{\theta}_i^1 = s_i^1 / x_{io}^1, \overline{\theta}_i^2 = s_i^2 / x_{io}^2, \overline{\theta}_i^{s1} = s_i^s / x_{io}^s, \overline{\theta}_i^{s2} = s_i^{s2} / x_{io}^s$, problem (16.19) can be written as

$$e_p = \max\left\{\sum_{i\in I_1} \overline{\theta}_i^1/|I_1| + \sum_{i\in I_2} \overline{\theta}_i^2/|I_2|\right.$$

$$\left. + \sum_{i\in I_s} [\overline{\theta}_i^{s2}/|I_s|) + L_i^1\gamma_i^1 - L_i^2\gamma_i^2]\right\}$$

subject to :

$$\begin{aligned}
&\sum_j \lambda_j^1 x_{ij}^1 - \lambda_{n+1}^1 x_{io}^1 + \overline{\theta}_i^1 x_{io}^1 \le 0,\ i\in I_1,\\
&\sum_j \lambda_j^1 x_{ij}^s - \lambda_{n+1}^1 x_{io}^s + \overline{\theta}_i^{s1} x_{io}^s \le 0,\ i\in I_s,\\
&\sum_j \lambda_j^2 x_{ij}^2 - \lambda_{n+1}^2 x_{io}^2 + \overline{\theta}_i^2 x_{io}^2 \le 0,\ i\in I_2,\\
&\sum_j \lambda_j^2 x_{ij}^s - \lambda_{n+1}^2 x_{io}^s + \overline{\theta}_i^{s2} x_{io}^s \le 0,\ i\in I_s, -\\
&\sum_j \lambda_j^1 y_{rj}^1 - \lambda_{n+1}^1 y_{ro}^1 \le 0,\ r\in R_1,\\
&-\sum_j \lambda_j^2 y_{rj}^2 + \lambda_{n+1} y_{ro}^2 \le 0,\ r\in R_2,\\
&\lambda_{n+1}^1 + \lambda_{n+1}^2 = 1,\\
&-\overline{\theta}_i^{s1}/|I_s| + \overline{\theta}_i^{s2}/|I_s| + \gamma_i^1 - \gamma_i^2 \le 0,\\
&\gamma_i^1, \gamma_i^2, \lambda_j^1, \lambda_j^2, \overline{\theta}_i^1, \overline{\theta}_i^2 \ge 0.
\end{aligned} \tag{16.20}$$

It can be seen that (16.20) is a direct generalization of (16.13). The equivalent of the R_1 measure given in (16.12) is $\overline{e}_p = 1 - e_p$.

It must be noted, of course that $e_p = e_d$ (the objective function value of (16.18)) is the maximum of the two components e_{1_o}, e_{2_0}, as per (16.2) and (16.3). the separate sales and service measures would fall out as part of the analysis.

16.5 Application to Bank Branches

To demonstrate the application of the additive structure, we again examine data on a sample of branches, with somewhat different outputs. Data on 20 branches is displayed in Table 16.4. The outputs chosen were:

Table 16.4 Sales and service outputs and inputs

Transit #	TOTMENUE	VISA	CAD	RSP	MORT	BPL	FSE	FSA	FSU	FOT
1	51,803	2,973	190,522	421	567	101	55.25	98.15	48.07	59.55
2	10,477	710	49,898	75	172	13	14.73	32.39	17.15	7.24
3	11,195	431	39,523	68	73	19	9.6	14.22	8.55	2.15
4	6,480	422	30,713	49	48	15	10.48	9.05	5.15	1
5	37,695	921	26,922	210	128	144	27.9	17.22	5.02	1.08
6	9,211	362	43,056	120	127	57	9.17	15.79	2.44	1
7	16,483	529	13,123	74	150	9	10.53	12.49	2.19	1.41
8	456	20	10,127	6	29	15	10.6	10	5.45	1.55
9	5,985	382	21,945	32	28	19	8.71	8.02	3.94	1
10	8,682	351	11,010	84	78	52	7.05	11.25	2.36	0.96
11	5,287	182	16,474	59	97	15	6.61	9.42	1.88	1
12	18,292	171	18,014	104	84	443	7.3	9.33	1.79	1.05
13	5,669	264	11,303	36	62	144	3.96	3.92	1	0.8
14	9,656	332	6,745	65	63	14	6.7	8.62	0.92	1.21
15	18,566	308	76,174	134	80	14	10.29	10.94	4.87	1.03
16	39,430	500	60,832	199	199	200	25.07	20.55	4.69	0.83
17	11,601	423	36,692	73	137	107	12.25	10.91	3.88	0.96
18	8,030	406	19,598	62	86	50	9	13.35	3.16	0.97
19	16,991	658	21,334	91	111	78	12.7	18.02	2.11	1.76
20	10,473	463	51,225	132	71	39	18.15	18.65	6.92	1

Service:
TOTEMU – total number of menu account transactions
VISA – number of Visa cash advances
CAD – number of commercial deposit transactions

Sales:
RSP – number of RSP account openings
MORT – number of mortgages transacted
BPL – number of variable rate consumer loans transacted

In the current example, inputs were restricted to personnel only. We have not included other operating expenses such as computers, rent, etc. Specifically, the inputs were:

Service: FSE – total number of full time equivalent service staff
Sales: FSA – total number of full time equivalent sales staff
Shared: FSU – total number of full time equivalent support staff
FST – total number of full time equivalent "other" staff.

As discussed earlier, in applying the models described herein, attention was paid to multiplier restrictions that reflect the relative weights to be placed on the various outputs. Specifically, the components of μ^1 and μ^{21} have been constrained in a ratio sense to obey time limits on branch transactions. For example, the specified time interval for a commercial deposit transaction is (in minutes) (1.2, 3.6); that for a VISA cash advance is (0.8, 2.5). To reflect these limits in the multipliers μ_2^1 and μ_3^1, we require $\frac{0.8}{3.6} \leq \mu_2^1/\mu_3^1 \leq \frac{2.5}{1.2}$. Similar restrictions have been applied to the components of μ^{21} to accommodate the time limits on the transaction portion of sales outputs.

No such detailed information was obtainable on the non-volume portion of the sales component. From interviews with branch consultants, it has been estimated that 30–50 % of the sales effort lies with the non-volume activity, and the remainder is the transaction or volume-related work. In general, this would imply that $\frac{3}{7} \leq \mu^{22} y_j^2/\mu^{21} y_j^2 \leq 1$ for each branch *k*. To simplify matters, we choose here to take a more restricted view, and constrain the ratio *for each product i* to be in this range. Specifically, $\frac{3}{7} \leq \mu_i^{22} y_{ij}^2/\mu_i^{21} y_{ij}^2 \leq 1$, implying that $\frac{3}{7} \leq \mu_i^{22}/\mu_i^{21} \leq 1$.

Table 16.4 displays the data on all inputs and outputs for a sample of 20 branches of the bank. The result from applying model (16.18), (augmented by the multiplier restrictions discussed above), are shown in Table 16.5. Recall that d represents the maximum inefficiency associated with the two components (sales and service). The corresponding efficiency measures e_s (sales) and e_T (service) are displayed. It is noted that $d = 1 - \min\{e_T, e_S\}$. As noted earlier, this model tends to force e_T and e_S together, and in a large percentage of the cases, the two measures are equal.

We have not directly addressed the issue of an aggregate measure of efficiency which should be some combination of the two separate measures. Arguably, this aggregate measure e_A should be some average of the component scores. A reasonable candidate for this might be of the form $e_A = \beta e_T + (1 - \beta) e_S$ where β is the proportion of total

Table 16.5 Results from model

Var1		2	3	4	5	6	7 8	8
α_1	0.742798	0.75	0.25	0.25	0.25	0.371086	0.746744	0.25
α_2	0.5	0.5	0.278917	0.5	0.25	0.5	0.5	0.5
d	70.14 %	85.13 %	40.00 %	80.24 %	0.00 %	23.59 %	94.27 %	34.77 %
ES	29.86 %	14.87 %	60.00 %	19.76 %	100.00 %	76.41 %	5.73 %	65.23 %
ET	29.86 %	14.87 %	60.00 %	19.76 %	100.00 %	76.41 %	5.73 %	65.23 %
Var	9	10	11	12	13	14	15	16
α_1	0.413508	0.507633	0.25	0.25	0.25	0.25	0.25	0.25
α_2	0.5	0.75	0.25	0.57965	0.417598	0.75	0.75	0.675252
d	29.79 %	23.65 %	66.80 %	0.00 %	0.00 %	0.00 %	0.00 %	0.00 %
ES	70.21 %	76.35 %	48.50 %	100 %	100 %	100 %	100 %	88.01 %
ET	70.21 %	76.35 %	33.20 %	100 %	100 %	100 %	100 %	100 %
Var	17	18	19	20	Average	SDev.		
α_1	0.25	0.75	0.75	0.25	40.16 %	21.62 %		
α_2	0.438533	0.75	0.75	0.75	54.45 %	17.12 %		
d	16.94 %	68.09 %	62.51 %	65.80 %	38.09 %	33.18 %		
ES	89.71 %	31.92 %	37.49 %	34.20 %	62.41 %	32.30 %		
ET	83.06 %	31.92 %	37.49 %	34.20 %	61.91 %	33.18 %		

resources consumed by the service component (dedicated service inputs together with shared inputs).

In the application of model (16.18), the splitting variables α_1 and α_2 have each been restricted to the range $.25 \leq \alpha \leq .75$. This range would need to be established by branch consultants in much the same manner that ranges on output multiplies might be set by way of time study estimates.

16.6 Conclusions

This chapter has examined model structures for dealing with multi-component efficiency measurement in a banking environment. The conventional DEA approach, as applied in bank related studies, has tended to concentrate on a single measure of performance for the DMU. Very often, however, there are multiple components or sub units within the DMU whose individual performance is required. The model provided herein provides a mechanism for developing multi-component measures.

References

Banker, R. D., Charnes, A., & Cooper, W. W. (1984). Some models for technical and scale efficiencies in data envelopment analysis. *Management Science, 30*, 1078–1092.

Beasley, J. E. (1995). Determining teaching and research efficiencies. *Journal of the Operational Research Society, 46*, 441–452.

Beasley, J. E., & Wong, Y.-H. B. (1990). Restricting weight flexibility in data envelopment analysis. *Journal of the Operational Research Society, 41*, 829–835.

Charnes, A. & Cooper, W.W. (1962). Programming with linear fractional functionals, *Naval Research Logistics Quarterly, 9*, 181–186.

Charnes, A., Cooper, W. W., & Rhodes, E. L. (1978). Measuring the efficiency of decision making units. *European Journal of Operational Research, 2*, 429–444.

Charnes, A., Cooper, W. W., Huang, Z. M., & Sun, D. B. (1990). Polyhedral cone-ratio DEA models with an illustrative application to large commercial banks. *Journal of Econometrics, 46*, 73–91.

Cook, W. D., & Hababou, M. (2001). Sales performance measurement in bank branches. *OMEGA, 29*, 299–307.

Cook, W. D., Hababou, M., & Tuenter, H. (2000). Multicomponent efficiency measurement and shared inputs in data envelopment analysis: An application to sales and service performance in bank branches. *Journal of Productivity Analysis, 14*, 209–224.

Cooper, W., Park, K., & Pastor, J. (1999). RAM: A range adjusted measure of inefficiency for use with additive models, and relations to other models and measures in DEA. *Journal of Productivity Analysis, 11*, 5–42.

Fare, R., & Grosskopf, S. (1996). Productivity and intermediate products: A frontier approach. *Economic Letters, 50*, 65–70.

Fare, R., & Lovell, C. A. K. (1978). Measuring the technical efficiency of production. *Journal of Economic Theory, 19*, 150–162.

Fare, R., & Primont, D. (1984). Efficiency measures for multiplant firms. *Operations Research Letters, 3*, 257–260.

Oral, M., & Yolalan, R. (1990). An empirical study on measuring operational efficiency and profitability of bank branches. *European Journal of Operational Research, 46*, 282–294.

Schaffnit, C., Rosen, D., & Paradi, J. C. (1997). Best practice analysis of bank branches: An application of data envelopment analysis in a large Canadian bank. *European Journal of Operational Research, 98*, 269–289.

Sherman, H. D., & Gold, F. (1985). Bank branch operating efficiency: Evaluation with data envelopment analysis. *Journal of Banking and Finance, 9*, 297–315.

Sherman, H. D., & Ladino, G. (1995). Managing bank productivity using data envelopment analysis (DEA). *Interfaces, 25*, 60–73.

Thompson, R. G., Langemeier, L. N., Lee, C. T., Lee, E., & Thrall, R. M. (1990). The role of multiplier bounds in efficiency analysis with application to Kansas farming. *Journal of Econometrics, 46*, 93–108.

Thrall, R. M. (1996). The lack of invariance of optimal dual solutions under translations. *Annals of Operations Research, 66*, 103–108.

Zieshang, K. (1984). An extended Farrell efficiency measure. *Journal of Economic Theory, 33*, 387–396.

Chapter 17
Evaluating Power Plant Efficiency: Hierarchical Models

Wade D. Cook, D. Chai, J. Doyle, and R. Green

Abstract In many efficiency-measurement settings there are identifiable groups or clusters of DMUs whose impacts should be captured in the analysis. In such problem settings at least two issues need to be considered. The first is that there may be both DMU-level and cluster-level factors each of which should be considered in their proper settings. The second issue is that we wish to identify both DMU-specific and DMU-cluster efficiency measures. In the current chapter we examine the problem of measuring efficiencies of a set of electric power plants, where each plant consists of a group of power units, hence clustering or grouping occurs naturally.

Keywords Hierarchy • Power plants • Components • Multi-level

This chapter is based upon Cook, W.D., D. Chai, J. Doyle, and R. Green 1998, Hierarchies and groups in DEA, *Journal of Productivity Analysis*, 10:177–198, with permission from Kluwer Academic Publishers, and Cook, W.D. and R. Green, 2003, Evaluating power plant efficiency: A hierarchical model, *Computers and Operations Research,* 32 (4), 813–823, with permission from Elsevier Science.

W.D. Cook (✉) • D. Chai
Schulich School of Business, York University, 4700 Keele Street, Toronto, ON M3J 1P3, Canada
e-mail: wcook@schulich.yorku.ca

J. Doyle
School of Business, Cardiff University, Aberconway Building, Colum Drive, Cardiff CF10 3EU, UK
e-mail: doyleJR@cardiff.ac.uk

R. Green
School of Management, University of Bath, Bath BA27AY, UK
e-mail: mnsrhg@bath.ac.uk

W.D. Cook and J. Zhu (eds.), *Data Envelopment Analysis*,
International Series in Operations Research & Management Science 208,
DOI 10.1007/978-1-4899-8068-7_17,

17.1 Introduction

In many problem settings that potentially lend themselves to analyses via DEA, there are identifiable groups or clusters of DMUs, whose impacts should be captured in the analysis. One form of grouping has been examined by Banker and Morey (1986), where the idea of categorical variables was discussed. Such variables allow for a comparison of any DMU with those in its own category and in those categories *below* it. Categorical variables generally apply in those situations where there is a natural nesting of the groups of DMUs. For example, in evaluating a set of banks, if the banks are arranged in increasing order according to the sizes of the towns or cities in which they are located, then categorical variables can be used to represent this size component, and banks in a given population category will be compared only to DMUs in this same category and to those in smaller population categories.

In many situations, however, where there is a grouping phenomenon present, categorical variables do not provide an appropriate structure for analysis. Consider the problem of evaluating DMUs such as hospitals in different parts of the country. Here, grouping may take several forms. First, in countries such as Canada or the United States, there may be jurisdictional considerations, e.g., state or provincial regulations can have budgetary or legislation implications for the hospitals. In Canada, for example, health care is under provincial rather than federal jurisdiction. Second, there may be different categories of medical units – extended care facilities, convalescent units, surgical units, and so on. Clearly, these DMUs do not form anything resembling a homogeneous set, making it necessary to address the group elements of the problem. At least two issues must be examined in the context of such problems:

Issue 1: There are both DMU (e.g., hospital) level factors and group (e.g., all extended care facilities versus all surgical facilities) level factors which should be dealt with in their proper settings;

Issue 2: We want not only to identify a measure of efficiency for each individual DMU (hospital), but also for each identified group of units. How do hospitals (as a group) in one jurisdiction compare, in an efficiency sense, to those in another jurisdiction? Do extended care facilities perform differently than surgical facilities?

In the following sections we examine the problem of efficiency evaluation when grouping of DMUs is a consideration. The discussion is based on the articles by Cook et al. (1998) and Cook and Green (2005). In Sect. 17.2 we present a problem setting where both individual DMU and group evaluation arise. The case illustrates two types of grouping – *hierarchical* grouping and grouping on *levels*. Section 17.3 presents appropriate model structures for evaluating group and individual DMU efficiency in a hierarchy. In particular we discuss a procedure for adjusting ratings of DMUs at any given level in a hierarchy to reflect ratings of groups of those units at levels higher up in the structure. In Sect. 17.4 we examine efficiency within groups on a level and develop a procedure for combining different efficiency ratings for a given DMU. In Sect. 17.5 the models are illustrated through an analysis of the application discussed in Sect. 17.2. In Sect. 17.6 the power plant evaluation problem is re-examined using the multicomponent concepts presented in Chap. 6. This arises from the need to deal with output shared among power units within a grouping. Section 17.7 illustrates the concepts using data similar to that found in Sect. 17.5. Conclusions and further directions follow in Sect. 17.8.

17.2 Hierarchical Structures: Power Plants

Ontario Hydro (now called Ontario Power Generation) is a crown corporation supplying electric power to both domestic and foreign markets in the northern USA. Two classes of power units or plants are managed under Hydro's jurisdiction – nuclear and thermal. While the number of nuclear units is relatively small, a total of 40 such units of varying ages, capacities, fuel types and so on are operated by the corporation. These latter will be the setting for the analysis of Sect. 17.5.

The standard measure of productivity used by management is the ratio of total annual expenditure (operating, maintenance and administration) to total energy produced, in megawatt hours per year. While it is the case that the total power production is a principal *output* of the operation, and is certainly the most convenient and readily available indicator of productive capability, management is interested in other, related indicators as well. What may be missing in this simplistic measure of productivity is a consideration of those factors that reflect management's skill. To a great extent, a power unit's efficiency measure should reflect the quality of maintenance that keeps it operating, and the abilities of management in charge of that maintenance. At least two types of other outputs should be considered, namely *outages* and *deratings*.

An outage is a situation in which a unit is shut down, hence it is not generating electric power. Types of outages include:

- planned outage, which is scheduled downtime (usually for major overhauls);
- maintenance outage, a form of scheduled down time, but for more minor, i.e., routine maintenance;
- forced outage, which is unscheduled and generally caused by equipment failure, environmental requirements, or other unforeseen incidents. There is generally some prior warning for this type of shutdown, and some delay is possible;
- sudden outage, which is a forced outage with no prior warning.

While it can be argued that operating hours essentially capture all forms of outages, it must be recognized that there is a difference between taking a unit out of service on a scheduled basis at non peak times, versus sudden brownouts or blackouts. The latter ignite public opinion, interrupt business operations, and generally reflect negatively on the organization. Thus, such outages should play a direct role in any measure of efficiency.

A derating is a *reduction* in unit capacity, where the operation may, for a number of reasons, operate at only a fraction (e.g., 75 % or 50 %) of its available (full) capacity. Breakdowns in coal belts, pulverizers or rollers (of which there are several operating in any plant) is a primary cause of such forced deratings. Environmental restrictions, in particular SO_2 emissions, can limit the extent to which a plant can operate at full capacity. Furthermore, such restrictions will often apply to a group of units (e.g., at a given geographical location).

As with outages, there are several forms of deratings, some of which are beyond the control of management and which have nothing to do with maintenance quality (e.g., grid or transmission network load restrictions), while others are a clear reflection of maintenance quality, such as equipment failures.

As with outputs, inputs should include several factors. In addition to expenditures, factors such as plant *age* and *available but not operating time* (ABNOT) should play a role as well. The latter factor (ABNOT) is the time during which the plant is able to operate, but for reasons beyond managements control (such as SO_2 restrictions), the plant is not running.

Grouping is a natural phenomenon here. Plants can be grouped by size or capacity, by geographical location, and so on. It is this necessity to view problems from a grouping and hierarchical perspective that we examine herein.

17.3 Models for Evaluating Plant Hierarchies

The power plant application discussed in the previous section provides an example of what might be termed a *pure hierarchy*. The basic DMU is the power unit. These 40 units are naturally clustered into eight plants.

17.3.1 The Two-Level Hierarchy

For simplicity of presentation in this subsection we assume there are only two levels in the hierarchy. Let the level 1 (power units) vectors of inputs and outputs be denoted $X(1)$, $Y(1)$ respectively, with $v(1)$, $\mu(1)$ representing the appropriate multipliers in the input orientation version of the CCR (Charnes et al. 1978) model.

In the normal case where we are interested only in a level 1 (power unit) analysis of efficiency, the "multiplier" form of the CCR model is:

$$\max \mu^T(1)Y_o(1) \tag{17.1a}$$

subject to:

$$v^T(1)X_o(1) = 1 \tag{17.1b}$$

$$\mu^T(1)Y_j(1) - \nu^T(1)X_j(1) \leq 0, \qquad j\varepsilon J \tag{17.1c}$$

$$\mu(1), \nu(1) \geq \varepsilon, \tag{17.1d}$$

where J is the set of DMUs under consideration. Suppose, however, that we want, in addition, to evaluate the relative efficiencies of the eight plants into which the 40 units are grouped. Clearly, one approach might be simply to evaluate each DMU relative to the entire set of 40 units as indicated above, (hence J would represent the entire set of power units), and use the average of the ratings for those units within any plant as representative of the standing of that plant. While it is difficult to argue that such an approach is wrong, it does possess some undesirable aspects. First, those factors that apply at the group level (level 2) are not represented (or at least not represented appropriately) in level 1. Second, and as indicated above, it would seem more appropriate at level 1 to evaluate a DMU relative to those DMUs in the same group only. In this case, J in (17.1c) above would refer to those units in a specific plant, whereupon, those factors which distinguish the groups (plants) can be omitted from the level 1 evaluation, and can more properly be applied at level 2. If this is done, then averaging within a plant does not help at all to understand the relative standings of the level 2 DMUs.

An alternative approach for evaluating efficiency at both levels 1 and 2, is to treat the level 2 groups themselves as decision making units, using a combination of the group-specific factors, and factors which emerge from level 1. The use of level 1 factors at level 2 may involve some form of aggregation as will be discussed in the next section.

For notational purposes define

K – the number of *groups* of level 1 DMUs, hence K is the number of DMUs at level 2;
k – a subscript representing a DMU at level 2;
j_k – a subscript representing a level 1 DMU that belongs to group k;
$Y_{kj_k}(1), X_{kj_k}(1)$ – level 1 outputs and inputs;
$Y_k^1(2)$, $X_k^1(2)$ – those level 2 outputs and inputs that are aggregates of factors that are used to evaluate level 1 DMUs;
$Y_k^2(2)$, $X_k^2(2)$ – those outputs and inputs used at level 2 that distinguish the K groups, and which were not used at level 1.

Let $\nu(1)$, $\mu(1)$ and $\nu(2)$, $\mu(2)$ denote the level 2 multipliers to be associated with $X_k^1(2)$, $Y_k^1(2)$ and $X_k^2(2)$, $Y_k^2(2)$ respectively. It is noted that $\nu(1)$, $\mu(1)$ are the same multipliers as used in level 1, as will be explained below.

In performing the analysis within a general model framework we make the following assumptions:

(a) When the "DMU" under consideration is a level 1 unit, we want to ensure that it is evaluated only relative to those units in the same group, hence DMUs in other groups should be excluded or *disengaged* from the constraint set;

(b) Level 2 DMUs (groups) should not interfere with, hence should be disengaged from, level 1 analyses;
(c) Level 1 DMUs should be included or *engaged* in the analysis of level 2 units.

Assumption (c) above is invoked with the argument that multipliers $\nu(1)$, $\mu(1)$ when applied at level 2 should also be feasible when applied to any level 1 units within the groups under consideration. Specifically, since the efficiency of a given group i at level 2 should be related to the efficiencies of that group's members, then the multipliers $\nu(1)$, $\mu(1)$ should be such that when applied to each member of the group, the ratio for that member should not exceed unity.

To accommodate the above considerations we propose the following general model. When applied to a level 2 DMU, the model would take the form:

$$\max e_o = \mu^T(1)Y_0^1(2) + \mu^T(2)Y_0^2(2) \tag{17.2a}$$

subject to:

$$v^T(1)X_0^1(2) + v^T(2)X_0^2(2) + Mw(2) = 1 \tag{17.2b}$$

$$\begin{aligned} &\mu^T(1)Y_k^1(2) + \mu^T(2)Y_k^2(2) - v^T(1)X_k^1(2) \\ &-v^T(2)X_k^2(2) - w(2) \le 0 \quad k = 1, \ldots K \end{aligned} \tag{17.2c}$$

$$\mu^T(1)Y_{kj_k}(1) - v^T(1)X_{kj_k}(1) - w_k(1) \le 0, j_k \varepsilon J_k, \quad k = 1, \ldots, K \tag{17.2d}$$

$$w(2) - w_k(1) \ge 0, \quad k = 1, \ldots, K \tag{17.2e}$$

$$\mu(1), \mu(2), v(1), v(2) \ge \varepsilon \tag{17.2f}$$

$$w_k(1), w(2) \ge 0, \quad \forall k \tag{17.2g}$$

When applied to a level 1 DMU, the model would take the form

$$\max e_0 = v^T(1)Y_0(1) \tag{17.2a$'$}$$

subject to:

$$v^T(1)X_0(1) + Mw_0(1) = 1 \tag{17.2b$'$}$$

(17.2c, 17.2d, 17.2e, 17.2f, and 17.2g)

Here M denotes a large positive number. In (17.2d), J_k denotes the index set of level 1 DMUs in group k (plant k). The notation $Y_0^1(2)$ in (17.2a) denotes the type 1 output (an aggregate of level 1 outputs) used at the second level and for a particular DMU $k =$ " 0. ". In (17.2a′) the notation $Y_0(1)$ denotes the output at level 1 for a particular DMU and group $(j_k, k) =$ " 0. " The variables $w_k(1)$, $w(2)$ are introduced to include or exclude certain DMUs from the analysis. In reference to the above discussion, these are referred to as *engagement variables*. It is noted that in (17.2b′), $w_0(1)$ refers to the particular level 2 groups $k =$ " 0 " in which the DMU under evaluation lies at the time. So, for example, all DMUs in a particular group k will be assigned the same variable $w_0(1)$.

In reference to assumptions (a), (b) above, we prove the following theorem:

Theorem 17.1 *In the evaluation of any level 1 DMU $j_{k_0} \varepsilon k_o$, only DMUs within the same group (k_0) as that DMU will be engaged. All other level 1 groups and all level 2 DMUs are disengaged.*

Proof Only the particular group $k =$ " 0 " in which the level 1 DMU under consideration lies, has its engagement variable ($w_0(1)$) involved in constraint (17.2b′). This variable will be forced to zero, otherwise the objective function value of (17.2a′) will equal zero. Furthermore, since all other engagement variables are free to assume the most favorable values possible (from the perspective of the DMU under evaluation), then all $w_k(1)$ (except for $w_0(1)$) and $w(2)$ will assume values large enough to render redundant all constraints in (17.2c), as well as all constraints in (17.2d) corresponding to those J_k, $k \neq$ " 0. " Since constraints (17.2e) are also redundant, the result follows.

Q.E.D.

From this theorem it follows as well that when a level 2 DMU is under evaluation, the engagement variable $w(2)$ will be forced to zero (hence *engaging* all level 2 DMUs). By virtue of constraints (17.2e), all $w_k(1) = 0$ as well, hence engaging all level 1 DMUs, thereby verifying assumption (c).

17.3.2 Efficiency Adjustments in a Hierarchy

In Sect. 17.5 we present an analysis of the efficiencies of power plants and groups of plants. One issue that arises in such multi level analyses has to do with adjustments in DMU efficiencies at one level to account for scores assigned at a higher level. Specifically, the scores achieved by individual DMUs (e.g., level 1) are measured only against others in the same group. To adjust these to reflect the standings of the groups themselves, it is necessary to merge the scores at these two levels in some reasonable manner. We describe a three step procedure to bring about the desired adjusted ratings.

Step 1: (Remove inter-group noise)
Scale the level 1 ratings by dividing each rating e_{kj_k} in group k by the average of the group k ratings. Specifically, define

$$f_{kj_k} = e_{kj_k}/\overline{e}_k \quad \text{where } \overline{e}_k = \left(\sum_{j_k \varepsilon J_k} e_{kj_k}\right) \Big/ |J_k|$$

where $|J_k|$ denotes the cardinality of J_k. Since the level 2 ratings are intended to account for any inter-group differences, this transformation is intended to remove any differences (noise) among the groups that are not level 2 – related. See Property 17.2 below and explanation following it.

Step 2: (Introduce level 2 adjustment)
Adjust the scaled ratings f_{kj_k} by multiplying them by the level 2 (group) ratings e_k. That is, define

$$g_{kj_k} = f_{kj_k} \times e_k.$$

Step 3: (Adjust to [0,1] scale)
Further adjust the step 2 ratings g_{kj_k} to ensure that the maximum level 1 rating is unity. Specifically, we want to adjust the g_{kj_k} ratings to the form

$$h_{kj_k} = g_{kj_k} \times R$$

where R is such that $h_{kj_k} \leq 1$, and $\max_{k,j_k}\{h_{kj_k}\} = 1$. Hence $R = \min_{k,j_k}\left\{1/g_{kj_k}\right\}$.

The final adjusted ratings therefore have two important properties:

Property 17.1 All level 1 ratings $h_{kj_k} \leq 1$, with at least one $h_{k_o j_{k_o}} = 1$.

Property 17.2 The averages of the ratings $\overline{h}_k$ within the K groups are such that $\frac{\overline{h}_{k_1}}{\overline{h}_{k_2}} = \frac{e_{k_1}}{e_{k_2}}$.

The latter property captures the fact that the final adjusted ratings not only represent the standing of DMUs (e.g., power units) within their own group k (plant), but also reflect their standing relative to DMUs in other groups. That is, if the rating e_{k_1} of one group k_1 is, for example, only 80 % of the rating e_{k_2} of another group k_2, then the averages for the DMUs in the two groups, namely $\overline{h}_{k_1}$ and $\overline{h}_{k_2}$, have this same property.

17.3.3 The Multi Level Hierarchy

The model (17.2a, 17.2b, 17.2c, 17.2d, 17.2e, 17.2f, and 17.2g) can be generalized to the case of an L-level hierarchy. We assume that the outputs and inputs used at any level ℓ are aggregates of $\ell - 1$ level factors together with any additional factors that distinguish the groups at the ℓ th level. We introduce the following notation:

ℓ – subscript representing a level in an L-level hierarchy;
K_ℓ – the number of DMUs at the ℓ th level;
k_ℓ – a subscript representing a DMU at level ℓ;
$j_{k\ell}$ – a subscript representing a DMU at level $\ell - 1$ that lies within a group k_ℓ (that is, within a DMU k_ℓ at the next level up in the hierarchy);
$J_{k\ell}$ – the subset of DMUs j_{k_ℓ} at level $\ell - 1$ that lie within group k_ℓ;

$\left\{Y^m_{k_\ell j_{k_\ell}}(\ell-1), X^m_{k_\ell j_{k_\ell}}(\ell-1)\right\}_{m=1}^{\ell-2}$ – those outputs and inputs used at level $\ell - 1$ that are aggregates of factors used for analysis of DMUs at lower levels $m \leq \ell - 2$. The subscript k_ℓ refers to the particular $\ell - 1$ level group (i.e. ℓ th level DMU), and j_{k_ℓ} to a DMU within that group;

$Y^{\ell-1}_{k_\ell j_{k_\ell}}(\ell-1), X^{\ell-1}_{k_\ell j_{k_\ell}}(\ell-1)$ – those outputs and inputs at level $\ell - 1$ that distinguish the DMUs at that level, and which were not used at any lower level;

$w_{k_\ell}(\ell-1)$ – denotes the engagement variables applicable at level $\ell - 1$. These distinguish the groups at this level.

$w(L)$ – denotes the engagement variable applicable at level L.

The model, when applied at the $\ell - 1$ level then takes the form:

$$\max c_o = \sum_{m=1}^{\ell-1} \mu^T(m) Y_0^m(\ell-1) \tag{17.3a}$$

$$\text{subject to}: \quad \sum_{m=1}^{\ell-1} \nu^T(m) X_0^m(\ell-1) + M w_0(\ell-1) = 1 \tag{17.3b}$$

$$\sum_{m=1}^{L} \mu^T(m) Y_j^m(L) - \sum_{m=1}^{L} \nu^T(m) X_j^m(L) - w(L) \leq 0, \quad j = 1, \ldots, K_L \tag{17.3c}$$

$$\sum_{m=1}^{\ell-1} \mu^T(m) Y_{k_\ell j_{k_\ell}}(\ell-1) - \sum_{m=1}^{\ell-1} \nu^T(m) X_{k_\ell j_{k_\ell}}(\ell-1) - w_{k_\ell}(\ell-1) \leq 0,$$
$$\ell = 2, \ldots, L, \quad k_\ell = 1, \ldots, K_\ell, j_{k_\ell} \varepsilon J_{k_\ell} \tag{17.3d}$$

$$w(L) - w_{k_L}(L-1) \geq 0, \quad k_L = 1, \ldots, K_L \tag{17.3e}$$

$$w_{k_\ell}(\ell-1) - w_{k_{\ell-1}}(\ell-2) \geq 0, \quad k_\ell = 1, \ldots, K_\ell, k_{\ell-1} \varepsilon I_{k_\ell} \tag{17.3f}$$

$$\mu(m), \nu(m) \geq \varepsilon, \quad m = 1, \ldots, \ell-1 \tag{17.3g}$$

where I_{k_ℓ} is comprised of those sets of DMUs at level $\ell - 2$ that make up the k_ℓ th set at level $\ell - 1$.

As with the 2-level problem discussed earlier, the engagement variables $w_{k_\ell}(\ell - 1)$ act to include or exclude sets of DMUs as the analysis proceeds. With regard to adjustments to ratings, a similar procedure could be applied here by starting at the top level L in the hierarchy to bring about alterations to the ratings at level L−1. Then, apply these adjusted ratings to alter the L−2 level ratings, and so on.

In this subsection we have examined the problem of evaluating DMUs and groups of DMUs which appear in the form of a hierarchy. In the following subsection this idea is extended to look at the alternative groupings of DMUs on the same level.

17.4 Grouping on Levels

The power plant application discussed above is a prime example of a pure hierarchy in that DMUs are grouped at each level according to a single attribute – in this case a jurisdictional or geographical attribute. In Sect. 17.5 we analyze the efficiency of the set of power plants and groupings thereof. In this case, the problem of efficiency evaluation seems to invite a 2-level analysis, in that plants can be grouped by a number of different attributes – capacity, geographical location, fuel type and so on. All these factors can be judged as level 2 attributes, although admittedly one can conceive of very complex mixes of these. One could, for example, group plants at the second level according to geographical location, then at a 3rd level group locations by capacity, assuming, of course, that only one capacity of plants exists at a given location. In the present example, this is not exactly the case. Of course, if at the third level we attempted to group by capacity, regardless of the location, then the hierarchical structure is destroyed. Groups at one level would be broken apart when going to the next level.

In the following subsection we will consider grouping only at one level (level 2 in the case of the power plants), and according to multiple attributes. If we wish to have plants at level 1 evaluated strictly within the groups that will form the DMUs at level 2, it would appear that multi attribute grouping implies simply replicating model (17.2a, 17.2b, 17.2c, 17.2d, 17.2e, 17.2f, and 17.2g) as many times as there are attributes. Suppose, for example, that we wish to group plants in two ways: (1) geographical and (2) according to capacity. The most practical approach would appear to be to run this model once for each type of grouping. This would lead to two sets of efficiency ratings. While an elaborate model with engagement variables can easily be formulated, there would seem to be no practical advantage in doing so.

17.4.1 Deriving an Aggregate Rating

The issue of grouping on a level according to a number of different attributes gives rise to the problem of how to derive some form of overall rating for a DMU. Suppose, for example, that plants are grouped by geographical location. A given plant j, when evaluated in a DEA manner, will be compared to other plants within the same group (at the same location). The number of other plants in that group and the efficiencies of those other plants will, of course, influence the score that j receives. When evaluated according to some other grouping attribute such as capacity, plant j will, in all likelihood, receive a different score. The problem then is how to view the aggregate or overall standing of j, given the different ratings for j that arise out of this multi attribute analysis.

One approach to this problem of deriving an overall efficiency measure is to introduce an importance multiplier on the i th attribute. To formalize this, assume there are I attributes, hence I different grouping types, and let e_{ij} denote the efficiency rating received by DMU j when viewed in terms of the grouping created by the i th attribute. Let α_i denote the weight or importance to be accorded attribute $i\varepsilon I$. The α_i may either be supplied weights or may need to be determined (discussed below). Using these multipliers, we define the aggregate efficiency of DMU j to be:

$$e_j = \sum_{i\varepsilon I} \alpha_i e_{ij}$$

In the event that the α_i are decision variables, there may or may not be information available as to appropriate values for these variables. In any event, and in the spirit of general DEA, one approach to deriving an aggregate rating for DMU j_o is to determine $\{\alpha_i\}$ through the optimization procedure

$$e^*_{j_o} = \max \quad e_{j_o} = \max \sum_{i=1}^{I} \alpha_i e_{ij_o} \tag{17.4a}$$

$$\text{subject to:}$$

$$\sum_{i=1}^{I} \alpha_i e_{ij} \leq 1, j\varepsilon J \tag{17.4b}$$

$$\alpha = (\alpha_i, \ldots \alpha_I)\varepsilon \Phi, \tag{17.4c}$$

where Φ defines the available information on the $\{\alpha_i\}$. Constraints (17.4b) bound the problem by requiring that the aggregate efficiency for each DMU not exceed 1.

One minimal set of restrictions on the α_i might be an ordinal ranking of the attributes. Suppose, for example, that the set of attributes consist of:

(a) geographical location,
(b) capacity,
(c) age,
(d) fuel type used.

Furthermore, assume that these attributes can be prioritized in order of importance to the organization (the utility company). With no loss of generality, assume that the most important attribute is geographical location, followed by capacity, then age, and finally fuel type. In notational terms, this would imply that $\alpha_1 > \alpha_2 > \alpha_3 > \alpha_4$. Introducing an infinitesimal ε, Φ may then be defined in this case by

$$\Phi = \left\{\alpha = (\alpha_1, \ldots, \alpha_I) \middle| \alpha_i - \alpha_{i+1} \geq \varepsilon, i = 1, \ldots, I - 1; \alpha_I \geq \varepsilon\right\} \tag{17.4d}$$

The idea of ordinal relations among multipliers in DEA was discussed in Ali et al. (1991) and Golany (1988). A somewhat similar structure appears in Cook, Kress and Seiford (1996) in the context of incorporating ordinal data within the DEA framework. Clearly, problem (17.4a, 17.4b, 17.4c, and 17.4d) is a set of J linear problems with each yielding a best or most efficient aggregate evaluation for the DMU j_o under consideration. One possible drawback to this approach is the fact that a different set of $\{\alpha_i\}$ will arise from each of the J optimizations. This, of course, can be a *general* criticism of the DEA approach.

17.4.2 A Common Set of Multipliers

If it is desirable to obtain a single or common set of multipliers $\{\alpha_i\}$, one approach to use in this particular instance is to determine the largest value of ε for which a feasible set of α_i exists. Specifically, solve the *single* optimization problem:

$$\varepsilon^* = \max\varepsilon \tag{17.4e}$$

subject to (17.4b, 17.4c, and 17.4d)

The set of α_i that are optimal in this problem provide a means of evaluating all DMUs on a common basis. The essence of this approach is that the minimum extent to which we distinguish or discriminate between the importance measures (α_i) attached to the various criteria is maximized.

17.4.3 Multiple Rankings of Attributes

In the above it is assumed that an overall *single* rank ordering of the attributes in I is at hand. This ordering is intended to express the relative importance of the various grouping mechanisms (geographical location, capacity, etc.). In some situations it may be necessary to ask the question "importance in what sense?" If environmental considerations are paramount, the above rank ordering which places geographical location first in importance may be appropriate. On the other hand, if new technology for powering the plants (new fuel types, e.g.,) is an issue, then the attribute 'fuel type' used may rank in first place. Therefore, multiple rankings of the attributes may be in order.

To formalize this concept, assume that Q *ranking vehicles* or mechanisms are to be considered. Let α_i^q denote the importance or weight to be given to attribute $i\varepsilon I$ when viewed from the perspective of ranking vehicle $q\varepsilon Q$. Furthermore, let the *decision variable* β_q represent the weight to be given to vehicle q. While various types of restrictions could be imposed on the β_q, we assume here that only positivity constraints are imposed, i.e. $\beta_q \geq \varepsilon$ for all q. If a rank ordering on the α_i^q is now imposed relative to each q, then Q feasible regions $\{\Phi_q\}_q=1^Q$ would be defined. Specifically, define

$$\Phi_q = \left\{\alpha^q = \left(\alpha_1^q, \alpha_2^q, \ldots, \alpha_I^q\right)\alpha_{i_\ell}^q - \alpha_{i_{\ell+1}}^q \geq \varepsilon, \ell = 1, \ldots, I-1; \alpha_{i_I}^q \geq \varepsilon\right\}, \tag{17.4f}$$

where $\alpha_{i_\ell}^q$ denotes the ℓ th ranked attribute from the point of view of the q th ranking vehicle. Following the logic of problem (17.4a, 17.4b, and 17.4c), an aggregate efficiency rating for DMU j_o could then be determined by solving the J problems:

$$e_{j_o}^* = \max e_{j_o} = \max \sum_{q=1}^{Q} \sum_{i=1}^{I} \beta_q \alpha_i^q e_{ij_o} \tag{17.5a}$$

subject to:

$$\sum_{q=1}^{Q} \sum_{i=1}^{I} \beta_q \alpha_i^q e_{ij} \leq 1, j\varepsilon J \tag{17.5b}$$

$$\alpha^q = \left(\alpha_1^q, \ldots, \alpha_I^q\right) \varepsilon \Phi_q, q = 1, \ldots, Q. \tag{17.5c}$$

$$\beta_q \geq \varepsilon, q = 1, \ldots, Q \tag{17.5d}$$

Problem (17.5a, 17.5b, 17.5c, and 17.5d), unlike the earlier single ranking vehicle formulation, is nonlinear with the product of the β_q and α_i^q. This formulation can be transformed to an equivalent linear structure, however, through a simple change of variables. That is, define

$$\delta_{qi} = \beta_q \alpha_i^q,$$

and note that the constraints $\alpha_{i_\ell}^q - \alpha_{i_{\ell+1}}^q \geq \varepsilon$ and $\alpha_{i_I}^q \geq \varepsilon$ can be replaced (through multiplication by β_q on both sides of the inequality) by $\delta_{qi_\ell} - \delta_{qi_{\ell+1}} \geq \varepsilon\beta_q$, and $\delta_{qi_I} \geq \varepsilon\beta_q$. Problem (17.5a, 17.5b, 17.5c, and 17.5d) is then equivalent to the linear problem:

$$e_{j_o}^* = \max e_{j_o} = \max \sum_{q=1}^{Q} \sum_{i=1}^{I} \delta_{qi} e_{ij_o} \tag{17.6a}$$

subject to:

$$\sum_{q=1}^{Q} \sum_{i=1}^{I} \delta_{qi} e_{ij} \leq 1, j\varepsilon J \tag{17.6b}$$

$$\delta_{qi_\ell} - \delta_{qi_{\ell+1}} - \varepsilon\beta_q \geq 0, \ell = 1, \ldots, I-1; q = 1, \ldots, Q \tag{17.6c}$$

$$\delta_{qi_I} - \varepsilon\beta_q \geq 0, q = 1, \ldots, Q. \tag{17.6d}$$

That is, given an optimal solution $(\delta^*_{iq}, \beta^*_q)$ to (17.6a, 17.6b, 17.6c, and 17.6d), then $\alpha_i^{q*} = \delta^*_{iq}/\beta^*_q$ and β^*_q constitute an optimal solution to (17.4f, 17.5a, 17.5b, 17.5c, and 17.5d), due to the fact that all δ^*_{iq}, β^*_q are strictly positive.

In certain situations the ranking vehicles referred to above may take the form of opinions offered by a set of Q voters (e.g. managers). That is, the relative importance of the I grouping attributes may be a matter of opinion, hence model (17.5a, 17.5b, 17.5c, and 17.5d) (and therefore (17.6a, 17.6b, 17.6c, and 17.6d)) is intended to derive a rating which takes into consideration the various opinions (rankings) offered.

Clearly the earlier comments regarding a common set of weights applies in the present situation as well.

In the following section an application is presented which illustrates some of the model structures presented in this and the previous section.

17.5 Efficiency Analysis of Power Plants: An Example

Earlier a description was given of a problem setting involving thermo generating plants, wherein it was argued that efficiency should be viewed in terms of a set of outputs and inputs. Table 17.1 shows the number of thermal units operating at each of eight locations, two of which (Plant 4(1) and Plant 7(1)) are each broken down into two groups for a total of ten groupings. Given also are the construction dates, fuel types and capacities in megawatt hours.

In the analyses of the plants, two levels were examined, namely, the individual power unit level (level 1) and a second level where plants are grouped in various ways. Two forms of analyses were carried out:

Table 17.1 Thermal plants

			Fuel	
Location	# units	Age range	Utilized	Size (MWH)
Plant 1	8	1971–1972	U.S. Bit. Coal & Western Cdn. Coal	500
Plant 2	8	1968	U.S. Bit. Coal	300
Plant 3	4	1970	U.S. Bit Coal	500
Plant 4 (1)	1	1964–1966	U.S. Bit Coal	100
Plant 4 (2)	2	1974–1975	Liquid Bit	150
Plant 5	4	1974	Oil	500
Plant 6	1	1978	Lignite Bit. Coal	200
Plant 7 (1)	4	1956	Gas/Coal	100
Plant 7 (2)	4	1960	Gas/Coal	200
Plant 8	4	1952	U.S. Bit. Coal	50

1. a hierarchical analysis at the two levels, where plants are grouped in level 2 by location;
2. analysis of efficiency on a level where, with different types of grouping, it is necessary to deal with several ratings for a given DMU.

Table 17.2 displays the raw data for the 40 plants under analysis.[1] Shown are three outputs and two inputs. These outputs and inputs are defined as follows:

Outputs

- OPER – a function of equivalent full capacity operating hours. This factor accounts for the fact that when operating at less than 100 % capacity (e.g. if the unit is derated to 50 % capacity), the operating hours during this period are prorated. To bring the scale of values for the units of measurement within the range of the scales used for other factors, we apply a scaling factor of $\frac{1}{10}$, i.e. OPER $= \frac{1}{10}\times$ full capacity operating hours.
- OUT – a function of the number of forced and sudden outages.
  OUT = N−K (# forced outages + # sudden outages). Sudden and forced outages, as unscheduled shutdowns of operations, are often consequences of equipment failure. Again, to bring scales into line we arbitrarily choose N = 200, K = 10.
- EQDER – a function of forced deratings caused by equipment failure.
  EQDER = N−K (# equipment related deratings), with N = 200 and K = 10 as above.

  Since on the output side, any measure used must be such that bigger is better, one cannot *directly* take outages as an output. To achieve the bigger is better condition, we subtract outages from some constant to create a proper scale measure. The value 200 has been chosen arbitrarily, but at the same time to yield "OUT" values that are in line with the scales used for other factors. Some sensitivity analyses were done relative to this parameter (200), and the particular value chosen was found to have very little effect on the final relative efficiency outcomes.

Inputs

- MAINT – the total maintenance expenditure (labor + materials) in thousands of dollars.

  Clearly, we could separate this into monetary inputs, but for purposes here we aggregate the two amounts into one figure.
- OCCUP – a function of total occupied hours, that is
  OCCUP $= \frac{1}{10}$ (Total hours available – available but not operating hours).

  In evaluating the ten level 2 DMUs (where, for example, the group of plants at Plant 1 is taken as a DMU), the averages of the level 1 DMUs make up the first three outputs and the first two inputs. For example, the average of the ten

[1] It is pointed out that this is sanitized data for illustration purposes only. It in no way reflects the actual operating positions of the various plants.

Table 17.2 Outputs and inputs for unit level analyses

		Outputs			Inputs	
Group	Unit	Oper	Out	Eqder	Maint	Occup
Plant 1	1	573	95	110	538	895
	2	560	138	120	290	770
	3	637	151	150	386	886
	4	685	139	160	290	760
	5	542	157	130	343	721
	6	520	100	120	470	810
	7	531	122	60	439	820
	8	511	135	160	293	888
Plant 2	1	521	102	93	440	771
	2	634	93	102	324	780
	3	610	86	75	378	825
	4	538	95	106	380	815
	5	591	116	119	241	880
	6	650	123	105	141	766
	7	621	107	91	355	823
	8	686	125	110	270	750
Plant 3	1	620	120	130	350	750
	2	550	81	95	630	770
	3	525	105	125	495	860
	4	580	125	106	345	800
Plant 4(1)	1	430	105	140	190	810
Plant 4(2)	1	560	110	105	280	770
	2	510	125	95	180	820
Plant 5	1	650	170	140	300	7,000
	2	550	120	120	275	800
	3	580	160	110	447	650
	4	640	110	130	370	720
Plant 6	1	480	95	125	228	880
Plant 7(1)	1	320	70	110	230	790
	2	250	60	110	220	790
	3	370	100	140	320	840
	4	280	90	100	280	810
Plant 7(2)	1	520	120	100	281	750
	2	430	100	140	302	850
	3	470	110	150	227	770
	4	410	80	110	254	825
Plant 8	1	475	100	120	179	750
	2	560	150	120	143	800
	3	510	120	110	114	750
	4	425	140	90	172	820

Plant 1 operating hours figures is 582. In addition to these aggregated figures, two further outputs, ENDER (a factor for environmental deratings) and planned capacity were used for the level 2 analyses. As well, a third input, average year of construction, was utilized. The data for the level 2 analyses is shown in Table 17.3.

Table 17.3 Group level data

	Outputs					Inputs		
Group	Oper.	Out.	Eqder	Ender	Cap	Maint.	Occup	Yr. const.
Plant 1	582	130	126	125	500	381	818	71
Plant 2	606	106	100	147	300	317	801	68
Plant 3	569	103	108	121	500	455	795	70
Plant 4(1)	430	105	140	111	100	190	810	65
Plant 4(2)	420	105	100	125	150	350	815	75
Plant 5	605	140	125	141	500	348	717	74
Plant 6	480	95	125	117	200	348	800	78
Plant 7(1)	305	80	115	110	100	263	808	56
Plant 7(2)	458	103	125	116	200	266	799	58
Plant 8	493	128	110	135	50	152	780	52

Table 17.4 Efficiency scores – hierarchical analysis (grouped by location)

(1)	(2)	(3)	(4)	(5)	(1)	(2)	(3)	(4)	(5)
Group	Unit	Group ratings	Unit ratings	Adjusted unit ratings	Group	Unit	Group ratings	Unit ratings	Adjusted unit ratings
Plant 1	1	100.0	70.8	70.8	Plant 4(1)	1	100.0	100.0	87.7
	2		99.1	99.1	Plant 4(2)	1	80.6	100.0	70.7
	3		86.0	86.0		2		100.0	70.7
	4		100.0	100.0	Plant 5	1	100.0	100.0	90.2
	5		100.0	100.0		2		93.5	84.3
	6		71.1	71.1		3		100.0	90.2
	7		76.1	76.1		4		95.7	86.3
	8		98.7	98.7	Plant 6	1	87.5	100.0	76.8
Plant 2	1	100.0	82.0	80.2	Plant 7(1)	1	87.1	100.0	76.4
	2		89.1	87.1		2		100.0	76.4
	3		80.7	78.9		3		100.0	76.4
	4		88.5	86.5		4		100.0	76.4
	5		95.1	93.0	Plant 7(2)	1	93.9	100.0	90.5
	6		100.0	97.8		2		84.5	76.4
	7		82.4	80.5		3		100.0	90.5
	8		100.0	97.8		4		79.6	72.0
Plant 3	1	100.0	100.0	94.9	Plant 8	1	100.0	100.0	89.7
	2		86.1	81.7		2		100.0	89.7
	3		83.7	79.4		3		100.0	89.7
	4		100.0	94.9		4		90.9	81.6

17.5.1 Hierarchical Analysis

Table 17.4 displays the outcomes from the hierarchical analysis. Here, power units have been grouped by location (Plant 1, Plant 2, . . ., Plant 8), and have been analyzed using the hierarchical DEA model (17.2a, 17.2b, 17.2c, 17.2d, 17.2e,

Table 17.5 Plant groupings by capacity and plant groupings by fuel type

Group	Capacity	Units included
Group 1	500 MWH	Plant 1, Plant 3, Plant 5
Group 2	200–300 MWH	Plant 2, Plant 6, Plant 7(2)
Group 3	<200 MWH	Plant 4(1), Plant 4(2), Plant 7(1), Plant 8
Group	Fuel type	Units included
Group 1	U.S. Bit. Coal	Plant 1, Plant 2, Plant 3, Plant 4(1), Plant 8
Group 2	Gas/Coal	Plant 7(1), Plant 7(2)
Group 3	Liquid Bit. Coal	Plant 4(2)
Group 4	Oil	Plant 5
Group 5	Lignite Bit. Coal	Plant 6

17.2f, and 17.2g) and (17.2a′, 17.2b′). The ten group ratings are shown under column (3). Column (4) provides the "within group" ratings of individual power units, i.e., those ratings achieved when units are compared only to the members of their own group. To obtain ratings whereby all 40 DMUs can be compared on a common basis, the suggested three-stage adjustment developed earlier has been applied to the column 4 figures. The resulting adjusted values are shown in column 5.

17.5.2 Grouping on Levels

In the above analyses, power units were grouped by location (e.g., the eight Plant 1 units formed one group). The within groups analyses resulted in the ratings shown in column 4 of Table 17.4. Two other types of groupings were then evaluated – by fuel type and by capacity. Table 17.5 specifies the memberships of the groups. When the within group analyses were carried out on the power units under these alternative groupings, ratings of units changed to reflect group membership. Table 17.6 displays power unit ratings under the different membership scenarios (columns (2),(3),(4)). The location scenario has been replicated here (from Table 17.4). To combine the three ratings for each power unit, model (17.4a, 17.4b, 17.4c, and 17.4d) and model (17.4e) with (17.4b, 17.4c, and 17.4d) were applied. The outcomes from these models are displayed under columns (5) and (6) respectively. In both instances the set Φ of (17.4d) is defined such that capacity is rated to be of highest importance, followed by location, then by fuel type, i.e.,

$$\text{Capacity} > \text{location} > \text{fuel}.$$

Although multiple rankings of attributes could clearly be applied to this example, such an analysis was not carried out here.

Table 17.6 Power unit ratings under different groupings

(1)		(2)	(3)	(4)	(5)	(6)
		Grouping by			Aggregate	Aggregate
Plant	Unit	Location	Capacity	Fuel	(District Wts.)	(Common Wts.)
1	1	70.8	68.8	69.6	69.8	69.6
	2	99.1	87.1	90.7	93.1	91.7
	3	86.0	82.0	86.0	84.7	84.0
	4	100.0	100.0	100.0	100.0	100.0
	5	100.0	90.0	100.0	96.7	95.0
	6	71.1	70.8	71.0	71.0	71.0
	7	76.1	69.6	76.1	73.9	72.8
	8	99.5	98.7	99.5	99.0	98.8
2	1	82.0	80.1	73.6	81.0	79.6
	2	89.1	88.9	87.4	89.0	88.7
	3	80.7	80.7	78.1	80.7	80.3
	4	88.5	80.4	72.1	84.4	81.7
	5	95.1	85.0	78.4	90.0	87.3
	6	100.0	100.0	100.0	100.0	100.0
	7	82.4	82.4	80.6	82.4	82.1
	8	100.0	100.0	100.0	100.0	100.0
3	1	100.0	89.0	90.7	94.5	92.9
	2	86.1	76.6	76.9	81.3	79.8
	3	83.7	68.9	68.9	76.3	73.8
	4	100.0	78.0	82.7	89.0	86.1
4(1)	1	100.0	100.0	100.0	100.0	100.0
4(2)	1	100.0	100.0	100.0	100.0	100.0
	2	100.0	88.1	100.0	96.0	94.0
5	1	100.0	100.0	100.0	100.0	100.0
	2	93.5	86.6	93.5	91.2	90.0
	3	100.0	100.0	100.0	100.0	100.0
	4	95.7	95.6	95.7	95.6	95.6
6	1	100.0	80.6	100.0	93.5	90.3
7(1)	1	100.0	80.5	72.3	90.2	85.6
	2	100.0	80.4	75.5	90.2	86.1
	3	100.0	96.2	85.4	98.1	95.7
	4	100.0	74.5	73.7	87.2	82.9
7(2)	1	100.0	95.8	100.0	98.6	97.9
	2	84.5	84.5	84.5	84.5	84.5
	3	100.0	100.0	100.0	100.0	100.0
	4	79.6	72.8	79.6	77.3	76.2
8	1	100.0	100.0	93.0	100.0	98.8
	2	100.0	100.0	100.0	100.0	100.0
	3	100.0	100.0	100.0	100.0	100.0
	4	90.9	90.9	89.4	90.9	90.7

17.6 Simultaneous Evaluation Across Levels

The model discussed above evaluates efficiencies at various levels in a hierarchy in a multi-stage fashion. Specifically, in stage 1, performance measures for power units within each plant are computed relative to their peers (within that plant's subset of units). In stage 2, the plants, at Level 2, are treated as DMUs, and requisite efficiency scores are computed there. Level 1 scores (for the power units) are then adjusted to reflect differences in efficiencies among the plants. In the hierarchical structure, DMUs at Level n have 2 types of inputs and outputs: (1) those consisting of aggregates of the corresponding factors at Level $n - 1$, and (2) additional measures that apply only at Level n.

In the current section we approach efficiency measurement at the various levels in this hierarchical structure by considering all levels simultaneously, and by directing the optimization at the highest level in the hierarchy. In the two-level setting, this means treating the plants at Level 2 as the DMUs, with the power units at Level 1 viewed as *components* of the DMUs. The complicating feature of this approach is the presence of plant-specific output factors which must be apportioned across the components in an equitable manner. The ideas used here are similar to those applied in Chap. 6 involving multi component efficiency in banking.

There appear to be at least two disadvantages of the two-stage approach discussed above. First, the measure applied (as suggested by the power authority) is simply related to the *frequency* of environmental deratings per year, as opposed to some function of the level of the SO_2 above or below the threshold. Arguably, it is the *quantity* of environmental damage that one may wish to capture as an output from the plant. Second, since the environmental variable only applies at the plant level, it is then the case in the hierarchical model that each power unit within that plant is *equally* penalized. Clearly, however, an individual power unit in a plant may contribute more or less toward the production of hazardous materials (e.g. SO_2) than is true for some other power unit. A power unit that is, for example, shut down for maintenance during peak pollution periods would not likely contribute as much to pollution accumulation as other units that were operating at full capacity during that time.

In this section, we present an augmented version of the DEA model that views both levels in the hierarchy simultaneously, generating performance measures for each plant and for the power units within those plants. Level 2 (plant level) variables are allocated across the level 1 power units. This is done in a manner consistent with any imposed constraints on the proportions of the output assigned to the various power units, and with the objective of maximizing the performance measure of the level 2 (plant) unit under consideration at any stage in the DEA model.

Consider the situation in which there are K power plants, with J_k power units within plant k. We define:

$Y_{kj_k} = (y_{kr_1j_k})$ – the R_1 – dimensional vector of outputs generated by power unit j_k in plant k.

$X_{kj_k} = (x_{kij_k})$ – the I – dimensional vector of inputs consumed by power unit j_k in plant k.

$Y_{ks} = (y_{kr_2s})$ – the R_2 – dimensional vector of outputs generated by plant k.

Let ν, μ, μ_s denote vectors of multipliers associated with X_{kj_k}, Y_{kj_k} and Y_{ks} respectively.

It is noted that in the current application, plant level (level 2) factors appear only on the output side. In the previous model, year of construction was taken as a level 2 input, but turned out to be relatively insignificant. While the model structure herein is easily extended to include both inputs and outputs, we restrict our attention only to such factors on the output side.

To facilitate model development, define the R_2 -dimensional decision vectors $\alpha^k_{j_k} = \left(\alpha^k_{r_2j_k}\right)$, where $\alpha^k_{r_2j_k}$ is the proportion of output y_{kr_2s} allocated to power unit j_k. As well, let Y_k, X_k denote the aggregates of the output vectors $\{Y_{kj_k}\}_{j_k}$ and input vectors $\{X_{kj_k}\}_{j_k}$, respectively. That is

$$Y_k = \sum_{j_k \varepsilon J_k} Y_{kj_k}, X_k = \sum_{j_k \varepsilon J_k} X_{kj_k}.$$

In this particular problem setting, aggregates derived in this manner make logical sense, although in some settings, sums of outputs may not be relevant.

The proportion $\alpha^k_{r_2j_k}$ of output y_{kr_2s} to be allocated to power unit j_k, may fall within certain logical bounds. Arguably, in the case that a given output r_2, is, for example, SO_2 emissions, the relative shares of this output allocated to two given units j_{k_1}, j_{k_2} could depend on a number of factors. These would include fuel types used, capacities in megawatt hours, operating hours, frequency of equipment failure deratings, etc. Since fuel type and capacity are fixed for units within the same plant, one can assume that $\alpha^k_{r_2j_k}$ is a function of factors such as operating hours. Reasonable bounds might take the form:

$$L_{j_k} \leq \alpha^k_{r_2j_k}/\alpha^k_{r_21} \leq U_{j_k}$$

Here, we assume that power unit #1 in plant k is taken as a standard, and other units j_k are compared to #1. L_{j_k} and U_{j_k} represent lower and upper limits respectively on the ratio of the proportions of output r_2 assigned to power units #1 and # j_k.

In the present two-level structure as described earlier, the plant (k) level performance measure (for any given set of multipliers (μ, μ_s, ν)) is given by:

$$e^k = [\mu Y_k + \mu_s Y_{ks}]/\nu X_k \tag{17.7}$$

Here, we distinguish between Y_k, the aggregate of level 1 (power unit) outputs, and Y_{ks}, the plant level (level 2) outputs that are to be allocated to the respective level 1 units. We can view Y_{ks} as a form of *shared* output (that is, shared among the power units). The *corresponding* j_k^{th} power unit performance ratio is given by:

$$e^k_{j_k} = \left[\mu Y_{kj_k} + \mu_s \alpha^k_{j_k} Y_{ks}\right]/\nu X_{kj_k} \tag{17.8}$$

We use here the notation $\alpha_{j_k}^k Y_s^k$ to denote the R_2 -dimensional vector $\left[\alpha_{1j_k}^k y_{k1s}, \alpha_{2j_k}^k y_{k2s}, \ldots, \alpha_{R_2 j_k}^k y_{kR_2 s}\right]^t$.

Property 17.3 The aggregate performance measure e^k of (17.7) is a convex combination of the J_k power unit measures $\left\{e_{j_k}^k\right\}_{j_{k=1}}^{J_k}$, defined in (17.8).

Property 17.4 A power plant k is efficient ($e^k = 1$) if and only if each power unit j_k within the plant is efficient ($e_{j_k}^k = 1$).

We now propose the following two-level variant of the standard CCR model:

$$\begin{aligned}
&\max \mathrm{e}^{\mathrm{o}} \\
&\text{subject to :} \\
&\mathrm{e}^{\mathrm{k}} \leq 1 && \text{all } k, \\
&e_{jk}^k \leq 1 && \text{all } k, j_k \varepsilon J_k \\
&L_{jk} \leq \alpha_{r_2 j_k}^k / \alpha_{r_2 1}^k \leq U_{jk} && \text{all } r_2, k, j_k \varepsilon J_k, \\
&\sum_{j_k \varepsilon J_k} \alpha_{r_2 j_k}^k = 1 && \text{all } r_2, k, \\
&\alpha_{r_2 j_k}, \mu_{r_1}, \mu_{sr_2}, v_i \geq 0, && \text{all } r_1, r_2, k, j_k i.
\end{aligned} \tag{17.9}$$

Problem (17.9) is nonlinear in two respects. First e^k and $e_{j_k}^k$ are linear fractional functionals. Second $e_{j_k}^k$ involves the product of variables $\mu_{sr_2} \alpha_{r_2 j_k}^k$. However, it can be shown that (17.9) is equivalent to a linear programming formulation, as given by the following theorem.

Theorem 17.2 *Problem (17.9) can be represented as a linear programming problem.*

Proof First it is noted that from Property (17.4), the constraints $e^k \leq 1$ are redundant, and can be removed from the problem. Make the change of variables

$$\gamma_{kr_2 j_k} = \mu_{sr_2} \cdot \alpha_{r_2 j_k}^k.$$

It is noted that the constraint set

$$L_{j_k} \leq \alpha_{r_2 j_k}^k / \alpha_{r_2 1}^k \leq U_{j_k}$$

becomes

$$Ł_{j_k} \alpha_{r_2 1}^k \leq \alpha_{r_2 j_k}^k \leq U_{j_k} \alpha_{r_2 1}^k$$

which, with multiplication through by μ_{sr_2}, becomes

$$L_{jk}\gamma_{r_2 1} \le \gamma_{kr_2 j_k} \le U_{j_k}\gamma_{r_2 1}.$$

As well, the convexity restriction

$$\sum_{j_k \varepsilon J_k} \alpha^k_{r_2 j_k} = 1$$

can be replaced by

$$\sum_{j_k \varepsilon J_k} \gamma^k_{r_2 j_k} = \mu_{sr_2}.$$

Following the standard conversion of Charnes and Cooper (1962) the linear fractional programming model (17.9) becomes

$$\begin{aligned}
&\max \mu Y^o + \mu_s Y^o_s, \\
&\text{subject to :} \\
&vX_o = 1, \\
&\mu Y_{kj_k} + \gamma^k_{j_k} Y_{ks} - vX_{kj_k} \le 0 \quad \text{all } k, j_k, \\
&L_{j_k}\gamma^k_{r_2 1} \le \gamma^k_{r_2 j_k} \le U_{j_k}\gamma^k_{r_2 1} \quad \text{all } r_2, k, j_k \\
&\gamma^k_{r_2 j_k}, \mu_r, \mu_{sr_2}, v_i \ge 0 \quad \text{all } r_1, r_2, k, j_k, i.
\end{aligned} \tag{17.10}$$

Clearly, problem (17.10) satisfies the necessary linearity property.

Q.E.D.

From the optimal solution of (17.10), one can compute $\hat{\alpha}^{\,k}_{\,r_2 j_k}$ from

$$\hat{\alpha}^{\,k}_{\,r_2 j_k} = \hat{\gamma}^{\,k}_{\,r_2 j_k} / \hat{\mu}_{\,sr_2}$$

In the following section, we apply model (17.10) to evaluate efficiencies of a set of power plants and corresponding power units.

17.7 Analysis of Efficiency: An Example

Considering again the data of Table 17.2, we can view the power plants (level 2 in the hierarchy) as aggregates of the units that comprise those plants. In this regard, the aggregates of all level 2 outputs and inputs can serve as level 2 factors. (As discussed previously, such aggregation may not be relevant in all cases, although it is so in this instance). In addition, there are factors that pertain primarily to the plant level only. The best example of such a factor in this situation is SO_2 emissions. The total environmental damage caused by a plant can be measured by the level (density of particulates) of SO_2 above some tolerable threshold, and multiplied by the number of hours that this phenomenon prevails during the year. Again, this factor falls into the more is worse category, as is true of the level 1 outputs, and was subtracted from the worst case value.

17.7.1 Proportional Split of Plant-Level Outputs

In the process of solving (17.10), the γ_{j_k} - variables (that give rise to the α_{j_k} - variables) are intended to split the shared output (SO_2) across the units in a plant, in a way that is most fair for that plant. If a particular power unit j_{k_1} in a plant is experiencing a higher degree of outages and equipment-related deratings than is true of the other units in that plant, then j_{k_1} should arguably be penalized with a smaller proportion of the environmental damage due to SO_2.

Unfortunately, the data is too coarse to be able to detect when a power unit was simultaneously experiencing equipment-related deratings, and environmental (SO_2) deratings. Clearly, if a power unit j_{k_1} was shut down for some reason on a given day when SO_2 emissions were high, the corresponding α_{jk_1} should be set to 0. To capture this idea we have imposed assurance region constraints, as per Thompson et al. (1990), of the form:

$$L_{j_k} \leq \alpha^k_{r_2 j_k} / \alpha^k_{r_2 1} \leq U_{j_k},$$

where we have numbered that power unit 1 as the unit whose total OUT + EQDER is lowest. (This is the power unit whose total number of hours of outage + equipment deratings is highest). The argument is that for plant $k, \alpha^k_{r_2 1}$ should be the lowest proportion among all units for that plant. We have then chosen $L_{j_k} = 1$ for all units j_k. Since it is unclear what the precise relationship is concerning the timing of non-environmental deratings and outages (as discussed above), we have chosen here to set all U_{j_k} equal to one another. We experimented with different values, and found that while the efficiency ratings of the various power units within a plant tended to decrease as U_{j_k} is lowered, their order (relative to one another) was quite stable. Table 17.7 displays the plant-level and associated power unit-level efficiency scores.

The advantage of viewing efficiency in this manner is that not only can one evaluate the performance of plants, but at the same time can uncover the extent to which each of the subunits (power units) within the plant is contributing to that performance. This permits management to identify which power units in a plant are under-performing, and which units could serve as benchmarks within that plant. Following Properties 17.3 and 17.4, it is noted that for efficient plants such as 4,5 and 6, all power units within these are efficient as well.

Table 17.7 Power plant and power unit ratings

Plant	Unit	Unit Rating	Plant Rating	Plant	Unit	Unit Rating	Plant Rating
1	1	0.64	0.833	4(1)	1	1	1
	2	1		4(2)	1	1	1
	3	0.99			2	1	
	4	1					
	5	1		5	1	1	1
	6	0.67			2	1	
	7	0.46			3	1	
	8	1			4	1	
2	1	0.58	0.861	6	1	1	1
	2	0.80					
	3	0.63		7(1)	1	0.87	0.84
	4	1		2	1		
	5	1		3	0.81		
	6	1		4	0.69		
	7	1					
	8	1		7(2)	1	0.82	0.935
					2	0.92	
3	1	0.82	0.793		3	1	
	2	0.91			4	1	
	3	0.79		8	1	0.99	0.997
	4	0.66			2	1	
					3	1	
					4	1	

17.8 Conclusions

This chapter has presented DEA-based models for evaluating the efficiency of a set of power plants, and corresponding power units as a hierarchical structure. In the earlier part of the chapter, hierarchical efficiency was viewed as a multi-stage process. In Sect. 17.7 however, hierarchical efficiency measurement is viewed at all levels simultaneously. This is accomplished by first defining the decision making units (DMUs), as the units at the highest level in the hierarchy (power plants in the current application). The elements lower down in the hierarchy are then viewed as components of the top level DMUs, and as such, have their efficiency evaluated as well.

A complicating feature of this latter structure is the presence of outputs at any level in the hierarchy that must be allocated among the components at the next stage down in that hierarchy. In the setting herein, this is accomplished by defining variables which provide for a split of such (plant-level) outputs among the power units within each plant. We demonstrate that this resulting non-linear model can be converted to a linear programming problem.

The developed models have been applied to 40 power generating units organized under eight plants. Sulphur dioxide (SO_2) emissions are generally regarded as a plant-level output which we wish to allocate to the power units under each plant. This allocation in practice could be a function of various factors including the percent downtime for scheduled maintenance, etc. The outcome of the efficiency evaluation is given in Table 17.7.

The application of DEA principles to hierarchical structures is an important area for research. Many organizational structures tend to exhibit such a profile. The ideas herein can potentially lend themselves to other areas of study, for example, supply chains. The ideas are also somewhat related, as well, to the concepts presented by Fare and Grosskopf (1996) regarding intermediate products, as well as structures studied in the network DEA model of Fare and Grosskopf (2000).

References

Ali, I., Cook, W., & Seiford, L. (1991). Strict vs weak ordinal relations for multipliers in data envelopment analysis. *Management Science, 37*(6), 733–738.

Banker, R., & Morey, R. (1986). The use of categorical variables in data envelopment analysis. *Management Science, 32*(12), 1613–1627.

Charnes, A., & Cooper, W. W. (1962). Programming with linear fractional functions. *Naval Research Logistics Quarterly, 9*, 181–186.

Charnes, A., Cooper, W. W., & Rhodes, E. L. (1978). Measuring the efficiency of decision making units. *European Journal of Operational Research, 2*(6), 429–444.

Cook, W. D., & Green, R. (2005). Evaluating power plant efficiency: A hierarchical model. *Computers and Operations Research, 32*(4), 813–823.

Cook, W. D., Kress, M., & Seiford, L. (1996). Data envelopment analysis in the presence of both quantitative and qualitative factors. *Journal of the Operational Research Society, 47*, 945–953.

Cook, W. D., Chai, D., Doyle, J., & Green, R. (1998). Hierarchies and groups in DEA. *Journal of Productivity Analysis, 10*, 177–198.

Färe, R., & Grosskopf, S. (1996). Productivity and intermediate products: A frontier approach. *Economic Letters, 50*(1), 65–70.

Färe, R., & Grosskopf, S. (2000). Network DEA. *Socio-Economic Planning Sciences, 34*, 35–49.

Golany, B. (1988). A note on including ordinal relations among multipliers in data envelopment analysis. *Management Science, 34*(8), 1029–1033.

Thompson, R., Langemeir, L., Lee, C., Lee, E., & Thrall, R. (1990). The role of multiplier bounds in efficiency analysis with application to Kansas farming, *Journal of Econometrics, 46*, 93–108.

Chapter 18
Multicomponent Efficiency Measurement and Core Business Identification in Multiplant Firms

Wade D. Cook and R.H. Green

Abstract As discussed in the previous chapters, the DMU may perform different types of functions. In that case it is desirable to derive a measure of performance not only at the DMU level but as well at the level of the particular functions within the DMU. In the current chapter we examine a set of manufacturing plants operating under a single umbrella, with the objective being to use the component or function measures to decide what might be considered as each plant's core business. It is proposed that this information can aid the company in any reorganization initiatives designed to capitalize on the strengths of each location (DMU).

Keywords Multicomponent • Core business • Shared inputs • Bundles

18.1 Introduction

The DEA model, developed by Charnes et al. (1978), provides a constant return to scale (CRS) methodology for evaluating the performance of a set of comparable decision making units (DMUs). In the usual setting, each DMU is evaluated in terms of a set of outputs that represent its accomplishments, and a set of inputs that represent the resources or circumstances at its disposal.

This chapter is based upon Cook, W. D. and R. H. Green, (2004), Multicomponent efficiency measurement and core business identification in multiplant firms: A DEA model, European Journal of Operational Research, 157, 540–551, with permission from Elsevier Publishing.

W.D. Cook (✉)
Schulich School of Business, York University, 4700 Keele Street, Toronto, ON M3J 1P3, Canada
e-mail: wcook@schulich.yorku.ca

R.H. Green
School of Management, University of Bath, Bath BA27AY, United Kingdom
e-mail: mnsrhg@bath.ac.uk

W.D. Cook and J. Zhu (eds.), *Data Envelopment Analysis*,
International Series in Operations Research & Management Science 208,
DOI 10.1007/978-1-4899-8068-7_18,

In some application areas, it has been recognized that the DMU may perform different *types* of functions. In such situations, it is desirable to derive a measure of performance, not only at the level of the DMU, but, as well, at the level of the particular function within the DMU. Cook and Roll (1993) were the first to examine the idea of partial efficiency measures, where the separate components of the DMU possess their own bundles of outputs and inputs. These bundles were assumed to be mutually exclusive of one another. Beasley (1995) examined both teaching and research components within a set of universities in the UK, and presented a nonlinear programming model for measuring DMU performance. A similar situation is encountered in Cook et al. (2000), and Cook and Hababou (2001), where sales and service components are evaluated within a set of bank branches. They discuss linear models for providing both overall performance of a branch, as well as separate component performance measures. In that context, as with Beasley (1995) the input is a shared resource to be allocated to two production units. The complicating feature in each of these problem settings, that was not present in Cook and Roll (1993), is the presence of *shared resources*. The existence of *shared resources* means that the usual DEA structure must be modified to provide for a splitting of those resources among the various components.

In the current chapter we examine a set of manufacturing plants operating under a single corporate umbrella, with the objective of identifying how well each plant performs in each of its components thus identifying what might be considered each plant's a core business. Here, each component consists of a group of products selected from the totality of products offered, according to the specific interests of the corporate decision maker. Unlike the aforementioned dual-component applications (e.g., sales and service components in a bank branch), these components may overlap. Examples are (1) those products made from rolled steel of given dimensions; (2) those products servicing the automotive industry, . . ., etc. This setting is clearly similar to those discussed above in that product groups are functions of the business, and, as will be seen, there are resources that are shared among those components. The models proposed here represent a departure from the earlier work of Beasley (1995), Cook and Roll (1993), Cook et al. (2000), and Cook and Hababou (2001), in two respects. First, we examine the extension of the earlier models to a multi-component (two or more) setting. Second, using this multi-component structure as a point of departure, we develop models for *identifying the most appropriate* product groupings for each plant (DMU).

Section 18.2 presents the problem setting in more detail. In Sect. 18.3, extensions of the models of Cook et al. (2000, 2001) and Beasley (1995) are presented. Multiple, and potentially overlapping components are considered. These models are appropriate where the issue is one of identifying overall performance, as well as isolating particular areas (components) where the plant can be improved. Section 18.4 extends this idea to those situations wherein the organization wishes to identify the segment of the business that is performing *best* in any given DMU. In this way, the *core business* of each plant can be isolated, thus aiding the company in any reorganization initiatives designed to capitalize on the strengths of each location. Section 18.5 discusses the application of these models in the plant setting described earlier. Conclusions are given in Sect. 18.6.

18.2 Multicomponent Efficiency Measurement and Core Business Identification

In this chapter we examine multi-component efficiency measurement from two perspectives. In the *first* situation, we make the assumption that the purpose of the performance assessment exercise is to determine an aggregate measure of efficiency, as well as measures for each of the separate components. Such evaluation will aid management in identifying the extent to which overall performance can be improved. As well, for specific business areas, the measures can point to those that are doing well, as well as those that require attention. Section 18.3 addresses this setting.

In the *second* situation, it is assumed that the organization wishes to go beyond simply identifying the level of performance of specific subunits of the business. Rather, it is desirable to identify the area(s) where DMUs are performing best, hence defining what might reasonably be regarded as each DMU's *core business*. A given DMU may then wish to focus its energies on this selected part of the operation, while de-emphasizing, or in some cases, even abandoning those portions of the business where it performs at a less than satisfactory level. This development is undertaken in Sect. 18.4.

To illustrate these ideas we examine a company with several plants that operate in the rolled steel industry. The company manufactures steel products, both of the *finished* variety that are sold on the open market, and semi-finished items that are custom-ordered, and sold to other manufacturers. These latter products can, for example, be items such as slit steel, used by other firms that manufacture steel doors and door frames. Other products, such as cylindrical bearings, are further along the value chain, and are purchased by companies that manufacture such consumer products as lawn mowers, or outboard motors for boats. Anticipating the detail given in Sect. 18.5, it is convenient to view the company's operations in terms of nine distinct products, and in conventional DEA terms each of these products would be considered an output. However, corporate management as well as being interested in the overall efficiency of each plant, is also interested in performance with respect to four overlapping groupings of these nine products. In what follows we will refer to a defined group of products, variously and interchangeably, as a component, subunit or segment. In some cases products are grouped to represent a particular market segment, e.g., automotive manufacturers who source certain products from the company. In other cases they are grouped to represent an internally meaningful segment of the operation, e.g., all products both semi-finished and finished, but pertaining to a certain size or quality of steel, or products made on particular machines.

In the section to follow, we present model structures for evaluating both the aggregate performance of each of a set of DMUs, as well as the performance of the separate subunits or components within a DMU's operation. For purposes of this development, we utilize the problem setting discussed herein as a backdrop.

18.3 Multicomponent Model Structures

The conventional model structure for evaluating the relative efficiency of each member of a set of DMUs is the DEA model of Charnes et al. (1978). Specifically, given an output vector $Y_k = (y_{1k}, y_{2k} \ldots, y_{Rk})$, and input vector $X_k = (x_{1k}, x_{2k}, \ldots, x_{Ik})$, for each of a set of n DMUs $k = 1, \ldots, n$, the constant returns to scale model is given by

$$\begin{aligned} &\max \ \mu_o Y_o / v_o X_o, \\ &\text{subject to:} \\ &\mu_o Y_k / v_o X_k \leq 1, \quad \text{all } k, \\ &\mu_o, v_o \geq \varepsilon. \end{aligned} \tag{18.1}$$

The structure in (18.1) presumes that one desires to measure the overall efficiency (e.g., operational efficiency) of each DMU, without consideration for the performance of subunits that may exist within the DMU. In the problem setting presented herein, we wish to provide for a more detailed performance evaluation, i.e., at the level of these subunits.

18.3.1 Multi-component Efficiency Measurement with Shared Inputs: Non-overlapping Subunits

Our point of departure for the discussion in this section, is the model structures of Cook et al. (2000), (see also Cook and Hababou 2001). There, the authors examine the problem of providing separate efficiency measures for both *sales* and *service* components of a set of bank branches for a major Canadian bank. Adopting the notation of Cook et al. (2000), and extending their model structure to "T" components, we have:

Parameters:

Y_k^t	= the R-dimensional vector of outputs included in the tth component of DMUk
R	= set of all outputs
R^t	= set of outputs generated by the tth component
X_k^t	= the I-dimensional vector of inputs dedicated to the tth component of DMUk
I	= set of all inputs
I^t	= set of inputs dedicated to the tth component
X_k^S	= the I^s-dimensional vector of inputs shared among the T components of DMU k
I^s	= set of shared inputs
L_i^t, U_i^t	= lower, upper limits on the portion of the ith shared resource, that can be assigned to the tth component of a DMU
T	= set of all components

Decision Variables:

μ_o^t	= vector of multipliers applied to outputs Y_o^t
ν_o^t	= vector of multipliers applied to inputs X_o^t
ν_o^{st}	= vector of multipliers applied to that portion of shared inputs X_o^S that are assigned to component t
α_o^t	= vector representing the proportion of shared inputs X_o^S allocated to the tth component

In the two-component problem addressed in Cook et al. (2000), the principal area of difficulty was the presence of shared inputs X_k^S.. Specifically, there are certain resources such as branch expenditure on computer technology and general branch staff, that are shared across the two components of the business. There is no well defined split of these resources across different functions, and the basic problem has to do with the allocation of these inputs among the components. To facilitate this, and at the same time extend the idea to the general case of T components, a decision vector α_k^t is introduced that permits the DMU k in question to apportion X_k^S among the T competing components. In Cook et al. (2000), this is done in a manner that optimizes the *aggregate* performance measure (of DMU "o") given by:

$$e_o^a = \sum_{t\varepsilon T} \mu_o^t Y_o^t \Big/ \left[\sum_{t\varepsilon T} \left(\nu_o^t X_o^t + \nu_o^{st} \left(\alpha_o^t X_o^S \right) \right) \right] \tag{18.2}$$

The component-specific performance measures e_o^t are given by:

$$e_o^t = \mu_o^t Y_o^t / \left(\nu_o^t X_o^t + \nu_o^{st} \left(\alpha_o^t X_o^S \right) \right) \tag{18.3}$$

It is pointed out that the notation $\alpha_o^t X_o^S$ represents the vector $\left(\alpha_{o1}^t x_{o1}^S, \alpha_{o2}^t x_{o2}^S, \ldots, \alpha_{oI^S}^t x_{oI^S}^S \right)$ of shared inputs allocated to component t by DMU "o".

In the discussion below, we distinguish between optimal performance measures and performance measures for a DMUk, evaluated in terms of the multipliers for a DMU "o" currently being considered. (Doyle and Green (1994) use the term *cross-evaluation* in this instance). For this purpose, we adopt the notation $\hat{e}_k^a, \hat{e}_k^t$ to denote the measures for DMUk that represent their optimal performance, while e_k^a, e_k^t denote performance relative to multipliers arising from the optimization of (some other) DMU "o".

The multi-component DEA model is given by:

$$\hat{e}_o^a = max e_o^a \quad \text{subject to} \tag{18.4a}$$

$$e_k^t \leq 1, \quad \text{all } t, k \tag{18.4b}$$

$$L_i^t \le \alpha_{oi}^t \le U_i^t \quad \text{all } t,\ i\ \varepsilon\ I^S, \tag{18.4c}$$

$$\sum_{t\varepsilon T} \alpha_{oi}^t = 1, \quad i\ \varepsilon I^S, \tag{18.4d}$$

$$\mu_o^t, \nu_o^t, \nu_o^{st} \ge \varepsilon, \quad \text{all } t. \tag{18.4e}$$

Here, the objective (18.4a) maximizes the overall performance measure for the DMU "o", in the spirit of the original DEA model of Charnes et al. (1978). Correspondingly, we restrict each component measure e_k^t by an upper bound of 1 in (18.4b). A permissible range on the proportion of the ith shared resource that can be allocated to the tth component by any DMU is given by (18.4c). Constraints (18.4d) specify that the proportional splits of any input i across the T components sum to unity. Finally, constraints (18.4e) restrict multipliers to be strictly greater than zero.

The limits L_i^t, U_i^t, on the proportions α_{oi}^t of the various inputs i to components t would need to be specified by the user. Such limits might generally arise from any information available at the plants regarding standard amounts of inputs i per unit of product in components t.

From the above discussion it is clear that problem (18.4) is a *restricted version* of problem (18.1). Specifically, any feasible solution to (18.4) is also feasible for (18.1). Problem (18.4) only permits multipliers which identify each component of the plant as a bona fide sub-DMU whose performance measure is captured at the same time as that of the entire plant. Problem (18.1), however, is focused purely at the plant level, with no recognition whatever of subunits.

Definition 18.1 A DMU "o" is said to be *efficient* if its aggregate score $\hat{e}_o^a = 1$.

Definition 18.2 A DMU "o" is said to be *efficient in its tth component* if $\hat{e}_o^t = 1$.

Theorem 18.1 *In model (18.4), the resulting aggregate performance measure $\hat{e}_k^a$ for any DMUk, does not exceed unity, i.e., $\hat{e}_k^a \le 1$.*

Proof If we define

$$\beta_k^t = \left[\nu_o^t X_k^t + \nu_o^{st}\left(\alpha_o^t X_k^s\right)\right] \Big/ \sum_{t\varepsilon T} \left(\nu_o^t X_k^t + \nu_o^{st}\left(\alpha_o^t X_k^s\right)\right),$$

then, the aggregate measure (in terms of the (μ_o, ν_o) multipliers), is given by

$$\hat{e}_k^a = \sum_{t\varepsilon T} \beta_k^t \hat{e}_k^t.$$

Hence, e_k^a is a convex combination of the component measures, and as such $e_k^a \le 1$.

Q.E.D.

Theorem 18.2 *In model (18.4), a DMU is efficient if and only if it is efficient in each of its components.*

Proof Case 1: Assume all component measures $\hat{e}_k^t = 1$.

By definition,

$$\hat{e}_k^a = \sum_{t\varepsilon T} \beta_k^t \hat{e}_k^t$$

from Theorem 18.1, and since $\sum_t \beta_k^t = 1$, it follows that $\hat{e}_k^a = 1$.

Case 2: Assume $\hat{e}_k^a = 1$. Then, if any $\hat{e}_k^t < 1$, it must be the case that

$$\hat{e}_k^a = \sum_{t\varepsilon T} \beta_k^t \hat{e}_k^t < 1,$$

as well, in contradiction.

Q.E.D.

We now examine multi-component performance measurement when overlaps can occur.

18.3.2 Multi-component Efficiency Measurement with Overlapping Subunits

The models presented above presume a set of subunits that are mutually exclusive. Arguably, in the bank branch setting of Cook and Hababou (2001), and Cook et al. (2000), sales and service components meet the mutual exclusivity requirement. In many settings this restriction may not hold, however, as is the case with the business components described later.

In the case where mutual exclusivity prevails, it is sufficient to subdivide a shared input among the set of components. That is, α_{oi}^t represents the portion of input i assigned to component t. It is not necessary to address how this portion α_{oi}^t is distributed among the outputs comprising component t. In case there is *overlap* among the components due to the existence of common outputs, the manner in which the proportions $\{\alpha_{oi}^t\}_t=1^T$ behave, is no longer clear. It is, for example, not true that $\sum_{t\varepsilon T} \alpha_{oi}^t = 1$, due to the overlap.

In recognition of the *overlap problem*, we need to be more exacting as to how the shared input i is assigned to outputs $r\varepsilon \mathfrak{R}$. Specifically, we define variables α_{oir} that denote the proportion of shared input x_{oi}^s (the ith component of vector X_o^s) that is allocated to output y_{or}. As well, let L_i^r, U_i^r, denote lower and upper bounds, respectively, on α_{oir}, and impose the constraint

$$\sum_{r \in \Re} \alpha_{oir} = 1.$$

The proportion α^t_{oi} of input i allocated to component t is then the sum of the proportions α_{oir} of i allocated to those outputs comprising t, i.e.

$$\alpha^t_{oi} = \sum_{t \varepsilon \Re^t} \alpha_{oir}.$$

For brevity in modelling, we henceforth denote the feasible set of $\alpha = (\alpha^t)$ by

$$\begin{aligned} \Lambda_o = \{\alpha_o = \left(\alpha^t_o\right) : (1)\, \alpha^t_{oi} = \sum_{r \varepsilon \Re^t} \alpha_{oir}; \\ (2)\, L^r_i \leq \alpha_{oir} \leq U^r_i; (3) \sum_{r \varepsilon \Re} \alpha_{oir} = 1, \\ \alpha_{oir} \geq 0, \quad \text{all}\ i \varepsilon \mathrm{I}^s, \ \text{all}\, t\}. \end{aligned}$$

The multi-component DEA model is then given by:

$$\begin{aligned} &\text{Max } e^a_o, \\ &\text{subject to} \end{aligned} \tag{18.5a}$$

$$e^t_k \leq 1, \quad \text{all } t, k, \tag{18.5b}$$

$$\alpha_o \varepsilon \Lambda_o, \tag{18.5c}$$

$$\mu^t_o, v^t_o, v^{st}_o \geq \varepsilon, \quad \text{all } t. \tag{18.5d}$$

It is noted that the objective function (18.5a) credits the DMU for producing an output y^t_{or} as many times as that output appears as a member of a component's output set. For example, an output y_{or}, contained in both components $t = 1$ and $t = 2$, (i.e., $y^1_{or_1} = y^2_{or_2}$), would appear in (18.5a) twice, as $\mu^1_{or_1} y_{or_1}$ and $\mu^2_{or_2} y_{or_2}$.

We point out, however, that, as in the case of non-overlapping subunits, it is also true here that problem (18.5) is simply a restricted version of problem (18.1), if we view the inputs X in (18.1) as all being shared inputs. This is captured by the following theorem.

Theorem 18.3 *Any feasible solution to problem (18.5) is feasible to (18.1).*

Proof Define the R-dimensional multiplier vector $U^t = (u^t_r)$ by

$$u^t_r = \begin{cases} \mu^t_r & \text{if product r is in component t} \\ 0 & \text{otherwise} \end{cases}$$

and let $U = \sum_{t\varepsilon T} U^t$. Letting Y denote the R-dimensional vector of all outputs as used in (18.1), it follows that

$$\sum_{t \varepsilon T} \mu^t Y^t = UY.$$

Similarly, one can replace the set of inputs $\{X^t\}$ by the I-dimensional vector $X(1) = (X^1, X^2, \ldots, X^T)$, and replace the set of "shared resource" vectors $\alpha^t X^s$ by the sum of these component shares to get X^s. Let $X = (X(1), X^s)$, the full vector of all inputs. Then, as with the output side, one can express the denominator of the performance measure as

$$\sum_{t \varepsilon T} [\nu^t X^t + \nu^{st}(\alpha^t X^s)] = VX,$$

where V is defined in terms of the ν^t, ν^{st} in a manner analogous to the definition of U in terms of $\{\mu^t\}$. Hence e_o^a in (18.5) can be written as

$$e_o^a = UY/VX.$$

Since it is true that each component measure $e_k^t \leq 1$, then it must also be true that the aggregate score $e_k^a \leq 1$ as well. Thus, any feasible solution to (18.5) is also feasible for (18.1).

Q.E.D.

Hence, the overlap of the components does not lead to inconsistencies in regard to problem (18.1). Defining the aggregate measure in this manner results in the following theorem. The Proof is analogous to those of Theorems 18.1 and 18.2, and is, therefore, omitted.

Theorem 18.4

(*a*) *The aggregate measure of efficiency given by* (*18.5a*) *does not exceed unity.*

(*b*) *A DMU will be aggregate-efficient, (the objective function* (*18.5a*) *will equal unity), if and only if it is efficient in each component measure.*

Model (18.5), thus, allows one to examine the performance of a DMU in each business area. As well, it provides an overall or aggregate measure of performance across all business components.

Because the orientation of model (18.5) is toward evaluation of the DMU at an aggregate level, with component measures arising only as a by-product, it can be argued that the individual subunits of the business may not be shown in their most favorable light. In some cases, the strategic intent of the organization might be to identify the core business for each DMU, the purpose being to focus the attention of the DMU toward the areas of the business at which it performs best. In the section to follow, we present model structures wherein the intention is to choose a core business component on behalf of each DMU. It should be pointed out that the identification of a core business component will not necessarily imply the immediate termination of all activities at a plant that are not included in that component. Rather, a DMU would initially continue to service all existing activities, possibly phasing out non-core activities as these are redistributed to where they are best accomplished over some time horizon.

18.4 Modelling Selection of Core Business Components

A typical problem setting would be one where each of a set of plants for a given company produces a full product line, for sale and distribution to customers. There can be a number of reasons why it is cost effective for a certain product line, for example, to be manufactured in particular locations, but not in others. Certain manufactured items may, for instance, require specialized and expensive equipment that the company might prefer to make available in only one location. Alternatively, certain customers (e.g. farmers) may be highly concentrated in one geographical area, meaning that a plant close to that concentration should produce products related to that customer group. As well, simple economies of scale may dictate that the production for a product be concentrated in only a few plants, or even a single plant.

The problem then is to identify which collection of products or product lines should be handled by any given plant, thus defining that plant's core business.

The conventional DEA model does not readily lend itself to resource allocation (i.e. allocation of shared inputs). The DEA approach focuses attention on the performance of a particular DMU "o". If the objective is to allocate components to DMUs (plants), and to divide shared resources among products (and thus among components), one needs to view this allocation process from the perspective of the entire *collection* of DMUs, simultaneously rather than from the conventional DEA perspective, i.e. iteratively, one DMU at a time.

To facilitate the allocation of components to DMUs, define the bivalent variables $\{d_k^t\}_{t=1}^T$, for each DMU k,

$$d_k^t = \begin{cases} 1 & \text{if component } t \text{ is assigned to DMU } k, \\ 0 & \text{otherwise.} \end{cases}$$

The aggregate performance (ratio) measure for the collection of DMUs, given an allocation defined by a chosen set of d_k^t values, can be expressed as:

$$\frac{\sum_k \left[\sum_t d_k^t \mu^t Y_k^t\right]}{\sum_k \left[\sum_t d_k^t \left(\nu^t X_k^t + \nu^{st}\left(\alpha_k^t X_k^s\right)\right)\right]}$$

The *optimal* assignment of components to DMUs, as defined by the d_k^t, is arguably that for which the ratio of aggregate output to aggregate input is maximized. The set of d_k^t for which this maximum occurs can be determined by solving the fractional programming problem:

$$\max\sum_k\left[\sum_t d_k^t\mu^t Y_k^t\right]\Big/\sum_k\left[\sum_t d_k^t\left(\nu^t X_k^t+\nu^{st}\left(\alpha_k^t X_k^s\right)\right)\right] \tag{18.6a}$$

subject to

$$\mu^t Y_k^t/\left(\nu^t X_k^t+\nu^{st}\left(\alpha_k^t X_k^s\right)\right)\le 1,\quad \text{all } k,t \tag{18.6b}$$

$$\alpha\,\varepsilon\,\Lambda_0, \tag{18.6c}$$

$$\sum_t d_k^t\ge 1,\quad \text{all } k, \tag{18.6d}$$

$$\sum_k d_k^t\ge 1,\quad \text{all } t, \tag{18.6e}$$

$$\mu^t,\nu^t,\nu^{st}\ge\varepsilon,\quad \text{all } t,\quad d_k^t\,\varepsilon\{0,1\},\quad \text{all } k,t. \tag{18.6f}$$

Constraints (18.6b) restrict the ratio of outputs to inputs in *any* component to not exceed unity. (18.6c) requires that the resource splitting variables satisfy conditions as defined earlier in Λ_υ. Constraints (18.6d) force each plant k to support *at least* one product group or component. Similarly, (18.6e) stipulates that each component must be produced at *one or more* of the plants.

It is conceivable that at the optimum, certain plants may be chosen to support several product groups, while other plants may service only one group.

Model (18.6a–f), assigns multipliers μ^t, ν^t, ν^{st} to each component t in each DMU k. While it is not the purpose of the model to measure the efficiency of the *entire* operation of each plant, the supplied (common set of) multipliers do in fact provide the basis for an efficiency score for each plant and the aggregate across all plants, should one want to extract these. That aggregate score clearly includes the contribution rendered by both core and non-core components of the plant. Admittedly, the set of multipliers is derived in a manner designed to display *core* components in their best light, and by implication, non-core components in a light less than best. Hence, non-core components may be represented in a disadvantageous manner. One might argue that this is appropriate since, over time, such non-core components will, in any event, be phased out. Thus, their estimated performance (by that stage) will be a non-issue. At the same time, the model does, in fact, recognize their existence, and the bounds $[L_i^r, U_i^r]$ appropriately force the allocation of shared resources across all components (both core and non-core). Thus, choice of these bounds by management affirms the continuing presence of non-core components in the operation.

Thus, the real purpose of the model is to single out those components of each plant on which that plant exhibits its best performance. It is these core components whose aggregate performance we wish to capture.

The implication of this is that when a set of plants exhibit inefficiency, it is often desirable to strive for *specialization*. The questions that management would like to answer are:

(1) In what parts of the operation should each plant specialize?
(2) If plant operations were reorganized to implement such specialization, what would be the anticipated performance of the resulting operation?
(3) How would each reorganized (future) plant perform?

Question 1: The purpose of the model is to extract those components at each plant that appear to be the ones in which the plant should specialize.

Question 2: While the model yields an aggregate performance across all core components in all plants, there is an implied measure of performance for each plant on a portion (core business portion) of that plant's operation. Specifically, using $\{\hat{d}_k^t\}_{t=1}^T$, for each k, the model yields a measure of performance for that subset of components in terms of the inputs that those components utilize, and the outputs generated by those components. This measure captures how the (reduced) plant *would* perform if non-core business elements were not present.

Question 3: In a reorganized structure, the essence of the model is that each plant would concentrate only on its core business activities. It is argued that if each plant were to scale up its operation such as to come to full capacity in its resource utilization, then it is hypothesized that the resulting output generated would be scaled up by the same factor.

To solve problem (18.6a–f), it can be shown that it is representable as a mixed integer linear programming problem. This is given by the following theorem.

Theorem 18.5 *Problem (18.6a–f) can be represented as a mixed integer (binary) linear problem.*

Proof Problem (18.6a–f) is equivalent to the mixed binary *nonlinear* programming model:

$$\max \sum_k \sum_t d_k^t \mu^t Y_k^t \tag{18.7a}$$

subject to

$$\sum_k \sum_t d_k^t \left(\nu^t X_k^t + \nu^{st}\left(\alpha_k^t X_k^s\right)\right) = 1 \tag{18.7b}$$

$$\mu^t Y_k^t - \left(\nu^t X_k^t + \nu^{st}\left(\alpha_k^t X_k^s\right)\right) \leq 0, \quad \text{all } k, t, \tag{18.7c}$$

$$\alpha \,\varepsilon\, \Lambda_o, \tag{18.7d}$$

$$\sum_t d_k^t \geq 1, \quad \text{all } k, \tag{18.7e}$$

$$\sum_k d_k^t \geq 1, \quad \text{all } t, \tag{18.7f}$$

$$\mu^t, \nu^t, \nu^{st} \geq \in, \quad \text{all } t, \; d_k^t \,\varepsilon\, \{0, 1\}, \quad \text{all } k, t. \tag{18.7g}$$

Make the change of variables:

$$\overline{\nu}^{st} = \nu^{st}\alpha^t, \mathrm{v}_k^{st} = d_k^t\overline{\nu}^{st}, \mathrm{v}_k^t = d_k^t\nu^t, u_k^t = d_k^t\mu^t.$$

It is noted that we can replace an expression such as $v_k^t = d_k^t\nu^t$ by the constraint set

$$\begin{aligned} v_k^t &\le Md_k^t, \\ \nu^t &\ge v_k^t, \\ \nu^t &\le v_k^t + M(1 - d_k^t), \end{aligned}$$

where M is a large positive number. Specifically, if $d_k^t = 0$, then $v_k^t = 0$; if $d_k^t = 1$, then $v_k^t = \nu^t$. A similar set of constraints can be imposed to replace the nonlinear expressions $u_k^t = d_k^t\mu^t$, and $v_k^{st} = d_k^t\upsilon^{st}$. Problem (18.7a–g) can then be written as the mixed binary linear programming model

$$\begin{aligned}
&\max \sum_k \sum_{t\varepsilon T} u_k^t Y_k^t, \\
&\text{subject to} \\
&\sum_k \sum_{t\varepsilon T} \left(v_k^t X_k^t + v_k^{st} X_k^s\right) = 1, \sum_{t\varepsilon T} \left[u_k^t Y_k^t - \left(v_k^t X_k^t + v_k^{st} X_k^s\right)\right] \le 0, \text{all } k, \\
&\upsilon_k^t \le \mathrm{M}d_k^t, \text{all t}, \\
&\nu^t \ge \mathrm{v}_k^t, \text{all t}, \\
&\nu^t \le \mathrm{v}_k^t + \mathrm{M}(1 - \mathrm{d}_k^t), \text{all t}, \\
&u_k^t \le \mathrm{Md}_k^t, \text{all t}, \\
&\mu^t \ge u_o^t, \text{all t}, \\
&\mu^t \le u_k^t + \mathrm{M}(1 - \mathrm{d}_k^t), \text{all t}, \\
&\upsilon_k^{st} \le \mathrm{Md}_k^t, \text{all t}, \\
&\overline{\nu}^t \ge \mathrm{v}_k^{st}, \text{all t}, \\
&\overline{\nu}^t \le \mathrm{v}_k^{st} + \mathrm{M}(1 - \mathrm{d}_k^t), \text{all k, t}, \\
&\alpha \varepsilon \Lambda, \\
&\sum_t \mathrm{d}_k^t \ge 1, \text{all k}, \\
&\sum_k \mathrm{d}_k^t \ge 1, \text{all t}, \\
&\overline{\nu}_i^{st} \ge \varepsilon \alpha_i^t, \text{all i, t}, \\
&\mu_r^t, \nu_i^t \ge \varepsilon, \ \text{all i, r, t}, \\
&u_{kr}^t, \mathrm{v}_{ki}^t \ge 0, \text{all i, r, k}, \\
&\upsilon_{ki}^{st} \ge 0, \text{all r, t, i} = 1, \ldots, \mathrm{I^S}, \\
&\mathrm{d}_k^t \varepsilon \{0, 1\}, \text{all k, t.}
\end{aligned} \tag{18.8}$$

This completes the proof.

Q.E.D.

There are clearly variations of this model where, for example, it may be pertinent for certain product groupings or components to be manufactured in only certain plants that are perhaps in the best possible position to handle them. This might be due to equipment capability, proximity of the market, and so on. Thus, for a given component t_o, we might require that $d_k^{to} = 0$, $k \varepsilon \overline{K}^{to}$, where K^{to} is the set of allowable plants for manufacturing component t_o, and $\overline{K}^{to}$ is its compliment.

In the section to follow, this model is used to allocate business components to ten plants within the company under study.

18.5 Application of Core Business Selection Model to a Set of Plants

In the problem studied, 10 plants currently operate under a single corporate umbrella, producing a variety of steel products including bearings, pipes and sheet steel of various sizes. Clearly, some of these products are of the finished goods variety (e.g. pipes), while others are semi-finished, becoming components in other manufactured items (bearings), or are sold to other plants for further manufacturing (sheet steel).

As indicated earlier, it is convenient to view each plant's business as consisting of various components. While it is the case that there can be a large number of products to consider (e.g. different sizes of circular bearings), here items have been grouped by management under a few major categories. For purposes of this study we present the operation of any plant as consisting of four (overlapping) components, defined by their outputs y_r^t, the number of units of output r in the tth component:

Component #1:

- All solid bearings (y_1^1)
- Circular bearings (automotive) (y_2^1)
- Sheet steel $\leq$4 ft in length (y_3^1).

Component #2:

- Solid bearings (automotive) y_1^2)
- Steel pipes $\leq$8 ft in length y_2^2)
- Sheet steel 4–8 ft in length (y_3^2)

Component #3:

- Steel pipes >8 ft in length (y_1^3)
- Sheet steel >8 ft in length (y_2^3)

Table 18.1 Outputs for four components

Plant	y_1^1	y_2^1	y_3^1	y_1^2	y_2^2	y_3^2	y_1^3	y_2^3	y_1^4	y_2^4	y_3^4	y_4^4
1	50	30	70	30	60	50	40	80	30	50	50	70
2	45	35	60	25	50	50	40	75	35	55	45	60
3	75	25	50	35	55	40	50	70	25	60	75	50
4	60	40	80	40	40	30	70	50	40	50	60	80
5	35	25	25	20	25	20	35	20	25	30	35	25
6	55	60	40	40	60	45	60	50	60	50	55	40
7	120	100	100	100	80	120	120	60	100	110	120	60
8	60	80	25	50	100	20	80	35	80	80	60	25
9	25	75	65	20	25	80	100	70	75	70	25	65
10	100	55	40	70	35	65	35	45	55	60	100	40

Component #4:

- Circular bearings (automotive) (y_1^4)
- Circular bearing (non-auto) (y_2^4)
- All solid bearings (y_3^4)
- Sheet steel ≤4 ft in length (y_4^4)

Table 18.1 displays the data for all outputs for the 10 plants considered.

The resources committed to the production of these product lines can be grouped under four headings, namely

- Shop labour (x_1)
- Machine labour (x_2)
- Steel splitting equipment (x_3)
- Lathes (x_4)

Shop labour and machine labour are measured in full time equivalent (FTE) staff. Both equipment variables are expressed in hundreds of hours of capacity available per month. Given the manner in which the four components have been defined, with the inherent overlap of products, all four of these inputs should be viewed as shared resources.

Table 18.2 shows the amounts of the four resources corresponding to each plant.

The connection between the shared inputs and the product outputs (y_r^t) is quite complex, and must be reflected through the α_{ir}. If a given input such as lathes (x_4) does not impact on a particular output such as sheet steel (≤4 ft) (y_3^1) then that particular variable α is set to zero. Figure 18.1 shows the input-to-output impact matrix.

In the figure, an "x" denotes the fact that the particular input contributes to the output shown. It must be noted as well, that when we have a product common to two or more components, the corresponding variables α_{ir} must be equated. For example, since sheet steel ≤4 ft is part of both components 1 and 4 (i.e., $y_3^1 = y_4^4$), then $\alpha_{1,3} = \alpha_{1,12}$.

Table 18.2 Shared resources

DMU	x_1	x_2	x_3	x_4
1	30	15	100	150
2	40	12	90	180
3	35	16	97	100
4	38	20	85	85
5	28	9	110	125
6	37	13	76	140
7	31	18	83	110
8	35	15	100	150
9	25	19	95	190
10	30	10	65	210

Fig. 18.1 Input versus output impact matrix

Input	y_1^1	y_2^1	y_3^1	y_1^2	y_2^2	y_3^2	y_1^3	y_2^3	y_1^4	y_2^4	y_3^4	y_4^4
x_1	—	—	x	—	x	x	x	x	—	—	—	x
x_2	x	x	—	x	—	—	—	—	x	x	x	—
x_3	—	—	x	—	x	x	x	x	—	—	—	x
x_4	x	x	—	x	—	—	—	—	x	x	x	—

For solution purposes we have restricted each α_{ir} to lie in the range 0.1–0.4. This means that for each shared input i, at least 10 % and not more than 40 % of that input would be dedicated to any given output r.. Although the decision on such bounds was difficult for management to pin down, the 0.1–0.4 range was deemed reasonable. As well, we impose both upper and lower limits on the numbers of plants to which any given component can be assigned. Specifically, we require for each component t:

$$1 \leq \sum_k d_k^t \leq 4.$$

Hence, at least one plant, and no more than four plants can be assigned component t.

Efficiency Results

Table 18.3 displays the optimal component assignment to plants. In summary:

Component #1 → Plants 5,7,10
Component #2 → Plants 6,8
Component #3 → Plants 1,3,9
Component #4 → Plants 2,4

The overall efficiency score corresponding to this assignment is 96.6 % (the value of objective function (18.7a)). Specifically, if plants are evaluated only on their core business components, their performance will be such that if viewed as a single entity, the aggregate score is 96.6 %. Table 18.4 displays both the current

Table 18.3 Assignment of components to plants

DMU	T_1	T_2	T_3	T_4
1	0	0	1	0
2	0	0	0	1
3	0	0	1	0
4	0	0	0	1
5	1	0	0	0
6	0	1	0	0
7	1	0	0	0
8	0	1	0	0
9	0	0	1	0
10	1	0	0	0

Table 18.4 Decomposition of DMU efficiency

	Assignment of components to plants				Partial efficiencies of component				
DMU	T_1	T_2	T_3	T_4	T_1	T_2	T_3	T_4	Aggregate efficiency
1	0	0	1	0	0.51	0.68	1.00	1.00	0.93
2	0	0	0	1	0.54	0.68	1.00	1.00	0.94
3	0	0	1	0	0.52	0.61	0.90	0.76	0.75
4	0	0	0	1	0.53	0.39	0.76	1.00	0.86
5	1	0	0	0	0.43	0.45	0.25	0.59	0.47
6	0	1	0	0	0.60	0.79	0.83	0.79	0.78
7	1	0	0	0	1.00	1.00	1.00	1.00	1.00
8	0	1	0	0	0.56	1.00	0.50	0.56	0.60
9	0	0	1	0	0.45	0.34	1.00	0.84	0.78
10	1	0	0	0	0.81	0.78	0.85	0.86	0.85

aggregate efficiencies for the 10 plants, as well as a decomposition of these scores into component efficiencies. For example, Plant #3 currently displays an overall performance score of 75 %. This is composed of partial efficiency scores of 52 %, 61 %, 90 % and 76 % for components 1, 2, 3 and 4, respectively. Recall that the measure of partial efficiency for a DMUk in its tth component is given by

$$e_k^t = \mu^t Y_k^t / \left[\nu^t X_k^t + \alpha_k^t \nu^{st} X_k^S \right].$$

It is noted, as well, that with the recommended component-to-plant assignments, plant #3 would be expected to have an efficiency of 90 % (up from 75 %), if it could ultimately phase out non-productive portions of its operation, and move its full emphasis to that part of the business defined by component #3. It must be emphasized that the component t_o assigned to a plant may not be the one whose partial efficiency is highest for that plant. Notice, for example, that component #2 is assigned to plant #6, with a partial efficiency of 79 %, yet component #3 actually performs better within that plant (at a partial efficiency of 83 %). This can occur because rather than minimizing a sum of efficiency ratios, we are optimizing the ratio of aggregate output (across all plants), to aggregate input.

18.6 Discussion

This chapter has examined the problem of identifying core business components for each of a set of comparable decision making units. In the context of a set of manufacturing plants, a modified version of the DEA model of Charnes et al. (1978) has been developed and demonstrated. Unlike conventional applications of DEA where the scope of the business (bundle of products produced) is assumed to remain fixed, the approach herein is intended to aid in making decisions pertaining to functional specialization in plants. An important by-product of the core-business selection process is the evaluation of efficiency of each component of the business as well as of the overall DMU. The result, as demonstrated by Table 18.4, is an efficiency profile that management can utilize in deciding where to aim for improvements and, as well, which components to de-emphasize or phase out.

We do not attempt to address issues relating to plant reorganization toward specialization. Rather, the model can aid management in choosing those (core) business activities to place within each plant. The logistics of restructuring and any change management considerations are beyond the scope of the current chapter.

One of the potential shortcomings of the model given here is the apparent absence of consideration of distribution costs on the input side. Specifically, in some settings, the choice of a particular plant as the location out of which a given component of the business will be operated, has distributional consequences. For example, manufacturing auto parts in a location remote from automobile plants (the customer) may be more costly than having them manufactured at a less efficient, but closer-to-market facility. In the application discussed herein, this issue was not highlighted as a major concern. Presumably, in situations where distribution is a major issue, one would need to augment the input bundle to include a provision for distribution costs.

References

Beasley, J. E. (1995). Determining teaching and research efficiencies. *Journal of the Operational Research Society, 46*, 4451–4452.

Charnes, A., Cooper, W. W., & Rhodes, E. L. (1978). Measuring the efficiency of decision making units. *European Journal of Operational Research, 2*, 429–444.

Cook, W. D., & Hababou, M. (2001). Sales performance measurement in bank branches. *OMEGA, 29*, 299–307.

Cook, W. D., & Roll, Y. (1993). Partial efficiencies in data envelopment analysis. *Socio-Economic Planning Sciences, 37*, 171–179.

Cook, W. D., Hababou, M., & Tuenter, H. J. H. (2000). Multi-component efficient measurement and shared inputs in data envelopment analysis: An application to sales and service performance in bank branches. *Journal of Productivity Analysis, 14*, 209–224.

Doyle, J. R., & Green, R. H. (1994). Efficiency and cross efficiency in DEA: Derivations, meanings and uses. *Journal of the Operational Research Society, 45*, 567–578.

Chapter 19
Two-Stage Network DEA with Bad Outputs

Hirofumi Fukuyama and William L. Weber

Abstract Conventional black-box DEA models allow producer performance to be measured for technologies where undesirable outputs are jointly produced by-products of desirable output production. These models allow for non-radial scaling of desirable outputs, undesirable outputs, and inputs and can account for slacks in the constraints that define the technology. We review some of these black-box performance measures and show how to measure performance in two-stage network models. In these kinds of network models inputs are used to produce intermediate outputs in a first stage and then, those intermediate outputs become inputs to a second stage where final desirable outputs and undesirable outputs are produced. The bias from using a black-box model when a network technology exists is examined as well as the bias from ignoring slacks in the constraints defining the network technology.

Keywords DEA • Two-stage network DEA • Network directional inefficiency • Network slacks-based inefficiency • Bad outputs • Black-box directional Russell inefficiency • Network directional Russell inefficiency • Budget-constrained inefficiency • Output loss indicator • Slack bias • Weakly efficient • Strongly efficient

H. Fukuyama (✉)
Faculty of Commerce, Fukuoka University, 8-19-1 Nanakuma, Jonan-Ku, Fukuoka 814-0180, Japan
e-mail: fukuyama@fukuoka-u.ac.jp

W.L. Weber
Department of Economics and Finance, Southeast Missouri State University, Cape Girardeau, MO 63701, USA
e-mail: wlweber@semo.edu

W.D. Cook and J. Zhu (eds.), *Data Envelopment Analysis*,
International Series in Operations Research & Management Science 208,
DOI 10.1007/978-1-4899-8068-7_19,

19.1 Introduction

In this chapter, we focus on the measurement of producer performance in two-stage network production models where undesirable outputs are jointly produced by-products of desirable output production. Fukuyama and Weber (2010) used a two-stage network model to analyze bank performance. The motive for their model was the disagreement among various researchers regarding whether deposits should be taken to be an output of banks or as an input (see Sealey and Lindley 1977; Hancock 1985; Barnett and Hahm 1994; Berger and Humphrey 1997). Instead, Fukuyama and Weber (2010) assumed that banks used labor, physical capital, and equity capital in a first stage to generate the intermediate output of deposits which become an input in a second stage where a portfolio of assets, including various kinds of loans and securities investments is produced. In addition, in the second stage of production some of the loans became non-performing: an undesirable by-product of producing loans.

Network models of production were developed by Färe and Grosskopf (1996, 2000). Various two-stage models where intermediate outputs are produced in a first stage and then used to produce final outputs in a second stage have been offered by Wang et al. (1997), Seiford and Zhu (1999), Chen and Zhu (2004), Sexton and Lewis (2003), Kao and Hwang (2008), Chen et al. (2009a, b) and Fukuyama and Weber (2010, 2012, 2013). In a non-network setting Färe et al. (1994) considered technologies with jointly produced undesirable outputs and the theory was extended by Färe et al. (2005), and Fukuyama and Weber (2008b). Murty et al. (2012) offer a two-stage model where pollution is generated in a first stage and then abated in a second stage of production. Two-stage network models are useful extensions to the first generation black-box models of DEA (data envelopment analysis), especially for measuring the performance of producers in industries that are vertically integrated. In the following sections we present the black-box DEA model and show how it can be extended to measure performance in a two-stage network DEA production process with undesirable outputs as by-products of desirable outputs. An Appendix provides some remarks on the synthetic data and estimates of each of the black-box and network directional distance functions that are presented in this paper.

19.2 Some Basics

19.2.1 Conceptual Black-Box and Network Technologies with Bad Outputs

The black-box DEA model (Charnes et al. 1978) assumes that inputs $\mathbf{x} \in \mathfrak{R}_+^N$ are used to produce desirable (good) outputs $\mathbf{y} \in \mathfrak{R}_+^M$ and undesirable (bad) outputs $\mathbf{b} \in \Re_+^L$. In this setting, the technology is defined by the black-box technology set:

$$BT = \{(\mathbf{x}, \mathbf{y}, \mathbf{b}) \ |\mathbf{x} \in \Re_+^N \text{ can produce } (\mathbf{y}, \mathbf{b}) \in \Re_+^{M+L}\}. \quad (19.1)$$

We assume that inputs, **x**, and final desirable outputs, **y**, are freely disposable and undesirable outputs are only weakly disposable. Free disposability of desirable outputs and inputs means that if $(\mathbf{x},\mathbf{y},\mathbf{b}) \in BT$ then for $(\mathbf{x}, -\mathbf{y}, \mathbf{b}) \leq (\mathbf{x}', -\mathbf{y}', \mathbf{b})$, $(\mathbf{x}',\mathbf{y}',\mathbf{b}) \in BT$. That is, it is possible to use more input or produce less desirable output and still remain in the black-box technology set. Weak disposability of undesirable outputs means that if $(\mathbf{x},\mathbf{y},\mathbf{b}) \in BT$ then for $0 \leq \theta \leq 1$, $(\mathbf{x}, \theta\mathbf{y}, \theta\mathbf{b}) \in BT$. Weak disposability implies an opportunity cost of reducing undesirable outputs; fewer desirable outputs must be produced in order to reduce the jointly produced bad outputs.

Two radial measures of performance are the Shephard (1953, 1970, 1974) input and output distance functions. The Shephard input distance function gives the maximum proportional contraction in inputs and the reciprocal of the Shephard output distance function gives the maximum proportional expansion in outputs. Non-radial projections of inputs and outputs to a production frontier are sometimes useful. Since we are considering jointly produced desirable and undesirable outputs we would like our performance measures to allow desirable outputs to be expanded while undesirable outputs and inputs are simultaneously contracted. One function that allows non-radial projections of output and inputs is the directional distance function. This distance function was developed by Chambers et al. (1996, 1998) as an extension of Luenberger's (1992, 1995) shortage and benefit functions. The directional distance function seeks the simultaneous maximum expansion in desirable outputs along the directional vector $\mathbf{g}^y = (g_1^y, \ldots, g_M^y) \in \Re_+^M$, contraction in undesirable outputs along the directional output vector $\mathbf{g}^b = (g_1^b, \ldots, g_L^b) \in \Re_+^L$, and contraction in inputs along the directional vector $\mathbf{g}^x = (g_1^x, \ldots, g_N^x) \in \Re_+^N$. To simplify notation we also use $\mathbf{g}^{yb} = (\mathbf{g}^y, \mathbf{g}^b) \in \Re_+^M \times \Re_+^L$ and $\mathbf{g} = (\mathbf{g}^x, \mathbf{g}^y, \mathbf{g}^b) \in \Re_+^N \times \Re_+^M \times \Re_+^L$.

Relative to (19.1), the black-box directional distance function is defined as

$$B\vec{D}(\mathbf{x}, \mathbf{y}, \mathbf{b}; \mathbf{g}) = \sup_{\beta} \{\beta | (\mathbf{x} - \beta\mathbf{g}^x, \mathbf{y} + \beta\mathbf{g}^y, \mathbf{b} - \beta\mathbf{g}^b) \in BT\}. \quad (19.2)$$

An observation of inputs and outputs is weakly efficient relative to the black-box technology if

$$(\mathbf{x}, \mathbf{y}, \mathbf{b}) \in BT \text{ and } (\mathbf{x} - \beta\mathbf{g}^x, y + \beta\mathbf{g}^y, b - \beta\mathbf{g}^b) \notin BT \text{ for } \beta > 0 \quad (19.3)$$

A producer is weakly efficient when $B\vec{D}(\mathbf{x}, \mathbf{y}, \mathbf{b}; \mathbf{g}) = 0$ and is inefficient when $B\vec{D}(\mathbf{x}, \mathbf{y}, \mathbf{b}; \mathbf{g}) > 0$ for the given directional vector, **g**. The interpretation of the directional distance function depends on the directional vector that is chosen.

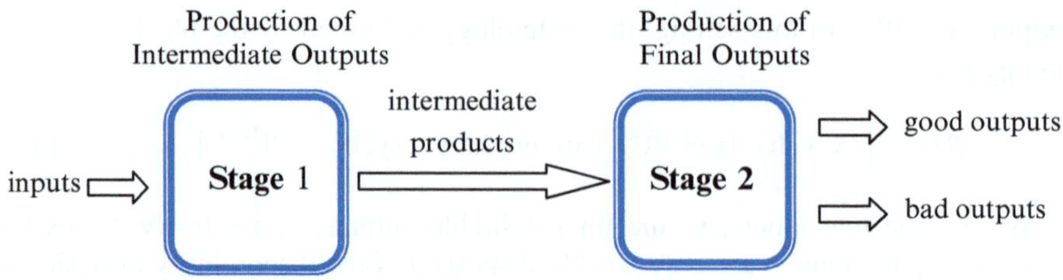

Fig. 19.1 Two-stage network production for DMUs

For instance, when $\mathbf{g} = (1,1, \ldots,1)$, the directional distance function gives the simultaneous maximum unit contraction in inputs and undesirable outputs and expansion in desirable outputs. An alternative directional vector, such as the producer's observed inputs and outputs, gives $\mathbf{g} = (\mathbf{g}^x,\mathbf{g}^y,\mathbf{g}^b) = (\mathbf{x},\mathbf{y},\mathbf{b})$. For this directional vector the directional distance function multiplied by 100 % gives the maximum percentage contraction in inputs and undesirable outputs and percentage expansion in desirable outputs. One or more of the components of the directional distance function can be set to zero. In such a case, the directional distance function holds the appropriate input or output constant and scales the remaining inputs and outputs to the frontier by the positive components of the directional vector. For instance, when $\mathbf{g}^x = 0$, the directional distance function holds inputs constant and seeks the maximum expansion in desirable outputs and contraction in undesirable outputs.

To move from the black-box technology to a two-stage network technology let $\mathbf{z} \in \Re_+^Q$ represent a vector of Q intermediate products. A two-stage network technology is defined as

$$NT = \left\{ (\mathbf{x}, \mathbf{y}, \mathbf{b}) \middle| \begin{array}{llll} \mathbf{x} \in \Re_+^N & \text{can produce} & \mathbf{z} \in \Re_+^Q & \text{(Stage 1)} \\ \mathbf{z} \in \Re_+^Q & \text{can produce} & (\mathbf{y},\ \mathbf{b}) \in \Re_+^{M+L} & \text{(Stage 2)} \end{array} \right\} \tag{19.4}$$

where the intermediate product vector $\mathbf{z}$ is determined endogenously.

Figure 19.1 depicts a two-stage network production technology. In the first stage of production a decision making unit (DMU) or firm employs exogenous inputs to produce intermediate outputs that become inputs to a second stage where desirable outputs are produced along with undesirable by-products. From here on, we use the term "network" to mean "two-stage network".

Using (19.4), the network directional distance function is written as

$$N\vec{D}(\mathbf{x}, \mathbf{y}, \mathbf{b}; \mathbf{g}) = \sup_{\beta} \{\beta | (\mathbf{x} - \beta\mathbf{g}^x, \mathbf{y} + \beta\mathbf{g}^y, \mathbf{b} - \beta\mathbf{g}^b) \in NT\}. \tag{19.5}$$

The network directional distance function expands desirable outputs and contracts undesirable outputs and inputs along the directional vector **g**.

Although intermediate outputs are produced in stage 1 and used as inputs in stage 2 of (19.4), these intermediate outputs are determined endogenously and so are not part of the network directional distance function.

We define an observation of inputs and outputs as weakly efficient for the network technology if

$$(\mathbf{x}, \mathbf{y}, \mathbf{b}) \in NT \text{ and } \left(\mathbf{x} - \beta\mathbf{g}^x, y + \beta\mathbf{g}^y, b - \beta\mathbf{g}^b\right) \notin NT \text{ for } \beta > 0. \tag{19.6}$$

That is, an observation is weakly efficient if it is not possible to simultaneously contract inputs and undesirable outputs and expand desirable outputs given the technology and directional vector. A producer is weakly efficient for the network technology if $N\vec{D}(\mathbf{x}, \mathbf{y}, \mathbf{b}; \mathbf{g}) = 0$ and inefficient if $N\vec{D}(\mathbf{x}, \mathbf{y}, \mathbf{b}; \mathbf{g}) > 0$.

19.2.2 The DEA Technology and Directional Inefficiency

In this section we follow the work of Fukuyama and Weber (2010) to develop DEA performance measures for a two-stage network technology. To estimate the distance functions given in (19.2) and (19.5) in DEA we assume there are $j = 1, \ldots, J$ decision making units (DMUs), each of which converts inputs $\mathbf{x}_j \in \Re_+^N$, into intermediate products $\mathbf{z}_j \in \Re_+^Q$ in a first stage of production and then uses the intermediate products in a second stage to produce final desirable outputs $\mathbf{y}_j \in \Re_+^M$ and bad (undesirable) outputs $\mathbf{b}_j \in \Re_+^L$.

Let $\mathbf{0}$ be an appropriate dimensional vector of zeros. The black-box DEA technology for (19.1) takes the form

$$BT = \left\{ (\mathbf{x}, \mathbf{y}, \mathbf{b}) \middle| \mathbf{x} \geq \sum_{j=1}^{J} \mathbf{x}_j\lambda_j;\ \ \mathbf{y} \leq \sum_{j=1}^{J} \mathbf{y}_j\lambda_j;\ \ \mathbf{b} = \sum_{j=1}^{J} \mathbf{b}_j\lambda_j;\ \ \boldsymbol{\lambda} \geq \mathbf{0} \right\} \tag{19.7}$$

where $\boldsymbol{\lambda}$ is a nonnegative intensity vector. Throughout this chapter, we assume all observed attributes of production are positive, i.e., $\mathbf{x}_j > \mathbf{0}$ ($\forall\, j$), $\mathbf{z}_j > \mathbf{0}$ ($\forall\, j$), $\mathbf{b}_j > \mathbf{0}$ ($\forall\, j$) and $\mathbf{y}_j > \mathbf{0}$ ($\forall\, j$). The black-box production possibility set (19.7) exhibits constant returns to scale. The directional distance function defined on (19.7) for DMU "o" is

$$\begin{aligned} B\vec{D}\left(\mathbf{x}_o, \mathbf{y}_o, \mathbf{b}_o; \mathbf{g}\right) = \max_{\beta, \boldsymbol{\lambda}, \mathbf{s}^x, \mathbf{s}^y} & \left\{ \beta \middle| \mathbf{x}_o \geq \sum_{j=1}^{J} \mathbf{x}_j\lambda_j + \beta\mathbf{g}^y + \mathbf{s}^x;\ \ \mathbf{y}_o \leq \sum_{j=1}^{J} \mathbf{y}_j\lambda_j - \beta\mathbf{g}^y - \mathbf{s}^y; \right. \\ & \left. \mathbf{b}_o = \sum_{j=1}^{J} \mathbf{b}_j\lambda_j + \beta\mathbf{g}^b;\ \ \boldsymbol{\lambda} \geq \mathbf{0};\ \ \mathbf{s}^x \geq \mathbf{0};\ \ \mathbf{s}^y \geq \mathbf{0} \right\} \end{aligned} \tag{19.8}$$

where $\mathbf{s}^x$ and $\mathbf{s}^y$ represent any remaining slack in the input and desirable output constraints that remains after the observed quantities ($\mathbf{x}_o$,$\mathbf{y}_o$,$\mathbf{b}_o$) have been scaled to the frontier along the scaling vector $\mathbf{g}$. The dual form of (19.8) is written as

$$\min_{\mathbf{v},\mathbf{u},\mathbf{w}} \left\{ \mathbf{v}\mathbf{x}_o - \mathbf{u}\mathbf{y}_o + \mathbf{r}\mathbf{b}_o \middle| \begin{array}{l} \mathbf{v}\mathbf{g}^x + \mathbf{u}\mathbf{g}^y + \mathbf{r}\mathbf{g}^b = 1; \ \ \mathbf{v}\mathbf{x}_j - \mathbf{u}\mathbf{y}_j + \mathbf{r}\mathbf{b}_j \geq \mathbf{0} \quad \forall j; \\ \mathbf{v} \geq \mathbf{0}; \ \ \mathbf{u} \geq \mathbf{0}; \ \ \mathbf{r} : free \end{array} \right\} \tag{19.9}$$

where $\mathbf{v}$, $\mathbf{u}$, and $\mathbf{r}$ are multipliers. Although Fukuyama and Weber (2009a) modeled (19.8) under the assumption of variable returns to scale, it is necessary to employ either constant or non-increasing returns to scale if the assumption of null jointness of bad outputs is to be satisfied.[1] See Fukuyama and Weber (2008a) for the implementation of non-increasing returns to scale. Shephard and Färe (1974) define and discuss the assumption of null jointness.

To incorporate the two-stage structure given in Fig. 19.1 in a DEA model we define the intensity vectors for the two stages as $\boldsymbol{\lambda}^1 = (\lambda_1^1, \ldots, \lambda_J^1) \in \Re_+^J$ and $\boldsymbol{\lambda}^2 = (\lambda_1^2, \ldots, \lambda_J^2) \in \Re_+^J$. The network production possibility set is

$$NT = \left\{ (\mathbf{x}, \mathbf{y}, \mathbf{b}) \middle| \begin{array}{l} \mathbf{x} \geq \sum_{j=1}^J \mathbf{x}_j \lambda_j^1; \ \ \mathbf{y} \leq \sum_{j=1}^J \mathbf{y}_j \lambda_j^2; \ \ \mathbf{b} = \sum_{j=1}^J \mathbf{b}_j \lambda_j^2; \\ \sum_{j=1}^J \mathbf{z}_j \lambda_j^1 \geq \hat{\mathbf{z}}; \ \ \sum_{j=1}^J \mathbf{z}_j \lambda_j^2 \leq \hat{\mathbf{z}}; \ \ \boldsymbol{\lambda}^1 \geq \mathbf{0}; \ \ \boldsymbol{\lambda}^2 \geq \mathbf{0}; \ \ \hat{\mathbf{z}} \geq \mathbf{0} \end{array} \right\} \tag{19.10}$$

where $\hat{\mathbf{z}}$ is endogenously determined in (19.10).

Using DEA, network directional inefficiency for DMU "o" takes the form:

$$\begin{aligned} & N\vec{D}(\mathbf{x}_o, \mathbf{y}_o, \mathbf{b}_o; \mathbf{g}) \\ & = \max_{\boldsymbol{\lambda}^1, \boldsymbol{\lambda}^2, \mathbf{s}^x, \mathbf{s}^y, \hat{\mathbf{z}}, \beta} \left\{ \beta \middle| \begin{array}{l} \mathbf{x}_o = \sum_{j=1}^J \mathbf{x}_j \lambda_j^1 + \beta \mathbf{g}^x + \mathbf{s}^x; \ \ \mathbf{y}_o = \sum_{j=1}^J \mathbf{y}_j \lambda_j^2 - \beta \mathbf{g}^y - \mathbf{s}^y; \\ \mathbf{b}_o = \sum_{j=1}^J \mathbf{b}_j \lambda_j^1 + \beta \mathbf{g}^b; \ \ \sum_{j=1}^J \mathbf{z}_j \lambda_j^1 \geq \hat{\mathbf{z}}; \ \ \sum_{j=1}^J \mathbf{z}_j \lambda_j^2 \leq \hat{\mathbf{z}}; \\ \boldsymbol{\lambda}^1 \geq \mathbf{0}; \ \ \boldsymbol{\lambda}^2 \geq \mathbf{0}; \ \ \mathbf{s}^x \geq \mathbf{0}; \ \ \mathbf{s}^y \geq \mathbf{0}; \ \ \hat{\mathbf{z}} \geq \mathbf{0}; \ \ \beta : free \end{array} \right\} \end{aligned} \tag{19.11}$$

[1] Null jointness means that if $\mathbf{b} = 0$ and $(\mathbf{x},\mathbf{y},\mathbf{b}) \in BT$ then $\mathbf{y} = 0$.

The network directional distance function projects the observed outputs and inputs of DMU$_o$ to the weakly efficient frontier. That is, $\left(\hat{\mathbf{z}}^*, \mathbf{x}_o - \beta^* \mathbf{g}^x, \mathbf{y}_o + \beta^* \mathbf{g}^y, \mathbf{b}_o - \beta^* \mathbf{g}^b\right) \in NT$ where the star (*) indicates an optimal solution to (19.11) and $\left(\hat{\mathbf{z}}^*, \mathbf{x}_o - \beta' \mathbf{g}^x, \mathbf{y}_o + \beta' \mathbf{g}^y, \mathbf{b}_o - \beta' \mathbf{g}^b\right) \notin NT$ for $\beta' > \beta^*$. Thus, a DMU must be a weakly efficient frontier point for both stages of production if $\hat{\mathbf{z}}^*$ is an optimal solution to (19.11). We note that the constraints on inputs and desirable outputs have slack variables added because of strong disposability. However, undesirable outputs only satisfy weak disposability and therefore have no slack variables. Although the projection is weakly efficient, it might leave positive slacks.

The dual form of (19.11) is

$$\min_{\mathbf{v},\mathbf{u},\mathbf{w}^1,\mathbf{w}^2,\mathbf{r}} \left\{ \mathbf{v}\mathbf{x}_o - \mathbf{u}\mathbf{y}_o + \mathbf{r}\mathbf{b}_o \,\middle|\, \begin{array}{l} \mathbf{v}\mathbf{g}^x + \mathbf{u}\mathbf{g}^y + \mathbf{r}\mathbf{g}^b = 1; \;\; \mathbf{v}\mathbf{x}_j - \mathbf{w}^1\mathbf{z}_j \geq \mathbf{0} \;\; \forall j; \\ \mathbf{w}^2\mathbf{z}_j - \mathbf{u}\mathbf{y}_j + \mathbf{r}\mathbf{b}_j \geq \mathbf{0} \;\; \forall j; \;\; \mathbf{w}^1 - \mathbf{w}^2 \geq \mathbf{0}; \\ \mathbf{v} \geq \mathbf{0}; \;\; \mathbf{w}^1 \geq \mathbf{0}; \;\; \mathbf{w}^2 \geq \mathbf{0}; \;\; \mathbf{u} \geq \mathbf{0}; \;\; \mathbf{r}: \mathit{free} \end{array} \right\}. \tag{19.12}$$

The non-negativity constraints $\hat{\mathbf{z}} \geq \mathbf{0}$ in (19.11) correspond to $\mathbf{w}^1 - \mathbf{w}^2 \geq \mathbf{0}$ in (19.12). Treating $\hat{\mathbf{z}}$ as free variables in (19.11) yields the corresponding dual form which can be obtained by replacing the inequality constraints $\mathbf{w}^1 - \mathbf{w}^2 \geq \mathbf{0}$ with equality constraints $\mathbf{w}^1 - \mathbf{w}^2 = \mathbf{0}$, but will not change the optimal objective values of (19.11) and (19.12).

Note that in (19.11) $\hat{\mathbf{z}}$ is positive. The first stage constraint, $\sum_{j=1}^{J} \mathbf{z}_j \lambda_j^1 \geq \hat{\mathbf{z}}$, and the second stage constraint, $\sum_{j=1}^{J} \mathbf{z}_j \lambda_j^2 \leq \hat{\mathbf{z}}$, leads to $\sum_{j=1}^{J} \mathbf{z}_j \lambda_j^1 \geq \hat{\mathbf{z}} \geq \sum_{j=1}^{J} \mathbf{z}_j \lambda_j^2 > 0$ because there exists a nonzero vector $(\boldsymbol{\lambda}^1, \boldsymbol{\lambda}^2)$ in (19.11) and $\mathbf{z}_j > 0$ ($\forall\, j$). This shows that a DMU must be a frontier point for both stages of production if $\hat{\mathbf{z}}^*$ is an optimal solution to (19.11). In their input-oriented Farrell measure, Kao and Hwang (2008) assumed equal shadow prices between the two stages. By allowing for different shadow prices, Chen et al. (2010) provided a modification but showed that their model is equivalent to Kao and Hwang's (2008). Hence, we assume a common shadow price vector $\mathbf{w} \in \Re_+^Q$ for the two divisions in (19.12).

The Farrell input efficiency measure ($F_I(\mathbf{x},\mathbf{y},\mathbf{b})$) is the reciprocal of Shephard's (1970) input distance function, $1/BD_I(\mathbf{x},\mathbf{y},\mathbf{b})$, and takes the form

$$F_I(\mathbf{x}, \mathbf{y}, \mathbf{b}) = \frac{1}{BD_I(\mathbf{x}, \mathbf{y}, \mathbf{b})} = \inf_{\theta} \{\theta | (\theta\mathbf{x}, \mathbf{y}, \mathbf{b}) \in BT\}. \tag{19.13}$$

Chen et al. (2010) estimated a network Farrell input efficiency model and implicitly used the equivalence in the following relations:

$$\sum_{j=1}^{J} \mathbf{z}_j\lambda_j^1 \geq \hat{\mathbf{z}} \text{ and } \sum_{j=1}^{J} \mathbf{z}_j\lambda_j^2 \leq \hat{\mathbf{z}} \Leftrightarrow \sum_{j=1}^{J} \mathbf{z}_j\lambda_j^1 - \sum_{j=1}^{J} \mathbf{z}_j\lambda_j^2 \geq \mathbf{0}$$
$$\Leftrightarrow \sum_{j=1}^{J} \mathbf{z}_j\left(\lambda_j^1 - \lambda_j^2\right) \geq \mathbf{0}. \tag{19.14}$$

It is easy to see that the relations in (19.14) hold under a directional model as well. Therefore, we replace the constraints, $\sum_{j=1}^{J} \mathbf{z}_j\lambda_j^1 \geq \hat{\mathbf{z}}$ and $\sum_{j=1}^{J} \mathbf{z}_j\lambda_j^2 \leq \hat{\mathbf{z}}$, by the constraints

$$\sum_{j=1}^{J} \mathbf{z}_j\left(\lambda_j^1 - \lambda_j^2\right) \geq \mathbf{0}. \tag{19.15}$$

The optimal intermediate outputs and intermediate inputs are obtained from (19.14) as $\sum_{j=1}^{J} \mathbf{z}_j\lambda_j^{1*}$ and $\sum_{j=1}^{J} \mathbf{z}_j\lambda_j^{2*}$. Equation (19.15) allows a portion of intermediate outputs to be consumed within the evaluated DMU so that the intermediate inputs do not necessarily have to equal the intermediate outputs. Hence, $\sum_{j=1}^{J} \mathbf{z}_j\lambda_j^{1*} > \sum_{j=1}^{J} \mathbf{z}_j\lambda_j^{2*}$ implies the existence of internal waste or internal consumption. If $\lambda_j^1 = \lambda_j^2$ ($\forall\, j$), the network directional distance function collapses to the black-box directional distance function (19.8).

In Table A.1 we report a synthetic data set where $J = 10$ producers use $N = 3$ inputs to produce $Q = 2$ intermediate outputs in a first stage of production. Then, in the second stage of production the $Q = 2$ intermediate outputs are used to produce $M = 3$ final outputs and $L = 2$ bad outputs. Table A.2 reports each of the performance measures that are presented in this paper. The estimates show that $N\vec{D}(\mathbf{x}_o, \mathbf{y}_o, \mathbf{b}_o; \mathbf{g}) \geq B\vec{D}(\mathbf{x}_o, \mathbf{y}_o, \mathbf{b}_o; \mathbf{g})$ since the two-stage process allows production possibilities to expand.

Table A.1 Synthetic data

j	x_1	x_2	x_3	z_1	z_2	y_1	y_2	y_3	b	p_1	p_2	p_3
1	8	10	1	12	5	5	5	12	3	1	1	1
2	7	9	2	14	6	7	7	19	2	1	1	1
3	9	5	4	16	7	15	10	17	3	1	1	1
4	4	14	5	20	3	3	15	10	4	1	1	1
5	9	4	6	15	6	2	12	18	3	1	1	1
6	6	12	7	12	7	4	8	19	5	1	1	1
7	10	12	8	25	8	12	10	21	5	1	1	1
8	12	6	9	22	8	15	5	25	4	1	1	1
9	15	5	8	22	9	14	7	15	3	1	1	1
10	14	7	1	20	9	9	9	13	4	1	1	1

Table A.2 Performance estimates

j	(1)	(2)	(3)	(4)	(5)	(6)	(7)	(8)
1	0	0.50482	0	1.94814	0	1.23397	1.23397	1.23397
2	0	0	0	0	0	0	0	0
3	0	0	0	0	0	0	0	0
4	0	0	0	0	0	0	0	0.18049
5	0	0	0	0.08333	0	0	0	0
6	0	0	0	1.38989	0	2.24679	0	2.24679
7	0.60889	1.00879	1.96581	1.98798	1.56439	1.63782	1.63782	1.63782
8	0	0	0	0	0	0.47475	0	0.47475
9	0.16667	0.16667	1.77778	1.77778	0.16667	0.16667	0.16667	0.16667
10	0	0.28906	0	2.47222	0	1.14634	1.14634	1.14634

(1) = $B\vec{D}(\mathbf{x}, \mathbf{y}, \mathbf{b}; \mathbf{g})$, (2) = $N\vec{D}(\mathbf{x}, \mathbf{y}, \mathbf{b}; \mathbf{g})$, (3) = $BSBI(\mathbf{x}_o,\mathbf{y}_o,\mathbf{b}_o;\mathbf{g})$, (4) = $NSBI(\mathbf{x}_o,\mathbf{y}_o,\mathbf{b}_o;\mathbf{g})$, (5) = $B\vec{D}_O(\mathbf{x}, \mathbf{y}, \mathbf{b}; \mathbf{g}^{yb})$, (6) = $BI\vec{D}_O(\mathbf{p}/\alpha, \mathbf{y}, \mathbf{b}; \mathbf{g}^{yb})$, (7) = $N\vec{D}_O(\mathbf{x}, \mathbf{y}, \mathbf{b}; \mathbf{g}^{yb})$, (8) = $NI\vec{D}_O(\mathbf{p}/\alpha, \mathbf{y}, \mathbf{b}; \mathbf{g}^{yb})$

19.3 Slacks-Based Inefficiency Measures

19.3.1 Black-Box Slacks-Based Inefficiency

Although the black-box and network directional distance functions scale inputs and outputs to the weakly efficient frontier the projection can still leave slacks in the constraints defining the DEA reference technology. When slacks exist it is possible to contract at least one of the inputs or expand at least one of the desirable outputs even though it is not possible to further contract all inputs and undesirable outputs and expand all desirable outputs along the directional vector **g**. Therefore, the directional distance function potentially underestimates the amount of producer inefficiency.

To account for slacks in the constraints defining the black-box DEA technology Tone (2001) introduced a slacks-based performance measure. In this section we extend Tone's method and incorporate undesirable outputs as part of the technology. We normalize the slacks in each constraint by the directional vector which allows the normalized slacks of inputs and outputs to be added. The black-box slacks-based inefficiency measure (*BSBI*) for DMU_o takes the form

$$BSBI(\mathbf{x}_o, \mathbf{y}_o, \mathbf{b}_o; \mathbf{g}) = \max_{\mathbf{s}^-, \mathbf{s}^+, \mathbf{s}^\#, \boldsymbol{\lambda}} \left\{ \frac{\frac{1}{N}\sum_{n=1}^{N} \frac{s_n^-}{g_n^x} + \frac{1}{M}\sum_{m=1}^{M} \frac{s_m^+}{g_m^y} + \frac{1}{L}\sum_{l=1}^{L} \frac{s_l^\#}{g_l^b}}{3} \middle| \begin{array}{l} \mathbf{x}_o = \sum_{j=1}^{J} \mathbf{x}_j \lambda_j + \mathbf{s}^-; \\ \mathbf{y}_o = \sum_{j=1}^{J} \mathbf{y}_j \lambda_j - \mathbf{s}^+; \\ \mathbf{b}_o = \sum_{j=1}^{J} \mathbf{b}_j \lambda_j + \mathbf{s}^\#; \\ \mathbf{s}^- \geq \mathbf{0};\ \mathbf{s}^+ \geq \mathbf{0};\ \mathbf{s}^\# \geq \mathbf{0};\ \boldsymbol{\lambda} \geq \mathbf{0} \end{array} \right\}. \tag{19.16}$$

In (19.16), the slack in each input constraint, s_n^- ($n = 1, \ldots, N$), is normalized by its corresponding component of $\mathbf{g}^x$. Similarly, slacks in the output constraints, s_m^+ ($m = 1, \ldots, M$), are normalized by $\mathbf{g}^y$, and slacks in the bad output constraints, $s_l^\#$ ($l = 1, \ldots, L$), are normalized by $\mathbf{g}^b$. Thus, $BSBI(\mathbf{x}_o,\mathbf{y}_o,\mathbf{b}_o;\mathbf{g})$ is independent of the units of measurement for inputs and outputs. The BSBI is a directional extension of Tone's (2001) slacks-based efficiency measure, or equivalently, Pastor et al. (1999) enhanced Russell graph efficiency measure.

Equivalent to (19.16) is the black-box directional Russell inefficiency measure, denoted by

$$BR(\mathbf{x}_o, \mathbf{y}_o, \mathbf{b}_o; \mathbf{g}) = \max \left\{ \frac{\frac{1}{N}\sum_{n=1}^{N} \xi_n^- + \frac{1}{M}\sum_{m=1}^{M} \xi_m^+ + \frac{1}{L}\sum_{l=1}^{L} \xi_l^\#}{3} \middle| \begin{array}{l} \mathbf{x}_o = \sum_{j=1}^{J} \mathbf{x}_j \lambda_j + \boldsymbol{\xi}^- \odot \mathbf{g}^x; \\ \mathbf{y}_o = \sum_{j=1}^{J} \mathbf{y}_j \lambda_j - \boldsymbol{\xi}^+ \odot \mathbf{g}^y; \\ \mathbf{b}_o = \sum_{j=1}^{J} \mathbf{b}_j \lambda_j + \boldsymbol{\xi}^\# \odot \mathbf{g}^b; \\ \boldsymbol{\xi}^- \geq \mathbf{0};\ \boldsymbol{\xi}^+ \geq \mathbf{0};\ \boldsymbol{\xi}^\# \geq \mathbf{0};\ \boldsymbol{\lambda} \geq \mathbf{0} \end{array} \right\}. \tag{19.17}$$

where $\boldsymbol{\xi}^- = (\xi_1^-, \ldots, \xi_N^-) \in \Re_+^N$, $\boldsymbol{\xi}^+ = (\xi_1^+, \ldots, \xi_M^+) \in \Re_+^M$ and $\boldsymbol{\xi}^\# = (\xi_1^\#, \ldots, \xi_L^\#) \in \Re_+^L$. The symbol '$\odot$' indicates component-wise multiplication. By setting $\xi_n^- = s_n^-/g_n^x$ ($\forall\ n$), $\xi_m^+ = s_m^+/g_m^y$ ($\forall\ m$) and $\xi_l^\# = s_l^\#/g_l^b$ ($\forall\ l$) in (19.17), we obtain the network SBI measure given in (19.16).

The dual formulation of (19.16) is

$$\min_{\mathbf{v},\mathbf{u},\mathbf{r}} \left\{ \mathbf{v}\mathbf{x}_o - \mathbf{u}\mathbf{y}_o + \mathbf{r}\mathbf{b}_o \;\middle|\; \begin{array}{l} \mathbf{v}\mathbf{x}_j - \mathbf{u}\mathbf{y}_j + \mathbf{r}\mathbf{b}_j \geq \mathbf{0} \quad \forall j \\ \mathbf{v} \geq \left[\dfrac{1}{3Ng_1^x}, \;\ldots, \dfrac{1}{3Ng_N^x}\right] \\ \mathbf{u} \geq \left[\dfrac{1}{3Mg_1^y}, \;\ldots, \dfrac{1}{3Mg_M^y}\right] \\ \mathbf{r} \geq \left[\dfrac{1}{3Lg_1^b}, \;\ldots, \dfrac{1}{3Lg_L^b}\right] \\ \mathbf{v} \geq \mathbf{0}, \;\; \mathbf{u} \geq \mathbf{0}, \;\; \mathbf{r} \geq \mathbf{0} \end{array} \right\}. \tag{19.18}$$

Hence, slacks-based inefficiency can be obtained by maximizing $\dfrac{\frac{1}{N}\sum_{n=1}^{N}\xi_n^- + \frac{1}{M}\sum_{m=1}^{M}\xi_m^+ + \frac{1}{L}\sum_{l=1}^{L}\xi_l^\#}{3}$ in (19.17) or minimizing $\mathbf{v}\mathbf{x}_o - \mathbf{u}\mathbf{y}_o + \mathbf{r}\mathbf{b}_o$ in (19.18). The inefficiency measures (19.16) and (19.17) are generalizations of the directional distance function given in (19.8). Fukuyama and Weber (2009a, 2010) discuss how the directional distance function in (19.8) is related to the slacks-based inefficiency measures given by (19.16), (19.17) and (19.18).

The black-box slacks-based inefficiency measure, $BSBI(\mathbf{x}_o,\mathbf{y}_o,\mathbf{b}_o;\mathbf{g})$, equals the average of mean input inefficiency, mean desirable output inefficiency, and mean undesirable output inefficiency. Furthermore, $BSBI(\mathbf{x}_o,\mathbf{y}_o,\mathbf{b}_o;\mathbf{g})$ collapses to the black-box directional distance function $(B\vec{D}(\mathbf{x}_o,\mathbf{y}_o,\mathbf{b}_o;\mathbf{g}))$ when the slacks-based projection and the directional projection are equal: $(\mathbf{x}^{BSBI},\mathbf{b}^{BSBI},\mathbf{y}^{BSBI}) = (\mathbf{x}^{B\vec{D}},\mathbf{b}^{B\vec{D}},\mathbf{y}^{B\vec{D}})$. When slacks exist $(\mathbf{x}^{BSBI},\mathbf{b}^{BSBI},-\mathbf{y}^{BSBI}) \lneq (\mathbf{x}^{B\vec{D}},\mathbf{b}^{B\vec{D}},-\mathbf{y}^{B\vec{D}})$ and $BSBI(\mathbf{x}_o,\mathbf{y}_o,\mathbf{b}_o;\mathbf{g}) > B\vec{D}(\mathbf{x}_o,\mathbf{y}_o,\mathbf{b}_o;\mathbf{g})$.

19.3.2 Network Slacks-Based Inefficiency

Adapting Tone and Tsutsui's (2009) ratio measure,[2] Fukuyama and Weber (2010) defined network slacks-based inefficiency (NSBI) as

[2] See Fukuyama and Mirdehghan (2012) for some discussion and analysis of the constraints associated with intermediate products.

$$NSBI(\mathbf{x}_o, \mathbf{y}_o, \mathbf{b}_o; \mathbf{g}) = \max_{\boldsymbol{\lambda}^1, \boldsymbol{\lambda}^2, \mathbf{s}^x, \mathbf{s}^y, \mathbf{s}^b, \hat{\mathbf{z}}} \left\{ \frac{\frac{1}{N}\sum_{n=1}^{N} \frac{s_n^-}{g_n^x} + \frac{1}{M}\sum_{m=1}^{M} \frac{s_m^+}{g_m^y} + \frac{1}{L}\sum_{l=1}^{L} \frac{s_l^\#}{g_l^b}}{3} \left| \begin{array}{l} \mathbf{x}_o = \sum_{j=1}^{J} \mathbf{x}_j \lambda_j^1 + \mathbf{s}^-; \\ \mathbf{y}_o = \sum_{j=1}^{J} \mathbf{y}_j \lambda_j^2 - \mathbf{s}^+; \\ \mathbf{b}_o = \sum_{j=1}^{J} \mathbf{b}_j \lambda_j^2 + \mathbf{s}^\#; \\ \sum_{j=1}^{J} \mathbf{z}_j \left(\lambda_j^1 - \lambda_j^2\right) \geq \mathbf{0}; \\ \boldsymbol{\lambda}^1 \geq \mathbf{0};\ \boldsymbol{\lambda}^2 \geq \mathbf{0};\ \mathbf{s}^- \geq \mathbf{0}; \\ \mathbf{s}^+ \geq \mathbf{0};\ \mathbf{s}^\# \geq \mathbf{0} \end{array} \right. \right\}. \tag{19.19}$$

Similar to $N\vec{D}(\mathbf{x}_o, \mathbf{y}_o, \mathbf{b}_o; \mathbf{g})$, slacks of intermediate products are not included in the objective function of (19.19), and only the slacks of the exogenous inputs, final outputs, and bad outputs are maximized. Equation (19.19) can be thought of as a directional slacks-based version of Färe and Grosskopf's (1996) network model depicted in Fig. 19.1. If $NSBI(\mathbf{x}_o,\mathbf{y}_o,\mathbf{b}_o;\mathbf{g}) = 0$, then DMU$_o$ is efficient. Values of $NSBI(\mathbf{x}_o,\mathbf{y}_o,\mathbf{b}_o;\mathbf{g}) > 0$ indicate inefficiency.

The dual of (19.19) is

$$\min_{\mathbf{v},\mathbf{w},\mathbf{u},\mathbf{r}} \left\{ \mathbf{v}\mathbf{x}_o - \mathbf{u}\mathbf{y}_o + \mathbf{r}\mathbf{b}_o \left| \begin{array}{l} \mathbf{v}\mathbf{x}_j - \mathbf{w}\mathbf{z}_j \geq \mathbf{0}\ \forall j;\ \mathbf{w}\mathbf{z}_j - \mathbf{u}\mathbf{y}_j + \mathbf{r}\mathbf{b}_j \geq \mathbf{0}\ \forall j; \\ \mathbf{v} \geq \left[\frac{1}{3Ng_1^x}, \ldots, \frac{1}{3Ng_N^x}\right];\ \mathbf{u} \geq \left[\frac{1}{3Mg_1^y}, \ldots, \frac{1}{3Mg_M^y}\right]; \\ \mathbf{r} \geq \left[\frac{1}{3Lg_1^b}, \ldots, \frac{1}{3Lg_L^b}\right];\ \mathbf{v} \geq \mathbf{0};\ \mathbf{u} \geq \mathbf{0};\ \mathbf{r} \geq \mathbf{0};\ \mathbf{w} \geq \mathbf{0}. \end{array} \right. \right\}. \tag{19.20}$$

The relation between the network directional distance function defined in (19.11) and network slacks-based inefficiency in (19.19) is similar to the one between the black-box slacks-based inefficiency measure (19.16) and the directional distance function (19.8). That is, $BSBI(\mathbf{x}, \mathbf{y}, \mathbf{b}; \mathbf{g}) \geq B\vec{D}(\mathbf{x}, \mathbf{y}, \mathbf{b}; \mathbf{g})$ and $NSBI(\mathbf{x}, \mathbf{y}, \mathbf{b}; \mathbf{g}) \geq N\vec{D}(\mathbf{x}, \mathbf{y}, \mathbf{b}; \mathbf{g})$.

Accounting for undesirable outputs, Färe and Grosskopf's (1996) input-based network efficiency measure takes the form

$$FG(\mathbf{x}_o, \mathbf{y}_o, \mathbf{b}_o) = \min_{\boldsymbol{\lambda}^1, \boldsymbol{\lambda}^2, \theta, \mathbf{s}^-, \mathbf{s}^+} \left\{ \theta \left| \begin{array}{l} \theta \mathbf{x}_o = \sum_{j=1}^{J} \mathbf{x}_j \lambda_j^1 + \mathbf{s}^-; \\ \mathbf{y}_o = \sum_{j=1}^{J} \mathbf{y}_j \lambda_j^2 - \mathbf{s}^+; \\ \mathbf{b}_o = \sum_{j=1}^{J} \mathbf{b}_j \lambda_j^2; \\ \sum_{j=1}^{J} \mathbf{z}_j \left(\lambda_j^1 - \lambda_j^2 \right) \geq \mathbf{0}; \\ \boldsymbol{\lambda}^1 \geq \mathbf{0};\ \boldsymbol{\lambda}^2 \geq \mathbf{0};\ \mathbf{s}^- \geq \mathbf{0};\ \mathbf{s}^+ \geq \mathbf{0} \end{array} \right. \right\}. \quad (19.21)$$

Similarly, Kao and Hwang's (2008) input-oriented efficiency measure can be modified to account for undesirable outputs as

$$KH(\mathbf{x}_o, \mathbf{y}_o, \mathbf{b}_o) = \max_{\mathbf{v}, \mathbf{u}, \mathbf{w}, \mathbf{r}} \left\{ \mathbf{u}\mathbf{y}_o - \mathbf{r}\mathbf{b}_o \left| \begin{array}{l} \mathbf{v}\mathbf{x}_o = 1 \\ \mathbf{w}\mathbf{z}_j - \mathbf{v}\mathbf{x}_j \leq \mathbf{0}\ \forall j, \\ \mathbf{u}\mathbf{y}_j - \mathbf{r}\mathbf{b}_j - \mathbf{w}\mathbf{z}_j \leq \mathbf{0}\ \forall j, \\ \mathbf{v} \geq \mathbf{0},\ \ \mathbf{w} \geq \mathbf{0},\ \ \mathbf{u} \geq \mathbf{0},\ \ \mathbf{r}\ \mathit{free} \end{array} \right. \right\}. \quad (19.22)$$

Since (19.21) and (19.22) are dual to each other they are equivalent models. This equivalence was established by Chen et al. (2010) for the case without bad outputs.

Table A.2 of the Appendix provides DEA estimates of $BSBI(\mathbf{x}_o,\mathbf{y}_o,\mathbf{b}_o;\mathbf{g})$ and $NSBI(\mathbf{x}_o,\mathbf{y}_o,\mathbf{b}_o;\mathbf{g})$ for the synthetic data set reported in Table A.1. Furthermore, the estimates verify that $BSBI(\mathbf{x}_o, \mathbf{y}_o, \mathbf{b}_o; \mathbf{g}) \geq B\vec{D}(\mathbf{x}_o, \mathbf{y}_o, \mathbf{b}_o; \mathbf{g})$ and $NSBI(\mathbf{x}_o, \mathbf{y}_o, \mathbf{b}_o; \mathbf{g}) \geq N\vec{D}(\mathbf{x}_o, \mathbf{y}_o, \mathbf{b}_o; \mathbf{g})$.

19.3.3 Decomposing Network Slacks-Based Inefficiency

In the context of the models presented in the previous sections two types of bias can arise when measuring the inefficiency of decision-making units. The first type of bias arises when there is a network production structure but inefficiency is estimated using one of the many standard black-box DEA models. The second type of bias arises when the inefficiency measure ignores potential input and output slacks.

We define a DMU as strongly efficient relative to the black-box technology if

$$(\mathbf{x}, \mathbf{y}, \mathbf{b}) \in BT \text{ and } \left(\mathbf{x} - \mathbf{s}_n^-, \mathbf{y} + \mathbf{s}_m^+, \mathbf{b} - \mathbf{s}_l^\#\right) \notin BT \text{ for } \mathbf{s}_n^- > 0, \mathbf{s}_m^+ > 0, \text{or } \mathbf{s}_l^\# > 0. \quad (19.23)$$

That is, an observation is strongly efficient if it is not possible to contract at least one of the inputs or undesirable outputs, or expand at least one of the desirable outputs given the technology. Similarly, a DMU is strongly efficient relative to the network technology if

$$(\mathbf{x}, \mathbf{y}, \mathbf{b}) \in NT \text{ and } \left(\mathbf{x} - \mathbf{s}_n^+, \mathbf{y} + \mathbf{s}_m^+, \mathbf{b} - \mathbf{s}_l^\#\right) \notin NT \text{ for } \mathbf{s}_n^- > 0, \mathbf{s}_m^+ > 0, \text{or } \mathbf{s}_l^\# > 0. \tag{19.24}$$

Network directional inefficiency, $N\vec{D}\left(\mathbf{x}_o, \mathbf{y}_o, \mathbf{b}_o; \mathbf{g}\right)$, contracts inputs and undesirable outputs, and expands desirable outputs to the weakly efficient frontier. However, the projected point of $N\vec{D}\left(\mathbf{x}_o, \mathbf{y}_o, \mathbf{b}_o; \mathbf{g}\right)$ can still leave slacks in the constraints that define the DEA technology. When slack exists, it is possible for a DMU to be on the weakly efficient frontier but able to expand at least one desirable output or contract at least one undesirable output or input. In contrast, $BSBI(\mathbf{x}_o,\mathbf{y}_o,\mathbf{b}_o;\mathbf{g})$ and $NSBI(\mathbf{x}_o,\mathbf{y}_o,\mathbf{b}_o;\mathbf{g})$ contract inputs and undesirable outputs and expand desirable outputs to the strongly efficient frontier, which is a subset of the weakly efficient frontier.

We define an indicator of slack bias for DMU_o as

$$SlackBias\left(\mathbf{x}_o, \mathbf{y}_o, \mathbf{b}_o; \mathbf{g}\right) = NSBI\left(\mathbf{x}_o, \mathbf{y}_o, \mathbf{b}_o; \mathbf{g}\right) - N\vec{D}\left(\mathbf{x}_o, \mathbf{y}_o, \mathbf{b}_o; \mathbf{g}\right). \tag{19.25}$$

Let the decision variables that are solutions to the network directional distance function defined in (19.11) be represented as β^*, $\mathbf{s}^{x*}$, $\mathbf{s}^{y*}$, and $\boldsymbol{\lambda}^*$. By setting $\mathbf{s}_n^- = \beta^*\mathbf{g}^x + \mathbf{s}^{x*}$, $\mathbf{s}_m^+ = \beta^*\mathbf{g}^y + \mathbf{s}^{y*}$ and $\mathbf{s}_l^\# = \beta^*\mathbf{g}^b$, and letting the intensity variables, λ_j^k in (19.19) equal the optimal intensity variables λ_j^{k*} from (19.11) for all j and $k = 1,2$, the objective function of (19.19) becomes

$$\begin{aligned}\frac{\frac{1}{N}\sum_{n=1}^{N}\frac{s_n^-}{g_n^x}+\frac{1}{M}\sum_{m=1}^{M}\frac{s_m^+}{g_m^y}+\frac{1}{L}\sum_{l=1}^{L}\frac{s_l^\#}{g_l^b}}{3} &= \frac{\frac{1}{N}\sum_{n=1}^{N}\frac{\beta^* g_n^x + s_n^{x*}}{g_n^x}+\frac{1}{M}\sum_{m=1}^{M}\frac{\beta^* g_m^y + s_m^{y*}}{g_m^y}+\frac{1}{L}\sum_{l=1}^{L}\frac{\beta^* g_l^b}{g_l^b}}{3} \\ &= \beta^* + \frac{\frac{1}{N}\sum_{n=1}^{N}\frac{s_n^{x*}}{g^x}+\frac{1}{M}\sum_{m=1}^{M}\frac{s_m^{y*}}{g_m^y}}{3}\end{aligned} \tag{19.26}$$

and all the constraints of (19.11) satisfy

$$\mathbf{x}_o = \sum_{j=1}^{J}\mathbf{x}_j\lambda_j^{1*} + \mathbf{s}^{-*}, \quad \mathbf{y}_o = \sum_{j=1}^{J}\mathbf{y}_j\lambda_j^{2*} - \mathbf{s}^{+*}, \quad \text{and } \mathbf{b}_o = \sum_{j=1}^{J}\mathbf{b}_j\lambda_j^{2*} + \mathbf{s}^{\#*}. \tag{19.27}$$

It follows that $NSBI(\mathbf{x}_o, \mathbf{y}_o, \mathbf{b}_o; \mathbf{g}) \geq \beta^* = N\vec{D}(\mathbf{x}_o, \mathbf{y}_o, \mathbf{b}_o; \mathbf{g})$. If $NSBI(\mathbf{x}_o,\mathbf{y}_o,\mathbf{b}_o;\mathbf{g})$ equals $N\vec{D}(\mathbf{x}_o, \mathbf{y}_o, \mathbf{b}_o; \mathbf{g})$, then $SlackBias(\mathbf{x}_o,\mathbf{y}_o,\mathbf{b}_o;\mathbf{g}) = 0$ and the network directional distance function projects the outputs and inputs to the strongly efficient frontier. When $NSBI(\mathbf{x}_o,\mathbf{y}_o,\mathbf{b}_o;\mathbf{g})$ is greater than $N\vec{D}(\mathbf{x}_o, \mathbf{y}_o, \mathbf{b}_o; \mathbf{g})$, there is at least one of the constraints defining the network technology where slack remains and $SlackBias(\mathbf{x}_o,\mathbf{y}_o,\mathbf{b}_o;\mathbf{g}) > 0$. Therefore, $SlackBias(\mathbf{x}_o,\mathbf{y}_o,\mathbf{b}_o;\mathbf{g})$ measures the amount that the network directional distance function underestimates inefficiency. A similar indicator of slack bias could be constructed for the black-box technology by estimating the amount that $BSBI(\mathbf{x}_o,\mathbf{y}_o,\mathbf{b}_o;\mathbf{g})$ is greater than $B\vec{D}(\mathbf{x}_o, \mathbf{y}_o, \mathbf{b}_o; \mathbf{g})$.

To investigate the second kind of bias-ignoring a network structure-let the optimal solution vector to the black-box slacks-based inefficiency, $BSBI(\mathbf{x}_o,\mathbf{y}_o,\mathbf{b}_o;\mathbf{g})$, be $(\mathbf{s}^{Bx-*},\mathbf{s}^{By+*},\mathbf{s}^{Bb*},\boldsymbol{\lambda}^*)$, where $\boldsymbol{\lambda}^* = (\lambda_1^*, \ldots \lambda_J^*)$ is the vector of intensity variables that defines the black-box technology. For the same DMU let the optimal solution vector to $NSBI(\mathbf{x}_o,\mathbf{y}_o,\mathbf{b}_o;\mathbf{g})$ be $(\mathbf{s}^{Nx-*},\mathbf{s}^{Nx+*},\mathbf{s}^{Nb*},\boldsymbol{\lambda}^{1*},\boldsymbol{\lambda}^{1*})$, where $\boldsymbol{\lambda}^{1*} = (\lambda_1^{1*}, \ldots,\lambda_J^{1*})$ and $\boldsymbol{\lambda}^{2*} = (\lambda_1^{2*}, \ldots,\lambda_J^{2*})$ are the optimal intensity vectors for stage 1 and stage 2 of the network technology. A feasible, but not necessarily optimal solution to $NSBI(\mathbf{x}_o,\mathbf{y}_o,\mathbf{b}_o;\mathbf{g})$ is found by choosing $\boldsymbol{\lambda}^1 = \boldsymbol{\lambda}^2 = \boldsymbol{\lambda}^*$. Since other choices of $\boldsymbol{\lambda}^1$ and $\boldsymbol{\lambda}^2$ that satisfy $\sum_{j=1}^{J} \mathbf{z}_j \left(\lambda_j^1 - \lambda_j^1\right) \geq \mathbf{0}$ are possible, $NSBI(\mathbf{x}_o,\mathbf{y}_o,\mathbf{b}_o;\mathbf{g})$ is greater than or equal to $BSBI(\mathbf{x}_o,\mathbf{y}_o,\mathbf{b}_o;\mathbf{g})$. We define the bias in the black-box estimate of inefficiency relative to the network estimate of inefficiency as

$$NetBias(\mathbf{x}_o, \mathbf{y}_o, \mathbf{b}_o; \mathbf{g}) = NSBI(\mathbf{x}_o, \mathbf{y}_o, \mathbf{b}_o; \mathbf{g}) - BSBI(\mathbf{x}_o, \mathbf{y}_o, \mathbf{b}_o; \mathbf{g}) \geq 0 \quad (19.28)$$

for $(\mathbf{x}_o,\mathbf{y}_o,\mathbf{b}_o) \in BT$. This indicator shows the bias in black-box slacks-based inefficiency that arises from ignoring an existing network structure.

From (19.25) and (19.28), we obtain the following network slacks-based inefficiency decompositions:

$$\begin{aligned} NSBI(\mathbf{x}_o, \mathbf{y}_o, \mathbf{b}_o; \mathbf{g}) &= SlackBias(\mathbf{x}_o, \mathbf{y}_o, \mathbf{b}_o; \mathbf{g}) + N\vec{D}(\mathbf{x}_o, \mathbf{y}_o, \mathbf{b}_o; \mathbf{g}) \\ NSBI(\mathbf{x}_o, \mathbf{y}_o, \mathbf{b}_o; \mathbf{g}) &= NetBias(\mathbf{x}_o, \mathbf{y}_o, \mathbf{b}_o; \mathbf{g}) + BSBI(\mathbf{x}_o, \mathbf{y}_o, \mathbf{b}_o; \mathbf{g}). \end{aligned} \quad (19.29)$$

For managers of a company, $SlackBias(\mathbf{x}_o,\mathbf{y}_o,\mathbf{b}_o;\mathbf{g})$ gives information on whether or not any input can be contracted or output expanded given a network structure. For example, bank managers who use benchmarking to evaluate their bank's performance might note that it is not possible to further expand outputs and simultaneously contract inputs and non-performing loans if $N\vec{D}(\mathbf{x}_o, \mathbf{y}_o, \mathbf{b}_o; \mathbf{g}) = 0$. On the other hand, if $SlackBias(\mathbf{x}_o,\mathbf{y}_o,\mathbf{b}_o;\mathbf{g}) > 0$, then the manager knows that even though they are on the weakly efficient frontier it is possible to expand at least one output or contract at least one input and this information can lead to further efficiency gains.

Similarly, an inefficiency measure that accounts for all slacks, such as *BSBI* ($\mathbf{x}_o$,$\mathbf{y}_o$,$\mathbf{b}_o$;$\mathbf{g}$), still provides no information on efficiency gains that might be realized by accounting for a network structure to the bank technology. For managers, positive values of *NetBias*($\mathbf{x}_o$,$\mathbf{y}_o$,$\mathbf{b}_o$;$\mathbf{g}$) tell them that they should look for efficiency gains by investigating the linkage between various divisions or stages of production within the bank and not focus solely on one specific output or input.

19.4 Budget-Constrained Inefficiency

19.4.1 Black-Box and Network Budget-Constrained Output Sets and Inefficiency

Sometimes producers have a fixed budget with which to hire inputs but have discretion over which inputs to hire. Different choices of inputs will yield different output possibility sets. Let the (direct) output possibility set be defined as

$$BP(\mathbf{x}) = \left\{(\mathbf{y},\mathbf{b})\middle|\mathbf{x} \in \Re^N_+ \text{ can produce } (\mathbf{y},\mathbf{b}) \in \Re^{M+L}_+\right\}. \tag{19.30}$$

The directional output distance function takes the form

$$B\vec{D}_O\left(\mathbf{x},\mathbf{y},\mathbf{b};\mathbf{g}^{yb}\right) = \sup\left\{\beta\middle|\left(\mathbf{y}+\beta\mathbf{g}^y,\mathbf{b}-\beta\mathbf{g}^b\right) \in BP(\mathbf{x})\right\}. \tag{19.31}$$

Following Shephard (1970, 1974), the budget-constrained (indirect) output set is defined as

$$BIP(\mathbf{p}/\alpha) = \left\{(\mathbf{y},\mathbf{b})\middle|(\mathbf{y},\mathbf{b}) \in BP(\mathbf{x}) \text{ and } \sum\nolimits_{n=1}^{N}(p_n/\alpha)x_n \leq 1\right\} \tag{19.32}$$

where input prices are represented by $\mathbf{p} = (p_1, \ldots, p_N) \in R^N_+$, $\alpha = \mathbf{px}$ is a positive budget used to hire inputs, and *BIP*($\mathbf{p}$/α) is the set of outputs that can be produced given the fixed budget and input prices. The budget-constrained output set can also be expressed as

$$BIP(\mathbf{p}/\alpha) = \{(\mathbf{y},\mathbf{b})|BC(\mathbf{p}/\alpha,\mathbf{y},\mathbf{b}) \leq 1\} \tag{19.33}$$

where

$$BC(\mathbf{p}/\alpha,\mathbf{y},\mathbf{b}) = \frac{1}{\alpha} \times \min_{\mathbf{x}} \ \{\mathbf{px}|(\mathbf{x},\mathbf{y},\ \mathbf{b}) \in BT, \quad \mathbf{x} \geq \mathbf{0}\}. \tag{19.34}$$

is the black-box cost efficiency measure. Färe and Primont (1995) showed that for $\sum_{n=1}^{N}(p_n/\alpha)x_n \leq 1$ that

$$BIP(\mathbf{p}/\alpha) = \underset{(\mathbf{p}/\alpha)\mathbf{x}\leq 1}{\cup} BP(\mathbf{x}). \tag{19.35}$$

Note also that $BP(\mathbf{x}) \subseteq BIP(\mathbf{p}/\alpha)$ for $\sum_{n=1}^{N}(p_n/\alpha)x_n \leq 1$. The relation given in (19.35) holds because there might be a different set of inputs that can be hired at the same cost as the observed inputs but be capable of producing even more output. Such a situation occurs when the observed input bundle is not allocatively efficient. Relative to the budget-constrained output set, $BIP(\mathbf{p}/\alpha)$, the budget-constrained directional output distance function takes the form

$$BI\vec{D}_O(\mathbf{p}/\alpha, \mathbf{y}, \mathbf{b}; \mathbf{g}^{yb}) = \sup\{\varphi | (\mathbf{y} + \varphi\mathbf{g}^y, \mathbf{b} - \varphi\mathbf{g}^b) \in BIP(\mathbf{p}/\alpha)\}. \tag{19.36}$$

Similar to the black-box direct output set (19.30) and budget-constrained output set (19.32), the network direct output set and network budget-constrained output set are defined as

$$NP(\mathbf{x}) = \{(\mathbf{y}, \mathbf{b}) | \mathbf{x} \text{ can produce } \mathbf{z} \text{ and } \mathbf{z} \text{ can produce } (\mathbf{y}, \mathbf{b})\} \text{ and} \tag{19.37}$$

$$\begin{aligned} NIP(\mathbf{p}/\alpha) &= \left\{(\mathbf{y}, \mathbf{b}) \middle| (\mathbf{y}, \mathbf{b}) \in NP(\mathbf{x}) \text{ and } \sum\nolimits_{n=1}^{N} (p_n/\alpha)x_n \leq 1\right\} \\ &= \{(\mathbf{y}, \mathbf{b}) | \; NC(\mathbf{p}/\alpha, \mathbf{y}, \mathbf{b}) \leq 1\} \end{aligned} \tag{19.38}$$

where

$$NC(\mathbf{p}/\alpha, \mathbf{y}, \mathbf{b}) = \frac{1}{\alpha} \times \min_{\mathbf{x}} \; \{\mathbf{px} | \; (\mathbf{x}, \mathbf{y}, \; \mathbf{b}) \in NT, \quad \mathbf{x} \geq \mathbf{0}\} \tag{19.39}$$

is the network cost efficiency measure (Fukuyama and Matousek 2011). The associated network directional output distance functions take the form

$$N\vec{D}_O(\mathbf{x}, \mathbf{y}, \mathbf{b}; \mathbf{g}^{yb}) = \sup\{\beta | (\mathbf{y} + \beta\mathbf{g}^y, \mathbf{b} - \beta\mathbf{g}^b) \in NP(\mathbf{x})\} \text{ and} \tag{19.40}$$

$$NI\vec{D}_O(\mathbf{p}/\alpha, \mathbf{y}, \mathbf{b}; \mathbf{g}^{yb}) = \sup\{\varphi | (\mathbf{y} + \varphi\mathbf{g}^y, \mathbf{b} - \varphi\mathbf{g}^b) \in NIP(\mathbf{p}/\alpha)\} \tag{19.41}$$

where subscript "O" in (19.40) and (19.41) indicates that performance is gauged for outputs holding inputs constant; i.e., $\mathbf{g}^x = 0$.

Gauging performance based on the network budget-constrained output set is often appropriate when DMUs do not seek to maximize profits or revenues. Such instances can occur for organizations in the public sector such as schools (Grosskopf et al. 1997) or non-profit or cooperative financial institutions such as Japanese credit cooperatives (Fukuyama et al. 1999) or in major professional sports leagues like the National Football League or National Basketball Association where each individual team faces a salary cap. In these cases the organizations may not seek to minimize costs, but instead seek to maximize desirable outputs and minimize undesirable outputs given their fixed budget. However, the organizations can choose the level and mix of inputs and in turn, performance depends on the outputs produced relative to maximum potential outputs. Comparing DMU efficiency relative to $BP(\mathbf{x})$ and $BIP(\mathbf{p}/\alpha)$ for the black-box model or $NP(\mathbf{x})$ and $NIP(\mathbf{p}/\alpha)$ for the network model provides an estimate of the lost output due to a misallocation of inputs, holding the total cost of production constant (Fukuyama and Weber 2009b, 2013).

19.4.2 DEA Implementation of Budget-Constrained Inefficiency

Estimates of the direct and budget-constrained directional output distance functions for the black-box and network technologies presented in the previous subsection can be obtained using DEA. The black-box output possibility set takes the form

$$BP(\mathbf{x}) = \left\{ (\mathbf{y}, \mathbf{b}) \middle| \sum_{j=1}^{J} \mathbf{x}_j\lambda_j \leq \mathbf{x};\ \sum_{j=1}^{J} \mathbf{y}_j\lambda_j \geq \mathbf{y};\ \sum_{j=1}^{J} \mathbf{b}_j\lambda_j = \mathbf{b};\quad \boldsymbol{\lambda} \geq \mathbf{0} \right\} \tag{19.42}$$

and the black-box budget-constrained output possibility set takes the form

$$BIP(\mathbf{p}/\alpha) = \left\{ (\mathbf{y}, \mathbf{b}) \middle| \sum_{j=1}^{J} \mathbf{x}_j\lambda_j \leq \mathbf{x};\ \sum_{j=1}^{J} \mathbf{y}_j\lambda_j \geq \mathbf{y};\ \sum_{j=1}^{J} \mathbf{b}_j\lambda_j = \mathbf{b};\ \sum_{n=1}^{N} p_n x_n \leq \alpha;\quad \boldsymbol{\lambda} \geq \mathbf{0} \right\} \tag{19.43}$$

The black-box directional output distance function defined on (19.42) is

$$\begin{aligned} B\vec{D}_O\left(\mathbf{x}_o, \mathbf{y}_o, \mathbf{b}_o; \mathbf{g}^{yb}\right) = \max_{\boldsymbol{\lambda},\ \beta} \Bigg\{ \beta \Bigg| & \sum_{j=1}^{J} \mathbf{x}_j\lambda_j \leq \mathbf{x}_o;\ \sum_{j=1}^{J} \mathbf{y}_j\lambda_j \geq \mathbf{y}_o + \beta\mathbf{g}^y; \\ & \sum_{j=1}^{J} \mathbf{b}_j\lambda_j = \mathbf{b}_o - \beta\mathbf{g}^b;\quad \boldsymbol{\lambda} \geq \mathbf{0};\quad \beta : free \Bigg\} \end{aligned} \tag{19.44}$$

and the budget-constrained directional output distance function defined on (19.43) is

$$\begin{aligned} BI\vec{D}_O\left(\mathbf{p}_o/\alpha_o, \mathbf{y}_o, \mathbf{b}_o\right) = \max_{\boldsymbol{\lambda},\ \mathbf{x},\ \beta} \Bigg\{ \beta \Bigg| & \sum_{j=1}^{J} \mathbf{x}_j\lambda_j \leq \mathbf{x};\ \mathbf{p}_o\mathbf{x} \leq \alpha_o;\ \sum_{j=1}^{J} \mathbf{y}_j\lambda_j \geq \mathbf{y}_o + \beta\mathbf{g}^y; \\ & \sum_{j=1}^{J} \mathbf{b}_j\lambda_j = \mathbf{b}_o - \beta\mathbf{g}^b;\quad \boldsymbol{\lambda} \geq \mathbf{0};\ \mathbf{x} \geq \mathbf{0};\quad \beta : free \Bigg\}. \end{aligned} \tag{19.45}$$

In (19.44), the inputs for DMU_o are taken as given and the desirable and undesirable outputs are projected to the frontier. In contrast, in (19.45), the inputs are optimally chosen so as to allow the maximum expansion in desirable outputs and contraction in undesirable outputs.

The DEA-based network output possibility set is represented by

$$NP(\mathbf{x}) = \left\{ (\mathbf{y}, \mathbf{b}) \middle| \begin{array}{l} \sum_{j=1}^{J} \mathbf{x}_j \lambda_j^1 \leq \mathbf{x}; \quad \sum_{j=1}^{J} \mathbf{y}_j \lambda_j^2 \geq \mathbf{y}; \quad \sum_{j=1}^{J} \mathbf{b}_j \lambda_j^2 = \mathbf{b}; \\ \sum_{j=1}^{J} \mathbf{z}_j \left(\lambda_j^1 - \lambda_j^2\right) \geq \mathbf{0}; \quad \boldsymbol{\lambda}^1 \geq \mathbf{0}; \quad \boldsymbol{\lambda}^2 \geq \mathbf{0} \end{array} \right\} \tag{19.46}$$

and the network budget-constrained output possibility set is

$$NIP(\mathbf{p}_o/\alpha_o) = \left\{ (\mathbf{y}, \mathbf{b}) \middle| \begin{array}{l} \sum_{j=1}^{J} \mathbf{x}_j \lambda_j^1 \leq \mathbf{x}, \ \ \mathbf{p}_o \mathbf{x} \leq \alpha_o; \quad \sum_{j=1}^{J} \mathbf{y}_j \lambda_j^2 \geq \mathbf{y}; \quad \sum_{j=1}^{J} \mathbf{b}_j \lambda_j^2 = \mathbf{b}; \\ \sum_{j=1}^{J} \mathbf{z}_j \left(\lambda_j^1 - \lambda_j^2\right) \geq \mathbf{0}; \quad \boldsymbol{\lambda}^1 \geq \mathbf{0}; \ \boldsymbol{\lambda}^2 \geq \mathbf{0}; \ \mathbf{x} \geq \mathbf{0} \end{array} \right\} \tag{19.47}$$

where input prices are allowed to vary across DMUs. We employ a network directional output distance function to measure technical inefficiency. This distance function measures the extra desirable output and contraction in undesirable output that can be produced for a given directional vector, $\mathbf{g}^{yb}$. The network directional output distance function is

$$\begin{aligned} & N\vec{D}_O\left(\mathbf{x}_o, \mathbf{y}_o, \mathbf{b}_o; \mathbf{g}^{yb}\right) \\ & = \max_{\boldsymbol{\lambda}^1, \boldsymbol{\lambda}^2, \beta} \left\{ \beta \middle| \begin{array}{l} \sum_{j=1}^{J} \mathbf{x}_j \lambda_j^1 \leq \mathbf{x}_o; \quad \sum_{j=1}^{J} \mathbf{y}_j \lambda_j^2 \geq \mathbf{y}_o + \beta \mathbf{g}^y; \quad \sum_{j=1}^{J} \mathbf{b}_j \lambda_j^2 = \mathbf{b}_o - \beta \mathbf{g}^b; \\ \sum_{j=1}^{J} \mathbf{z}_j \left(\lambda_j^1 - \lambda_j^2\right) \geq \mathbf{0}; \quad \boldsymbol{\lambda}^1 \geq \mathbf{0}; \ \boldsymbol{\lambda}^2 \geq \mathbf{0}; \quad \beta : free \end{array} \right\}. \end{aligned} \tag{19.48}$$

This distance function measures output technical inefficiency. If $N\vec{D}_O\left(\mathbf{x}_o, \mathbf{y}_o, \mathbf{b}_o; \mathbf{g}^{yb}\right) = 0$ and $(\mathbf{y}_o, \mathbf{b}_o)$ is not dominated by any other activities in $NP(\mathbf{x}_o)$, then DMU_o is weakly efficient relative to the output possibility set $NP(\mathbf{x}_o)$. Values of $N\vec{D}_O\left(\mathbf{x}_o, \mathbf{y}_o, \mathbf{b}_o; \mathbf{g}^{yb}\right) > 0$ indicate inefficiency. The dual to (19.48) is

$$\min_{\mathbf{v},\mathbf{w},\mathbf{u},\mathbf{r}} \left\{ \mathbf{v}\mathbf{x}_o - \mathbf{u}\mathbf{y}_o + \mathbf{r}\mathbf{b}_o \left| \begin{array}{l} \mathbf{u}\mathbf{g}^y + \mathbf{r}\mathbf{g}^b = 1; \;\; \mathbf{v}\mathbf{x}_j - \mathbf{w}\mathbf{z}_j \geq \mathbf{0} \;\; \forall j; \\ \mathbf{w}\mathbf{z}_j - \mathbf{u}\mathbf{y}_j + \mathbf{r}\mathbf{b}_j \geq \mathbf{0} \;\; \forall j; \\ \mathbf{v} \geq \mathbf{0}; \;\; \mathbf{w} \geq \mathbf{0}; \;\; \mathbf{u} \geq \mathbf{0}; \;\; \mathbf{r} : free \end{array} \right. \right\}. \tag{19.49}$$

If DMU$_o$ is allowed to choose inputs and intermediate outputs, performance can be measured by the network budget constrained directional output distance function defined on (19.47) as:

$$NI\vec{D}_O\left(\mathbf{p}_o/\alpha_o, \mathbf{y}_o, \mathbf{b}_o; \mathbf{g}^{yb}\right) = \max_{\boldsymbol{\lambda}^1, \boldsymbol{\lambda}^2, \mathbf{x}, \varphi} \left\{ \varphi \left| \begin{array}{l} \sum_{j=1}^{J} \mathbf{x}_j \lambda_j^1 \leq \mathbf{x}, \;\; \mathbf{p}_o \mathbf{x} \leq \alpha_o; \;\; \sum_{j=1}^{J} \mathbf{y}_j \lambda_j^2 \geq \mathbf{y}_o + \varphi \mathbf{g}^y; \\ \sum_{j=1}^{J} \mathbf{b}_j \lambda_j^2 = \mathbf{b}_o - \varphi \mathbf{g}^b; \\ \sum_{j=1}^{J} \mathbf{z}_j \left(\lambda_j^1 - \lambda_j^2\right) \geq \mathbf{0}; \;\; \boldsymbol{\lambda}^1 \geq \mathbf{0}; \;\; \boldsymbol{\lambda}^2 \geq \mathbf{0}; \\ \mathbf{x} \geq \mathbf{0}; \;\; \varphi : free \end{array} \right. \right\}, \tag{19.50}$$

which is based on the budget-constrained (indirect) network directional output distance function. If $NI\vec{D}_O\left(\mathbf{p}_o/\alpha_o, \mathbf{y}_o, \mathbf{b}_o; \mathbf{g}^{yb}\right) = 0$ and ($\mathbf{y}_o$,$\mathbf{b}_o$) is not dominated by any other activities in $NIP(\mathbf{p}_o/\alpha_o)$, then DMU$_o$ produces on the budget-constrained frontier of (19.47) and is weakly efficient for the directional vector $\mathbf{g}^{yb}$. A DMU is inefficient if $NI\vec{D}_O\left(\mathbf{p}_o^x/\alpha_o, \mathbf{y}_o, \mathbf{b}_o; \mathbf{g}^{yb}\right) > 0$, with larger values indicating greater inefficiency. The dual to (19.50) is

$$\min_{\mathbf{v},\mathbf{w},\mathbf{u},\mathbf{r},\zeta} \left\{ \zeta - \mathbf{u}\mathbf{y}_o + \mathbf{r}\mathbf{b}_o \left| \begin{array}{l} \mathbf{u}\mathbf{g}^y + \mathbf{r}\mathbf{g}^b = 1; \;\; \mathbf{v}\mathbf{x}_j - \mathbf{w}\mathbf{z}_j \geq \mathbf{0} \;\; \forall j; \\ \mathbf{w}\mathbf{z}_j - \mathbf{u}\mathbf{y}_j + \mathbf{r}\mathbf{b}_j \geq \mathbf{0} \;\; \forall j; \;\; - v_n + \zeta(p_n/\alpha_o) \geq 0 \;\; \forall n; \\ \mathbf{v} \geq \mathbf{0}; \;\; \mathbf{w} \geq \mathbf{0}; \;\; \mathbf{u} \geq \mathbf{0}; \;\; \mathbf{r} : free; \;\; \zeta \geq 0 \end{array} \right. \right\}. \tag{19.51}$$

Since the network budget-constrained output set allows inputs to be chosen given a budget it allows more alternative output vectors to be produced so $NIP(\mathbf{p}_o/\alpha_o) \supseteq NP(\mathbf{x}_o)$ for the vector ($\mathbf{p}_o/\alpha_o$, $\mathbf{x}_o$) satisfying $\sum_{n=1}^{N} (p_{no}/\alpha_o) x_n \leq 1$. Therefore it is also the case that

$$NI\vec{D}_O\left(\mathbf{p}_o/\alpha_o, \mathbf{x}_o, \mathbf{y}_o, \mathbf{b}_o; \mathbf{g}^{yb}\right) \geq N\vec{D}_O\left(\mathbf{x}_o, \mathbf{y}_o; \mathbf{g}^{yb}\right). \tag{19.52}$$

Let the difference between the left and right-hand sides of (19.52) represent the loss of output due to a misallocation of inputs. The network output loss indicator equals

$$NOL(\mathbf{x}_o, \mathbf{y}_o, \mathbf{b}_o; \mathbf{g}^{yb}) = NI\vec{D}_O(\mathbf{p}_o/\alpha_o, \mathbf{y}_o, \mathbf{b}_o; \mathbf{g}^{yb}) - N\vec{D}_O(\mathbf{x}_o, \mathbf{y}_o, \mathbf{b}_o; \mathbf{g}^{yb}) \quad (19.53)$$

which can be arranged as

$$\begin{aligned} NI\vec{D}_O(\mathbf{p}_o/\alpha_o, \mathbf{y}_o, \mathbf{b}_o; \mathbf{g}^{yb}) = {} & NOL(\mathbf{p}_o/\alpha_o, \mathbf{x}_o, \mathbf{y}_o, \mathbf{b}_o; \mathbf{g}^{yb}) \\ & + N\vec{D}_O(\mathbf{x}_o, \mathbf{y}_o, \mathbf{b}_o; \mathbf{g}^{yb}) \end{aligned} \quad (19.54)$$

Therefore, overall network inefficiency consists of two parts: technical inefficiency caused by the failure to maximize final outputs given inputs, $N\vec{D}_O(\mathbf{x}_o, \mathbf{y}_o, \mathbf{b}_o; \mathbf{g}^{yb})$, and the lost output due to a misallocation of inputs given the budget and input prices, $NOL(\mathbf{p}_o/\alpha_o, \mathbf{x}_o, \mathbf{y}_o, \mathbf{b}_o; \mathbf{g}^{yb})$.

Table A.2 provides DEA estimates of the black-box and network budget constrained directional distance functions. In addition the estimates show that $I\vec{D}_O(\mathbf{p}_o/\alpha_o, \mathbf{x}_o, \mathbf{y}_o, \mathbf{b}_o; \mathbf{g}^{yb}) \geq B\vec{D}_O(\mathbf{x}_o, \mathbf{y}_o, \mathbf{b}_o; \mathbf{g}^{yb})$ and that $NI\vec{D}_O(\mathbf{p}_o/\alpha_o, \mathbf{x}_o, \mathbf{y}_o, \mathbf{b}_o; \mathbf{g}^{yb}) \geq N\vec{D}_O(\mathbf{x}_o, \mathbf{y}_o, \mathbf{b}_o; \mathbf{g}^{yb})$ since the black-box and network output sets are contained within the respected budget-constrained black-box technology and budget-constrained network technology.

Slacks-based measures of performance evaluated relative to the direct and budget constrained output possibility sets can also be constructed. We leave that as an exercise for the reader.

19.5 Summary and Conclusions

Black-box models of a production technology assume that inputs enter and outputs emerge from a metaphorical black-box. Two-stage network models are a way of partially opening the black-box. In network models, producers use inputs in a first stage of production to produce an intermediate output that becomes an input used to produce final outputs in the second stage. We employed directional distance functions to measure the performance of producers for both kinds of technologies. Directional distance functions can be easily estimated using DEA and can allow for non-radial projections of inputs and outputs to the frontier technology. Directional distance functions are also particularly useful when undesirable outputs such as pollution or bad loans are jointly produced as by-products along with desirable outputs. The models we develop all assumed that these undesirable outputs are part of the final outputs produced in stage 2.

Recent work by Akther et al. (2013) has extended the network models presented in this chapter to a dynamic framework where decisions made in one period impact future period's production possibilities. Specifically, Akther et al. (2013) allowed the undesirable outputs produced in stage 2 in one period to become an undesirable input to stage 1 in a subsequent period. In the context of production by banks, such a model is useful because non-performing loans become a drag on the fiscal position of banks in future periods and regulatory requirements generally require larger amounts of financial equity capital to offset those non-performing loans. Thus, producers might want to consider how to allocate a fixed amount of input between various production periods so as to maximize the size of a dynamic output possibility set. Färe et al. (2012) have considered these kinds of time substitution opportunities.

Acknowledgement This research is partially supported by the Grants-in-aid for scientific research, fundamental research (B) 19310098 and (C) 23510165, the Japan Society for the Promotion of Science.

Appendix

The synthetic data set in Table A.1 is used to estimate the black-box and two-stage network performance indicators presented in the paper. In the first stage $N = 3$ inputs are used to produce $Q = 2$ intermediate outputs. In the second stage the $Q = 2$ intermediate outputs become inputs to produce $M = 3$ desirable outputs and $L = 1$ undesirable output. We choose a directional vector of $g = (g_1^x, g_2^x, g_3^x, g_1^y, g_2^y, g_3^y, g_1^b) = (1,1,1,1,1,1,1)$ to estimate each of the black-box and network performance indicators given in (19.8), (19.11), (19.16), and (19.19). We choose $g = (g_1^y, g_2^y, g_3^y, g_1^b) = (1,1,1,1)$ for the black-box and network directional output distance functions given by (19.44), (19.45), (19.48), and (19.50). Estimates were obtained using GAMS (Generalized Algebraic Modeling System) with the Minos solver.

References

Akther, S., Fukuyama, H., & Weber, W. L. (2013). Estimating two-stage network slacks-based inefficiency: An application to Bangladesh banking. *Omega: The International Journal of Management Science, 41*, 88–96.

Barnett, W., & Hahm, J.-H. (1994). Financial firm production of monetary services: A generalized symmetric Barnett variable profit function approach. *Journal of Business and Economics Statistics, 12*, 33–46.

Berger, A. N., & Humphrey, D. B. (1997). Efficiency of financial institutions: International survey and directions for future research. *European Journal of Operational Research, 98*(2), 175–212.

Chambers, R. G., Chung, Y., & Färe, R. (1996). Benefit and distance functions. *Journal of Economics Theory, 70*, 407–419.

Chambers, R. G., Chung, Y., & Färe, R. (1998). Profit, directional distance functions and Nerlovian efficiency. *Journal of Optimization Theory and Applications, 98*(2), 351–364.

Charnes, A., Cooper, W. W., & Rhodes, E. (1978). Measuring the efficiency of decision making units. *European Journal of Operational Research, 2*, 429–444.

Chen, Y., Cook, W. D., Li, N., & Zhu, J. (2009a). Additive efficiency decomposition in two-stage DEA. *European Journal of Operational Research, 196*, 1170–1176.

Chen, Y., Liang, L., & Zhu, J. (2009b). Equivalence in two-stage DEA approaches. *European Journal of Operational Research, 193*(2), 600–604.

Chen, Y., Cook, W. D., & Zhu, J. (2010). Deriving the DEA frontier for two-stage processes. *European Journal of Operational Research, 202*(1), 138–142.

Chen, Y., & Zhu, J. (2004). Measuring information technology's indirect impact on firm performance. *Information Technology & Management Journal, 5*(1–2), 9–22.

Färe, R., & Grosskopf, S. (1996). Productivity and intermediate products: A frontier approach. *Economics Letters, 50*, 65–70.

Färe, R., & Grosskopf, S. (2000). Network DEA. *Socio-Economic Planning Sciences, 34*, 35–49.

Färe, R., & Primont, D. (1995). *Multi-output production and duality: Theory and applications.* Norwell: Kluwer Academic.

Färe, R., Grosskopf, S., Lovell, C. A. K., & Pasurka, C. (1994). Multilateral productivity comparisons when some outputs are undesirable: A non-parametric approach. *Review of Economics and Statistics, 71*(1), 90–98.

Färe, R., Grosskopf, S., Noh, D.-W., & Weber, W. (2005). Characteristics of a polluting technology: Theory and practice. *Journal of Econometrics, 126*, 469–492.

Färe, R., Grosskopf, S., Margaritis, D., & Weber, W. L. (2012). Technological change and timing reductions in greenhouse gas emissions. *Journal of Productivity Analysis, 37*, 205–216.

Fukuyama, H., & Matousek, R. (2011). Efficiency of Turkish banking: Two-stage system variable returns to scale model. *Journal of International Financial Markets, Institutions & Money, 21*(1), 75–91.

Fukuyama, H., & Mirdehghan, S. M. (2012). Identifying the efficiency status in network DEA. *European Journal of Operational Research, 220*(1), 85–92.

Fukuyama, H., & Weber, W. L. (2008a). Japanese banking inefficiency and shadow pricing. *Mathematical and Computer Modelling, 71*(11/12), 1854–1867.

Fukuyama, H., & Weber, W. L. (2008b). Estimating inefficiency, technological change and shadow prices of problem loans for regional banks and Shinkin Banks in Japan. *The Open Management Journal, 1*, 1–11.

Fukuyama, H., & Weber, W. L. (2009a). A directional slacks-based measure of technical inefficiency. *Socio-Economic Planning Sciences, 43*(4), 274–287.

Fukuyama, H., & Weber, W. L. (2009b). Estimating indirect allocative inefficiency and productivity change. *Journal of the Operational Research Society, 60*(11), 1594–1608.

Fukuyama, H., & Weber, W. L. (2010). A slacks-based inefficiency measure for a two-stage system with bad outputs. *Omega: International Journal of Management Science, 38*, 398–409.

Fukuyama, H., & Weber, W. L. (2012). Estimating two-stage technology inefficiency. *International Journal of Operations Research and Information Systems, 3*(2), 1–23.

Fukuyama, H., & Weber, W. L. (2013, in press). Network performance of Japanese credit cooperatives, 2004–2007. *International Journal of Information Technology & Decision Making.*

Fukuyama, H., Guerra, R., & Weber, W. L. (1999). Efficiency and ownership: Evidence from Japanese credit cooperatives. *Journal of Economics and Business, 51*(6), 473–487.

Grosskopf, S., Hayes, K., Taylor, L., & Weber, W. L. (1997). Budget constrained frontier measures of fiscal equality and efficiency in schooling. *Review of Economics and Statistics, 79*, 116–124.

Hancock, D. (1985). The financial firm: Production with monetary and nonmonetary goods. *Journal of Political Economy, 93*, 859–880.

Kao, C., & Hwang, S. N. (2008). Efficiency decomposition in two-stage data envelopment analysis: An application to non-life insurance companies in Taiwan. *European Journal of Operational Research, 185*(1), 418–429.

Luenberger, D. G. (1992). Benefit functions and duality. *Journal of Mathematical Economics, 21*, 461–481.

Luenberger, D. G. (1995). *Microeconomic theory.* New York: McGraw-Hill.

Murty, S., Russell, R. R., & Levkoff, S. B. (2012). On modeling pollution-generating technologies. *Journal of Environmental Economics and Management, 64*(1), 117–135.

Pastor, J. T., Ruiz, J. L., & Sirvent, I. (1999). An enhanced Russell graph efficiency measure. *European Journal of Operational Research, 115*, 596–607.

Sealey, C., & Lindley, J. (1977). Inputs, outputs, and a theory of production and cost at depository financial institutions. *Journal of Finance, 32*, 1251–1266.

Seiford, L. M., & Zhu, J. (1999). Profitability and marketability of the top 55 US commercial banks. *Management Science, 45*(9), 1270–1288.

Sexton, T. R., & Lewis, H. F. (2003). Two-stage DEA: An application to major league baseball. *Journal of Productivity Analysis, 19*, 227–249.

Shephard, R. W. (1953). *Cost and production functions*. Princeton: Princeton University Press.

Shephard, R. W. (1970). *Theory of cost and production functions*. Princeton: Princeton University Press.

Shephard, R. W. (1974). *Indirect production functions*. Meisenheim Am Glan: Verlag Anton Hain.

Shephard, R. W., & Färe, R. (1974). The law of diminishing returns. *Zeitschrift für Nationalökonomie, 34*, 69–90.

Tone, K. (2001). A slacks-based measure of efficiency in data envelopment analysis. *European Journal of Operational Research, 130*, 498–509.

Tone, K., & Tsutsui, M. (2009). Network DEA: A slacks-based measure approach. *European Journal of Operational Research, 197*, 243–252.

Wang, C. H., Gopal, R., & Zionts, S. (1997). Use of data envelopment analysis in assessing information technology impact on firm performance. *Annals of Operations Research, 3*, 191–213.

Chapter 20
Performance Measurement of Major League Baseball Teams Using Network DEA

Herbert F. Lewis

Abstract Data envelopment analysis (DEA) has been extensively applied to measure the performance of individual athletes and teams in a variety of sports as well as to analyze nations competing in the Olympics. Most of the models presented in the literature are single-stage DEA models which treat the underlying process of converting inputs into outputs as a "black box." In many situations, analysts are interested in investigating the sources of inefficiency within the organization in order to improve organizational performance. To accomplish this, researchers have developed two-stage and network DEA methodologies.

In this chapter, we model an MLB team as comprised of a front office operation which consumes money in the form of player salaries to acquire offensive and defensive talent and an on-field operation which uses the talent to outscore opponents and win games. We present a network DEA methodology to measure performance of the front office operation, the on-field operation, and the overall team. Finally, we conduct two industry-wide studies of Major League Baseball which utilize the network DEA methodology.

Keywords Two-stage DEA • Network DEA • Major League Baseball • Efficiency measurement in sports

20.1 Introduction

Baseball is a sport in which two teams, each consisting of nine players, compete on a field referred to as a baseball diamond due to its shape. Each team takes turns on offense (batting) and defense (pitching and playing the field). Traditionally, the visiting team begins on offense. The batting team sends its players one at a time to

H.F. Lewis (✉)
College of Business, Stony Brook University, Stony Brook, NY 11794-3775, USA
e-mail: Herbert.Lewis@StonyBrook.edu

W.D. Cook and J. Zhu (eds.), *Data Envelopment Analysis*,
International Series in Operations Research & Management Science 208,
DOI 10.1007/978-1-4899-8068-7_20,

try to hit a hard ball (thrown by a defensive player called a pitcher) with a wooden bat. For a batter to be successful, he must safely arrive at a base, which he can accomplish in several ways. Once the batter arrives safely at a base, he becomes a base runner. A base runner scores a run by advancing four bases and touching home plate. A base runner can advance along the bases by the actions of future batters or by stealing bases.

The defense tries to prevent offensive players from advancing around the bases, which it can accomplish in several ways. Each success by the defense records one out; when three outs are recorded, the teams switch roles (the fielding team becomes the batting team and the batting team becomes the fielding team). When both teams have batted, they have completed one inning. A game consists of nine innings. The winning team is the team that has scored the most runs by the end of the game. If there is a tie at the end of nine innings, the game continues until one team has more runs than the other does at the end of an inning. See Lorimer (2002) for a more extensive discussion on baseball.

Major League Baseball (MLB) is a professional baseball league in the United States and Canada. MLB is made up of two leagues: the National League (NL) and the American League (AL). From 1901 until the early 1960s, each league consisted of eight teams. At this time, each league began to expand. By 1969, each league was comprised of 12 teams, making it necessary to split each league into two divisions. Expansion continued and in 1994, each league further split into three divisions. Currently, MLB is comprised of 30 teams. There are 15 teams in the NL and 15 in the AL. Each of the three divisions within each league contains five teams.

The leagues play under essentially identical rules with one major exception: since the early 1970s, the American League allows the use of a designated hitter who bats in place of the pitcher. This potentially leads to generally greater offensive production in the AL because pitchers are commonly poor batters and designated hitters are often very good offensively.

Prior to 1961, each team played 154 regular season intra-league games. Since then, each team plays 162 regular season games. Until 1997, these games were all intra-league. Since then, each team plays roughly 144 intra-league games and 18 interleague games.

Major League Baseball has become a multi-billion dollar industry with many individual player salaries in the tens of millions of dollars. With so much money at stake, it is important for MLB teams to manage resources efficiently. Thus, in this chapter, we present a model framework for measuring the performance of MLB teams and use it to perform industry-wide analyses of Major League Baseball. Our model framework utilizes recent extensions to the *data envelopment analysis* (DEA) methodology: namely, *two-stage DEA* and *network DEA*. DEA is a linear programming-based methodology that is widely used to evaluate relative efficiency of decision making units (DMUs) in situations in which there are multiple inputs and multiple outputs. Its mathematical development can be traced to Charnes et al. (1978), who built on the work of Farrell (1957) and others.

The remainder of this chapter is organized as follows. The next section surveys the application of DEA in baseball and other sports. In Sect. 20.3, we briefly describe two-stage DEA and network DEA and review the related literature. In Sect. 20.4, we present our network DEA model framework for measuring the efficiency of MLB teams. Section 20.5 presents two previously published MLB industry-wide studies that apply the two-stage and network DEA methodology. Finally, we present concluding remarks in Sect. 20.6.

20.2 DEA in Baseball and Other Sports

DEA has been extensively applied to measure the performance of individual athletes and teams in baseball and other sports as well as nations in the Olympics. In this section we summarize the literature.

20.2.1 DEA in Baseball

Howard and Miller (1993) use DEA to identify underpaid, equitably paid, and overpaid MLB players. For each of the 433 players in the study, stolen bases, games played, at-bats, runs scored, hits, doubles, triples, home runs, runs batted in, batting average, put outs, assists, errors, fielding average, and years in the league are used as the inputs to the DEA model. Player salary is the output of the DEA model. A separate analysis is performed for each position. A reference set for each player is provided from which an equitable salary can be determined.

Mazur (1994) measures efficiency of MLB batters, pitchers, and teams during the 1986, 1987, and 1988 seasons. The author performs separate analyses for each league in each season. The model for batters uses standardized batting average, standardized number of home runs, and standardized number of runs batted in for batters having at least 200 at bats in a given season. These measures define the triple crown frontier (TCF). The model for pitchers uses standardized earned run average, standardized hits to innings pitched ratio, and standardized base on balls to strike-outs ratio for pitchers having at least 100 innings pitched in a given season. These measures define the pitching dominance frontier (PDF). A TCF efficiency score is determined for each batter and team and a PDF efficiency score is determined for each pitcher and team in each season. Regression models for each league and season suggest that a team's TCF efficiency score and a team's PDF efficiency score are significant indicators of its winning percentage.

Anderson and Sharp (1997) present a radial input-oriented CCR DEA model (Charnes et al. 1978) for measuring performance of MLB batters called the Composite Batter Index (CBI). Their model uses one input (plate appearances) and five outputs (dominance transformations of walks, singles, doubles, triples, and home runs). The authors compute CBI scores for players in both the American League and the National League from 1901 to 1993 resulting in 186 analyses. Players with fewer than 350 at-bats with one team in a given season are omitted from the analysis. Historical results indicate that batting has matured over the decades.

Specifically, league-wide CBI scores have increased over time. In addition, the proportion of players with low CBI scores has increased and the proportion of players with high CBI scores has increased over the study period. Finally, the authors develop and test a method for reducing the effect of noise in DEA. Thus allowing the CBI score to estimate a player's skill as opposed to his productivity.

Sueyoshi et al. (1999) present a goal programming model to rank Japanese baseball players in the Central League during the 1995 season. The goal program utilizes the offensive earned-run average (OERA) index (Cover and Keilers 1977) and results from a slack-adjusted DEA. The DEA model uses at-bats and double plays as input measures and singles, doubles, triples, home runs, runs batted in, steals, sacrifices, and walks as output measures. The authors compare the player rankings resulting from the OERA index, the DEA model, and the goal program.

Einolf (2004) applies two BCC DEA models (Banker et al. 1984) to measure efficiency of teams in MLB from 1985 to 2001 and in the National Football League (NFL) from 1981 to 2000. The model for MLB team efficiency has two inputs (total salary paid to position players and total salary paid to pitchers) and three outputs (team wins, team batting average, and team earned-run average). Similarly, the model for NFL team efficiency has two inputs (total salary paid to offensive players and total salary paid to defensive players) and three outputs (team wins, team offensive yards per attempt, and team defensive yards per attempt). The author uses the DEA results to compare the leagues and concludes that, on average, MLB teams are less efficient than NFL teams. MLB teams in large markets tend to spend more and tend to be less efficient than those in small markets. A second conclusion is that, on average, NFL teams became more efficient after the salary cap was introduced.

Hadley and Ruggiero (2006) apply two BCC DEA models (Banker et al. 1984) to determine the contract zone for arbitration-eligible MLB players. One DEA model reflects the player's point of view, measuring worth relative to players who earn more and have relatively lower performance. The other model reflects the owner's point of view, measuring worth relative to players who earn less and have relatively higher performance. A double frontier is generated based on these two models. The authors demonstrate the approach on position players eligible for arbitration between the 2001 and 2002 seasons. They use the contract zone determined by the DEA models and the player's final arbitrated salary to calculate each player's Relative Contract Position (RCP). The RCP is a measure of whether the final arbitrated salary is favorable to the player (RCP close to 1) or to the owner (RCP close to 0). Finally, a tobit regression indicates that player performance is the only significant independent variable in predicting RCP. Player characteristics (race and position), team characteristics (winning percentage and market size), and whether a player is a free agent or arbitration-eligible are unrelated to RCP.

Volz (2009) uses an output oriented BCC DEA model (Banker et al. 1984) and survival time analysis to analyze the effect of minority status on managerial survival in MLB over the period from 1985 to 2006. Team position player salaries, team pitching salaries, and average salary of all other in-division teams are used as the inputs to the DEA model and regular season winning percentage is used as the output of the DEA model. The efficiency scores computed by the DEA are included as covariates in the survival time analysis. The author concludes that on average, minorities are 9.6 % more likely to return the following season. In addition, managerial survival is independent of winning percentage.

20.2.2 DEA in Other Sports

DEA has been used to measure individual and team performance in other sports such as basketball, soccer, and European football. DEA has also been used to measure efficiency of athletes in non-team sports such as golf and tennis. In addition, DEA has been applied to evaluate efficiency of nations competing in the Olympics.

20.2.2.1 Basketball

Fizel and D'Itri (1997, 1999) apply DEA to measure the efficiency of coaches in NCAA Division I college basketball from 1984 to 1991. The DEA models use player talent and opposition power as the inputs and winning percentage as the output. In these studies, the authors examine the importance of team effectiveness (winning percentage) and managerial efficiency on hiring and firing of coaches. Results indicate that, although hiring and firing of coaches is often based on team effectiveness, managerial efficiency may be a better measure when making these decisions.

Cooper et al. (2009) use the two-step procedure for the selection of weights proposed in Cooper et al. (2007) to measure effectiveness of basketball players in the Spanish Premier League. They focus on player outputs such as points scored and percentage of free throw successes and leave out such things as player salaries and other inputs.

20.2.2.2 Soccer and European Football

Haas (2003a) investigates the efficiency of 20 English Premier League clubs during the 2000/2001 season using DEA. The input variables are wage bills for players and coaches and the output variables are points awarded and total revenues. Population of each club's home town is introduced in the model as a site characteristic. The author finds that efficiency and club effectiveness are unrelated. The sensitivity of results is analyzed with regard to different model specifications and variable combinations. In all models at least 25 % of the clubs are on the efficient frontier.

Haas (2003b) applies DEA to measure the technical and scale efficiencies of teams in Major League Soccer (MLS) during the 2000 season. This study uses the same inputs and outputs as in Haas (2003a). Absolute number of spectators is also included as an output. The author finds that efficiency scores are highly correlated with league performance and that the largest part of team inefficiency can be explained by scale inefficiency as opposed to technical inefficiency.

Haas et al. (2004) study the efficiency of football teams in the German Bundeslign during the 1999/2000 season using DEA. The input variables and output variables are the same as those in Haas (2003a). In addition, average stadium utilization is included as an output variable in the model. Findings indicate that efficiency scores are not correlated with effectiveness in the league. Medium-sized and small-sized teams tend to outperform large-sized teams. The authors also decompose the sources of inefficiency into technical inefficiency and scale inefficiency.

Espitia-Escuer and GarcÍa-CebriÁn (2004) use DEA to measure the efficiency of teams in the Spanish First Division from 1998 to 2001. The number of players used, attacking moves, the minutes of possession of the ball, and the shots and headers are the input variables and the number of points achieved is the output variable. The authors conclude that the efficient teams do not always correspond with those that finished highest in the league at the end of the season.

Espitia-Escuer and GarcÍa-CebriÁn (2006) use an output oriented DEA model to evaluate the performance of Spanish First-Division soccer teams between the years 1998 and 2005. The authors use the same inputs and output as in Espitia-Escuer and GarcÍa-CebriÁn (2004). The main finding is that the final league position of a team depends more on its efficient use of resources than on its potential.

Barros and Leach (2006) apply an input oriented DEA model to panel data on English Premier League Football Clubs in the years 1998/1999 to 2002/2003. The authors measure three outputs (points obtained in the season, attendance and turnovers) and four inputs (number of players, wages, net assets, and stadium facilities expenditures). The main conclusion is that the clubs display equivalent managerial skills, but they do not display equivalent scale efficiency.

García-Sánchez (2007) present a three-stage DEA model to measure performance of teams in the Spanish Professional Football League during the 2004/2005 season. The first stage consumes offensive talent (attacking moves, passes to the penalty area and shots at goal) and defensive talent (ball recovery and goalkeeper's actions) as inputs and produces goals scored by the team and the inverse of goals scored by the opposing teams as outputs. The outputs from the first stage determine the inputs to the second stage. The second stage outputs reflect the final ranking of the team. Finally, the third stage input is determined from the output of the second stage and the output is the number of spectators who attended the team's home games. Site characteristics related to province population and stadium size are considered in the third stage of the model. Results indicate that technical inefficiency of the defense is greater than that of the offense. In addition, teams with the most experience are more effective than those with little experience.

Guzmán and Morrow (2007) use an input oriented DEA to measure the efficiency of clubs in the English Premier League from 1997/1998 to 2002/2003. The authors consider two inputs (directors' remuneration and general expenses) and two outputs (points won in a season and total revenue for the corresponding financial year). A second study is performed using the Malmquist productivity index (Malmquist 1953) to measure the change in productivity over the study period.

Results indicate that clubs which were successful on the field achieved relatively low efficiency scores, while other clubs that enjoyed less success on the field were relatively more efficient. In addition, there was little evidence that teams improved their productivity over time.

Boscá et al. (2009) analyze the performance of Italian and Spanish football clubs using DEA during the 2000/2001, 2001/2002, and 2002/2003 seasons. The authors select goals scored as the offensive output, goals conceded as the defensive output, four offensive inputs (shots-on-goal, attacking plays made by the team, passes into the opposing team's centre area, and minutes of possession) and four defensive inputs (the inverse of shots-on-goal made by the opposing team, the inverse of attacking plays made by the opposing team, the inverse of passes to the centre area made by the opposing team, and the inverse of minutes of possession by the opposing team). Results indicate that the Spanish league is more homogeneous and competitive than the Italian league. In addition, to improve competitiveness in the Italian league, it is more important to improve defensive, rather than offensive, efficiency. On the other hand, to improve the ranking in the Spanish league, the best strategy is to improve offensive efficiency when playing at home and then to improve offensive efficiency when playing away from home.

González-Gómez and Picazo-Tadeo (2010) use DEA to measure performance of Spanish professional football teams at competition level (League, King's Cup, and European competitions) from season 2001/2002 to season 2006/2007 and use the results as a proxy of fan satisfaction. The DEA model has three outputs (the points obtained in the league at the end of each season, the number of rounds played in the King's Cup, and the number of matches played in European competitions) and four inputs (the number of players in each season, the average number of spectators per match, the number of seasons played in the First Division, and the trophies in national and international competitions).

20.2.2.3 The Olympics

Lozano et al. (2002) present a variable returns-to-scale DEA model to measure performance of nations competing in five summer Olympic games (from 1984 to 2000). The authors use two inputs (GNP and population of the country under consideration) and three outputs (the numbers of gold, silver, and bronze medals won by the country under consideration). Weights are used to differentiate between the value associated with each medal type.

Churilov and Flitman (2006) use DEA to generate a ranking of the nations that participated in the Sydney 2000 summer games. Their goal is "to design an objective impartial system of analysis of the Olympics results which the majority of participating countries would agree upon as a measuring tool without significant bias." The inputs to the DEA model are population of the country under consideration, its GDP per capita (in U.S. dollars), its disability adjusted life expectancy, and its index of equality of child survival. The model consists of four outputs determined from utility functions on the numbers of gold, silver, and bronze medals won by the country under consideration.

Li et al. (2008) use a variable returns-to-scale context-dependent assurance region DEA model (Cook and Zhu 2008) to "fairly" rank the performance of 78 different nations that participated in six summer Olympics (from 1984 to 2004). Nations are classified into four groups based on wealth. This classification is used to impose the assurance region restrictions. Inputs to the model include population of the country under consideration and its GDP per capita (in U.S. dollars). Outputs are the numbers of gold, silver, and bronze medals won by the country under consideration.

Wu et al. (2009) use cross efficiency evaluation (Sexton et al. 1986) to measure performance of nations that competed in six summer Olympic games (from 1984 to 2004). The authors use the same inputs and outputs as in Li et al. (2008) and weight the outputs as in Lozano et al. (2002).

Wu et al. (2010) use an integer-valued DEA model to evaluate efficiency of nations involved in the Beijing Olympics. The inputs and outputs are the same as those in Li et al. (2008). In this analysis, the target outputs (number of gold, silver, and bronze medals) determined from the DEA must be integer values.

20.2.2.4 Golf and Tennis

Fried et al. (2004) use DEA to measure the efficiency of golfers on the PGA, LPGA, and SPGA tours during the 1998 season. For each golfer, a performance under pressure index and an athletic ability performance index are determined.

Ruiz et al. (2013) use DEA to measure efficiency of professional tennis players. The authors provide an index of the overall performance of players by aggregating the Association of Tennis Professionals (ATP) statistics and compare the results to the ATP rankings.

20.3 Two-Stage and Network DEA

With the exception of García-Sánchez (2007), who presents a three-stage DEA model for teams in the Spanish Professional Football League, the DEA models discussed in the previous section are all variations of the standard single-stage DEA model. Such models treat the production process in which inputs are converted into outputs as a "black box" and provide little insight as to the sources of inefficiency. These single-stage DEA models are appropriate in many situations including when the objective of the study is to rank DMUs based on performance. However, in many other situations, analysts and DMU managers seek more detailed information to assist them in improving managerial performance.

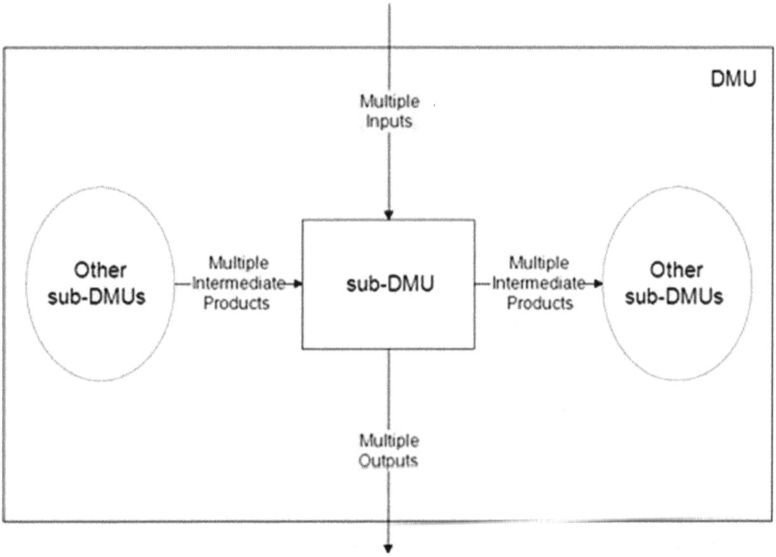

Fig. 20.1 Internal structure of a DMU in a network DEA model

20.3.1 Two-Stage and Network DEA Methodology

To address this issue, researchers have proposed two-stage and network DEA models. In network DEA, each DMU is comprised of two or more sub-DMUs. Each resource consumed by a sub-DMU either enters the DMU from outside (input to the DMU) or is produced by another sub-DMU (intermediate product). Each product produced by a sub-DMU either exits the DMU (output of the DMU) or is consumed by another sub-DMU (intermediate product). Figure 20.1 shows the internal structure of a DMU in a network DEA model. The DMU is a directed acyclic graph in which the nodes correspond to sub-DMUs and the arcs correspond to inputs to the DMU, outputs from the DMU, or intermediate products from one sub-DMU to another.

In this chapter, we apply the two-stage methodology and network DEA methodology proposed by Sexton and Lewis (2003) and Lewis and Sexton (2004a), respectively. The methodologies allow the analyst to measure the efficiency of each sub-DMU as well as the efficiency of the DMU itself. To measure the efficiency of a given sub-DMU, solve a standard single-stage DEA model for the sub-DMU. To evaluate the DMU-level efficiency use the directed acyclic structure of the underlying graph to identify a partial order of the sub-DMUs. Resolve the DEA model for each sub-DMU in accordance with the partial order, assuming that all sub-DMUs that precede the sub-DMU under analysis are efficient. Then, for an input (output) oriented model, the DMU-level efficiency (inverse efficiency) is the largest (smallest) of the ratios, computed for each input (output) of what could have been consumed (produced) to what was actually consumed (produced).

20.3.2 Two-Stage and Network DEA Literature

Over the past two decades, several papers have been published on the theory, methodology, and application of two-stage and network DEA. Färe and Whittaker (1995) apply an input oriented two-stage DEA model to study relative efficiency of dairy production. Seiford and Zhu (1999) evaluate the performance of 55 U.S. commercial banks using a two-stage network DEA model. In another study, Färe and Grosskopf (2000) present a network DEA model for the Swedish Institute for Health Economics. Zhu (2000) applies two-stage network DEA to develop a multi-factor financial performance model to examine Fortune Global 500 companies. Castelli et al. (2001) describe a DEA-like model that evaluates the efficiencies of each of a number of interdependent sub-DMUs within a larger DMU. Their analysis assesses sub-DMU efficiency relative to other sub-DMUs within the same DMU. Chen and Zhu (2004) develop an efficiency model that identifies the efficient frontier of a two-stage production process linked by intermediate measures. They illustrate the approach on a set of firms in the banking industry. Yang (2006) creates a two-stage DEA model to provide managerial insights for the Canadian life and health insurance industry. Chen et al. (2006) contend that two-stage DEA with a single intermediate product can behave as a parametric linear model. They develop a nonlinear DEA model to evaluate the impact of information technology on multiple stages of a business operation along with information on how to distribute IT-related resources so that efficiency is achieved. Färe et al. (2007) survey network DEA models and present three network DEA examples. Liang et al. (2008) examine and extend the two-stage DEA model using game theory concepts. They also investigate the relationship among non-cooperative, centralized, and standard DEA approaches. Kao and Hwang (2008) develop a two-stage DEA model and apply it to measure efficiency of non-life insurance companies in Taiwan. Chen et al. (2009a) develop an additive efficiency decomposition approach to generalize the two-stage DEA model presented by Kao and Hwang (2008). Chen et al. (2009b) examine the relationship and equivalence between the two-stage DEA approaches of Chen and Zhu (2004) and Kao and Hwang (2008). Tone and Tsutsui (2009) present a slacks-based measure approach to network DEA that applies to differing model orientations. They demonstrate their methodology by measuring the efficiency of electric power companies. Chen et al. (2010) develop an approach for determining the frontier points for inefficient DMUs within the framework of two-stage DEA. Tone and Tsutsui (2010) present a dynamic slacks-based measure model that can evaluate the overall efficiency of the DMUs as well as the efficiencies of the individual sub-DMUs in a network DEA. In a survey paper, Cook et al. (2010) review and classify several two-stage network DEA structures. In many of these models, the first stage processes the DMU's inputs into intermediate products and the second stage converts the intermediate products into outputs. Lewis and Mazvancheryl (2011) develop a network DEA model to measure the efficiency of the customer satisfaction process and apply it to the automobile industry. Holod and Lewis (2011) present a two-stage DEA model to measure efficiency of bank holding companies which resolves a long time dilemma by treating deposits as an intermediate product as opposed to an input or an output to the process. Mallikarjun et al. (2013) study the relationship between efficiency and government subsidization of the U.S. commuter rail system using an unoriented network DEA model.

20.4 Network DEA Model for a Major League Baseball Team

Sexton and Lewis (2003) present a sequential two-stage DEA model for measuring the efficiency of MLB teams. Each MLB team consists of a front office operation and an on-field operation. The methodology provides efficiency scores for the front office operation and the on-field operation as well as the overall organization. The two-stage methodology is then extended in Lewis and Sexton (2004a) to a network DEA model which allows for efficiency measurement of organizations with more complex internal structures. The network DEA model further divides the front office operation and on-field operation of an MLB team. The two-stage and network DEA methodologies allow for constant or variable returns-to-scale processes and permit input oriented or output oriented models. In addition, Lewis et al. (2013) present an unoriented two-stage DEA methodology and apply it to measure efficiency of MLB teams during the 2009 season.

Figure 20.2 presents our network representation of an MLB team. The front office operation consumes money in the form of position player and pitcher salaries to acquire offensive and defensive talent. The on-field operation uses this talent to score runs and to prevent the team's opponents from scoring runs in order to win games.

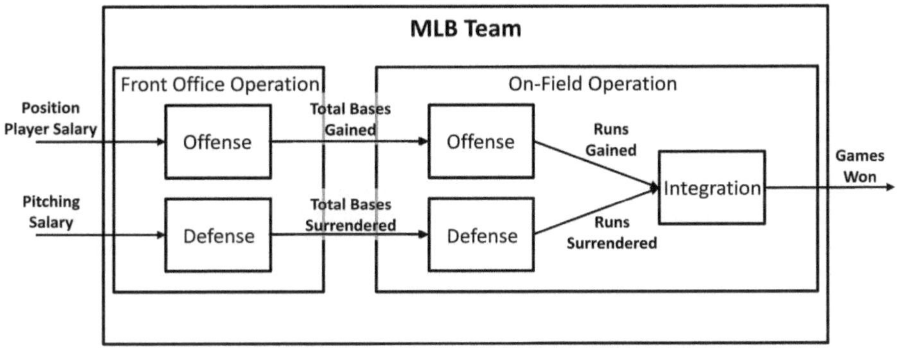

Fig. 20.2 Network model of an MLB team consisting of a front office operation and an on-field operation

20.4.1 Inputs, Intermediate Products, and Outputs

Total player salary (*TPS*) of a team in a season consists of position player salaries (*POS*) which reflect the offense and pitching salaries (*PIT*) which reflect the defense. Offensive talent can be measured by total bases gained (*TBG*) by the team in a season. MLB uses a statistic called total bases (*TB*) to measure offensive performance. Specifically, MLB's definition of total bases for a team in a season is $TB = S + 2D + 3T + 4HR$ where S is the number of singles, D is the number of doubles, T is the number of triples, and HR is the number of home runs hit by the team. We extend this definition by adding BB, the number of walks received by the team and E, the number of fielding errors committed by the opposing team. Thus, $TBG = TB + BB + E$. We point out that, with the exception of the relatively rare hit by pitch and catcher, fielder, or umpire interference, our definition of *TBG* includes all the ways in which a batter can reach first base without an out being recorded. We recognize that not every error results in the batter reaching first base. However, each error results in at least one runner (and in many cases the batter) advancing at least one base. We elect to model errors as the approximate equivalent of singles and walks. Defensive talent can be measured by total bases surrendered (*TBS*) to the team's opponents in a season. We define *TBS* identically to *TBG* except that the summands refer to the number of such hits and walks surrendered by the team, and the number of fielding errors committed by the team, in the given season. Runs gained (*RG*) is the number of runs scored by the team in a season. Runs surrendered (*RS*) is the number of runs scored by the team's opponents in a season. We note that *TBS* and *RS* are "reverse quantities," in the sense that larger values correspond to less, rather than more, defensive contribution. We use the methodology developed by Lewis and Sexton (2004b) to incorporate reverse quantities in our models. The output of the process is games won (*GW*) by the team in a season. We note that various inputs, intermediate products, and outputs may be aggregated or disaggregated and sub-DMUs may be split or combined depending on the analyst's preferences and the data available.

20.4.2 Model Orientation and Returns-to-Scale

We select an output orientation for each MLB team as well as its front office operation and its on-field operation because we feel that the appropriate improvement for an inefficient team is to increase the number of games it wins rather than decrease its total player salary. This orientation is consistent with each team's long-term goal of qualifying for post-season play. The input orientation would imply that all teams seek to hold its games won at current levels, an assumption that, we believe, contradicts the fundamental competitive nature of baseball teams. We recognize that individual teams may make economic decisions to spend less (or more) on player salaries. However, we see this as a scale change, not evidence of an input orientation.

We select a variable returns-to-scale model for the front office operation because of the "threshold" nature in which player salaries result in offensive and defensive production. At very low levels of player salary, we expect the marginal return to be less than the average return. Low budget teams will tend to sign weaker players and yet must conform to minimum salary levels set by the Major League Baseball Players Association contract with MLB. Below a certain threshold, therefore, we expect non-increasing returns-to-scale. Eventually, as player salary increases beyond this threshold, the team is better able to sign superior players who contribute significantly on the field. Here we expect non-decreasing returns-to-scale. At very high salary levels, we again expect the marginal return to be less than the average return. Superstar players who command the highest salaries are unlikely to provide offensive and defensive performance commensurate with their salaries. Above a second threshold, therefore, we expect non-increasing returns-to-scale.

We select a variable returns-to-scale model for the on-field operation because of the "threshold" nature in which offensive and defensive performance combine to win games. At very low levels of total bases gained and total bases surrendered, we would expect the marginal return to be less than the average return. Weak teams are likely to lose many games by several runs and therefore experience only a small increase in games won for a given increase in offensive and defensive performance. Below a certain threshold, therefore, we expect non-increasing returns-to-scale. Eventually, as performance increases beyond this threshold, the average margin of loss diminishes and the marginal return increases as the team begins to win some close games that they would otherwise have lost. Here we expect non-decreasing returns-to-scale. At very high levels of total bases gained and total bases surrendered, we again expect the marginal return to be less than the average return. Strong teams are likely to win many games by several runs and would therefore experience only a small increase in games won for a given increase in offensive and defensive performance. Above a second threshold, therefore, we expect non-increasing returns-to-scale. In addition, the limit on the number of games that a team can win – it cannot win more than it plays – must lead eventually to non-increasing returns-to-scale. Given our selection of variable returns-to-scale in both the front office operation and the on-field operation, we select a variable returns-to-scale model for the MLB organization.

20.4.3 Network DEA Model Formulation

Let POS_j be the total salary remunerated to position players by team j in a season, PIT_j be the total salary remunerated to pitchers by team j in a season, TBG_j be the total bases gained by team j in a season, TBS_j be the total bases surrendered by team j in a season, RG_j be the runs gained by team j in a season, RS_j be the runs surrendered by team j in a season, and GW_j be the games won by team j in a season. Define θ_{1k} to be the inverse efficiency of the front office offense for team k, ε_{2k} to be the efficiency of the front office defense for team k, θ_{3k} to be the inverse efficiency of the on-field offense for team k, ε_{4k} to be the efficiency of the on-field defense for team k, and θ_{5k} to be the inverse efficiency of the on-field integration for team k. Further, define λ_{1j} to be the weight placed on the front office offense of team j by team k, λ_{2j} to be the weight placed on the front office defense of team j by team k, λ_{3j} to be the weight placed on the on-field offense of team j by team k, λ_{4j} to be the weight placed on the on-field defense of team j by team k, and λ_{5j} to be the weight placed on the on-field integration of team j by team k.

The output oriented variable returns-to-scale model for the front office offense is:

$$
\begin{aligned}
&\text{Max } \theta_{1k} \\
&\text{s.t.} \\
&\quad \sum_{j=1}^{n} \lambda_{1j} POS_j \leq POS_k \\
&\quad \sum_{j=1}^{n} \lambda_{1j} TBG_j \geq \theta_{1k} TBG_k \\
&\quad \sum_{j=1}^{n} \lambda_{1j} = 1 \\
&\quad \lambda_{1j} \geq 0; \quad j = 1, 2, \ldots, n \\
&\quad \theta_{1k} \geq 0
\end{aligned}
$$

The output oriented variable returns-to-scale model for the front office defense is:

$$
\begin{aligned}
&\text{Min } \varepsilon_{2k} \\
&\text{s.t.} \\
&\quad \sum_{j=1}^{n} \lambda_{2j} PIT_j \leq PIT_k \\
&\quad \sum_{j=1}^{n} \lambda_{2j} TBS_j \leq \varepsilon_{2k} TBS_k \\
&\quad \sum_{j=1}^{n} \lambda_{2j} = 1 \\
&\quad \lambda_{2j} \geq 0; \quad j = 1, 2, \ldots, n \\
&\quad \varepsilon_{2k} \geq 0
\end{aligned}
$$

The output oriented variable returns-to-scale model for the on-field offense is:

$$\begin{aligned}
&\text{Max } \theta_{3k}\\
&\text{s.t.}\\
&\quad \sum_{j=1}^{n} \lambda_{3j} TBG_j \leq TBG_k\\
&\quad \sum_{j=1}^{n} \lambda_{3j} RG_j \geq \theta_{3k} RG_k\\
&\quad \sum_{j=1}^{n} \lambda_{3j} = 1\\
&\quad \lambda_{3j} \geq 0; \quad j = 1, 2, \ldots, n\\
&\quad \theta_{3k} \geq 0
\end{aligned}$$

The output oriented variable returns-to-scale model for the on-field defense is:

$$\begin{aligned}
&\text{Min } \varepsilon_{4k}\\
&\text{s.t.}\\
&\quad \sum_{j=1}^{n} \lambda_{4j} TBS_j \geq TBS_k\\
&\quad \sum_{j=1}^{n} \lambda_{4j} RS_j \leq \varepsilon_{4k} RS_k\\
&\quad \sum_{j=1}^{n} \lambda_{4j} = 1\\
&\quad \lambda_{4j} \geq 0; \quad j = 1, 2, \ldots, n\\
&\quad \varepsilon_{4k} \geq 0
\end{aligned}$$

The output oriented variable returns-to-scale model for the on-field integration is:

$$\begin{aligned}
&\text{Max } \theta_{5k}\\
&\text{s.t.}\\
&\quad \sum_{j=1}^{n} \lambda_{5j} RG_j \leq RG_k\\
&\quad \sum_{j=1}^{n} \lambda_{5j} RS_j \geq RS_k\\
&\quad \sum_{j=1}^{n} \lambda_{5j} GW_j \geq \theta_{5k} GW_k\\
&\quad \sum_{j=1}^{n} \lambda_{5j} = 1\\
&\quad \lambda_{5j} \geq 0; \quad j = 1, 2, \ldots, n\\
&\quad \theta_{5k} \geq 0
\end{aligned}$$

To determine the organizational inverse efficiency for team k, we use the network DEA methodology presented in Lewis and Sexton (2004a). Let $TBG_k^* = \sum_{j=1}^{n} \lambda_{1j}^* TBG_j$

and $TBS_k^* = \sum_{j=1}^{n} \lambda_{2j}^* TBS_j$ where λ_{1j}^* and λ_{2j}^* are the optimal weights obtained when solving the front office offense model for team k and the front office defense model for team k, respectively. We next resolve the on-field offense model for team k using TBG_k^* as the RHS of the first constraint and resolve the on-field defense model for team k using TBS_k^* as the RHS of the first constraint.

$$
\begin{aligned}
&\text{Max } \theta_{3k} \\
&\text{s.t.} \\
&\quad \sum_{j=1}^{n} \lambda_{3j} TBG_j \le TBG_k^* \\
&\quad \sum_{j=1}^{n} \lambda_{3j} RG_j \ge \theta_{3k} RG_k \\
&\quad \sum_{j=1}^{n} \lambda_{3j} = 1 \\
&\quad \lambda_{3j} \ge 0; \quad j = 1, 2, \ldots, n \\
&\quad \theta_{3k} \ge 0
\end{aligned}
$$

$$
\begin{aligned}
&\text{Min } \varepsilon_{4k} \\
&\text{s.t.} \\
&\quad \sum_{j=1}^{n} \lambda_{4j} TBS_j \ge TBS_k^* \\
&\quad \sum_{j=1}^{n} \lambda_{4j} RS_j \le \varepsilon_{4k} RS_k \\
&\quad \sum_{j=1}^{n} \lambda_{4j} = 1 \\
&\quad \lambda_{4j} \ge 0; \quad j = 1, 2, \ldots, n \\
&\quad \varepsilon_{4k} \ge 0
\end{aligned}
$$

Let ${}^*RG_k^* = \sum_{j=1}^{n} {}^*\lambda_{3j}{}_{3j}^{*} RG_j$ where ${}^*\lambda_{3j}^*$ are the optimal weights obtained when solving the on-field offense model for team k, assuming the front office offense is efficient and ${}^*RS_k^* = \sum_{j=1}^{n}{}^* \lambda_{4j}^* RS_j$ where ${}^*\lambda_{4j}^*$ are the optimal weights obtained when solving the on-field defense model for team k, assuming the front office defense is efficient. We next resolve the on-field integration model for team k using ${}^*RG_k^*$ and ${}^*RS_k^*$ as the RHS of the first and second constraints, respectively.

$$
\begin{aligned}
&\text{Max } \theta_{5k} \\
&\text{s.t.} \\
&\quad \sum_{j=1}^{n} \lambda_{5j} RG_j \leq {}^*RG_k^* \\
&\quad \sum_{j=1}^{n} \lambda_{5j} RS_j \geq {}^*RS_k^* \\
&\quad \sum_{j=1}^{n} \lambda_{5j} GW_j \geq \theta_{5k} GW_k \\
&\quad \sum_{j=1}^{n} \lambda_{5j} = 1 \\
&\quad \lambda_{5j} \geq 0; \quad j = 1, 2, \ldots, n \\
&\quad \theta_{5k} \geq 0
\end{aligned}
$$

Finally, let ${}^*GW_k^* = \sum_{j=1}^{n} {}^*\lambda_{5j}{}_{5j}^* GW_j$ where $^*\lambda_{5j}^*$ are the optimal weights obtained when solving the on-field integration model for team k, assuming the front office offense, the front office defense, the on-field offense, and the on-field defense are all efficient. The organizational (overall team) inverse efficiency for team k is $\theta_k = {}^*GW_k^*/GW_k$.

20.4.4 Extension to Other Team Sports

The model can be applied to measure the performance of teams in other sports. In football, for example, total player salary (*TPS*) of a team in a season consists of offensive team salary (*OS*), defensive team salary (*DS*), and special team salary (*SS*). Offensive talent can be measured by total yards gained (*TYG*) by the team in a season. *TYG* is the sum of passing and rushing yards gained (while on offense), kickoff and punt return yards gained (while on special teams), interception and fumble return yards gained (while on defense) and penalty yards gained (while on offense or special teams) by the team in a season. Defensive talent can be measured by total yards surrendered (*TYS*) to the team's opponents in a season. *TYS* is the sum of passing and rushing yards surrendered (while on defense), kickoff and punt return yards surrendered (while on special teams), interception and fumble return yards surrendered (while on offense) and penalty yards surrendered (while on defense or special teams) to the team's opponents in a season. Points gained (*PG*) is calculated from the number of touchdowns, extra points, two-point conversions, field goals, and safeties scored by the team in a season. Points surrendered (*PS*) is calculated from the number of touchdowns, extra points, two-point conversions, field goals, and safeties scored by the team's opponents in a season. The output of the process is games won (*GW*) by the team in a season.

20.5 Two Studies of MLB Using Two-Stage and Network DEA

In this section, we present two published studies which apply two-stage and network DEA models to measure MLB team efficiency. The first study (Lewis et al. 2007) is published in the *Journal of Sports Economics*. The second study (Lewis et al. 2009) is published in the *European Journal of Operational Research*.

20.5.1 Player Salaries, Organizational Efficiency, and Competitiveness in MLB

In this study published in the *Journal of Sports Economics* (Lewis et al. 2007), we use a two-stage DEA model as part of a larger analysis to determine the minimum total player salary required for a team to be competitive for each season and count the number of teams that are noncompetitive due to low total player salary in each season. Next, we determine the salary at which a team is overspending on total player salary for each season and count the number of teams that overspend on total player salary in each season. Finally, we examine the relationship between market size, efficiency, and competitiveness. The study period is the non-strike seasons from 1985 to 2002.

20.5.1.1 Motivation and Research Questions

MLB, unlike other business enterprises, depends on stiff competition for economic survival. Baseball is entertainment; tight division races, unpredictable playoff series, and the periodic emergence of new champions enhance the entertainment value of the sport, ensuring the league's future fan base. However, while individual teams need the league to succeed, winning is the key to their economic success. Winning increases fan interest, brings more people to the ballpark, improves television ratings, and bolsters sales of team-related merchandise, all of which add to the team's prosperity.

Baseball entered the era of free agency on December 23, 1975, and player salaries have since grown to extraordinary levels. In 1975, the average player salary was $44,676; in 2002, it was $2,384,779, an average annual growth rate of nearly 16 % per year (nearly 11 % per year adjusted for inflation) for 27 years. During this period, MLB grew by 25 %, expanding from 24 to 30 teams. Some teams, notably those located in larger markets and those possessing greater financial resources, found it easier than other teams to sign free agents to high-salary, multi-year contracts, thereby cornering the market on the most talented players and threatening the competitive balance on the field.

In July 2000, the *Commissioner's Blue Ribbon Panel on Baseball Economics* (Levin et al. 2000) reported on the revenue disparities in MLB. The Panel found that these disparities were affecting competition, that the disparities were becoming worse, and that the limited revenue sharing and payroll taxes approved in the 1996 labor agreement with the players were having little effect. Moreover, the Panel concluded that the cost of trying to be competitive was raising ticket and concession prices, jeopardizing MLB's position as the affordable family spectator sport. The Panel's recommendations included greater revenue sharing and a competitive balance tax, both of which are part of the 2002 labor agreement with the players.

In 2002, the total player salary for the New York Yankees was $125.93 million while that of the Tampa Bay Devil Rays was $34.38 million. With one team's total player salary equal to 3.66 times that of another team, it is reasonable to ask whether the team with the lower salary can effectively compete with the team with the higher salary, and the extent to which market size influences competitiveness. More specifically, we pose the following research questions for the study period:

1. How much does a team need to spend on total player salary to be competitive?
2. What is the maximum total player salary that a team can pay without overspending?
3. How many teams are noncompetitive due to low total player salary?
4. How many teams are overspending on total player salary?
5. How does noncompetitiveness due to low total player salary relate to market size?
6. How does overspending on total player salary relate to market size?

20.5.1.2 Study Methodology

We present an overview of the study methodology in Fig. 20.3. In a given season, we apply two-stage DEA to measure the relative efficiency of each MLB team. We use a logistic regression model to classify teams as competitive versus noncompetitive. For each season, we use the Gini index to determine the minimum total player salary to be competitive and the maximum total player salary without overspending. Finally, we model the transitions of teams among the competitive and noncompetitive states according to a Markov process.

On page 5 of the report (Levin et al. 2000), the Commissioner's Blue Ribbon Panel defines *competitive balance* as the state in which "... every well-run club has a regularly recurring reasonable hope of reaching post-season play." Our analysis entails parsing this statement into operational definitions of "well-run" and "reasonable hope of reaching post-season play."

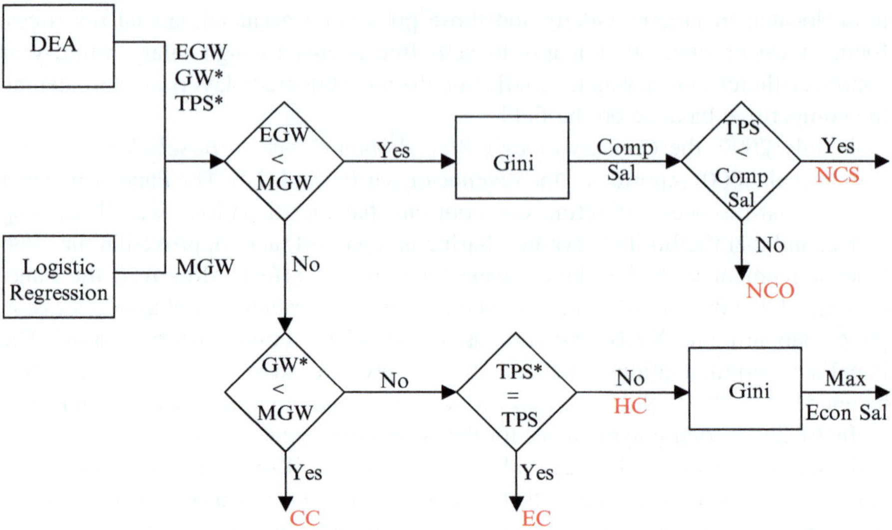

Fig. 20.3 An overview of the methodology used in this study to classify MLB teams

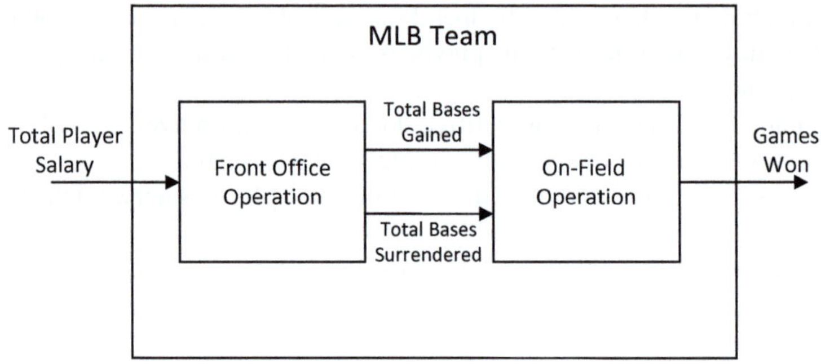

Fig. 20.4 Sequential two-stage model of an MLB team consisting of a front office operation and an on-field operation

Efficiency Measurement

To define "well-run," we turn to the theory of productive efficiency in the management science and economics literature. We apply the two-stage DEA methodology described in Sexton and Lewis (2003) to compute the efficiency of every MLB team in the study period relative to the frontier created by all other teams in the same season. The two-stage production model is presented in Fig. 20.4.

Define λ_{1j} to be the weight placed on the front office operation of team *j* by the front office operation of team *k*, λ_{2j} to be the weight placed on the on-field operation of team *j* by the on-field operation of team *k*, λ_j to be the weight placed on the team *j* by team *k* when determining the organizational inverse efficiency of team *k*, ε_{1k} to be the efficiency of the front office operation of team k, θ_{1k} to be the inverse efficiency of the front office operation of team k, θ_{2k} to be the inverse efficiency of the on-field operation of team k, and θ_k to be the organizational inverse efficiency of team k.

First, we solve the following DEA model to determine the front office inverse efficiency of team k:

$$
\begin{aligned}
&\text{Max } \theta_{1k} \\
&\text{s.t.} \\
&\quad \sum_{j=1}^{n} \lambda_{1j} TPS_j \leq TPS_k \\
&\quad \sum_{j=1}^{n} \lambda_{1j} TBG_j \geq \theta_{1k} TBG_k \\
&\quad \sum_{j=1}^{n} \lambda_{1j} TBS_j \leq \varepsilon_{1k} TBS_k \\
&\quad \theta_{1k} c_{1k} = 1 \\
&\quad \sum_{j=1}^{n} \lambda_{1j} = 1 \\
&\quad \lambda_{1j} \geq 0; \quad j = 1, 2, \ldots, n \\
&\quad \theta_{1k}, \varepsilon_{1k} \geq 0
\end{aligned}
$$

Next, we solve the following DEA model to determine the on-field inverse efficiency of team k:

$$
\begin{aligned}
&\text{Max } \theta_{2k} \\
&\text{s.t.} \\
&\quad \sum_{j=1}^{n} \lambda_{2j} TBG_j \leq TBG_k \\
&\quad \sum_{j=1}^{n} \lambda_{2j} TBS_j \geq TBS_k \\
&\quad \sum_{j=1}^{n} \lambda_{2j} GW_j \geq \theta_{2k} GW_k \\
&\quad \sum_{j=1}^{n} \lambda_{2j} = 1 \\
&\quad \lambda_{2j} \geq 0; \quad j = 1, 2, \ldots, n \\
&\quad \theta_{2k} \geq 0
\end{aligned}
$$

Let $TBG_k^* = \sum_{j=1}^{n} \lambda_{1j}^* TBG_j$ and $TBS_k^* = \sum_{j=1}^{n} \lambda_{1j}^* TBS_j$ where λ_{1j}^* are the optimal weights obtained when solving the front office model for team k. Then, we solve the following DEA model to determine the organizational inverse efficiency for team k using TBG_k^* and TBS_k^* as the RHS of the first and second constraints, respectively:

$$\begin{aligned}
&\text{Max } \theta_k \\
&\text{s.t.} \\
&\sum_{j=1}^{n} \lambda_j TBG_j \le TBG_k^* \\
&\sum_{j=1}^{n} \lambda_j TBS_j \ge TBS_k^* \\
&\sum_{j=1}^{n} \lambda_j GW_j \ge \theta_k GW_k \\
&\sum_{j=1}^{n} \lambda_j = 1 \\
&\lambda_j \ge 0; \quad j = 1, 2, \ldots, n \\
&\theta_k \ge 0
\end{aligned}$$

We compute the number of games each team would have won had it been efficient, i.e., the team's efficient games won (*EGW*), using the formula $EGW = {}^*GW_k^* = \sum_{j=1}^{n} {}^*\lambda_j^* GW_j$, where $*\lambda_j^*$ are the optimal weights obtained when solving the organizational model for team k, assuming the front office is efficient. The organizational (overall team) inverse efficiency for team k is $\theta_k = {}^*GW_k^*/GW_k$.

Logistic Regression

We interpret the phrase "reasonable hope of reaching post-season play" to mean that a team must have at least the same probability of reaching post-season play as it would have if all teams in its league were equally talented. We refer to this probability as the team's *balanced probability*. The balanced probability for a given team in a given season depends on the playoff qualification condition in effect. Before 1969, the two leagues had no division structure and only the league champions qualified for post-season play. Thus, the balanced probability for a team before 1969 depended on only the number of teams in its league. Between 1969 and 1993, each league consisted of two divisions and each division winner qualified for post-season play. During this period, the balanced probability for a team depended on only the number of teams in the team's division. Since 1994, each league consists of three divisions. Between 1994 and 2011, each division winner qualified for post-season play, as does the "wild card" team, which is the non-division winner with the highest winning percentage in the league. Thus, between 1994 and 2011, the balanced probability for a team depends on both the number of teams in its division and the number of teams in its league.

For each of the playoff qualification conditions, we compute its balanced probability. We interpret each probability as the minimum probability of qualifying for post-season play that a team must achieve to be competitive under its playoff qualification condition.

Next, we construct a logistic regression model for a team's probability of qualifying for post-season play. We use data for the seasons 1903 through 2002 (except 1904, when there was no post-season, and the strike seasons 1981, 1994, and 1995). For each team in each season, we use *GW* as the independent variable and a binary indicator variable equal to 1 if the team qualified for post-season play in that season, or equal to zero if it did not qualify. We also include indicator variables that identify the playoff qualification condition that applied to the league and division in which the team played in that season. Therefore, the logistic regression computes a team's probability of qualifying for post-season play given its number of games won and the playoff qualification condition that applied to the league and division in which the team played in that season.

We use the logistic regression model to compute *MGW*, the minimum number of games a team must win to be competitive under each playoff qualification condition. Thus, a team is competitive if and only if it would have won at least *MGW* had it been efficient. In other words, we say that a team is competitive if and only if $EGW \geq MGW$.

Gini Index

We then determine the minimum total player salary needed to be competitive in each season, which we call the *competitive salary* for that season. To do this, within each season, we sort the teams according to total player salary from low to high and use the Gini index to identify a total player salary that partitions the teams into two sets, one of which consists primarily of competitive teams and one of which consists primarily of noncompetitive teams. The competitive salary in that season is the total player salary of the lowest paid team in the primarily competitive set.

We now partition the noncompetitive teams into two groups:

- **Noncompetitive Due to Low Total Player Salary (NCS)**: A noncompetitive team is *noncompetitive due to low total player salary* if its total player salary is less than the competitive salary.
- **Noncompetitive for Other Reasons (NCO)**: A noncompetitive team is *noncompetitive for other reasons* if its total player salary is greater than the competitive salary.

Next, we analyze the competitive teams. In order to do this, we need to provide more definitions. Define GW^* to be the number of games that an efficient on-field operation would have won given the actual performance of the front office. We note that $GW \leq GW^* \leq EGW$. We obtain GW^* from the DEA of the on-field operation of the two-stage model. Let TPS^* be the total player salary of the efficient front office operation. We note that $TPS^* \leq TPS$. We obtain TPS^* from the DEA of the front office operation.

We now partition the competitive teams into three groups:

- **Conditionally Competitive (CC)**: A competitive team is *conditionally competitive* if $GW^* < MGW$. The team is spending enough money on total player salary but inefficiency in the front office has resulted in insufficient player performance to win enough games to achieve the balanced probability of qualifying for post-season play. The front office must become more efficient for this to happen. We note that a conditionally competitive team may be overspending on player salaries if $TPS^* < TPS$.
- **Economically Competitive (EC)**: A competitive team is *economically competitive* if $GW^* \geq MGW$ and $TPS^* = TPS$. The team has sufficient player performance on the field to achieve the balanced probability of qualifying for post-season play. Moreover, there is no evidence that the team is overspending on total player salary.
- **Hypercompetitive (HC)**: A competitive team is *hypercompetitive* if $GW^* \geq MGW$ and $TPS^* < TPS$. The team has sufficient player performance on the field to achieve the balanced probability of qualifying for post-season play. However, there is evidence that the team is overspending on total player salary.

We use the Gini index again, this time to determine the value of total player salary that partitions hypercompetitive teams from other teams. We call this value of total player salary the *hypercompetitive salary* for the given season.

Markov Analysis

Finally, we model the transitions of teams among these five states (NCS, NCO, CC, EC, and HC) according to a Markov process. We test the five row distributions for statistical independence and compute the steady-state probabilities and the mean first passage times from each state to each other state.

20.5.1.3 Data for the Study

We obtain market size data from the *United States Census Bureau* and *Statistics Canada*. We extract player salary data from the *USA Today Website*. We gather games won, whether the team qualified for post-season play, and the team performance data required to compute total bases gained and total bases surrendered from the *Baseball Archive Database* and the *Major League Baseball Official Website*.

We were unable to find data on the number of opposition errors, which is required in the calculation of total bases gained. We estimated this number for each team in each season by subtracting the team's own errors committed from the total committed in that team's league and dividing by one less than the number of teams in the league. This approximation ignores the minor effects of interleague play and the somewhat different schedules played by different teams, and assumes that teams are equally likely to commit errors against each team they play. In addition, we were unable to find data to support MLB's definition of total bases in the calculation of total bases surrendered for seasons prior to 1999. We estimated this quantity by identifying the relationship between total hits and total bases using regression analysis.

20.5.1.4 Study Results

We apply the two-stage DEA model to measure the efficiency of the front office operation, the on-field operation, and the overall organization of each team in each season of the study period and explore the relationship between efficiency and competitiveness. Next, we determine the competitive salary and hypercompetitive salary for each season in the study period and classify teams as competitive and noncompetitive. Finally, we examine how market size relates to efficiency and competitiveness.

Efficiency, Wins, and Competitiveness

Figure 20.5 illustrates the relationship between *EGW* and *GW* for all teams in the study period as determined by the DEA. The teams that lie along the line defined by $EGW = GW$ are organizationally efficient. All organizationally inefficient teams lie above this line. Different symbols indicate whether the team was noncompetitive due to low total player salary, noncompetitive due to other reasons, conditionally competitive, economically competitive, or hypercompetitive.

Table 20.1 shows, for each playoff qualification condition, the probability that a team would qualify for the playoffs if every team in its league or division were equally talented. For example, consider a team playing in a four-team division within a 14-team league with a wild card. This team has a balanced probability of 0.318 of qualifying for post-season play. Figure 20.6 shows the logistic regression model for this condition. The model indicates that a team playing under this condition must win at least 86.1 games to have a probability of qualifying for the playoffs equal to or greater than 0.318. Thus, under this playoff qualification condition, $MGW = 86.1$. Similar analyses lead to the *MGW* values shown in Table 20.1.

Competitive and Hypercompetitive Salary

Figure 20.7 shows the relationship between *TPS* and *EGW* for the 2000 season. Similar relationships hold in all other seasons in the study period. Three teams were noncompetitive in 2000, when the *MGW* was 85.6 in the American League East and Central, 86.1 in the American League West, 88.2 in the National League East and

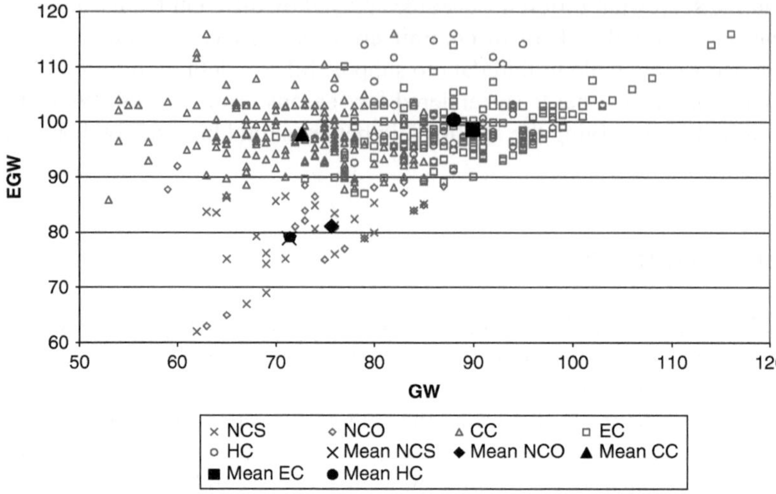

Fig. 20.5 The relationship between efficient games won and games won for all teams in the study period

Table 20.1 The balanced probability of a team qualifying for the playoffs given its playoff qualification condition and the minimum number of games a team needs to win to have a probability of qualifying for the playoffs at least as large as the balanced probability

Number of teams in league	Number of teams in division	Wild card	Balanced probability	MGW
10	–	No	0.100	92.8
8	–	No	0.125	88.9
14	7	No	0.143	89.5
12	6	No	0.167	88.2
16	6	Yes	0.231	87.3
16	5	Yes	0.262	88.2
14	5	Yes	0.272	85.6
14	4	Yes	0.318	86.1

West, and 87.3 in the National League Central. They were the Minnesota Twins, the Florida Marlins, and the Houston Astros – their efficient games won were 74.3, 79.0, and 81.1, respectively. The Gini index analysis indicates that the two teams with the lowest total player salaries (the Minnesota Twins and the Florida Marlins) were noncompetitive due to low total player salary. The lowest total player salary in the primarily competitive group is \$23.13 million, belonging to the Kansas City Royals. Thus, the competitive salary in 2000 was \$23.13 million.

We cannot explain why the Houston Astros were noncompetitive in 2000 other than to say that it was not due to low total player salary. However, we point out that 2000 was the Astros' first season in their new ballpark, one with dramatically

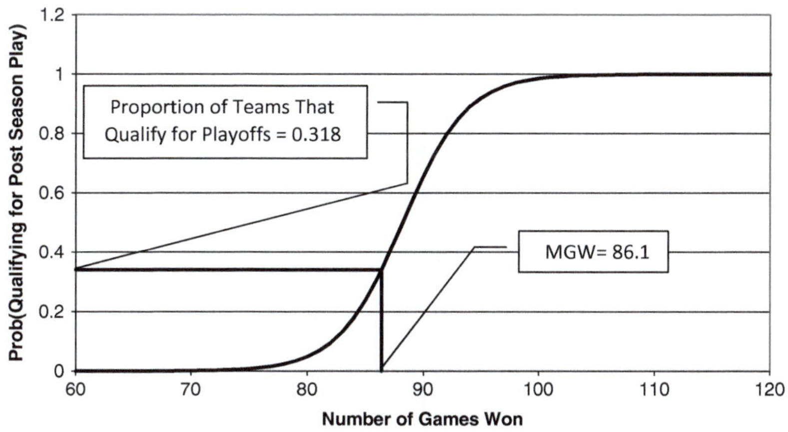

Fig. 20.6 The logistic regression model represents the probability that a team in a four-team division within a 14-team league, with a wild card, qualifies for the playoffs

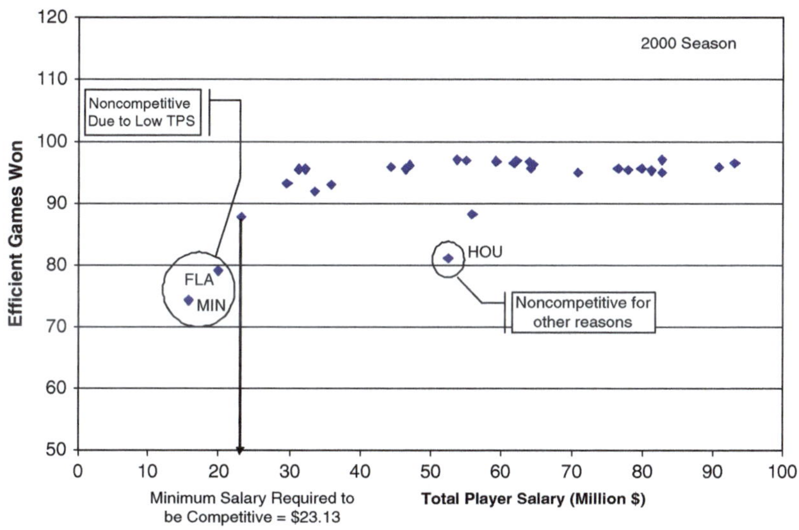

Fig. 20.7 The relationship between total player salary and efficient games won in the year 2000

different playing conditions than those found in the Astrodome. We also note that the Astros were competitive in both 1999 and 2001.

Table 20.2 and Fig. 20.8 show the competitive and hypercompetitive salary along with the team minima, mean, and maxima salaries for the non-strike seasons between 1985 and 2002. We find that the competitive salary ranges from \$6.19 million in 1985 to \$38.67 million in 2002, an average annual growth rate of 10.7 %

Table 20.2 The minimum, maximum, and mean total player salaries for the non-strike seasons between 1985 and 2002 along with the minimum total player salary to be competitive and hypercompetitive salaries

Season	1985	1986	1987	1988	1989	1990	1991	1992	1993	1994	1995	1996	1997	1998	1999	2000	2001	2002
Minimum	50.10	4.26	3.56	5.41	7.19	9.50	10.72	8.46	8.83			16.26	10.77	9.16	16.18	15.65	24.13	34.38
Competitive	6.19	9.19	4.09	6.00	9.09	12.66	14.05	13.49	18.20			23.02	17.27	22.73	22.20	23.13	24.13	38.67
Mean	10.64	11.46	10.79	11.63	13.82	17.20	22.95	30.04	30.38			33.94	40.25	40.34	47.37	56.20	65.48	67.49
Hypercompetitive	9.81	12.31	11.47	15.27	19.68	20.52	25.96	35.86	28.85			49.70	53.45	63.46	73.59	77.88	74.72	125.93
Maximum	16.20	19.64	20.01	19.44	20.27	23.57	37.28	44.46	45.75			54.71	62.24	68.99	85.03	92.94	112.29	125.93

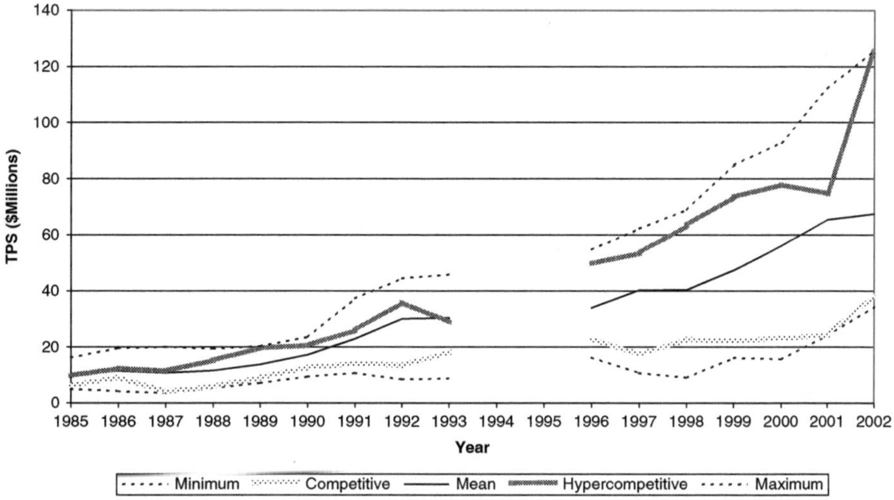

Fig. 20.8 The minimum, maximum, and mean total player salaries for the non-strike seasons between 1985 and 2002 along with the minimum total player salary to be competitive and hypercompetitive salaries

per year, adjusted for inflation. The hypercompetitive salary ranges from $9.81 million in 1985 to $125.9 million in 2002, an average annual growth rate of 12.6 % per year, adjusted for inflation. Interestingly, we observe that the team minimum, mean, and maximum salaries have risen at nearly the same average annual percentage rate, namely 11.2 % for the minimum, 10.8 % for the mean, and 12.1 % for the maximum. This suggests that, over the study period, it has not become relatively more costly to be competitive in MLB.

Moreover, the competitive salary has remained low relative to the mean total player salary in each season. As Fig. 20.9 shows, the ratio of the competitive salary in a given season to the minimum total player salary in the same season has remained stable around its mean of 1.5. Therefore, a rule of thumb is that a team's total player salary must be at least 50 % larger than the lowest total player salary in a given season to be competitive. The least squares regression line in Fig. 20.9 has a slope that is very nearly zero (0.0012 per year).

Classifying Teams as Competitive or Noncompetitive

We find that, in each season, there were between zero and four teams that were noncompetitive due to low total player salary, as shown in Fig. 20.10. We conclude that, in each season in the study period except for 2001, there existed teams that were noncompetitive due to low total player salary and that the number of such teams was relatively small. As Fig. 20.10 also shows, there were between zero and eight teams that were noncompetitive for reasons other than salary. These are teams

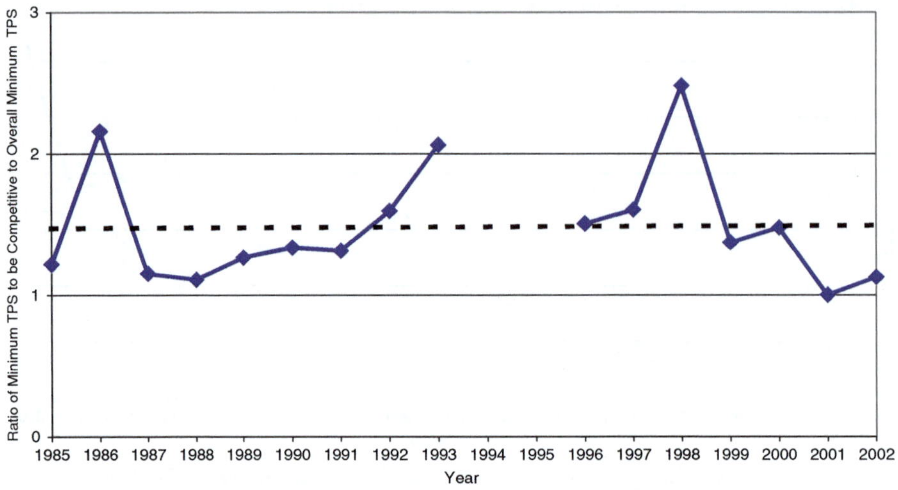

Fig. 20.9 The ratio of the minimum total player salary to be competitive to the overall minimum total player salary has remained stable around a mean of approximately 1.5 between 1985 and 2002

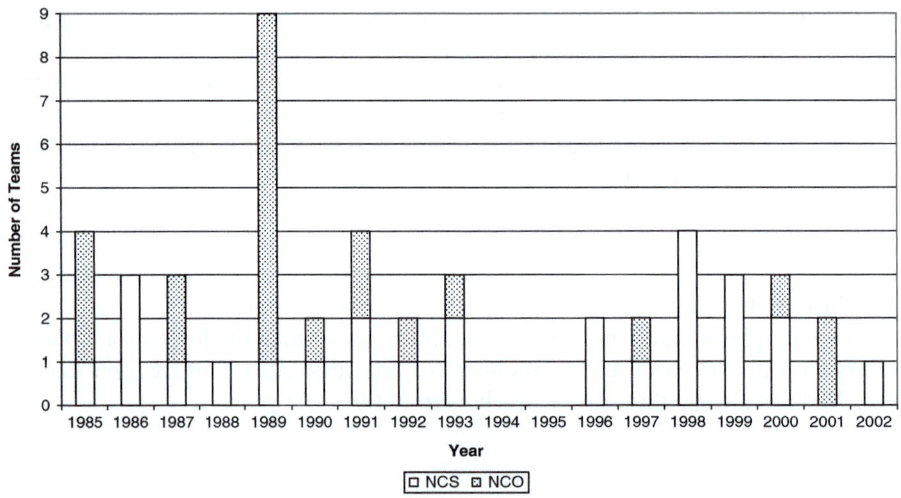

Fig. 20.10 The number of teams which are noncompetitive due to low total player salary and noncompetitive for other reasons in each non-strike season

whose total player salaries exceeded the minimum required to be competitive but had $EGW < MGW$. We cannot say why these teams are noncompetitive. During the study period, 49 of 442 teams (11.1 %) have been noncompetitive. Of these 49 teams, 27 (55.1 %) were noncompetitive due to low total player salary, while 22 (44.9 %) were noncompetitive for other reasons.

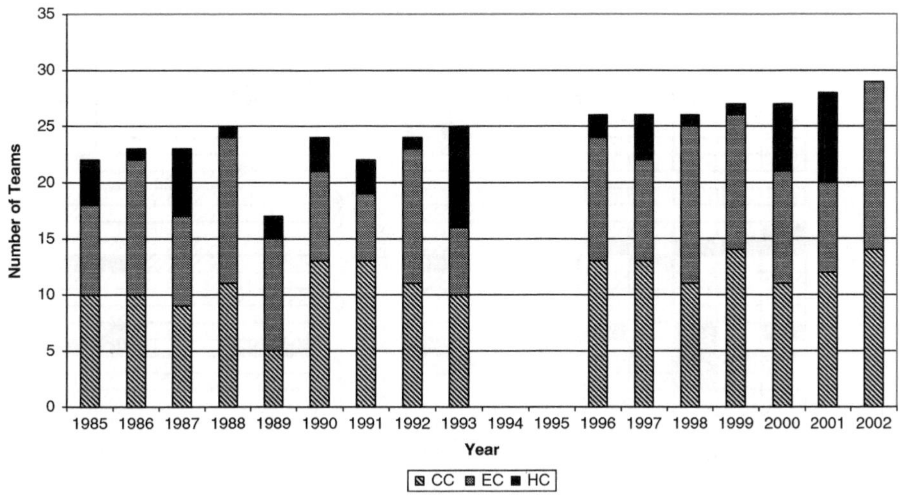

Fig. 20.11 The number of conditionally competitive, economically competitive, and hypercompetitive teams in each non-strike season

We find that, in each season, there were between zero and nine teams that were hypercompetitive, as shown in Fig. 20.11. During the study period, 95 of the 442 teams (21.5 %) overspent on player salaries.

We find that 51 of these 95 teams (54.7 %) were hypercompetitive, 41 (43.2 %) were conditionally competitive, and 2 (2.1 %) were noncompetitive for other reasons. We also find that, in each season, there were between 6 and 14 economically competitive teams, and that there were between 5 and 14 conditionally competitive teams. None of these categories demonstrate significant trends over time.

Figure 20.12 displays the competitive status of each team in each season during the study period. We observe that 16 teams have never been noncompetitive due to low total player salary during the study period. Note that only the Minnesota Twins have been noncompetitive due to low total player salary in four of the 18 seasons in the study period, and no team has been noncompetitive due to low total player salary more often. The Cleveland Indians and the Montreal Expos were each noncompetitive due to low total player salary three times. The Seattle Mariners (1985–1986), the Cleveland Indians (1992–1993), the Pittsburgh Pirates (1997–1998), the Montreal Expos (1998–1999), the Minnesota Twins (1986–1987 and 1999–2000), and the Florida Marlins (1999–2000) were noncompetitive due to low total player salary for two consecutive seasons. Thus, there is no evidence that being noncompetitive due to low total player salary is a chronic condition.

We observe that two teams (the Anaheim/California Angels and the Milwaukee Brewers) were conditionally competitive 12 times during the 16 seasons analyzed. In addition, three teams (the Chicago Cubs, the Kansas City Royals, and the San Diego Padres) were conditionally competitive 11 times, while the Philadelphia

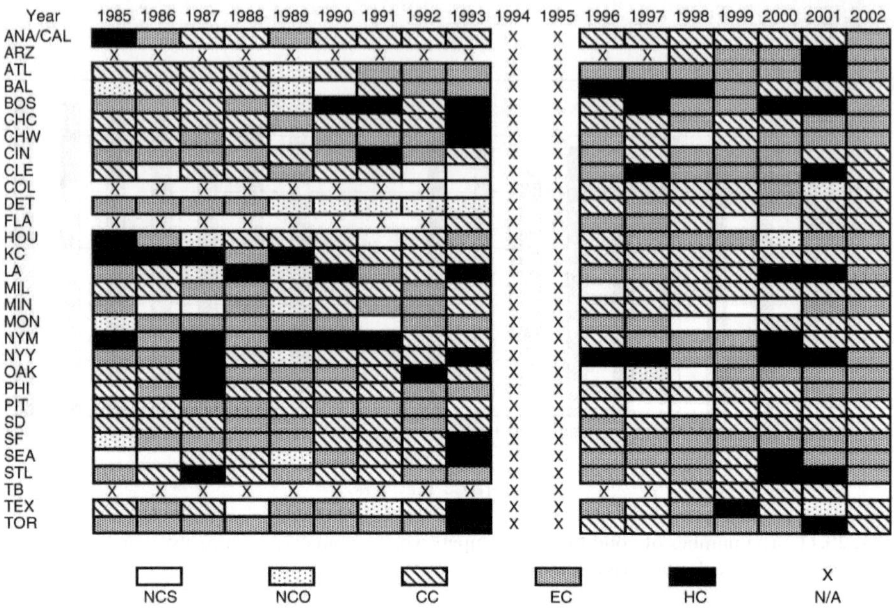

Fig. 20.12 The competitive status of each team in each season during the study period

Phillies and the Pittsburgh Pirates were conditionally competitive 10 and 9 times, respectively. Thus, this competitive status, in which the team has spent sufficient money on player salary but the front office has failed to produce sufficient talent on the field to be competitive, has been a persistent problem in these seven franchises.

We find that one team (the Toronto Blue Jays) was economically competitive in 11 seasons, while two teams (the Cincinnati Reds and the San Francisco Giants) were economically competitive in 10 seasons and two teams (the Atlanta Braves and the Montreal Expos) were economically competitive in 9 seasons. These franchises consistently paid sufficient player salaries to be competitive, and their front offices used the money to place sufficient talent on the field.

Three teams (the Boston Red Sox, the New York Mets, and the New York Yankees) were hypercompetitive six times, while the Los Angeles Dodgers and the Kansas City Royals were hypercompetitive five and four times, respectively. Moreover, nine teams have never been hypercompetitive and another nine teams have been hypercompetitive only once.

Markov Analysis

We model the transition of teams among the five states according to a Markov process. Ignoring transitions that spanned the strike seasons, the estimated transition matrix is

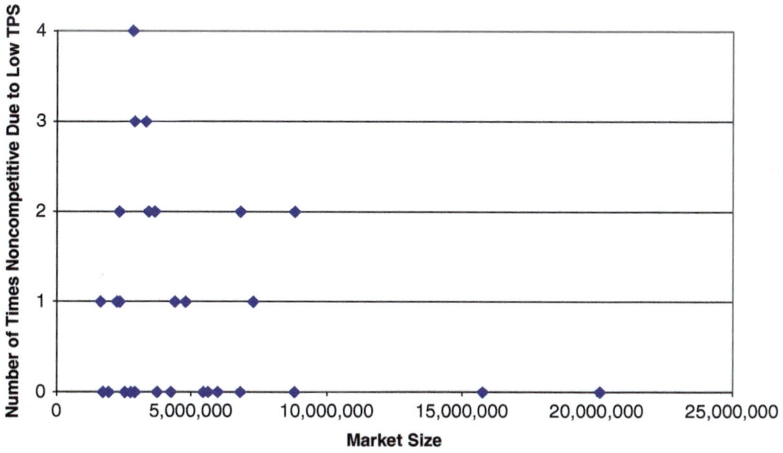

Fig. 20.13 The number of times that a team has been noncompetitive due to low total player salary between 1985 and 2002 is negatively related to the size of the market in which it plays

$$\mathbf{P} = \begin{pmatrix} 0.304 & 0.044 & 0.391 & 0.261 & 0 \\ 0.095 & 0.191 & 0.381 & 0.191 & 0.143 \\ 0.064 & 0.039 & 0.564 & 0.244 & 0.090 \\ 0.021 & 0.050 & 0.270 & 0.539 & 0.121 \\ 0 & 0.023 & 0.256 & 0.442 & 0.279 \end{pmatrix}$$

where the order of the states is NCS, NCO, CC, EC, and HC. A chi-square test shows that the probability distributions in the rows are significantly different ($\chi^2 = 92.75$ with df $= 16$, $P < 0.00005$). The steady-state probabilities associated with this transition matrix are $\boldsymbol{\pi} = (0.055 \quad 0.048 \quad 0.397 \quad 0.378 \quad 0.122)$.

The matrix of mean first passage times is

$$\mathbf{M} = \begin{pmatrix} 18.3 & 24.1 & 3.0 & 3.8 & 11.5 \\ 23.4 & 20.6 & 3.1 & 4.0 & 9.7 \\ 24.3 & 24.3 & 2.5 & 3.8 & 10.3 \\ 25.8 & 23.9 & 3.6 & 2.6 & 9.8 \\ 26.6 & 24.7 & 3.7 & 2.9 & 8.2 \end{pmatrix}$$

Market Size, Efficiency, and Competitiveness

Figure 20.13 shows the number of times each team was noncompetitive due to low total player salary versus the team's market size, defined as the population of the team's metropolitan area according to the 2000 U.S. census and the 2001 Canadian census. We find evidence that the number of times that a team has been noncompetitive due to low total player salary between 1985 and 2002 is negatively related to the size of the market in which it plays ($P = 0.0464$ in a

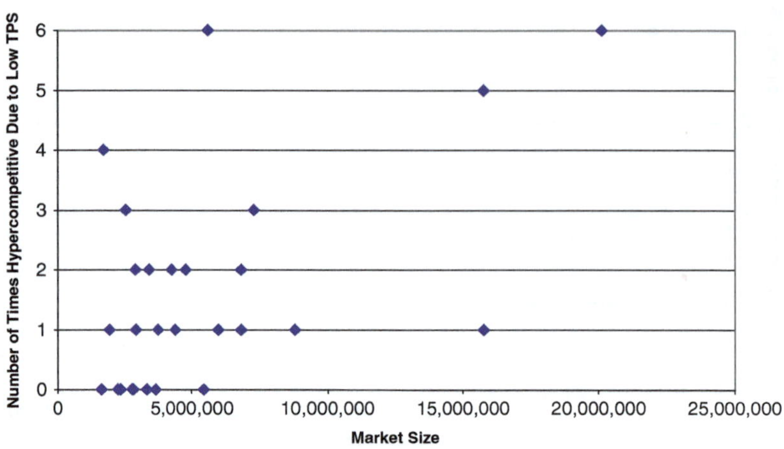

Fig. 20.14 The number of times that a team has been hypercompetitive between 1985 and 2002 is positively related to the size of the market in which it plays

Poisson regression). The teams that played in markets below five million people were NCS in 8.4 % of their seasons while teams that played in markets above five million people were NCS in 3.1 % of their seasons ($P = 0.027$). We note that, while the four teams in the two largest markets – two in New York and two in Los Angeles/Anaheim – were never noncompetitive due to low total player salary, the Chicago White Sox, who play in the third largest market, were noncompetitive due to low total player salary in two seasons (1989 and 1998). In addition, we see that seven of the 18 teams that play in markets below five million people have not been noncompetitive due to low total player salary in the study period.

Figure 20.14 shows the number of times each team was hypercompetitive versus the team's market size. We find evidence that the number of times that a team has been hypercompetitive between 1985 and 2002 is positively related to the size of the market in which it plays ($P < 0.00005$ in a Poisson regression). The teams that played in markets above five million people were hypercompetitive in 17.2 % of their seasons while teams that played in markets below five million people were hypercompetitive in 7.6 % of their seasons ($P = 0.0027$). Of the 18 teams with market size below five million, eight have never been hypercompetitive and four have been hypercompetitive once. Of the 12 teams with market size above five million, 11 have been hypercompetitive at least once, including the Boston Red Sox, the New York Mets, and the New York Yankees six times each and the Los Angeles Dodgers five times.

Figure 20.15 shows the relationship between efficient games won and market size for MLB teams during the study period. The regression line shown in Fig. 20.15 has a slope of 2.262 games per 10 million people (P-value = 0.0019), suggesting that an efficient New York team, with market size approximately equal to 20.13 million, would win roughly four more games in a season than would an efficient Milwaukee team, with market size equal to 1.65 million.

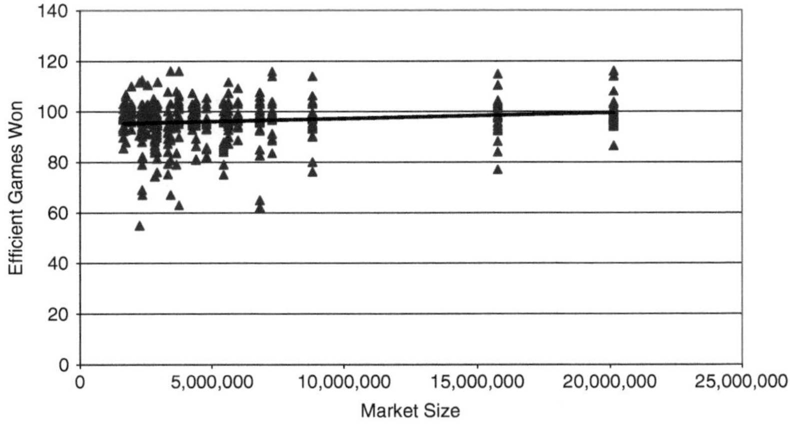

Fig. 20.15 Efficient games won versus market size for MLB teams between 1985 and 2002 excluding the strike seasons of 1994 and 1995

20.5.1.5 Conclusions of the Study

In this section, we summarize our results by responding to each research question.

How much does a team need to spend on total player salary to be competitive?
The competitive salary ranges from $6.19 million in 1985 to $38.67 million in 2002, an average annual growth rate of 10.7 % per year, adjusted for inflation. The team minimum, mean, and maximum salaries have risen at nearly the same average annual percentage rate. This suggests that, over the study period, it has not become relatively more costly to be competitive in MLB. Moreover, the competitive salary has remained low relative to the mean total players salary in each season. We find that the ratio of the competitive salary to the minimum total player salary has remained stable around its mean of 1.5.

What is the maximum total player salary that a team can pay without overspending?
The hypercompetitive salary is $125.9 million in 2002, up from $9.81 million in 1985. This is an average annual increase of 12.6 % (adjusted for inflation) per year. It is increasing over time as a percentage of maximum total player salary.

How many teams are noncompetitive due to low total player salary?
We find that, in each season, there were between zero and four teams that were noncompetitive due to low total player salary. We conclude that, in each season in the study period except for 2001, there existed teams that were noncompetitive due to low total player salary and that the number of such teams was relatively small. There were between zero and eight teams that were noncompetitive for other reasons. The Markov analysis suggests that, in any given season, 5.5 % of the teams (1.65 out of 30 teams) will be noncompetitive due to low total player salary, while another 4.8 % of the teams will be noncompetitive for other reasons (1.44 out of 30 teams).

How many teams are overspending on total player salary?
We also find that between zero and nine teams are hypercompetitive in a given season. During the study period, 95 of 442 teams (21.5 %) have been overspending on total player salary. Of these overspending teams, 52 (54.7 %) were hypercompetitive, 41 (43.2 %) were conditionally competitive, and two (2.1 %) were noncompetitive due to other reasons. The Markov analysis suggests that, in a given season, 12.2 % (3.66 out of 30 teams) of the teams will be hypercompetitive.

How does noncompetitiveness due to low total player salary relate to market size?
We find evidence that the number of times that a team has been noncompetitive due to low total player salary between 1985 and 2002 is negatively related to the size of the market in which it plays. However, we see that seven of the 18 teams that play in markets below five million people have not been noncompetitive due to low total player salary in the study period. Six of the 18 teams have been noncompetitive due to low total player salary more than once in this period. While the four teams in the two largest markets were never noncompetitive due to low total player salary, the Chicago White Sox, who play in the third largest market, were noncompetitive due to low total player salary in two seasons.

The size of the team's market relates to the number of games it can win if it is efficient. An efficient New York team, playing in the largest market, can expect to win roughly four more games per season than an efficient Milwaukee team, playing in the smallest market.

How does overspending on total player salary relate to market size?
Large market teams are more likely to be hypercompetitive than small market teams. Of the 18 teams with market size less than five million, eight have never been hypercompetitive, while four have been hypercompetitive only once. Meanwhile, of the 12 teams with market size greater than five million, 11 have been hypercompetitive at least once.

20.5.2 Organizational Capability, Efficiency, and Effectiveness in MLB

In this study published in the *European Journal of Operational Research* (Lewis et al. 2009), we use a network DEA model as part of a larger analysis to explore the relative contributions of team capability and managerial efficiency to team effectiveness in the context of Major League Baseball. We analyze every MLB team over the past century to capture long-term, persistent relationships. We perform separate analyses of regular season effectiveness and post-season effectiveness. The study period for the regular season analysis is from 1901 through 2002 (excluding the strike-shortened seasons in 1981, 1994, and 1995), during which there are 1934 observations. The study period for the post-season analysis is from 1903 through 2002 (excluding 1904 when there was no post-season play and the strike-shortened seasons in 1981, 1994, and 1995), during which there are 282 observations.

20.5.2.1 Motivation and Research Questions

To be effective, organizations need capabilities relevant to their missions and they must manage those capabilities efficiently. Without adequate talent, even a well-managed organization will fail to achieve its goals. Similarly, the inefficient utilization of resources will cause a well-equipped organization to fail. Of course, a powerfully equipped organization can compensate for managerial inefficiencies more easily than can a marginally equipped organization.

We anticipate that the relative contributions of capability and managerial efficiency are significant factors in organizational resource allocation decisions. Capability will be relatively more important in industries in which labor is highly paid. Examples of such industries include high-tech manufacturing, universities, hospitals, and professional sports. Efficiency will be relatively more important in industries in which labor is inexpensive. Examples of such industries include low-tech manufacturing, fast-food restaurant chains, janitorial services, and retail services.

MLB team owners, general managers, scouts, field managers, and coaches acquire, develop, and manage talent. Knowing the relative impact of talent and efficient use of that talent on team effectiveness can greatly enhance decisions both on and off the field. In this context, we pose the following research questions for the regular season and post-season study periods, respectively:

1. How much does team capability and managerial efficiency contribute to regular season effectiveness in MLB?
2. How much does team capability and managerial efficiency contribute to post-season effectiveness in MLB?

20.5.2.2 Study Methodology

We present mathematical models to measure regular season team capability, regular season team efficiency, regular season team effectiveness, and post-season team effectiveness. We then use weighted linear regression to evaluate the contributions of regular season team capability and regular season team efficiency to the variation in regular season and post-season team effectiveness.

Measuring Regular Season Capability

We measure the organizational capability of an MLB team during the regular season using offensive and defensive measures based on a variant of MLB's definition of total bases. We refer to these measures as *O-Capability* and *D-Capability*.

The capability of a team depends on its ability to get players on base and its ability to prevent its opponent's players from reaching base. We measure offensive capability in any season as $(TB_{Off} + BB_{Off} + E_{Off})/GP$ where TB_{Off} is the team's total bases gained on offense, BB_{Off} is the number of walks received by the team, E_{Off} is the number of fielding errors committed by the opposing team, and GP is the number of games played by the team. This approach to measuring offensive capability assumes constant returns-to-scale, i.e., that the sum in the numerator is proportional to the number of games played.

Defensive capability is defined identically except that the terms refer to the number of total bases and walks surrendered by the team, and number of fielding errors committed by the team, in the given season. Thus, we measure defensive capability in any season as $(TB_{Def} + BB_{Def} + E_{Def})/GP$ where TB_{Def} is the team's total bases surrendered on defense, BB_{Def} is the number of walks surrendered by the team, and E_{Def} is the number of fielding errors committed by the team. Observe that *D-Capability* has the property that larger numerical values are representative of less rather than more defensive capability. Thus, *D-Capability* is a reverse quantity. Note that we also assume constant returns-to-scale for defensive capability.

Measuring Regular Season Efficiency

To measure efficiency of an MLB team, we use the network DEA model developed by Lewis and Sexton (2004a). Figure 20.16 shows our network representation of the on-field operation of an MLB team.

The on-field operation of an MLB team is comprised of three sub-DMUs. The offense sub-DMU consumes offensive contributions (TB_{Off}, BB_{Off}, and E_{Off}) to produce runs gained on offense (R_{Off}), the defense sub-DMU consumes defensive contributions (TB_{Def}, BB_{Def}, and E_{Def}) to prevent runs surrendered on defense (R_{Def}), and the integration sub-DMU consumes R_{Off} and R_{Def} to produce games won (GW). Note that TB_{Def}, BB_{Def}, E_{Def} and R_{Def} are reverse quantities.

We use four efficiency scores to evaluate managerial performance. The first – the *O-Efficiency* – measures the efficiency of the offense sub-DMU. A team increases its *O-Efficiency* by clustering its hits, walks, and the errors committed by the opposing team, by stealing more bases and taking extra bases on hits, and by leaving fewer runners on base. The DEA model for *O-Efficiency* is

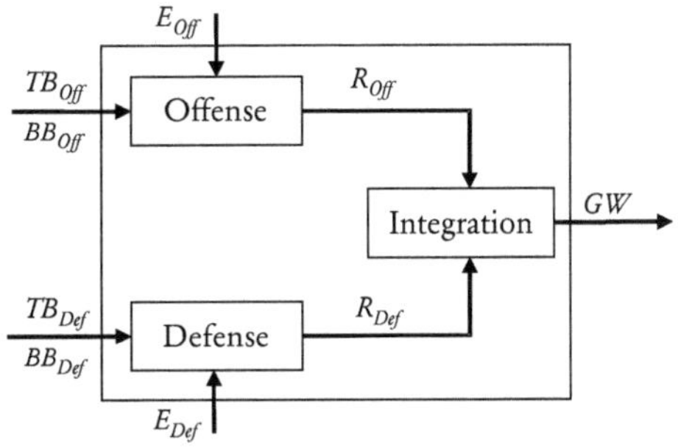

Fig. 20.16 Network model of the on-field operation of an MLB team

$$
\begin{aligned}
&\text{Max } \theta_{1k} \\
&\text{s.t.} \\
&\quad \sum_{j=1}^{n} \lambda_{1j} TB_{Off\,j} \leq TB_{Off\,k} \\
&\quad \sum_{j=1}^{n} \lambda_{1j} BB_{Off\,j} \leq BB_{Off\,k} \\
&\quad \sum_{j=1}^{n} \lambda_{1j} E_{Off\,j} \leq E_{Off\,k} \\
&\quad \sum_{j=1}^{n} \lambda_{1j} R_{Off\,j} \geq \theta_{1k} R_{Off\,k} \\
&\quad \sum_{j=1}^{n} \lambda_{1j} = 1 \\
&\quad \lambda_{1j} \geq 0; \quad j = 1, 2, \ldots, n \\
&\quad \theta_{1k} \geq 0
\end{aligned}
$$

In this model, λ_{1j} represent the weight that DMU k places on DMU j when measuring the efficiency of the offense sub-DMU and θ_{1k} is the inverse efficiency of the offense sub-DMU.

The second efficiency score – the *D-Efficiency* – measures the efficiency of the defense sub-DMU. A team increases its *D-Efficiency* by scattering the hits, walks, and the errors it commits, by preventing stolen bases and extra bases on hits, and by leaving more opposition runners on base. The DEA model for *D-Efficiency* is

$$
\begin{aligned}
&\text{Min } \varepsilon_{2k}\\
&\text{s.t.}\\
&\quad \sum_{j=1}^{n} \lambda_{2j} TB_{Def\,j} \geq TB_{Def\,k}\\
&\quad \sum_{j=1}^{n} \lambda_{2j} BB_{Def\,j} \geq BB_{Def\,k}\\
&\quad \sum_{j=1}^{n} \lambda_{2j} E_{Def\,j} \geq E_{Def\,k}\\
&\quad \sum_{j=1}^{n} \lambda_{2j} R_{Def\,j} \leq \varepsilon_{2k} R_{Def\,k}\\
&\quad \sum_{j=1}^{n} \lambda_{2j} = 1\\
&\quad \lambda_{2j} \geq 0; \quad j = 1, 2, \ldots, n\\
&\quad \varepsilon_{2k} \geq 0
\end{aligned}
$$

In this model, λ_{2j} represent the weight that DMU k places on DMU j when measuring the efficiency of the defense sub-DMU and ε_{2k} is the efficiency of the defense sub-DMU.

The third efficiency score – the *W-Efficiency* – measures the efficiency of the integration sub-DMU. A team increases its *W-Efficiency* by winning more close games. The DEA model for *W-Efficiency* is

$$
\begin{aligned}
&\text{Max } \theta_{3k}\\
&\text{s.t.}\\
&\quad \sum_{j=1}^{n} \lambda_{3j} R_{Off\,j} \leq R_{Off\,k}\\
&\quad \sum_{j=1}^{n} \lambda_{3j} R_{Def\,j} \geq R_{Def\,k}\\
&\quad \sum_{j=1}^{n} \lambda_{3j} GW_{j} \geq \theta_{3k} GW_{k}\\
&\quad \sum_{j=1}^{n} \lambda_{3j} = 1\\
&\quad \lambda_{3j} \geq 0; \quad j = 1, 2, \ldots, n\\
&\quad \theta_{3k} \geq 0
\end{aligned}
$$

In this model, λ_{3j} represent the weight that DMU k places on DMU j when measuring the efficiency of the integration sub-DMU and θ_{3k} is the inverse efficiency of the integration sub-DMU.

The fourth efficiency score – the *F-Efficiency* – measures the efficiency of the entire DMU. The *F-Efficiency* is computed as the efficiency of the integration sub-DMU using the optimal values R^*_{Off} and R^*_{Def} produced by the offense sub-DMU and the defense sub-DMU, respectively. The DEA model for *F-Efficiency* is

$$
\begin{aligned}
&\text{Max } \theta_k \\
&\text{s.t.} \\
&\quad \sum_{j=1}^{n} \lambda_j R_{Off\,j} \le R^*_{Off\,k} \\
&\quad \sum_{j=1}^{n} \lambda_j R_{Def\,j} \ge R^*_{Def\,k} \\
&\quad \sum_{j=1}^{n} \lambda_j GW_j \ge \theta_k GW_k \\
&\quad \sum_{j=1}^{n} \lambda_j = 1 \\
&\quad \lambda_j \ge 0; \quad j = 1, 2, \ldots, n \\
&\quad \theta_k \ge 0
\end{aligned}
$$

In this model, λ_j represent the weight that DMU k places on DMU j when measuring the efficiency of the entire DMU and θ_k is the inverse efficiency of the entire DMU.

All of the DEA models assume variable returns-to-scale, an output orientation, and use a common frontier over teams in all seasons in the study period. We justify the use of a common frontier for all seasons based on the observation that the process by which MLB teams convert inputs into outputs has remained essentially unchanged throughout the study period. While it may be true that offensive and defensive capabilities have evolved during the study period, the variable returns-to-scale assumption neither rewards nor penalizes a team based on the season in which it played. During this period, there were 1934 observations. All the models except for the *O-Efficiency* model involve reverse quantities, which we incorporate using the methodology presented in Lewis and Sexton (2004b).

Measuring Regular Season Effectiveness

The number of games an MLB team wins in a given season is a measure of its effectiveness. However, during the study period, not all teams played the same number of regular season games. Therefore, we define the regular season organizational effectiveness of an MLB team as the team's winning percentage during the season, defined as $WPct = GW/GP$.

Table 20.3 The first column shows the possible seven-game post-season series outcomes, ranked from best to worst for Team A. The second column shows the probabilities of the possible seven-game post-season series outcomes, where p is the probability that Team A wins any given game, and $q = 1 - p$. The third column shows the seven-game post-season series effectiveness of Team A for each possible series outcome

Team A	Probability	Post-season series effectiveness
Wins in 4	p^4	1
Wins in 5	$4p^4q$	$q^4(1 + 4p + 10p^2 + 20p^3) + p^4(20q^3 + 10q^2 + 4q)$
Wins in 6	$10p^4q^2$	$q^4(1 + 4p + 10p^2 + 20p^3) + p^4(20q^3 + 10q^2)$
Wins in 7	$20p^4q^3$	$q^4(1 + 4p + 10p^2 + 20p^3) + 20p^4q^3$
Loses in 7	$20p^3q^4$	$q^4(1 + 4p + 10p^2 + 20p^3)$
Loses in 6	$10p^2q^4$	$q^4(1 + 4p + 10p^2)$
Loses in 5	$4pq^4$	$q^4(1 + 4p)$
Loses in 4	q^4	q^4

Analyzing Regular Season Effectiveness

We use *WPct* as the dependent variable in a weighted linear regression with *O-Capability*, *D-Capability*, *O-Efficiency*, *D-Efficiency*, *W-Efficiency*, and *F-Efficiency* as the independent variables and regular season games played as the weights. We evaluate the contributions of each independent variable to the variation in regular season effectiveness in three ways. First, we compute the coefficient of partial determination for each independent variable. Second, we compute the R^2 contribution of a given independent variable. Third, we compare (absolute) regression coefficients between the two capability measures and within the efficiency measures.

Measuring Post-season Effectiveness

We measure a team's post-season series effectiveness as the probability that the team would have performed at least as well as it actually did. For a given team A, we rank the series outcomes from best to worst. For example, in a best-of-seven game series, the ranked outcomes for team A are shown in the first column of Table 20.3. Next, we determine the probability that team A wins a given game versus an opposing team B. Let α be the regular season winning percentage of team A and β be the regular season winning percentage of team B. Then, the probability that team A wins a given game is $p = \alpha/(\alpha + \beta)$ and the probability that team B wins a given game is $q = 1 - p = \beta/(\alpha + \beta)$. The second column of Table 20.3 shows the probability distribution for team A in a best-of-seven series. We measure the post-season series effectiveness of team A as the sum of the probabilities from the worst outcome for team A to the outcome that occurred. The third column of Table 20.3 shows these values.

Prior to 1969, we measure a team's post-season effectiveness as its World Series effectiveness. Since 1969, we measure a team's post-season effectiveness as a weighted average of its individual series effectiveness measures, using the maximum series lengths as the weights.

Analyzing Post-season Effectiveness

We examine how capability and efficiency during the regular season relate to post-season effectiveness. We use post-season effectiveness as the dependent variable in a weighted linear regression with (regular season) *O-Capability*, *D-Capability*, *O-Efficiency*, *D-Efficiency*, *W-Efficiency*, and *F-Efficiency* as the independent variables and post-season games played as the weights. We evaluate the contributions of each independent variable to the variation in post-season effectiveness using the coefficient of partial determination, the R^2 contribution, and the (absolute) regression coefficients.

20.5.2.3 Data for the Study

We obtained games won, post-season records, and team performance data (such as runs, total bases, walks, and errors) from the *Baseball Archive Database* and the *Major League Baseball Official Website*. We were unable to find data on the number of opposition errors and the number of opposition total bases for seasons prior to 1999. We estimated these quantities as described in Sect. 20.5.1.3.

20.5.2.4 Study Results

In this section, we present summary statistics of our capability, efficiency, and effectiveness measures of all MLB teams and post-season teams. We also perform a series of hypothesis tests to compare the capability, efficiency, and effectiveness measures of post-season and non-post-season teams. In addition, we report the results of our regular season and post-season regression analyses.

Summary Statistics of Capability, Efficiency, and Effectiveness Measures

Table 20.4 presents descriptive statistics of regular season capability for all regular season teams and post-season teams. Figures 20.17 and 20.18 are histograms of *O-Capability* for all regular season teams and post-season teams, respectively, while Figs. 20.19 and 20.20 are histograms of *D-Capability* for all regular season teams and post-season teams, respectively. On average, a regular season team gains (and surrenders) 17.24 total bases per game. We note that a typical post-season team gains 18.40 total bases and surrenders 16.42 total bases during a regular

Table 20.4 Descriptive statistics of regular season *O-Capability* and *D-Capability* for all regular season and post-season teams

Variable	Teams	N	Mean	SD	Minimum	1st quartile	Median	3rd quartile	Maximum
Regular season	All	1,934	17.24	1.66	12.50	16.11	17.25	18.31	23.10
O-Capability	Post-season	282	18.40	1.62	13.88	17.36	18.44	19.47	23.10
Regular season	All	1,934	17.24	1.63	12.34	16.20	17.25	18.30	24.13
D-Capability	Post-season	282	16.42	1.40	12.34	15.52	16.53	17.25	20.22

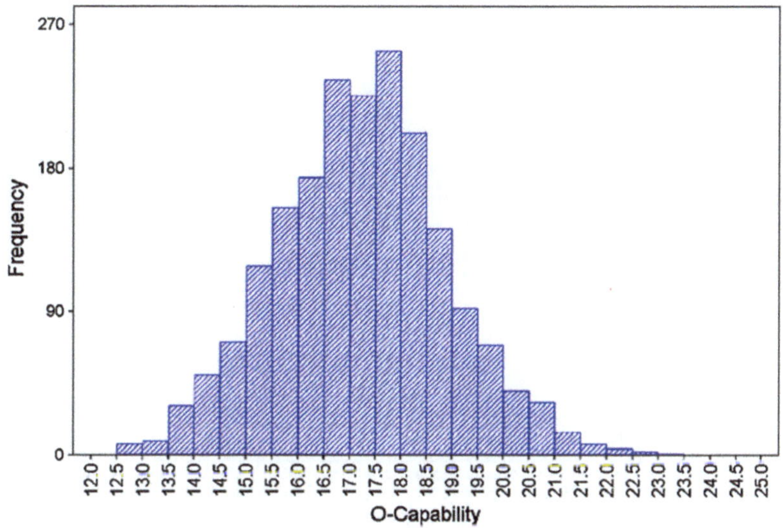

Fig. 20.17 Histogram of regular season *O-Capability* for all regular season teams

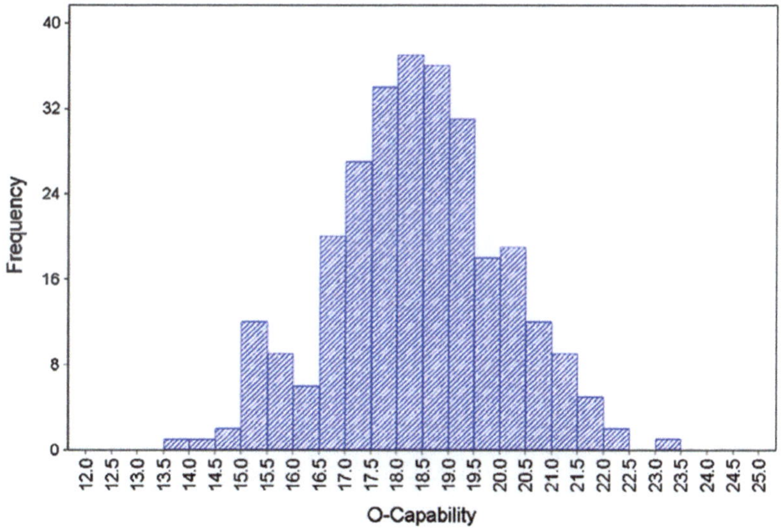

Fig. 20.18 Histogram of regular season *O-Capability* for post-season teams

season game. Thus, teams that achieve the post-season are typically above average offensively and defensively.

Table 20.5 presents descriptive statistics of regular season efficiency for all regular season teams and post-season teams. Figures 20.21 and 20.22 are

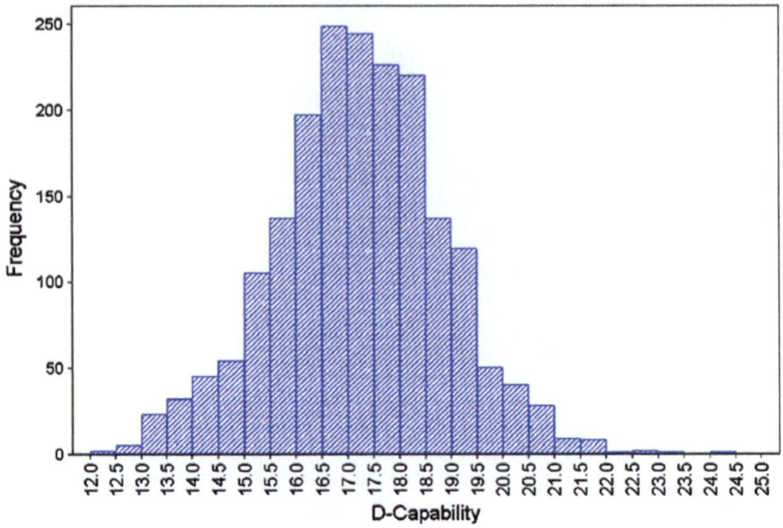

Fig. 20.19 Histogram of regular season *D-Capability* for all regular season teams

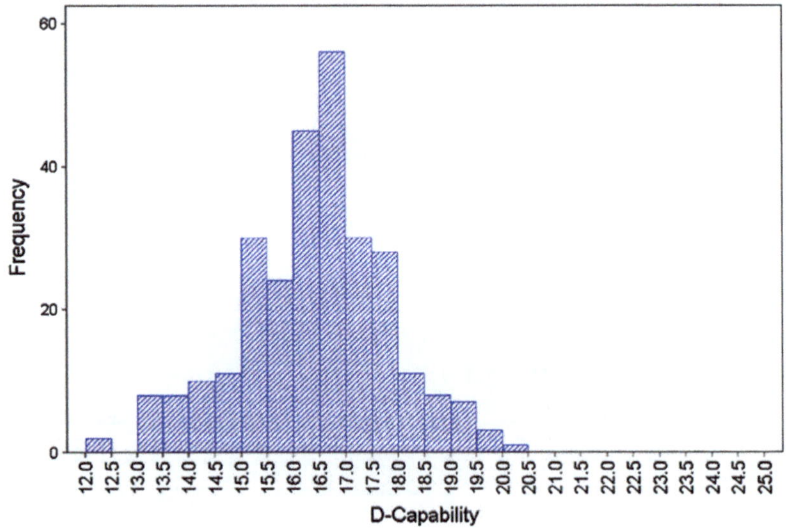

Fig. 20.20 Histogram of regular season *D-Capability* for post-season teams

histograms of *O-Efficiency* for all regular season teams and post-season teams, respectively. Figures 20.23 and 20.24 are histograms of *D-Efficiency* for all regular season teams and post-season teams, respectively. Figures 20.25 and 20.26 are histograms of *W-Efficiency* for all regular season teams and post-season teams,

Table 20.5 Descriptive statistics of regular season *O-Efficiency*, *D-Efficiency*, *W-Efficiency*, and *F-Efficiency* for all regular season and post-season teams

Variable	Teams	N	Mean	SD	Minimum	1st quartile	Median	3rd quartile	Maximum
Regular season	All	1,934	1.13	0.06	1	1.09	1.13	1.17	1.38
O-Efficiency	Post-season	282	1.09	0.05	1	1.06	1.10	1.12	1.22
Regular season	All	1,934	0.90	0.05	0.74	0.87	0.90	0.93	1
D-Efficiency	Post-season	282	0.92	0.04	0.78	0.89	0.92	0.95	1
Regular season	All	1,934	1.16	0.09	1	1.10	1.15	1.21	1.62
W-Efficiency	Post-season	282	1.09	0.04	1	1.06	1.09	1.12	1.23
Regular season	All	1,934	1.36	0.18	1	1.23	1.33	1.45	2.46
F-Efficiency	Post-season	282	1.17	0.06	1	1.13	1.17	1.21	1.33

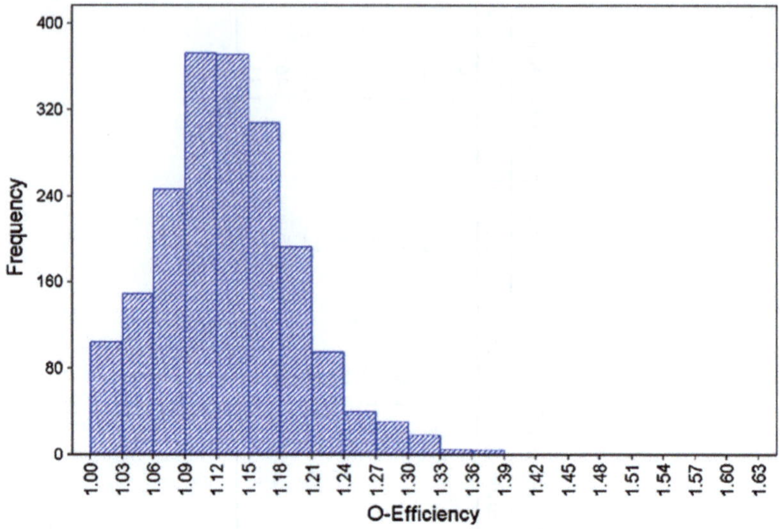

Fig. 20.21 Histogram of regular season *O-Efficiency* for all regular season teams

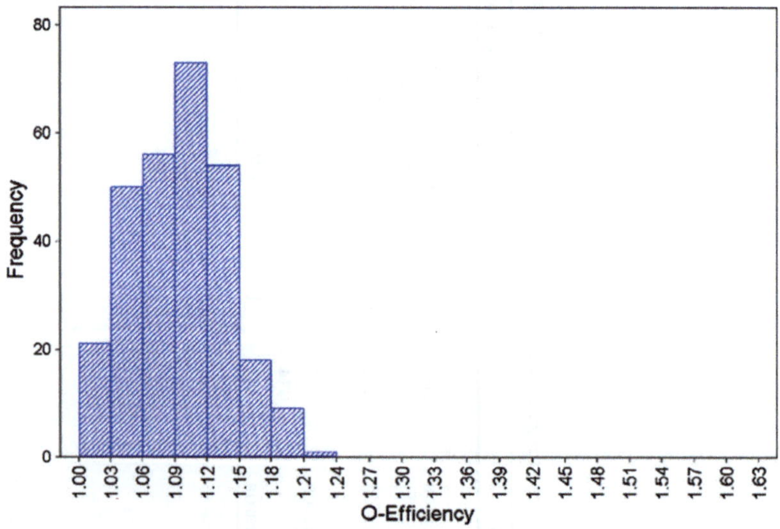

Fig. 20.22 Histogram of regular season *O-Efficiency* for post-season teams

respectively. Figures 20.27 and 20.28 are histograms of F-*Efficiency* for all regular season teams and post-season teams, respectively. On average, a regular season team should be able to increase its runs gained by 13 % (given its total bases gained), decrease its runs surrendered by 10 % (given its total bases surrendered), and win 16 % more games (given its runs gained and runs surrendered). Overall, a

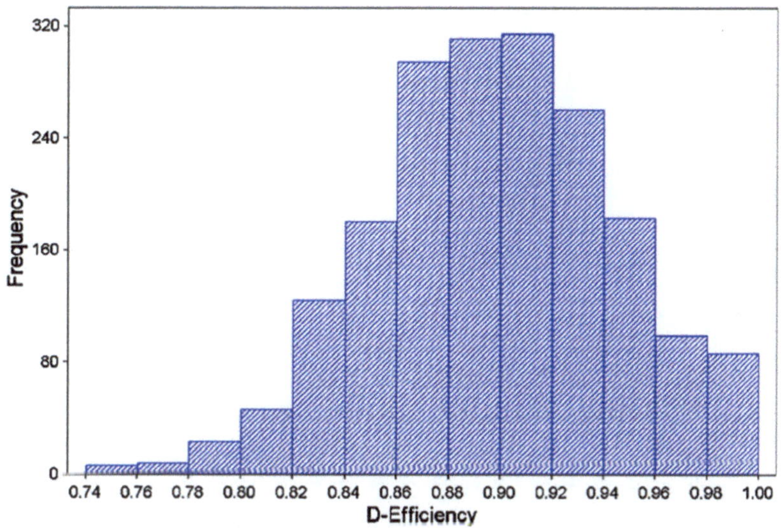

Fig. 20.23 Histogram of regular season *D-Efficiency* for all regular season teams

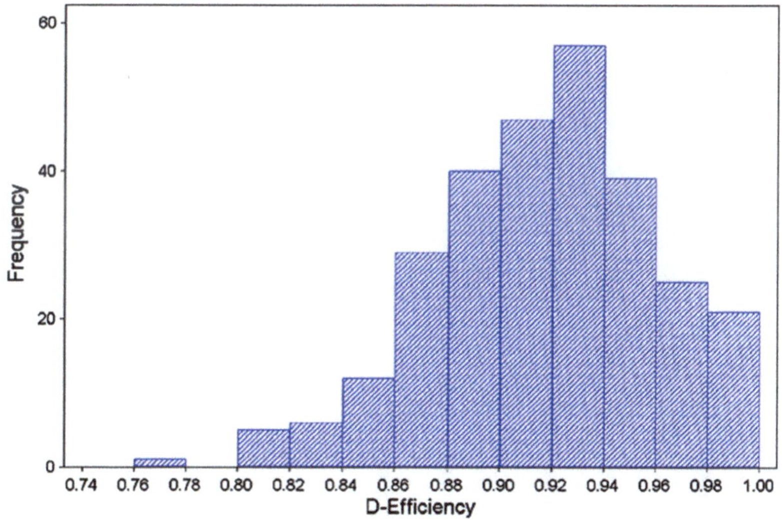

Fig. 20.24 Histogram of regular season *D-Efficiency* for post-season teams

typical regular season team should be able to win 36 % more games if it were efficient in the offense, defense, and integration sub-DMUs. Typical post-season teams demonstrate above average efficiency. On average, a post-season team should be able to increase its runs gained by 9 % (given its total bases gained),

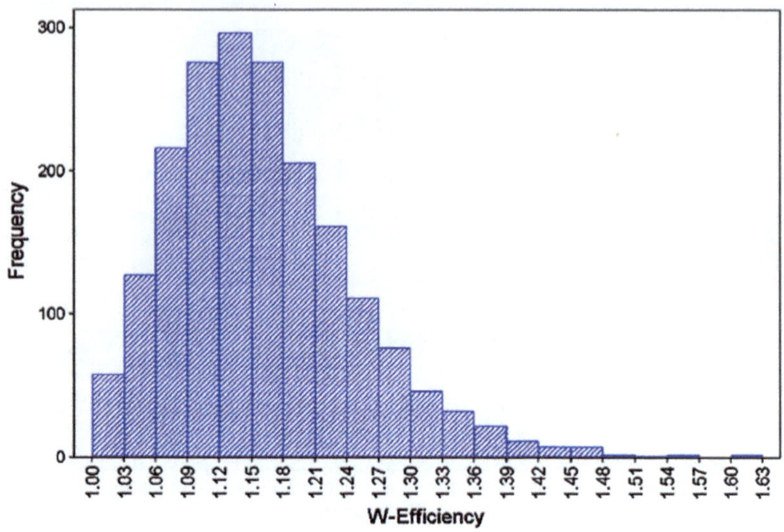

Fig. 20.25 Histogram of regular season *W-Efficiency* for all regular season teams

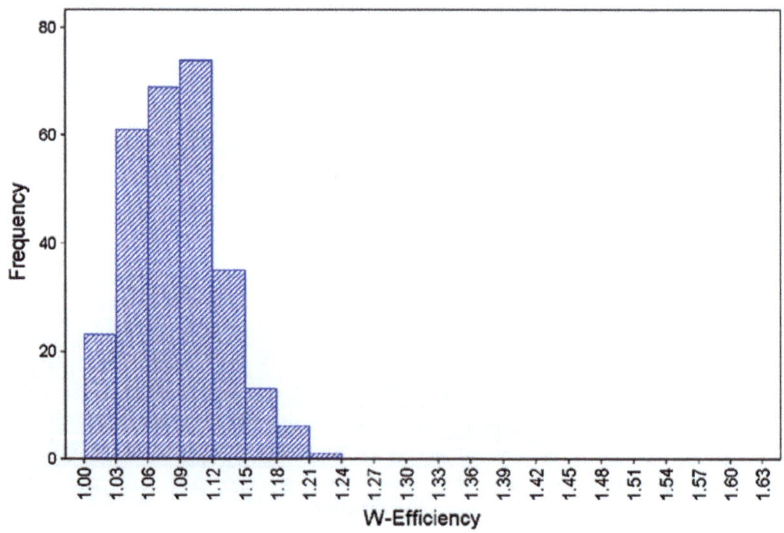

Fig. 20.26 Histogram of regular season *W-Efficiency* for post-season teams

decrease its runs surrendered by 8 % (given its total bases surrendered), and win 9 % more games (given its runs gained and runs surrendered). Overall, a typical post-season team should be able to win 17 % more games if it were efficient in the offense, defense, and integration sub-DMUs.

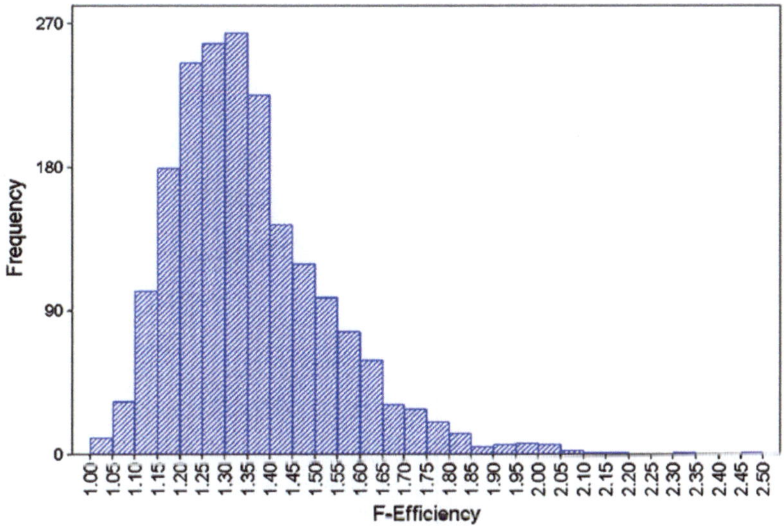

Fig. 20.27 Histogram of regular season *F-Efficiency* for all regular season teams

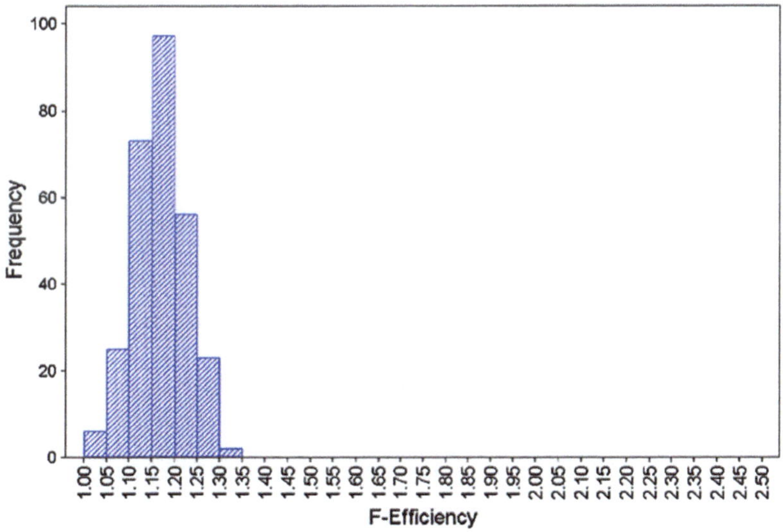

Fig. 20.28 Histogram of regular season *F-Efficiency* for post-season teams

Table 20.6 presents descriptive statistics of regular season effectiveness for all regular season teams and post-season teams and post-season effectiveness for all post-season teams. Figures 20.29 and 20.30 are histograms of regular season effectiveness for all regular season teams and postseason teams, respectively, and

Table 20.6 Descriptive statistics of regular season effectiveness for all regular season and post-season teams and post-season effectiveness for post-season teams

Variable	Teams	N	Mean	SD	Minimum	1st quartile	Median	3rd quartile	Maximum
Regular season *WPct*	All	1,934	0.50	0.08	0.23	0.44	0.50	0.56	0.75
	Post-season	282	0.61	0.04	0.51	0.58	0.60	0.63	0.75
Division series effectiveness	Post-season	56	0.58	0.33	0.11	0.30	0.60	0.88	1
League series effectiveness	Post-season	124	0.58	0.30	0.08	0.31	0.58	0.86	1
World series effectiveness	Post-season	192	0.57	0.29	0.05	0.34	0.58	0.81	1
Post-season effectiveness	Post-season	282	0.52	0.27	0.05	0.31	0.51	0.74	1

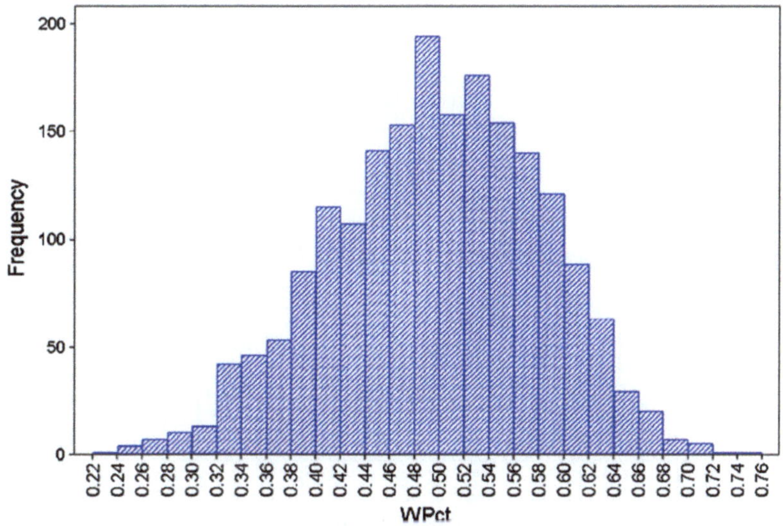

Fig. 20.29 Histogram of regular season effectiveness for all regular season teams

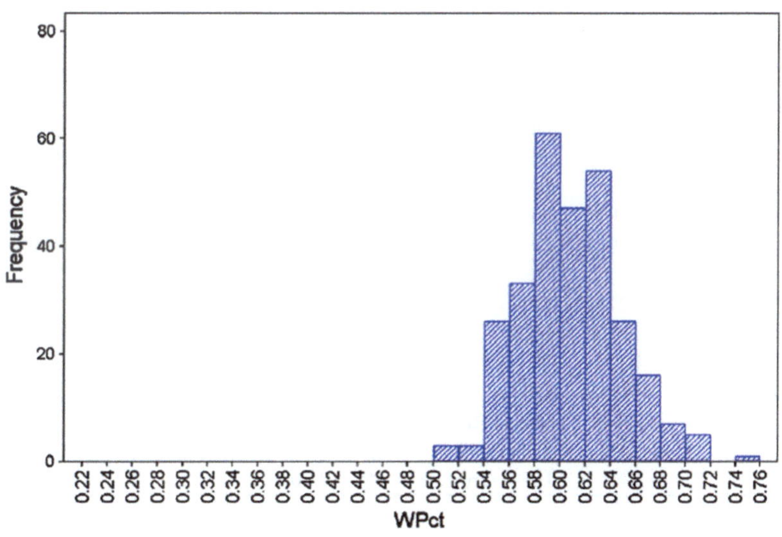

Fig. 20.30 Histogram of regular season effectiveness for post-season teams

Fig. 20.31 is a histogram of post-season effectiveness for post-season teams. We see that no team has won fewer than 23 % or more than 75 % of its regular season games. A typical post-season team wins 61 % of its regular season games and each post-season team has won at least 51 % of its regular season games.

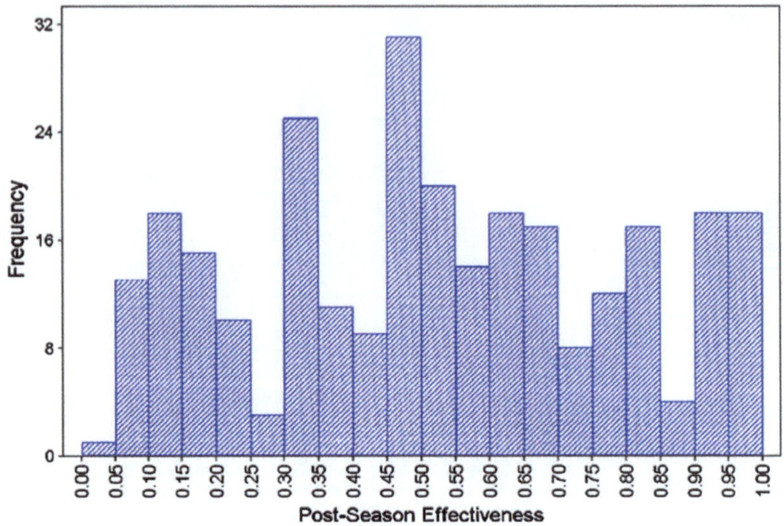

Fig. 20.31 Histogram of post-season effectiveness for post-season teams

Table 20.7 Results of hypothesis tests on post-season and non-post-season teams

Variable	Two-sample t test			Wilcoxon rank-sum test	
O-Capability	t = 13.23	DF = 1932	P < 0.00005	Z = 12.318	P < 0.00005
	F = 1.04	DF = 281,1651	P = 0.3371		
D-Capability	t = −10.37	DF = 423.4	P < 0.00005	Z = −9.518	P < 0.00005
	F = 1.37	DF = 1651,281	P = 0.0005		
O-Efficiency	t = −13.74	DF = 495.1	P < 0.00005	Z = −11.055	P < 0.00005
	F = 1.99	DF = 1651,281	P < 0.00005		
D-Efficiency	t = 8	DF = 406.2	P < 0.00005	Z = 7.623	P < 0.00005
	F = 1.21	DF = 1651,281	P = 0.0212		
W-Efficiency	t = −26.26	DF = 751.4	P < 0.00005	Z = −17.582	P < 0.00005
	F = 4.12	DF = 1651,281	P < 0.00005		
F-Efficiency	t = −41.07	DF = 1293.6	P < 0.00005	Z = −23.408	P < 0.00005
	F = 9.04	DF = 1651,281	P < 0.00005		
WPct	t = 42.97	DF = 672.7	P < 0.00005	Z = 24.134	P < 0.00005
	F = 3.47	DF = 1651,281	P < 0.00005		

Hypothesis Tests on Capability, Efficiency, and Effectiveness Measures

We perform a series of hypothesis tests to determine differences in capability, efficiency, and effectiveness between post-season and non-post-season teams. Table 20.7 presents the results of both a two-sample t-test and a Wilcoxon rank-sum test for each variable. Table 20.7 also shows the results of the F-tests to determine whether to assume equal or unequal variances when performing the t-tests. The results of the F-tests indicate that we should assume unequal variances

Table 20.8 Regression output for the regular season analysis

Weighted least squares linear regression of winning percentage					
Weighting variable: Games played					
Predictor variables	**Coefficient**	**Std error**	**T**	**P**	
Constant	1.36968	0.01867	73.36	0.0000	
O-Capability	0.03537	4.283E-04	82.57	0.0000	
D-Capability	−0.04247	3.907E-04	−108.71	0.0000	
O-Efficiency	−0.26566	0.00935	−28.42	0.0000	
D-Efficiency	−0.13465	0.00957	−14.06	0.0000	
W-Efficiency	−0.25583	0.00686	−37.28	0.0000	
R^2	0.9245	Resid. Mean Square (MSE)			0.08455
Adjusted R^2	0.9243	Standard deviation			0.29077
Source	DF	SS	MS	F	P
Regression	5	1,996.91	399.383	4,723.70	0.0000
Residual	1,928	163.01	0.085		
Total	1,933	2,159.92			

for all the t-tests except for the one involving *O-Capability*. We summarize the hypothesis tests for each variable below:

- *O-Capability*: Post-season teams on average gain more total bases than do non-post-season teams.
- *D-Capability*: Post-season teams on average surrender fewer total bases than do non-post-season teams.
- *O-Efficiency*: Post-season teams have on average lower inverse efficiency scores at the offense sub-DMU than do non-post-season teams.
- *D-Efficiency*: Post-season teams have on average higher efficiency scores at the defense sub-DMU than do non-post-season teams.
- *W-Efficiency*: Post-season teams have on average lower inverse efficiency scores at the integration sub-DMU than do non-post-season teams.
- *F-Efficiency*: Post-season teams have on average lower organizational inverse efficiency scores than do non-post-season teams.
- *WPct*: Post-season teams have on average higher regular season winning percentages than do non-post-season teams.

Regular Season Regression Analysis

The sample size for the regression model is 1934 regular season teams. We omitted *F-Efficiency* from the model because of its high colinearity with the other three efficiency scores. Table 20.8 shows the resulting regression model.

We observe that all five independent variables are highly statistically significant and that all five coefficients have the expected sign. Recall that *D-Capability* is a reverse quantity and that larger values of *O-Efficiency*, *D-Efficiency*, and *W-Efficiency* indicate greater potential to increase output and therefore signify

Table 20.9 Coefficient of partial determination and the R^2 contribution of each capability measure and each efficiency measure in the regular season analysis

Variable	Coefficient of partial determination	R^2 contribution
O-Capability	0.780	0.267
D-Capability	0.860	0.463
O-Efficiency	0.295	0.032
D-Efficiency	0.093	0.008
W-Efficiency	0.419	0.054

lower efficiency. The model explains over 92 % of the variation in *WPct*, indicating that omitted factors and random variation account for no more than 8 % of the variation in regular season effectiveness. Table 20.9 shows the coefficients of partial determination and the R^2 contribution of each independent variable.

The results show that capability contributes more to regular season effectiveness than does efficiency. Specifically, the coefficients of partial determination of *O-Capability* and *D-Capability* are 0.780 and 0.860, respectively, while those of *O-Efficiency*, *D-Efficiency*, and *W-Efficiency* are 0.295, 0.093, and 0.419, respectively. We observe a similar pattern in the R^2 contributions.

Defensive capability appears to be more important than offensive capability, as indicated by the higher coefficient of partial determination, higher R^2 contribution, and larger (absolute) regression coefficient for *D-Capability* relative to *O-Capability* ($t = 12.25$, $df = 3866$, $P < 0.001$). However, good management apparently can enhance offense more than it can enhance defense, as indicated by the higher coefficient of partial determination, higher R^2 contribution, and larger (absolute) regression coefficient for *O-Efficiency* relative to *D-Efficiency* ($t = 9.79$, $df = 3866$, $P < 0.001$). The coefficients of partial determination and the R^2 contributions suggest that *W-efficiency* contributes somewhat more to regular season effectiveness than does *O-Efficiency*, although the (absolute) regression coefficients of *W-Efficiency* and *O-Efficiency* are nearly equal.

Post-season Regression Analysis

The sample size for the regression model is 282 post-season teams. We find that a team's post-season performance is virtually unrelated to offensive and defensive capabilities and that only overall efficiency on the field (a combination of *O-Efficiency*, *D-Efficiency*, and *W-Efficiency*) has even the slightest relationship to post-season performance. Overall efficiency on the field can account for just over 1 % of post-season performance, suggesting that nearly 99 % of post-season success is attributable to chance and other unidentified factors.

20.5.2.5 Conclusions of the Study

In this section, we summarize our results by responding to each research question.

How much does team capability and managerial efficiency contribute to regular season effectiveness in MLB?

We conclude that both capability and efficiency are significant contributors to regular season effectiveness in MLB. However, capability is more important than efficiency. This supports our speculation that capability is more important than efficiency in industries where labor is highly paid. Moreover, we conclude that defensive capability contributes more to regular season effectiveness than does offensive capability. This supports an organizational strategy that places greater emphasis on defense (primarily pitching) relative to offense (primarily hitting).

Among the three efficiency measures, we conclude that the team's ability to win close games, as indicated by its *W-Efficiency*, has the greatest contribution to regular season effectiveness. This suggests that managers who employ effective strategies late in the game, such as pinch-hitting and relief pitching, can significantly influence the team's overall effectiveness.

We also find that the team's *O-Efficiency* has greater influence on its regular season effectiveness than does its *D-Efficiency*. We speculate that this may be because the offense typically has greater control over the tactics that increase *O-Efficiency* relative to the control that the defense has over the tactics that increase *D-Efficiency*. For example, the offense decides when to try to steal a base, when to attempt a hit-and-run play, and when a runner seeks to advance an extra base on a hit. There are few tactics that the defense can employ, such as pitching out and having the pitcher keep runners close to their bases, leaving the defense in a generally reactive position. Thus, the defense tends to rely more on capability – the ability of the pitcher and the catcher to prevent stolen bases and the throwing abilities of the outfielders – than on efficiency.

How much does team capability and managerial efficiency contribute to post-season effectiveness in MLB?

We conclude that regular season capabilities and efficiencies are poor predictors of post-season effectiveness. Thus, post-season success is overwhelmingly determined by chance in that even talented and well-managed teams have little relative advantage in post-season play. We believe that this is due primarily to two factors. First, opposing teams in the post-season are likely also to be talented and well managed, nullifying any relative advantage. Second, post-season series are short – either five or seven games in almost all cases – so that an inferior team maintains a significant chance of winning the series with the help of a few lucky bounces.

20.6 Conclusion

Data envelopment analysis has been extensively applied to measure the performance of individual athletes and teams in a variety of sports as well as to analyze nations competing in the Olympics. Most of the models presented in the literature are single-stage DEA models which treat the underlying process of converting inputs into outputs as a "black box." This approach is appropriate in many situations including when the purpose of the analysis is to rank decision making units (individual athletes, teams, or nations).

In other situations, analysts are interested in investigating the sources of inefficiency within the organization in order to improve organizational performance. For example, the owner of a sports team may be interested in evaluating the efficiency of various organizational sub-processes under the control of different administrators (general managers, talent scouts, on-field managers, or coaches) in order to make personnel decisions. To accomplish this, researchers have developed two-stage and network DEA methodologies.

In this chapter, we model an MLB team as comprised of a front office operation which consumes money in the form of player salaries to acquire offensive and defensive talent and an on-field operation which uses the talent to outscore opponents and win games. We present a network DEA methodology to measure performance of the front office operation (offense and defense), the on-field operation (offense, defense, and integration), and the overall team. We utilize the methodology in two industry-wide studies of Major League Baseball.

In the first study, we use a two-stage DEA model as part of a larger analysis to determine the minimum total player salary required for a team to be competitive, to count the number of teams that are noncompetitive due to low total player salary, to determine the hypercompetitive salary, to count the number of hypercompetitive teams, and to examine the relationship between market size, efficiency, and competitiveness. In order to address these issues, we need to classify the MLB teams as noncompetitive due to low total player salary, noncompetitive for other reasons, conditionally competitive, economically competitive, and hypercompetitive. The classification process utilizes the front office, on-field, and overall team efficiency scores obtained from the two-stage DEA model.

In the second study, we use a network DEA model as part of a larger analysis to explore the relative contributions of team capability and managerial efficiency to team effectiveness during the regular season and the post-season. In order to build the model for team effectiveness, we utilize the *O-Efficiency*, *D-Efficiency*, *W-Efficiency*, and *F-Efficiency* scores determined from the network DEA model.

We emphasize that each of these studies requires the results obtained from the two-stage and network DEA models in order to perform the analysis. A single-stage DEA model does not provide the in-depth information the analyst needs to address the research questions.

References

Anderson, T., & Sharp, G. (1997). A new measure of baseball batters using DEA. *Annals of Operations Research, 73*, 141–155.

Banker, R. D., Charnes, A., & Cooper, W. W. (1984). Some models for the estimation of technical and scale inefficiencies in data envelopment analysis. *Management Science, 30*, 1078–1092.

Barros, C. P., & Leach, S. (2006). Performance evaluation of the English Premier Football League with data envelopment analysis. *Applied Economics, 38*(12), 1449–1458.

Baseball Archive Database. http://www.seanlahman.com/baseball-archive/

Boscá, J. E., Liern, V., Martínez, A., & Sala, R. (2009). Increasing offensive or defensive efficiency? An analysis of Italian and Spanish football. *Omega, 37*(1), 63–78.

Castelli, C., Pesenti, R., & Ukovich, W. (2001). DEA-like models for efficiency evaluations of specialized and interdependent units. *European Journal of Operational Research, 132*, 274–286.

Charnes, A., Cooper, W. W., & Rhodes, E. (1978). Measuring the efficiency of decision making units. *European Journal of Operational Research, 2*(6), 429–444.

Chen, Y., & Zhu, J. (2004). Measuring information technology's indirect impact on firm performance. *Information Technology and Management Journal, 5*(1–2), 9–22.

Chen, Y., Liang, L., Yang, F., & Zhu, J. (2006). Evaluation of information technology investment: A data envelopment analysis approach. *Computers and Operations Research, 33*(5), 1368–1379.

Chen, Y., Cook, W. D., Li, N., & Zhu, J. (2009a). Additive efficiency decomposition in two-stage DEA. *European Journal of Operational Research, 196*(3), 1170–1176.

Chen, Y., Liang, L., & Zhu, J. (2009b). Equivalence in two-stage DEA approaches. *European Journal of Operational Research, 193*(2), 600–604.

Chen, Y., Cook, W. D., & Zhu, J. (2010). Deriving the DEA frontier for two-stage processes. *European Journal of Operational Research, 202*(1), 138–142.

Churilov, L., & Flitman, A. (2006). Towards fair ranking of Olympics achievements: The case of Sydney 2000. *Computers and Operations Research, 33*(7), 2057–2082.

Cook, W. D., & Zhu, J. (2008). CAR-DEA: Context-dependent assurance regions in DEA. *Operations Research, 56*(1), 169–178.

Cook, W. D., Liang, L., & Zhu, J. (2010). Measuring performance of two-stage network structures by DEA: A review and future perspective. *Omega, 38*, 423–430.

Cooper, W. W., Ruiz, J. L., & Sirvent, I. (2007). Choosing weights from alternative optimal solutions of dual multiplier models in DEA. *European Journal of Operational Research, 180* (1), 443–458.

Cooper, W. W., Ruiz, J. L., & Sirvent, I. (2009). Selecting non-zero weights to evaluate effectiveness of basketball players with DEA. *European Journal of Operational Research, 195*(2), 563–574.

Cover, T. M., & Keilers, C. W. (1977). An offensive earned-run average for baseball. *Operations Research, 25*, 729–740.

Einolf, K. W. (2004). Is winning everything? A data envelopment analysis of Major League Baseball and the National Football League. *Journal of Sports Economics, 5*(2), 127–151.

Espitia-Escuer, M., & García-Cebrián, L. I. (2004). Measuring the efficiency of Spanish First-Division soccer teams. *Journal of Sports Economics, 5*(4), 329–346.

Espitia-Escuer, M., & García-Cebrián, L. I. (2006). Performance in sports teams: Results and potential in the professional soccer league in Spain. *Management Decision, 44*(8), 1020–1030.

Färe, R., & Grosskopf, S. (2000). Network DEA. *Socio-Economic Planning Sciences, 3*, 249–267.

Färe, R., & Whittaker, G. (1995). An intermediate input model of dairy production using complex survey data. *Journal of Agriculture Economics, 46*, 201–213.

Färe, R., Grosskopf, S., & Whittaker, G. (2007). Network DEA. In J. Zhu & W. D. Cook (Eds.), *Modeling data irregularities and structural complexities in data envelopment analysis* (pp. 209–240). New York: Springer.

Farrell, M. J. (1957). The measurement of productive efficiency. *Journal of the Royal Statistical Society, 120*(3), 253–290.

Fizel, J. L., & D'Itri, M. P. (1997). Managerial efficiency, managerial succession and organizational performance. *Managerial and Decision Economics, 18*(4), 295–308.

Fizel, J. L., & D'Itri, M. P. (1999). Firing and hiring of managers: Does efficiency matter? *Journal of Management, 25*(4), 567–585.

Fried, H. O., Lambrinos, J., & Tyner, J. (2004). Evaluating the performance of professional golfers on the PGA, LPGA and SPGA tours. *European Journal of Operational Research, 154*(2), 548–561.

García-Sánchez, I. M. (2007). Efficiency and effectiveness of Spanish football teams: A three-stage-DEA approach. *Central European Journal of Operations Research, 15*(1), 21–45.

González-Gómez, F., & Picazo-Tadeo, A. J. (2010). Can we be satisfied with our football team? Evidence from Spanish professional football. *Journal of Sports Economics, 11*(4), 418–442.

Guzmán, I., & Morrow, S. (2007). Measuring efficiency and productivity in professional football teams: Evidence from the English Premier League. *Central European Journal of Operations Research, 15*(4), 309–328.

Haas, D. J. (2003a). Productive efficiency of English football teams – A data envelopment analysis approach. *Managerial and Decision Economics, 24*(5), 403–410.

Haas, D. J. (2003b). Technical efficiency in the Major League soccer. *Journal of Sports Economics, 4*(3), 203–215.

Haas, D. J., Kocher, M. G., & Sutter, M. (2004). Measuring efficiency of German football teams by data envelopment analysis. *Central European Journal of Operations Research, 12*(3), 251–268.

Hadley, L., & Ruggiero, J. (2006). Final-offer arbitration in Major League Baseball: A nonparametric analysis. *Annals of Operations Research, 145*(1), 201–209.

Holod, D., & Lewis, H. F. (2011). Resolving the deposit dilemma: A new DEA bank efficiency model. *Journal of Banking and Finance, 35*, 2801–2810.

Howard, L. W., & Miller, J. L. (1993). Fair pay for fair play: Estimating pay equity in professional baseball with data envelopment analysis. *Academy of Management Journal, 36*(4), 882–894.

Kao, C., & Hwang, S. N. (2008). Efficiency decomposition in two-stage data envelopment analysis: An application to non-life insurance companies in Taiwan. *European Journal of Operational Research, 185*(1), 418–429.

Levin, R. C., Mitchell, G. J., Volcker, P. A., & Will, G. F. (2000). *The report of the independent members of the Commissioner's Blue Ribbon Panel on Baseball Economics*. http://mlb.mlb.com/mlb/downloads/blue_ribbon.pdf

Lewis, H. F., & Mazvancheryl, S. K. (2011). A model for efficiency analysis of the customer satisfaction process. *Innovative Marketing, 7*, 33–45.

Lewis, H. F., & Sexton, T. R. (2004a). Network DEA: Efficiency analysis of organizations with complex internal structure. *Computers and Operations Research, 31*(9), 1365–1410.

Lewis, H. F., & Sexton, T. R. (2004b). Data envelopment analysis with reverse inputs and outputs. *Journal of Productivity Analysis, 21*(2), 113–132.

Lewis, H. F., Sexton, T. R., & Lock, K. A. (2007). Player salaries, organizational efficiency, and competitiveness in Major League Baseball. *Journal of Sports Economics, 8*(3), 266–294.

Lewis, H. F., Lock, K. A., & Sexton, T. R. (2009). Organizational capability, efficiency, and effectiveness in Major League Baseball: 1901–2002. *European Journal of Operational Research, 197*(2), 731–740.

Lewis, H. F., Mallikarjun, S., & Sexton, T. R. (2013). Unoriented two-stage DEA: The case of the oscillating intermediate products. *European Journal of Operational Research, 229*, 529–539.

Li, Y., Liang, L., Chen, Y., & Morita, H. (2008). Models for measuring and benchmarking Olympics achievements. *Omega, 36*(6), 933–940.

Liang, L., Cook, W. D., & Zhu, J. (2008). DEA models for two-stage processes: Game approach and efficiency decomposition. *Naval Research Logistics, 55*(7), 643–653.

Lorimer, L. (2002). *Baseball desk reference* (pp. 60–69). New York: Dorling Kindersley.

Lozano, S., Villa, G., Guerrero, F., & Cortés, P. (2002). Measuring the performance of nations at the Summer Olympics using data envelopment analysis. *Journal of the Operational Research Society, 53*(5), 501–511.

Major League Baseball Official Website. http://mlb.mlb.com/home

Mallikarjun, S., Lewis, H. F., & Sexton, T. R. (2013). Operational performance of U.S. public rail transit and implications for public policy. *Socio-Economic Planning Sciences*, (forthcoming).

Malmquist, S. (1953). Index numbers and indifference surfaces. *Trabajos de Estadística, 4*, 209–242.

Mazur, M. J. (1994). Evaluating the relative efficiency of baseball players. In A. Charnes, W. W. Cooper, A. Y. Lewin, & L. M. Seiford (Eds.), *Data envelopment analysts: Theory, methodology, and application* (pp. 369–391). Boston: Kluwer Academic.
Ruiz, J. L., Pastor, D., & Pastor, J. T. (2013). Assessing professional tennis players using data envelopment analysis (DEA). *Journal of Sports Economics, 14*(3), 276–302.
Seiford, L. M., & Zhu, J. (1999). Profitability and marketability of the top 55 U.S. commercial banks. *Management Science, 45*(9), 1270–1288.
Sexton, T. R., & Lewis, H. F. (2003). Two-stage DEA: An application to Major League Baseball. *Journal of Productivity Analysis, 19*(2–3), 227–249.
Sexton, T. R., Silkman, R. H., & Hogan, A. J. (1986). Data envelopment analysis: Critique and extensions. In R. H. Silkman (Ed.), *Measuring efficiency: An assessment of data envelopment analysis* (pp. 73–105). San Francisco: Jossey-Bass.
Statistics Canada. http://www.statcan.gc.ca/start-debut-eng.html
Sueyoshi, T., Ohnishi, K., & Kinase, Y. (1999). A benchmark approach for baseball evaluation. *European Journal of Operational Research, 115*(3), 429–448.
Tone, K., & Tsutsui, M. (2009). Network DEA: A slacks-based measure approach. *European Journal of Operational Research, 197*, 243–252.
Tone, K., & Tsutsui, M. (2010). Dynamic DEA: A slacks-based measure approach. *Omega, 38*, 145–156.
United States Census Bureau. http://www.census.gov/
USA Today. http://www.usatoday.com/sports/mlb/salaries/
Volz, B. (2009). Minority status and managerial survival in Major League Baseball. *Journal of Sports Economics, 10*(5), 522–542.
Wu, J., Liang, L., & Yang, F. (2009). Achievement and benchmarking of countries at the Summer Olympics using cross efficiency evaluation method. *European Journal of Operational Research, 197*(2), 722–730.
Wu, J., Zhou, Z., & Liang, L. (2010). Measuring the performance of nations at the Beijing Summer Olympics using an integer-valued DEA model. *Journal of Sports Economics, 11*(5), 549–566.
Yang, Z. (2006). A two-stage DEA model to evaluate the overall performance of Canadian life and health insurance companies. *Mathematical and Computer Modeling, 43*(7–8), 910–991.
Zhu, J. (2000). Multi-factor performance measure model with an application to Fortune 500 companies. *European Journal of Operational Research, 123*(1), 105–124.

Chapter 21
Production and Marketing Efficiencies of the U.S. Airline Industry: A Two-Stage Network DEA Approach

Wen-Min Lu, Shiu-Wan Hung, Qian Long Kweh, Wei-Kang Wang, and En-Tzu Lu

Abstract This chapter presents an application of a two-stage network data envelopment analysis (DEA) for examining the performance of 30 U.S. airline companies. The airline industry is a subject of concern because the industry is a major contributor to a country's or even global economic development. Although a number of studies have explored airline performance using DEA, relatively few studies have applied a two-stage DEA model. The current chapter examines production efficiency and marketing efficiency through an additive two-stage network DEA model. This approach allows the black-box of the performance measurement process to be assessed, thus, providing a new direction in measuring airline performance. The chapter includes a managerial decision-making matrix and makes suggestions to help airline managers improve performance for airlines. In addition, a regression analysis of the effect of corporate governance mechanisms on airlines performance is conducted. Given the volatility of growth in the airline industry, it is expected that we will see more research related to performance management in the industry.

W.-M. Lu
Department of Financial Management, National Defense University, No. 70, Section 2, Zhongyang North Road, Beitou, Taipei 112, Taiwan
e-mail: wenmin.lu@gmail.com

S.-W. Hung (✉)
Department of Business Administration, National Central University, No. 300, Jung-Da Road, Jung-Li City, Tao-Yuan 320, Taiwan
e-mail: shiuwan@mgt.ncu.edu.tw

Q.L. Kweh
Department of Accounting, College of Business Management and Accounting, Universiti Tenaga Nasional, Sultan Haji Ahmad Shah Campus, Muadzam Shah, Pahang 26700, Malaysia
e-mail: qlkweh@uniten.edu.my

W.-K. Wang • E.-T. Lu
Department of Accounting, Yuan Ze University, 135 Yuan-Tung Road, Chung-Li, Taiwan
e-mail: jameswang@saturn.yzu.edu.tw; karrylu1220@gmail.com

W.D. Cook and J. Zhu (eds.), *Data Envelopment Analysis*, International Series in Operations Research & Management Science 208, DOI 10.1007/978-1-4899-8068-7_21, © Springer Science+Business Media New York 2014

Keywords Two-stage data envelopment analysis (DEA) • Truncated regression • Corporate governance • Managerial decision-making matrix

21.1 Introduction

Since the United Stated Congress passed the Airline Deregulation Act in 1978, the air transport market has seen significant changes. This Open Skies policy allowed low-cost carriers (LCCs) to enter the air transport market. LCCs use a simple type of aircraft, secondary airports and simplified routes to reduce their operating costs, which allows them to provide lower fares to customers. The rapid expansion of LCCs has caused traditional airlines to confront fierce competition. Rising labor costs and volatile fuel prices affect all airlines. Competition in the airline industry is at an all-time high, challenging providers to reduce costs while improving quality. In this environment, the ability to attract new customers while retaining existing ones through superior customer service is not only a key competitive differentiator, but a necessity. Obstacles met in the search for flight information can diminish customers' perceptions of an airline's capability, decrease the opportunity for future revenue, and open the door for other carriers to win their business. In today's highly competitive market, airlines are deploying a range of innovations in terms of customer service and support in order to improve operating performance. The focus has moved from attempts to characterize performance in terms of a simple indicator, e.g., revenues, to a multi-dimensional systems perspective.

The Data Envelopment Analysis (DEA) is a linear programming based technique that converts multiple output and input measures into a single comprehensive measure of performance. This is attained through the construction of an empirical-based production or resource conversion frontier, and by the identification of peer groups. The philosophy behind DEA is predicated on the fact that a frontier transformation function empirically captures the underlying process defining firms' production activities. The application of DEA is strongly supported in the multitude of empirical analysis methods in different fields of profit (Seiford 1997; Zhu 2000; Gattoufi et al. 2004; Cooper et al. 2006). DEA has also been widely applied in evaluating airline performance (Sengupta 1999; Barbot et al. 2008; Barros and Peypoch 2009). The traditional DEA model is based on one-stage activities, which neglect intermediate measures or linking activities (Fare and Whittaker 1995; Chen and Zhu 2004; Tone and Tsutsui 2009). This study establishes a two-stage DEA model to overcome the shortcomings of the traditional one-stage DEA. While production efficiency indicates the relative efficiency of a firm in the production process, marketing efficiency reflects the relative performance of a firm in the marketing process. This study evaluates the relative efficiency of airlines in the US, in response to the changing nature of the airline market.

Corporate governance is a multi-faceted subject. An important theme of corporate governance is the nature and extent of accountability of particular individuals in the organization and mechanisms that try to reduce or eliminate the principal-agent problem. A related, but separate, thread of discussions focus on the impact of a corporate governance system on economic efficiency, with a strong emphasis on shareholders' welfare; this aspect is particularly present in contemporary public debates and developments in regulatory policy. Since the failures of well-known companies such as Enron, WorldCom, Tyco and Merck, academics and practitioners have shown increasing interest in corporate governance. Corporate governance is the set of processes, customs, policies, laws, and institutions affecting the way a corporation (or company) is directed, administered or controlled. Corporate governance also includes the relationships among the many stakeholders involved and the goals for which the corporation is governed. In contemporary business corporations, the main external stakeholder groups are shareholders, debt holders, trade creditors, suppliers, customers and the communities affected by the corporations' activities. Internal stakeholders include the boards of directors, executives and other employees. Chiang and Lin (2007), Bennedsen et al. (2008), Carline et al. (2009) and Sueyoshi et al. (2010) demonstrate that corporate governance is correlated with organizational performance. Gompers et al. (2003) illustrate that good governance positively affects a firm's performance. Several governance factors may affect the performance of airlines. To explore the impact of exogenous factors on corporate performance, Simar and Wilson (2007) verify that truncated regression is more appropriate than Tobit regression.

This study adopts bootstrapped DEA scores with truncated regression to analyze the relationship between corporate governance and airline performance. The significant difference between the present study and the studies mentioned above is that the former adopts a two-stage DEA to explore airline performance and addresses production efficiency and marketing efficiency in order to better understand intermediate measures or linking activities. Additionally, this study uses a managerial decision-making matrix in order to help airline managers rapidly improve corporate efficiency or strategies. Finally, this study uses truncated regression in order to analyze the relationship between corporate governance and performance and to guide managers toward competitiveness in the airline industry. The important contributions of this study include: (1) developing an innovative two-stage production process that includes production efficiency and marketing efficiency in order to assess the operating performance of airlines; (2) implementing truncated regression (Simar and Wilson 2007) in order to investigate whether or not corporate governance affects airline performance; (3) integrating production efficiency and marketing efficiency in order to address managerial decision-making. As a result, management could use the managerial decision-making matrix to set up improvable strategies.

The remainder of this chapter is organized as follows: Sect. 21.2 presents a literature review; Sect. 21.3 describes the research design, including our two-stage DEA methodology, truncated regression, collection of the sample data and the criteria for variables to evaluate performance; Sect. 21.4 presents empirical data and analyzes the results; and Sect. 21.5 presents the conclusion.

21.2 Operating Performance Measurement Approaches in the Airline Industry

In today's globalized world, air transportation systems play the role of providing a service to connect virtually all countries around the world. Over the last decade, the airline industry has experienced the fastest growth and has contributed to the creation of economies worldwide. However, the industry was greatly affected by economic challenges such as increased competition and volatility. The September 11 attacks in 2001 and the severe acute respiratory syndrome (SARS) in 2002 and 2003 caused the 2001/2003 aviation crisis, and the recent global financial crisis of 2007–2009 caused the 2008/2009 aviation crisis (Franke and John 2011). During these periods, even the most lucrative airlines lost money and this implies that airline managers have to take extra caution in choosing information that reflects their operating performance (Gramani 2012). Therefore, it is no surprise that there has been a recent spate of interest in the study of airline performance as researchers have applied an arsenal of tools to evaluate airline performance.

21.2.1 Uni-dimensional Measures

There is a long-standing debate over which measures reflect operating performance well. Among performance measurement technique, Francis et al. (2005) document that benchmarking is the most used method in the aviation industry. In the academic field, researchers have used various measures as indicators of the operating performance of airlines, including ratio analysis on accounting-based performance and market-based performance.

The traditional ratio measures are simple and easy to understand with each indicator providing a single dimensional measure of operating performance. Various ratios have been used in prior studies (see for example, Feng and Wang 2000). An example of a market-based ratio measure used is Tobin's Q. Lee et al. (2013) used approximate q, which is calculated as the summation of a firm's market value, liquidating value of outstanding preferred shares, and value of short-term liabilities. Raghavan and Rhoades (2005) employed an operating profit margin to indicate the profitability of the U.S. airlines. In examining the performance of international airlines, Backx et al. (2002) used several ratios, such as return on sales, return on assets, and employee productivity. Despite the prevalent use of ratio measures, Chuang et al. (2008) argue that the use of traditional performance measures like the Sharpe ratio could provide wrong information to investors regarding the stock performances of airlines.

Other measures that have been previously used are aggregate measures (Gorin and Belobaba 2004), activity-based costing (Lin 2012), an integrated approach of an analytic hierarchy process (AHP) and fuzzy TOPSIS method (Aydogan 2011), benchmarking (Francis et al. 2002), a multi-attribute decision making model (Chang and Yeh 2001) and even newly-developed explanatory frameworks (Tan and Rae 2009).

21.2.2 Data Envelopment Analysis

Prior studies document that DEA is superior to traditional methods, such as ratio analysis (Lu et al. 2014). Zhu (2000) argues that the single output to input financial ratios may not characterize the financial performance of a company that is complex in nature. Instead of using mere ratios or other individual financial variables, various attributes can be accommodated so that possible interactions between them can be captured in order to derive the efficiency scores of decision making units (DMUs) under DEA (Yeh 1996; Homburg 2001; Biener and Eling 2011; Lu and Hung 2011; Premachandra et al. 2011; Fang et al. 2013; Matthews 2013; Yang and Morita 2013). This means that DEA provides additional information through the computed efficiency scores of financial measures aggregation as compared to such financial ratios as return on assets that are uni-dimensional and have subjective problematic interpretations (Feroz et al. 2003). Moreover, comparing multiple inputs and outputs of DMUs for measuring relative DMU's efficiency allows for the identification of benchmarking. Other advantages of DEA include the identification of sources and the amount of inefficiency in each input and each output for a DMU as well as its ability for benchmarking purposes (Cooper et al. 2006).

Contemporary research in the aviation industry has applied DEA in order to evaluate organization performance. Through a rigorous analysis, Liu et al. (2013b) provide a summary of five major DEA application areas, among which is transportation. In air transportation, Schefczyk (1993) is the leading article measuring the operational performance of airlines. Using data from 15 airlines, the author uses DEA to analyze the operational performance of airlines and concludes with an analysis of the strategic factors of high profitability and performance in the airline industry.

Sengupta (1999) evaluates the performance of seven major airlines between 1988 and 1994 by using the DEA method. The results of this study showed that techniques and the allocation efficiency of the airlines changed significantly during this period. Scheraga (2004) investigates whether relative operational efficiency implied superior financial mobility. He used DEA to derive efficiency scores for 38 airlines in North America, Europe, Asia and the Middle East, and found that relative operational efficiency did not inherently imply superior financial mobility.

Chiou and Chen (2006) employ DEA to evaluate 15 Taiwanese domestic air routes from three perspectives proposed by Fielding et al. (1978). The results of the DEA model suggest that ten routes were relatively cost efficient, five routes were relatively cost effective and four routes were relatively service effective. The study also performed clustering analysis to categorize the routes into four clusters.

Based on the characteristics of each route, the authors addressed directions for improvement.

Barbot et al. (2008) use DEA and total factor productivity (TFP) to analyze the efficiency and productivity of the 49 member airlines of IATA. The study found that low-cost carriers perform more efficiently than full-service carriers, and larger airlines are more efficient than smaller ones. With respect to geographic areas, the author noted that the European and American carriers were more effective than airlines in Asia Pacific and China/North Asia. The results of the DEA analysis illustrate that efficiency and effectiveness are not always correlated. The results of the TFP analysis show that the airlines operating within more homogeneous and regulatory structured areas, like North America, are more uniform in their productivity.

Greer (2008) uses DEA and the Malmquist productivity index to examine changes in the productivity of the major U.S. passenger airlines from 2000 to 2004. The study suggests that there was a significant improvement in the productivity of the carriers during this period. Most of the productivity improvements came from the efficiency laggards' catching up with efficiency leaders in the industry. Barros and Peypoch (2009) apply DEA to evaluate the efficiency of 27 airlines in the Association of European Airlines (AEA), from 2000 to 2005. The study found that almost all European airlines operate at a high level of pure technical efficiency and scale efficiency. In the second stage, the study used bootstrapped truncated regression and noted that population and network alliances are the most important influences on the efficiency of airlines.

Hong and Zhang (2010) use DEA to analyze the operations of 29 airlines from 1998–2002 in order to explore whether a high degree of cargo business improves the operational efficiency of mixed passenger/cargo airlines. It was found that airlines with a high degree of cargo business are significantly more efficient than ones with a low degree of cargo business. Moreover, the authors found no statistically significant difference between airlines with similar degrees of cargo business.

Merkert and Hensher (2011) evaluate key determinants of 58 passenger airlines' efficiency using bootstrapped DEA scores in the first stage and partially bootstrapped random effects Tobit regressions in the second stage. They show that airline size and fleet mix characteristics have positive impacts on technical, allocative and cost efficiencies. Although the age of an airline's fleet does have a positive impact on its allocative and cost efficiency, it has no significant impact on its technical efficiency, which means that an old fleet can possibly achieve higher efficiency than a young fleet. Merkert and Williams (2013) also apply a two-stage approach to measure the efficiency of 18 European public service obligations airlines. In the first stage, they document a bootstrapped DEA analysis to compute the technical efficiency of the sample airlines. In the second stage, they regress the first-stage DEA efficiency scores against explanatory variables (determinants of airline performance).

Wang et al. (2011) also apply a two-stage approach, where they use DEA to assess the operating performance of 30 airlines in the U.S. and regress the first-stage DEA efficiency scores against corporate governance mechanisms. They assess competitive advantages of airlines through efficiency decomposition, cluster analysis, and multidimensional scaling. Their truncated regression analysis shows that airline performance is related to their characteristics and corporate governance mechanisms. Studying 17 European airlines, Lozano and Gutiérrez (2011) employ a multi-objective DEA approach to determine the trade-offs among environmental impact, fleet cost and operating cost, and a Slacks-Based Measure (SBM) of DEA to assess the technical efficiency of each airline. Their results show that approximately 50 % of the airlines are technically inefficient.

The study of Gramani (2012), which is closely related to our study, applies a two-phase data envelopment analysis approach to examine the operational and financial performances for Brazilian and American airlines from 1997 to 2006. The intermediate input in their study is the inverted efficiency scores obtained in the first phase. In contrast, our study is set in a different research setting in which we establish a two-stage DEA model with intermediate measures in a single implementation. This is consistent with that of Zhu (2011), who examined the performance of fleet maintenance in the first stage and the performance of revenue generation in the second stage through an application of the centralized model developed by Liang et al. (2008).

In summary, prior studies have considered different inputs and outputs with no unanimous orientation. That is, input-oriented, output-oriented, and non-oriented models have been applied in prior studies. With respect to the assumption on returns to scale, they assume either constant returns to scale (CRS) or variable returns to scale (VRS). Both radial and non-radial methods have been employed. A drawback of traditional DEA models is that they ignore intermediate measures or linking activities. To overcome this problem, an integrated additive two-stage DEA model is discussed in the next section.

21.3 Research Design

From the influential study of Charnes et al. (1978), the first outcome of note is the definition of DEA: a "measure of the efficiency of any decision making unit (DMU) is obtained as the maximum of a ratio of weighted outputs to weighted inputs subject to the condition that the similar ratios for every DMU be less than or equal to unity" (Charnes et al. 1978). DEA, a widely used linear-programming-based composite tool, is a non-parametric mathematical technique for measuring the relative efficiency of DMUs, in particular the efficiency of the DMUs in transforming inputs into outputs. Particularly, DEA first establishes an "efficient frontier" formed by a set of DMUs that exhibit best practices and then assigns the efficiency level to other non-frontier units according to their distance from the efficient frontier. Put differently, a company is technically efficient if it cannot improve any of its inputs or outputs without reducing some of its other inputs or outputs (Cooper et al. 2004).

DEA has been in existence for more than 30 years. Since the publication of the seminal paper by Charnes et al. (1978), the number of DEA-related published research articles has exceeded 4,000 entries up to year 2007 (Emrouznejad et al. 2008) and has accumulated approximately 4,500 papers in the ISI Web of Science database up through the year 2009 (Liu et al. 2013a). The usefulness of DEA is evident in its wide acceptance, in which it has been widely applied to various industrial and non-industrial contexts, such as transportation, banking, education, etc. in the academic field. Furthermore, almost no one in the DEA research community is cognizant of the development and application of DEA to real world scenarios in countries such as China and UAE (Liu et al. 2013b). That is, DEA has become a widely used tool for evaluating corporate performance in the study of management and related disciplines (Seiford and Zhu 1999; Zhu 2000; Banker et al. 2004; Wang 2005; Lu and Hung 2010; Sueyoshi and Goto 2010; Chang et al. 2008).

Taken together, applying DEA, a multi-factor performance measurement model, to measure corporate performance is more advantageous than traditional performance measures and can better capture managerial efficiency in managing organizational resources. Those publications noted earlier are definitely great examples to illustrate the pros of using DEA to address real world problems or corporate performance.

However, the traditional DEA model, which is based on a one-stage approach ignores intermediate measures or linking activities (Chen and Zhu 2004; Tone and Tsutsui 2009). That is, the conventional DEA model treats each DMU as a “black box” where only initial inputs and final outputs are assessed for efficiency measurement (Chen and Yan 2011). A two-stage DEA model is able to overcome the shortcoming of the traditional one-stage DEA model.

21.3.1 Two-Stage Transformation

Evaluating organizational performance is a complex process that cannot take into account just one criterion or just one dimension. A number of studies have applied DEA, which converts multiple inputs and outputs into a single efficiency score in order to evaluate the performance of organizations (Seiford 1997; Gattoufi et al. 2004; Emrouznejad et al. 2008).

One disadvantage of traditional DEA models is that they neglect intermediate measures or linking activities (Fare and Whittaker 1995; Chen and Zhu 2004; Tone and Tsutsui 2009). In order to adequately evaluate the operating performance of airlines, this study proposes a two-stage production process that includes production efficiency and marketing efficiency, as shown schematically in Fig. 21.1.

In each stage, input and output variables are chosen based on the literature from studies on the field of aviation (Schefczyk 1993; Charnes et al. 1996; Ray and Hu 1997; Alam et al. 1998; Sengupta 1999; Scheraga 2004; Barbot et al. 2008; Greer 2008). In the first stage, each airline uses six inputs to produce two outputs, which are then used as inputs in the second stage to produce two further outputs. The input, intermediate and output variables used in this study are defined as follows.

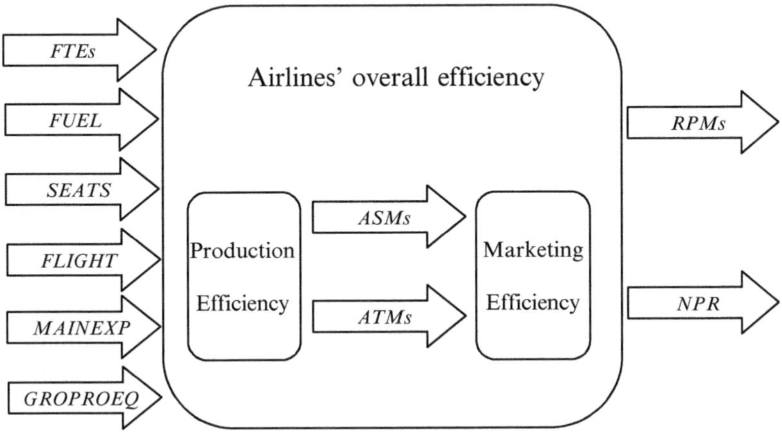

Fig. 21.1 Two-stage production processes for airline

Input variables:

Employees (FTEs): The number of full-time equivalent employees (FTEs)
Fuel Consumed (FUEL): The total gallons of fuel consumed during the current period
Seating Capacity (SEATS): The total number of seats in all aircraft
Flight Equipment (FLIGHT): The cost of flight equipment
Maintain Expense (MAINEXP): The maintenance, materials and repairs expenses in the income statement
Ground Property & Equipment (GROPROEQ): The cost of equipment and property minus that of flight equipment

Intermediate variables:

Available Seat Miles (ASMs): The total number of seat miles that were available to passengers (i.e. aircraft miles flown times the number of seats available for revenue passenger use)
Available Ton Miles (ATMs): The sum of the products obtained from the number of tons available to carry revenue load passengers, freight and mail on each flight stage multiplied by miles flown

Output variables:

Revenue Passenger Mile's (RPMs): The scheduled revenue miles flown by passengers (i.e. revenue passengers carried times miles flown)
Non-Passenger Revenue (NPR): The total amount of passenger revenue subtracted from gross sales

21.3.2 Data Collection and Descriptive Statistics

The data were extracted from the Compustat database, the Bureau of Transportation Statistics (BTS), annual reports, and proxy statements published in 2006. The study first selected 36 airlines listed on the US stock exchanges, including 26 US airlines and 10 American Depositary Receipt (ADR) airlines. Each airline is treated as a DMU in the DEA analysis. The samples included an additional consolidated statement from a different source, while airlines that were missing data were eliminated. As a result, the final sample list contained 30 airlines, including 21 US airlines and 9 ADR airlines. We examined the efficiency of the airlines with the two-stage production process, in which all of the outputs from the first stage were used as inputs to the second stage. Furthermore, we explored whether characteristics of corporate governance affected airline performance.

Descriptive statistics for the 30-airline sample are provided in Table 21.1. Panels B and C in Table 21.1 show the correlation coefficients for inputs and outputs in the two stages. The results reveal a significantly positive relation between inputs and outputs. The data set satisfies the assumption of isotonicity, that is, an increase in any input should not result in a decrease in any output. Besides, DEA requires that the number of DMUs be at least twice the total number of input and output variables (Golany and Roll 1989). In this study there are six inputs and two outputs in stage one, with two inputs and two outputs in stage two. Each stage meets the criterion, i.e., $30 > 2(6 + 2)$ in the first stage and $30 > 2(2 + 2)$ in the second stage. The DEA model of this study is thus deemed valid.

21.3.3 Additive Efficiency Decomposition in Two-Stage DEA

A number of studies (Seiford 1997; Gattoufi et al. 2004; Emrouznejad et al. 2008) have employed two-stage processes to evaluate the operating efficiency of peer organizations using different DEA models. In general, DEA models can be sub-divided into four categories: separate DEA models (SDEA; e.g., Karlaftis 2004; Chiou and Chen 2006), separate two-stage DEA models (STDEA; e.g., Seiford and Zhu 1999; Keh et al. 2006), network DEA models (NDEA; e.g., Yu and Lin 2008; Yu 2008; Kao 2009; Tone and Tsutsui 2009; Cook et al. 2009), and integrated two-stage DEA models (ITDEA; e.g., Kao and Hwang 2008; Chen et al. 2009; Cook et al. 2010). The SDEA cannot conduct the two-stage efficiency, with intermediate measures, in a single implementation. Hence, because of intermediate measures, the performance improvement of one stage affects the efficiency status of the other.

The lack of interrelated performance among different stages in SDEA may be solved by the NDEA or ITDEA models. However, due to the complexity of the modeling, the scale economy and slack values for each DMU are hard to compute using the NDEA model proposed by Yu and Lin (2008), Yu (2008) and Kao (2009), which is only applicable to the case of constant returns to scale. The NDEA model

Table 21.1 Descriptive statistics and correlation coefficients for the 30 airlines

Variables	Mean	Std. dev.	Median	Minimum	Maximum
Panel A: Efficiency of airlines					
Inputs					
FTEs (Employee)	24.75	26.99	13.06	0.37	103.05
FUEL (Gallons 000s)	789,702.17	885,551.19	264,196.00	2,551.00	3,178,000.00
SEATS (Seats)	34,403.67	34,089.43	18,733.50	190.00	132,718.00
FLIGHT (US$ 000s)	5,074,473.43	6,134,591.70	2,238,455.50	847.00	22,913,000.00
MAINEXP (US$ 000s)	357,400.10	352,232.28	218,975.50	3,785.00	1,157,104.00
GROPROEQ (US$ 000s)	948,654.63	1,302,488.84	261,790.00	1,724.00	4,578,824.00
Intermediate					
ASMs (000s)	51,510,127.83	54,875,526.86	24,504,735.00	83,829.00	187,575,000.00
ATMs (000s)	8,311,688.83	8,342,892.92	3,801,467.00	32,808.00	28,904,778.00
Outputs					
RPMs (000s)	40,344,345.87	43,982,663.61	18,424,492.50	32,656.00	149,426,000.00
NPR (US$ million)	1,320.44	2,247.04	319.17	7.84	10,485.02

Panel B: Correlation coefficients for inputs and outputs in the First-Stage

	FTEs	*FUEL*	*SEATS*	*FLIGHT*	*MAINEXP*	*GROPROEQ*	*ASMs*	*ATMs*
FTEs	1.000							
FUEL	0.905	1.000						
SEATS	0.820	0.919	1.000					
FLIGHT	0.875	0.891	0.923	1.000				
MAINEXP	0.935	0.912	0.858	0.866	1.000			
GROPROEQ	0.835	0.815	0.908	0.943	0.820	1.000		
ASMs	0.917	0.969	0.950	0.933	0.933	0.882	1.000	
ATMs	0.633	0.763	0.784	0.724	0.668	0.666	0.766	1.000

(continued)

Table 21.1 (continued)

Variables	Mean	Std. dev.	Median	Minimum	Maximum
Panel C: Correlation coefficients for inputs and outputs in the Second-Stage					
	ASMs	*ATMs*	*RPMS*	*NPR*	
ASMs	1.000				
ATMs	0.766	1.000			
RPMS	0.999	0.763	1.000		
NPR	0.550	0.414	0.547	1.000	

Note:
Input variables:
FTEs = the number of full-time equivalent employees (FTEs)
FUEL = the total amount of gallons of fuel consumed during the current period
SEATS = the total number of seats in all the aircraft
FLIGHT = the cost of flight equipment
MAINEXP = the maintenance, materials and repairs expenses in the income statement
GROPROEQ = the cost of equipment and property minus that of flight equipment
Intermediate variables:
ASMs = the total number of seat miles that were available to passengers (i.e. aircraft miles flown times by the number of seats available for revenue passenger use)
ATMs = the sum of the products obtained from the number of tons available for the carrying of revenue load passengers, freight and mail on each flight stage multiplied by miles flown
Output variables:
RPMs = the scheduled revenue miles flown by passengers (i.e. revenue passengers carried times by miles flown)
NPR = the total amount of passenger revenue subtracted from gross sales

(Tone and Tsutsui 2009) does not show the relative importance or contribution of the performances of individual stages to the overall performance of the entire process. The ITDEA model proposed by Chen et al. (2009) and Cook et al. (2010) can be applied to both technologies of variable and constant returns to scale and represents the relative importance or contribution of the performances of individual stages to the overall performance of the entire process. One reasonable choice of the weights for each stage is by the relative size of each stage.

Therefore, this study adopts the additive efficiency decomposition of Chen et al. (2009) and Cook et al. (2010) by establishing a two-stage DEA model with intermediate measures in a single implementation. Consider the two-stage production process presented in Fig. 21.1. Assume we have n airlines and that airline j uses m inputs (x_{ij}, $i = 1, \ldots, m$) to produce d outputs (z_{pj}, $p = 1, \ldots, d$) in the first stage; these are then used as inputs in the second stage to produce s outputs (y_{rj}, $r = 1, \ldots, s$). The efficiency measure for stages 1–2 of the process under VRS for an observed airline becomes β_1 and β_2.

$$\beta_1 = \left(\sum_{p=1}^{d} \eta_p z_{po} + \kappa^A\right) / \left(\sum_{i=1}^{m} v_i x_{io}\right) \tag{21.1}$$

$$\beta_2 = \left(\sum_{r=1}^{s} u_r y_{ro} + \kappa^B\right) / \left(\sum_{p=1}^{d} \eta_p z_{po}\right). \tag{21.2}$$

The overall efficiency measure of the two-stage process can reasonably be represented as a convex linear combination of the two stage-level measures, namely,

$$\theta = w_1\beta_1 + w_2\beta_2 \quad \text{where } w_1 + w_2 = 1.$$

The weights (w_1 and w_2) are intended to represent the relative importance or contribution of the performances of individual stages to the overall performance of the two stage process. The weights (w_1 and w_2) in each stage are determined based on the relative size of that stage. To be more specific, ($\sum_{i=1}^{m} v_i x_{io}$ + $\sum_{i=1}^{m} \eta_p z_{po}$) represents the total size of, or total amount of, resources consumed by the two stage process. Assume that w_1 and w_2 are defined as the proportion of the total input used at each stage, then

$$w_1 = \left(\sum_{i=1}^{m} v_i x_{io}\right) / \left(\sum_{i=1}^{m} v_i x_{io} + \sum_{p=1}^{d} \eta_p z_{po}\right) \text{ and}$$
$$w_2 = \left(\sum_{p=1}^{d} \eta_p z_{po}\right) / \left(\sum_{i=1}^{m} v_i x_{io} + \sum_{p=1}^{d} \eta_p z_{po}\right).$$

Thus, the overall efficiency θ is in the form

$$\theta = \left(\sum_{p=1}^{d} \eta_p z_{po} + \sum_{r=1}^{s} u_r y_{ro} + \kappa^A + \kappa^B\right) / \left(\sum_{i=1}^{m} v_i x_{io} + \sum_{p=1}^{d} \eta_p z_{po}\right) \tag{21.3}$$

The overall efficiency θ of the two stage process can be optimized, subject to the restrictions that the individual measures (β_1 and β_2) must not exceed unity in the linear programming format, after making the usual Charnes and Cooper transformation (1962). The linear programming problem of additive efficiency decomposition in the two-stage DEA under the VRS model is as follows.

$$\begin{aligned} & Max \sum_{p=1}^{d} \eta_p z_{po} + \sum_{r=1}^{s} u_r y_{ro} + \kappa^A + \kappa^B \\ & \text{subject to} \\ & \sum_{i=1}^{m} v_i x_{io} + \sum_{p=1}^{d} \eta_p z_{po} = 1 \\ & \sum_{p=1}^{d} \eta_p z_{pj} + \kappa^A - \sum_{i=1}^{m} v_i x_{ij} \leq 0 \\ & \sum_{r=1}^{s} u_r y_{rj} + \kappa^B - \sum_{p=1}^{d} \eta_p z_{pj} \leq 0 \\ & v_i, \eta_p, u_r \geq 0; \kappa^A, \kappa^B \text{ free in sign}; \quad j = 1, 2, \ldots, n. \end{aligned} \tag{21.4}$$

If $\kappa^A = \kappa^B = 0$ in (21.4), then the technology is said to exhibit CRS. The dual model of (21.4) under the VRS model is as follows:

$$\begin{aligned} & Min \ \ \theta_k \\ & \text{subject to} \\ & \sum_{j=1}^{n} \lambda_{jk} x_{ij} \leq \theta_k x_{ik}, & i = 1, \ldots, m, \\ & \sum_{j=1}^{n} \mu_{jk} z_{pj} - \sum_{j=1}^{n} \lambda_{jk} z_{pj} \leq \theta_k z_{pk} - z_{pk}, & p = 1, \ldots, d, \\ & \sum_{j=1}^{n} \mu_{jk} y_{rj} \geq y_{rk}, & r = 1, \ldots, s, \\ & \sum_{j=1}^{n} \lambda_{jk} = 1, \\ & \sum_{j=1}^{n} \mu_{jk} = 1, \\ & \lambda_{jk} \geq 0; \quad j = 1, 2, \ldots, n, \\ & \mu_{jk} \geq 0; \quad j = 1, 2, \ldots, n. \end{aligned} \tag{21.5}$$

The solution based on formula (21.5), λ_{jk} can be utilized to determine whether unit j is a peer of the observed unit k in the first stage. If it is zero, then unit j is not a peer, otherwise, λ_{jk} serves as an indication of how much unit j is to be learned by the observed unit k. The larger λ_{jk} which is the stronger unit j, is related to the observed unit. μ_{jk} plays the same role in the second stage.

Once we obtain the overall efficiency, models similar to (21.4) can be developed to determine the efficiency of each stage. Specifically, assuming the pre-emptive priority of stage 1, the following model determines that stage's efficiency TE_k^1, while maintaining the overall efficiency score at θ_k calculated from (21.6),

$$
\begin{aligned}
& TE_k^1 = \sum\nolimits_{d=1}^{D} \pi_d z_{dk} + \kappa^A \\
& s.t. \\
& \quad \sum\nolimits_{d=1}^{D} \pi_d z_{dj} + \kappa^A - \sum\nolimits_{i=1}^{m} w_i x_{ij} \le 0 \\
& \quad \sum\nolimits_{r=1}^{s} \mu_r y_{rj} + \kappa^B - \sum\nolimits_{d=1}^{D} \pi_d z_{dj} \le 0 \\
& \quad (1-\theta_k) \sum\nolimits_{d=1}^{D} \pi_d z_{dk} + \sum\nolimits_{r=1}^{s} \mu_r y_{rk} + \kappa^A + \kappa^B = \theta_k \\
& \quad \sum\nolimits_{i=1}^{m} w_i x_{ik} = 1 \\
& \quad \pi_d, \mu_r, w_i \ge 0, \quad j = 1, \ldots, n, \\
& \quad \kappa^A, \kappa^B \text{ free in sign.}
\end{aligned}
\tag{21.6}
$$

Similarly, if stage 2 is to be given pre-emptive priority, the following model determines the efficiency TE_k^2 for that stage, while

$$
\begin{aligned}
& TE_k^2 = \sum\nolimits_{r=1}^{s} \mu_r y_{rk} + \gamma_B \\
& s.t. \\
& \quad \sum\nolimits_{d=1}^{D} \pi_d z_{dj} + \kappa^A - \sum\nolimits_{i=1}^{m} w_i x_{ij} \le 0 \\
& \quad \sum\nolimits_{r=1}^{s} \mu_r y_{rj} + \kappa^B - \sum\nolimits_{d=1}^{D} \pi_d z_{dj} \le 0 \\
& \quad \sum\nolimits_{d=1}^{D} \pi_d z_{dk} + \sum\nolimits_{r=1}^{s} \mu_r y_{rk} - \theta_k \sum\nolimits_{i=1}^{m} w_i x_{ik} + \kappa^A + \kappa^B = \theta_k \\
& \quad \sum\nolimits_{d=1}^{D} \pi_d z_{dk} = 1 \\
& \quad \pi_d, \mu_r, w_i \ge 0, \quad j = 1, \ldots, n, \\
& \quad \kappa^A, \kappa^B \text{ free in sign.}
\end{aligned}
\tag{21.7}
$$

21.3.4 Truncated Regression

In the DEA literature, Tobit regression has been used to investigate whether performance is affected by exogenous factors. This chapter assumes and tests the regression condition as:

$$TE_j = \alpha + X_j \boldsymbol{\beta} + \varepsilon_j, \quad j = 1, \ldots, n. \tag{21.8}$$

In (21.8), α is the intercept, ε_j is the residual value, and X_j is a vector of observation-specific variables for airline j that is expected to be related to the airline's efficiency score, represented by TE_j. Nevertheless, Simar and Wilson (2007) illustrate that Tobit regression is inappropriate to analyze the efficiency score under DEA. They also developed a truncated-regression model with bootstraps instead of the Tobit model, and illustrate satisfactory performance in Monte Carlo experiments.

This study follows the approach of Simar and Wilson (2007) by adopting the exogenous factors (corporate governance proxy variables) that would affect the performance of airlines. It is noted that the distribution of ε_j is restricted by the condition $\varepsilon_j \geq 1 - \alpha - X_j\boldsymbol{\beta}$ in (21.8). This study modifies (21.8) and assumes that the distribution before truncation is truncated normal with zero mean, unknown variance, and truncation point, which are determined by different conditions. Equation 21.9 is the result after modification.

$$\hat{TE}_j \approx \alpha + X_j\beta + \varepsilon_j, \quad j = 1, \ldots, n, \tag{21.9}$$

where $\varepsilon_j \sim N(0, \sigma_\varepsilon^2)$, such that $\varepsilon_j \geq 1 - \alpha - X_j\boldsymbol{\beta}$, $j = 1, \ldots, n$. This study uses the regression process of parametric bootstrapping to estimate parameters $(\boldsymbol{\beta}, \sigma_\varepsilon^2)$, estimates (21.8) by maximizing the corresponding likelihood function and gives heed to $(\boldsymbol{\beta}, \sigma_\varepsilon^2)$. Readers not familiar with the details of the estimation algorithm are referred to Simar and Wilson (2007).

21.4 Empirical Results

This section presents the empirical results of the two-stage DEA approach and truncated regression. In the first phase, we examine the production and marketing performances. Combining production efficiency and marketing efficiency, we construct a decision-making matrix to identify the relative positions of our sample airlines. In the second phase, we further help airline managers to improve their operating performance through regression analysis. That is, we regress production efficiency and marketing efficiency (dependent variables), respectively, on a number of corporate governance mechanisms and control variables (explanatory variables).

21.4.1 Measuring Production and Marketing Performances

Based on the controllable aspect of a manager, this study adopts additive efficiency decomposition (Chen et al. 2009; Cook et al. 2010) under the assumption of input minimization (also known as input orientation) to measure the operating performance of the multi-stage production of airlines, with intermediate measures, in a single implementation model. One opting for DEA analysis should choose either the CRS or VRS model. As Avkiran (2001) suggests, the way to choose between CRS and VRS is to run the performance models under each assumption and compare the efficiency scores. In this study, a Wilcoxon Matched Pairs Test is applied to perform the evaluation. The mean of the paired differences between CRS and VRS scores are not significantly greater than zero, thus supporting the CRS

Table 21.2 Additive efficiency decomposition for airlines

		Additive efficiency decomposition				
Classification	Company name	Production	Marketing	w_1	w_2	OTE
Full-service carriers						
	AIR FRANCE-KLM-ADR	1.000	1.000	0.500	0.500	1.000
	AMR CORP/DE	0.808	0.924	0.553	0.447	0.860
	BRITISH AIRWAYS PLC-ADR	1.000	0.893	0.500	0.500	0.947
	CHINA EASTERN AIRLINES-ADR	0.755	0.832	0.570	0.430	0.788
	CHINA SOUTHN AIRLS LTD-ADR	1.000	0.848	0.500	0.500	0.924
	CONTINENTAL AIRLS INC-CL B	1.000	0.942	0.500	0.500	0.971
	COPA HOLDINGS SA	1.000	0.846	0.500	0.500	0.923
	DELTA AIR LINES	0.885	0.909	0.530	0.470	0.897
	DEUTSCHE LUFTHANSA AG ADR	0.722	0.958	0.581	0.419	0.821
	GREAT LAKES AVIATION LTD	0.356	0.705	0.737	0.263	0.448
	HAWAIIAN HOLDINGS	1.000	1.000	0.500	0.500	1.000
	LAN AIRLINES SA-ADR	0.741	0.907	0.574	0.426	0.812
	MESA AIR GROUP	0.726	0.866	0.579	0.421	0.785
	NORTHWEST AIRLINES	0.804	1.000	0.554	0.446	0.892
	PINNACLE AIRLINES	1.000	0.897	0.500	0.500	0.948
	REPUBLIC AIRWAYS HLDGS	1.000	0.935	0.500	0.500	0.967
	SKYWEST	0.843	0.906	0.543	0.457	0.872
	TAM SA-ADR	1.000	0.858	0.500	0.500	0.929
	UAL	1.000	0.955	0.500	0.500	0.977
	Average efficiency score	0.876	**0.904**			0.882
Low-cost carriers						
	AIRTRAN HOLDINGS	0.964	0.843	0.509	0.491	0.905
	ALASKA AIR GROUP	0.839	0.886	0.544	0.456	0.860
	ALLEGIANT TRAVEL	1.000	0.959	0.500	0.500	0.980
	EXPRESSJET HOLDINGS	0.889	1.000	0.529	0.471	0.941
	FRONTIER AIRLINES HOLDINGS	1.000	0.871	0.500	0.500	0.935
	GOL LINHAS AEREAS INTEL-ADR	1.000	0.847	0.500	0.500	0.923
	JETBLUE AIRWAYS	1.000	0.955	0.500	0.500	0.978
	MAIR HOLDINGS	1.000	0.659	0.500	0.500	0.829
	RYANAIR HOLDINGS PLC-ADR	1.000	0.904	0.500	0.500	0.952
	SOUTHWEST AIRLINES	0.877	0.846	0.533	0.467	0.863
	US AIRWAYS GROUP	0.984	0.940	0.504	0.496	0.962
	Average efficiency score	**0.959**	0.883			**0.921**
	Total average efficiency score	0.907	0.896			0.896

Note:
w_1 represents the relative importance of production efficiency to overall performance
w_2 represents the relative importance of marketing efficiency to overall performance
OTE = $w_1$production efficiency + $w_2$marketing efficiency

assumption in the efficiency assessment. The 'Production Efficiency,' 'Marketing Efficiency' and the relative importance of individual stages to the overall performance of the entire process are presented in Table 21.2.

DEA efficiency scores do not measure the productive efficiencies of the decision-making units in an absolute sense. Instead, they measure their efficiencies relative to 'the efficient', meaning the best existing empirical practice and decision-making units in the dataset, i.e. those that receive a DEA efficiency score of 1. This does not preclude the existence of some inefficiency among the most efficient decision-making units, so the efficient ones need not be completely efficient in an absolute sense. DEA is a frontier analysis, where the boundary of the production possibilities set (sometimes called the "efficiency frontier") is specified by linear combinations of the input–output vectors of the efficient firms. Firms that are not found to be efficient receive efficiency scores of less than one (in an input-oriented DEA model), with the extent to which their scores fall short of one measures how inefficient they are relative to the efficient firms. An average efficiency score for the decision-making units in a dataset essentially measures how inefficient, on average, the firms in the dataset are compared to the efficient firms.

The score of relative efficiency ranges from 0 to 1. An airline with the score of one is relatively efficient; otherwise, one with a score of less than 1 is relatively inefficient. Table 21.2 shows that the mean scores of production efficiency and marketing efficiency are 0.907 and 0.896, respectively. This finding indicates that, in the area of production efficiency, there are smaller differences in the relative efficiencies of the carriers than there are in their marketing efficiencies. This result suggests that the policy-makers in these airlines should focus first on improving marketing strategies and then proceed to improving their revenue passenger miles and non-passenger revenue.

To determine whether differences exist in various operating characteristics, including carrier type (either full-service carriers or low-cost carriers) for production and marketing efficiencies, a non-parametric statistical analysis (Mann–Whitney test) is used (Brockett and Golany 1996) for unknown distribution scores. Table 21.2 shows that the low-cost carriers have higher production efficiency than the full-service ones, with scores of 0.959 and 0.876 respectively. However, the full-service carriers have higher marketing efficiency than the low-cost ones, 0.904 and 0.883 respectively. Most low-cost carriers do not carry cargo or provide other services. Their main source of revenue comes from passengers. We speculate that the marketing inefficiency of low-cost carriers is due to lower non-passenger revenue. Due to the small sample size, the result of the Mann–Whitney test shows no significant difference at the 5 % level.

21.4.2 Managerial Decision-Making Matrix

To identify the relative positions of the 30 airlines, we constructed a decision-making matrix by combining production efficiency and marketing efficiency to help

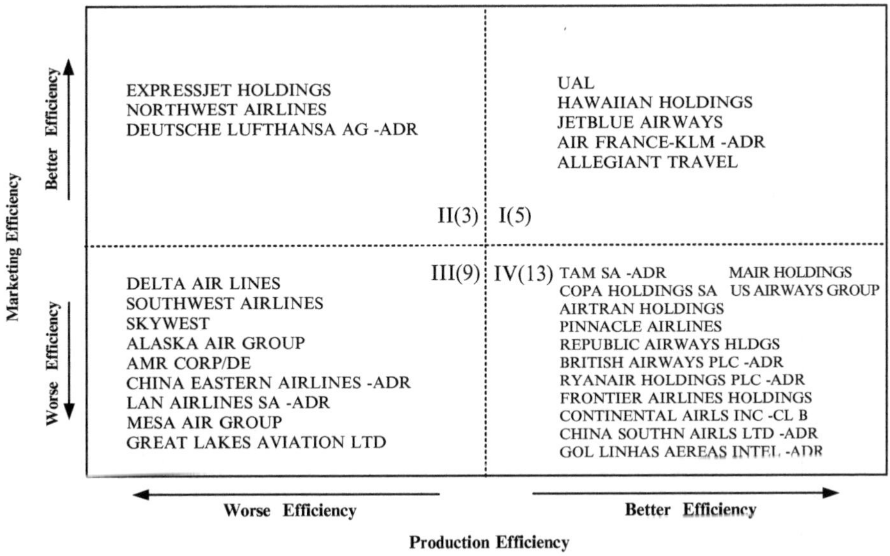

Fig. 21.2 Managerial decision-making matrix for airlines

airline managers and boards of directors improve corporate efficiency. The horizontal axis of the matrix measures production efficiency, while the vertical axis of the matrix measures marketing efficiency. A large value indicates better marketing efficiency. In contrast, a small value indicates a lower marketing efficiency.

Each airline is classified into a quadrant by examining (1) whether the production efficiency is equal to or less than 0.95, and (2) whether the marketing efficiency is greater than or smaller than 0.95. The decision-making matrix, shown in Fig. 21.2, is divided into four quadrants, according to the importance and urgency of the decision-making process. In order to find information indicating by how much and in what areas an inefficient airline needs to improve, a non-zero slack analysis was used to find targets and potential improvements for the inefficient airlines. Such analysis can identify marginal contributions in efficiency ratings with an additional decrease in specific input amounts. Table 21.3 reports the results of our slack analysis. Based on the results shown in Table 21.3, the inefficient DELTA AIR LINES, as an example, can decrease its number of employees (FTEs) by 10.35 %, its fuel consumed (FUEL) by 10.35 %, its seating capacity (SEATS) by 10.35 %, its flight equipment (FLIGHT) by 10.35 %, its maintenance expenses (MAINEXP) by 31.6 %, its ground property and equipment (GROPROEQ) by 56.95 %, its available seat miles (ASMs) by 10.35 % and its available ton miles (ATMs) by 53.37 %, so as to be as efficient as its peer group. This result suggests that DELTA AIR LINES is seriously over-utilizing operational efficiency and should enhance its management's ability operate. The total potential improvement also indicates that the inefficient DELTA AIR LINES has the greatest potential to

Table 21.3 Potential improvements for inefficient airlines

		Potential improvement							
Company name	Zone	FTEs	FUEL	SEATS	FLIGHT	MAINEXP	GROPROEQ	ASMs	ATMs
AIR FRANCE-KLM-ADR	I	0.00 %	0.00 %	0.00 %	0.00 %	0.00 %	0.00 %	0.00 %	0.00 %
AIRTRAN HOLDINGS	IV	−9.54 %	−9.54 %	−9.54 %	−9.54 %	−9.54 %	−9.54 %	−9.54 %	−15.09 %
ALASKA AIR GROUP	III	−13.97 %	−13.97 %	−13.97 %	−13.97 %	−38.99 %	−33.05 %	−13.97 %	−13.97 %
ALLEGIANT TRAVEL	I	−2.04 %	−2.04 %	−2.04 %	−2.04 %	−2.04 %	−2.04 %	−2.04 %	−2.04 %
AMR CORP/DE	III	−13.99 %	−13.99 %	−13.99 %	−13.99 %	−13.99 %	−40.33 %	−13.99 %	−13.99 %
BRITISH AIRWAYS PLC-ADR	IV	−5.33 %	−5.33 %	−5.33 %	−5.58 %	−5.92 %	−6.04 %	−5.33 %	−5.33 %
CHINA EASTERN AIRLINES-ADR	III	−42.44 %	−21.18 %	−21.18 %	−27.48 %	−21.18 %	−21.18 %	−21.18 %	−226.13 %
CHINA SOUTHN AIRLS LTD-ADR	IV	−7.62 %	−7.62 %	−7.62 %	−7.62 %	−7.62 %	−7.62 %	−7.62 %	−7.62 %
CONTINENTAL AIRLS INC-CL B	IV	−2.91 %	−2.91 %	−2.91 %	−2.91 %	−2.91 %	−12.25 %	−2.91 %	−2.91 %
COPA HOLDINGS SA	IV	−7.71 %	−7.71 %	−7.71 %	−7.71 %	−7.71 %	−7.71 %	−7.71 %	−80.02 %
DELTA AIR LINES	III	−10.35 %	−10.35 %	−10.35 %	−10.35 %	−31.60 %	−56.95 %	−10.35 %	−53.37 %
DEUTSCHE LUFTHANSA AG-ADR	II	−17.90 %	−18.97 %	−17.90 %	−17.90 %	−17.90 %	−34.81 %	−17.90 %	−76.37 %
EXPRESSJET HOLDINGS	II	−5.86 %	−30.12 %	−5.86 %	−5.86 %	−34.53 %	−39.30 %	−5.86 %	−310.92 %
FRONTIER AIRLINES HOLDINGS	IV	−6.47 %	−6.47 %	−6.47 %	−6.47 %	−6.47 %	−6.47 %	−6.47 %	−46.84 %
GOL LINHAS AEREAS INTEL-ADR	IV	−7.67 %	−7.67 %	−7.67 %	−7.67 %	−7.67 %	−7.67 %	−7.67 %	−46.34 %
GREAT LAKES AVIATION LTD	III	−90.50 %	−55.22 %	−55.84 %	−56.85 %	−90.36 %	−55.22 %	−55.22 %	−541.31 %
HAWAIIAN HOLDINGS	I	0.00 %	0.00 %	0.00 %	0.00 %	0.00 %	0.00 %	0.00 %	0.00 %
JETBLUE AIRWAYS	I	−4.21 %	−3.52 %	−2.25 %	−2.25 %	−2.25 %	−7.95 %	−2.25 %	−2.25 %
LAN AIRLINES SA-ADR	III	−28.40 %	−30.71 %	−18.79 %	−18.79 %	−18.79 %	−18.79 %	−18.79 %	−72.17 %
MAIR HOLDINGS	IV	−70.52 %	−17.07 %	−43.14 %	−17.07 %	−74.99 %	−65.98 %	−17.07 %	−17.07 %
MESA AIR GROUP	III	−31.83 %	−35.42 %	−21.50 %	−21.50 %	−73.27 %	−21.50 %	−21.50 %	−212.82 %
NORTHWEST AIRLINES	II	−10.85 %	−10.85 %	−10.85 %	−17.21 %	−44.53 %	−54.00 %	−10.85 %	−81.84 %
PINNACLE AIRLINES	IV	−5.17 %	−5.17 %	−5.17 %	−5.17 %	−5.17 %	−5.17 %	−5.17 %	−5.17 %
REPUBLIC AIRWAYS HLDGS	IV	−5.71 %	−3.76 %	−3.27 %	−3.27 %	−17.20 %	−3.27 %	−3.27 %	−3.27 %
RYANAIR HOLDINGS PLC-ADR	IV	−4.80 %	−4.80 %	−4.80 %	−4.80 %	−4.80 %	−4.80 %	−4.80 %	−73.25 %
SKYWEST	III	−66.10 %	−12.82 %	−24.37 %	−12.82 %	−78.82 %	−12.82 %	−12.82 %	−559.14 %
SOUTHWEST AIRLINES	III	−13.74 %	−13.74 %	−13.74 %	−13.81 %	−39.35 %	−13.74 %	−13.74 %	−42.29 %
TAM SA -ADR	IV	−7.08 %	−7.08 %	−7.08 %	−7.08 %	−7.08 %	−7.08 %	−7.08 %	−78.60 %
UAL	I	−2.27 %	−3.36 %	−2.27 %	−2.27 %	−3.06 %	−3.57 %	−2.27 %	−2.27 %
US AIRWAYS GROUP	IV	−3.79 %	−10.34 %	−3.79 %	−3.79 %	−3.79 %	−22.52 %	−3.79 %	−21.26 %

decrease their inefficiency. Therefore, managers should expect to spend most of their efforts in this area.

The purpose of this study is to understand the utilization of resources and the decision-making orientation of each airline. We also propose a number of methods to improve the airlines' efficiency. The airlines located in the four zones are described as follows.

Zone I: The airlines in this zone demonstrate higher efficiency in both production and marketing than airlines in other zones. There are five airlines in this zone: HAWAIIAN HOLDINGS, UAL, JETBLUE AIRWAYS, AIR FRANCE-KLM-ADR, and ALLEGIANT TRAVEL. In both stages, these airlines were found to be superior to other airlines, and are regarded as benchmarks because of their outstanding efficiency. If they manage and control resources effectively, they will be able to maintain their leading position.

Zone II: There are three airlines in this zone: EXPRESSJET HOLDINGS, NORTHWEST AIRLINES and DEUTSCHE LUFTHANSA AG-ADR. They were found to be not as efficient as the airlines in Zone I. Despite their relatively inferior production efficiency, they performed remarkably in marketing. These airlines should improve their ability to reallocate ASMs and ATMs in order to achieve more effective outcomes in the production process.

Zone III: This zone contains nine airlines: DELTA AIR LINES, SOUTHWEST AIRLINES, SKYWEST, ALASKA AIR GROUP, AMR CORP/DE, CHINA EASTERN AIRLINES-ADR, LAN AIRLINES SA-ADR, MESA AIR GROUP and GREAT LAKES AVIATION LTD. Both their production efficiency and marketing efficiency were found to be inferior, with China Eastern Airlines bearing the lowest score. All of these airlines should attempt to increase both production efficiency and marketing efficiency. In addition to enhancing managerial capabilities and reorganizing resources, these airlines should concentrate on substantive issues and effective strategies.

Zone IV: There are 13 airlines in this zone, MAIR HOLDINGS, CONTINENTAL AIRLS INC-CL B, REPUBLIC AIRWAYS HLDGS, COPA HOLDINGS SA, CHINA SOUTHN AIRLS LTD-ADR, BRITISH AIRWAYS PLC-ADR, GOL LINHAS AEREAS INTEL-ADR, RYANAIR HOLDINGS PLC-ADR, TAM SA-ADR, PINNACLE AIRLINES, FRONTIER AIRLINES HOLDINGS, US AIRWAYS GROUP and AIRTRAN HOLDINGS. They had better production efficiency, but lower marketing efficiency. This suggests that all policy-makers in these airlines should focus first on improving marketing strategies and then proceed to improving their revenue passenger miles and non-passenger revenue.

To summarize, we find that almost all variables are maximized in Zone I. Thus, we can say that these airlines use resources efficiently. One input resource might be further reduced in Zone II. Boards and management could focus on how to reduce maintenance costs through communication and discussion. For example, if they leased newer aircraft, maintenance expenses could be reduced. In Zone III, the output resources are relatively smaller than those in other zones. Thus, managers could revise their strategies in order to increase output resources (ASMs, ATMs and RPMs). They could learn about management strategies from those airlines with the best practices in Zone I. Non-passenger revenue is the smallest in Zone VI. Boards and management could improve their strategy by adding other non-passenger services. If these services were incorporated, then these airlines could move up to Zone I.

21.4.3 The Relationship Between the Airlines' Performance and Corporate Governance

Jensen (1993) and Chiang and Lin (2007) point out that large board sizes can lead to some problems, such as coordination and communication, allow the CEO to control the board easily and so give rise to certain agency related problems. However, more directors in the board can allow for more specialists from different fields and therefore higher-quality decision-making. Furthermore, resource dependent theory demonstrates that board size is associated with a firm's ability to acquire key resources from outside (Zahra and Pearce 1989; Xie et al. 2003). Committees are established by the board. There are four common types of committees: audit, nominating, remuneration and executive. Vafeas (1999) found that committees and a firm's performance were negatively related, but the relation was not significant. In terms of boards' internal functions, one of their major tasks is deciding the frequency of meetings (Vafeas 1999). Andres and Vallelado (2008) demonstrate that meetings provide board members with ways to discuss and exchange ideas about how they wish to monitor managers. Jensen (1993), however, indicates that board meetings were not necessarily helpful to performance.

With respect to composition and independence, there can be two types of directors: executive and non-executive. Non-executive directors' major duties include monitoring, disciplining and advising managers; therefore, they can reduce conflicts of interest between insiders and shareholders (Harris and Raviv 2008). Andres and Vallelado (2008) point out that an appropriate, not excessive, number of non-executive directors would be more efficient in monitoring and advising functions and thus would improve performance. Nevertheless, Yermack (1996) shows that firms with a high percentage of non-executive directors have inferior performance. Baliga et al. (1996) and Bhagat and Bolton (2008) showed that when a firm separates the functions of the CEO and the chairman, performance is better than those with CEO duality. Nevertheless, Jensen (1993) observes that in CEO duality, the CEO can control information more effectively than the other board members and so can impede monitoring. Sonnenfeld (2002) argues that the average age of directors can be used as a proxy for experience. Older directors have more professional experience in firms and industries and so board quality can be promoted. However, Stathopoulos et al. (2004) suggests that older directors are less effective in ensuring firm performance. With respect to managerial ownership, Jensen and Mecking (1976) point out that when managerial ownership increases, the interests of managers and shareholders become more similar. Therefore, managerial ownership is significantly positively related to performance.

To summarize, prior studies have rarely explored whether the characteristics of corporate governance affect an airline's performance. Therefore, in this section, we will explore whether corporate governance affects airline performance.

For the dependent variable of the study, we apply the efficiency results from the two-stage DEA in the first part. For independent variables, we use *Board size* (*BOASIZE*), *Committee* (*COMNUM*), *Meetings* (*MEETYEA*), *Non-executive director* (*NEXDIR*), *CEO duality* (*CEODUALITY*), *Directors' age* (*DIRAGE*) and *Executive officers ownership* (*EXEOWN*) to represent corporate governance. As proposed by Backx et al. (2002) and Barros and Peypoch (2009), this study uses five control variables.

Corporate governance variable

Board size (*BOASIZE*)*:* The number of directors on the board including executive and non-executive directors

Committees (*COMNUM*)*:* The number of committees established by the board, for instance, auditing, nominating, remuneration and executive committees.

Meetings (*MEETYEA*)*:* The annual number of board meetings for each airline

Non-executive director (*NEXDIR*)*:* The number of independent non-executive directors on the board

CEO duality (*CEODUALITY*)*:* A dummy variable for airlines, which equals 1 if the CEO is also chairman of the board, and 0 otherwise

Directors' age (*DIRAGE*): The average age of the board directors

Executive officers ownership (*EXEOWN*)*:* The percent of the firm's outstanding shares owned by the executive officers

Control variable

Average age of aircraft (AVGAGE): The average age of all aircraft

Average aircraft size (AVGSIZE): The average number of seats on the aircraft

Average stage length (AVGSTAGE): The average distance flown on each segment of every route

Dummy International (INTER_DUM): A dummy variable for airlines, which equals 1 if the airline has international flights and 0 otherwise

Dummy low cost carrier (LCC_DUM): A dummy variable for airlines, which equals 1 if the airline is a low-cost carrier and 0 otherwise

To explore whether characteristics of corporate governances affect an airline's performance, we estimate the truncated-regression model as follows:

$$\begin{aligned} TE_i = \alpha &+ \beta_1 BOASIZE_i + \beta_2 COMNUM_i + \beta_3 MEETYEA_i + \beta_4 NEXDIR_i \\ &+ \beta_5 CEODUALITY_i + \beta_6 DIRAGE_i + \beta_7 EXEOWN_i + \delta_1 AVGAGE_i \\ &+ \delta_2 AVGSIZE_i + \delta_3 AVGSTAGE_i + \delta_4 INTER\_DUM_i + \delta_5 LLC\_DUM_i + \varepsilon_i \end{aligned}$$

TE represents the empirical result of the efficiency score obtained from the production efficiency or marketing efficiency of the two-stage DEA

Table 21.4 Results of truncated regression by using the heteroskedasticity-robust standard error

Variable	Production efficiency	Marketing efficiency
Intercept	1.9850*	2.2265**
BOASIZE	−0.0638**	−0.0063
COMNUM	0.0358*	0.0331
MEETYEA	−0.0053	0.0034
NEXDIR	0.0119	0.0479**
CEODUALITY	0.0352	0.0629*
DIRAGE	0.0026	−0.0010
EXEOWN	0.0005	0.0165*
AVGAGE(years)	−0.0364	−0.0229
AVGSIZE	0.0041**	−0.0028*
AVGSTAGE(miles)	0.0004**	−0.0002
INTER_DUM	−0.0169	0.2779**
LCC_DUM	0.0547	0.0737
Adjusted R-squared	0.2956	0.3125
Variance	2.880	2.8312

Note:
*, ** and ***, indicates the statistical significance at the 10 %, 5 %, and 1 % levels, respectively
PE = the efficiency score obtained from the weighted average of the two-stage DEA efficiency score
BOASIZE = the number of board members
COMNUM = the number of committees established by the board
MEETYEA = the number of meetings held by the board per year
NEXDIR = the ratio of non-executive board members to board size, which is also a measure of board independence
CEODUALITY = a dummy variable, which is equal to 1 if the CEO is also the chairman of the board and 0 otherwise
DIRAGE = the average age of board directors
EXEOWN = the percentage of outstanding shares owned by the executive officers
AVGAGE = the average age of all aircraft
AVGSIZE = the average number of seats on each aircraft
AVGSTAGE = the average distance flown on each segment of every route
INTER_DUM = a dummy variable for airlines, which is equal to 1 if the airline has international scheduled flights and 0 otherwise
LCC_DUM = a dummy variable for airlines, which is equal to 1 if the airline is a low-cost carrier and 0 otherwise

efficiency scores. This study also performs a simulation test, which includes 3,000 experimental observations, in order to confirm the fitness of the truncated-regression model. To enhance the robustness of our empirical results, the following additional analyses were completed. First, we used the variance inflation factors' diagnostics (Neter et al. 1985) for collinearity analysis. No evidence of collinearity between independent variables was found in our regression models. Next, the White test (White 1980) was used to check the heteroskedasticity of the residuals, and evidence of heteroskedasticity could be found. Finally, the heteroskedasticity-robust standard error was used to construct a heteroskedasticity-robust t statistic.

Table 21.4 presents the regression results of the separate efficiency by using the heteroskedasticity-robust standard error.

In terms of production efficiency, the results of a truncated-regression suggested that the independent variables *BOASIZE* and *COMNUM* are significantly related to airline performance at 5 % levels. They also indicated that *BOASIZE* and *MEETYEA* have an inverse relation with airline performance, while *COMNUM* has a positive significant relationship with airline performance.

The results indicate that smaller boards lead to better airline performance. Provan (1980) found that board size is positively related to performance and argued that board size is related to the company's ability to acquire key resources such as budgets, external funding and leverage. Zahra and Pearce (1989) and Kiel and Nicholson (2003) propose that larger boards are likely to be heterogeneous in their industrial background and expertise, which improves the company's decision making and thus enhances its performance. Xie et al. (2003) reported that larger boards can mitigate earnings management. The results of the current study are contrary to their findings. We speculate that too many directors on the board may lead to greater personnel compensation, which in turn makes it more difficult to integrate management decisions. With respect to *COMNUM*, the results indicate a significantly positive relationship with production performance. Vafeas (1999) found a negative but not significant relationship between performance and the number of committees. The more committees established by the board (for instance, auditing, nominating, remuneration and executive committees), the tighter control will be.

The study shows that *MEETYEA* has no significant negative relation with production performance, while *CEODUALITY* showed no significant relation to performance. When the CEO is also the chairman of the board, serious agency problems can arise, and monitoring by the board can be reduced (Fama and Jensen 1983). Baliga et al. (1996) showed that a firm with separate CEO and chairman of the board positions performed better than a firm with CEO duality. *DIRAGE* is a proxy for experience, and older directors may be familiar with the firm and industries. The results show that older directors have more experience and contribute to the airline's performance. *EXEOWN* had positive relations with performance. This is consistent with the findings of Jensen and Mecking (1976) in which they show a positive relation between performance and stock ownership by executive officers.

Regarding marketing efficiency, the truncated-regression analysis shows that the independent variables *NEXDIR* and *EXEOWN* are significantly positively related to airline performance at 5 % levels. Non-executive directors should scrutinize the performance of management in meeting agreed goals and objectives, in monitoring, and where necessary, removing, senior management, and in succession planning. Directors may be recruited for their ability to offer support and advice in specialized areas such as marketing, product development or financial restructuring. On the other hand, *BOASIZE* and *DIRAGE* were found to be negatively related to airline performance. Board size refers to the number of directors serving in a firm's board. Large boards may destroy corporate value. The result here suggests that some boards may be larger than optimal and that it may be worthwhile for some airlines to reevaluate their optimal board size. Besides, a high level of executive officer ownership reflects too many chiefs in an airline, which in turn may spoil the company.

While previous studies have discussed the influence of control variables, this study shows that some control variables are significant. *AVGSIZE* and *AVGSTAGE* are significant in the production model, and *AVGAGE*, *AVGSIZE* and INTER DUM are significant in the marketing model. These airlines should increase the average number of seats on an aircraft and the average distance flown on each segment of every route to improve production efficiency, and decrease the average number of seats on an aircraft and the average distance flown on each segment of every route and increase international flights in order to improve marketing performance.

Finally, it should be noted that the use of variables measured in monetary terms in arriving at the DEA efficiency scores may render the scores somewhat questionable as measures of the relative production and marketing efficiencies of the carriers in the dataset. Variables such as the FLIGHT, MAINEXP, GROPROEQ and NPR are measured in United States dollars, while the remaining variables are measured in their physical units, which is how they should be measured when evaluating the relative productive efficiency of a decision-making unit. Different carriers in the dataset pay different prices for each unit of flight equipment, for their maintenance services, and for each item in their ground property equipment, and receive different prices for each ton of freight and mail they transport, especially since the carriers used in the dataset are based in different countries (France, Germany, China, Brazil, the United States, etc.). The currency exchange rate from one country to another may lead to different operational costs for firms in different geographical regions, and so the relative DEA efficiency scores may not actually track differences in the relative efficiencies of the carriers. This will be the topic of our future research.

21.5 Conclusion

While transport industries have become increasingly important in the global economy, issues in the aviation industry are especially important for a large, free market economy like the United States, because they can influence both global economic development and international politics. Although the efficiency of the aviation industry has been widely discussed in previous literature, and the DEA technique is frequently used to evaluate efficiency, there are still some important points not previously explored. As a research topic, the issue of corporate governance in the aviation industry has rarely been investigated. From the perspective of research methods, the problem with the traditional DEA model is the concept of a one-stage process, which neglects intermediate measures or linking activities. The concept of a two-stage process has been applied infrequently at best in previous studies of the aviation industry. Therefore, this chapter aimed to establish a two-stage DEA to measure efficiency, to discuss corporate governance and to evaluate the benchmarks of airlines from a more complete viewpoint. The results of this study can provide United States airlines with insights into resource allocation and competitive advantage, and can help them to improve their strategic decision-making, specifically regarding operational styles, under fierce competition in the aviation industry.

The findings can briefly be described as follows. First, the 30 airlines researched had an average production efficiency of 63 % and an average marketing efficiency of 33 %. This suggests that managers should focus first on improving the inefficient allocation of resources in production and then their marketing efficiency. Secondly, low-cost carriers are more efficient, on average, than full-service ones in production. This finding is consistent with the findings of Barbot et al. (2008). On the other hand, full-service carriers are more efficient, on average, than low-cost carriers in marketing. Thirdly, we can state that corporate governance influences firm performance. The results of truncated regression on board size, average age of directors, and percentage of outstanding shares owned by executive officers all show significant, positive relations to performance. Number of committees and CEO duality both present significant negative relations with performance. This means that these airlines can modify corporate governance to strengthen their efficiency and competitiveness. Finally, we used the managerial decision-making matrix to find benchmark airlines in order to help managers improve corporate performance.

Our findings can provide guidelines for coping with corporate governance issues in the aviation industry. Future research might use Malmquist productivity change index techniques to examine long-term variance in airline performance. It could also prevent the results from being affected by external, short-term factors. Such an approach would allow a dynamic view of the multidimensional performance of airlines. It is also hoped that the models and methods implemented in this study can help to bring about related research in other industries.

References

Alam, I., Semenick, M., & Sickles, C. (1998). The relationship between stock market returns and technical efficiency innovations: Evidence from the US airline industry. *Journal of Productivity Analysis, 9*(1), 35–51.

Andres, P. D., & Vallelado, E. (2008). Corporate governance in banking: The role of the board of directors. *Journal of Banking & Finance, 32*(12), 2570–2580.

Avkiran, N. (2001). Investigating technical and scale efficiencies of Australian universities through data envelopment analysis. *Socio-Economic Planning Sciences, 35*(1), 57–80.

Aydogan, E. K. (2011). Performance measurement model for Turkish aviation firms using the rough-AHP and TOPSIS methods under fuzzy environment. *Expert Systems with Applications, 38*(4), 3992–3998.

Backx, M., Carney, M., & Gedajlovic, E. (2002). Public, private and mixed ownership and the performance of international airlines. *Journal of Air Transport Management, 8*(4), 213–220.

Baliga, B., Moyer, R., & Rao, R. (1996). CEO duality and firm performance: What's the fuss? *Strategic Management Journal, 17*(1), 41–53.

Banker, R. D., Janakiraman, S., & Natarajan, R. (2004). Analysis of trends in technical and allocative efficiency: An application to Texas public school districts. *European Journal of Operational Research, 154*(2), 477–491.

Barbot, C., Costa, A., & Sochirca, E. (2008). Airlines performance in the new market context: A comparative productivity and efficiency analysis. *Journal of Air Transport Management, 14*(5), 270–274.

Barros, C. P., & Peypoch, N. (2009). An evaluation of European airlines' operational performance. *International Journal of Production Economics, 122*(2), 525–533.

Bennedsen, M., Kongsted, H. C., & Nielsen, K. M. (2008). The causal effect of board size in the performance of small and medium-sized firms. *Journal of Banking & Finance, 32*(6), 1098–1109.

Bhagat, S., & Bolton, B. (2008). Corporate governance and firm performance. *Journal of Corporate Finance, 14*(3), 257–273.

Biener, C., & Eling, M. (2011). The performance of microinsurance programs: A data envelopment analysis. *The Journal of Risk and Insurance, 78*(1), 83–115.

Brockett, P. L., & Golany, B. (1996). Using rank statistics for determining programmatic efficiency differences in data envelopment analysis. *Management Science, 42*(3), 466–472.

Carline, N. F., Linn, S. C., & Yadav, P. K. (2009). Operating performance changes associated with corporate mergers and the role of corporate governance. *Journal of Banking & Finance, 33*(10), 1829–1841.

Chang, Y.-H., & Yeh, C.-H. (2001). Evaluating airline competitiveness using multiattribute decision making. *Omega International Journal of Management Sciences, 29*(5), 405–415.

Chang, D. S., Kuo, Y. C., & Chen, T. Y. (2008). Productivity measurement of the manufacturing process for outsourcing decisions: The case of a Taiwanese printed circuit board manufacturer. *International Journal of Production Research, 46*(24), 6981–6995.

Charnes, A., & Cooper, W. W. (1962). Programming with fractional functionals. *Naval Research Logistics Quarterly, 9*(3–4), 181–186.

Charnes, A., Cooper, W. W., & Rhodes, E. (1978). Measuring the efficiency of decision making units. *European Journal of Operational Research, 2*, 429–444.

Charnes, A., Galleegos, A., & Li, H. (1996). Robustly efficient parametric frontiers via multiplicative DEA for domestic and international operations of the Latin American airline industry. *European Journal of Operational Research, 88*(3), 525–536.

Chen, C., & Yan, H. (2011). Network DEA model for supply chain performance evaluation. *European Journal of Operational Research, 213*(1), 147–155.

Chen, Y., & Zhu, J. (2004). Measuring information technology's indirect impact on firm performance. *Information Technology & Management Journal, 5*(1–2), 9–22.

Chen, Y., Cook, W. D., Li, N., & Zhu, J. (2009). Additive efficiency decomposition in two-stage DEA. *European Journal of Operational Research, 196*(3), 1170–1176.

Chiang, M. H., & Lin, J. H. (2007). The relationship between corporate governance and firm productivity: Evidence from Taiwan's manufacturing firms. *Corporate Governance, 15*(5), 768–779.

Chiou, Y. U., & Chen, Y. H. (2006). Route-based performance evaluation of Taiwanese domestic airlines using data envelopment analysis. *Transportation Research Part E, 42*(2), 116–127.

Chuang, I. Y., Chiu, Y.-C., & Edward Wang, C. (2008). The performance of Asian airlines in the recent financial turmoil based on VaR and modified Sharpe ratio. *Journal of Air Transport Management, 14*(5), 257–262.

Cook, W. D., Liang, L., & Zhu, J. (2009). Measuring performance of two-stage network structures by DEA: A review and future perspective. *Omega International Journal of Management Sciences, 38*(6), 423–430.

Cook, W. D., Zhu, J., Yang, F., & Bi, G.-B. (2010). Network DEA: Additive efficiency decomposition. *European Journal of Operational Research, 207*(2), 1122–1129.

Cooper, W. W., Seiford, L. M., & Zhu, J. (2004). Data envelopment analysis: History, models and interpretations. In W. W. Cooper, L. M. Seiford, & J. Zhu (Eds.), *Handbook on data envelopment analysis*. Boston: Kluwer Academic.

Cooper, W. W., Seiford, L. M., & Tone, K. (2006). *Introduction to data envelopment analysis and its uses: With DEA-solver software and references*. New York: Springer.

Emrouznejad, A., Parker, B. R., & Tavares, G. (2008). Evaluation of research in efficiency and productivity: A survey and analysis of the first 30 years of scholarly literature in DEA. *Socio-Economic Planning Sciences, 42*(3), 151–157.

Fama, E. F., & Jensen, M. C. (1983). Separation of ownership and control. *Journal of Law and Economics, 26*(2), 301–324.

Fang, H.-H., Lee, H.-S., Hwang, S.-N., & Chung, C.-C. (2013). A slacks-based measure of super-efficiency in data envelopment analysis: An alternative approach. *Omega International Journal of Management Sciences, 41*(4), 731–734.

Fare, R., & Whittaker, G. (1995). An intermediate input model of dairy production using complex survey data. *Journal of Agricultural Economics, 46*(2), 201–213.

Feng, C.-M., & Wang, R.-T. (2000). Performance evaluation for airlines including the consideration of financial ratios. *Journal of Air Transport Management, 6*(3), 133–142.

Feroz, E. H., Kim, S., & Raab, R. L. (2003). Financial statement analysis: A data envelopment analysis approach. *Journal of the Operational Research Society, 54*(1), 48–58.

Fielding, G. J., Glauthier, R. E., & Lave, C. A. (1978). Performance evaluation for bus transit. *Transportation Research Part A, 19*(1), 73–82.

Francis, G., Humphreys, I., & Fry, J. (2002). The benchmarking of airport performance. *Journal of Air Transport Management, 8*(4), 239–247.

Francis, G., Humphreys, I., & Fry, J. (2005). The nature and prevalence of the use of performance measurement techniques by airlines. *Journal of Air Transport Management, 11*(4), 207–217.

Franke, M., & John, F. (2011). What comes next after recession? – Airline industry scenarios and potential end games. *Journal of Air Transport Management, 17*(1), 19–26.

Gattoufi, S., Oral, M., & Reisman, A. (2004). Data Envelopment Analysis literature: A bibliography update (1996–2001). *Socio-Economics Planning Sciences, 38*(2–3), 122–159.

Golany, B., & Roll, Y. (1989). An application procedure for DEA. *Omega International Journal of Management Sciences, 17*(3), 237–250.

Gompers, P. A., Ishii, J. L., & Metrick, A. (2003). Corporate governance and equity prices. *Quarterly Journal of Economics, 118*(1), 107–155.

Gorin, T., & Belobaba, P. (2004). Impacts of entry in airline markets: Effects of revenue management on traditional measures of airline performance. *Journal of Air Transport Management, 10*(4), 257–268.

Gramani, M. C. N. (2012). Efficiency decomposition approach: A cross-country airline analysis. *Expert Systems with Applications, 39*(5), 5815–5819.

Greer, M. R. (2008). Nothing focuses the mind on productivity quite like the fear of liquidation: Changes in airline productivity in the United States, 2000–2004. *Transportation Research Part A, 42*(2), 414–426.

Harris, M., & Raviv, A. (2008). A theory of board control and size. *Review of Financial Studies, 21*(4), 1797–1832.

Homburg, C. (2001). Using data envelopment analysis to benchmark activities. *International Journal of Production Economics, 73*(1), 51–58.

Hong, S., & Zhang, A. (2010). An efficiency study of airlines and air cargo/passenger divisions: A DEA approach. *World Review of Intermodal Transportation Research, 3*(1–2), 137–149.

Jensen, M. C. (1993). The modern industrial revolution, exit, and the failure of internal control systems. *Journal of Finance, 48*(3), 831–880.

Jensen, M. C., & Meckling, W. H. (1976). Theory of the firm: Managerial behavior, agency costs, and ownership structure. *Journal of Financial Economics, 3*(4), 305–360.

Kao, C. (2009). Efficiency decomposition in network data envelopment analysis: A relational model. *European Journal of Operational Research, 192*(3), 949–962.

Kao, C., & Hwang, S. N. (2008). Efficiency decomposition in two-stage data envelopment analysis: An application to non-life insurance companies in Taiwan. *European Journal of Operational Research, 185*(1), 418–429.
Karlaftis, M. G. (2004). A DEA approach for evaluating the efficiency and effectiveness of urban transit systems. *European Journal of Operational Research, 152*(2), 354–364.
Keh, H. T., Chu, S., & Xu, J. (2006). Efficiency, effectiveness and productivity of marketing in service. *European Journal of Operational Research, 170*(1), 265–276.
Kiel, G. C., & Nicholson, G. J. (2003). Board composition and corporate performance: How the Australian experience informs contrasting theories of corporate governance. *Corporate Governance, 11*(3), 189–205.
Lee, S., Seo, K., & Sharma, A. (2013). Corporate social responsibility and firm performance in the airline industry: The moderating role of oil prices. *Tourism Management, 38*, 20–30.
Liang, L., Cook, W. D., & Zhu, J. (2008). DEA models for two-stage processes: Game approach and efficiency decomposition. *Naval Research Logistics, 55*(7), 643–653.
Lin, W.-C. (2012). Financial performance and customer service: An examination using activity-based costing of 38 international airlines. *Journal of Air Transport Management, 19*, 13–15.
Liu, J. S., Lu, L. Y. Y., Lu, W.-M., & Lin, B. J. Y. (2013a). Data envelopment analysis 1978–2010: A citation-based literature survey. *Omega International Journal of Management Sciences, 41*(1), 3–15.
Liu, J. S., Lu, L. Y. Y., Lu, W.-M., & Lin, B. J. Y. (2013b). A survey of DEA applications. *Omega International Journal of Management Sciences, 41*(5), 893–902.
Lozano, S., & Gutiérrez, E. (2011). A multiobjective approach to fleet, fuel and operating cost efficiency of European airlines. *Computers & Industrial Engineering, 61*(3), 473–481.
Lu, W.-M., & Hung, S.-W. (2010). Assessing the performance of a vertically disintegrated chain by the DEA approach – A case study of Taiwanese semiconductor firms. *International Journal of Production Research, 48*(4), 1155–1170.
Lu, W.-M., & Hung, S.-W. (2011). Exploring the operating efficiency of Technology Development Programs by an intellectual capital perspective – A case study of Taiwan. *Technovation, 31*(8), 374–383.
Lu, W.-M., Wang, W.-K., & Kweh, Q. L. (2014). Intellectual capital and performance in the Chinese life insurance industry. *Omega International Journal of Management Sciences, 42*(1), 65–74.
Matthews, K. (2013). Risk management and managerial efficiency in Chinese banks: A network DEA framework. *Omega International Journal of Management Sciences, 41*(2), 207–215.
Meckling, W. H. (1976). Theory of the firm: Managerial behavior, agency costs and ownership structure. *Journal of Financial Economics, 3*(4), 305–360.
Merkert, R., & Hensher, D. A. (2011). The impact of strategic management and fleet planning on airline efficiency – A random effects Tobit model based on DEA efficiency scores. *Transportation Research Part A: Policy and Practice, 45*(7), 686–695.
Merkert, R., & Williams, G. (2013). Determinants of European PSO airline efficiency – Evidence from a semi-parametric approach. *Journal of Air Transport Management, 29*, 11–16.
Neter, J., Wasserman, W., & Kutner, M. H. (1985). *Applied linear statistical models: Regression, analysis of variance, and experimental designs*. Homewood: R.D. Irwin.
Premachandra, I. M., Chen, Y., & Watson, J. (2011). DEA as a tool for predicting corporate failure and success: A case of bankruptcy assessment. *Omega International Journal of Management Sciences, 39*(6), 620–626.
Provan, K. G. (1980). Board power and organizational effectiveness among human service agencies. *The Academy of Management Journal, 23*(2), 221–236.
Raghavan, S., & Rhoades, D. L. (2005). Revisiting the relationship between profitability and air carrier safety in the US airline industry. *Journal of Air Transport Management, 11*(4), 283–290.
Ray, S. C., & Hu, X. (1997). On the technically efficient organization of an industry: A study of U.S. airlines. *Journal of Productivity Analysis, 8*(1), 5–18.

Schefczyk, M. (1993). Operational performance of airlines: An extension of traditional measurement paradigms. *Strategic Management Journal, 14*(4), 301–317.

Scheraga, C. A. (2004). Operational efficiency versus financial mobility in the global airline industry: A data envelopment and Tobit analysis. *Transportation Research Part A, 38*(5), 383–404.

Seiford, L. M. (1997). A bibliography for Data Envelopment Analysis (1978–1996). *Annals of Operations Research, 73*, 393–438.

Seiford, L. M., & Zhu, J. (1999). Profitability and marketability of the top 55 U.S. commercial banks. *Management Science, 45*(9), 1270–1288.

Sengupta, J. K. (1999). A dynamic efficiency model using data envelopment analysis. *International Journal of Production Economics, 62*(3), 209–218.

Simar, L., & Wilson, P. (2007). Estimation and inference in two-stage, semi-parametric models of production processes. *Journal of Econometrics, 136*(1), 31–64.

Sonnenfeld, J. A. (2002). What makes great boards great. *Harvard Business Review*, RO209H.

Stathopoulos, K., Espenlaub, S., & Walker, M. (2004). U.K. executive compensation practices: New economy versus old economy. *Journal of Management Accounting Research, 16*, 57–92.

Sueyoshi, T., & Goto, M. (2010). Measurement of a linkage among environmental, operational, and financial performance in Japanese manufacturing firms: A use of data envelopment analysis with strong complementary slackness condition. *European Journal of Operational Research, 207*(3), 1742–1753.

Sueyoshi, T., Goto, M., & Omi, Y. (2010). Corporate governance and firm performance: Evidence from Japanese manufacturing industries after the lost decade. *European Journal of Operational Research, 203*(3), 724–736.

Tan, K. H., & Rae, R. H. (2009). Uncovering the links between regulation and performance measurement. *International Journal of Production Economics, 122*(1), 449–457.

Tone, K., & Tsutsui, M. (2009). Network DEA: A slacks-based measure approach. *European Journal of Operational Research, 197*(1), 243–252.

Vafeas, N. (1999). Board meeting frequency and firm performance. *Journal of Financial Economics, 53*(1), 113–142.

Wang, W.-K. (2005). A knowledge-based decision support system for measuring the performance of government real estate investment. *Expert Systems with Applications, 29*(4), 901–912.

Wang, W.-K., Lu, W.-M., & Tsai, C.-J. (2011). The relationship between airline performance and corporate governance amongst US Listed companies. *Journal of Air Transport Management, 17*(2), 148–152.

White, H. (1980). A heteroskedasticity-consistent covariance matrix estimator and direct test for heteroskedasticity. *Econometrica, 4*, 817–838.

Xie, B., Davidson, W. N., & DaDalt, P. J. (2003). Earnings management and corporate governance: The role of the board and the audit committee. *Journal of Corporate Finance, 9*(3), 295–316.

Yang, X., & Morita, H. (2013). Efficiency improvement from multiple perspectives: An application to Japanese banking industry. *Omega International Journal of Management Sciences, 41*(3), 501–509.

Yeh, Q. J. (1996). The application of data envelopment analysis in conjunction with financial ratios for bank performance evaluation. *The Journal of the Operational Research Society, 47*(8), 980–988.

Yermack, D. (1996). Higher market valuation of companies with a small board of directors. *Journal of Financial Economics, 40*(1), 185–211.

Yu, M. M. (2008). Assessing the technical efficiency, service effectiveness, and technical effectiveness of the world's railways through NDEA analysis. *Transportation Research Part A, 42*(10), 1283–1294.

Yu, M. M., & Lin, E. T. J. (2008). Efficiency and effectiveness in railway performance using a multi-activity network DEA model. *Omega International Journal of Management Sciences, 36*(6), 1005–1017.

Zahra, S. A., & Pearce, J. A. (1989). Boards of directors and corporate financial performance: A review and Integrated model. *Journal of Management, 15*(2), 291–334.

Zhu, J. (2000). Multi-factor performance measure model with an application to Fortune 500 companies. *European Journal of Operational Research, 123*(1), 105–124.

Zhu, J. (2011). Airlines performance via two-stage network DEA approach. *The Business and Economics Research Journal, 4*(2), 260–269.

Chapter 22
Network Representations of Efficiency Analysis for Engineering Systems: Examples, Issues and Research Opportunities

Konstantinos (Kostas) Triantis

Abstract Network efficiency models depict internal production/service processes, and/or alternative perspectives, and/or different time periods. Researchers in the efficiency measurement field are investigating and applying these models in a variety of ways. However, in very few instances are these representations focused on engineering systems. This chapter presents two very distinct network efficiency models that are applied to engineering systems. The first uses the radial and slacks based network DEA models to assess the efficiency performance of a downtown space reservation system (DSRS). This system has been designed as an approach to mitigate traffic congestion in an urban downtown area. The implementation of the network DEA models identify the determinants of efficiency performance for the agency operating the DSRS, for the traveler using the DSRS and for the community where the DSRS resides. The second example pertains to asset management and more specifically to highway maintenance management. An alternative network efficiency representation is used where a system dynamics modeling approach provides a way to study dynamic efficiency performance and assess highway maintenance policies. Through these examples, issues pertaining to opening the production black box to evaluate internal processes, the validity of the axiomatic foundations of DEA for the network models, the relevance of the structure of the network models in terms of suggesting resulting system behaviors, temporal and dynamic efficiency performance associated with the network efficiency models are discussed suggesting future research directions.

This chapter is based in part on work supported by the National Science Foundation, while working at the Foundation. Any opinion, finding, and conclusions and recommendations expressed in this paper are those of the author and do not necessarily reflect the views of the National Science Foundation.

K. (Kostas) Triantis (✉)
Grado Department of Industrial and Systems Engineering, Northern Virginia Center, System Performance Laboratory (SPL), Virginia Tech, Falls Church, VA, USA
e-mail: triantis@vt.edu

W.D. Cook and J. Zhu (eds.), *Data Envelopment Analysis*, International Series in Operations Research & Management Science 208, DOI 10.1007/978-1-4899-8068-7_22,

Keywords Network DEA • Dynamic efficiency • Demand based traffic congestion mitigation • Highway maintenance management

22.1 Introduction and Context

Even though efficiency measurement and improvement has been a significant area of scholarly research, engineers have not extensively used the efficiency measurement paradigm to evaluate system performance and design engineering systems (Triantis 2011, 2013) even though there are notable exceptions in the literature (e.g., Cooper et al. 1992).

There are number of issues that provide challenges and opportunities for this type of research. Typically, the focus of the analysis for engineering systems is at the very micro level. This suggests that the "input/output transformation box" typically considered in efficiency analyses needs to be "opened" and studied in detail. This requires access and understanding of the underlying technologies, processes, information exchange, organizational settings, and social/behavioral considerations. On a very fundamental level, from an engineering perspective, a first understanding of what needs to be measured lies in having a full appreciation and knowledge of the physical and engineering relationships that govern these systems.

However, engineering systems are not designed, built and operated in a vacuum. There are organizations and design teams that are tasked to do so by exchanging important information and making decisions. Given, this reality, understanding the interdependencies between system performance and the organizational entities that are responsible for these engineering systems is paramount. This suggests that we need a deeper appreciation and integration of the social/behavioral and information sciences in our measurement analyses and thinking.

While the efficiency literature is based on an axiomatic framework (Vaneman and Triantis 2003) the engineering design literature does not enjoy a similar axiomatic foundation. What this means is that while in efficiency measurement we rely on production theory, in engineering design the theory is still evolving while at the same time engineering design literature borrows knowledge and representations from organizational, optimization, decision, and probabilistic theoretical frameworks among others. At the end, we need to make sure that the axiomatic framework on which efficiency evaluation methods are based on are relevant for the specific systems that we are evaluating and designing.

An important consideration is when in the system life-cycle performance assessment is being conducted. Most efficiency studies rely on an ex-post assessment where historical performance is analyzed. For engineering systems, the evaluation of performance during design (conceptual, preliminary and detailed) (Blanchard and Fabrycky 2010) is just as important as the assessment of performance during operational phases. This performance evaluation during the design phase requires the identification of the production possibility space or the design possibility space and to make sure that the axiomatic framework underlying efficiency measurement still holds for the various systems being considered.

Related to the previous point, i.e., as to when in the system life cycle the performance analysis is conducted, are the temporal and dynamic considerations of system performance (Fallah-Fini et al. 2013a). In this chapter we borrow the dynamic concepts presented by Sterman (2000) that help describe the dynamic characteristics of systems. More specifically the concepts that we take into account is the consideration of causation, feedback mechanisms, delays, and non-linear relationships. In our dynamic representations and models the structure of the system leads to the observed dynamic system behaviors and the resulting system performance.

In this chapter, we build on the concept of networks in efficiency analysis and how this concept can be used to address some of the issues described in this Section. More specifically, in Sect. 22.2, we employ the notion of network DEA (both radial network DEA (Färe and Grosskopf 2000) and slacks based network DEA (Tone and Tsutsui 2009)) in the context of a transportation system that has yet to be designed, i.e., the downtown space reservation system (Zhao et al. 2010a, 2011). The DSRS uses concepts from transportation engineering and efficiency measurement and combines system optimization, neural networks, traffic micro-simulation and network DEA approaches.

Additionally, in Sect. 22.3 we provide a description of a different type of network that considers the dynamics of highway maintenance (Fallah-Fini et al. 2010). The highway maintenance example combines concepts from highway deterioration and efficiency measurement and combines system dynamics simulation, optimization and efficiency measurement. In both of these sections we do not replicate the details of the mathematical formulations, models and discussions in the papers that have already been published or are under review. However, we do wish to highlight some of the issues that have been briefly described in this section and focus on future research opportunities (Sect. 22.4).

22.2 The Downtown Space Reservation System (DSRS)[1]

22.2.1 The Initial DSRS Conceptualization

In transportation engineering, congestion analysis is a continuing research concern. Travel demand based approaches attempt to reduce congestion by defining, evaluating, and implementing congestion mitigation strategies. An experimental travel demand management approach that has yet to be implemented is the downtown space reservation system (DSRS) (Zhao et al. 2010a) whose main objective is to mitigate downtown traffic congestion.

Within the DSRS, travelers who want to drive to an urban downtown area have to reserve their time slots in advance before embarking on their trips.

[1] This section is adopted from Zhao et al. (2010a, b, 2011).

The transportation agency who operates the DSRS, allocates time slots to travelers based on the availability of the road network capacity. Only the travelers who get permission from the transportation agency can drive in the downtown area during the requested time period. This system is analogous to the idea of making reservations in advance to secure a seat on an airline for a trip taking into account carrier capacity. In the case of the DSRS, the traveler is securing a time slot to visit the downtown area taking into account road capacity.

The proposed DSRS consists of two modules, an offline optimization module and an online decision making module (based on a neural network approach). In the offline module, an optimization model is solved based on historical travel information. Two objectives are included in the optimization problem, i.e., the total number of travelers that the transportation system handles during a certain time period and the revenue obtained from the downtown space reservation system. From a travel demand mitigation point of view, the mobility of people is improved by restraining the excessive amount of automobiles entering the downtown area. From an economic point of view, revenue is maximized. It is assumed that this revenue can be used to finance public transportation systems.

In order to take into account the stochastic variations in travel demand, a neural network approach was used to construct the online module. Assuming that we have hundreds of historical demand scenarios, we obtained optimal solutions for each scenario (using the CPLEX platform). Given that artificial neural networks have the capability to "learn from experience" (Teodorović and Vukadinović 1998), they can be taught from the historical demand scenarios and the derived optimal solutions. From this learning process, the system is able to recognize a situation characterized by the number of reservations that already have been made for each vehicle class during each time period and the corresponding revenue generated from the reservations. Therefore, when a new request comes in, the neural network can rely on this historical information to provide a real time decision. In addition, new requests become historical information and the system can be updated at predetermined time intervals.

22.2.2 The Micro-simulation Evaluation

From a system performance measurement point of view, there were a number of challenges. First is the fact that there were no ex-post data to use for the performance analysis. The DSRS has been proposed but not yet implemented so there were no available historical operational data. Second, the level of aggregation at which the performance analysis could take place needed to be decided.

Initially, Zhao et al. (2010b) used a microscopic traffic simulation approach executed in VISSIM to evaluate the DSRS. The microscopic traffic simulation emulated the physics of traffic flow at a microscopic level. The simulation was conducted for a revised road network representing downtown Boise, Idaho. The issues that were tested in the simulation included: whether the DSRS improves traffic performance when compared without the DSRS; how the DSRS performs compared with a reservation system that uses the First Come First Serve (FCFS) principle; how specific DSRS parameters (such as, the relative importance of traveler throughput versus revenue generation) influence the transportation network system performance. The performance parameters that were outputs from the simulation included typical engineering variables such as: average delay time per vehicle, average speed, total travel time, total vehicle miles, total delay time, fuel usage and costs, and emissions.

22.2.3 Social Welfare Evaluation for the Three Perspectives

Yet in addition to the standard transportation engineering measures of performance one is faced with issues related to performance from the agency's/provider's, travelers'/users' and community's points of view where the social welfare impact of the DSRS needs to be considered. These three perspectives represent important stakeholders in the transportation system and their interactions determine the overall performance of the transportation system. By social welfare we are suggesting that above and beyond the engineering traffic flow impacts of the DSRS, one needs to take into account the impact of the system as far as the operational issues associated with the transportation agency that is providing the service, the quality of service that is experienced by the travelers using the system and the sustainability issues insofar as the community is concerned.

At the initial stages of the DSRS development (Zhao et al. 2010a), the design of the system was anchored around the objective of mitigating congestion. This means that the original optimization model, which is at the core of the DSRS, was not formulated to consider multiple stakeholder perspectives (agency, traveler, community). In order to consider these multiple perspectives, the network DEA approach was considered as a potentially viable approach. Nevertheless, there were a number of issues that needed to be resolved as the following section suggests.

22.2.4 The Network DEA Approach

22.2.4.1 Assumptions and Considerations

As stated in the Sect. 22.1, the production possibility space or in this case the design possibility space needed to be defined. This required the determination of the decision making units and the definition of the inputs and outputs associated with each of the three perspectives (agency, traveler, and community). Additionally, it was assumed that the production axioms (Vaneman and Triantis 2003) governing the inputs and outputs held as part of the associated service processes that are part of the three perspectives. The three processes in the context of this research are the service provision process of the transportation agency, the service consumption process of the travelers, and the environmental impact process for the community as a function of automobile travel in the downtown area.

An understanding of each of these processes provided the requisite background to define the input and output variables that were used in the network DEA model that linked the three perspectives. It was assumed that the system was being evaluated at the beginning of the system life-cycle. The dynamics associated with the various processes of the DSRS and the impact of organizational/behavioral/ information issues (e.g., the way the transportation agency would implement the DSRS) were not considered as part of the measurement evaluation. An open issue that still remains unresolved was how to account for the fact that the DSRS is to be used by many travelers. The network DEA approach assumed that for the travelers' perspective we would consider average values associated with multiple travelers for the input and output variables that were defined for the traveler perspective. This aggregation issue however, requires further investigation in the future.

The essence of the network DEA efficiency measurement approach was to compare and contrast various instances (scenarios) that occur in the transportation network under the execution of the Downtown Space Reservation System (DSRS). The scenarios constituted the production possibility set for our analysis. In other words, the scenarios generated by the traffic micro-simulation constituted the decision making units (28 in total). In this context, the data that were used were viewed as ex-ante versus ex-post data.

The scenarios varied in terms of the total demand level (i.e. number of vehicles per control period), the reservation policies (i.e. the weights assigned to the traveler throughput and revenue in the objective function of the optimization model (Zhao et al. 2010a)) and the inherent stochastic behavior of the traffic assignment and the traffic flow in the simulation. The demand level varied from 6,000 to 7,000 (vehicles/control period) and was chosen according to the transportation network size of the traffic simulation model. The relative importance (and consequently weights of the DSRS optimization model) associated with the traveler throughput and revenue was arbitrarily assigned due to the lack of practical references. Operational costs were assumed constant for all of the 28 DMUs. The data from the simulation model were complemented with revenue data from the original optimization model of the DSRS.

22.2.4.2 The Network DEA Approach: The Initial and Final Representations[2]

The idea behind the network DEA formulation is that users and community stakeholders are likely to be outcome oriented whereas providers are output oriented. Furthermore, we assumed that users are more focused on their mobility and

[2] The mathematical formulations for the radial and slacks-based network models are described in Zhao et al. 2011.

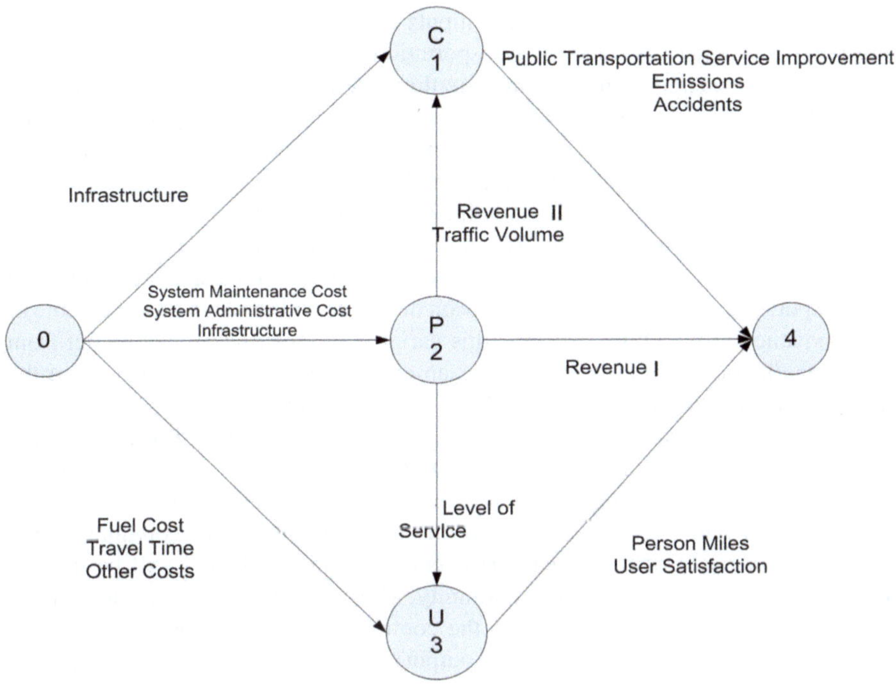

Fig. 22.1 Three perspectives of the performance network structure (initial conceptualization) (Reprinted with permission from Zhao et al. 2011)

this was reflected with the travel time related measures. It was assumed that transportation agencies are mostly concerned with system efficiency and effectiveness, which is reflected by the revenue, the level of service, and the vehicle miles traveled. Last but not least, the community typically is concerned with environmental and safety issues that are associated with traffic. Therefore, sustainability oriented measures (such as, emissions) are more appropriate to reflect their interests.

The network of Fig. 22.1 represents the conceptual underlying structure of the DSRS transportation system with respect to the three perspectives and their interrelationships. This initial conceptualization was arrived at from input from transportation engineers. The network consists of five nodes. Node 0 and node 4 are dummy nodes. The purpose of these nodes is to distribute inputs to and collect outputs from the intermediate nodes (nodes 1, 2 and 3). Therefore, the performance of the network reflects the interrelationship among the three perspectives captured by nodes 1, 2, and 3. Node 1 represents the community's perspective that is directly impacted by the transportation system. Node 2 represents the viewpoint of the transportation service provider whereas node 3 is the transportation user's perspective. The connection between nodes is directed, indicating the material transformation from inputs to outputs.

From the agency's perspective, the inputs to the transportation system include different operational costs and the transportation system infrastructure. The operational costs considered in this research are the system maintenance and administrative costs that the transportation agency wishes to minimize. It is also assumed that the agency makes decisions on whether to improve the transportation infrastructure, so it is considered as an input to the agency node 2. The outputs from the agency node 2 include revenue (Revenue I and II in Fig. 22.1), traffic volume, and level of service (LOS). While collecting revenue (Revenue I) to maintain the transportation system is in itself an objective for the agency, revenue (Revenue II) is also collected as a final output. It is assumed that traffic flow on the roads will result in traffic volume as a consequence of the DSRS and thus this variable is considered as an output from the agency's node 2. LOS is included as an output for node 2, because one of the agency's goals is to provide a certain LOS to the traveler.

From the community's point of view, the inputs are the infrastructure, the revenue (Revenue I) from executing the DSRS and the traffic volume. Infrastructure is imposed in the community's territory, so it is viewed as an input for node 1. The traffic volume will result in emissions and accidents for the community, and we assume that part of the revenue (Revenue I) from the DSRS will be used to improve the transit system in the community. Thus, the traffic volume and revenue (Revenue I) are included as inputs to the community node, and emissions (undesirable output), accidents (undesirable output) and public transportation improvements (desirable output) are the outputs.

From the travelers' perspective, the fuel cost, travel time and other costs including the reservation fee spent on the trips are considered as inputs by most travelers. These costs are direct costs. Since node 3 reflects the traveler's perspective, the measurement of the output is considered to be person miles rather than vehicle miles therefore the outputs from node 3 are person miles traveled and user satisfaction. Among all the variables in the representation of Fig. 22.1, there are two types of inputs/outputs – intermediate inputs/outputs and initial inputs/final outputs. The final outputs are the outputs that are accumulated in node 4, such as emissions, accidents, and person miles. The intermediate outputs, LOS, traffic volume and revenue, are the outputs from agency's node 2 and they are also the inputs to nodes 1 and 3.

Given the data from the micro-simulation model and from the original DSRS formulation the network DEA representation that was finally executed is represented by Fig. 22.2. Travel time, vehicle miles, average speed, fuel costs, emissions and personal miles (calculated from total vehicle miles and average occupancy) was obtained from the micro-simulation whereas revenue was obtained from the DSRS optimization. Radial network DEA and slacks based network DEA models were computed for the network (both input and output orientations) whereas the Banker et al. (1984) efficiency scores were computed for each of the nodes (Fig. 22.3). The reason why the efficiency scores for each of the individual nodes were computed was to determine the differences in performance evaluation using the network DEA and the DEA individual formulations.

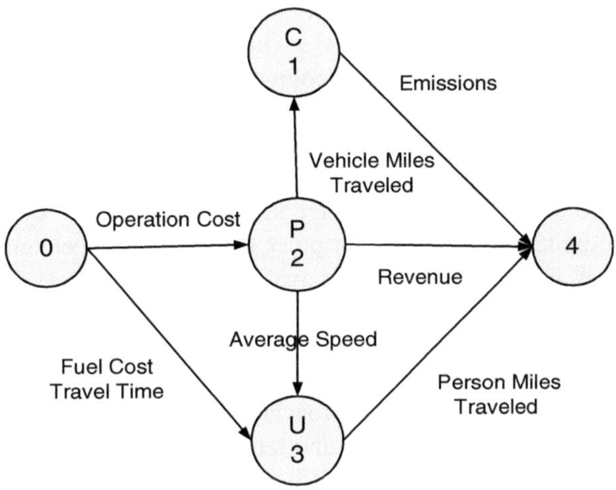

Fig. 22.2 Three perspectives of the performance network structure (final representation) (Reprinted with permission from Zhao et al. 2011)

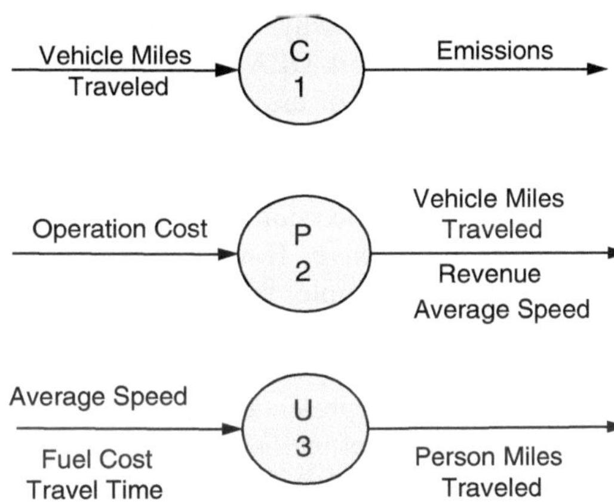

Fig. 22.3 Single node representations for each perspective (Reprinted with permission from Zhao et al. 2011)

22.2.5 Network DEA: Conclusions from the Example

The differences between the radial network and slacks based models for this example lie in the way the efficiency scores are computed. The radial and the slacks based model network DEA approaches provide different performance assessments. For example, when considering both approaches, the node that dominates is different given that the radial network efficiency score focuses primarily on the relatively efficient node in the network and ignores the inferior performance of the other nodes. Whereas, the slacks based network measure considers the average performance of all nodes. According to this information, the decision maker may be inclined to focus on very different interventions so as to improve system performance. For instance, based on the results from our example, the radial network DEA will lead the decision maker to focus more on the agency's perspective, while the slacks based network model will lead decision maker to focus on the traveler.

One of the core assumptions is that the network DEA structure is representative of the underlying processes (in this example, transportation processes (from an agency's and traveler's points of view) and community processes (in terms of community resilience). The network structure assumes two things. The first is that the input and output variables considered for each node are accurate representations of the service transformation processes. The second is that the interactions (co-dependencies) between nodes are reflected by the intermediate inputs and outputs. This means that other forms of co-dependencies (physical, informational and behavioral) are not at explicitly considered.

This research uses a combination of a DSRS optimization, a neural network, a micro simulation evaluation and a network DEA approaches. What this suggests is that when evaluating alternative system designs it is reasonable to combine analytical with simulation approaches. In our example, the DSRS optimization model itself could not convey important information, such as which traffic flow conditions are best suited for the DSRS, whether the design of the system meets stakeholders' requirements, and how the DSRS influences agency, traveler, and community performance. This is why it was necessary to additionally execute the micro-simulation and network DEA approaches.

Simulation is one of the most popular tools used by transportation engineers. It has been used to test and analyze the DSRS (Zhao et al. 2010b). The simulation model provides various transportation measures (e.g. travel time, average delay, etc.) and helps the decision maker appreciate the system from a transportation engineering perspective. However, additional performance aspects need to be considered. For example, the simulation does not directly tell us how various input/output variables affect overall system performance and whether social welfare goals are met by key stakeholder entities. The network DEA approach provides a single index as representative of the overall system efficiency and identifies appropriate sources of inefficiency across the various perspectives.

Therefore, the DSRS optimization model represents the system that was designed; the simulation and the network DEA models are the supporting approaches that provide an assessment of this system design. The two evaluation approaches are complementary. The simulation approach supports the network DEA model by providing data, and the network DEA performance measurement complements the simulation model by taking into account the key perspectives that are impacted by the potential implementation of the DSRS.

Returning to the discussion of Sect. 22.1, the network DEA approach helps the decision maker understand the system that is being evaluated by opening the DEA transformation "black box". This enables decision makers to locate the sources of inefficiency more precisely. For example, if the network DEA model shows that inefficiency is linked mainly to the traveler, the decision maker might improve the traveler throughput via a pricing policy adjustment. Additionally, we assumed that the axiomatic framework on which DEA is based holds for each of the three nodes of our example and for the network as a whole. Nevertheless, we did not consider the dynamic characteristics of the DSRS system which brings us to the next topic in this chapter.

22.3 Dynamic Representations of Performance Measurement Networks

While the static network DEA performance models for engineering systems provide an initial understanding of the determinants of efficiency performance within these systems, they do not consider the dynamic characteristics of these systems. In terms of dynamic network performance models, the efficiency research community has chosen to approach this issue using two distinct and separate directions.

The first simply extends network DEA formulations to include time (see for example, Tone and Tsutsui 2013). An alternative approach is to model the system's dynamic behavior explicitly (for example, using either system dynamics (Vaneman and Triantis 2007) or agent based modeling (Dougherty et al. 2013)) and then include efficiency concepts as a way to assess system performance. This latter direction allows for the explicit consideration of causation, feedback mechanisms, delays, and non-linear relationships whereas the former direction introduces temporal variations of efficiency measures explicitly. As described in the example of Sect. 22.2, the structure of network DEA model does not suggest anything in terms of the resulting system behavior whereas, the structure of system dynamics or agent based models result in various forms of system behaviors once one executes the simulations. On the other hand, the ways to measure efficiency performance with the system dynamics (Fallah-Fini et al. 2013b) or agent based models (Dougherty et al. 2013) are not straightforward. Consequently, one can view both directions as complementary since they address alternative representations of dynamics and efficiency measurement.

In order to complement the discussion of Sect. 22.2 we offer an example of a system dynamics representation of a system where performance assessment is an important objective. In the example that follows we use system dynamics to explicitly consider highway deterioration and renewal and briefly describe how efficiency measurement considerations are incorporated. The modeling details are described in Fallah-Fini et al. (2010, 2012, 2013b) which we do not replicate in this brief overview.

22.3.1 Infrastructure Management: Obtaining an Optimum Strategy for Road Maintenance[3]

For the highly challenged U.S. road infrastructure, major budgetary restrictions at the State and Federal levels and the significant growth in traffic demand have led to a continual pressure to improve the performance of highway maintenance practices. This has led to a series of analyses (Fallah-Fini et al. 2010, 2012, 2013b) that have

[3] This section is adopted from Fallah-Fini et al. (2010, 2012).

attempted to assess the privatization of road maintenance operations by state Departments of Transportation (DOTs). The research findings of these studies have indicated that road agencies should use hybrid contracting approaches that include best practices of both traditional (public) and performance-based (private) highway maintenance contracting. The analyses used an empirical dataset of pavement condition and maintenance expenditures over the years 2002–2008 corresponding to 17 miles of interstate highway that lay in one of the counties in the state of Virginia, USA. The data allowed for the calibration of the developed system dynamics models.

In the dynamic efficiency measurement model (Fallah-Fini et al. 2013b) the performance of highway maintenance operations was evaluated where the inter-temporal dependences between consumption of inputs (i.e., maintenance budget) and realization of outputs (i.e., improvement in road condition) were explicitly captured. We built on a micro representation of pavement deterioration and renewal (Fallah-Fini et al. 2010, 2012) and studied the impact of the allocation of scarce maintenance budgets over time. We introduced a concept of efficiency that contrasts the optimized budget allocations to the actual ones. The policies that were found through the optimization showed that road authorities should give higher priorities to preventive maintenance than corrective maintenance.

Initially, in order to establish the basic model we identified key maintenance dynamics associated with road maintenance and then we represented the deterioration and renewal processes of road maintenance using a physical understanding of these processes at the pavement level. The deterioration and maintenance dynamics can be summarized in two major feedback loops (See the causal loop in Fig. 22.4). The pavement condition deteriorates as a function of traffic load and environmental conditions. The balancing loop B1 (Maintenance Fix) depicts how the maintenance operations performed by road agencies bring the road condition towards desired conditions by reducing the road area under distress. On the other hand, the reinforcing loop R1 (Accelerated Deterioration) depicts the effect of a budget shortfall on delaying maintenance and the further deterioration of pavement conditions. This initial qualitative representation of the deterioration and maintenance dynamics served as the input to the physical simulation model. For details of the simulation model refer to Fallah-Fini et al. (2010, 2013b).

The conceptualization of the dynamic evolution of road condition over time expands the dynamic representation in network DEA introduced by Färe et al. (2007). We assume that part of the highway network at period t is affected by a set of deterioration factors such as climate conditions, traffic load, etc. Based on the condition of the road, appropriate maintenance operations are performed and the road evolves to a new condition at the end of period t. The new road condition is used as an input at the start of period $t + 1$ when the road section goes under a similar transformation process. This means that the maintenance treatments during period t affect the road condition at the end of period t which is the starting point for period $t + 1$. Thus, the required maintenance operations during period $t + 1$ (and consequently the road condition at the end of period $t + 1$) depend on the maintenance operations/inputs that have been performed in a stream of previous periods.

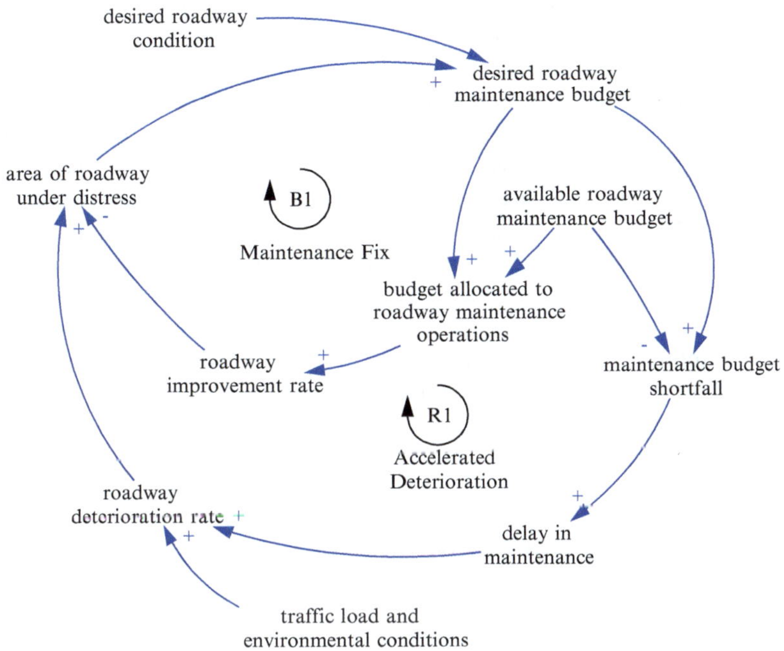

Fig. 22.4 The highway deterioration and maintenance causal loop diagram (Adopted from Fallah-Fini et al. 2010)

In such a setting, any "static" network DEA efficiency measurement framework that ignores the inter-temporal effects of inputs and managerial decisions for future streams of outputs (i.e., future road conditions) is likely to be unrealistic. The premise of this research is that successful evaluation and improvement of the performance of road maintenance practices requires a long-term perspective that takes into account the dynamics of road deterioration and maintenance.

The pavement engineering literature was studied to understand and capture the physics of the pavement deterioration (Huang 2004). Within the System Dynamics (SD) framework (Sterman 2000), the physically based dynamics were investigated in conjunction with macro-level maintenance operations. This combination allowed for constructing a simulation model that is grounded in the physics of road operations (i.e., the pavement distress generation and propagation, the effects of aging, the effect of deferred maintenance), that considers environmental conditions (the load in terms of vehicles and climate conditions) and material delays, that incorporates managerial factors (i.e., budget constraints, priorities in terms of the type of maintenance action (preventive, corrective and restorative), the thresholds associated with each type of maintenance and the actual amount of funds allocated to conduct maintenance for each road section). When executing the simulation model one can observe an adjustment path of the road condition to a new condition at the end of the simulation that is affected by the physics of deterioration, environmental conditions (traffic demand and climate conditions) and maintenance policies.

In defining and measuring system efficiency, we compared the actual road condition adjustment path (change in the state of the system) to a benchmark that represents the expected road condition adjustment path under an optimal budget allocation strategy over time. This concept is an augmentation of the output oriented concept of efficiency. To find the benchmark, we introduced a payoff representation that is a function of the state of the system at time t. For highway maintenance, the main objectives of the road authorities are to improve the condition of the highway network and maximize drivers' utilities while minimizing the costs. Thus, as an example, the payoff representation could be defined as the drivers' utilities at any point of time as a function of the condition of the road network state minus the maintenance costs. Then, starting from an arbitrary state at time t_0, the infinite horizon optimal adjustment path for the road condition can be constructed by following the optimal maintenance decisions obtained from solving an optimization problem (Fallah-Fini et al. 2013b) that maximizes driver utilities.

22.4 Conclusions and Future Research

The two examples presented in this chapter suggest that we have only scratched the surface in terms of obtaining viable network efficiency representations of engineering systems. The challenges and opportunities summarized in Sect. 22.1 remain. More specifically, micro performance representations of systems as a function of implementing the network DEA approach are opening the production "black box". In so doing, the identification of important processes that impact system performance are studied. This allows for an expanded exploration of the determinants of efficiency performance both within and between nodes and processes that are fundamental for the network DEA approach. In this sense, there is an opportunity to contribute to theory by experimentally discovering determinants of efficiency performance for a number of systems and applications.

Furthermore, we still have a limited understanding of how the structure of the efficiency network relates to which nodes and determinants of efficiency performance are important. In the case of the DSRS, the computational approach (radial versus slacks based efficiency determination) suggested that different nodes are important (i.e. the agency's versus the traveler's). This does not assist decision makers to arrive at consistent performance improvement interventions.

In terms of understanding and measuring dynamic efficiency (Fallah-Fini et al. 2013a) of engineering systems, there are potentially two distinct directions. The first simply extends network DEA formulations to include time (see for example, Tone and Tsutsui 2013). An alternative approach is to model the system's dynamic behavior explicitly (for example, using either system dynamics (Vaneman and Triantis 2007) or agent based modeling (Dougherty et al. 2013)) and then include efficiency concepts as a way to assess system performance. As suggested by the highway maintenance example of Sect. 22.3, one can view both directions as complementary since they address alternative representations of dynamics and efficiency measurement.

As argued in Sect. 22.1, engineering systems are not designed, built and operated in a vacuum. There are organizations and design teams that are tasked to do so by exchanging important information and making decisions. This suggests that we need a deeper appreciation and the integration of the social/behavioral and information sciences in our efficiency analyses and thinking. While the efficiency literature relies primarily on economic and operations research thinking it is the contention of the author that understanding of efficiency performance will be incomplete without the input from the social and behavioral sciences (sociology, psychology, cognitive sciences, decision sciences, etc.) computer science (cyber physical systems), and engineering (control systems, environmental engineering, electrical engineering, structural engineering, etc.). This suggests an even expanded inter-disciplinary approach to efficiency measurement. Network representations such as network DEA offer a viable vehicle for realizing this inter-disciplinary perspective. However, all of this is contingent on consistently revisiting the axiomatic framework on which efficiency analysis is based on.

References

Banker, R. D., Charnes, A., & Cooper, W. W. (1984). Some models for estimating technical and scale efficiencies in data envelopment analysis. *Management Science, 30*(9), 1078–1092.

Blanchard, B., & Fabrycky, W. (2010). *Systems engineering and analysis* (5th ed.). Prentice Hall International Series in Industrial & Systems Engineering. Upper Saddle River, New Jersey 07458.

Cooper, W. W., Sinha, K. K., & Sullivan, R. S. (1992). Measuring complexity in high-technology manufacturing: Indexes for evaluation. *Interfaces, 4*(22), 38–48.

Dougherty, F., Ambler, N., & Triantis, K. (2013). *Complex adaptive systems approach for productive efficiency analysis – Part I: Conceptualization* (Working Paper). System Performance Laboratory, Virginia Tech.

Fallah-Fini, S., Rahmandad, H., Triantis, K., & de la Garza, J. (2010). Optimizing highway maintenance policies and practices: Operational and dynamic considerations. *System Dynamics Review: Special Issue on Transportation, 26*(3), 216–238.

Fallah-Fini, S., Triantis, K., de la Garza, J., & Seaver, B. (2012). Measuring the efficiency of highway maintenance contracting strategies: A bootstrapped non-parametric meta-frontier approach. *European Journal of Operational Research, 219*(1), 134–145.

Fallah-Fini, S., Triantis, K., & Johnson, A. (2013a, forthcoming). Efficiency performance: State-of-the-art. *Journal of Productivity Analysis.*

Fallah-Fini, S., Triantis, K., Rahmandad, H., & de la Garza, J. (2013b). Measuring dynamic efficiency of highway maintenance operations. *Omega*. Under review.

Färe, R., & Grosskopf, S. (2000). Network DEA. *Socio-Economic Planning Sciences, 34*, 35–49.

Färe, R., Grosskopf, S., & Whittaker, G. (2007). Network DEA. In J. Zhu & W. D. Cook (Eds.), *Modeling data structures irregularities and structural complexities in data envelopment analysis* (pp. 209–240). New York: Springer.

Huang, Y. (2004). *Pavement analysis and design*. Upper Saddle River: Pearson/Prentice Hall.

Sterman, J. (2000). *Business dynamics*. Boston: McGraw-Hill.

Teodorović, D., & Vukadinović, K. (1998). *Traffic control and transport planning: A fuzzy sets and neural networks approach*. Boston/Dordrecht/London: Kluwer Academic.

Tone, K., & Tsutsui, M. (2009). Network DEA: A slacks based measure approach. *European Journal of Operational Research, 197*(1), 243–252.

Tone, K., & Tsutsui, M. (2013). *Dynamic DEA with network structure: A slacks based measure approach*. Workshop on DNDEA. Graduate Institute for Policy Studies, 7-22-1 Roppongi, Minato-ku, Tokyo 106–8677.

Triantis, K. (2011). Chapter 14: Engineering applications of DEA. In *Handbook of data envelopment analysis*. Kluwer, pp. 363–402, 101 Philip Drive, Norwell, MA 02061, U.S.A.

Triantis, K. (2013). Engineering design and performance measurement: Issues and future research opportunities. *International Journal of Data Envelopment Analysis*. Under review.

Vaneman, K., & Triantis, K. (2003). The dynamic production axioms and system dynamics behaviors: The foundation for future integration. *Journal of Productivity Analysis, 19*, 93–113.

Vaneman, W., & Triantis, K. (2007). Evaluating the productive efficiency of dynamical systems. *IEEE Transactions on Engineering Management, 54*(3), 600–611.

Zhao, Y., Triantis, K., Teodorović, D., & Edara, P. (2010a). A travel demand management strategy: The downtown space reservation system. *European Journal of Operational Research, 205*(3), 584–594.

Zhao, Y., Triantis, K., & Edara, P. (2010b). Evaluation of travel demand strategies: A microscopic traffic simulation approach. *Transportation, 37*(3), 549–571.

Zhao, Y., Triantis, K., Murray-Tuite, P., & Edara, P. (2011). Performance measurement of a transportation network with a downtown space reservation system: A network-DEA approach. *Transportation Research Part E, 47*(6), 1140–1159.

Index

W.D. Cook and J. Zhu (eds.), *Data Envelopment Analysis*,
International Series in Operations Research & Management Science 208,
DOI 10.1007/978-1-4899-8068-7, © Springer Science+Business Media New York 2014

E

N

U

V

MIX
Papier aus verantwortungsvollen Quellen
Paper from responsible sources
FSC® C105338

If you have any concerns about our products,
you can contact us on
ProductSafety@springernature.com

In case Publisher is established outside the EU,
the EU authorized representative is:
Springer Nature Customer Service Center GmbH
Europaplatz 3, 69115 Heidelberg, Germany

Printed by Libri Plureos GmbH
in Hamburg, Germany